www.brookscole.com

brookscole.com is the World Wide Web site for Thomson Brooks/Cole and is your direct source to dozens of online resources.

At *brookscole.com* you can find out about supplements, demonstration software, and student resources. You can also send email to many of our authors and preview new publications and exciting new technologies.

brookscole.com
Changing the way the world learns®

OCEANOGRAPHY

AN INVITATION TO MARINE SCIENCE

FIFTH EDITION

TOM GARRISON

Orange Coast College
University of Southern California

THOMSON

BROOKS/COLE

Australia • Canada • Mexico • Singapore • Spain • United Kingdom • United States

THOMSON

BROOKS/COLE

Earth Science Editor: *Keith Dodson*
Development Editor: *Mary Arbogast*
Assistant Editor: *Carol Benedict*
Editorial Assistant: *Melissa Newt*
Technology Project Manager: *Erika Yeoman Saler*
Marketing Manager: *Kelley McAllister*
Marketing Assistant: *Leyla Jowza*
Advertising Project Manager: *Nathaniel Bergson-Michelson*
Project Manager, Editorial Production: *Teri Hyde*
Print/Media Buyer: *Rebecca Cross*
Permissions Editor: *Joohee Lee*

Production Service: *Joan Keyes, Dovetail Production Services*
Text Designer: *Joan Keyes, Dovetail Production Services*
Photo Researchers: *Myrna Engler, Tom Garrison*
Copy Editor: *Carol Reitz*
Illustrators: *Precision Graphics*
Cover Designer: *Irene Morris*
Cover Image: *Ben Osborne Photography Ltd.*
Cover Printer: *Quebecor World Dubuque*
Compositor: *Graphic World, Inc.*
Printer: *Quebecor World Dubuque*

For more information about our products, contact us at:
Thomson Learning Academic Resource Center
1-800-423-0563

For permission to use material from this text, contact us by:
Phone: 1-800-730-2214
Fax: 1-800-730-2215
Web: http://www.thomsonrights.com

Library of Congress Control Number:

Student Edition ISBN: 0-534-40887-7
Advantage Series Student Edition: 0-534-40533-9
Instructor's Edition ISBN: 0-534-40890-7
International Student Edition ISBN 0-534-40993-8

Brooks/Cole—Thomson Learning
10 Davis Drive
Belmont, CA 94002
USA

Asia
Thomson Learning
5 Shenton Way #01-01
UIC Building
Singapore 068808

Australia/New Zealand
Thomson Learning
102 Dodds Street
Southbank, Victoria 3006
Australia

Canada
Nelson
1120 Birchmount Road
Toronto, Ontario M1K 5G4
Canada

Europe/Middle East/Africa
Thomson Learning
High Holborn House
50/51 Bedford Row
London WC1R 4LR
United Kingdom

Latin America
Thomson Learning
Seneca, 53
Colonia Polanco
11560 Mexico D.F.
Mexico

To my family and my students:
My hope for the future

About the Author

Tom Garrison (Ph.D., University of Southern California) is a professor in the Marine Science Department at Orange Coast College in Costa Mesa, California, one of the largest undergraduate marine science departments in the United States. Dr. Garrison also holds an adjunct professorship at the University of Southern California. He has been named Outstanding Marine Educator by the National Marine Technology Society, is a member of the COSEE staff, writes a regular column for the journal *Oceanography*, and was a winner of the prestigious Salgo-Noren Foundation Award for Excellence in College Teaching. Garrison was an Emmy Award team participant as writer and science advisor for the PBS syndicated *Oceanus* television series, and writer and science advisor for *The Endless Voyage*, a set of television programs in oceanography completed in 2003. His widely used textbooks in oceanography and marine science are the college market's best sellers.

His interest in the ocean dates from his earliest memories. As he grew up with a U.S. Navy admiral as a dad, the subject was hard to avoid! He had the good fortune to meet great teachers who supported and encouraged this interest. Years as a midshipman and commissioned naval officer continued the marine emphasis; graduate school and 30+ years of teaching have allowed him to pass his oceanic enthusiasm to more than 65,000 students.

Dr. Garrison is married, has two children (a daughter who teaches 4th grade; a son who is an executive in international trade), a son-in-law, a new granddaughter with another on the way, and a very patient cadre of teaching assistants. He and his family reside in Newport Beach, California.

John Wilson Cramer IV

Brief Contents

Contents

Preface

This book was written to provide an *interesting*, clear, current, and reasonably comprehensive overview of the marine sciences. It was designed for college and university students who are curious about Earth's largest feature but who may have little formal background in science. Oceanography is broadly interdisciplinary; students are invited to see the *connections* between astronomy, economics, physics, chemistry, history, meteorology, geology, and ecology—areas of study they once considered separate. It's no surprise that oceanography courses have become increasingly popular!

Students bring a natural enthusiasm to their study of this field. Even the most indifferent reader will perk up when presented with stories of encounters with huge waves, photos of giant squids, tales of exploration under the best and worst of circumstances, evidence that vast chunks of Earth's surface slowly move, news of Earth's past battering by asteroids, micrographs of glistening diatoms, and data showing the growing economic importance of seafood and marine materials. If pure spectacle is required to generate an initial interest in the study of science, oceanography wins hands down!

In the end, however, it is subtlety that triumphs. Studying the ocean reinstills in us the sense of wonder we all felt as children when we first encountered the natural world. There is much to tell. The story of the ocean is a story of change and chance—its history is written in the rocks, the water, and the genes of the millions of organisms that have evolved here.

The Fifth Edition

My aim in writing this book was to produce a text that would enhance students' natural enthusiasm for the ocean. My students have been involved in this book from the very beginning—indeed, it was their request for a readable, engaging, and thorough text that initiated the project a long time ago. During the 20 years I have been writing textbooks, my enthusiasm for oceanic knowledge has increased (if that is possible), forcing my patient reviewers and editors to weed out an excessive number of exclamation points. But enthusiasm does shine through. One student reading the final manuscript of the first edition commented, "At last, a textbook that does not read like stereo instructions." Good!

This new edition builds on its predecessors. As before, a great many students have participated alongside professional marine scientists in the writing and reviewing process. In response to their recommendations, as well as those of instruc-

tors who have adopted the book and the many specialists and reviewers who contributed suggestions for strengthening the earlier editions, I have:

- Emphasized the *process* of science throughout. Underlying assumptions and limitations are discussed as appropriate in all sections of the book.
- Rearranged the chapters on the physics and chemistry of seawater to present the physical properties before discussing chemistry.
- Rearranged the chapters on primary productivity and trophic relationships for more logical flow.
- Enhanced the visual program for increased clarity and accuracy, adding and modifying illustrations to make ideas easier to grasp.
- Expanded discussions of physical oceanography, marine geology, primary productivity, environmental issues, and the ocean's contribution to world economics.
- Covered communities more thoroughly, emphasizing community ecology in the sections on marine biology.
- Modified every chapter to reflect current thought and recent research.
- Generated a new Web site specific to this textbook. In addition to keying more than 3,000 active sites to specific points in the text, the Web site now includes an integrated learning system, *OceanographyNow* that is described in more detail below.
- Refined the pedagogy to include summary grids, "Chapter in Perspective" sections to place material in context, and analytical review questions.

The Plan

The plan of the book is straightforward: We begin with a brief look at the history of marine science (with additional historical information sprinkled throughout later chapters). Because all matter on Earth except hydrogen and some helium was generated in stars, our story of the ocean starts with stars. Have oceans evolved elsewhere? The theories of Earth structure and plate tectonics are presented next, as a base on which to build the explanation of bottom features that follows. A survey of ocean physics and chemistry prepares us for discussions of atmospheric circulation, classical physical oceanography, and

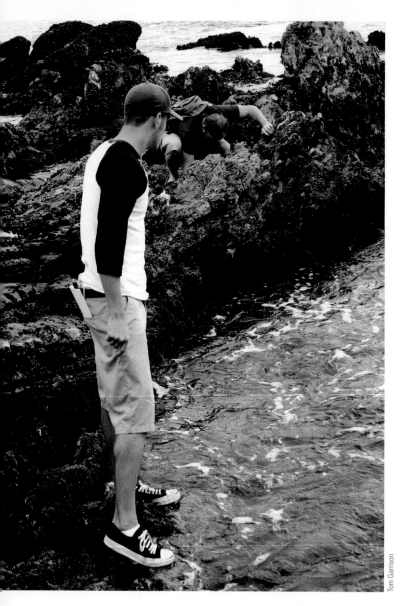

Tom Garrison

Go into the natural world whenever the opportunity arises. If you don't live near a coast, just walk outside and look around! Air, water, and Earth are intimately intertwined. The more you look at weather, water, life, rocks, and time, the clearer the connections will become.

coastal processes. Our look at marine biology begins with an overview of the problems and benefits of living in seawater, continues with a discussion of the production and consumption of food, and ends with taxonomic *and* ecological surveys of marine organisms. The last chapters treat marine resources and environmental concerns.

Again, *connections* between disciplines are emphasized throughout. Marine science draws on several fields of study, integrating the work of specialists into a unified whole. For example, a geologist studying the composition of marine sediments on the deep seabed must be aware of the biology and life histories of the organisms in the water above, the chemistry that affects the shells and skeletons of the creatures as they fall to the ocean floor, the physics of particle settling and water density and ocean currents, and the age and underlying geol-

ogy of the study area. This book is organized to make those connections from the beginning.

Organization and Pedagogy

A broad view of marine science is presented in 18 chapters, each free-standing (or nearly so) to allow an instructor to assign chapters in any order he or she finds appropriate. Each chapter begins with a **vignette,** a short illustrated tale, observation, or sea story to whet your appetite for the material to come. Some vignettes spotlight scientists at work; others describe the experiences of people or animals in the sea. Each vignette ends with a brief overview of the chapter's high points and a few advance organizers for what's to come.

The chapters are written in an **engaging style.** Terms are defined and principles developed in a straightforward manner. Some of the more complex ideas are initially outlined in broad brushstrokes; then the same concepts are discussed again in greater depth after you have a clear view of the overall situation. When appropriate to their meanings, the derivations of words are shown. **Measurements** are given in both metric (SI) and American systems. At the request of a great many students, the units are written out (that is, we write *kilometer* rather than *km*) to avoid ambiguity and for ease of reading.

The photos, charts, graphs, and paintings in the **extensive illustration program** have been chosen for their utility, clarity, and beauty. **Boxes** in each chapter present commentaries of special interest on unique topics or controversies. **Internet icons** are provided at nearly all of the subheads, boxes, and essays, indicating that text-specific Web links provide additional information.

Concluding each chapter is a **Questions from Students** section. These questions are ones that students have asked me over the years. This material is an important extension of the chapters and occasionally contains key words and illustrations. Each chapter ends with an array of study materials for students, beginning with a new feature, **Chapter in Perspective,** that combines a narrative review with a brief look ahead at the next chapter. A chart of **Key Ideas** follows, each including references to critical illustrations and pages. Important **Terms and Concepts to Remember** are listed next; these are also defined in an extensive **Glossary** in the back of the book. **Study Questions** (review, critical thinking, and analytical) are also included in each chapter; writing the answers to these questions will cement your understanding of the concepts presented. The **annotated bibliography** at the book's Web site will be helpful when you wish to know more about a particular topic.

Appendixes will help you master measurements and conversions, geological time, latitude and longitude, chart projections, the mathematics of the Coriolis effect and tidal forces, the taxonomy of marine organisms, and the periodic table of elements. In case you'd like to join us in our life's work, the last appendix discusses **jobs in marine science.**

The book has been thoroughly **student-tested.** You need not feel intimidated by the concepts—this material has been mastered by students just like you. Read slowly and go step by step through any parts that give you trouble. Your predecessors have found the ideas presented here to be useful, inspiring, and applicable to their lives. Best of all, they have found the subject to be *interesting!*

The Web Site

OceanographyNow is available through the text's Companion Web Site at http://oceanography.brookscole.com/garrison5e. In addition to **OceanographyNow**, students using the Web site will have access to maps, Web links, Internet and *InfoTrac®College Edition* exercises, learning objectives, discussion questions, chapter outlines, and much more.

What does an integrated learning system do?
OceanographyNow and the new edition of this text were developed in concert to enhance one another and provide you with a seamless, integrated learning system. As you work through the text, you will see notes that direct you to the dynamic, media-enhanced activities and helpful tutorials in **OceanographyNow**.

How does it work?
Students log on to **OceanographyNow** from the book's Companion Web Site at http://oceanography.brookscole.com/garrison5e by using the free PIN code packaged with this text. You'll immediately notice the system's simple, browser-based format—as easy to use as surfing the Web. A simple click of the mouse gives you the freedom to enter and explore the system at any point. You can build a complete, personalized learning plan for yourself by taking advantage of all three of **OceanographyNow**'s powerful components:

1. **What I Know** A *Pre-Test* is your first step. Each chapter's *Pre-Test* includes approximately 20 multiple-choice questions, and it helps you identify any conceptual weaknesses. The questions concentrate on the most common student mistakes, and the test's response will help you zero in on the error. Your results may be e-mailed to your professor so that both of you can work together on the best learning strategy. After completing the *Pre-Test*, you are presented with a detailed *Personalized Learning Plan* that outlines the concepts that you need to review, identifying where the concepts can be found in the text as well as in the program.

2. **What I Need to Learn** Working from your Personalized Learning Plan, you will encounter activities and movies that function together to help you master key concepts including a remarkable bank of animated figures and animations that vividly demonstrate oceanographic processes. Animated from three-dimensional models of *Active Figures* in this and other Brooks/Cole's earth science texts, each interactive student exploration demonstrates a key concept and is paired with conceptual questions to help you focus on what is happening in the model.

3. **What I've Learned** An optional, 10-question *Post-Test* ensures that you have mastered the concepts in each chapter. As with the *Pre-Test*, students' results may be e-mailed to your instructor.

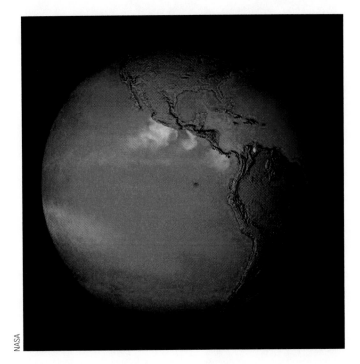

NASA

The world ocean swirls with heat. Without the ocean's thermostatic effects, and its ability to move heat from near the equator to the polar regions, Earth would be a dramatically different (and uninhabitable) place.

Suggestions for Using This Book

1. *Begin with a preview.* Scout the territory ahead—read the vignette that begins each chapter, flip through the assigned pages reading only the headings and subheadings, look at the figures and read captions that catch your attention.

2. *Keep a pen and paper handy.* Jot down a few questions—any questions—that this quick glance stimulates. *Why* is the deep ocean cold if the inside of Earth is so hot? *What* makes storm conditions like those seen in 1997–98? *Where* did sea salt come from? *Will* global warming actually be a problem? *Does* anybody still hunt whales? Writing questions will help you focus when you start studying.

3. *Now read in small but concentrated doses.* Each chapter is written in a sequence and tells a story. The logical progression of ideas is going somewhere. Find and follow the organization of the chapter. Stop occasionally to review what you've learned. Flip back and forth to review and preview.

4. *Strive to be actively engaged!* Write marginal notes, underline occasional passages (underlining whole sections is seldom useful), write more questions, draw on the diagrams, check off subjects as you master them, make flashcards while you read (if you find them helpful), *use your book!*

5. *Monitor your understanding.* If you start at the beginning of the chapter you will have little trouble understanding the concepts as they unfold. But if you find yourself at the bottom of the page having only scanned (rather than understood) the material, stop there and start that part again.

NASA/ORBIMAGE

The Gulf Stream, the world's fastest ocean current, jets northeastward into the open Atlantic at Cape Hatteras. Water evaporating from the warm sea surface has condensed into clouds. Warmth transported by the Gulf Stream greatly influences Europe's weather and climate.

Look ahead to see where we're going. Remember, students have been here before, and I have listened to their comments to make the material as clear as I can. This book was written for you.

6. *Check out the study tools on the book's Web site.* OceanographyNow provides media-enhanced activities and helpful tutorials. It allows you to develop a personal learning plan and focus on the concepts you most need to master.

7. *Use the Internet sites.* We have placed more than 1,500 Internet sites through the text for you to explore. They provide expanded information and different ways of looking at the material in the text. If you have access to a computer, fire it up and scan the designated site as you read the associated passage of the book.

8. *Enjoy the journey.* Your instructor would be glad to share his or her understanding and appreciation of marine science with you—you have only to ask. Students, instructors, and authors all work together toward a common goal: an appreciation of the beauty and interrelationships a growing understanding of the ocean can provide.

Acknowledgments

Jack Carey at Thomson Learning, the grand master of college textbook publishing, willed the first edition of this book into being. His suggestions have been combined with those of more than 850 undergraduate students and 134 reviewers to contribute to my continuously growing understanding of marine science. Donald Lovejoy, Stanley Ulanski, Richard Yuretich, Ronald Johnson, and John Mylroie deserve special recognition for many years of patient direction. For this fifth edition, I have especially depended on the expert advice of Ronald E. Johnson, Old Dominion University; Suzanne O'Connell, Wesleyan University; James H. Meyers, Winona State University; Jennifer Latimer, Indiana Purdue University at Indianapolis; Lloyd W. Kitazono, California State University Maritime Academy; Richard Strickland, University of Washington; Edward Laine, Bowdoin College; James E. Mackin, Prairie View A&M University; Mark Boardman, Miami University; Susan Libes, Coastal Carolina University; Raphael Kudela, University of California at Santa Cruz; Maria-Serena Poli, Eastern Michigan University; Randy Reed, Shasta College; Geno Pawlak, University of Hawaii at Honolulu; Arne Winguth, University of Wisconsin at Madison; and Steve P. Lund at my alma mater, the University of Southern California.

My long-suffering departmental colleagues Dennis Kelly, Jay Yett, and Robert Profeta again should be awarded medals for putting up with me, answering hundreds of my questions, and being so forbearing through the book's lengthy gestation period. Thanks again to our dean, Stanley Johnson, and our college president, Gene Farrell, for supporting this project and encouraging our faculty to teach, conduct research, and be involved in community service. Our past and present department teaching assistants deserve praise as well, especially Timothy Riddle, the Internet wizard responsible—among many other things—for the Web site and extensive links.

Yet another round of gold medals should go to my family for being patient (well, *relatively* patient) during those years of days and nights when Dad was holed up in his dark reference-littered cave, listening to really loud Glenn Gould Bach recordings, working late on The Book. Thank you, Marsha, Jeanne, Greg, Grace, and John, for your love and understanding.

The people who provided pictures and drawings have worked miracles to obtain the remarkable images in these pages. To mention just a few: Gerald Kuhn sent classics taken by his late SIO colleague Francis Shepard, Vincent Courtillot of the University of Paris contributed the remarkable photo of the Aden Rift, Catherine Devine at Cornell provided time-lapse graphics of tsunami propagation, Jian Lin of Woods Hole sent a photo of Iceland's Thingvellir Graben, Robert Headland of the Scott Polar Research Institute in Cambridge searched out prints of polar subjects, Charles Hollister at Woods Hole kindly provided seafloor photos from his important books, Andreas Rechnitzer and Don Walsh recalled their exciting days with *Trieste*, and Bruce Hall, Pat Mason, Ron Romanoski, Ted Delaca, William Cochlan, Christopher Ralling, Mark McMahon, John Shelton, Alistair Black, Howard Spero, Eric Bender, Ken-ichi Inoue, and Norman Cole contributed beautiful slides. Seran Gibbard provided the highest-resolution images yet made of the surface of Titan, and Michael Malin forwarded truly beautiful images of erosion on Mars. Herbert Kawainui Kane again allowed us to reprint his magnificent paintings of Hawaiian subjects. Deborah Day and Cindy Clark at the Scripps Institution, Jutta Voss-Diestelkamp at the Alfred Wegener Institut in Bremerhaven, and David Taylor at the Centre for Maritime Research in Greenwich dug through their archives one more time. Don Dixon, William Hartmann, Ron Miller, and William Kaufmann provided paintings, Dan Burton sent photos, and Andrew

Goodwillie printed customized charts. Wim van Egmond contributed striking photomicrographs of diatoms, forams, and copepods. Bill Haxby at Lamont provided truly beautiful seabed scans. Karen Riedel helped with DSDP core images. James Ingle offered me a desk at Stanford whenever I needed it. NOAA, NASA, USGS, the Smithsonian Institution, the Royal Geographical Society, the U.S. Navy, and the U.S. Coast Guard came through time and again, as did private organizations like Alcoa Aluminum, Cunard, Shell Oil, The Maersk Line, Grumman Aviation, Breitling-SA, CNN, Associated Press, Oakley, and the *Los Angeles Times.* The Woods Hole team was also generous: Philip Richardson, William Schmitz, Larry Madin, Susumu Honjo, Doug Webb, James Broda, Albert Bradley, and Kathy Patterson all provided photographs, diagrams, and advice. Individuals with special expertise have also been willing to share: Hank Brandli processed satellite digital images of storms, Peter Sloss at the National Geophysical Data Center helped me sort through computer-generated seabed images, Maria Zuber at M.I.T. took time away from guiding the current crop of Mars landers to provide a provocative image of river outflow patterns on that planet, Dr. Hans-Ulrich Schmincke contributed his photograph of an ophiolite formation on Cyprus, Dr. A. Rubino of the University of Hamburg sent his images of internal waves, Kevin Arrigo at Stanford shared his CFC-related phytoplankton research, Steven Grand of the University of Texas provided a descending deep-slab image, Hans-Peter Bunge of Princeton patiently explained mantle–core dynamics, Michael Gentry again mined the archives of the Johnson Space Center for Earth images, Jurrie van der Woulde at JPL helped with images of oceans elsewhere, John Maxtone-Graham of New York's Seaport Museum found me a rogue wave picture, Ed Ricketts, Jr., contributed a portrait of his father, and Professor Lynton Land of the University of Texas sent a rare photo of a turbidity current. Dave Sandwell at Scripps shared his astonishing satellite-generated imagery of the seabed. Rick Grigg at the University of Hawaii encouraged me to tackle some tricky bits of wave physics. Without their inestimable goodwill, a project like this would not be possible.

The Brooks/Cole team performed the customary miracles. The charge was again led by Joan Keyes, perpetually calm production editor and champion of every known means of digital communication. The text was polished by Mary Arbogast, a good friend and the best developmental editor in the Orion arm of the galaxy, and Carol Reitz, the copy editor who saved me from many errors. Myrna Engler worked tirelessly on photo research and permissions, and Joohee Lee tackled permissions. Production Project Manager Teri Hyde kept us all running in the same direction. And Keith Dodson, my excellent editor, gave birth to this book and his first child simultaneously. What skill!

My thanks to all.

A Goal and a Gift

The goal of all this effort: *To allow you to gain an oceanic perspective.* "Perspective" means being able to view things in terms of their relative importance or relationship to one another. An oceanic perspective lets you see this misnamed planet in a new light and helps you plan for its future. You will see that water, continents, seafloors, sunlight, storms, seaweeds, and society are connected in subtle and beautiful ways.

The ocean's greatest gift to humanity is intellectual—the constant challenge its restless mass presents. Let yourself be swept into this book and the class it accompanies. Give yourself time to ponder: "Meditation and water are wedded forever," wrote Herman Melville in *Moby Dick.* Take pleasure in the natural world. Ask questions of your instructors and TAs, read some of the references, try your hand at the questions at the ends of the chapters.

Be optimistic. Take pleasure in the natural world. Please write to me when you find errors or if you have comments. Above all, *enjoy yourself!*

Tom Garrison
Orange Coast College
University of Southern California
tgarriso@mail.occ.cccd.edu

Reviewers

ERNEST ANGINO, *University of Kansas*
M. A. ARTHUR, *Pennsylvania State University*
HENRY A. BART, *La Salle University*
STEVEN R. BENHAM, *Pacific Lutheran University*
LATSY BEST, *Palm Beach Community College*
EDWARD BEUTHER, *Franklin and Marshall College*
WILLIAM L. BILODEAU, *California Lutheran University*
MARK BOARDMAN, *Miami University*
JULIE BRIGHAM-GRETTE, *University of Massachusetts at Amherst*
LAURIE BROWN, *University of Massachusetts*
KEITH A. BRUGGER, *University of Minnesota, Morris*
ZANNA CHASE, *Lamont-Doherty Earth Observatory, Columbia University*
KARL M. CHAUFF, *St. Louis University*
WILLIAM COCHLAN, *San Francisco State University*
JAMES E. COURT, *City College of San Francisco*
RICHARD DAME, *University of South Carolina, Columbia*
DAVID DARBY, *University of Minnesota at Duluth*
ROBERT J. FELLER, *University of South Carolina, Columbia*
L. KENNETH FINK, JR., *University of Maine*
BRUCE FOUKE, *University of Illinois, Urbana*
DIRK FRANKENBURG, *University of North Carolina, Chapel Hill*
ROBERT R. GIVEN, *Marymount College, Rancho Palos Verdes*
WILLIAM GLEN, *U.S. Geological Survey*
KAREN GROVE, *San Francisco State University*
BARRON HALEY, *West Valley College*
JACK C. HALL, *University of North Carolina at Wilmington*
WILLIAM HAMNER, *University of California, Los Angeles*
WILLIAM B. HARRISON III, *Western Michigan University*
DAVID HASTINGS, *University of British Columbia*
TED HERMAN, *West Valley College*
JOSEPH HOLLIDAY, *El Camino College*
ANDREA HUVARD, *California Lutheran University*
JAMES C. INGLE, JR., *Stanford University*
RONALD E. JOHNSON, *Old Dominion University*
SCOTT D. KING, *Purdue University*
LLOYD W. KITAZONO, *California State University Maritime Academy*
JOHN A. KLASIK, *California Polytechnic University, Pomona*
ERNEST C. KNOWLES, *North Carolina State University, Raleigh*
EUGENE KOZLOFF, *Friday Harbor, WA*
MICHELLE KOMINZ, *Western Michigan University*
LAWRENCE KRISSEK, *Ohio State University*
RAPHAEL KUDELA, *University of California at Santa Cruz*
ALBERT M. KUDO, *University of New Mexico*
FRANK T. KYTE, *University of California, Los Angeles*
EDWARD LAINE, *Bowdoin College*
LYNTON S. LAND, *University of Texas, Austin*
JENNIFER LATIMER, *Indiana Purdue University at Indianapolis*
RICHARD W. LATON, *Western Michigan University*
RUTH LEBOW, *University of California, Los Angeles Extension*
DOUGLAS R. LEVIN, *Bryant College*
LARRY LEYMAN, *Fullerton College*

SUSAN LIBES, *Coastal Carolina University*
TIMOTHY LINCOLN, *Albion College*
DONALD L. LOVEJOY, *Palm Beach Atlantic College*
STEVE P. LUND, *University of Southern California*
MICHAEL LYLE, *Tidewater Community College*
JAMES E. MACKIN, *Prairie View A&M University*
DAVID C. MARTIN, *Centralia College*
ELLEN MARTIN, *University of Florida*
BRIAN MCADOO, *Vasser College*
JAMES MCWHORTER, *Miami-Dade Community College, Kendall Campus*
CHRIS METZLER, *Mira Costa College*
JAMES H. MEYERS, *Winona State University*
RICHARD W. MURRAY, *Boston University*
JOHN MYLROIE, *Mississippi State University*
CONRAD NEWMAN, *University of North Carolina, Chapel Hill*
SUZANNE O'CONNELL, *Wesleyan University*
JAMES G. OGG, *Purdue University*
B. L. OOSTDAM, *Millersville University*
GENO PAWLAK, *University of Hawaii at Honolulu*
JAN PECHENIK, *Tufts University*
BERNARD PIPKIN, *University of Southern California*
MARK PLUNKETT, *Bellevue Community College*
K. M. POHOPIEN, *Covina, CA*
MARIA-SERENA POLI, *Eastern Michigan University*
RANDY REED, *Shasta College*
RICHARD G. ROSE, *West Valley College*
JUNE R. P. ROSS, *Western Washington University*
WENDY L. RYAN, *Kutztown University*
ROBERT J. SAGER, *Pierce College, Washington*
ROBERT F. SCHMALZ, *Pennsylvania State University*
DON SEAVY, *Olympic College*
SAM SHABB, *Highline Community College*
WILLIAM G. SIESSER, *Vanderbilt University*
RALPH SMITH, *University of California, Berkeley*
SCOTT W. SNYDER, *East Carolina University*
MORRIS L. SOTONOFF, *Chicago State University*
JAMES F. STRATTON, *Eastern Illinois University*
RICHARD STRICKLAND, *University of Washington*
KENT SYVERSON, *University of Wisconsin*
J. COTTER THARIN, *Hope College*
STANLEY ULANSKI, *James Madison University*
J. J. VALENCIC, *Saddleback College*
RAYMOND E. WALDNER, *Palm Beach Atlantic College*
JILL M. WHITMAN, *Pacific Lutheran University*
P. KELLY WILLIAMS, *University of Dayton*
ARNE WINGUTH, *University of Wisconsin at Madison*
BERT WOODLAND, *Centralia College*
JOHN H. WORMUTH, *Texas A&M University*
RICHARD YURETICH, *University of Massachusetts at Amherst*
MEL ZUCKER, *Skyline College*

CHAPTER 1

A view of Earth. Beneath a very thin atmosphere, most of its surface is covered by a liquid-water ocean averaging 3,796 meters (12,451 feet) deep. This beautiful satellite image of the northeastern Atlantic from Iceland to central Spain shows storms over southern England, vapor trails left by airliners off the French coast, and a turquoise blue bloom of plankton in the North Sea.

NASA/GSFC

KNOWING THE OCEAN WORLD

Oceanography⬡Now™

This icon, appearing throughout the book, indicates an opportunity to explore interactive tutorials, animations, or practice problems available on the OceanographyNow Web site at http://oceanography.brookscole.com/garrison5e

Science and the Story of the Ocean

Think of oceanography as the story of the ocean. In this chapter the main character—the world ocean—is introduced in broad brush strokes. We begin our investigation of the ocean with an overview of the process of science and the history of marine science.

Like other sciences, marine science is based in curiosity. In particular, the question How do we know? is vital to an understanding of the physical world. We arrive at tentative explanations for the features and processes of things we can see, feel, touch, and hear by using the scientific method, a systematic way of asking and answering questions about the natural world. As you read this chapter—and the rest of the book— keep in mind the scientific logic that underpins the objects and ideas you're learning about. It's *always* good to ask questions!

The process of asking and answering questions about the ocean has occupied us since humans first encountered a large body of salty water. In the beginning marine studies had only practical aims: travel, trade, warfare, and finding food more easily. Later the search for new knowledge became a goal in itself. The first part of the historical overview that follows focuses on the mastery of marine knowledge for voyaging; it describes some of the voyagers and their journeys, the inventions that made their adventures possible, and some of the discoveries they made. The second part covers voyaging for science, the rise of oceanographic institutions, and the present thrust of ocean science.

As you read you'll notice a subtle progression—from the involvement of whole human populations, to the contributions of extraordinary individuals, to epic scientific expeditions, to the rise of national and academic centers of learning. Oceanographers now spend more time ashore than at sea; some work exclusively with data sent by space-based sensors or robot sea probes.

Scientific oceanography began with an unlikely adventure that had its origin over a few beers in an off-campus Scottish pub. At the University of Edinburgh a brilliant professor and an enthusiastic student imagined how the (literally) unfathomable ocean might be explored. Their conversations resulted in the invention of the word *oceanography* and the launching, in 1872, of the *Challenger* expedition, the longest and most important voyage of oceanic exploration ever completed.

The oceanographers aboard HMS *Challenger*—and marine scientists today—would agree with Isaac Newton's observation that history is important: "If I have seen farther, it is because I have stood on the shoulders of giants."

For Internet sites related to topics in the text, go to the Web site and click on the numbered icon.

An Ocean World

Ours is not a particularly large planet, and it is not unusual in overall composition. Its sun and its position in the galaxy are unremarkable. What *is* extraordinary is the liquid-water ocean dominating its cloud-shrouded surface. This brilliant blue ocean affects and moderates temperature and dramatically influences weather. Its creatures directly provide about 2% of humanity's food. From beneath its floor is pumped about one-third of the world's petroleum and natural gas. The ocean borders most of Earth's largest cities; nearly half of the planet's 6 billion human inhabitants live within 150 kilometers (240 miles) of a coastline.[1] The ocean is a primary shipping and communication route and a major recreational resource. The dry land on which nearly all of human history has unfolded is hardly visible from space; nearly three-quarters of the planet is covered by water.

The **ocean**[2] may be defined as the vast body of saline water that occupies the depressions of Earth's surface. More than

---

[1] Throughout this book metric measurements precede American measurements. For a quick review of metric (SI) units and their abbreviations, please turn to Appendix I.

[2] When an important new term is introduced and defined, it is printed in **boldfaced type**. These terms are listed at the end of the chapter and defined in the Glossary.

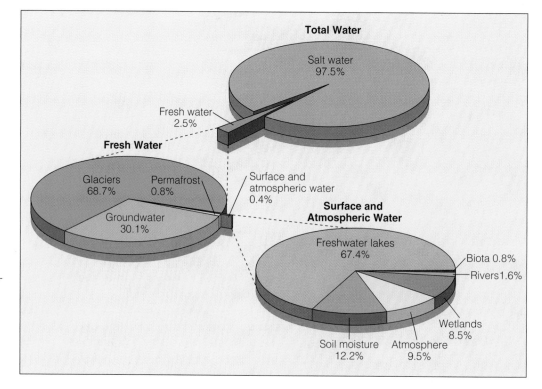

FIGURE 1.1 The relative amounts of water in various locations on or near Earth's surface. More than 97% of the water lies in the ocean. Of all water at Earth's surface, ice on land contains about 1.7%, groundwater 0.8%, rivers and lakes 0.007%, and the atmosphere 0.001%.

97% of the water on or near Earth's surface is contained in the ocean; less than 3% is held in land ice, groundwater, and all the freshwater lakes and rivers (**Figure 1.1**).

Traditionally we have divided the ocean into artificial compartments called *oceans* and *seas* using the boundaries of continents and imaginary lines such as the equator. In fact, there are few dependable natural divisions, only one great mass of water. The Pacific and Atlantic Oceans, the Mediterranean and Baltic Seas, so named for our convenience, are in reality only temporary features of a single **world ocean.** In this book I refer to the ocean *as a single entity*, with subtly different characteristics at different locations but with very few natural partitions. Such a view emphasizes the interdependence of ocean and land, life and water, atmospheric and oceanic circulation, and natural and man-made environments.

On a *human* scale the ocean is impressively large. It covers 361 million square kilometers (139 million square miles) of Earth's surface. The average depth of the ocean is about 3,796 meters (12,451 feet), the volume of seawater is 1.37 billion cubic kilometers (329 million cubic miles), the average temperature is a cool 3.9°C (39°F). The ocean's mass is a staggering 141 billion billion metric tons. If Earth's contours were leveled to a smooth ball, the ocean would cover it to a depth of 2,686 meters (8,810 feet). The average land elevation is only 840 meters (2,772 feet), but the average ocean depth is $4\frac{1}{2}$ times greater!

On a *planetary* scale, however, the ocean is insignificant. Its average depth is a tiny fraction of Earth's radius—the blue ink representing the ocean on an 8-inch paper globe is proportionally thicker. The ocean accounts for only slightly more than 0.02% of Earth's mass, or 0.13% of its volume. Much more water is trapped within Earth's hot interior than there is in its ocean and atmosphere. Some characteristics of the world ocean are summarized in **Figure 1.2.**

Marine Science, Oceanography, and the Nature of Science

Marine science (or **oceanography**) is the process of discovering unifying principles in data obtained from the ocean, its associated life-forms, and the bordering lands. It draws on several disciplines, integrating the fields of geology, physics, biology, chemistry, and engineering as they apply to the ocean and its surroundings. Nearly all marine scientists specialize in one

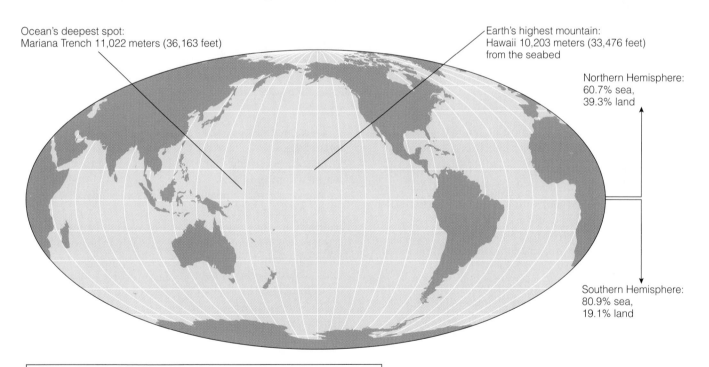

Ocean's deepest spot:
Mariana Trench 11,022 meters (36,163 feet)

Earth's highest mountain:
Hawaii 10,203 meters (33,476 feet)
from the seabed

Northern Hemisphere:
60.7% sea,
39.3% land

Southern Hemisphere:
80.9% sea,
19.1% land

Some Statistics for the World Ocean
Area: 361,100,000 square kilometers (139,400,000 square miles)
Mass: 141 billion billion metric tons (155 billion billion tons)
Volume: 1,370,000,000 cubic kilometers (329,000,000 cubic miles)
Average depth: 3,796 meters (12,451 feet)
Average temperature: 3.9°C (39.0°F)
Average salinity: 34,482 grams per kilogram (0.56 ounce per pound), 3.4%
Average land elevation: 840 meters (2,772 feet)
Age: About 4 billion years
Future: Uncertain

FIGURE 1.2 The proportion of sea versus land, shown on an equal-area projection of Earth. (An equal-area projection is a map drawn to represent areas in their correct relative proportions.) The average depth of the ocean is $4\frac{1}{2}$ times greater than the average land elevation. Note the extent of the Pacific Ocean, Earth's most prominent single feature.

FIGURE 1.3 Sunset on an ocean world. A liquid-water ocean moderates temperature and dramatically influences weather, nurtures life, and provides crucial natural resources.

area of research, but they also must be familiar with related specialties and appreciate the linkages between them.

- *Marine geologists* focus on questions such as the composition of the inner Earth, the mobility of the crust, the characteristics of seafloor sediments, and the history of Earth's ocean, continents, and climate. Some of their work touches on areas of intense scientific and public concern, including earthquake prediction and the distribution of valuable resources.

- *Physical oceanographers* study and observe wave dynamics, currents, and ocean–atmosphere interaction. Their predictions of long-term climate trends are becoming increasingly important as pollutants change Earth's atmosphere.

- *Marine biologists* work with the nature and distribution of marine organisms, the impact of oceanic and atmospheric pollutants on the organisms, the isolation of disease-fighting drugs from marine species, and the yields of fisheries.

- *Chemical oceanographers* study the ocean's dissolved solids and gases, and their relationships to the geology and biology of the ocean as a whole.

- *Marine engineers* design and build oil platforms, ships, harbors, and other structures that enable us to use the ocean wisely.

Other marine specialists study the techniques of weather forecasting, ways to increase the safety of navigation, methods to generate electricity, and much more. **Figure 1.4** shows marine scientists in action.[3]

Marine scientists today are asking some critical questions about the origin of the ocean, the age of its basins, and the nature of the life-forms it has nurtured. We are fortunate to live at a time when scientific study may be able to answer some of those questions. **Science** is a systematic *process* of asking questions about the observable world by gathering and then studying information (data), but the information itself is not science. Science interprets raw information by constructing a general explanation with which the information is compatible.

Scientists start with a question—a desire to understand something they have observed or measured. They then form a tentative explanation for the observation or measurement. This explanation is called a working **hypothesis,** a speculation about the natural world that can be tested and verified or disproved by further observations and controlled experiments. (An **experiment** is a test that simplifies observation in nature or in the laboratory by manipulating or controlling the conditions under which observations are made.) A hypothesis that is

---

[3] Would you like to join us? Appendix IX discusses careers in oceanography.

b

FIGURE 1.4 Doing marine science is sometimes anxious, sometimes routine, and always interesting. (**a**) Deploying a data-gathering buoy from the slippery deck of a rolling ship takes courage and timing. (**b**) Routine maintenance and inspections are part of every marine scientist's day.

Scripps Institution of Oceanography

Tom Garrison

a

consistently supported by observation or experiment is advanced to the status of a **theory,** a statement that explains the observations. The largest constructs, known as **laws,** summarize experimental observations. Laws are principles explaining events in nature that have been observed to occur with unvarying uniformity under the same conditions. A law summarizes observations; a theory provides an explanation for the observations.

Theories and laws in science do not arise fully formed or all at once. Scientific thought progresses as a continuing chain of questioning, testing, and matching theories to observations. A theory is strengthened if new facts support it. If not, the theory is modified or a new explanation is sought. The power of science lies in the ability of the process to operate *in reverse*—that is, in the use of a theory or law to make predictions and anticipate new facts to be observed.

This procedure, often called the **scientific method,** is an orderly process by which theories are verified or rejected. It is based on the assumption that nature "plays fair"—that the rules governing natural phenomena do not change capriciously as our powers of questioning and observing improve. We believe that the answers to our questions about nature are *ultimately knowable.*

There is no one "scientific method." Some researchers observe, describe, and report on some subject and leave it to others to hypothesize. Scientists don't have one single method in common; the general method they employ is a critical attitude about being *shown* rather than being *told* and taking a logical approach to problem solving. **Figure 1.5** summarizes the main points of the scientific method. Note that the process is circular—new theories and laws always suggest new questions.

Nothing is ever proven absolutely true by the scientific method. Theories may change as our knowledge and powers of observation change; thus all scientific understanding is tentative. Science is neither a democratic process nor a popularity contest. The conclusions about the natural world that we reach by the process of science may not always be comfortable, easily understood, or immediately embraced, but if those conclusions consistently match observations, they may be considered true.

This book describes some of the results of the scientific process as it has been applied to the world ocean. It presents facts, interpretations of facts, examples, stories, and some of the crucial discoveries that have led to our present understanding of the ocean and the world on which it formed. As the

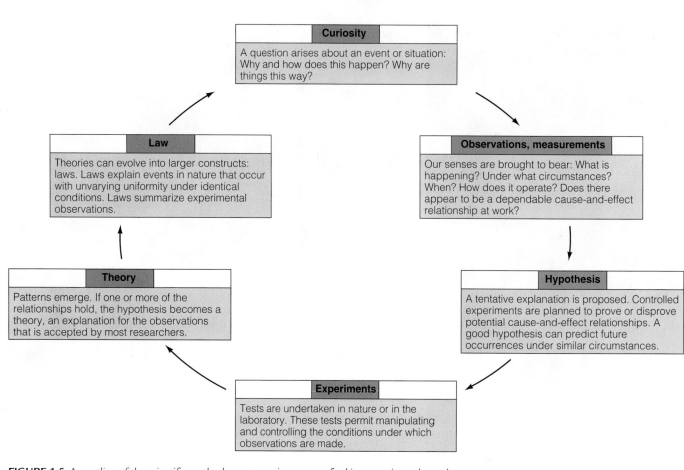

FIGURE 1.5 An outline of the scientific method, a systematic process of asking questions about the observable world and then testing the answers to those questions. There is not a single "scientific method." Science rests on a critical attitude about being shown rather than being told and then a logical approach to problem solving. Application of the scientific method leads us to truth based on the observations and measurements that have been made—a work in progress, never completed. Indeed, the formulation of new theories and laws always leads to more questions. The external world, not internal conviction, must be the testing ground for scientific beliefs.

results of science change, so will the ideas and interpretations presented in books like this one.

Early Voyaging and Discovery

The gradual development of the processes of scientific inquiry and the progress of human history are wonderfully intertwined. It has taken a long time for humans to appreciate the nature of the world, but we're a restless and inquisitive lot, and despite the ocean's great size, we have populated nearly every inhabitable place. This fact was aptly illustrated when European explorers set out to "discover" the world only to be met by native peoples at almost every landfall! Clearly the ocean did not prevent the spread of humanity. The early history of marine science is closely associated with the history of voyaging.

Ocean transportation offers people the benefits of mobility and greater access to food supplies. Any coastal culture skilled at raft building or small boat navigation would have economic and nutritional advantages over less skilled competitors. From the earliest period of human history, then, understanding and appreciating the ocean and its life-forms benefited those patient enough to learn.

The first direct evidence we have of **voyaging**, traveling on the ocean for a specific purpose, comes from records of trade in the Mediterranean Sea. The Egyptians organized shipborne commerce on the Nile River, but the first regular ocean traders were probably the Cretans, or the Phoenicians who inherited maritime supremacy in the Mediterranean after the Cretan civilizations were destroyed by earthquakes and political instability around 1200 B.C. Skilled sailors, the Phoenicians carried their wares through the Strait of Gibraltar to markets as distant as Britain and the west coast of Africa. Given the simple ships they used, this was quite an achievement.

The Greeks began to explore outside the Mediterranean into the Atlantic Ocean around 900–700 B.C. (**Figure 1.6**). Early Greek seafarers noticed a current running from north to south beyond Gibraltar. Believing that only rivers had currents, they decided that this great mass of water, too wide to see across, was part of an immense flowing river. The Greek name

FIGURE 1.6 A Greek ship from about 500 B.C. Such ships were used for trade and to continue the exploration of the Atlantic outside the Mediterranean.

for this river was *okeanos*. Our word *ocean* is derived from *oceanus*, a Latin variant of that root. Phoenician sailors were also very much at home in this "river," but like the Greeks they rarely ventured out of sight of land.

As they went about their business, early mariners began to record information to make their voyages easier and safer—the location of rocks in a harbor, landmarks and the sailing times between them, the direction of currents. These first **cartographers** (chart makers) were probably Mediterranean traders who made routine journeys from producing areas to markets. Their first charts (from about 800 B.C.) were drawn to jog their memory for obvious features along the route. Today's **charts** are graphic representations that primarily depict water and water-related information. (*Maps* primarily represent land.) For more on maps and charts, please see Appendix IV.

In this early time other cultures also traveled on the ocean. The Chinese began to engineer an extensive system of inland waterways, some of which connected with the Pacific Ocean to make long-distance transport of goods more convenient. The Polynesian peoples had been moving easily among islands off the coasts of Southeast Asia and Indonesia since 3000 B.C. and were beginning to settle the mid-Pacific islands. Though none of these civilizations had contact with the others, each developed methods of charting and navigation. All these early travelers were skilled at telling direction by the stars and by the position of the rising or setting sun.

Curiosity and commerce encouraged adventurous people to undertake ever more ambitious voyages. But these voyages were possible only with the coordination of astronomical direction finding (and knowledge of the shape and size of Earth), advanced shipbuilding technology, accurate graphic charts (not just written descriptions), and, perhaps most important, a growing understanding of the ocean itself. Marine science, the organized study of the ocean, began with the technical studies of voyagers.

Science for Voyaging

Progress in applied marine science began at the **Library of Alexandria** in Egypt. Founded in the third century B.C. by Alexander the Great, the library constituted history's greatest accumulation of ancient writings. The library and the adjacent museum could be considered the first university in the world. Scholars worked and researched there, and students came from around the Mediterranean to study. Written knowledge of all kinds—characteristics of nations, trade, natural wonders, artistic achievements, tourist sights, investment opportunities, and other items of interest to seafarers—was warehoused around its leafy courtyards. When any ship entered the harbor, the books (actually scrolls) it contained were by law removed and copied; the *copies* were returned to the owner and the originals kept for the library. Caravans arriving overland were also searched. Manuscripts that described the Mediterranean coast were of great interest. Traders quickly realized the competitive benefit of this information.

Yet marine science was only one of the library's many research areas. For 600 years it was the greatest repository of wisdom of all kinds and the most influential institution of higher learning in the ancient world. Here, perhaps, was the first instance of cooperation between a university and the commercial community, a partnership that has paid dividends for both science and business ever since.

Alexandrian Contributions

The second librarian at Alexandria (from 235 B.C. until 192 B.C.) was the Greek astronomer, philosopher, and poet **Eratosthenes of Cyrene.** This remarkable man was the first to calculate the circumference of Earth. The Greek Pythagoreans had realized Earth was spherical by the sixth century B.C., but Eratosthenes was the first to estimate its true size.

Eratosthenes had heard from travelers returning from Syene (now Aswan, site of the great Nile dam) that at noon on the longest day of the year the sun shone directly onto the waters of a deep vertical well. In Alexandria he noticed that a vertical pole cast a slight shadow on that day. He measured the shadow angle and found it to be a bit more than $7°$, about $\frac{1}{50}$ of a circle. He correctly assumed that the sun is a great distance from Earth, which means that the sun's rays would approach Syene and Alexandria in essentially parallel lines. If the sun were directly overhead at Syene but not directly overhead at Alexandria, then Earth's surface would have to be curved. But what was the planet's *circumference*?

By studying the reports of camel caravan traders, he estimated the distance from Alexandria to Syene at about 785 kilometers (491 miles). Eratosthenes now had the two pieces of information needed to derive the circumference of Earth by geometry. **Figure 1.7** shows his method. The precise size of the units of length (*stadia*) Eratosthenes used is thought to have been 555 meters (607 yards), and historians estimate that his calculation, made in about 230 B.C., was accurate to within about 8% of the true value. Within a few hundred years most people in the West who had contact with the library or its scholars knew Earth's approximate size.

Even without the contributions of Eratosthenes, the significance of the Library of Alexandria to marine science is immense. In Alexandria traders, explorers, scholars, and students had a place to conduct research and exchange information (and rumors) about the seas. Library researchers invented the astronomical, geometric, and mathematical base for **celestial navigation,** the technique of finding one's position on Earth by reference to the apparent positions of heavenly bodies.

Cartography flourished. The first workable charts that represented a spherical surface on a flat sheet were developed by Alexandrian scholars. Latitude and longitude, systems of imaginary lines dividing the surface of Earth, were invented by Eratosthenes. **Latitude** lines were drawn parallel to the equator, and **longitude** lines ran from pole to pole (**Box 1.1**). Eratosthenes placed the lines through prominent landmarks and important places to create a convenient, though irregular, grid. Our present regular grid of latitude and longitude was invented by Hipparchus (c. 165–127 B.C.), a librarian who divided the surface of Earth into 360°. A later Egyptian–Greek, Claudius Ptolemy (A.D. 90–168), *oriented* charts by placing east to the right and north at the top. Ptolemy's division of degrees into minutes and seconds of arc is still used by navigators.[4]

---

[4] For more information on latitude and longitude, please see Appendix III.

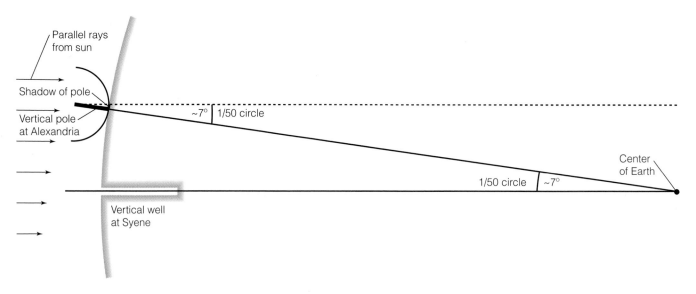

FIGURE 1.7 A diagram showing Eratosthenes's method for calculating the circumference of Earth. As described in the text, he used simple geometric reasoning based on the assumptions that Earth is spherical and that the sun is very far away. He was able to discover the circumference of Earth to within about 8% of its true value. This knowledge was available more than 1,700 years before Columbus began his voyages. (The diagram is not drawn to scale.)

Labels in figure: Parallel rays from sun; Shadow of pole; Vertical pole at Alexandria; Vertical well at Syene; ~7° 1/50 circle; 1/50 circle ~7°; Center of Earth

BOX 1.1
Latitude and Longitude

A sphere has no edges, no beginnings or ends, so what should we use as a frame of reference for positioning and navigation? The question was first successfully addressed by geographers at the Library at Alexandria in Egypt. In the third century B.C. Eratosthenes drew latitude and longitude lines through important places (**Figure a**). The Alexandrian perception of the world is reflected in the size of the continents and the central position of Alexandria.

A later Alexandrian scholar divided Earth into an orderly grid based on 360 increments, or "degrees" (*degre* = step). The equator was a natural dividing point for the north–south (latitude) positioning grid, but there was no natural dividing point for the east–west (longitude) grid. Not surprisingly, Alexandria was arbitrarily selected as the first "zero longitude" and a regular grid laid out east and west of that city.

The general scheme has withstood the test of time, but there has been controversy. Though use of the equator as "zero latitude" has never been in question, each seafaring country wanted the prestige of having the world's longitude centered on its capital. For centuries maritime nations issued charts with their own longitude "zeros." After much political disagreement, nations agreed in 1884 that the Greenwich meridian near London would be the world's "zero longitude" (**Figures b–d**). Given the accuracy of that meridian's known position and the long history and success of British navigation and timekeeping, Greenwich was an excellence choice.

To read more on latitude, longitude, time, and navigation, please turn to Appendix III.

a

(**a**) The world, according to a chart from the third century B.C. attributed to Eratosthenes. He drew latitude and longitude lines through important places rather than spacing them at regular intervals as we do today.

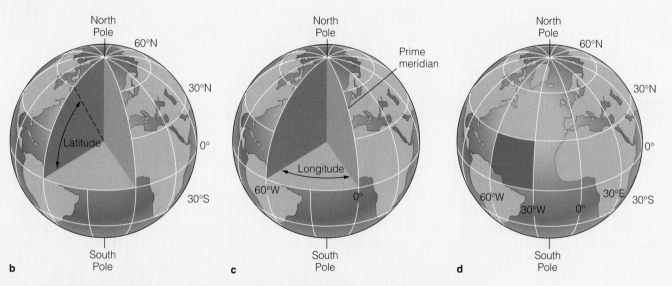

b **c** **d**

Latitude and longitude are measured as angles between lines drawn from the center of Earth to the surface. (**b**) Latitude is measured as the angle between a line from Earth's center to the equator and a line from Earth's center to the measurement point. (**c**) Longitude is measured as the angle between a line from Earth's center to the measurement point and a line from Earth's center to the prime (or Greenwich) meridian, which is a line drawn from the North Pole to the South Pole passing through Greenwich, England. (**d**) Lines of latitude are always the same distance apart, while the distance between two lines of longitude varies with latitude.

FIGURE 1.8 The modern library of Alexandria. Sponsors of the new *Bibliotheca Alexandrina,* opened in the spring of 2002, hope it will become "a lighthouse of knowledge to the whole world." The goal of this conference center and storehouse is not to restore the past but to revive the ancient library's questing spirit.

Ptolemy also introduced an "improvement" to Eratosthenes's surprisingly accurate estimate of Earth's circumference. Unfortunately Ptolemy wrongly depended on flawed calculations of the effects of atmospheric refraction. He publicized an estimate of the size of Earth that was too small—about 70% of the true value. This error, coupled with his mistake of overestimating the size of Asia, greatly reduced the apparent width of the unknown part of the world between the Orient and Europe. More than 1,500 years later these mistakes made it possible for Columbus to convince people he could reach Asia by sailing west.

Though it weathered the dissolution of Alexander's empire, the Library of Alexandria did not survive the subsequent period of Roman rule. The last librarian was Hypatia, the first notable woman mathematician, philosopher, and scientist. In Alexandria she was a symbol of science and knowledge, concepts the early Christians identified with pagan practices. The mission of the library, as personified by the last librarian, antagonized the governors and citizens of Alexandria. After years of rising tensions, in A.D. 415 a mob brutally murdered Hypatia and burned the library with all its contents. Most of the community of scholars dispersed, and Alexandria ceased to be a center of learning in the ancient world. The academic loss was incalculable, and trade suffered because shipowners no longer had a clearinghouse for updating the nautical charts and information they had come to depend on. All that remains of the library today is a remnant of an underground storage room. We will never know the true extent and influence of its collection of more than 700,000 irreplaceable scrolls.

Western intellectual development slackened during the so-called Dark Ages that followed the fall of the Roman Empire in A.D. 476. For almost 1,000 years, until the European Renaissance, much of the progress in medicine, astronomy, philoso-phy, mathematics, and other vital fields of human endeavor was made by the Arabs or imported by them from Asia. For example, the Arabs used the Chinese-invented compass for navigating caravans over seas of sand, and their understanding of the Indian Ocean's periodic winds—the monsoons—allowed an Arabian navigator to guide Vasco da Gama from East Africa to India in 1498. Earlier, at the height of the Dark Ages, Vikings raided and explored to the south and west. And half a world away the Polynesians continued some of the most extraordinary voyages in history.

Voyages of the Oceanian Peoples

In the history of human migration, no voyaging saga is more inspiring than that of the **Polynesian** colonizations, the peopling of the central and eastern Pacific islands. A profound knowledge of the sea was required for these voyages, and the story of the Polynesians is a high point in our chronology of marine science applied to travel by sea.

The Polynesians are one of four cultures that inhabited some 10,000 islands scattered across nearly 26 million square kilometers (10 million square miles) of open Pacific Ocean (**Figure 1.9**). The Southeast Asian or Indonesian ancestors of the Oceanian peoples, as these cultures are collectively called, spread eastward in the distant past. Although experts differ in their estimates, there is some consensus that by 30,000 years ago New Guinea was populated by these wanderers and that by 20,000 years ago the Philippines were occupied. By between 900 and 800 B.C. the so-called cradle of Polynesia—Tonga, Samoa, the Marquesas, and the Society Islands—was settled. Oceanian navigators may already have been using shells attached to a bamboo grid to represent the positions of their islands. (A Micronesian stick chart from recent times is shown in **Figure 1.10**.)

FIGURE 1.9 The Polynesian Triangle. Ancestors of the Polynesians had spread from Southeast Asia or Indonesia to New Guinea and the Philippines by about 20,000 years ago. The mid-Pacific islands have been colonized for about 2,500 years, but the explosive dispersion that led to the settlement of Hawaii occurred about A.D. 450–600. Arrows show a possible direction and order of settlement.

For a long and evidently prosperous period the Polynesians spread from island to island until the easily accessible islands had been colonized. Eventually, however, overpopulation and depletion of resources became a problem. Politics, intertribal tensions, and religious strife shook society. Groups of people scattered in all directions from some of the "cradle" islands during a period of explosive dispersion. Between A.D. 300 and 600 Polynesians successfully colonized nearly every inhabitable island within the vast triangular area shown in Figure 1.9. Easter Island was found against prevailing winds and currents,

and the remote islands of Hawaii were discovered and occupied. These were among the last places on Earth to be populated.

How did these risky voyages into unexplored territory come about? Religious warfare may have been the strongest stimulus to colonization. If the losers of a religious war were banished from the home islands under penalty of death, their only hope for survival was to reach a distant and hospitable new land.

Seafaring had been a long tradition in the home islands, but long trips called for radical new technology. Great dual-hulled

Tom Garrison

FIGURE 1.10 A modern Micronesian stick chart. Knots or shells tied at the junctions between bamboo sticks represent islands. Straight strips represent patterns of regular waves; bent strips depict waves curving around islands.

sailing ships, some capable of transporting as many as 100 people, were designed and built. New navigation techniques were perfected that depended on the positions of stars barely visible to the north. New ways of storing food, water, and seeds were devised. Whole populations left their home islands in fleets designed especially for long-distance discovery. In some cases fire was nurtured on board in case of landfall on an island that lacked volcanic flame. But a new island was only a possibility, a dream. Their gods may have promised the voyagers safe deliverance to new lands, but how many fleets set out from the troubled homelands only to fall victim to storms, thirst, or other dangers?

Yet in that anxious time the Polynesians practiced and perfected their seafaring knowledge. To a skilled navigator, a change in the rhythmic set of waves against the hull could indicate an island out of sight over the horizon. The flight tracks of birds at dusk could suggest the direction of land. The positions of the stars told stories, as did the distant clouds over an unseen island. The smell of the water, or its temperature, or salinity, or color, conveyed information—as did the direction of the wind relative to the sun, and the type of marine life clustering near the boat. The sunrise colors, the sunset colors, the

hue of the moon—every nuance had meaning, every detail had been passed in ritual from father to son. The greatest Polynesian minds were navigators, and reaching Hawaii was their greatest achievement.

Of all the islands colonized by the Polynesians, Hawaii is farthest away, across an ocean whose guide stars were completely unknown to the southern navigators. The Hawaiian Islands are isolated in the northern Pacific. There are no islands of any significance for more than 3,200 kilometers (2,000 miles) to the south. Moreover Hawaii lies beyond the equatorial doldrums, a hot and often windless stretch across which these pioneers must somehow have paddled. And yet some fortunate and knowledgeable people colonized Hawaii sometime between A.D. 450 and 600. Try to imagine their feelings of relief and justification upon reaching a promised paradise under a new night sky. Think of that first approach to the high islands of Hawaii, the first unlimited drink of fresh water, the first solid Earth after months of uncertainty (**Figure 1.11**)!

Within a hundred years of their first arrival, Hawaiian navigators were routinely piloting vessels on regular return trips to the Marquesas and the Society Islands (Tahiti and others). Some of the trips were undertaken to import needed food species to the newly found islands, but others were made to recruit new citizens and leaders to "green-clad Hawaii."

At a time when seafarers of other civilizations sailed beside the comforting bulk of a charted coast, Polynesians looked to the open sea for sustenance, deliverance, and hope. Their great knowledge of the ocean protected them.

Meanwhile, Back in Europe . . .

The Dark Ages were periodically punctuated by the raids of **Vikings,** bands of Scandinavian adventurers and treasure seekers whose remarkably fast, strong, and stable ships enabled them to row up rivers faster than a horse and rider could spread warning (**Figure 1.12**). Danish and Norwegian Vikings swept down the coast of Europe; they methodically pillaged Paris, robbed monasteries in Ireland, and looted Britain. The Swedish Vikings foraged as far away as Kiev and Constantinople! In A.D. 859 Vikings spent a week or so ashore in Morocco, rounding up prisoners for sale as slaves or to hold for ransom. Sixty-two Viking ships participated, a spectacular display of technology, sea power, seamanship, and navigation.

At first the Europeans were powerless against these marauders, but eventually the need for common defense overcame provincial hostility and xenophobia. One of the causes of the Renaissance in Europe may have been the experience of banding together for protection against these northern raiders.

As the French, Irish, and British defenses became more effective, the Norwegian Vikings began to look west. Iceland and Greenland had been discovered by ships blown off course during storms. Iceland was colonized by about A.D. 700; Greenland by 995. In an early voyage from Norway to Greenland in 986, a commuter named Bjarni Herjulfsson was blown past his goal by unfavorable winds. For about five days he sailed up and down the coast of a new land (which was, in fact, North America) without landing or making charts. His sketchy reports

FIGURE 1.11 The discovery of Hawaii: "Looking anew at the clouds we saw a sight difficult to comprehend. What had appeared as an unusual cloud formation was now revealed as the peak of a gigantic mountain, a mountain of unbelievable size, a white mountain—a pillar that seemed to support the sky! We watched in wonder until nightfall. Then to the south of that mountain a dull red glow lighted the underside of the lifting clouds, revealing the shape of another mountain. It brightened as the night darkened. That mountain seemed to be burning! No one slept that night. Our two ships thrashed along in the night wind, and the dreadful red beacon lighted our way." (By kind permission of the author and artist, Herbert Kauainui Kane.)

Bjørn Landstrøm

FIGURE 1.12 A Viking ship from around A.D. 900. This painting is a reconstruction drawn from a ship found in 1880 at the bottom of a Norwegian fjord. Sturdy and intended for travel over open water, the ship is 23.3 meters (76 feet) long and 5.25 meters (17 feet) wide. Oarsmen sat on loose benches or chests, and holes for the oars could be covered by small disks so that water would not flow in when the ship heeled under sail.

FIGURE 1.13 The treasure ship, largest in a vast Chinese fleet whose purpose was to show kindness to people of distant places. The fleet sailed the Pacific and Indian Oceans between 1405 and 1433.

kindled a real-estate fever; Leif, son of Eric the Red, purchased Bjarni's ship and returned. His party found salmon-filled lakes, vines and grapes, and fodder for cattle in what was probably the northeastern tip of Newfoundland. With a bit of advertising overstatement, he called the place *Vinland* ("wine-land").

By A.D. 1000 the Norwegians had colonized Vinland. The settlements were modest, and at first relationships with the natives were encouraging. Unlike the Spanish who landed in the New World 500 years later, the Norwegian colonists tried to cooperate with the locals, to learn from them, and to help them in a mutual pact of assistance. Unfortunately misunderstandings arose, battles ensued, and the weather turned colder. The colony had to be abandoned in 1020. The Norwegians lacked the numbers, the weapons, and the trading goods to make the colony a success.

Chinese Contributions

As the Dark Ages distracted Europeans, Chinese navigators became more skilled, and their vessels grew larger and more seaworthy. They then set out to explore the other side of the world. Between 1405 and 1433 Admiral Zheng He (pronounced *jung huh*) commanded the greatest fleet the world had ever known. At least 317 ships and 37,000 men undertook seven missions to explore the Indian Ocean, Indonesia, and around the tip of Africa into the Atlantic. Their aim: to display the wealth and power of the young Ming dynasty and to show

kindness to people of distant places. The largest ship in the fleet, with nine masts and a length of 134 meters (440 feet), was a huge treasure ship carrying objects of the finest materials and craftsmanship (**Figure 1.13**). The mission of the fleet was not to accumulate such treasure but to give it away! Indeed the primary purpose of these expeditions was to convince all nations with which the fleet had contact that China was the only truly civilized state and beyond any imaginable need for knowledge or assistance.

Many technical innovations had been required to make such an ambitious undertaking possible. In addition to the compass, the Chinese invented the central rudder, watertight compartments, and sophisticated sails on multiple masts, all of which were critically important for the successful operation of large sailing vessels. Until Europeans adopted the rudder around 1100, long-distance voyaging in a Western ship large enough to be stable in rough seas was usually difficult. Early Mediterranean traders and, later, the Polynesians and the Vikings had used specialized steering oars held against the right side (steer-board = starboard side) of their boats. Though this system worked well in protected waters, the small area of the steering oar (and the exposed position of the steersman) made it difficult to hold a course on long ocean passages. The centrally mounted submerged rudder solved that problem. Also, dividing the ship into separate compartments below the waterline meant that flooding due to hull damage could be confined to a relatively small area of the ship, and the vessel could then be repaired and saved from sinking. Because sails provided the power to move, advances in sail design could drastically influence the success of any voyage. The Chinese fitted their trapezoidal or triangular sails with battens (pieces of bamboo inserted into stitched seams running the width of the sail) and placed the sails on multiple masts. The sails resembled venetian blinds covered with cloth. It was not necessary for Chinese sailors to climb the masts to unfurl the sails every time the wind changed; everything could be done from the deck with windlasses and lines. The shape of the sails made it easier to sail close to the wind in confined seaways.

Despite enjoying these technical advances, the Chinese intentionally abandoned oceanic exploration in 1433. The political winds had changed, and the cost of the "reverse tribute" system was judged too great. In all, until late in the twentieth century, the Chinese made very few contributions to our understanding of the ocean. Still, their voyaging technology filtered into the West and made subsequent discoveries possible.

The Age of Discovery: From Prince Henry to Magellan

Jolted by internal awakening and external reality, Renaissance Europeans set out to explore the world by sea. They did not undertake exploration for its own sake, however; any voyage had to have a material goal. Trade between East and West had long been dependent on arduous and insecure desert caravan routes through the central Asian and Arabian deserts. This commerce was cut off in 1453 when the Turks captured Constantinople, and an alternative ocean route was needed.

A European visionary who thought ocean exploration held the key to great wealth and successful trade was **Prince Henry the Navigator,** third son of the royal family of Portugal (**Figure 1.14**). Prince Henry established a center at Sagres for the study of marine science and navigation ". . . through all the watery roads." Although he personally was not well traveled (he went to sea only twice in his life), captains under his patronage explored from 1451 to 1470, compiling detailed charts wherever they went. Henry's explorers pushed south into the

unknown and opened the west coast of Africa to commerce. He sent out small, maneuverable ships designed for voyages of discovery and manned by well-trained crews. For navigation his mariners used the **compass**—an instrument (invented in China in the fourth century B.C.) that points to magnetic north. Although Arab traders had brought the compass from China in the twelfth century, navigators still considered it a magical tool. They concealed the compass in a special box (predecessor of today's binnacle) and consulted it out of view of the crew. Henry's students knew Earth was round, but because of the errors publicized by Claudius Ptolemy, they were wrong in their estimation of its size.

A master mariner (and skilled salesman), **Christopher Columbus** "discovered" the New World quite by accident. Native Americans had been living on the continent for about 11,000 years, and the Norwegian Vikings had made about two dozen visits to a functioning colony on the continent 500 years before his noisy arrival; yet Columbus gets the credit. Why? Because his interesting souvenirs, exaggerated stories, inaccurate charts, and promises of vast wealth excited the imagination of royal courts. Columbus made North America a media event without ever sighting it!

Columbus wasn't trying to discover new lands. His intention was to pioneer a sea route to the rich and fabled lands of the East, made famous more than 200 years earlier in the overland travels of Marco Polo. As "Admiral of the Ocean Sea," Columbus was to have a financial interest in the trade routes he blazed. He was familiar with Prince Henry's work and, like all other competent contemporary navigators, knew Earth was spherical. By sailing west he believed he could come close to his eastern destination, the latitude of which he thought he knew. Because of wishful thinking and dependence on Ptolemy's data, however, Columbus made the *smallest* estimate of Earth's size by any navigator in modern history; he assumed Earth to be only about half its actual size!

Not surprisingly, Columbus mistook the New World for his goal of India or Japan. He thought that the notable absence of wealthy cities and well-dressed inhabitants resulted from striking the coast too far north or south of his desired latitude. He made three more trips to the New World but went to his grave believing that he had found islands off the coast of Asia. He never saw the mainland of North America and never realized the size and configuration of the continents whose future he had so profoundly changed.

Other explorers quickly followed, and Columbus's error was soon corrected. Charts drawn as early as 1507 included the New World (**Figure 1.15**). Such charts perhaps inspired **Ferdinand Magellan** (**Figure 1.16a**), a Portuguese navigator in the service of Spain, to believe that he could open a westerly trade route to the Orient. Unfortunately the chart makers estimated the Americas and the Pacific Ocean to be much smaller than they actually are. (Compare Figure 1.15 with Magellan's route, shown in **Figure 1.16b.**) In the Philippines Magellan was killed, and his men decided to continue sailing west around the world. Only 34 of the original crew of 260 survived, returning to Spain three years after they had set out. But they had proved it was possible to circumnavigate the globe.

FIGURE 1.14 Prince Henry of Portugal, the Navigator. In the mid-1400s Henry established a center at Sagres for the study of marine science and navigation ". . . through all the watery roads."

FIGURE 1.15 The Waldseemüller Map, published in 1507—the first map to name America and to show the New World as separate from Asia. The deep gores are designed to form a globe about 10 centimeters (4 inches) in diameter. This is an image of the only known copy to survive of the 1,000 printed from 12 wood blocks. It was purchased in 2003 by the U.S. Library of Congress for US$10 million.

The Magellan expedition's return to Spain in 1522 marks the end of the Age of European Discovery. An unpleasant era of exploitation of the human and natural resources of the Americas followed. Native empires were destroyed, and objects of priceless cultural value were melted into coin to fund European warfare and greed.

Voyaging for Science

British sea power arose after the Age of Discovery to compete with the colonial aspirations of France and Spain. Sailing ships require dependable supply and repair stations, especially in remote areas. The great powers sent out expeditions to claim appropriate locations, preferably inhabited by friendly natives eager to help provision ships half a globe from home. The French sent Admiral de Bougainville into the South Pacific in the mid-1760s. His 1768 claim for France of what is now called French Polynesia opened the area to the powerful European nations. The British followed immediately.

James Cook

Scientific oceanography begins with the departure from Plymouth Harbor in 1768 of HMS *Endeavour* under the capable command of **James Cook** of the British Royal Navy (**Figure 1.17**). An intelligent and patient leader, Cook was also a skillful navigator, cartographer, writer, artist, diplomat, sailor, scientist, and dietitian. The primary reason for the voyage was to assert the British presence in the South Seas, but the expedi-

tion had numerous scientific goals as well. First Cook conveyed several members of the Royal Society (a scientific research group) to Tahiti to observe the transit of Venus across the disk of the sun. Their measurements verified calculations of planetary orbits made earlier by Edmund Halley (later of comet fame) and others. Then Cook turned south into unknown territory to search for a hypothetical southern continent, which some philosophers believed had to exist to balance the landmass of the Northern Hemisphere. Cook and his men found and charted New Zealand, mapped Australia's Great Barrier Reef, marked the positions of numerous small islands, made notes on the natural history and human habitation of these distant places, and initiated friendly relations with many chiefs. Cook survived an epidemic of dysentery contracted by the ship's company while ashore in Batavia (Djakarta) and sailed home to England, completing an around-the-world voyage in 1771. Because of his insistence on cleanliness and ventilation, and because his provisions included cress, sauerkraut, and citrus extracts, his sailors avoided scurvy—a vitamin-C deficiency disease that for centuries had decimated crews on long voyages.

The Admiralty was deeply impressed. Cook was promoted to the rank of commander and in 1772 was given command of the ships *Resolution* and *Adventure*, in which he embarked on one of the great voyages in scientific history.[5] On this second voyage Cook charted Tonga and Easter Islands and discovered

---

[5] Sailing master in *Resolution* was the 21-year-old William Bligh, later the object of the famous mutiny aboard HMS *Bounty*.

New Caledonia in the Pacific and South Georgia in the Atlantic. He was first to circumnavigate the world at high latitudes. Though he sailed to 71°S latitude, he never sighted Antarctica. He returned home again in 1775.

Posted to the rank of captain, Cook set off in 1776 on his third, and last, expedition in *Resolution* and *Discovery*. His commission was to find a northwest passage around Canada and Alaska, or a northeast passage above Siberia. He "discovered" the Hawaiian Islands (Hawaiians were there to greet him, of course, as shown in **Figure 1.18**) and charted the west coast of North America. After searching unsuccessfully for a passage across the top of the world, Cook retraced his route to Hawaii to provision his ships for departure home. On 14 February 1779, after an elaborate farewell dinner with the chief of the island of Hawaii, Cook and his officers prepared to return to *Resolution*, anchored in Kealakekua Bay. The Englishmen somehow angered the Hawaiians and were beset by the crowd. Cook, among others, was killed in the fracas.

Cook deserves to be considered a scientist as well as an explorer because of the accuracy, thoroughness, and completeness in his descriptions. He and the scientists aboard took samples of marine life, land plants and animals, the ocean floor, and geological formations; they also reported the characteristics of these samples in their logbooks and journals. Cook's navigation was outstanding, and his charts of the Pacific were accurate enough to be used by the Allies in World War II invasions of the Pacific islands. He drew accurate conclusions, did not exaggerate his findings, and opened friendly diplomatic relations with many native populations. Cook recorded and successfully interpreted events in natural history, anthropology, and oceanography. Unlike many captains of his day, he cared for his men. He was a thoughtful and clear writer. This

a

FIGURE 1.16 (**a**) Ferdinand Magellan, a Portuguese navigator in service to Spain whose expedition was first to circumnavigate the world. (**b**) Track of the Magellan expedition, the first voyage around the world. Magellan himself did not survive the voyage; only 34 out of 260 sailors managed to return after three years of dangerous travel.

b

Magellan killed
27 April 1521

Voyage began 20 September 1519
Voyage ended 6 September 1522

Rio de Janeiro
15 December 1519

Straits of Magellan
15 October 1520

FIGURE 1.17 Captain James Cook, Royal Navy, painted in 1776 by Nathaniel Dance, shortly before embarking on his third, and fatal, voyage. Cook is shown as a fully matured, self-confident captain who has twice circled the globe, penetrated into the Antarctic, and charted coastlines from Newfoundland to New Zealand.

first marine scientist peacefully changed the map of the world more than any explorer or scientist in history.

The Longitude Problem

How did Cook (or Columbus, or any ocean explorer) know where he was? Unless explorers could record position accurately on a chart, exploration was essentially useless. They could not find their way home, nor could they or anyone else find the way back to the lands they had discovered.

At night Columbus and his European predecessors used the stars to find latitude and, as a consequence, knew their position north or south of home. You can do this, too. In the Northern Hemisphere take a simple protractor and measure the angle between the horizon, your eye, and the north polar star. The protractor reads approximately in degrees of latitude. To find the Indies, for example, Columbus dropped south to a line of latitude and followed it west. But to pinpoint a location you need both latitude *and* the east–west position of longitude. (For more about latitude, longitude, and their use in marine navigation, see Appendix III.)

You can find longitude with a clock. First determine local noon by observing the path of the shadow of a vertical shaft—it is shortest at noon—and set your clock accordingly. After

FIGURE 1.18 First contact! Captain James Cook, commanding HMS *Resolution* off the Hawaiian island of Kaua´i, wrote in 1778: "It required but little address to get them to come along side, but we could not prevail upon any one to come on board; they exchanged a few fish they had in the canoes for anything we offered them, but valued nails, or iron above every other thing; the only weapons they had were a few stones in some of the canoes and these they threw overboard when they found they were not wanted."

traveling some distance to the west, you will notice that noon according to your *clock* no longer marks the time when the shadow of the *shaft* is shortest at your new location. If "clock" noon occurs 3 hours before "shaft" noon, you can do some simple math to see how far west of your starting point you have come. Earth turns toward the east, making one rotation of 360° in 24 hours, so its rotation rate is 15° per hour (360°/24 hours = 15°/hour). The 3-hour difference between "clock" noon and "shaft" noon puts you 45° west of your point of origin (3 × 15° = 45°). The more accurate the clock (and the measurement of the shaft's shadow), the more accurate your estimate of westward position.

The time method described here would work in theory, but in Columbus's time—and for many years afterward—no clocks were accurate enough to make this calculation practical after a few days at sea. Indeed they were governed by pendulums, which are useless in a rolling ship.

The key to the longitude problem was to invent a sturdy clock that ran at a constant rate under any circumstance, even the changeable conditions of a ship at sea. How such clocks were invented is a fascinating story.

In 1707 a British battle fleet commanded by Sir Clowdisley Shovell ran aground in the Scilly Isles because they had lost track of their longitude after many weeks at sea; four ships and 2,000 men were lost. This was the most serious in a long string of English maritime disasters due to inaccurate estimation of longitude. In 1714, after lengthy study, the British government offered a £20,000 prize (equivalent to more than US$12 million in modern currency) for a method of determining longitude to within $\frac{1}{2}°$ after 30 days at sea.

In 1728 **John Harrison,** a Yorkshire cabinetmaker, began working on a clock that would be accurate enough to determine longitude. His radical new timepiece, called a **chronometer,** was governed not by a pendulum but by a spring escapement. His first version (**Figure 1.19a**) was tested at sea in 1736, and Harrison was awarded £500 as encouragement to continue his efforts. Over the next 25 years he built three more clocks, culminating in 1760 in his Number Four (**Figure 1.19b**), perhaps the most famous timekeeper in the world.

A sea trial of Number Four was begun in HMS *Deptford* in 1761. Harrison, too old and infirm to accompany the chronometer, sent his son and collaborator to tend the instrument. *Deptford* crossed the Atlantic from England to Jamaica and made a near-perfect landfall. After taking the clock's known error rate into account—its "rate of going" was $2\frac{2}{3}$ seconds a day—the clock was found to be only 5 seconds slow. This would have meant an error in longitude of only 2.3 kilometers (1.4 miles), an astonishing achievement by then-current standards of long-distance navigation.

Technically Harrison had won the prize—Number Four had more than met the criteria set by the British Board of Longitude—but he was granted only part of the promised reward. Understandably the officials would not hand over the money until it had been determined that the clock's secrets could be applied to quantity production. Just as understandably Harrison did not wish his life's work compromised without compensation. He feared (correctly, as it turned out) that once the

FIGURE 1.19 The first chronometers. (**a**) Harrison's Number One timekeeper, a clock built to prove principles but unfit for use at sea. (**b**) The Number Four timekeeper, which won the £20,000 award offered by the British Board of Longitude. Both are functioning and are on display at the National Maritime Museum in Greenwich, England.

clockwork was examined by a competent watchmaker, his ideas would be copied. Finally, in 1769, a single copy was made; its success clinched Harrison's achievement. Captain Cook took a copy of Harrison's fourth chronometer on his last two voyages, but Harrison received the balance of the prize only in 1773 (when he was 80) and then only through the direct intervention of King George III.

All four of Harrison's chronometers are on view functioning in Britain's National Maritime Museum at Greenwich, in eastern London. Greenwich is an ideal site for the museum; in 1884 the Greenwich meridian, a longitude line at the naval observatory there, became "longitude zero" for the world (**Figure 1.20**). Not since Eratosthenes's selection of Alexandria as the first "longitude zero" had Western nations recognized a common base for positioning.

The Sampling Problem

Marine science advances by the analysis of samples. The chronometer permitted investigators to determine the precise location at which they collected samples of water, bottom sediments, and marine life; but first they had to overcome the difficulty of actually *obtaining* a sample. Sampling of floor sediments or bottom water is not an easy task in the deep ocean. The line used to suspend the sampling device snakes back and forth as currents strike it, and the weight of the line itself makes

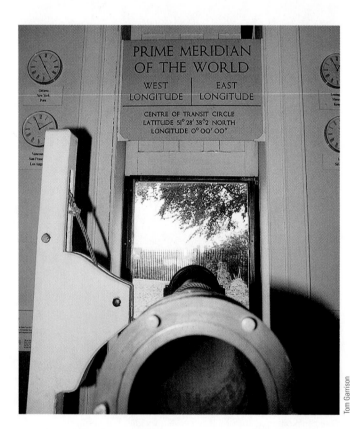

FIGURE 1.20 Looking north along the telescope in the prime meridian transit circle at the Royal Observatory in Greenwich, England. Zero longitude is defined as the line passing north and south through the center of the instrument.

it difficult to tell when the sampler has hit bottom. Deploying and recovering the line are laborious and time consuming, and sometimes the sampling device does not work properly. Early bottom-sampling devices (such as those used by Cook) were simple wax-covered lead weights lowered to shallow bottoms to pick up sediments and test the suitability of anchorages. Later devices took deep-water samples, extracted cores from the sediments, grabbed samples of the bottom, or scooped biological specimens from the ocean floor.

The first researchers to attack the deep-sampling problem successfully were British explorers Sir John Ross and his nephew Sir James Clark Ross. During an expedition to scout the Northwest Passage in 1818, Sir John Ross obtained a bottom sample from 1,919 meters (6,294 feet) near Greenland by using a clamping sampler to trap the specimen. Sir James Clark Ross, discoverer of the Ross Sea and the area of Antarctica known as Victoria Land, obtained depth **soundings** (depth measurements) of 4,433 meters (14,545 feet) and 4,893 meters (16,054 feet) in the South Atlantic.

Sampling techniques improved through the century. Using a sounding method perfected in the late 1840s by a U.S. Navy midshipman, American commodore Matthew Maury used a long lightweight line and lead weight to discover the Mid-Atlantic Ridge, an important hidden range of mountains. Fridtjof Nansen perfected the deep-water sampling bottle that bears his name near the end of the century. Even today, in spite of modern advances, deep sampling remains difficult. Remotely operated vehicles can work at great depths and return samples and pictures to the surface, but their electronic complexity makes them delicate and expensive to operate.

The First Scientific Expeditions

Great as Cook's contributions undoubtedly were, his three voyages (and those of the Rosses) were not purely scientific expeditions. These men were British naval officers engaged in Crown business, concerned with charting, foreign relations, and natural phenomena as they applied to Royal Navy matters. The first genuine *only-for-science* expedition may well have been the British *Challenger* expedition of 1872–76, but the United States got into the act first with a hybrid expedition in 1838.

The United States Exploring Expedition

After a ten-year argument over its potential merits, the **United States Exploring Expedition** was launched in 1838. It was primarily a naval expedition, but its captain was somewhat freer in maneuvering orders than Cook had been. The work of the scientists aboard the flagship USS *Vincennes* and the expedition's five other vessels helped establish the natural sciences as reputable professions in America. Had it not been for the combative and disagreeable personality of its leader, Lieutenant Charles Wilkes (**Figure 1.21**), this expedition might have

FIGURE 1.21 Lt. Charles Wilkes soon after his return from the United States Exploring Expedition in 1842. Wilkes commanded the largest number of ships sent on an exploring expedition since Chinese Admiral Zheng He's in 1431.

United States Naval Academy Museum

become as famous as those of Cook or the later *Challenger* voyage.

The expedition departed on a four-year circumnavigation. Its goals included showing the flag, whale scouting, mineral gathering, charting, observing, and pure exploration. One unusual goal was to disprove a peculiar theory that Earth was hollow and could be entered through huge holes at either pole.

Wilkes's team explored and charted a large sector of the east Antarctic coast and made observations that confirmed the landmass as a continent. A map of the Oregon Territory produced in 1841, one of 241 maps and charts drawn by members of the expedition, proved especially valuable when connected to the map of the Rocky Mountains prepared the following year by Captain John C. Fremont. Hawaii was thoroughly explored, and Wilkes led an ascent of Mauna Loa, one of the two highest peaks of Hawaii's largest island. James Dwight Dana, the expedition's brilliant geologist, confirmed Charles Darwin's hypothesis of coral atoll formation (about which more will be found in Chapter 12). The expedition returned with many scientific specimens and artifacts, which formed the nucleus of the collection of the newly established Smithsonian Institution in Washington, D.C. No evidence of polar holes was found!

Upon their return in 1842, Wilkes and his "scientifics" prepared a final report totaling 19 volumes of maps, text, and illus-

trations. The report is a landmark in the history of American scientific achievement.

The Work of Matthew Maury

At about the time the Wilkes expedition returned, **Matthew Maury (Figure 1.22)**, a Virginian and fellow U.S. naval officer, became interested in exploiting winds and currents for commercial and naval purposes. After being crippled in a stagecoach accident, in 1842 Maury was given charge of the Navy's Depot of Charts and Instruments. There he studied a huge and neglected treasure trove of ships' logs, with their many regular readings of temperature and wind direction. By 1847 Maury had assembled much of this information into coherent wind and current charts. Maury began to issue these charts free to mariners in exchange for logs of their own new voyages.

Slowly a picture of planetary winds and currents began to emerge. Maury himself was a compiler, not a scientist, and he was vitally interested in promoting maritime commerce. His understanding of currents built on the work of **Benjamin Franklin.** Nearly a hundred years earlier Franklin had noticed the peculiar fact that the fastest ships were not always the fastest ships; that is, hull speed did not always correlate with out-and-return time on the European run. Franklin's cousin, a Nantucket merchant named Tim Folger, noted Franklin's puzzlement and provided him with a rough chart of the

FIGURE 1.22 Matthew Fontaine Maury, perhaps the first person for whom oceanography was a full-time occupation. This photograph was probably taken in 1853.

U.S. Naval Observatory Library. Artist: Beverly Stautz

"Gulph Stream" that he (Folger) had worked out. By staying within the stream on the outbound leg and adding its speed to their own, and by avoiding it on their return, captains could traverse the Atlantic much more quickly. It was Franklin who published, in 1769, the first chart of any current (**Figure 1.23**).

But Maury was the first person to sense the worldwide pattern of surface winds and currents. Based on his analysis, he produced a set of directions for sailing great distances more efficiently. Maury's sailing directions quickly attracted worldwide notice: He had shortened the passage for vessels traveling from the American East Coast to Rio de Janeiro by 10 days, and to Australia by 20. His work became famous in 1849 during the California gold rush; his directions made it possible to save 30 days on the voyage around Cape Horn to California. Applicable U.S. charts still carry the inscription "Founded on the researches of M.F.M. while serving as a lieutenant in the U.S. Navy." His crowning achievement, *The Physical*

FIGURE 1.23 Benjamin Franklin's 1769 chart of the Gulf Stream system. His cousin, Tim Folger, discovered that Yankee whalers had learned to use the Gulf Stream to their advantage. Others, especially English shipowners, were slower to learn. Folger, himself a sea captain, wrote that Nantucket whalers ". . . in crossing it have sometimes met and spoke with those packets who were in the middle of and stemming it. We have informed them that they were stemming a current that was against them to the value of three miles an hour and advised them to cross it, but they were too wise to be counseled by simple American fishermen."

Geography of the Seas, a book explaining his discoveries, was published in 1855.

Maury, considered by many to be the father of physical oceanography, was perhaps the first man to undertake the systematic study of the ocean as a full-time occupation.

Charles Darwin and HMS *Beagle*

Though he was later to become famous for his explanation of biological evolution through natural selection, **Charles Darwin**'s first contributions to science were in the field of marine biology (**Figure 1.24**). Darwin became interested in natural history during his student days in Cambridge, where he was studying to become a minister of the Church of England. Because of the influence of John Henslow, a professor of botany, Darwin was selected in 1831 to accompany HMS *Beagle* on a voyage to South America and some Pacific islands as an unpaid naturalist. Nearly all of Darwin's subsequent work stemmed directly from observations he made on this $4\frac{1}{2}$-year voyage. In his first major work, *Structure and Distribution of Coral Reefs* (1842), Darwin correctly argued that atolls and reefs such as those he observed could have resulted only from subsidence of the seafloor, with corals growing upward as their bases dropped. His monographs on barnacle biology, volcanic islands, and fossils are overshadowed by his remarkable work *On the Origin of Species*, published in 1859. More on the contributions of Charles Darwin may be found in Chapter 13 on pages 322 and 325.

The *Challenger* Expedition

The first sailing expedition devoted completely to marine science was conceived by Charles Wyville Thomson, a professor of natural history at Scotland's University of Edinburgh, and his Canadian-born student, John Murray. Stimulated by their own curiosity and by the inspiration of Charles Darwin's voyage in HMS *Beagle*, they convinced the Royal Society and the

British government to provide a Royal Navy ship and trained crew for a prolonged and arduous voyage of exploration across the oceans of the world. Thomson and Murray even coined a word for their enterprise: **oceanography.** Though the term literally implies only marking or charting, it has come to mean the science of the ocean. Prime Minister Gladstone's administration and the Royal Society agreed to the endeavor provided that a proportion of any financial gain from discoveries was handed over to the Crown. This arranged, the scientists made their plans.

HMS *Challenger*, a 2,306-ton steam corvette (**Figure 1.25**), set sail on 21 December 1872 on a four-year voyage around the world, covering 127,600 kilometers (79,300 miles). Although the captain was a Royal Navy officer, the six-man scientific staff directed the course of the voyage. *Challenger*'s track is shown in **Figure 1.26**.

One important mission of the ***Challenger* expedition** was to investigate Edinburgh professor Edward Forbes's contention that life below 549 meters (1,800 feet) was impossible because of high pressure and lack of light. The steam winch on board

FIGURE 1.24 Charles Darwin late in life pictured between a marine iguana (left) and a land iguana (right), its presumed evolutionary antecedent. Darwin studied marine iguanas during his voyage aboard HMS *Beagle*.

The Royal Geographic Society, London

FIGURE 1.25 Lt. Pelham Aldrich, first lieutenant of HMS *Challenger*, kept a detailed journal of the *Challenger* expedition. With accuracy and humor he kept this record in good weather and bad, and he had the patience and skill to include watercolors of the most exciting events. This is part of the first page of his journal.

FIGURE 1.26 HMS *Challenger*'s track from December 1872 to May 1876. The *Challenger* expedition remains the longest continuous oceanographic survey on record.

Archive, Scripps Institute of Oceanography

FIGURE 1.27 Emptying a trawl net, an engraving from the *Challenger Report*.

made deep sampling practical, and samples from depths as great as 8,185 meters (26,850 feet) were collected off the Philippines. Through the course of 492 deep soundings with mechanical grabs and nets at 362 stations (including 133 dredgings), Forbes was proved resoundingly wrong. With each hoist, animals new to science were strewn on the deck; in all, staff biologists discovered 4,717 new species! **Figure 1.27** shows one of the large trawl nets used in making some of these discoveries.

The scientists also took salinity, temperature, and water-density measurements during these soundings. Each reading contributed to a growing picture of the physical structure of the deep ocean. The crew completed at least 151 open-water trawls and stored 77 samples of seawater for detailed analysis ashore. The expedition collected new information on ocean currents, meteorology, and the distribution of sediments. The locations and profiles of coral reefs were charted. Thousands of pounds of specimens were brought to British museums for study. Manganese nodules, brown lumps of mineral-rich sediments, were discovered on the seabed, sparking interest in deep-sea mining.

This first pure oceanographic investigation was an unqualified success. The discovery of life in the depths of the oceans stimulated the new science of marine biology. The scope, accuracy, thoroughness, and attractive presentation of the researchers' written reports made this expedition a high point in scientific publication. The *Challenger Report*, the record of the expedition, was published between 1880 and 1895 by Sir John Murray in a well-written and magnificently illustrated 50-volume set; it is still used today. Indeed it was the *Report*, rather than the voyage itself, that provided the foundation for the new science of oceanography. The expedition's many financial spin-offs indicated that pure research was a good investment, and the British government realized quick profits from the exploitation of newly discovered mineral deposits on islands. The *Challenger* expedition remains history's longest continuous scientific oceanographic expedition.

With successes like these, the pace of exploration accelerated. American naturalist Alexander Agassiz, sailing in 1877 in the U.S. Coast and Geodetic Survey ship *Blake*, collected data corroborating the *Challenger* material at 355 deep-sea stations. The distribution of manganese nodules was found to be widespread. Further work by Agassiz and his students around the turn of the twentieth century in the survey ship *Albatross* helped train a generation of influential American marine biologists. In 1886 the Russians entered the field of marine exploration with the three-year cruise of *Vitiaz* under the leadership of S. O. Makarov; their main contribution was a careful analysis of the salinity and temperature of North Pacific waters.

The Rise of Modern Sea Power

Marine science is also applied to military interests. **Sea power** is the means by which a nation extends its military capacity onto the ocean. History has been greatly influenced by sea power—for example, the defeat of the Persian fleet by the Greeks at Salamis in 480 B.C. and the triumph of British admiral Horatio Nelson over French forces at Trafalgar in 1805 led to eras of cultural and economic supremacy by both nations.

In 1892 **Alfred Thayer Mahan,** an American naval officer and historian, published *The Influence of Sea Power upon History, 1660–1783.* Based on his studies of the rise and fall of nation-states, this book had profound consequences for the development of the modern world. Mahan stressed the interdependence of military and commercial control of seaborne commerce, and the ability of safe lines of transportation and communication to influence the outcomes of conflicts. Coming at a time of unprecedented technological improvements in shipbuilding, Mahan's work was read avidly in Great Britain, Germany, and the United States. For better or worse, the naval hardware, strategy, and tactics of the last century's greatest wars—along with their outcomes—were influenced by his clear analysis.

Voyages for Science in the Twentieth Century

In the twentieth century oceanographic voyages became more technically ambitious and expensive. Scientist-explorers sought out and investigated places that once had been too difficult to reach. Though the deep-ocean floor was coming into reach, it was the forbidding polar ocean that attracted their first attentions.

Polar Exploration

Polar oceanography began with the pioneering efforts of Fridtjof Nansen (**Figure 1.28a**). Nansen courageously allowed his specially designed ship *Fram* to be trapped in the Arctic ice, where he and his crew of 13 drifted with the pack for nearly four years (1893–96), exploring to 85°57'N, a record for

the time. The 1,650-kilometer (1,025-mile) drift of *Fram* proved that no Arctic continent existed. Nansen's studies of the drift, of meteorological and oceanographic conditions, of life at high latitudes, and of deep sounding and sampling techniques form the underpinnings of modern polar science.

Living up to her name—*Fram* means "forward" in Norwegian—Nansen's ship continued to play a pivotal role in exploration (**Figure 1.28b**). In 1910 Roald Amundsen, a student of Nansen's, set out in the sturdy little vessel for the coast of Antarctica, the first leg of a journey to the South Pole. Nansen himself settled down to a long and distinguished career as an oceanographer, inventor, zoologist, artist, statesman, and professor. He was awarded the Nobel Peace Prize in 1922 for his unstinting work in worldwide humanitarian causes.

Scientific curiosity, national pride, new ideas in shipbuilding, advances in nutrition, and great personal courage led in the early years of the twentieth century to the golden age of polar exploration. After some heroic attempts by a number of explorers to reach the poles, an American naval officer, Robert E. Peary, accompanied by his African American assistant Matthew Henson and four Inuit (Eskimos), reached the vicinity of the North Pole in April 1909. A party of five men led by Roald Amundsen of Norway achieved the South Pole in December 1911.

Peary's and Amundsen's scientific contributions were limited to a few meteorological observations, but around the same time another Antarctic explorer, Captain Robert Falcon Scott of the British Royal Navy, led a scientific expedition to the South Pole. Chief scientist of the British expedition was Dr. Edward A. Wilson, a Cambridge-trained physician and naturalist with a special interest in ornithology. In 1911, before the assault on the pole, Wilson and three other scientists conducted the first extensive scientific investigations at high latitude. Laboring in a cramped expedition hut at Cape Evans, the researchers tended continuous-reading barometers and temperature gauges. They noted the direction of winds at the surface (and aloft by observing the smoke plume from Mt. Erebus, a nearby active volcano), measured atmospheric radioactivity, and recorded displays of auroras. They also made weekly observations of the strength and direction of Earth's magnetic field, towed for plankton, trapped fish beneath the ice, made soundings through cracks in the pack ice, analyzed water samples from various depths, collected samples of life from various freshwater lakes and pools ashore, explored nearby dry valleys and glaciers, and analyzed the ecological relationships among species. Ice received their full attention; using methods pioneered by Nansen, the team made the first detailed on-site studies of the formation, deposition, temperature, and thermal conductivity of sea ice.

Although the expedition was a scientific success, the polar journey itself was a tragic failure. Under Scott's leadership, members of the expedition arrived at the pole four weeks after Amundsen, but they died on the return trip to base camp.

Modern technology has eased the burden of high-latitude travel. In 1958, under the command of Captain William Anderson, the U.S. nuclear submarine *Nautilus* sailed beneath

FIGURE 1.28 (**a**) Fridtjof Nansen, pioneering Norwegian oceanographer and polar explorer, looking every inch the Viking. In 1908 Nansen became the first professor of oceanography, a post created for him at Christiania University. (**b**) Nansen's 123-foot schooner *Fram* ("forward"). With 13 men, *Fram* sailed on 22 June 1893 to the high Arctic with the specific purpose of being frozen into the ice. *Fram* was designed to slip up and out of the frozen ocean, and drifted with the pack ice to within about 4° of the North Pole. The whole harrowing adventure took nearly four years. The ship's 1,650-kilometer (1,025-mile) drift proved no Arctic continent existed beneath the ice. Living conditions aboard can be sensed from this recently discovered photograph. (**c**) A very fast, immensely strong, silent nuclear submarine makes an ideal platform for oceanographic research. Scientists aboard USS *Hawkbill* prepare to take water and ice samples at the North Pole in the summer of 1998. The 1998 and 1999 expeditions aboard USS *Hawkbill* made use of a sub-bottom profiler to provide the first images of the shallow strata of the Arctic Ocean floor.

the North Pole during a submerged transit beneath the Arctic pack from Point Barrow, Alaska, to the Norwegian Sea.

At the request of oceanographic researchers, the U.S. Navy initiated a series of research cruises in the northern polar ocean. Called SCICEX (for Scientific Ice Exercise) and funded by governmental agencies, the cruises began in 1993 and continued through 2000. **Figure 1.28c** shows USS *Hawk-* *bill*, a fast and powerful *Sturgeon*-class submarine hardened for surfacing through ice, during operations in the Arctic. The 1998 and 1999 expeditions aboard USS *Hawkbill* made use of a sub-bottom profiler to provide the first images of the shallow strata of the Arctic Ocean floor. The Navy hosted annual 45- to 60-day research cruises aboard *Sturgeon*-class submarines through 2000.

BOX 1.2

Marine Archaeology

Much can be gained from the study of underwater artifacts. Objects protected beneath soft marine sediment can age more gracefully than objects exposed to eroding water, winds, or frost. As you would expect, study of well-preserved objects from the seabed can provide insight into ship construction or the layout of a town, but there is always more to learn—information about a culture and people in the form of personal objects, fragments of clothing, remnants of cargo, even scraps of written material or paintings. The resting place of the RMS *Titanic* was discovered in 1985, and artifacts brought ashore for study have stimulated our imagination (and provided the story line for one of the most profitable movies ever made).

With luck, underwater archaeology can also solve mysteries. The loss of the CSS *Hunley* in February 1864 has bewildered generations of sailors and historians. The first submarine to sink a ship in battle, *Hunley* was also the first application of stealth technology to naval warfare. Designed to destroy Union ships blockading Charleston harbor during the American Civil War, the unstable 12-meter- (39-foot-) long sub was powered by eight sailors turning a crank attached to the propeller. The crew had enough air for 150 minutes—long enough to attach a bomb to the hull of the USS *Housatonic* and send it to the bottom. Their mission accomplished, the captain turned *Hunley* toward shore. They flashed the "success" signal as they returned—one blue light from a lantern. Then they disappeared.

No one knew *Hunley*'s location until 1995 when a research team located the hull using side-scan solar. In 1999 *Hunley* was raised on two massive trusses fitted with nylon slings and inflated foam pillows (**Figure a**). The trusses and sub were loaded aboard a barge and taken to the Charleston Naval Base for a restoration process expected to last about ten years. The privately funded *Hunley* Foundation expects the cost of recovery and conservation to exceed US$10 million.

When they began their work, the archaeologists found that currents had filled the hull with fine sediment. By December 2001 they had removed the bodies of the captain and eight crewmen and recovered a number of artifacts (**Figure b**), including the captain's logbook. The technical sophistication of the submarine has surprised researchers; the steering mechanism and hull shape incorporated design elements that were not seen again until the German U-boats of World War I. *Hunley* will make its final resting place in a museum in South Carolina.

1.24

a

b

(**a**) The confederate submarine *Hunley* is raised by crane on 8 August 2000 at the former Charleston Naval Base in North Charleston, South Carolina. *Hunley*, which sank with its crew of nine on 17 February 1864 after blowing up the Union blockade ship *Housatonic* with a charge of black powder, was lifted from its watery grave about 4 miles off Sullivans Island, secured on a barge, and transported to a conservation lab at the old base. (**b**) The chief archaeologist for the excavation discovers a gold piece in *Hunley*'s wreckage. It probably saved the captain's life by blocking a bullet in an earlier battle on land.

FIGURE 1.29 R/V *Chikyu* ("Earth") at its launch on 18 January 2002. Beginning in 2005 *Chikyu* will be the lead vessel in the International Ocean Drilling Program.

Other Twentieth-Century Voyages

In 1925 the German ***Meteor* expedition**, which crisscrossed the South Atlantic for two years, introduced modern optical and electronic equipment to oceanographic investigation. Its most important innovation was use of an **echo sounder**, a device that bounces sound waves off the ocean bottom, to study the depth and contour of the seafloor (see Figure 4.2). The echo sounder revealed to *Meteor* scientists a varied and often extremely rugged bottom profile rather than the flat floor they had anticipated.

Atlantis, launched in 1931, was the first U.S. research ship built specifically for ocean studies. Investigations by her scientists confirmed Matthew Maury's findings of a mid-Atlantic ridge and helped discover its extent. In 1937 the 32-meter (104-foot) schooner *E. W. Scripps*, under the direction of Harald Sverdrup, began a wide-ranging program of chemical, biological, and geophysical exploration off the coast of southern California. These voyages led to publication in 1942 of *The Oceans*, the first modern reference work on all phases of marine science.

In October 1951 a new HMS *Challenger* began a two-year voyage that would make precise depth measurements in the Atlantic, Pacific, and Indian Oceans and in the Mediterranean Sea. With echo sounders, measurements that would have taken the crew of the first *Challenger* nearly 4 hours to complete could be made in seconds. *Challenger II*'s scientists discovered the deepest part of the ocean's deepest trench, naming

it Challenger Deep in honor of their famous predecessor. In 1960 U.S. Navy Lieutenant Don Walsh and Jacques Piccard descended into the Challenger Deep in *Trieste*, a Swiss-designed, blimplike bathyscaphe. (That dramatic story may be read in Box 4.1.)

But in many ways the last voyage to be discussed here is the most portentous of all. In 1968 the drilling ship *Glomar Challenger* set out to test a controversial hypothesis about the history of the ocean floor. It was capable of drilling into the ocean bottom beneath more than 6,000 meters (20,000 feet) of water and recovering samples of seafloor sediments. These long and revealing plugs of seabed rock provided confirming evidence for seafloor spreading and plate tectonics. (The wonderful details are found in Chapter 3.) In 1985 deep-sea drilling duties were taken over by the much larger and more technologically advanced ship *JOIDES Resolution*.[6] Beginning in October 2003, deep-drilling responsibility will be passed to the Integrated Ocean Drilling Program (IODP), an international research consortium that will operate a successor to *JOIDES Resolution* and an even larger drillship, R/V *Chikyu* ("Earth") (**Figure 1.29**). The new Japanese ship, fully operational in 2007, will contain equipment capable of drilling cores as much as 11 kilometers (7 miles) long! The vessel will have

---

[6] *JOIDES* stands for Joint Oceanographic Institutions for Deep Earth Sampling.

equipment to control any flows of oil or gas, so it can safely drill deep into sedimentary basins on continental margins considered unsafe for *JOIDES Resolution*. This ship will house one of the most completely equipped geological laboratories ever put to sea.

The Rise of Oceanographic Institutions

The demands of scientific oceanography have become greater than the capability of any single voyage. Oceanographic institutions, agencies, and consortia evolved in part to ensure continuity of effort. The first of these coordinating bodies was founded by Prince Albert I of Monaco, who endowed his country's oceanographic laboratory and museum in 1906 (**Figure 1.30**). The most famous alumnus of Albert's Musée

Musée océanographique de Monaco, photo F.P.

FIGURE 1.30 Scientists and technicians at work in the first autonomous oceanographic institution. Founded by Prince Albert I of Monaco in 1908, the Institut Océanographique continues research and public education programs in Monaco and Paris.

Océanographique is Jacques Cousteau, co-inventor in 1943 of the scuba underwater breathing system. Monaco also became the site of the International Hydrographic Bureau, founded in 1921 as an association of maritime nations. This bureau published one of the first general charts of the ocean showing bottom contours.

A consortium of Japanese industries and governmental agencies established the Japan Marine Science and Technology Center (JAMSTEC) in 1971. In 1989 JAMSTEC launched *Shinkai 6500*, now the deepest-diving manned submersible. *Kaiko*, a remotely controlled robot sister that became fully operational in 1995, is the deepest-diving vehicle presently in service (see the Chapter 4 opener). JAMSTEC is a major contributor to the IODP program.

In the United States the three preeminent oceanographic institutions are the Woods Hole Oceanographic Institution on Cape Cod, founded in 1930 (and associated with the Massachusetts Institute of Technology and the neighboring Marine Biological Laboratory, founded in 1888); the Scripps Institution of Oceanography, founded in La Jolla, California, and affiliated with the University of California in 1912 (**Figure 1.31**), and the Lamont–Doherty Earth Observatory of Columbia University, founded in 1949.[7]

The U.S. government has been active in oceanographic research. Within the Department of the Navy are the Office of Naval Research, the Office of the Oceanographer of the Navy, the Naval Oceanic and Atmospheric Research Laboratory, and the Naval Ocean Systems Command. These agencies are responsible for oceanographic research related to national defense. The National Oceanic and Atmospheric Administration (NOAA), founded within the Department of Commerce in 1970, seeks to facilitate commercial uses of the ocean. NOAA includes the National Ocean Service, the National Weather Service, the National Marine Fisheries Service, and the Office of Sea Grant.

Satellite Oceanography

The National Aeronautics and Space Administration (NASA), organized in 1958, has become an important institutional contributor to marine science. For four months in 1978 NASA's *Seasat*, the first oceanographic satellite, beamed oceanographic data to Earth. More recent contributions have been made by satellites beaming radar signals off the sea surface to determine wave height, variations in sea-surface contour and temperature, and other information of interest to marine scientists.

The first of a new generation of oceanographic satellites was launched in 1992 as a joint effort of NASA and the Centre National d'Études Spatiales (the French space agency). The centerpiece of **TOPEX/Poseidon**, as the project is known, is a satellite orbiting 1,336 kilometers (835 miles) above Earth in an orbit that allows coverage of 95% of the ice-free ocean every ten days. The satellite's *TOPography EXperiment* uses a

---

[7] There are, of course, other prominent institutions involved in studying the ocean. Some thoughts on how a student might enter the field appear in Appendix IX, "Working in Marine Science."

a

b

FIGURE 1.31 (a) Woods Hole Oceanographic Institution, Woods Hole, Massachusetts. Marine science has been an important part of this small Cape Cod fishing community since Spencer Fullerton Baird, then assistant secretary of the Smithsonian Institution, established the U.S. Commission of Fish and Fisheries there in 1871. The Marine Biological Laboratory was founded in 1888, the Oceanographic Institution in 1930. (b) Scripps Institution of Oceanography, La Jolla, California. Begun in 1892 as a portable laboratory-in-a-tent, Scripps was founded by William Ritter, a biologist at the University of California. Its first permanent buildings were erected in 1905 on a site purchased with funds donated by philanthropic newspaper owner E. W. Scripps and his sister, Ellen.

positioning device that allows researchers to determine its position to within 1 centimeter ($\frac{1}{2}$ inch) of Earth's center. The radars aboard can then determine the height of the sea surface with unprecedented accuracy. Other experiments in this five-year program include sensing water vapor over the ocean, determining the precise location of ocean currents, and determining wind speed and direction.

Jason-1, NASA's ambitious follow-on to *TOPEX/Poseidon*, was launched in December 2001. Now flying 1 minute and 370 kilometers (230 miles) ahead of *TOPEX/Poseidon* on an identical ground track, its primary task is to monitor global climate interactions between the sea and the atmosphere. A five-year mission is planned.

SEASTAR, launched by NASA in 1997, carries a color scanner called SeaWiFS (sea-viewing wide-field-of-view sensor). This device measures the distribution of chlorophyll at the ocean surface, a measure of marine productivity.

Aqua, one of three of NASA's next generation of Earth-observing satellites, was launched into polar orbit on 4 May 2002 (**Figure 1.32**). It is the centerpiece of a project named for the large amount of information that it will be collecting about Earth's water cycle, including evaporation from the oceans, water vapor in the atmosphere, phytoplankton and dissolved organic matter in the oceans, and air, land, and water temper-

atures. *Aqua* will soon be joined by sisters *Terra* and *Aura* to monitor Earth and air.

A satellite system you can use every day? The U.S. Department of Defense has built the **GPS (Global Positioning System)**, a "constellation" of 24 satellites (21 active and 3 spare) in orbit 16,900 kilometers (10,600 miles) above Earth. The satellites are spaced so that at least four of them are above the horizon from any point on Earth. Each satellite contains a computer, an atomic clock, and a radio transmitter. On the ground every GPS receiver contains a computer that calculates its own geographic position using information from at least three of the satellites. The longitude and latitude it reports are accurate to less than 1 meter (39.37 inches), depending on the type of equipment used. Hand held GPS receivers can be purchased for less than US$70. The use of the GPS in marine navigation and positioning has revolutionized data collection at sea.

Satellite oceanography is an important frontier, and discoveries made by satellites are discussed in later chapters.

Alphabet Oceanography

Oceanography has become big science. Governments and institutions now cooperate in funding research to describe large-scale marine processes, not just to describe the nature of

FIGURE 1.32 *Aqua,* one of three of NASA's next generation of Earth-observing satellites, was launched into polar orbit on 4 May 2002. It is the centerpiece of a project named for the large amount of information that it will be collecting about Earth's water cycle, including evaporation from the oceans, water vapor in the atmosphere, phytoplankton and dissolved organic matter in the oceans, and air, land, and water temperatures.

small attributes of the ocean. The largest programs are known by their acronyms—words formed from the first letters of their titles. Here are a few of the most important:

- CLIVAR. The CLImate VARiability and Predictibility program continues the work of TOGA (the Tropical Ocean and Global Atmosphere program), which officially ended in 1994, and WOCE (the World Ocean Circulation Experiment), which ended in 1998. It is scheduled to become the largest scientific program ever attempted by physical oceanographers. CLIVAR's three goals are: (1) to study seasonal climate variability and the dynamics of the global ocean–atmosphere–land system, (2) to study climate variability and predictability in the 10- to 100-year ranges, and (3) to detect human-induced changes to atmospheric temperature and circulation.

- CoOP. The interdisciplinary Coastal Ocean Processes program seeks to increase understanding of processes that occur at the edges of ocean basins. Continental shelves are a special focus for the investigations, which have biological and physical components.

- GLOBEC. Global Ocean Ecosystem Dynamics will increase our understanding of the causes of variations in the populations of marine organisms resulting from global climate change. Though focused mainly on zooplankton biology, the program will also address issues of biological diversity.

- JGOFS. Scientists in the Joint Global Ocean Flux Study have concentrated on the ocean's chemical, physical, and

biological processes to increase understanding of the ocean's carbon cycle. Their goals are to determine what processes control the movement of carbon between the ocean and the atmosphere and to improve our ability to make global-scale predictions of the likely response of the ocean and atmosphere to human activities.

- IODP. The successor to ODP (the Ocean Drilling Program), the Integrated Ocean Drilling Program is an international consortium that will use advanced technology and a new vessel (R/V *Chikyu,* see again Figure 1.29) to drill into the deep seabed. IODP's goal is to discover the geological histories of the ocean basins and their margins. ODP's discoveries figure prominently in the history of the ocean floor discussed in Chapters 3 and 4, and IODP will continue to make important contributions.

- RIDGE. The Ridge Interdisciplinary Global Experiments (and its international component Inter-RIDGE) are directed at the dynamics of mid-ocean spreading centers. As you will learn in the next chapter, ridges are the boundaries along which new oceanic crust is created. Remotely piloted research vehicles are playing an important role in these studies.

- WCRP. The World Climate Research Program encompasses studies of Earth's physical climate system. Its goals are to provide quantitative answers to the questions being raised about natural climate variability and to establish a basis for predictions of global and regional climatic variations.

Table 1.1 Time Line for the History of Marine Science

Date	Event	Date	Event
4000 B.C.	Egyptian trade on Nile.	1758	Carolus Linnaeus publishes tenth edition of *Systema Naturae*, in which biological nomenclature is formalized (see **Chapter 13**).
3800 B.C.	First maps showing water (river charts).		
1200 B.C.	Phoenicians trade from Mediterranean to Britain and West Africa.	1760	John Harrison's Number Four chronometer (see **Chapter 1**).
1000 B.C.	Polynesians first inhabit Tonga, Samoa.	1768	James Cook's first voyage of discovery (see **Chapter 1**).
900 B.C.	Greeks first use the term *okeanos*, root of our word *ocean*.	1769	Benjamin Franklin publishes first chart showing an ocean current (see **Chapter 1**).
800 B.C.	First graphic aids to marine navigation.	1779	James Cook dies in Hawaii.
600 B.C.	Greek Pythagoreans assume a spherical Earth.	1818	John Ross takes first deep-water and sediment samples.
325 B.C.	Pytheas voyages to Britain, links tides to movement of the moon (see **Chapter 11**); Chinese invent the compass.	1831	Charles Darwin departs on five-year voyage aboard HMS *Beagle* (see **Chapter 13**).
300 B.C.	Library founded at Alexandria.	1835	Gaspard Coriolis publishes first papers on an object's horizontal motion across Earth's surface (see **Chapter 8**).
230 B.C.	Eratosthenes calculates circumference of Earth, invents latitude and longitude.	1836	William Harvey devises a taxonomy of seaweeds (see **Chapter 14**).
127 B.C.	Hipparchus arranges latitude and longitude in regular grid by degrees.	1838	Departure of the United States Exploring Expedition.
A.D. 150	Claudius Ptolemy errs in estimating Earth's circumference.	1847	Hans Christian Oersted observes plankton (see **Chapter 14**).
A.D. 415	Alexandrian library destroyed.	1855	Matthew Maury publishes *Physical Geography of the Seas* (see **Chapter 1**).
A.D. 500	Hawaii colonized by Polynesians.		
A.D. 780	Viking raids begin.	1859	Darwin's *Origin of Species* published (see **Chapter 13**).
1000	Norwegian colonies in North America.	1872	Departure of *Challenger* expedition.
1460	Prince Henry the Navigator dies.	1877	Alexander Agassiz begins research in *Blake*.
1492	Columbus's first voyage.	1880	William Dittmar determines major salts in seawater (see **Chapter 7**).
1522	Magellan's crew completes first circumnavigation.		
1609	Hugo Grotius publishes *Mare Liberum*, the foundation for all modern law of the sea (see **Chapter 18**).	1888	Marine Biological Laboratory founded at Woods Hole, Massachusetts.
1687	Isaac Newton's publication of *Principia Mathematica*, which includes an explanation of the operation of gravity (see **Chapter 11**).	1890	Alfred Thayer Mahan completes *The Influence of Sea Power upon History*.
		1891	Sir John Murray and Alphonse Renard classify marine sediments (see **Chapter 5**).
1742	Anders Celsius invents the centigrade temperature scale (see **Chapter 6**).	1893	Fridtjof Nansen in Arctic in *Fram* (see **Chapter 1**).

History in Progress

As we have seen, hundreds of marine scientists and their students are using an impressive array of equipment to probe the world ocean. *Marine science is by necessity a field science:* Ships and distant research stations are essential to its progress. The business of operating the ships and staffing the research stations is costly and sometimes dangerous, yet these researchers are willing to meet the daily challenge. They feel a sense of continuity within their separate specialties because oceanographers must be familiar with the scientific literature, the written history of their fields. History is not an abstract area of interest, but rather a part of their daily lives. Some of the important milestones in the history of marine science are listed in **Table 1.1**.

One cannot help but feel that the Alexandrian scholars would have appreciated the 2,000-year effort to understand the ocean. Application of the scientific method to oceanographic studies has yielded many benefits in that time, not the least of which is the satisfaction of knowing a small fraction of the story of Earth and its ocean. The lure of voyaging has been behind many of these discoveries. Scientists in the oceanographic research ships, laboratories, and libraries of the world go on collecting knowledge today. With luck and support, their efforts will continue into the distant future.

Questions from Students[8]

1. Where did the word *ocean* come from?

Ocean derives from the Greek *okeanos* (oceanus), a word that means "outer sea" (in contrast to the Greeks' "inner sea," the

[8] Each chapter ends with a few questions students have asked me after a lecture or reading assignment. These questions and their answers may be interesting to you, too.

Table 1.1   Time Line for the History of Marine Science (continued)

Date	Event	Date	Event
1900	Richard D. Oldham identifies P and S waves on seismograph (see **Chapter 3**).	1969	Santa Barbara, California, oil well blowout captures national attention (see **Chapter 18**).
1906	Prince Albert I of Monaco establishes the Musée Océanographique.	1970	National Oceanic and Atmospheric Administration (NOAA) established.
1907	Bertram Boltwood calculates age of Earth by radioactive decay (see **Chapter 3**).	1970	John Tuzo Wilson writes brief history of the tectonic revolution in geology in *Scientific American* (see **Chapter 3**).
1911	Roald Amundsen first at South Pole.		
1912	Alfred Wegener's Frankfurt lectures on continental drift (see **Chapter 3**).	1974	Project FAMOUS (French-American Mid-Ocean Undersea Study) maps and samples the Mid-Atlantic Ridge, a zone of seafloor spreading (see **Chapter 3**).
1912	Scripps Institution allied with the University of California.	1977	*Alvin* finds hydrothermal vents in the Galápagos rift (see **Chapters 3, 4,** and **16**).
1918	Vilhelm Bjerknes formulates theory of atmospheric fronts (see **Chapter 8**).	1978	*Seasat,* the first satellite dedicated to ocean studies, is launched.
1921	International Hydrographic Bureau founded.	1985	*JOIDES Resolution* replaces *Glomar Challenger* in Deep Sea Drilling Project (see **Chapters 1** and **3**).
1925	Departure of *Meteor* expedition; first echo sounder in operation (see **Chapters 1** and **3**).	1985	R. D. Ballard locates wreck of *Titanic.*
1930	Woods Hole Oceanographic Institution founded.	1987	Observations of supernova 1987A confirm theories of the origin of elements (see **Chapter 2**).
1931	*Atlantis* launched.		
1937	*E. W. Scripps* launched.	1991	JOI researchers bore to a depth of 2 kilometers (1.24 miles) beneath the seafloor near the Galápagos Islands (see **Chapter 3**).
1942	*The Oceans,* first modern reference text, published.		
1943	Jacques Cousteau and Emile Gagnan invent the scuba regulator and tank combination, the "aqualung."	1992	U.S.–French *TOPEX/Poseidon* satellite launched.
1949	Maurice Ewing forms the Lamont–Doherty Geological Observatory (see **Chapters 1** and **3**).	1995	*Keiko,* a small remotely controlled Japanese submersible, sets a new depth record: 10,978 meters (36,008 feet) in the Challenger Deep.
1958	U.S. nuclear submarine Nautilus makes first submerged transit of the Arctic ice pack, passes through North Pole (see **Chapter 1**).	1998	*Galileo* spacecraft finds possible evidence of an ocean on Jupiter's moon Europa (see **Chapter 2**).
1960	Bathyscaphe *Trieste* carrying Jacques Piccard and Don Walsh reaches bottom of deepest trench at 10,915 meters (35,801 feet) (see **Chapter 4**).	2000	*Mars Global Surveyor* photographs channels perhaps carved by flowing water (see **Chapter 2**).
		2002	R/V *Chikyu,* lead ship of the Inegrated Ocean Drilling Program, is launched (see **Chapters 1** and **3**).
1962	Rachel Carson's book *Silent Spring* initiates the U.S. environmental movement.	2003	Inauguration of the Integrated Ocean Drilling Program (IODP) (see **Chapters 3** and **4**).
1968	*Glomar Challenger* returns first cores, indicating the age of Earth's crust. The cores support theories of plate tectonics (see **Chapters 3** and **5**).	2004	Mars *Rover* explores Gusev crater to seek evidence of water.

Mediterranean). The term connotes a large moving river. Okeanos was also a mythical Greek titan who was god of the sea before Poseidon and father of the Oceanids, or ocean nymphs. The later Latin name for the ocean was *oceanus*. The Latin word evolved into the Middle English term *occean*, in which the double *c* was pronounced like a *k*. This later became the English word *ocean* (which, until this century, was pronounced with a soft *c*, as in the word *celery*). Perhaps the best name for Earth would be Oceanus, but so far I've not had much luck in changing the name!

2. You wrote that "Nothing is ever proven absolutely true by the scientific method." What good is it then? Can't we depend on the process of science?

One philosopher of science has described truth as a liquid: It flows around ideas and is hard to grasp. The progressive improvement in our understanding of nature is subject to the limitations inherent in our observations. As our observations become more accurate, so do our conclusions about the natural world. But because observation (and interpretation of observation) is never perfect, truth can never be absolute. In the 1920s, for example, astronomers assumed that the universe was limited to our own Milky Way galaxy. Observations made with a large new telescope on Mt. Wilson in California by Harlow Shapley and Edwin Hubble allowed them to measure more distant objects. Galaxies were discovered in profusion, "like grains of sand on a beach," in Shapley's words. Thus truth changed its shape. In a more directly marine example, one of the *Challenger* expedition's responsibilities was investigating Edinburgh professor Edward Forbes's contention that life below about 550 meters (1,800 feet) was impossible. Consistent sampling below that depth was not practical until the advent of *Challenger*'s steam winch. Sure enough, when observations were made, life was there and truth changed. We learn as we go. We depend on the underlying assumption that nature "plays fair"—that is, is consistent and does not capriciously change the rules as our powers of observation grow. What we have learned so far is of inestimable practical and aesthetic value, and we have only scratched the surface.

3. How do we know when people reached certain locations? How do researchers know New Guinea was inhabited by 30,000 years ago, the tip of South America by 11,400 years ago, and Hawaii by A.D. 450–600?

Researchers use various methods to date the human artifacts they find. The **radiometric dating** technique, for example, depends on the slow and predictable decrease in radioactivity of naturally radioactive materials. One commonly used form of radiometric dating measures the amount of radioactive carbon in once-living material such as wood or bone. A small amount of the carbon dioxide in the atmosphere contains carbon atoms that have been made radioactive by natural processes. This carbon dioxide is made into glucose and then into cellulose (and wood) by plants. The analysis assumes a fixed ratio of radioactive to nonradioactive carbon dioxide in the air through time. By knowing how the present proportion of radioactive carbon in wood differs from the presumed initial proportion, scientists know how long ago the wood ceased taking up CO_2 and therefore when it stopped living. Animal remains may also be dated this way because animals derive nutrients from photosynthesizing plants. Organic material up to 40,000 years old—including human bones, tools with wood or bone pieces, and food remains—can all be dated using this method. More on the topic of radiometric dating can be found in Box 3.2.

4. It would be difficult for humans to walk from Siberia to Alaska today. How was it possible in the past?

The great migrations to the Americas took place at the end of the last ice age, about 13,000 years ago. At that time the large amount of water trapped as ice on the continents caused sea level to fall 100 to 125 meters (300 to 400 feet) lower than it is today. The lower sea level exposed land in the Bering Sea and the Aleutian arc between Siberia and Alaska. Lower sea level, combined with the jam of pack ice against the islands themselves, made a passage for migrating game. People followed the animals for food. Both ended up in the "New World" in the process.

5. What's this about a chronometer not having to keep perfect time? I thought you had to know exactly what time it is to be able to calculate your longitude.

Yes, you need accurate time. But a chronometer is valuable not because it necessarily keeps perfect time, but because it loses or gains time at a constant, known rate. Each day the navigator multiplies the number of seconds the clock is known to gain (or lose) by the number of days since the clock was last set—and then adds the total to the time shown on the chronometer's face to obtain the real time. The value of a chronometer lies entirely in its consistency.

6. If Columbus was unsuccessful in his attempt to sail around the world and Magellan died without completing his circumnavigation, who was the first captain to complete the trip?

Sir Francis Drake of England was the first captain to sail his own ship around the world. His expedition, begun in 1577, lasted three years. In the eastern Pacific he raided Spanish merchantmen and captured a fortune in gold, silver, coins, and precious stones. He was the first European to sight the west coast of what is now Canada, and he claimed California for Queen Elizabeth I. His transit of the Pacific lasted 68 days, and on his return to England he concluded spice trade negotiations with various heads of state; some of the agreements remain in effect to this day! On 26 September 1580 he returned to England a wealthy man, was knighted by the Queen, and went on in 1588 to help defeat the Spanish Armada.

Little was known of Drake's explorations until relatively recently. The trade and geographic information he brought home to England was considered so valuable that it was given the highest security classification—thus few people saw it or benefited from it!

7. Speaking of James Cook, what was his motivation for those extraordinary voyages?

One could say simply that he was a serving Royal Navy Officer and was ordered to go. But Beaglehole, Hough, and other biographers suggest the story is more complex. How did a relatively unschooled man become leader of one of the first scientific oceanographic expeditions? Cook had the usual attributes of a successful person—intelligence, strength of character, meeting the right people at the right time, health, focus, luck—but he also had a driving intellectual curiosity and rare (for that era) tolerance and respect for alien cultures. As Hough (1994) writes, "Cook stood out like a diamond amidst junk jewelery. . . ." It was no surprise that the Lords of the Admiralty settled upon this unique man to lead the adventure.

And the atmosphere of that adventure? Another excerpt from Hough's 1994 biography is instructive: ". . . we can still stand back in wonder at the dimensions of these orders, at the sheer effrontery of asking nearly one hundred men to embark in a wooden vessel, scarcely more than one hundred feet in length, and dependent for its mobility upon the whim of winds, and currents and tides, to sail to the other side of a world only dimly charted, there to carry out an exacting observation, and then to discover a completely uncharted continent, survey its coastline and take note of its characteristics. Perhaps an even greater wonder was that men of experience and wisdom were prepared to take on this undertaking with pleasure and excitement."

8. Did Columbus discover North America?

No. He never saw North America.

9. You wrote that the future of oceanography lies in the big institutions. Is there a place for individual initiative in marine science?

Always. Every adventure begins with a person sitting quietly nurturing an idea. The notion may seem crazy at first, or it may seem impossible to prove or disprove, but the idea won't go away. An individual shares the idea with colleagues. If a research consensus is reached, plans are made, grants are proposed and funded, data flow. But the trail always begins with one person and his or her idea.

Chapter in Perspective

In this chapter you learned that science is a process of asking and answering questions. Decisions are made based on observable information, not on preconceived ideas or submission to authority. There is no "absolute truth" in science; explanations of natural phenomena will vary as our powers of observation change. The scientific process depends on an open mind, the collection of information, consistent experimentation, and the assumption that nature "plays fair"—that is, that the rules governing natural processes don't capriciously change.

Science and exploration have gone hand in hand. Voyaging for necessity evolved into voyaging for scientific and geographic discovery. The transition to scientific oceanography was complete when the *Challenger Report* was published in 1895. The rise of the great oceanographic institutions quickly followed, and those institutions and their funding agencies today mark our path into the future.

In the next chapter you'll learn of our place in the universe and about the origin of Earth and ocean. Earth is about 4.6 billion years old (that's 4,600 million years). The ocean formed early in Earth's history, and life followed almost immediately thereafter. Earth is density stratified—that is, its inner layers are heavier than its outer layers. Indeed the origin and interaction of those layers occupy our attention for the next few chapters.

Key Ideas

✓ The world ocean may be considered as a single entity with temporary partitions. It covers about 71% of Earth's surface.
Review Figure 1.2, pages 2–3

✓ Marine science—oceanography—is the process of discovering unifying principles in data obtained from the ocean.
Review page 2

✓ Science is the process of asking questions about the observable world and then testing the answers to those questions. The external world, not internal conviction, is the testing ground for scientific beliefs.
Review Figure 1.5, pages 4–6

✓ The ocean did not prevent humans from occupying nearly every place on Earth that could sustain them.
Review page 6

✓ The origins of marine science lie in voyaging—traveling on the ocean for a purpose. Technological advances made during voyaging and later marine exploration led to the rise of scientific oceanography.
Review pages 7–12

✓ Earth's shape and circumference were accurately estimated around 230 B.C. at the Library of Alexandria, Egypt. The use of latitude and longitude for positioning and navigation also began there.
Review Figure 1.7 and Box 1.1, pages 8 and 9

✓ Christopher Columbus did not discover North America, and he did not sail around the world.
Review page 15

✓ Captain James Cook, Royal Navy, was perhaps the first ocean observer careful enough to be considered a marine scientist.
Review Figure 1.17, pages 16–18

✓ The *Challenger* expedition was the first purely scientific voyage of oceanic exploration.
Review Figures 1.25–1.27, pages 23–25

✓ Polar studies greatly advanced marine science at the beginning of the twentieth century.
Review Figure 1.28, pages 25–26

✓ Nearly all research is conducted not by individuals but by teams of specialists working in large, nationally funded oceanographic or military institutions.
Review Figures 1.30 and 1.31, pages 29 and 30

✓ The tools of modern oceanography include satellites, manned and remotely controlled vehicles, and computer modeling.
Review Figure 1.32, pages 29–32

Terms and Concepts to Remember

Study Questions

Reviewing What You've Learned

1. Which hemisphere contains the greatest percentage of ocean? Is most of Earth's water in the ocean?
2. Which is greater: the average depth of the ocean or the average height of the continents above sea level?
3. Can the scientific method be applied to speculations about the natural world that are not subject to test or observation?
4. Why do we say there's one world ocean? What about the Pacific and Atlantic Oceans, the Baltic and Mediterranean Seas?
5. What were the stimuli to Polynesian colonization? How were the long voyages accomplished? How were Polynesian voyages different from (and similar to) those of the Vikings?
6. What were the main stimuli to European voyages of exploration during the Age of Discovery? Why did it end?
7. What were the oceanographic contributions of Prince Henry the Navigator, Benjamin Franklin, Matthew Maury, and Charles Darwin?
8. How can you find your approximate latitude and longitude at sea?
9. What were the goals and results of the United States Exploring Expedition? What U.S. institution greatly benefited from its efforts?
10. What was the first purely scientific oceanographic expedition, and what were some of its accomplishments? What contributions did the earlier, hybrid expeditions make?

Thinking Critically

1. How could you convince a ten-year-old that Earth is round? What evidence would a child offer that it's flat? How can you counter those objections?
2. How did the Library of Alexandria contribute to the development of marine science? What happened to most of the information accumulated there? Why do you suppose the residents of Alexandria became hostile to the librarians and the many achievements of the library?
3. How did Eratosthenes calculate the approximate size of Earth? Which of his assumptions was the "shakiest"?
4. If Columbus didn't discover North America, then who did?
5. Sketch briefly the major developments in marine science since 1900. Do individuals, separate voyages, or institutions figure most prominently in this history?
6. **InfoTrac College Edition Project** Navigation technology has become highly sophisticated in the past 30 years.

Ships' crews now rely on satellite transmissions and computer systems to tell them their location, often with accuracy within a few meters. Companies are promoting small Global Positioning System (GPS) devices for use by automobile travelers and hikers. What are the pros and cons of increased reliance on such devices? Research this question using InfoTrac College Edition.

Thinking Analytically

1. Imagine that you set your watch at local noon in Kansas City on Monday and then fly to the coast on Tuesday. You stick a pole into the ground on a sunny day at the beach, wait until its shadow is shortest, and look at your watch. The watch says 10:00 A.M. Are you on the East Coast or the West Coast? What is the difference in longitude? (Hint: 360° divided by 24 hours is 15°. The sun moves through the sky at a rate of 15° per hour.)
2. Look at Box 1.1, figure **b**. Provide a rough estimate of the latitude and longitude of your home.
3. Magellan's crew kept very careful records of their circumnavigation, yet when they returned home, they were one day "off." Why? Had they gained a day or lost a day?
4. Replicate Eratosthenes's measurement of the diameter of Earth. Try this technique: Contact a friend who lives about 800 kilometers (500 miles) north or south of you (a distance comparable to the distance between Alexandria and Syene). Drive a tall pole into the ground at each location. Make sure the poles are vertical (using a weight on a string). Watch around noon, and when the pole casts the shortest shadow, measure the sun's angle of inclination from the shadow cast. Can you take it from there?

Oceanography Now™

Assess your understanding of this chapter's topics with additional quizzing and conceptual-based problems at http://oceanography.brookscole.com/garrison5e

Study and Research on the Internet

Assess your understanding of this chapter's topics with additional quizzing and conceptual-based problems at:

http://oceanography.brookscole.com/garrison5e

- **InfoTrac College Edition Research** leads you to critical thinking exercises that use InfoTrac College Edition® as a research tool.
- **Regional InfoTrac College Edition** articles are organized into East Coast, West Coast, and Gulf Coast regions, allowing you to study oceanography on a more local level.

For more readings, go to InfoTrac College Edition, your online research library, at:

http://infotrac.thomsonlearning.com

CHAPTER 2

GFSC/NASA

The Milky Way shines over the ocean horizon. Stars and sea share a common origin.

ORIGINS

Stars and Seas

To understand the ocean, we need to understand how it formed and evolved through time. Since the world ocean is the largest feature of Earth's surface, it should not be surprising that we believe the origin of the ocean is linked to the origin of Earth. The origin of Earth is linked to that of the solar system and the galaxies.

The origin of Earth and the ocean is a long and wonderful story—one we've only recently come to know. As you read this chapter, you may be startled to discover that most of the atoms that make up Earth, its ocean, and its inhabitants were formed within stars billions of years ago. Stars spend their lives changing hydrogen and helium into heavier elements. As they die, some stars eject these elements into space during cataclysmic explosions. The sun and the planets, including Earth, condensed from a cloud of dust and gas enriched by the recycled remnants of exploded stars.

Our ocean is not a direct remnant of that cloud, however. Most of the ocean formed later, as water vapor trapped in Earth's outer layers escaped to the surface through volcanic activity during the planet's youth. The vapor cooled and condensed to form an ocean. Comets may have delivered additional water to the new planet's surface. Life originated in the ocean soon after, developing and flourishing in the ocean for more than 3 billion years before venturing onto the unwelcoming continents. Life and Earth have grown old together.

The Origin of Earth

We have always wondered about our origins—how Earth was formed, how the ocean arose, and how life came to be. In the last 50 years researchers using the scientific method have determined a tentative age for the ocean, Earth, and the universe. They have developed hypotheses about how matter is assembled, how stars and planets are formed, and even how life may have arisen. Many of the details are still sketchy, of course, but these hypotheses have predicted some important recent discoveries in subatomic physics and molecular biology. Perhaps the most dramatic twentieth-century discoveries in natural science have been those dealing with the origin and history of the universe.

The universe apparently had a beginning. The **big bang**, as that event is modestly named, occurred about 14 billion years ago. All of the mass and energy of the universe are thought to have been concentrated at a geometric point at the beginning of space and time, the moment when the expansion of the universe began. We don't know what initiated the expansion, but it continues today and will probably continue for billions of years, perhaps forever.

The very early universe was unimaginably hot, but as it expanded it cooled. About a million years after the big bang, temperatures fell enough to permit the formation of atoms from the energy and particles that had predominated up to that time. Most of these atoms were hydrogen, then as now the most abundant form of matter in the universe. About a billion years after the big bang, this matter began to congeal into the first galaxies and stars.

Galaxies and Stars

A **galaxy** is a huge rotating aggregation of stars, dust, gas, and other debris held together by gravity. Our galaxy is named the **Milky Way galaxy** (from Greek *galaktos* = milk).[1] A galaxy very similar to our own is shown in **Figure 2.1**.

The **stars** that make up a galaxy are massive spheres of incandescent gases. They are usually intermingled with diffuse clouds of gas and debris. In spiral galaxies like the Milky Way, the stars are arrayed in curved arms radiating from the galactic center. Other galaxies are elliptical or irregular in shape. Our

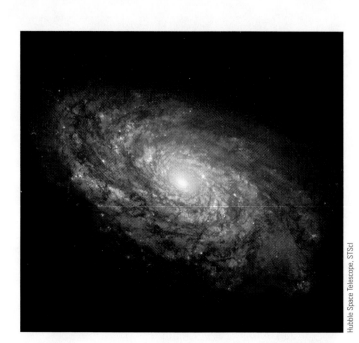

FIGURE 2.1 This spiral galaxy in the constellation Coma Berenices is very similar in size and structure to our own Milky Way galaxy. The galaxy, photographed in 1999 by the Hubble Space Telescope, is 62 million light-years away and about 56,000 light-years in diameter. The stars in the foreground are in our galaxy. Note the clouds of dust and gas that obscure our view of some parts of the spiral arms. Planets and oceans are made of such material.

Hubble Space Telescope, STScI

part of the Milky Way is populated with many stars, but distances within a galaxy are so huge that the star nearest the sun is about 42 trillion kilometers (26 trillion miles) away. Astronomers tell us there are perhaps 100 billion galaxies in the universe and 100 billion stars in each galaxy. Imagine more stars in the Milky Way than grains of sand on a beach![2]

Our sun is a typical star (**Figure 2.2**). The sun and its family of planets, called the **solar system**, is located about three-fourths of the way out from the galaxy's center, in a spiral arm. We orbit the galaxy's brilliant core, taking about 230 million years to make one orbit—even though we are moving at about 280 kilometers per second (half a million miles an hour). Earth has made about 20 circuits of the galaxy since the ocean formed.

---

[1] Because they can be useful as well as interesting, the derivations of words are sometimes included.

[2] In July 2003 astronomers announced that their survey of the total number of stars in the known universe had reached 70 sextillion—about ten times more stars than grains of sand on all the world's beaches and deserts!

FIGURE 2.2 A filament of hot gas erupts from the face of our sun. Like all normal stars, our sun is powered by nuclear fusion—the welding together of small atoms to make larger ones. These violent reactions generate the heat, light, matter, and radiation that pour from stars into space. The entire Earth could easily fit into this filament's outstretched arms.

TRACE, NASA

The Lives of Stars

As we will see, most of the substance of Earth and its ocean was formed by stars. Stars form in **nebulae**, large, diffuse clouds of dust and gas within galaxies. With the aid of telescopes and infrared sensing satellites, astronomers have observed such clouds in our own and other galaxies. They have seen stars in different stages of development and have inferred a sequence in which these stages occur. The **condensation theory**, a theory based on this inference, explains how stars and planets are believed to form.

The life of a star begins when a diffuse area of a spinning nebula starts to shrink and heat up under the influence of its own weak gravity. Gradually the cloudlike sphere flattens and condenses at the center into a knot of gases called a **protostar** (*protos* = first). The original diameter of the protostar may be many times the diameter of our solar system, but gravitational energy causes it to contract, and the compression raises its internal temperature. When the protostar reaches a temperature of about 10 million degrees Celsius (18 million degrees Fahrenheit), nuclear fusion begins; that is, hydrogen atoms begin to fuse together to form helium, a process that liberates even more energy. This rapid release of energy, which marks the transition from *protostar* to *star*, stops the young star's shrinkage. (The process is shown in the top half of **Figure 2.3**.)

After fusion reactions begin, the star becomes stable—neither shrinking nor expanding, and burning its hydrogen fuel at a steady rate. Over a long and productive life, the star converts a large percentage of its hydrogen to atoms as heavy as carbon or oxygen. This stable phase does not last forever, though. The life history and death of a star depend on its initial mass. When a medium-mass star (like our sun) begins to

consume carbon and oxygen atoms, its energy output slowly rises and its body swells to a stage aptly named *red giant* by astronomers. The dying giant slowly pulsates, incinerating its planets and throwing off concentric shells of light gas enriched with these heavy elements. But most of the harvest of carbon and oxygen is forever trapped in the cooling ember at the star's heart.

Stars much more massive than the sun have shorter but more interesting lives. They too fuse hydrogen to form atoms as heavy as carbon and oxygen; but because they are larger and hotter, their internal nuclear reactions consume hydrogen at a much faster rate. In addition, higher core temperatures permit the formation of atoms up to the mass of iron.

The dying phase of a massive star's life begins when its core—depleted of hydrogen—collapses in on itself. This rapid compression causes the star's internal temperature to soar. When the infalling material can no longer be compressed, the energy of the inward fall is converted to a cataclysmic expansion called a **supernova**. The explosive release of energy in a supernova is so sudden that the star is blown to bits and its shattered mass is accelerated outward at nearly the speed of light. The explosion lasts only about 30 seconds, but in that short time the nuclear forces holding apart individual atomic nuclei are overcome—and atoms heavier than iron are formed. The gold of your rings, the mercury in a thermometer, and the uranium in nuclear power plants were all created during such a brief and stupendous flash. The atoms produced by a star through millions of years of orderly fusion, *and* the heavy atoms generated in a few moments of unimaginable chaos, are sprayed into space (**Figure 2.4**). Every chemical element heavier than hydrogen—most of the atoms that make up the planets, the ocean, and living creatures—was manufactured by the stars.

The Formation of the Solar System

Earth and its ocean are the indirect result of a supernova explosion. The thin cloud, or **solar nebula**, from which our sun and its planets formed was probably struck by the shock wave and some of the matter of an expanding supernova remnant. Indeed the turbulence of the encounter may have caused the condensation of our solar system to begin. The solar nebula was affected in at least two important ways: First, the shock wave caused the condensing mass to spin; second, the nebula absorbed some of the heavy atoms from the passing supernova remnant. In other words, a massive star had to live its life (constructing elements in the process) and then undergo explosive disintegration in order to seed heavy elements back into the nebular nursery of dust and gas from which our solar system arose. The planets are made mostly of matter assembled in a star (or stars) that disappeared billions of years ago. We are made of that stardust. Our bones and brains are composed of ancient atoms formed by stellar fusion long before the solar system existed.

By about 5 billion years ago, the solar nebula was a rotating, disk-shaped mass of about 75% hydrogen, 23% helium, and 2% other material (including heavier elements, gases, dust, and ice). Like a spinning skater bringing in his or her arms, the

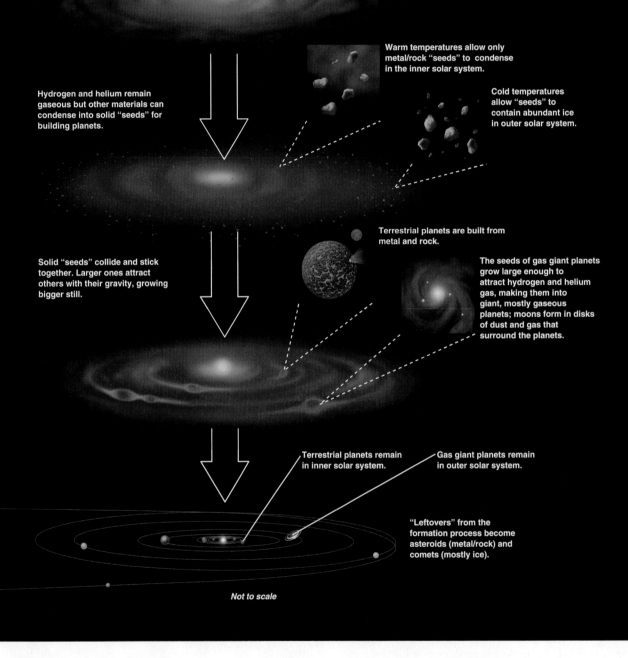

A nebula (a large, diffuse gas cloud of gas and dust) contracts under gravity. As it contracts, the nebula heats, flattens, and spins faster, becoming a spinning disk of dust and gas.

Star will be born in center.

Planets will form in disk.

Warm temperatures allow only metal/rock "seeds" to condense in the inner solar system.

Hydrogen and helium remain gaseous but other materials can condense into solid "seeds" for building planets.

Cold temperatures allow "seeds" to contain abundant ice in outer solar system.

Terrestrial planets are built from metal and rock.

Solid "seeds" collide and stick together. Larger ones attract others with their gravity, growing bigger still.

The seeds of gas giant planets grow large enough to attract hydrogen and helium gas, making them into giant, mostly gaseous planets; moons form in disks of dust and gas that surround the planets.

Terrestrial planets remain in inner solar system.

Gas giant planets remain in outer solar system.

"Leftovers" from the formation process become asteroids (metal/rock) and comets (mostly ice).

Not to scale

FIGURE 2.3 The origin of a solar system in the spiral arm of a galaxy. Our sun and its family of planets were formed in this way about 5 billion years ago.

European Southern Observatory

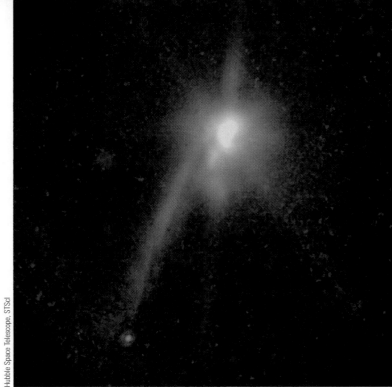

Hubble Space Telescope, STScI

FIGURE 2.4 A photograph of the Crab Nebula, the remains of a massive star whose supernova stage was witnessed—and position marked—by Chinese astronomers in A.D. 1054. In a few tens of thousands of years it will have fully dispersed its wealth of heavy elements into the surrounding galactic gas.

FIGURE 2.5 The first image of a possible planet outside our solar system. It is the small, bright object to the lower left, at the end of the filament of dust and gas extending from a newly formed sun.

nebula spun faster as it condensed. Material concentrated near its center became the protosun. Much of the outer material eventually became **planets**, the smaller bodies that orbit a star and do not shine by their own light.

Look again at Figure 2.3. The new planets formed in the disk of dust and debris surrounding the young sun through a process known as **accretion**—the clumping of small particles into large masses. Bigger clumps with stronger gravity pulled in most of the condensing matter. Near the protosun, where temperatures were highest, the first materials to solidify were substances with high boiling points, mainly metals and certain rocky minerals. The planet Mercury, closest to the sun, is mostly iron because iron is a solid at high temperatures. Somewhat farther out, in the cooler regions, magnesium, silicon, water, and oxygen condensed. Methane and ammonia accumulated in the frigid outer zones. Earth's array of water, silicon-oxygen compounds, and metals results from its middle position within that accreting cloud. The gas giant planets of the outer solar system—Jupiter, Saturn, Uranus, and Neptune—are composed mostly of methane and ammonia ices because those gases can congeal only at cold temperatures.

The period of accretion lasted perhaps 30 to 50 million years. The protosun became a star—our sun—when its internal temperature was high enough to fuse atoms of hydrogen into helium. The violence of these nuclear reactions sent a solar wind of radiation sweeping past the inner planets, clearing the area of excess particles and ending the period of rapid accretion. Gases like those we now see on the giant outer plan-

ets may once have surrounded the inner planets, but this rush of solar energy and particles stripped them away.

This process might not be rare. As you'll learn in Box 2.1, more than 100 planets have been discovered orbiting other stars. One of them is shown in **Figure 2.5**.

Earth and Ocean

The young Earth, formed by the accretion of cold particles, was probably homogeneous throughout. Then, during the accretion phase, Earth's surface was heated by the impact of asteroids, comets, and other falling debris. This heat, combined with gravitational compression and heat from the decay of radioactive elements accumulating within the newly assembled planet, caused Earth to partially melt. Gravity pulled most of the iron and nickel inward to form the planet's core. The sinking iron released huge amounts of gravitational energy, which, through friction, heated Earth even more. At the same time a slush of lighter minerals—silicon, magnesium, aluminum, and oxygen-bonded compounds—rose toward the surface, forming Earth's crust (**Figure 2.6**). This important process, called **density stratification**, lasted perhaps 100 million years.[3]

---

[3] **Density** is an expression of the relative heaviness of a substance; it is defined as the mass per unit volume, usually expressed in grams per cubic centimeter (g/cm^3). The density of pure water is $1\ g/cm^3$. Granite rock is about 2.7 times denser, at $2.7\ g/cm^3$.

a

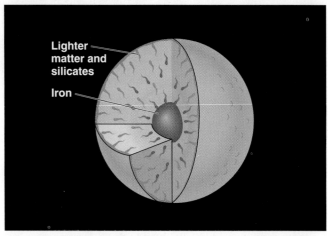

Lighter matter and silicates

Iron

b

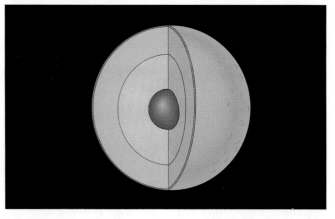

c

FIGURE 2.6 A representation of the formation of Earth. (**a**) The planet grew by the aggregation of particles. Meteors and asteroids bombarded the surface, heating the new planet and adding to its growing mass. At the time, Earth was composed of a homogeneous mixture of materials. (**b**) Earth lost volume because of gravitational compression. High temperatures in the interior turned the inner Earth into a semisolid mass; dense iron (red drops) fell toward the center to form the core, while less dense silicates moved outward. Friction generated by this movement heated Earth even more. (**c**) The result of *density stratification* is evident in the formation of the inner and outer core, the mantle, and the crust.

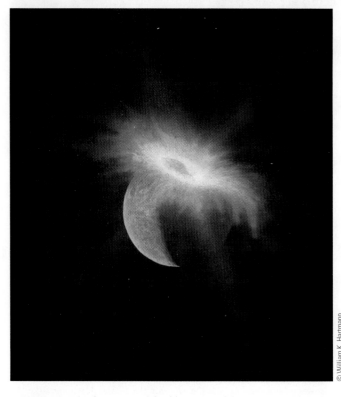

© William K. Hartmann

FIGURE 2.7 The first stage of the formation of the moon. A planetary body somewhat larger than Mars smashed into the young Earth about 4.4 billion years ago. The rocky mantle of the impactor was ejected to form a ring of debris around Earth, and its metallic core fell into Earth's core and joined with it. Rocks brought from the lunar surface by Apollo astronauts suggest the ejected material condensed soon after to become our moon.

Then Earth began to cool. Its first surface is thought to have formed about 4.6 billion years ago. That surface did not remain undisturbed for long. A planetary body somewhat larger than Mars smashed into the young Earth and broke apart (**Figure 2.7**). The metallic core fell into Earth's core and joined with it, while the rocky mantle was ejected to form a ring of debris around Earth. The debris began condensing soon after and became our moon. The newly formed moon, still glowing from heat generated by the kinetic energy of infalling objects, is depicted in **Figure 2.8**. Could a similar cataclysm happen today? The issue is addressed in Box 13.1.

Radiation from the energetic young sun had stripped away our planet's outermost layer of gases, its first atmosphere; but soon gases that had been trapped inside the forming planet burped to the surface to form a second atmosphere. This volcanic venting of volatile substances—including water vapor—is called **outgassing** (**Figure 2.9a**). As the hot vapors rose, they condensed into clouds in the cool upper atmosphere. Though most of Earth's water was present in the solar nebula during the accretion phase, recent research suggests that a barrage of icy comets colliding with Earth may also have contributed a portion of the accumulating mass of water, this ocean-to-be (**Figure 2.9b**)

FIGURE 2.8 Within a thousand years of the giant impact, our moon (foreground) was forming. In this painting the sky is dominated by a red-hot Earth, recently reshaped and melted by the moon-forming impact. The ring of debris will eventually fall to Earth or be captured by the still-growing molten moon.

a   b

FIGURE 2.9 Sources of the ocean. (**a**) Outgassing. Volcanic gases emitted by fissures add water vapor, carbon dioxide, nitrogen, and other gases to the atmosphere. Volcanism was a major factor in altering Earth's original atmosphere; later the action of photosynthetic bacteria and plants was another. (**b**) Comets may have delivered some of Earth's surface water. Intense bombardment of the early Earth by large bodies—comets and asteroids—probably lasted until about 3.8 billion years ago.

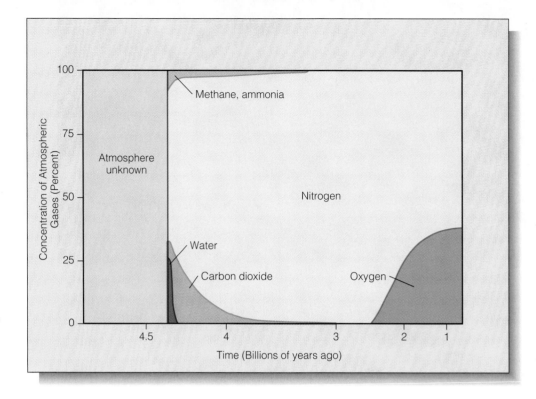

FIGURE 2.10 The evolution of our atmosphere. The early atmosphere had high concentrations of water and carbon dioxide with traces of methane and ammonia. Much of the carbon dioxide dissolved in seawater to form carbonic acid, then combined with crustal rocks. Nitrogen became the dominant gas. After the emergence of photosynthetic organisms, oxygen began to accumulate in the atmosphere. C.J. Allegre and S.H. Schneider. "The Evolution of the Earth," Scientific American, October 1994. reprinted by permission of Ian Worpole.

Earth's surface was so hot that no water could collect there and no sunlight could penetrate the thick clouds. (A visitor approaching from space 4.5 billion years ago would have seen a vapor-shrouded sphere blanketed by lightning-stroked clouds.) After millions of years the upper clouds cooled enough for some of the outgassed water to form droplets. Hot rains fell toward Earth, only to boil back into the clouds again. As the surface became cooler, water collected in basins and began to dissolve minerals from the rocks. Some of the water evaporated, cooled, and fell again. The world ocean was gradually accumulating.

These heavy rains may have lasted about 20 million years. Large amounts of water vapor and other gases continued to escape through volcanic vents during that time and for millions of years thereafter. The ocean grew deeper. Evidence suggests that Earth's crust grew thicker as well, perhaps in part from chemical reaction with oceanic compounds.

The physical expanse and distribution of the early ocean are matters of some controversy. Most researchers hold that masses of rock have always protruded through the ocean surface to form continents. However, some recent studies suggest that water may have covered Earth's entire surface for some 200 million years before the continents emerged. Although most of the ocean was in place about 4 billion years ago, ocean formation continues very slowly even today: About 0.1 cubic kilometer (0.025 cubic mile) of new water is added to the ocean each year, mostly as steam flowing from volcanic vents and in the form of microscopic cometary fragments.

The composition of the early atmosphere (often called the *primitive atmosphere*) was much different from today's (**Figure 2.10**). Geochemists believe it may have been rich in carbon dioxide, nitrogen, and water vapor, with traces of ammonia and methane. Beginning about 3.5 billion years ago, this mixture began a gradual alteration to its present composition, mostly nitrogen and oxygen. At first this change was brought about by carbon dioxide dissolving in seawater to form carbonic acid, then combining with crustal rocks. The chemical breakup of water vapor by sunlight high in the atmosphere also played a role. Then about 1.5 billion years later, the ancestors of today's green plants produced—by photosynthesis—enough oxygen to oxidize minerals dissolved in the ocean and surface sediments. Oxygen began to accumulate in the atmosphere. (This monumental event in Earth's history is called the *oxygen revolution*. You'll find more about it in Chapter 15.)

Oceanography Now™ | Click Oceanography
Interactive to view an animation on the earth's layers.

The Origin of Life

Life, at least as we know it, would be inconceivable without large quantities of water. Water has the ability to retain heat, moderate temperature, dissolve many chemicals, and suspend nutrients and wastes. These characteristics make it a mobile stage for the intricate biochemical reactions that allowed life to begin and prosper on Earth.

Life on Earth is formed from aggregations of a few basic kinds of carbon compounds. Where did the carbon compounds come from? There is growing consensus that most of the organic (that is, carbon-containing) materials in these compounds were transported to Earth by the comets, asteroids, meteors, and interplanetary dust particles that crashed into our planet during its birth.

FIGURE 2.11 An environment for biosynthesis? Weak sunlight and unstable conditions on Earth's surface may have favored the origin of life on mineral surfaces near deep-ocean hydrothermal vents similar to the one shown here.

FIGURE 2.12 Fossil of a bacteria-like organism (with an artist's reconstruction) that photosynthesized and released oxygen into the atmosphere. Among the oldest fossils ever discovered, this microscopic filament from northwestern Australia is about 3.5 billion years old.

The young ocean was a thin broth of organic and inorganic compounds in solution.

In laboratory experiments mixtures of dissolved compounds and gases thought to be similar to Earth's early atmosphere have been exposed to light, heat, and electrical sparks. These energized mixtures produce simple sugars and most of the biologically important amino acids. They even produce small proteins and nucleotides (components of the molecules that transmit genetic information between generations). The main chemical requirement seems to be the absence (or near absence) of free oxygen, a compound that can disrupt any unprotected large molecule.

Did *life* form in these experiments? No, the compounds that formed are only building blocks of life. But the experiments do tell us something about the commonality and unity of life on Earth. The facts that these crucial compounds can be synthesized so easily and are present in virtually all living forms are probably not coincidental. Those compounds are "permitted" by physical laws and by the chemical composition of this planet. The experiment also underscores the special role of water in life processes. The fact that all life, from a jellyfish to a dusty desert weed, depends on saline water within its cells to dissolve and transport chemicals is certainly significant. It strongly suggests that simple, self-replicating—living—molecules arose somewhere in the early ocean. It also suggests that all life on Earth is of common origin and ancestry.

The early steps in the evolution of living organisms from simple organic building blocks, a process known as **biosynthesis**, are still speculative. Planetary scientists suggest that the sun was faint in its youth. It put out so little heat—about 30% less than today—that the ocean may have been frozen to a

depth of around 300 meters (1,000 feet). The ice would have formed a blanket that kept most of the ocean fluid and relatively warm. Periodic fiery impacts by asteroids, comets, and meteor swarms could have thawed the ice, but between batterings it would have reformed. In 2002 chemists Jeffrey Bada and Antonio Lazcano suggested that organic material may have formed and then been trapped beneath the ice—protected from the atmosphere, which contained chemical compounds capable of shattering the complex molecules. The first living molecules might have arisen deep below the layers of surface ice, on clays or pyrite crystals at cool mineral-rich seeps on the ocean floor (**Figure 2.11**).

A similar biosynthesis cannot occur today. Living things have changed the conditions in the ocean and atmosphere, and those changes are not consistent with any new origin of life. For one thing, green plants have filled the atmosphere with oxygen. For another, some of this oxygen (as ozone) now blocks most of the dangerous wavelengths of light from reaching the surface of the ocean. And finally, the many tiny organisms present today would gladly scavenge any large organic molecules as food.

How long ago might life have begun? The oldest fossils yet found, from northwestern Australia, are between 3.4 and 3.5 billion years old (**Figure 2.12**). They are remnants of fairly complex bacteria-like organisms, indicating that life must have originated even earlier, probably only a few hundred million years after a stable ocean formed. Evidence of an even more ancient beginning has been found in the form of carbon-based residues in some of the oldest rocks on Earth, from Akilia Island near Greenland. These 3.85-billion-year-old specks of carbon bear a chemical fingerprint that researchers feel could only have come from a living organism. Life and Earth have grown old together; each has greatly influenced the other.

The Distant Future of Earth?

Our descendants may enjoy another 5 billion years of Earth as we know it today. But then our sun, like any other star, will begin to die. The sun is not massive enough to become a supernova, but after a billion-year cooling period the re-energized sun's red giant phase will engulf the inner planets. Its fiery atmosphere will expand to a radius greater than the orbit of Earth. The ocean and atmosphere, all evidence of life, the crust, and perhaps the whole planet will be recycled into component atoms and hurled by shock waves into space (**Figure 2.13**). Our successors, if any, will have perished or fled to safer worlds. Its fuel exhausted and its energies spent, the sun will cool to a glowing ember and ultimately to a dark cinder. Perhaps a new system of star and planets will form from the debris of our remains.

The history of past and future Earth is shown as a time line in **Figure 2.14**.

FIGURE 2.13 The end of a solar system? The glowing gas in this beautiful nebula once formed the outer layers of a sunlike star that exploded only 10,000 years ago. The inner loops are being ejected by a strong wind of particles from the remnant central star. If planets orbited this star, their shattered remnants are contained in the outward-rushing orange filaments at the periphery. Perhaps 6 billion years from now, observers 5,000 light-years away would see a similar sight as our sun passes the end of its life.

FIGURE 2.14 Earth's history and future. Note that with the exception of the last column, each column indicates an expansion of part of the column to its left.

BOX 2.1

Comparative Oceanography

Planets with liquid on or near their surfaces may not be as rare as once thought. Water itself is not scarce. In the solar system, for example, Jupiter has hundreds of times more water than Earth does, nearly all of it in the form of ice. In 1998 ice was discovered in deep craters near our moon's south pole. Astronomers have even located water molecules drifting free in space. But *liquid* water is unexpected.

Consider the conditions necessary for a large permanent ocean of liquid water to form on a planet. An ocean world must move in a nearly circular orbit around a stable star. The distance of the planet from the star must be just right to provide a temperature environment in which water is liquid. Unlike most stars, a water planet's sun must not be a double or multiple star, or the orbital year would have irregular periods of intense heat and cold. The materials that accreted to form the planet must have included both water and substances capable of forming a solid crust. The planet must be large enough that its gravity will keep the atmosphere and ocean from drifting off into space.

Although it might be a bit premature to consider "comparative oceanography" as a career choice, researchers are increasingly certain that liquid water exists (or existed quite recently) on at least three other bodies in our solar system. We can begin to compare and contrast them.

Europa and Ganymede

The spacecraft *Galileo* passed close to Europa—a moon of Jupiter—in early 1997. Photos sent to Earth revealed a cracked, icy crust covering what appears to be a slushy mix of ice and water (**Figures a** and **b**). The jigsaw-puzzle pattern of ice pieces appears to have formed when the ice crust cracked apart, moved slightly, and then froze together again. In another pass early in 2000, *Galileo* detected a distinctive magnetic field, the signature of a salty liquid-water ocean below the ice.

The volume of this ocean is astonishing. Though Europa is slightly smaller than our own moon, the depth of the ocean averages about 160 kilometers (100 miles). The amount of water in its ocean is perhaps 40 times that of Earth's! Europa's ocean is probably kept liquid by heat escaping Europa's interior and by gravitational friction of tidal forces generated by Jupiter itself. Though the surface of the ice is about 8 kilometers (5 miles) thick and as cold as the surface of Jupiter, the liquid interior of the ocean, cradled deep in rocky basins, may be warm enough to sustain life. No continents emerge from this alien sea. A mission to the surface of Europa is being planned for launch in 2008.

Ganymede, Jupiter's largest satellite, was surveyed by *Galileo* in May 2000; the photographs showed structures

a   b

(**a**) A photograph of the icy surface of Europa—a moon of Jupiter—taken by the *Galileo* spacecraft on 20 February 1997. Jupiter's gravitational pull twists Europa, cracking the ice crust and warming the interior. In some areas the ice has broken into large pieces that have shifted away from one another. These can fit together like pieces of a jigsaw puzzle (**b**). This suggests the ice crust is lubricated by slush or liquid water.

(**d**) The edge of the north polar cap of Mars, photographed in April 2002, shows layers of water ice over darker layers of sand. The image is about 5 kilometers (3 miles) across.

(**c**) Mars color-coded by elevation. The blue areas are the farthest below "sea level," with red and brown indicating the highest elevations. A liquid ocean may have covered the blue areas at some time in the past. Note especially the prominent outflow channels and streamline features at the edges of the uplands.

strikingly similar to those on Europa. Again, magnetometer data suggested a salty ocean beneath a moving, icy crust.

Mars

Europa and Ganymede may have icy oceans now, but Mars, a much nearer neighbor, may have had an ocean in the distant past. An ocean could have occupied the low places of the northern hemisphere of Mars between 3.2 billion and 1.2 billion years ago when conditions were warmer (**Figure c**). Current models suggest that early in its history Mars had a thick, carbon dioxide–rich atmosphere, much like the atmosphere of early Earth. Carbon dioxide is a "greenhouse" gas; it traps the sun's heat like the glass panels of a greenhouse. The atmosphere kept Mars warm and allowed water to flow freely. In

2001 cameras aboard *Mars Global Surveyor* sent photos from the surface; they may show evidence that water once flowed from rock layers below the surface onto the bottom of a crater, leaving sediments that look suspiciously marine.

Where is the water now? Over the eons, rocks on the Martian surface absorbed the carbon dioxide and the atmosphere grew thin and cold. The ocean disappeared, its water binding to rocks or freezing beneath the planet's surface. Mars has become much colder in the past billion years, perhaps because of the loss of greenhouse gases in the atmosphere. If a large quantity of water is present today, most of it probably lies at the poles (**Figure d**) or beneath the surface in the form of permafrost.

Could wet pockets exist today? In June 2002 Michael Malin released photographs from *Mars Global Surveyor* that showed clear evidence of recently flowing water (**Figure e**). Images of crater walls show channels that end in fan-shaped "aprons"—debris fields similar to the alluvial fans deposited by water flowing rapidly over arid landscapes on Earth. The surfaces of these aprons are not marked by craters or the dark dust that covers nearly all Martian features. These gullies appear to be young.

How might liquid water form channels on such a cold planet? Warmed by volcanic heat within Mars, water might seep from the cliffs, then freeze to form ice dams at the tops of these gullies. Stressed by growing pressure from the warm water rising behind it, the dams might give way in a gully-forming flood. Confirmation must await a visit by planetary oceanographers.

Titan

Must an ocean consist of liquid water? Hydrocarbons have been seen on the surface of Titan, Saturn's largest moon. In 1999 scientists using the huge Keck Telescope in Hawaii de-

Malin Space Science Systems/MGS/JPL/NASA

(e) Young gullies on Mars. These valleys, possibly caused by flowing water, are about 3 kilometers (2 miles) wide and line a south-facing wall of a large crater. Researchers suggest these furrows formed between 10,000 and 1 million years ago. Their recent formation is indicated by the lack of surface craters.

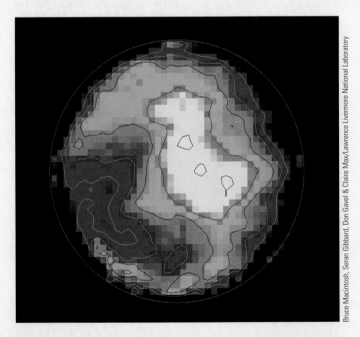

Bruce Macintosh, Seran Gibbard, Don Gavel & Claire Max/Lawrence Livermore National Laboratory

(f) In 1999 astronomers were able to peer through the haze on Saturn's moon Titan and glimpse an ocean of liquefied natural gas. The surface temperature is a chilly −180°C (−215°F).

© Don Davis

(g) An artist's conception of the surface of Titan showing an ocean of liquid methane and ethane. The continent in the foreground may consist of water ice.

tected cold, dark, infrared-absorbing organic matter surrounding a bright area about the size of Australia (**Figure f**). The evidence suggests a cold liquid ocean of methane, ethane, and other hydrocarbons (**Figure g**). The bright "continent" is probably made of water ice. More information may be obtained in 2004 by the *Cassini* spacecraft when it drops a probe into Titan's surface.

Extrasolar Planets

More than 100 planets had been discovered outside our solar system by the fall of 2003, and new planets are being discovered at the rate of about two per month! Most of these planets were found by watching the wobbling path a star takes through space when influenced by the gravity of a massive companion planet. In June 1996 astronomers at the University of Arizona announced the discovery of water vapor, methane, and ammonia in the atmosphere of one of the largest of these planets. The high temperature of its surface (as high as 700°C, or 1,300°F) would prevent the formation of a liquid-water ocean, but smaller and cooler planets with similar atmospheres have been found around two stars nearer the sun. Researchers at San Francisco State University have estimated surface temperatures on these planets at 85°C (185°F) and 44°C (112°F). These planets could have water in both vapor and liquid form—rain and oceans are possible.

Life and Oceans?

Could the presence of oxygen be a clue to the existence—or past existence—of life or of an ocean on a planet or moon? Since CO_2 is the "normal" composition for the atmosphere of a terrestrial planet, a large quantity of atmospheric oxygen would be unexpected—after all, oxygen is one of the most reactive of gases. Any oxygen in an atmosphere is likely to react with other materials. The red, rust-colored rocks typical of Mars almost certainly resulted from the oxidation (rusting) of iron-containing minerals. If a planet is found with lots of free oxygen, something is probably replenishing that oxygen.

That "something" is probably life. The action of photosynthetic organisms (including plants) produces excess oxygen. Without photosynthesis, Earth's atmosphere would be all but oxygen-free. As noted in this chapter, life, at least on Earth, almost certainly originated in the ocean. If the atmospheres of distant planets contain significant quantities of oxygen, an ocean and life might be possible. Scientists are close to technologies that will allow them to detect chemical signatures in the atmospheres of planets orbiting other stars.

2.9

Questions from Students

1. How do scientists know how old Earth, the solar system, and the galaxy are? How can they calculate the age of the universe?

The age estimates presented in this chapter are derived from interlocking data obtained by many researchers using different sources. One source is meteorites, chunks of rock and metal formed at about the same time as the sun and planets and out of the same cloud. Many have fallen to Earth in recent times. We know from signs of radiation within these objects how long it has been since they were formed. That information, combined with the rate of radioactive decay of unstable atoms in meteorites, moon rocks, and the oldest rocks on Earth, allows astronomers to make reasonably accurate estimates of how long ago these objects formed. By the way, there is essentially no evidence to support the contention that Earth is between 6,000 and 10,000 years old.

As for the age of the universe itself, by April 2002 astronomers had obtained very accurate measurements of its rate of expansion. By calculating backward, they found the universe would have begun its expansion between 13 and 14 billion years ago.

2. Life appears to have arisen on Earth soon after the formation of a stable surface. Could life have formed on other planets?

We have no evidence, direct or indirect, of life on other planets around our sun or elsewhere in the universe. Yet it seems provincial to assume that life could have arisen only here. The formation of organic molecules from simple chemicals receiving energy from lightning, heat, ultraviolet light, and other sources may be quite common, and increasing complexity in these compounds may be a universal phenomenon.

3. Would life on other planets resemble life on Earth?

Organisms elsewhere might be very different. Recall that life on this planet probably arose in the ocean, and all life-forms here carry an ocean of sorts within their bodies. On a planet without water, the organisms would be much different.

For example, on a hypothetical planet with an ammonia ocean, life would not have a structure of cells surrounded by lipid membranes. Lipid membranes are the sheets of fatty molecules that keep the inside of a cell separate from the environment, and ammonia prevents them from forming. Without membranes, cells as we know them are not possible. Notwithstanding this argument, life need not be confined to planets with water. Other life-forms may exist, based on other "brews."

4. What does it mean when someone says we're made of "starstuff"?

The late Carl Sagan popularized that term. His intention was to emphasize the interrelationship of the living and nonliving worlds—to indicate that life is an extension of stellar physics. Interestingly, three of the four most abundant elements—other than hydrogen and helium—are carbon, nitrogen, and oxygen. These elements are the primary ingredients of living organisms on Earth. Since stars are found all over the universe, and these elements appear to be easily constructed, the stuff of life must be common throughout the universe. Whether life itself has evolved elsewhere is a different question—we'll have to wait and see.

5. How far away are exploding stars? What if a star exploded in our neighborhood of the galaxy? Would we notice anything?

Of course it depends on what you mean by "neighborhood," but the outcome of a nearby event could be ugly. Look again at Figures 2.4 and 2.13. The first figure shows a cataclysmic supernova—an explosion capable of yielding such heavy elements as mercury, gold, and uranium. The second figure shows the end of a star like our sun, a relatively tame event as such things go. The intense bursts of gamma rays and X-rays from a huge supernova (a hypernova) could sterilize everything in part of a galaxy's spiral arm; nothing alive based on water and proteins would survive. The radiation from the disintegration of a sunlike star would be less catastrophic. Astronomers have detected gamma-ray bursts since the 1960s, but only in 2003 was a gamma burst directly associated with the first light from a hypernova. Fortunately the event happened in a distant galaxy.

Chapter in Perspective

In this chapter you learned that most of the atoms that make up Earth, its ocean, and its inhabitants were formed within stars billions of years ago. Stars spend their lives changing hydrogen and helium into heavier elements. As they die, some stars eject the elements into space during cataclysmic explosions. The sun and planets, including Earth, condensed from a cloud of dust and gas enriched by the recycled remnants of exploded stars.

Earth formed by accretion, the clumping of small particles into a large mass. The mass heated as it grew, eventually melting. The heavy iron and nickel crashed toward Earth's center to become its core; the lighter silicates and aluminum compounds rose to the surface to form a crust. The earth became *density stratified*—that is, layered by density.

The ocean formed as soon as Earth was cool enough for water to remain liquid. Life followed soon thereafter.

In the next chapter you will learn about Earth's inner layers—layers that are density stratified. You'll find these layers to be heavier and hotter as depth increases, and you'll learn how we know what's inside our planet even though we've never been past the outermost layer. As you'll see, today's earthquakes and volcanoes, and the slow movement of continents, are all remnants of our distant cosmological past.

Key Ideas

✓ The objects comprising our solar system condensed about 5 billion years ago from a thin cloud that had been enriched by heavy elements made in exploding stars.
Review Figure 2.3 and 2.4, pages 40–41

✓ Earth is *density stratified*. During its formation heavy materials fell inward to form the core, and lighter substances rose to become the outer layers. Ocean and atmosphere are the least dense of these layers.
Review Figure 2.6, page 42

✓ Earth first had a solid surface about 4.6 billion years ago. The ocean formed from clouds of steam and water vapor when Earth's surface became cool enough to allow liquid water to rest on the surface.
Review Figure 2.9, page 43

✓ Life probably arose on Earth shortly after its formation, about 4 billion years ago. Life may have arisen in the deep ocean.
Review pages 44–46

✓ Earth's atmosphere has changed substantially through time. The most dramatic change began about 2 billion years ago when photosynthetic organisms began producing large quantities of oxygen.
Review Figure 2.10, page 44

✓ Other planets and moons may have—or may have had—oceans.
Review Box 2.1, pages 47–50

Terms and Concepts to Remember

accretion, 41	nebula, 39
big bang, 38	outgassing, 42
biosynthesis, 45	planet, 41
condensation theory, 39	protostar, 39
density, 41	solar nebula, 39
density stratification, 41	solar system, 38
galaxy, 38	star, 38
Milky Way galaxy, 38	supernova, 38–39

Study Questions

Reviewing What You've Learned

1. How old is Earth? On what are those estimates based?
2. What is the relationship between galaxies and stars? What things make up galaxies?
3. What element makes up most of the detectable mass in the universe?
4. Briefly trace the life of a typical star.
5. How are planets and stars related?
6. How are light elements converted into heavy ones?
7. Will all stars end their lives as supernovas? What happens to the heavy elements made by small stars?
8. What is density stratification? What does it have to do with the present structure of Earth?
9. Are the ocean and present atmosphere "leftovers" from the original atmosphere of Earth?
10. What is biosynthesis? Where do researchers think it might have occurred on our planet? Could it happen again today?

Thinking Critically

1. Marine biologists sometimes say that all life-forms on Earth, even desert lizards and alpine plants, are marine. Why?
2. Where did Earth's surface water come from?
3. Do you think water planets are common in the galaxy? What about planets in general?
4. How might Earth be different if an ocean had not formed on its surface?
5. How do we know what happened so long ago?
6. Considering what must happen to form them, do you think ocean worlds are relatively abundant in the galaxy? Why or why not?
7. Could water planets form around multiple-star systems? Stars with greatly variable output?
8. **InfoTrac College Edition Project** Life as we know it on Earth apparently requires water in some form. The planet Mars appears not to have liquid water on its surface, but evidence suggests that it may have had water at one time. Could Mars have also hosted Earth-like life? Might life still exist there? Do some research using InfoTrac College Edition to support your answer.

Thinking Analytically

1. A light-year is the distance light can travel in one year. Light travels at 300,000 kilometers (186,000 miles) per second. Commercial television broadcasting began in 1939. Television signals travel at the speed of light. How far away would a space probe have to be before it could no longer detect those signals?
2. Density is mass per unit volume. Granite rock weighs about 2.6 g/cm³; water weighs about 1.0 g/cm³. Knowing their sizes, how might you determine whether Europa or Ganymede is hiding a large liquid-water ocean beneath an icy crust?

3. Can you think of any way an astronomer could detect a large planet orbiting a star without actually seeing the planet? (Hint: How would the star move as the planet orbits it?)

Oceanography **Now**™

Assess your understanding of this chapter's topics with additional quizzing and conceptual-based problems at http://oceanography.brookscole.com/garrison5e

Study and Research on the Internet

Assess your understanding of this chapter's topics with additional quizzing and conceptual-based problems at:

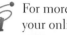 http://oceanography.brookscole.com/garrison5e

- **InfoTrac College Edition Research** leads you to critical thinking exercises that use InfoTrac College Edition® as a research tool.
- **Regional InfoTrac College Edition** articles are organized into East Coast, West Coast, and Gulf Coast regions, allowing you to study oceanography on a more local level.

For more readings, go to InfoTrac College Edition, your online research library, at:

http://infotrac.thomsonlearning.com

CHAPTER 3

A huge cauldron appeared about two days after the 1996 eruption began beneath Iceland's Vatnajôkull glacier. This part of the glacier was 400-600 meters (1,300-2,000 feet) thick, but its foundation had been melted by rapidly rising lava. On 5 November enough water had accumulated to float the glacier off its floor. Water then burst from beneath the glacier, causing a catastrophic flood. Heat from within Earth powers volcanic eruptions, causes earthquakes, and moves great slabs of Earth's surface.

EARTH STRUCTURE AND PLATE TECTONICS

Fire and Ice

This chapter describes the inner structure of Earth and how that structure forms the ocean floor and the continents. You may be surprised to discover that the rigid, brittle surface of our planet floats on a hot, deformable layer of partially melted rock. Over long spans of time, movement in and below this layer carries continents across the face of Earth, splits ocean basins, and shatters cities. Much of what you'll learn in the next three chapters depends on an understanding of this movement and its consequences.

Nowhere are the consequences more consistently apparent than in Iceland. The 253,000 inhabitants of this island country live on a part of the ocean floor that rises above sea level just south of the Arctic Circle. The seabed is pushed up by a long plume of hot material rising from near Earth's core. Maintaining civilization on top of a rising shaft of molten rock is not without its challenges; it is estimated that the volcanoes of Iceland have produced one-third of the total lava that has flowed onto dry land since the year 1500! With more hot springs and vapor vents than any other country in the world, Iceland is the most geologically active place on Earth.

What happens when a volcano erupts beneath a thick ice cap? On 30 September 1996 an eruption began beneath the Vatnajôkull glacier in central Iceland. About 36 hours later, the eruption had broken through the ice (see **figure**). The interaction of lava and ice assured a violent outcome; rhythmic eruptions beginning on 2 October threw ash and smoke more than 4,600 meters (15,000 feet) into the air. Meltwater flooded into the crater and surrounding area, and by 5 November enough water had accumulated to float the glacier off its floor. Water then burst from beneath the glacier, causing a catastrophic flood, the largest recorded in the twentieth century (55,000 cubic meters per second!).

Because of advance warning, there was no loss of life from the eruption or subsequent flood. Damage to roads, bridges, and power lines was considerable, however. The citizens of Iceland tap steam from underground reservoirs to generate electricity and heat community swimming pools, and their steam-heated greenhouses provide fresh vegetables year-round, but those benefits must be weighed alongside the occasional bouts of geological terror!

As you read this chapter you should be especially alert to the idea of *density*. How does density affect the way Earth's inner layers are stacked? How can such heavy layers move? Does their density affect their movement? How did people figure out what's inside our planet without having been there or having deep samples to study?

A Layered Earth

As we saw in Chapter 2, Earth was formed by accretion from a cloud of dust, gas, ices, and stellar debris. Gravity later sorted the components by density, separating Earth into layers (see again Figure 2.6). Because each deeper layer is denser than the layer above, we say that Earth is **density stratified.** Remember, **density** is an expression of the relative heaviness of a substance; it is defined as the mass per unit volume, usually expressed in grams per cubic centimeter (g/cm³). The density of pure water is 1 g/cm³. Granite rock is about 2.7 times denser, at 2.7 g/cm³.

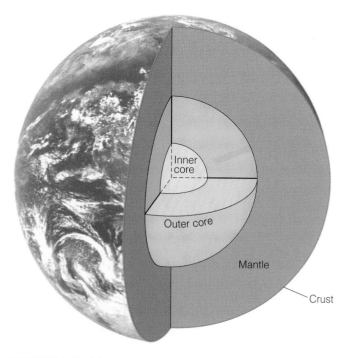

FIGURE 3.1 Earth in cross section, showing the major internal layers.

It might seem easy to satisfy our curiosity about the nature of inner Earth by digging or drilling for samples. The deepest hole drilled so far was bored by researchers on the Kola Peninsula in Russia. In 1992 they reached a depth of 12,063 meters (7.5 miles), where temperature (245°C, 473°F) and pressure squeezed the hole closed. Drilling has also been conducted at sea. The oceanic drilling record is held by *JOIDES Resolution*. In 1991, after 12 years of intermittent effort, a drill aboard the ship penetrated 2 kilometers (1.2 miles) of seafloor beneath 2.5 kilometers (1.6 miles) of seawater.

No matter where investigators drill, the samples of rock they recover are not exceptionally dense. The deepest probes have penetrated less than 1/500 of the radius of Earth, however. Should we assume that Earth consists of lightweight (less dense) rock all the way through?

Thanks to studies of Earth's orbit begun in the late 1700s, we know Earth's total mass. This mass is much greater than would have been predicted from even the deepest rocks ever collected. Earth's interior must therefore contain heavier (more dense) substances than the rocks from the deep drill holes. We might expect materials from the interior to get denser gradually with increasing depth, but geologists have shown that the density of the materials increases abruptly at specific depths. Thus they are convinced that Earth has *distinct interior layers*, somewhat resembling the inside of an onion (**Figure 3.1**).

The Evidence for Layering

Scientists have known since the mid-1800s that low-frequency waves can travel through the interior of Earth. The forces that cause **earthquakes** generate low-frequency waves called **seismic waves** (*seismos* = earthquake). Some of these waves radiate through Earth, reflecting or bending as they travel, and eventually reappear at the surface. Careful study of the time and location of their arrival at the surface, along with changes in the frequency and strength of the waves themselves, has revealed information about the nature of Earth's interior. (We use the same kind of analysis to select a ripe watermelon. If we tap the outside and hear a *tick*, we suspect the melon isn't ripe. A *thunk* indicates a winner.)

Primary and Secondary Seismic Waves

Two kinds of seismic waves are particularly useful for analyzing Earth's interior structure. One kind of wave, the **P wave** (or primary wave), is a compressional wave similar in behavior to a sound wave. Rapidly pushing and pulling a very flexible spring (like a Slinky) generates P waves. The **S wave** (or secondary wave) is a transverse wave like that seen in a rope shaken side to side. Both kinds of seismic waves are shown in **Figure 3.2**.

P waves and S waves are generated simultaneously at the source of an earthquake. P waves travel through Earth nearly twice as fast as S waves, so they arrive first at a distant **seismograph** (*seismos* = earthquake, *graphos* = to write), an instrument that senses and records earthquakes. Liquids are unable to transmit the side-to-side S waves, but they do propagate compressional P waves. Solid rock transmits both kinds of waves. Analysis of the characteristics of seismic waves returning to Earth's surface after passage through the interior suggests which parts of the interior are solid, liquid, or partially melted.

FIGURE 3.2 P waves (primary waves) are compressional waves like those seen in a Slinky that is alternately stretched and compressed. S waves (secondary waves) are side-to-side waves like those seen in a shaken rope. Both kinds of waves are associated with earthquakes. (Source: From *The Earth: An Introduction to Physical Geology,* 4/e by Edward J. Tarbuck and Frederick K. Lutgens. © 1993 Macmillan College Publishing. Reprinted by permission of Pearson Education, Inc. Upper Saddle River, NJ 07458.)

Shadow Zones

In 1900 the English geologist Richard Oldham first identified P and S waves on a seismograph. If Earth were perfectly homogeneous, seismic waves would travel at constant speeds from an earthquake, and their paths through the interior would be straight lines (**Figure 3.3a**). Oldham's investigations, however, showed that seismic waves were arriving *earlier* than expected at seismographs far from the quake. This meant that the waves must have traveled *faster* as they went down into Earth. They must also have been refracted—bent back toward the surface (**Figure 3.3b**).[1] Oldham reasoned that the waves were being influenced by passage through areas of Earth with different density and elastic properties than those seen at the surface. This reasoning showed that Earth is not homogeneous and that its properties vary with depth.

In 1906 Oldham made a critical discovery: No S waves survived deep passage through Earth (**Figure 3.3c**). Oldham deduced that a dense fluid structure, or core, must exist within Earth to absorb the S waves. He further predicted that a **shadow zone**, a wide band from which S waves were absent, would be found on the side of Earth opposite the location of an earthquake. The existence of the liquid core and shadow zone was verified by seismographic analysis in 1914.

What about the P waves—the compressional waves that can pass through liquid? Oldham found that P waves arrived at a seismograph farthest away from an earthquake (that is, on the opposite side of the globe) much more slowly than expected. They had been deflected but not stopped by Earth's core (**Figure 3.3d**). Working from this information, later researchers were able to calculate that the core–mantle boundary is about 2,900 kilometers (1,800 miles) below the surface.

More sensitive seismographs were developed in the 1930s. In 1935 the Danish seismologist Inge Lehmann suggested that the very faint, very low-frequency P waves discovered opposite earthquake sites had speeded up as they passed through an inner core, indicating that it was a solid. Measurements of subtle differences in the pull of gravity, plus a more accurate estimate of Earth's mass (derived from precise timings of the orbits of artificial satellites), gave further clues to the layered structure of Earth. By the early 1960s another new generation of sensitive seismographs stood ready to provide geologists with an even better understanding of Earth's inner configuration. To confirm their theories, scientists needed data from a very large earthquake.

Alaska, 27 March 1964: The "Experiment" That Confirmed a Theory

They did not have to wait long. On the last Friday of March 1964, at 5:36 P.M. one of the largest earthquakes ever recorded struck 144 kilometers (90 miles) east of Anchorage, Alaska

---

[1] The principle of *refraction* is explained and illustrated in Figure 6.18.

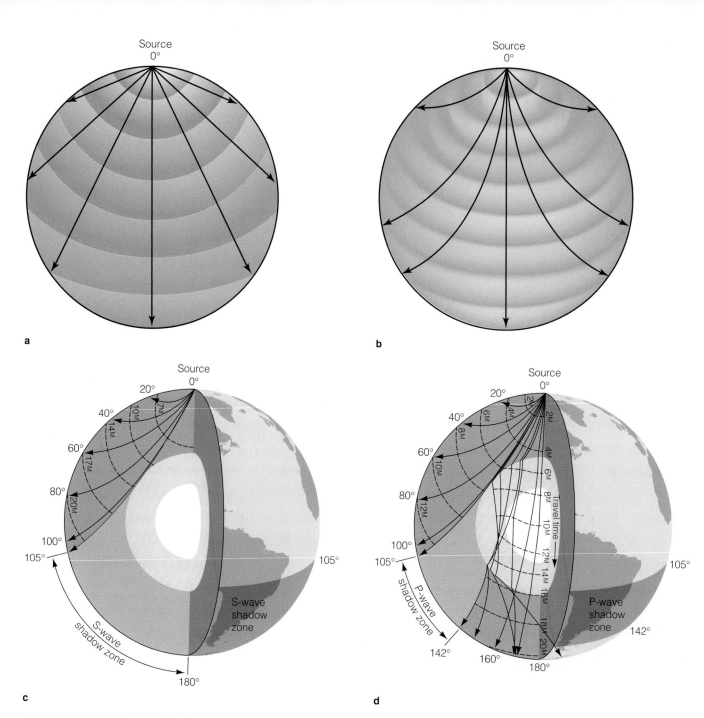

ACTIVE **FIGURE 3.3** How earthquakes contributed to our model of the layered Earth. (**a**) Earthquake waves passing through a homogeneous planet would not be reflected or refracted (bent). The waves would follow linear paths. (**b**) In a planet that becomes gradually denser and more rigid with depth, the waves would bend along evenly curved paths. (**c**) Earth has a liquid outer core through which the side-to-side S waves cannot penetrate, creating a large "shadow zone" between 105° and 180° from an earthquake's source. (Note that M is time in minutes.) (**d**) P waves (compressional waves) can penetrate the liquid outer core but are bent in transit. A P-wave shadow zone forms between 105° and 142° from an earthquake's source. Very sensitive seismographs can sometimes detect weak P-wave signals reflecting off the solid inner core (see Box 3.1). (Source for **a, b**: From *The Earth's Dynamic Systems*, 9/e by Hamblin and Christiansen. Reprinted by permission of Pearson Education, Inc., Upper Saddle River, NJ 07458.) (Source for **c** and **d**: From *Understanding Earth*, 3/e by Frank Press and Raymond Siever. © 1994, 1998, and 2001 by W. H. Freeman and Company. Used with permission.)

Oceanography⊗Now™ | At the Active Figures link at http://oceanography.brookscole.com/garrison5e you can view how earthquakes contributed to the model of the layered earth.

BOX 3.1   Echoes from the Core

By carefully studying low-frequency pulses of energy generated by the forces that cause earthquakes, researchers have been able to discover a good deal about the nature of Earth's interior. In one of the twentieth century's nicest pieces of geological detective work, scientists discovered the concentric inner and outer cores within our planet.

In 1906 British geologist Richard Oldham published a paper showing that the earthquake-related low-frequency pulses called S waves do not penetrate straight through Earth. He also found that P waves arrive at measuring instruments farthest away from an earthquake (that is, on the opposite side of the globe) much more slowly than expected. Oldham thought that a dense fluid core must exist within Earth to slow and absorb the waves. He further predicted that a shadow zone, a wide band from which seismic waves would be nearly absent, would encircle the side of Earth opposite the location of an earthquake. Such a shadow zone could be formed by reflection and refraction of waves by a liquid core. The existence of the shadow zones and liquid core was verified in 1914.

A more subtle feature was discovered by Inge Lehmann, a Danish seismologist. Denmark is well situated for recording waves that pass through the core of Earth from large earthquakes around the southern Pacific rim. By 1935 a new generation of very sensitive detection instruments was available, and Lehmann used them to sense the very faint, very low-frequency waves that reach Earth's surface opposite an earthquake. The waves had speeded up as they passed through an inner core, indicating that it was a solid. Delicate secondary reflections suggested some of the characteristics of this material. Lehmann had discovered a rigid white-hot mass about the size of the moon suspended within Oldham's liquid core.

Excellent teachers had prepared both researchers to learn from the ghostly echoes of the cores. Although Oldham's educational background was fairly typically for a Victorian gentlemen of the era, Inge Lehmann's personal history was not. Educated at the first coeducational school in Denmark, Lehmann decided to pursue a scientific career, partly because "no difference between the intellect of boys and girls was recognized [at the school], a fact that brought me disappointment later in life when I had to realize that this was not the general attitude." A few years after graduating from the University of Copenhagen in mathematics and physical sciences, Lehmann was appointed chief of the seismological department of the Royal Danish Geodetic Institute in Copenhagen, a post she held until her retirement in 1953. She passed away in 1993 at the age of 105.

3.7

National Portrait Gallery, London

Richard Oldham, discoverer of Earth's core.

© National Survey & Cadastre, Denmark

Inge Lehmann, discoverer of Earth's inner core.

Sverdlosk

Traces from Lehmann's 1936 paper showing the arrival in Russia of seismic waves from a New Zealand earthquake on 16 June 1929. The core wave arrivals are marked by arrows. Source: B. Bolt, "Earthquakes and Geological Discovery," Scientific American Books, ©1993. (Reprinted with permission of W. H. Freeman & Co.)

FIGURE 3.4 Much of the Anchorage suburb of Turnagain Heights was destroyed by landslides and ground liquefaction in the 27 March 1964 Alaska earthquake. The earthquake's magnitude was 9.2—the highest ever recorded in the United States. Seismic waves associated with the earthquake passed through Earth and emerged at distant locations, carrying information about Earth's interior.

Wilbur Garrett/NGS Image Collection

(**Figure 3.4**). The release of energy tore the surface of Earth for 800 kilometers (500 miles) between the small port of Cordova in the east and Kodiak Island in the west. In some places the vertical movement of the crust was 3.7 meters (12 feet); one small island was lifted 11.6 meters (38 feet). Horizontal movement caused the greatest damage: 65,000 square kilometers (25,000 square miles) of land abruptly moved west. In $4\frac{1}{2}$ minutes of violent shaking, Anchorage had moved sideways 2 meters (6.6 feet), and the town of Seward had moved 14 meters (46 feet)! A seismic (earthquake-generated) sea wave destroyed two harbors. More than 75% of the state's commerce was disrupted, and thousands of people were made homeless. Damage exceeded $750 million, and, considering the violence of the earthquake, it is a wonder that only 115 lives were lost.

Seismic stations over much of the world saw extraordinarily large P waves arrive from Alaska. Many of the 800 seismographs online worldwide were physically damaged. The arrival times of the P and S waves at each station were carefully noted. When correlated with the frequency, intensity, and phase characteristics of the waves, this information helped to confirm the models of Earth layering. The shaken citizens of Anchorage probably didn't derive much comfort from the knowledge gained from "their" earthquake.

Earth's Inner Structure

Although researchers have never directly collected samples from below the outermost layer of Earth, they have indirect evidence about the chemical composition, density, temperature, and thickness of each layer of Earth's interior. This evidence was pieced together from measurements of earthquake shocks, volcanic gases, and variations in the pull of gravity.

Classifying the Layers by Composition

The first useful scientific classification of Earth's interior emphasized chemical composition. Geologists named the layers *crust*, *mantle*, and *core*.

The **crust** is the thin, relatively lightweight outermost layer. It accounts for only 0.4% of Earth's total mass and less than 1% of its volume. The crust beneath the ocean differs in thickness, composition, and age from the crust of the continents. The thin **oceanic crust** is mostly **basalt**—a heavy, dark-colored rock composed largely of oxygen, silicon, magnesium, and iron. Its density is about 2.9 g/cm³. By contrast, the most common material in the thicker **continental crust** is **granite**, a familiar light-colored rock composed mainly of oxygen, silicon, and aluminum. Its density is about 2.7 g/cm³.

The **mantle,** the layer beneath the crust, comprises 68% of Earth's mass and 83% of its volume. Mantle materials are thought to contain mainly silicon and oxygen with iron and magnesium. The mantle's average density is about 4.5 g/cm³. It is about 2,900 kilometers (1,800 miles) thick.

The **core,** Earth's innermost layer, consists mainly of iron (90%) and nickel, along with silicon, sulfur, and heavy elements. Its average density is about 13 g/cm³, and its radius is approximately 3,470 kilometers (2,160 miles). The core accounts for about 31.5% of Earth's mass and about 16% of its volume.

Classifying the Layers by Physical Properties

Subdividing Earth on the basis of chemical composition does not always reflect the *physical* properties and behavior of rock materials in these layers. Different conditions of temperature and pressure prevail at different depths, and these conditions influence the physical properties of the materials subjected to them.

Slabs of Earth's relatively cool and solid crust and upper mantle float—and move—over the hotter, pliable (deformable) mantle layer directly below. Physical properties are more important than chemical ones in determining this movement, so geologists have devised a classification based on physical properties. These are shown in **Figure 3.5:**

- The **lithosphere** (*lithos* = rock) is Earth's cool, rigid outer layer—100–200 kilometers (60–125 miles) thick. It is composed of the continental and oceanic crusts *and* the uppermost cool and rigid portion of the mantle.

- The **asthenosphere** (*asthenes* = weak) is the hot, partially melted, slowly flowing layer of upper mantle below the lithosphere extending to a depth of 350–650 kilometers (220–400 miles).

- The **lower mantle** extends to the core. The asthenosphere and the mantle below the asthenosphere (the lower mantle) have a similar chemical composition. Although it is hotter, the mantle below the asthenosphere does not melt because of rapidly increasing pressure. As a result it is more dense and flows more slowly.

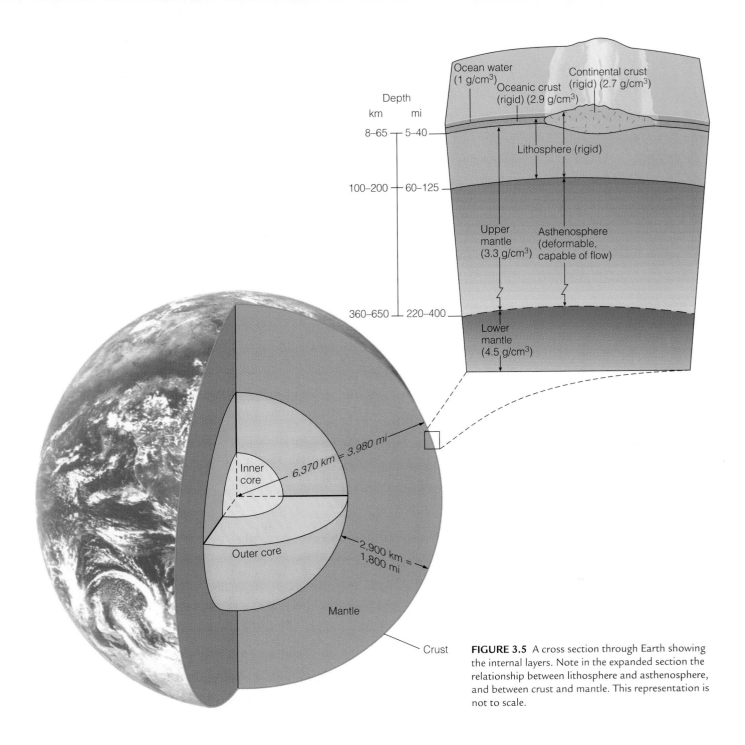

FIGURE 3.5 A cross section through Earth showing the internal layers. Note in the expanded section the relationship between lithosphere and asthenosphere, and between crust and mantle. This representation is not to scale.

■ The **core** has two parts. The outer core is a dense, viscous liquid. The inner core is a solid with a maximum density about 16 g/cm³, nearly six times the density of granite rock. Both parts are very hot, with an average temperature of about 5,500°C (9,900°F). Recent evidence indicates that the inner core may be as hot as 6,600°C (12,000°F) at its center—hotter than the surface of the sun!

Figure 3.5 shows the lithosphere and asthenosphere in detail. *Note that the rigid sandwich of crust and upper mantle—the lithosphere—floats on (and is supported by) the denser deformable asthenosphere.* Note also that the structure of

oceanic lithosphere differs from that of continental lithosphere. Because the thick granitic continental crust is not exceptionally dense, it can project above sea level. In contrast, the thin, dense basaltic oceanic crust is almost always submerged.

Isostatic Equilibrium

Why do large regions of continental crust stand high above sea level? If the asthenosphere is partially melted and deformable, why don't mountains sink because of their

a

b

FIGURE 3.6 The principle of buoyancy. (a) A ship sinks until it displaces a volume of water equal in weight to the weight of the ship and its cargo. (b) Icebergs sink into water so that the same proportion of their volume (about 90%) is submerged. The more massive the iceberg, the greater this volume is. The large iceberg rides higher but also extends to a greater depth than the small one.

mass and disappear? Another look at Figure 3.5 will help to explain the situation. The mountainous parts of continents have "roots" extending into the asthenosphere. The continental crust and the rest of the lithosphere "float" on the denser asthenosphere. The situation involves buoyancy, the principle that explains why ships and icebergs float.

Buoyancy is the ability of an object to float in a fluid by displacing a volume of that fluid equal in weight to the floating object's own weight. *A steel ship floats because it displaces a volume of water equal in weight to its own weight plus the weight of its cargo.* An empty containership displaces a smaller volume of water than the same ship fully loaded (**Figure 3.6a**). The water that supports the ship is not *strong* in the mechanical sense; water does not support a ship the same way a steel bridge supports the weight of a car. Buoyancy, rather than mechanical strength, supports the ship and its cargo.

The buoyancy of an object depends on both its density *and* its mass. Think of floating objects of different *sizes* but the same *density*—icebergs, for example. All icebergs float with about 10% of their volume exposed above sea level. Because the submerged portion (the "root") of an iceberg is nine times larger than its exposed portion, larger icebergs ride higher and extend to greater depths than smaller ones (**Figure 3.6b**). An iceberg 10 meters (33 feet) high will stick out of the water about 1 meter (3 feet), but a 100-meter- (330-foot-) high iceberg will stand 10 meters (33 feet) above sea level. *For an iceberg to stand high in the water, it must have a deep submerged portion to support it.*

Now think of continents. Any part of a continent that projects above sea level must be supported in the same way. Consider the continent that contains Mt. Everest, the highest of Earth's mountains at 8.84 kilometers (29,007 feet) above sea level. Mt. Everest and its neighboring peaks are not supported by the *mechanical* strength of the materials within Earth; nothing in our world is that strong. Over a long period, and under the tremendous weight of the overlying crust, the asthenosphere behaves like a dense, viscous, slowly moving fluid. *The continent's mountains float high above sea level because the lithosphere gradually sinks into the deformable asthenosphere until it has displaced a volume of asthenosphere equal in mass to the mountains' mass.* The mountains stand at great height, nearly in balance with their subterranean underpinnings but susceptible to rising or falling as erosion or crustal stresses dictate. Lower regions are supported by shallower "roots." In a slow-motion version of a ship floating in water, the entire continent stands in **isostatic equilibrium** (*isos* = equal, *stasis* = standing).

What happens when a mountain erodes? In much the same way a ship rises when cargo is removed, Earth's crust will rise in response to the reduced load. Ancient mountains that have undergone millions of years of erosion often expose rocks that were once embedded deep within their roots. This kind of isostatic readjustment results in the thinning of the continental crust beneath the mountains and subsidence beneath areas of deposited sediments. This process is shown in **Figure 3.7.**

Unlike the asthenosphere on which the lithosphere floats, crustal rock does not slowly flow at normal surface temperatures. A ship or iceberg reacts to any small change in weight with a correspondingly small change in vertical position in the water, but an area of continent or ocean floor cannot react to every small weight change because the underlying rock is *not* liquid, the deformation does not occur rapidly, and the edges of the continent or seabed are mechanically bound to adjacent crustal masses. When the force of uplift or downbending exceeds the mechanical strength of the adjacent rock, the rock will fracture along a plane of weakness—a **fault.** The adjacent crustal fragments will move vertically in relation to each other. This sudden adjustment of the crust to isostatic forces by fracturing, or faulting, is one cause of earthquakes. The lithosphere does not always behave like a rigid, brittle solid, however. Where forces are applied slowly enough, some of this material may deform without breaking.

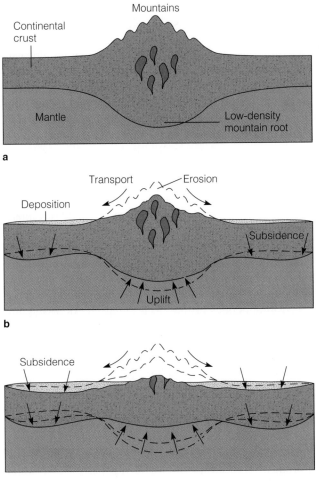

Mountains

Continental crust

Mantle

Low-density mountain root

a

Transport — Erosion

Deposition

Subsidence

Uplift

b

Subsidence

c

ACTIVE **FIGURE 3.7** Erosion and isostatic readjustment can cause continental crust to become thinner in mountainous regions. As mountains are eroded over time (**a–c**), isostatic uplift causes their roots to rise. The same thing happens when a ship is unloaded or an iceberg melts. Further erosion exposes rocks that were once embedded deep within the peaks. Deposition of sediments away from the mountains often causes nearby crust to sink.

Oceanography⊜Now™ | At the Active Figures link at http://oceanography.brookscole.com/garrison5e you can view the process of erosion and isotatic readjustment.

Toward a New Understanding of Earth

Earth's layered internal structure—its brittle lithosphere floating on the hot, viscous asthenosphere—ensures that its surface will be geologically active. Earthquakes and volcanoes attest to the commotion within. But how do internal layering and heat contribute to mountain building, the arrangement of continents, the nature of the seafloors, and the wealth of seemingly randomly distributed geological features found everywhere? Are there patterns and order in this apparent chaos? Let's see how we reached our current answers to these questions.

Sources of Internal Heat

Heat from within Earth keeps the asthenosphere pliable and the lithosphere in motion. Studies of the sources and effects of Earth's internal heat have contributed to our understanding of the planet's internal construction. In the late 1800s the British mathematician and physicist William Thomson (Lord Kelvin) calculated that Earth was about 80 million years old, an estimate based on the rate at which the planet would have cooled from an original molten mass. Geologists tried to explain mountain building based on Lord Kelvin's assumption of progressive cooling. In their drying-fruit model, Earth was considered to have shrunk as it cooled. Mountains were thought to be shrinkage wrinkles—like those seen as a grape becomes a raisin—and earthquakes were thought to be caused by jerks during this wrinkling. Earth's true age was not then known, and the "wrinkle" theory depended on rapid cooling of a relatively young Earth. When further research showed that Earth is about 4.6 billion years old (see Chapter 2), calculations of heat flow indicated that our planet should have cooled almost completely by now. Earthquakes, active volcanoes, and hot springs clearly indicated it had not! There must be other sources of heat energy besides the trapped ancient heat of formation. Better explanations for mountain building and earthquakes were needed.

An important source of heat that was not recognized in Lord Kelvin's time is **radioactive decay.** Though most atoms are stable and do not change, some forms of elements are unstable and give off heat when their nuclei break apart (decay). Radioactive particles are ejected in the process. As we also saw in Chapter 2, radioactive decay within the newly formed Earth released heat that contributed to the melting of the original mass. Most of the melted iron sank toward the core, releasing huge amounts of energy. This residual heat combines with the much greater heat given off by the continuous decay of radioactive elements within the crust and upper mantle (primarily potassium, uranium, and thorium).

Some of Earth's internal heat journeys toward the surface by **conduction,** a process analogous to the slow migration of heat along a skillet's handle. Some heat also rises by **convection** in the asthenosphere and mantle. Convection occurs when a fluid or semisolid is heated, expands and becomes less dense, and rises. (It is convection that causes air to rise over a warm radiator—see, for example, Figure 8.7.)

So, even after 4.6 billion years, heat continues to flow out from within Earth. As we shall see, this heat, not raisinlike global shrinkage, builds mountains and volcanoes, causes earthquakes, moves continents, and shapes ocean basins.

Oceanography⊜Now™ | Click Oceanography Interactive to view an animation of absolute dating through geological times.

The Age Debate

Geologists read Earth's history in its surface rocks and features. In the past the effort to understand geological processes was especially hindered by an incomplete understanding of Earth's age. Advances in geology had to wait—quite literally—for the time to be right.[2]

As you saw in Chapter 2, interlinking lines of evidence now tell us that the 4.6-billion-year age of Earth can be accepted with confidence. But before appropriate tools were developed, researchers were frustrated by tradition and seemingly contradictory data. At the end of the eighteenth century most European natural scientists believed in a young Earth, one that had formed only about 6,000 years ago. This age had been determined not by an analysis of rocks, but through the genealogy of the Bible's Old Testament. Imaginative reading of the book of Genesis in 1654 had convinced Irish bishop James Ussher that the Creation had taken place on 26 October 4004 B.C.

James Hutton—a Scottish physician with an interest in geology—decided that the biblical account of the Creation was incorrect because it implied that the landscape was mostly stable and unchanging. Hutton had measured geological changes: the rate at which stream beds eroded, the distribution of sediments by rivers, and the patterns of rocks in the Scottish countryside. From his observations, Hutton concluded that the rate of geological change today is not greatly different from the rate of change in the past. His principle of **uniformitarianism,** formalized in 1788, suggested that all of Earth's geological features and history could be explained by processes identical to ones acting today, and that these processes must have been at work for a very long time (**Figure 3.8**).

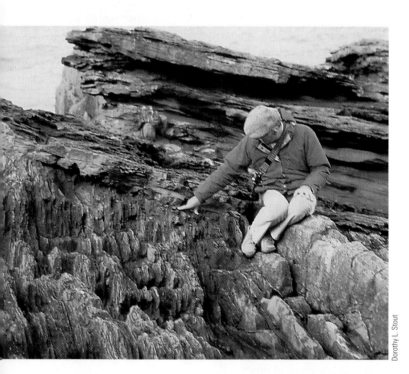

FIGURE 3.8 Rocks at Siccar Point, Scotland, that helped convince James Hutton of Earth's great age. Both layers of rock, now at right angles, were laid down horizontally on the seafloor at different times. After the first layer was deposited, earth movements tilted it vertically and uplifted it to form mountains. Erosion wore down those mountains. When the lower layer was once again submerged, the upper layers were deposited horizontally on top. Now both sets of rock layers have been uplifted and eroded. Clearly this process required time—and lots of it.

A few scientists agreed with Hutton. His detractors, however, asked some painful questions: If Earth is very old, and if erosive forces have continued uniformly through time, why isn't Earth's surface eroded flat? Why isn't the ocean brimming with sediment? What post-Creation forces could build mountains?

Believers in another school of thought, **catastrophism,** were able to answer these objections by interpreting the biblical account of the Creation literally. The catastrophists were convinced that Earth was very young and that the biblical flood was responsible for the misleading appearance of Earth's great age. The flood, they maintained, had folded and exposed strata, toppled mountains, filled shallow ocean basins with sediments, and caused many plants and animals to become extinct. This hypothesis had the additional benefit of explaining how seashell fossils could be present on mountaintops.

A further complication was introduced in the late 1850s when Charles Darwin and Alfred Russell Wallace proposed a rational mechanism—natural selection—by which new kinds of living things might come about. Because natural selection required long periods of time to generate the overwhelming variety of life-forms on Earth, biological evidence also suggested an ancient Earth. The arguments intensified.

Scientists attempting to prove or disprove uniformitarianism, catastrophism, and evolution made a wealth of new discoveries during the last half of the nineteenth century. As we've seen, they improved the seismograph and discovered long-distance earthquake (seismic) waves. Others probed the ocean floors, drew more accurate charts, collected mineral samples from great heights and depths, measured the flow of heat from within Earth, and identified patterns in the worldwide distribution of fossils. All the theories were reassessed. *The evidence from these explorations convinced most researchers that Earth was truly of great age.* The stage was now set for a revolution in geology, the development of what we know today as the theory of plate tectonics. The first steps toward the theory were tentative, however, and some of its proponents were dismissed as lunatics.

A Puzzling Fit 3.15

As Leonardo da Vinci noticed on early charts, in some regions the continents look as if they would fit together like jigsaw-puzzle pieces if the intervening ocean were removed. In 1620 Francis Bacon also wrote of a "certain correspondence" between shorelines on either side of the South Atlantic. In 1885 Edward Suess, a respected German scientist, elaborated on suggestions that the Southern Hemisphere's continents might once have been a single large landmass. He based his belief in part on the similarities of fossils found on these continents, especially fossils of the fern *Glossopteris.* Suess was not taken seriously by his colleagues because he could not explain how the continents had moved. Still, the fit of South America to Africa was striking (**Figure 3.9**).

As they probed the submerged edges of the continents, marine scientists found that the ocean bottom nearly always

---

[2] For information on geological time and its divisions, please see Appendix II.

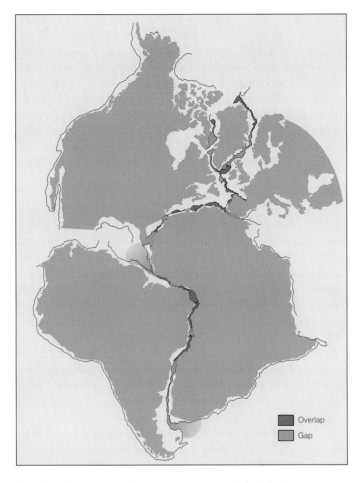

FIGURE 3.9 The fit of all the continents around the Atlantic at a water depth of about 137 meters (450 feet), as calculated by Sir Edward Bullard at the University of Cambridge in the 1960s. This graphic was an effective stimulus to the tectonic revolution.

Overlap
Gap

FIGURE 3.10 Alfred Lothar Wegener writes in his journal at the beginning of what would be his last expedition in Greenland in 1930. His remarkable book *The Origin of Continents and Oceans* was published in 1915. In it he outlined interdisciplinary evidence for his theory of continental drift.

sloped gradually out to sea for some distance and then dropped steeply to the deep-ocean floor. They realized that these shelflike continental edges were extensions of the continents themselves. In the few locations where they had measurements, researchers found that the fit between South America and Africa, impressive at the shoreline, was even better along the submerged edges of the continents.

Though so accurate a fit almost certainly could *not* have occurred by chance, no one had yet proposed a mechanism that could separate whole continents into moving pieces. If such a mechanism did exist, its gradual operation would surely require a great deal of time.

Oceanography☙Now™ | Click Oceanography Interactive to view an animation of changing Earth through geological times.

Continental Drift

Into the fray stepped **Alfred Wegener**, a busy German meteorologist and polar explorer (**Figure 3.10**). In a lecture in 1912 he proposed a startling and original theory, **continental drift.**

Wegener suggested that all Earth's land had once been joined into a single supercontinent surrounded by an ocean. He called the landmass **Pangaea** (*pan* = all, *gaea* = Earth, land) and the surrounding ocean **Panthalassa** (*pan* = all, *thalassa* = ocean). Wegener thought Pangaea had broken into pieces about 200 million years ago. Since then, he said, the pieces had moved to their present positions and were still moving.

Wegener's evidence included the apparent shoreline fit of continents across the North and South Atlantic and new information on offshore contours obtained by contemporary oceanographic expeditions. He pointed to Suess's *Glossopteris* fossils; to areas of erosion apparently caused by the same glacier in tropical areas now widely separated (South Africa, India, and Australia); and to Ernest Shackleton's 1908 discovery of coal, the fossilized remains of tropical plants, in frigid Antarctica. Wegener even suggested that volcanic activity was powered by the friction of continental movement.

Unlike anyone before him, Wegener also proposed a mechanism to account for the hypothetical drift. He believed that the heavy continents were slung toward the equator on the spinning Earth by a centrifugal effect. This inertia, coupled with the tidal drag on the continents from the combined effects of sun and moon, would account for the phenomenon of drifting continents, he thought.

Wegener was dismissed as a crank. His detractors claimed, with some justification, that he had carefully selected only those data supporting his hypothesis, ignoring contrary evidence. Where, for instance, were the wakes or tracks through old seabed that the migrating continents would leave? But a few geologists sided with Wegener. These "drifters" were hesitant to embrace the centrifugal force theory, yet they were unable to propose an alternative power source that could move the massive granitic continents.

The greatest block to the acceptance of continental drift was geologists' view of Earth's mantle. The available evidence seemed to suggest that a deep, solid mantle supported the crust and mountains *mechanically* (not *isostatically*) from below. Drift would be impossible with this kind of subterranean construction. A few perceptive seismic researchers then noticed that the upper mantle reacted to earthquake waves as if it were a deformable mass, not a rigid solid. Perhaps such a layer would resemble a slug of iron heated in a blacksmith's forge; it would deform with pressure and even flow slowly. Established geologists dismissed this interpretation, however, saying that the mountains would simply fall over or sink without rigid underpinnings. By 1926 the "drifters" were in full retreat. When Wegener died on an expedition across Greenland in 1930, his theory was already in eclipse.

The Idea Transformed

The theory of continental drift refused to die; those neatly fitted continents provided a haunting reminder of Wegener to anyone looking at an Atlantic chart. In 1935 a Japanese scientist, Kiyoo Wadati, speculated that earthquakes and volcanoes near Japan might be associated with continental drift. In 1940 seismologist Hugo Benioff plotted the locations of deep earthquakes at the edges of the Pacific. His charts revealed the true extent of the **Pacific Ring of Fire,** a circle of violent geological activity surrounding much of the Pacific Ocean. Seismographs were now beginning to reveal a worldwide pattern of earthquakes and volcanoes. Deep earthquakes did not occur randomly over Earth's surface but were concentrated in zones that extended in lines along Earth's surface.

Benioff, Wadati, and others wondered what could cause such an orderly pattern of deep earthquakes. Many of the lines corresponded with a worldwide system of oceanic ridges, the first of which was plotted in 1925 by oceanographers aboard the research ship *Meteor* working in the middle of the North Atlantic. **Figure 3.11** is a plot of about 10,000 earthquakes. Notice the odd pattern they form—almost as if Earth's lithosphere is divided into sections! Benioff's sensitive seismographs also began to gather strong evidence for a deformable, nonrigid layer in the upper mantle. Could the continents somehow be sliding on that layer?

Other seemingly unrelated bits of information were accumulating. **Radiometric dating** of rocks was perfected after World

FIGURE 3.11 Seismic events worldwide, January 1977 through December 1986. The locations of about 10,000 earthquakes are colored red, green, and blue to represent event depths of 0–70 kilometers, 70–300 kilometers, and below 300 kilometers, respectively.

War II (**Box 3.2**). This technique is based on the discovery that unstable, naturally radioactive elements lose particles from their nuclei and ultimately change into new stable elements. The radioactive decay occurs at a predictable rate, and measuring the ratio of radioactive to stable atoms in a sample provides its age. To the surprise of many geologists, the maximum age of the ocean floor and its overlying sediments was radiometrically dated to less than 200 million years, only about 4% of the age of Earth. The centers of the continents are *much* older; some parts of the continental crust are more than 4 billion years old, about 90% of the age of Earth. *Why was oceanic crust so young?*

Attention had turned to the deep-ocean floors, the complex profiles of which were now being revealed by **echo sounders,** devices that measure depth by bouncing high-frequency sound waves off the bottom (see Figure 4.4). In particular, scientists aboard the Lamont–Doherty Geological Observatory deep-sea research vessel *Vema* (a converted three-masted schooner) invented deep survey techniques as they went. After World War II they probed the bottom with powerful echo sounders and looked beneath sediments with reflected pressure waves generated by surplus Navy depth charges dropped gingerly overboard. The overall shape of the Mid-Atlantic Ridge was slowly revealed. The ridge's conformance to shorelines on either side of the Atlantic raised many eyebrows (**Figure 3.12**). Ocean floor sediments were shown to be thickest at the edge of the Atlantic and thinnest near this mid-ocean ridge.

Vema and other research ships also compiled more complete, accurate charts of the submerged edges of continents. At Cambridge University Sir Edward Bullard used a computer to process these data to achieve the best possible fit of the continental jigsaw-puzzle pieces around the Atlantic. The fit was astonishingly good (see again Figure 3.9).

Mantle studies were keeping pace. The first links in the Worldwide Standardized Seismograph Network, begun during the International Geophysical Year in 1957, were beginning to report data from seismic waves reflected and refracted through the planet's inner layers. This information verified the existence of a layer in the upper mantle that caused a decrease in the velocity of seismic waves. This finding strongly suggested that the layer was deformable. Perhaps the lithosphere was isostatically balanced in this partially melted layer, and perhaps continents could move around in it *if* a suitable power source existed.

Oceanography ⊜ Now™ | Click Oceanography Interactive to view an animation of distribution of volcanoes through volcanism.

Oceanography ⊜ Now™ | Click Oceanography Interactive to view an animation of plate locations through plate tectonics.

The Breakthrough: From Seafloor Spreading to Plate Tectonics

In 1960 Professor Harry Hess of Princeton University and Robert Dietz of Scripps Institution of Oceanography proposed a radical idea to explain the features of the ocean floor and the "fit" of the

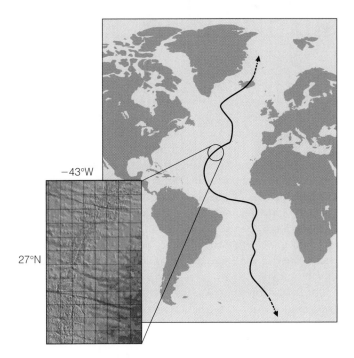

FIGURE 3.12 The Mid-Atlantic Ridge, showing its conformance to the coastlines of the adjacent continents. The inset shows a detail of the ridge generated from side-scan sonar data—the central rift is clearly visible. Red and orange indicate the crest of the ridge; dark blue, the deeper seabed on either side.

continents. They suggested that new seafloor develops at the Mid-Atlantic Ridge (and the other newly discovered ocean ridges) and then spreads outward from this line of origin. Continents would be carried along by the same forces that cause the ocean to grow. This motion could be powered by **convection currents,** slow-flowing circuits of material within the mantle.

Seafloor spreading, as the new hypothesis was called, pulled many loose ends together. If the mid-ocean ridges were **spreading centers** and sources of new ocean floor rising from the asthenosphere, then they should be hot. They were; indeed they were found to be lines of volcanoes! If the new oceanic crust cooled as it moved from the spreading center, then it should shrink in volume and become denser, and the ocean should be deeper farther from the spreading center. It was. Sediments at the edges of the ocean basin should be thicker than those near the spreading centers. They were, and they were also older.

Did this mean that Earth was continuously expanding? Because there was no evidence for a growing Earth, the creation of new crust at spreading centers would have to be balanced by the destruction of crust somewhere else. Then researchers discovered that the crust plunges down into the mantle along the periphery of the Pacific. The process is known as **subduction,** and these areas are called **subduction zones** (or Wadati–Benioff zones in honor of their discoverers). The zones of concentrated earthquakes (see again Figure 3.11) were found in regions of crustal formation (spreading centers) and crustal destruction (subduction zones).

In 1965 the ideas of continental drift and seafloor spreading were integrated into the overriding concept of **plate tectonics** (*tekton* = builder; *architect* has the same root), primarily by the

BOX 3.2   Absolute and Relative Dating

Radioactive decay is the process by which unstable atomic nuclei break apart. As we have seen, radioactive decay is accompanied by the release of heat, some of which warms Earth's interior and helps drive the processes of plate tectonics. Although it is impossible to predict exactly when any one unstable nucleus in a sample will decay, it is possible to discover the time required for one-half of all the unstable nuclei in a sample to decay. This time is called the **half-life.** Every radioactive element has its own unique half-life. For example, one of the radioactive forms of uranium has a half-life of 4.5 billion years, and a radioactive form of potassium has a half-life of 8.4 billion years. During each half-life, one-half of the remaining amount of the radioactive element decays to become a different element.

Radiometric dating is the process of determining the age of rocks by observing the ratio of unstable radioactive elements to stable decay products. Geologists consider radiometric dating a form of **absolute dating** because the age of a rock that contains a radioactive element may be determined with an accuracy of 1 to 2% of its actual age. The **figure** shows how samples may be dated by this means. Using radiometric dating, researchers have identified small zircon grains from western Australian sandstone that are 4.2 billion years old. The zircons were probably eroded from nearby continental rocks and deposited by rivers. (Older crust is now unidentifiable, having been altered and converted into other rocks by geological processes.)

Relative dating is a method of dating a sample by comparing its position to the positions of other samples. Younger sediments are typically laid down over older deposits; events are placed in their proper sequence. If a group of rocks or fossilized remains contains no radioactive elements, researchers can determine whether the sample is older or younger than a different sample close by, but not the actual age of the assemblages. The two methods of dating can work together to determine the age of materials.

Oceanography⬡Now™ | Click Oceanography Interactive to view an animation of relative dating and absolute dating through geological times.

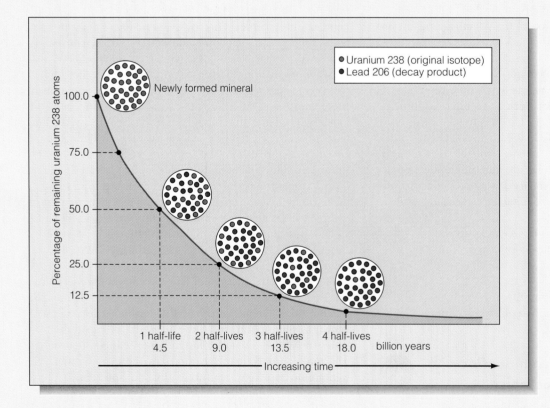

The rate of radioactive decay of a radioactive form of uranium (^{238}U) into lead. The half-life of ^{238}U is 4.5 billion years. During each half-life, one-half of the remaining amount of the radioactive element decays to become a different element. The assumption is made that the system has remained closed; no radioactive atoms or decay products have been added or removed from the sample. Ages obtained from absolute and relative dating of continental rocks and rocks and sediments from the seabeds coincide with ages predicted by plate tectonics theory.

a   b

Lithosphere

A plate is the cooled surface layer of a convection current in upper mantle.

FIGURE 3.13 Convection. (**a**) As a pot of water heats to boiling, the heated water rises; it falls again as it cools near the surface. (**b**) A tectonic plate is the cooled surface layer of a convection current in the upper mantle. Plate movement is caused by the plate sliding off the raised ridges along a spreading center, and by its cool and dense leading edge being pulled downward by gravity into the mantle. (Source: From *Understanding Earth*, 3/e by Frank Press and Raymond Siever. ©1994, 1998, and 2001 by W. H. Freeman and Company. Used with permission.)

work of **John Tuzo Wilson,** a geophysicist at the University of Toronto. In this theory Earth's outer layer consists of about a dozen separate major lithospheric **plates** floating on the asthenosphere. When heated from below, the deformable asthenosphere expands, becomes less dense, and rises. It turns aside when it reaches the lithosphere, lifting and cracking the

crust to form the plate edges. The newly forming pair of plates (one on each side of the spreading center) slide down the swelling ridges; they diverge from the spreading center (**Figure 3.13**). New seabed forms in the area of divergence. The large plates include both continental and oceanic crust. The major plates jostle about like huge slabs of ice on a warming lake. Plate movement is slow in human terms, averaging about 5 centimeters (2 inches) a year. The plates interact at converging, diverging, or sideways-moving boundaries, sometimes forcing one another below the surface or wrinkling into mountains.

Plate movement appears to be caused by two forces acting in nearly equal proportion:

■ Plates form and slide off the raised ridges of the spreading centers.

■ Plates are pulled downward into the mantle by their cool, dense leading edges.

We now know that *through the great expanse of geological time, this slow movement remakes the surface of Earth, expands and splits continents, and forms and destroys ocean basins.* The less dense, ancient granitic continents ride high in the lithospheric plates, rafting on the slowly moving asthenosphere below. This process has progressed since Earth's crust first cooled and solidified. **Figure 3.14** presents an overview of the tectonic system.

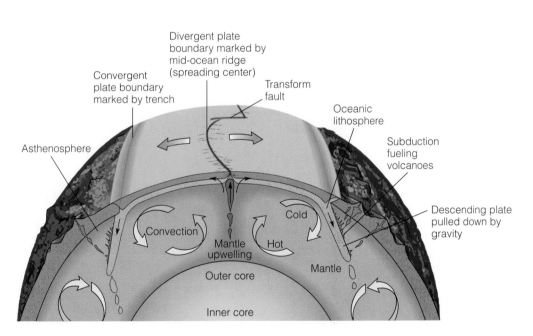

ACTIVE FIGURE 3.14 The tectonic system is powered by heat. Some parts of the mantle are warmer than others, and convection currents form when warm mantle material rises and cool material falls. Above the mantle floats the cool, rigid lithosphere, which is fragmented into plates. Plate movement is powered by gravity: The plates slide down the ridges at the places of their formation; their dense, cool leading edges are pulled back into the mantle. Plates may move away from one another (along the ocean ridges), toward one another (at subduction zones or areas of mountain building), or past one another (as at California's San Andreas Fault). Smaller localized convection currents form cylindrical plumes that rise to the surface to form hot spots (like the Hawaiian Islands). Note that the whole mantle is involved in thermal convection currents.

Oceanography≋Now™ | At the Active Figures link at http://oceanography.brookscole.com/garrison5e you can view an overview of the tectonic system.

Literally and figuratively, it all fits; a cooling, shrinking, raisinlike wrinkling is no longer needed to explain Earth's surface features. This twentieth-century understanding of the ever-changing nature of Earth has given fresh meaning to historian Will Durant's warning: "Civilization exists by geological consent, subject to change without notice."

After a series of raucous scientific meetings in 1966 and 1967, the revolution in geology entered a period of rapid consolidation. In 1968 *Glomar Challenger* drilled her first deep-ocean crustal cores and provided the confirmation of plate tectonics. Researchers found supporting data from many sources that tended to confirm Wilson's surprising synthesis. Every scientist had to reexamine his or her specialty in light of this new information. Zoologists found new explanations for the unusual animals of Australia. Biologists discovered a new cause of the isolation required for the formation of new species by natural selection. Paleontologists found an explanation for similar fossils on different continents. Resource specialists could at last explain why coal deposits were buried in Antarctica. Some geologists were pleased; some were skeptical. All were eager to explore further to prove or disprove this new theory.

This historical overview aims to convey the sense of discovery and excitement surrounding the twentieth-century revolution in geology. Now we can investigate the workings of plate tectonics in more detail.

Plate Boundaries

The lithospheric plates and their margins are shown in **Figure 3.15.** The plates float on a dense, deformable asthenosphere and are free to move relative to one another. Plates interact with neighboring plates along their mutual boundaries. In **Figure 3.16** movement of Plate A to the left (west) requires it to slide along its north and south margins. An overlap is produced in front (to the west), and a gap is created behind (to the east). Different places on the margins of Plate A experience separation and extension, convergence and compression, and transverse movement (shear).

The three types of plate boundaries that result from these interactions are called *divergent*, *convergent*, and *transform boundaries*, depending on their movement. **Figure 3.17** provides an overview of the plate boundary interactions we will study next.

Oceanography⬡Now™ | Click Oceanography Interactive to view an animation on plate boundaries through plate tectonics.

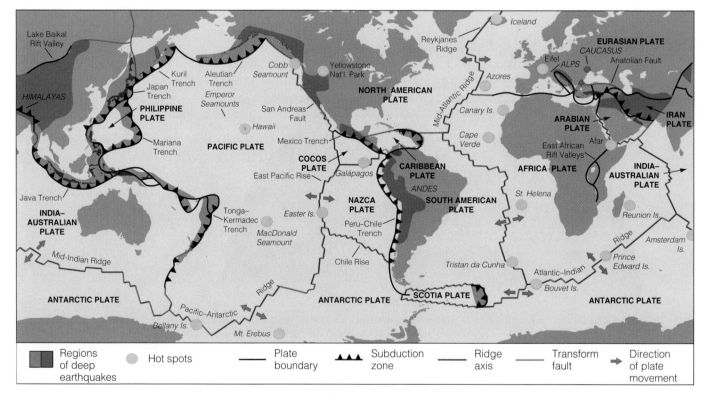

FIGURE 3.15 The major lithospheric plates, showing their directions of relative movement and the locations of the principal hot spots. Note the correspondence of plate boundaries and earthquake locations—compare this figure to Figure 3.11. Most of the million or so earthquakes and volcanic events each year occur along plate boundaries.

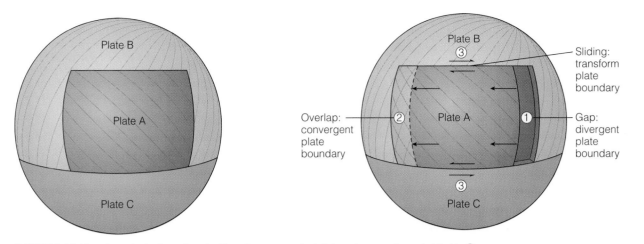

FIGURE 3.16 Plate boundaries in action. As Plate A moves to the left (west), a gap forms behind it ① and an overlap with Plate B forms in front ②. Sliding occurs along the top and bottom sides ③. The margins of Plate A experience the three types of interactions: at ①, extension characteristic of divergent boundaries; at ②, compression characteristic of convergent boundaries; and at ③, the shear characteristic of transform plate boundaries. Figure 3.17 shows these kinds of boundaries in more detail.

a

b

ACTIVE FIGURE 3.17 (a) Plate boundaries on the surface of a sphere. The divergent ①, convergent ②, and transform ③ margins match those of the margins of Plate A in Figure 3.16. (b) Extension of divergent boundaries causes splitting and rifting ①, compression at convergent boundaries produces buckling and shortening ②, and translation at transform boundaries causes shear ③.

Oceanography Now™ | At the Active Figures link at http://oceanography.brookscole.com/garrison5e you can view the three types of plate boundaries.

a — Warping, stretching of Pangaea

Lithosphere { Continental crust

b — Formation of rift valley

c — South America   Young Atlantic   Africa

FIGURE 3.18 A model for the formation of a new plate boundary: the breakup of Pangaea and the formation of the Atlantic beginning about 210 million years ago. (**a**) As the lithosphere began to crack, a rift formed beneath the continent, and molten basalt from the asthenosphere began to rise. (**b**) As the rift continued to open, the two new continents were separated by a growing ocean basin. Volcanoes and earthquakes occurred along the active rift area, which is now the mid-ocean ridge. The East African Rift Valley, though not yet submerged, currently resembles this stage. (**c**) A new ocean basin (shown in green) forms beneath a new ocean. The Red Sea (**d**) currently resembles this stage. Note the remarkable sawtooth configuration of the peaks on the horizon and their similarity to the diagram. (**e**) The South Atlantic in cross section, showing the mid-ocean ridge in the middle of the growing basin.

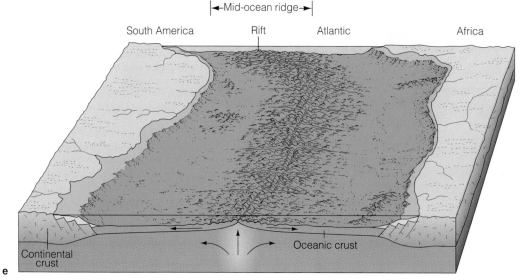

d

V. Courtillot

e — |◄—Mid-ocean ridge—►|

South America   Rift   Atlantic   Africa

Continental crust   Oceanic crust

Divergent Plate Boundaries: Forming Ocean Basins

The spreading center at the Mid-Atlantic Ridge is a **divergent plate boundary**, a line along which two plates are moving apart. Oceanic crust forms along divergent plate boundaries. The formation of the Atlantic is shown in **Figure 3.18**. Heat from the lower crust and mantle accumulates beneath the continents. About 210 million years ago this heat caused the asthenosphere to expand and rise, lifting and fracturing the lighter, solid lithosphere above. The plate and its embedded continent split in two, and a new plate boundary—a rift valley—formed between the pieces. As the broken plate separated at this new spreading center, molten rock called **magma** rose into the crustal fractures. (Magma is called *lava* when found above ground.) Some of the magma solidified in the fractures; some erupted from volcanoes on the seafloor. Together these processes produced new oceanic crust. The South Atlantic, a large new ocean basin that formed between the diverging plates when the rift became deep enough for water to collect, is shown in **Figure 3.18e**. A long mid-ocean ridge divided by a central rift valley traverses the ocean floor roughly equidistant from the shorelines in both the North and South Atlantic, terminating north of Iceland.

Plate divergence is not confined to the Atlantic, nor has it been limited to the last 200 million years. As may also be seen in Figure 3.15, the Mid-Atlantic Ridge has counterparts in the Pacific and Indian Oceans. The Pacific floor, for example, diverges along the East Pacific Rise and the Pacific Antarctic Ridge, spreading centers that form the eastern and southern boundaries of the great Pacific Plate. In East Africa rift valleys have formed relatively recently as plate divergence begins to separate another continent. As happened in the Red Sea (**Figure 3.18d**), the ocean will invade when the rift becomes deep enough.

Figure 3.19 shows how divergence has formed other ocean basins. About 20 cubic kilometers (4.8 cubic miles) of new ocean crust forms each year.

Convergent Plate Boundaries: Recycling Crust, Building Island Arcs and Continents

Because Earth is not getting larger, divergence in one place must be offset by convergence in another. Oceanic crust is destroyed at **convergent plate boundaries,** regions of violent geological activity where plates are pushing together. South America, embedded in the westward-moving South American Plate, encounters the Pacific's Nazca Plate as it moves eastward. The relatively thick and light continental lithosphere of South America rides up and over the heavier oceanic lithosphere of the Nazca Plate, which is subducted along the deep trench that parallels the west coast of South America. **Figure 3.20** is a cross section through these plates.

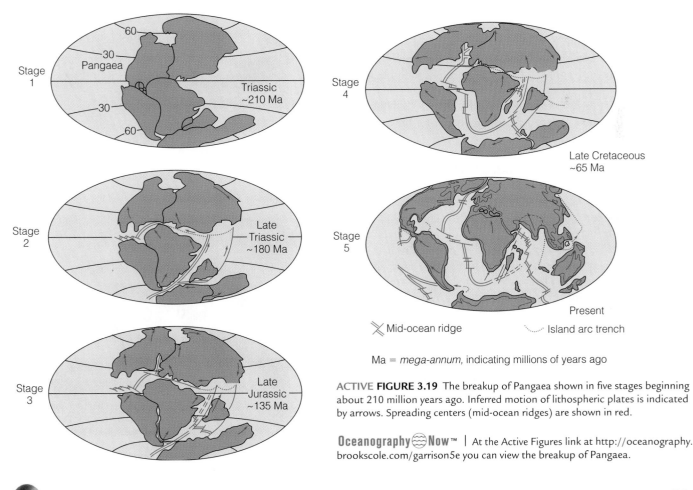

ACTIVE FIGURE 3.19 The breakup of Pangaea shown in five stages beginning about 210 million years ago. Inferred motion of lithospheric plates is indicated by arrows. Spreading centers (mid-ocean ridges) are shown in red.

Oceanography ⬯Now™ | At the Active Figures link at http://oceanography. brookscole.com/garrison5e you can view the breakup of Pangaea.

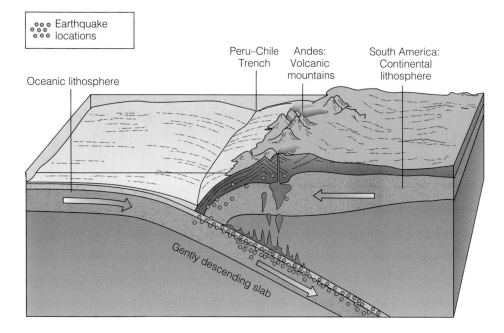

Oceanic lithosphere

Peru–Chile Trench

Andes: Volcanic mountains

South America: Continental lithosphere

Gently descending slab

ACTIVE FIGURE 3.20 A cross section through the west coast of South America, showing the convergence of a continental plate and an oceanic plate. The subducting oceanic plate becomes more dense as it descends, its downward slide propelled by gravity. Starting at a depth of about 100 kilometers (60 miles), heat drives water and other volatile components from the subducted sediments into the overlying mantle, lowering its melting point. Masses of the melted material, rich in water and carbon dioxide, rise to power Andean volcanoes.

Oceanography〰Now™ | At the Active Figures link at http://oceanography. brookscole.com/garrison5e you can view a cross section of a subducting oceanic plate near the west coast of South America.

The subducting plate's periodic downward lurches cause earthquakes. Some of the oceanic crust and its sediments will melt as the plate plunges downward, forming a magma rich in water and carbon dioxide. In places this magma rises to the surface through overlying layers and causes volcanic eruptions. The active volcanoes of Central America and South America's Andes Mountains are a product of this activity, as are the area's numerous earthquakes. The North American Cascade volcanoes, including Mount St. Helens, result from similar processes. Most of the subducted crust mixes with the mantle. As shown in **Figure 3.21,** some of it continues downward through the mantle, eventually reaching the core–mantle boundary 2,800 kilometers (1,700 miles) beneath the surface! Subduction at converging oceanic plates was responsible for the great Alaska earthquake of 1964. Plate convergence (and divergence) is faster in the Pacific than in the Atlantic, in a few places reaching a rate of 18 centimeters (7 inches) a year. You can now clearly see the source of the Pacific Ring of Fire.

In the previous example continental crust met oceanic crust. What happens when two *oceanic* plates converge? One of the colliding plates will usually be older, and therefore cooler and denser, than the other. Pulled by gravity, this heavier plate will slip steeply below the lighter one into the asthenosphere. The ocean bottom is distorted in these areas to form deep trenches, the ocean's greatest depths. Water and carbon dioxide trapped with the melting rock of the subducting plate rise into the overlying mantle, lowering its melting point. This fluid mix of magma and subducted material forms a relatively light magma that powers vigorous volcanoes, but the volcanoes emerge from the seafloor rather than from a continent. These volcanoes appear in patterns of curves on the overriding oceanic crust; when they emerge above sea level they form curving arcs of islands (**Figure 3.22**).

Convergent margins are vast "continent factories" where materials from the surface descend and are heated, com-

CENTRAL AMERICA

Trench

W   E

Core–mantle boundary

JAPAN   Trench

W   E

Core–mantle boundary

FIGURE 3.21 Vertical slices through Earth's mantle beneath Central America and Japan showing the distribution of warmer (red) and colder (blue) material. The configuration of colder material suggests that the subducting slabs beneath both areas have penetrated to the core–mantle boundary, which is at a depth of about 2,900 kilometers (1,800 miles).

a

b

FIGURE 3.22 (a) The formation of an island arc along a trench as two oceanic plates converge. The volcanic islands form as masses of magma reach the seafloor. The Japanese islands were formed in this way. (b) The distribution of shallow, intermediate, and deep earthquakes for part of the Pacific Ring of Fire in the vicinity of the Japan Trench. Note that earthquakes occur on only one side of the trench, the side on which the plate subducts. The site of the catastrophic 1995 Kobe subduction earthquake is marked.

pressed, partially liquefied, separated, mixed with surrounding materials, and recycled to the surface. Relatively light continental crust is the main product, and it is produced at a rate of about 1 cubic kilometer (0.24 cubic mile) per year. Some geophysicists believe all of Earth's continental crust may have originated from granitic rock produced in this way. The island arcs may have coalesced to form larger and larger continental masses.

Two plates bearing continental crust can also converge. The most spectacular example of such a collision, between the India–Australian and Eurasian Plates some 45 million years ago, formed the Himalayas. Because both plates are of approximately equal density, neither plate edge is being subducted; instead, both are compressed, folded, and uplifted, as **Figure 3.23** shows. The lofty top of Mt. Everest is made of rock formed from sediments deposited long ago in a shallow sea!

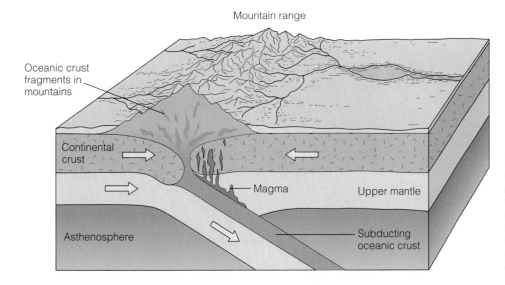

FIGURE 3.23 A cross section through southern China, showing the convergence of two continental plates. Neither plate is dense enough to subduct; instead, their compression and folding uplift the plate edges to form the Himalayas. Notice the massive supporting "root" beneath the emergent mountain needed for isostatic equilibrium.

Transform Plate Boundaries: Fracturing Crust

Remember, movement of lithospheric plates over the mantle is occurring on the surface of a sphere, not on a plane. The axis of spreading is not a smoothly curving line, but rather a jagged trace abruptly offset by numerous faults. These features are called **transform faults** (see **Figure 3.24,** and look for these features on the mid-ocean rides of Figure 3.15). Transform faults are so named because the relative plate motion is changed, or transformed, along them. We will discuss transform faults in more detail in Chapter 4 in our discussion of the mid-ocean ridge system (see, for example, Figure 4.24), but the concept is important in our discussion of plate boundaries because lithospheric plates shear laterally past one another at **transform plate boundaries.** Crust is neither produced nor destroyed at this type of junction.

The potential for earthquakes at transform plate boundaries can be great as the plate edges slip past each other. The eastern boundary of the Pacific Plate is a long transform fault system. As you can see in Figure 3.15, California's San Andreas Fault is merely the most famous of the many faults marking the junction between the Pacific and North American Plates (**Figure 3.25**). The Pacific Plate moves steadily, but its movement is stored elastically at the North American Plate boundary until friction is overcome. Then the Pacific Plate lurches in abrupt jerks to the northwest along much of its shared border with the North American Plate, an area that includes the major population centers of California. These jerks cause California's famous earthquakes. Because of this movement, coastal southwestern California is gradually sliding north along the rest of North America; some 50 million years from now, it will encounter the Aleutian Trench.

Oceanography🌐Now™ | Click Oceanography Interactive to view an animation of earthquakes in space and time and seismic risk, USA through earthquakes and tsunami.

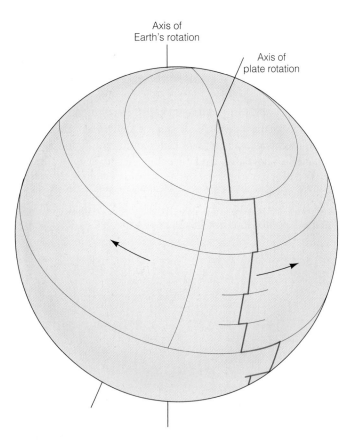

FIGURE 3.24 Transform faults (red) form because the axis of seafloor spreading on the surface of a sphere cannot follow a smoothly curving line. The motion of two diverging lithospheric plates (arrows) rotates about an imaginary axis extending through Earth. The plates shear laterally past one another at transform faults. Look for these features worldwide on Figures 3.15 and 3.25.

FIGURE 3.25 A long transform plate boundary, which includes California's San Andreas Fault.

Table 3.1   Characteristics of Plate Boundaries

Plate Boundary		Plate Movement	Seafloor	Events Observed	Example Locations
Divergent plate boundaries	Ocean–ocean	Apart	Forms by seafloor spreading.	Ridge forms at spreading center. Ocean basin expands, plate area increases. Many small volcanoes and/or shallow earthquakes.	Mid-Atlantic Ridge, East Pacific Rise
	Continent–continent		New ocean basin may form as continent splits.	Continent spreads, central rift collapses, ocean fills basin.	East African Rift Valley, Red Sea
Convergent plate boundaries	Ocean–continent	Together	Destroyed at subduction zones.	Dense oceanic lithosphere plunges beneath less dense continental. Earthquakes trace path of down-moving plate as it descends into asthenosphere. A trench forms. Subducted plate partially melts. Magma rises to form continental volcanoes.	Western South America, Cascade Mountains in western United States
	Ocean–ocean			Older, cooler, denser crust slips beneath less dense crust. Strong quakes. Deep trench forms in arc shape. Subducted plate heats in upper mantle, magma rises to form curving chains of volcanic islands.	Aleutians, Marianas
	Continent–continent		Closure of Ocean Basins.	Collision between masses of granitic continental lithosphere. Neither mass is subducted. Plate edges are compressed, folded, uplifted; one may move beneath the other.	Himalayas, Alps
Transform plate boundaries		Past each other	Neither created nor destroyed.	A line (fault) along which lithospheric plates move past each other. Strong earthquakes along fault.	San Andreas Fault; South Island, New Zealand
				Transform faults across spreading center.	Mid-ocean ridges

Plate Interactions—A Summary

There are, then, two kinds of plate divergences:

- Divergent oceanic crust (such as in the Mid-Atlantic)
- Divergent continental crust (as in the Rift Valley of East Africa)

And there are three kinds of plate convergences:

- Oceanic crust toward continental crust (west coast of South America)
- Oceanic crust toward oceanic crust (northern Pacific)
- Continental crust toward continental crust (Himalayan Mountains)

Transform boundaries mark the locations at which crustal plates move past one another (San Andreas Fault).

Each of these movements produces a distinct topography, and each zone contains potential dangers for its human inhabitants.

The characteristics of plate boundaries are summarized in **Table 3.1.**

The Confirmation of Plate Tectonics

The theory of plate tectonics has had the same effect on geology that the theory of evolution has had on biology. In each case a catalog of seemingly unrelated facts was unified by a powerful central idea. As we will see in this section, many discoveries contributed to our present understanding of plate tectonics, but the most compelling evidence is locked within the floors of the young ocean basins themselves.

Oceanography☸Now™ | Click Oceanography Interactive to view an animation of plate boundaries through plate tectonics.

Paleomagnetism

A compass needle points toward the magnetic north pole because it aligns with Earth's magnetic field (**Figure 3.26**). Tiny particles of an iron-bearing magnetic mineral called magnetite occur naturally in basaltic magma. When this magma erupts at mid-ocean ridges, it cools to form solid rock. The magnetic minerals act like miniature compass needles. As they

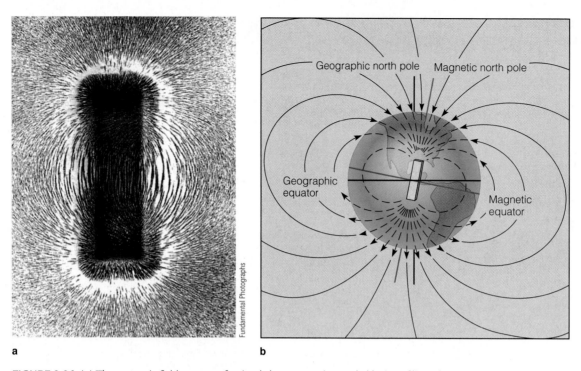

a

b

FIGURE 3.26 (**a**) The magnetic field pattern of a simple bar magnet is revealed by iron filings that align themselves along the lines of magnetic force. (**b**) Earth's magnetic field is thought to arise from fluid motions and electric currents in the outer core. The likely energy source is heat from the solid inner core causing convection currents in the outer core. These flowing currents of liquid iron, coupled with the rotation of Earth, form the magnetic field in a process analogous to the way a power-station generator produces electricity. The axis of Earth's magnetic field is tilted about 11° from the axis of the geographic North and South Poles. Compass needles point to magnetic north, not to geographic (true) north.

cool below their **Curie point** (or Curie temperature)—about 580°C (1080°F)—to form new seafloor, the magnetic minerals' magnetic fields align with Earth's magnetic field. Thus the orientation of Earth's magnetic field at that particular time becomes frozen in the rock as it solidifies. Any later change in the strength or direction of Earth's magnetic field will not significantly change the characteristics of the field trapped within the solid rocks. **Figure 3.27** shows the process. The "fossil," or remanent, magnetic field of a rock is known as **paleomagnetism** (*palaios* = ancient).

A **magnetometer** measures the amount and direction of residual magnetism in a rock sample. In the late 1950s geophysicists towed sensitive magnetometers just above the ocean floor to detect the weak magnetism frozen in the rocks. When plotted on charts, the data revealed a pattern of symmetrical magnetic stripes or bands on both sides of a spreading center (**Figure 3.28a**). The magnetized minerals contained in the rocks of some stripes add to Earth's present magnetic orientation to enhance the strength of the local magnetic field, but the magnetism in rocks in adjacent stripes weakens it. What could cause such a pattern?

In 1963 geologists Drummond Matthews, Frederick Vine, and Lawrence Morley proposed a clever interpretation. They knew similar magnetic patterns had been found in layered lava flows on land that had been independently dated by other means. They also knew that Earth's magnetic field reverses at

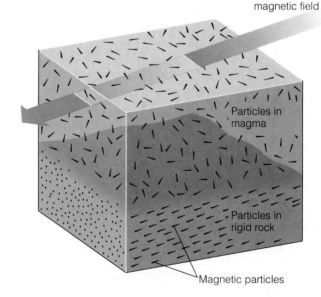

FIGURE 3.27 Particles of iron-bearing magnetite occur naturally in basaltic magma. As this rock cools to form new seabed, the magnetic particles cool past their Curie point, "locking" their magnetic orientation to that of Earth's prevailing magnetic field. If Earth's magnetic field changes direction later, the "locked" particles will not respond, but magnetic particles in any new (hot) magma above will orient to the new field direction. (Source: From *Understanding Earth,* 3/e by Frank Press and Raymond Siever. © 1994, 1998, and 2001 by W.H. Freeman and Company. Used with permission.)

= Normal magnetic polarity

= Reversed magnetic polarity

= Direction of plate movement

Ma = mega-annum, millions of years ago

b

ACTIVE FIGURE 3.28 Patterns of paleomagnetism and their explanation by plate tectonic theory. (a) When scientists conducted a magnetic survey of a spreading center, the Mid-Atlantic Ridge, they found bands of weaker and stronger magnetic fields locked in the rocks. (b) The molten rocks forming at the spreading center take on the polarity of the planet when they are cooling and then move slowly in both directions from the center. When Earth's magnetic field reverses, the polarity of newly formed rocks changes, creating symmetrical bands of opposite polarity.

Oceanography ⌷Now™ | At the Active Figures link at http://oceanography.brookscole.com/garrison5e you can view the patterns of paleomagnetism.

irregular intervals of a few hundred thousand years. In a time of reversal a compass needle would point south instead of north, and any particles of magnetic material falling below their Curie points in fresh seafloor basalt at a spreading center would be imprinted with the reversed field. The alternating magnetic stripes represent rocks with alternating magnetic polarity—one band having normal polarity (magnetized in the same direction as today's magnetic field direction), and the next band having reversed polarity (opposite from today's direction). These researchers realized that the pattern of alternating weak and strong magnetic fields was symmetrical because freshly magnetized rocks born at the ridge are spread apart and carried away from the ridge by plate movement (**Figure 3.28b**).

By 1974 scientists had compiled charts showing the paleomagnetic orientation—and the age—of the seafloors of the eastern Pacific and the Atlantic for about the last 200 million years (**Figure 3.29**). Plate tectonics beautifully explains these patterns, and the patterns themselves are among the most compelling of all arguments for the theory.

Paleomagnetic data have recently been used to measure spreading rates, to calibrate the geological time scale, and to reconstruct continents. Paleomagnetism has been among the most productive specialties in geology for the past three decades, and other lines of paleomagnetic investigation have also shed light on the process of plate tectonics.

Apparent Polar Wandering

Geologists already knew that periodic magnetic field reversals were not the only unusual feature of Earth's magnetic field. A plot of the apparent position of the north magnetic pole, as measured by the magnetic orientation of rocks in North America, South America, Europe, and Africa, showed that the magnetic pole seemed to have moved. It appeared to have migrated to its present position from a point much farther south, in the Pacific Ocean.

Because *actual* pole wandering was a very remote scientific possibility and having more than one north magnetic pole at the same time would be impossible, the pattern strongly suggested that the poles had probably stayed put and the continents had moved in relation to the pole and to one another.

Geophysicists examined this possibility. If these continents had once been united and had drifted over Earth's surface together, they should have identical paths of apparent polar wandering, terminating in coincidence at the North Pole's present position (**Figure 3.30**). Armed with this information, geologists quickly noticed a fault through Scotland's Caledonian Mountains joined with the Cabot Fault extending from Newfoundland to Boston. This and other evidence strongly suggested that the continents have indeed drifted apart, carrying their magnetized rocks with them.

Mantle Plumes and Hot Spots

Mantle plumes are continent-sized columns of superheated mantle that originate at the core–mantle boundary. The largest known plume, called a **superplume,** is now lifting all of Africa.

■ Pleistocene to Recent (0–1.6 Ma)	■ Paleocene (58–66 Ma)
□ Pliocene (1.6–5 Ma)	□ Late Cretaceous (66–88 Ma)
■ Miocene (5–24 Ma)	■ Middle Cretaceous (88–118 Ma)
■ Oligocene (24–37 Ma)	■ Early Cretaceous (118–144 Ma)
■ Eocene (37–58 Ma)	■ Late Jurassic (144–161 Ma)

Ma = mega-annum, millions of years ago

FIGURE 3.29 The age of the ocean floors. The colors represent seafloor spreading over the last 200 million years as revealed by paleomagnetic patterns. Note especially the relative symmetry of the Atlantic basin in contrast with the asymmetrical Pacific, where the spreading center is located close to the eastern margin and intersects the coast of California.

Its relatively sharp edges extend from Scotland to the Indian Ocean, and from the Mid-Atlantic Ridge to the Red Sea. The center of Africa is fracturing and spreading, and what will eventually be new seabed is forming rapidly in the East African rift valleys.

Plumes and superplumes are conduits for heat from Earth's core. Current research suggests that the heat in the asthenosphere that powers plate tectonics is resupplied from the core by superplumes. Indeed, in the not-so-distant past, superplume heat may have been responsible for some of the most dramatic events on Earth's surface. A huge outpouring of Earth's interior occurred over much of present-day India about 65 million years ago. The Indian subcontinent was deluged with more than a million cubic kilometers of lava! The stacks of lava are known as the Deccan Traps. If distributed evenly, this cataclysmic series of eruptions would have covered Earth's surface with a layer of lava 3 meters (10 feet) thick! Similar mega-eruptions happened about 17 million years ago in what is now the U.S. Pacific Northwest, and 248 million years ago in

Siberia. The atmospheric effects of such tremendous upheavals almost certainly led to mass extinctions.

Hot spots are one of the surface expressions of plumes of magma rising from relatively stationary sources of heat in the mantle (**Figure 3.31**). Hot spots are not always located at plate boundaries, and no one knows why their source of heat is localized or what anchors them in place. As lithospheric plates slide over these fixed locations, the plates are weakened from below by rising heat and magma. A volcano can form over the hot spot; but because the plate is moving, the volcano is carried away from its source of magma after a few million years and becomes inactive. It is replaced at the hot spot by a new volcano a short distance away. A chain of volcanoes and volcanic islands results (**Figure 3.32**).

Figure 3.33 shows the most famous of these "assembly line" chains, which extends from the old eroded volcanoes of the Emperor Seamounts to the still-growing island of Hawaii. In fact, the abrupt bend in the chain was caused by a change in the direction of movement of the Pacific Plate, from largely

a

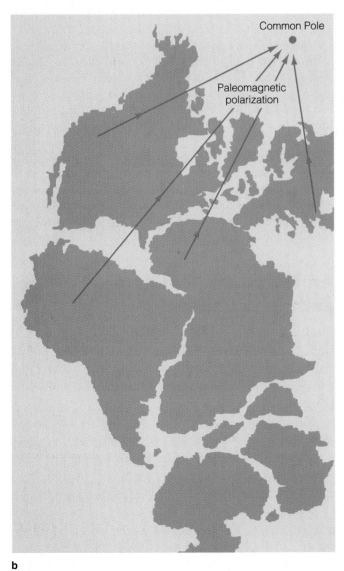

Common Pole

Paleomagnetic polarization

b

northward to more westward, about 40 million years ago. The next Hawaiian island that will come into being—already named Loihi—is building on the ocean floor at the southeastern end of the chain. Now about 1,000 meters (3,300 feet) beneath the surface, Loihi will break the surface about 30,000 years from now.

There are other hot spots in the Pacific. The island chains formed by their activity also jog in the Hawaiian pattern, indicating that they are positioned on the same lithospheric plate. Chains of undersea volcanoes in the Atlantic, centered on the Mid-Atlantic Ridge, suggest that a similar process is at work there. Hot spots can exist beneath continental crust as well; Yellowstone National Park is believed to be over a hot spot beneath the westward-moving North American Plate. Look once again at Figure 3.15 to see the locations of hot spots around the world. The configuration and length of all these chains of volcanoes and geothermal sites are consistent with the theory of plate tectonics.

 Oceanography ⬡ Now™ | Click Oceanography Interactive to view an animation of distribution of volcanoes through volcanism.

The Age and Distribution of Sediments

If the ocean basins are genuinely ancient, and if the processes that produce sediments have been operating for most or all of that time, then both the thickness and age of sediments on the ocean floor should be great. They are not. The young spreading ridges are almost free of sediment, and the oldest edges of the basins support layers of sediment 15 to 20 times thinner than the age of the ocean itself would suggest. The oldest sediments of the ocean basins are rarely more than 180 million years old (see Figure 5.23). This is because sediments are subducted at a plate's leading edge.

The Oceanic Ridges

The location and configuration of the oceanic ridges are clear evidence of past events. The volcanic nature of ridge islands like Iceland, the shape of the longitudinal rifts splitting the ridge

FIGURE 3.30 Polar wandering? (**a**) The magnetic properties of rocks in North America, South America, Europe, and Africa suggest that the north magnetic pole apparently migrated to its present position from much farther south, possibly from a point in the Pacific Ocean. The paleomagnetic evidence also suggests that, if the continents had remained fixed in place, Earth would have had different magnetic poles at the same time—an impossibility. (**b**) The problem can be solved if the pole has remained fixed but the continents have moved. For example, if Europe and North America were previously joined, the paleomagnetic fields preserved in their rocks would indicate a single pole location until they drifted apart. Evidence from South America and Africa supports this interpretation. (Source: From *Earth's Dynamic Systems*, 8/e by Hamblin & Christiansen, © 1998. Reprinted by permission of Pearson Education, Inc., Upper Saddle River, NJ 07458.)

Depth
km

300

650

1,000

1,450

1,900

2,350

2,800

170 160 150 140

Depth
mi

186

403

620

899

1,178

1,457

1,736

West Longitude   Degrees North Latitude

FIGURE 3.31 Imaging a mantle plume as individual slices down to the core–mantle boundary. Maps of P-wave velocity anomalies at different depths beneath the Hawaiian Islands show a vast area of unusually high temperature (red and yellow areas) that extends up from the core–mantle boundary (about 2,800 km or 1,750 miles, from the surface). This area of high temperature is a mantle plume. Activity atop this plume powers the Hawaiian volcanoes, and plumes like this are thought to bring to the asthenosphere much of the heat needed to power plate tectonics. The position of the Hawaiian Islands is shown on each slice for orientation, but of course the islands are only at Earth's surface. (Source: After R. Montelli, "Seismic Tomography Beyond Ray Theory," Ph.D. thesis, Princeton University, 2003. Used with permission.)

tops, and the sinking of the seabed as new oceanic crust cools and travels outward are all consistent with the theory of plate tectonics. The distribution of transform faults and fracture zones along the oceanic ridges (features you'll learn about in Chapter 4) also supports plate tectonics theory, as does on-the-spot geological observations made by researchers in deep submersibles.

Terranes

3.31

Buoyant continental and oceanic plateaus (submerged small fragments of continents), island arcs, and fragments of granitic rock and sediments can be rafted along with a plate and scraped off onto a continent when the plate is subducted. This process is similar to what happens when a sharp knife is scraped across a table top to remove pieces of cool candle wax. The wax accumulates and wrinkles on the knife blade in the same way landmasses and ocean sediments accumulate against the face of a continent as the lithosphere in which they are embedded reaches a plate boundary. Plateaus, isolated segments of seafloor, ocean ridges, ancient island arcs, and parts of continental crust that are squeezed and sheared onto the face of a continent are called **terranes**. The thickness and low density of terranes prevent their subduction. A simplified account of terrane accumulation is diagrammed in **Figure 3.34.**

Terranes are surprisingly common. New England, much of the Pacific Northwest of North America, and most of Alaska appear to be composed of this sort of crazy-quilt assemblage of material, some of which has evidently arrived from thousands of kilometers away. For example, western Canada's Vancouver Island may have moved north some 3,500 kilometers (2,200 miles) in the last 75 million years.

Curiously, terranes can also contain fragments of dense oceanic crust. Roughly 0.001% of oceanic lithosphere is not subducted, but rather *obducted*—scraped off—onto the edges of continents. The heavy wrinkled rocks contain pillow basalts and material derived from the upper mantle. Called **ophiolites** (*ophion* = snake) because of their sinuous shape, these assemblages can contain metallic ores similar to those known to exist at the mid-ocean-ridge spreading centers (**Figure 3.35**). Ophiolites are found on all continents, and as might be expected, those near the edges of the present continents are generally younger than those embedded deep within continental interiors. Some of these ancient reminders of past episodes of plate tectonics are more than 1.2 billion years old.

An Overview, Problems, and Implications

The theory of plate tectonics reveals much about the nature of Earth's surface. **Figure 3.36** summarizes surface tectonic activity.

In case you think geophysicists have all the answers, however, consider just a few of the theory's unsolved problems:

- Why should long *lines* of asthenosphere be any warmer than adjacent areas?

- Is the increasing density of the leading edge of a subducting plate more important than the plate's sliding off the swollen mid-ocean ridge in making the plate move?

- Why do mantle plumes form? What causes a superplume? How long do plumes last?

- How far do most plates descend? Recent evidence suggests that much of the material spans the entire mantle, reaching the edge of the outer core.

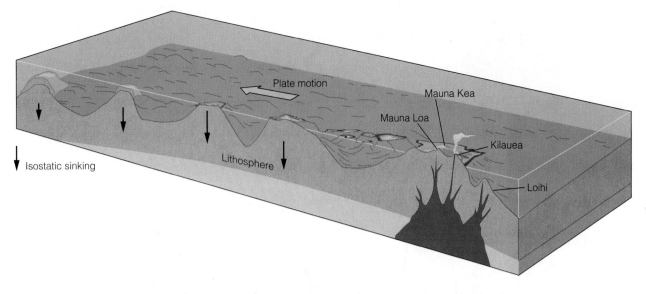

ACTIVE **FIGURE 3.32** Formation of a volcanic island chain as an oceanic plate moves over a stationary mantle plume and hot spot. The age of the islands increases toward the left. New islands will continue to form over the hot spot. In this example, showing the formation of the Hawaiian Islands, Loihi is such a newly forming island.

Oceanography ⊜Now™ | At the Active Figures link at http://oceanography.brookscole.com/garrison5e you can view the formation of a volcanic island chain.

FIGURE 3.33 The Hawaiian chain, islands formed one by one as the Pacific Plate slid over a hot spot. The oldest known member of the chain, the Meiji Seamount, formed about 70 million years ago (70 Ma), and the bend in the chain shows that the plate changed direction about 40 million years ago. The island of Hawaii still has active volcanism, but the next island in the chain, Loihi, has begun building on the ocean floor. The upper arrow shows the initial direction of plate motion; the lower arrow shows the present direction.

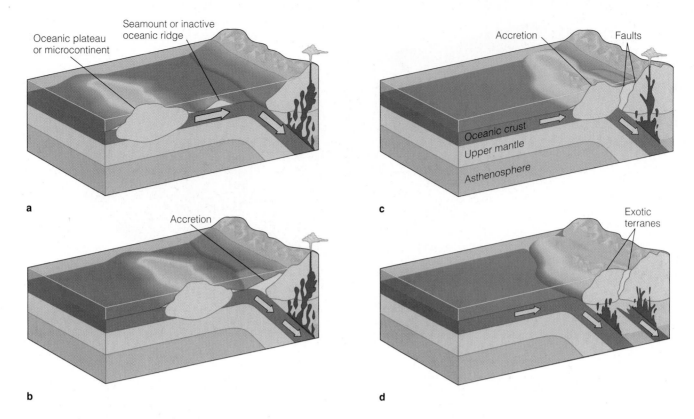

a

b

c

d

ACTIVE FIGURE 3.34 Terrane formation. Oceanic plateaus usually composed of relatively low density rock are not subducted into the trench with the oceanic plate. Instead, they are "scraped off," causing uplifting and mountain building as they strike a continent (**a–d**). Though rare, assemblages of subducting oceanic lithosphere known as ophiolites can also be scraped off (obducted) onto the edges of continents.

Oceanography 🌀 Now™ | At the Active Figures link at http://oceanography.brookscole.com/garrison5e you can view terrane formation.

FIGURE 3.35 Ophiolites—dense, mineral-rich assemblages named for their serpent-like shape.

- Has seafloor spreading always been a feature of Earth's surface? Has a previously thin crust become thicker with time, permitting plates to function in the ways described here?

- There is evidence of tectonic movement prior to the breakup of Pangaea. Will the process continue indefinitely, or are there cycles within cycles?

Though there is clearly much to learn, plate tectonics is already an especially powerful predictive theory. Discoveries and insights made by the researchers mentioned in this chapter, and hundreds of others, have borne out the intuition of Alfred Wegener. Our understanding of the process will evolve as more data become available, but there seems very little chance that geologists will ever return to the dominant pre-1960 view of a stable and motionless crust. Plate tectonic theory shows us the picture of an actively cycling Earth and an ever-changing surface, with *a single world ocean* changing shape and shifting position as the plates slowly move.

The configuration of the ocean basins—discussed in the next chapter—is the result of plate tectonic activity. The variety of these features will make more sense now that you are armed with an understanding of the theory.

Stage 1: Embryonic
Motion: Rift flank uplift, rift valley subsidence
Features: Complex system of rift valleys and lakes on continent
Example: East African Rift Valleys

Stage 2: Juvenile
Motion: Divergence (spreading)
Features: Narrow sea with matching coasts. Oceanic ridge formed.
Example: Red Sea

Stage 3: Mature
Motion: Divergence (spreading)
Features: Ocean basin with continental margins. Ocean continues to widen at oceanic ridge.
Example: Atlantic Ocean, Arctic Ocean

Stage 4: Declining
Motion: Convergence (subduction)
Features: Subduction begins. Island arcs and trenches form around basin edge.
Example: Pacific Ocean

Stage 5: Terminal
Motion: Convergence, collision, and uplift
Features: Oceanic ridge subducted. Narrow, irregular seas with young mountains.
Example: Mediterranean Sea

Stage 6: Suturing
Motion: Convergence and uplift
Features: Mountains form as two continental crust masses collide, are compressed, and override.
Example: India–Eurasia collision, Himalaya Mountains

ACTIVE FIGURE 3.36 The Wilson cycle, named in honor of John Tuzo Wilson's synthesis of plate tectonics. Over great spans of time, ocean floors form and are destroyed. Mountains erode, sediments subduct, and continents rebuild. Seawater moves from basin to basin.

Oceanography Now™ | At the Active Figures link at http://oceanography.brookscole.com/garrison5e you can view surface tectonic activity.

Questions from Students

1. How far beneath Earth's surface have people actually ventured?

Gold miners in South Africa have excavated ore 3,777 meters (12,389 feet, 2.35 miles) beneath the surface, where the rock temperature is 55°C (131°F). The miners work in pairs, one chipping the rock and the other aiming cold air at his partner. After a time they trade places.

2. What is the difference between crust and lithosphere? Between lithosphere and asthenosphere?

Lithosphere includes crust (oceanic and continental) and rigid upper mantle down to the asthenosphere. The velocity of seismic waves in the crust is much different from that in the mantle. This suggests differences in chemical composition or crystal structure, or both. The lithosphere and asthenosphere have different physical characteristics: The lithosphere is generally rigid, but the asthenosphere is capable of slow movement. The asthenosphere and lithosphere also transmit seismic waves at different speeds.

3. What are the most abundant elements inside Earth?

You may be surprised to learn that oxygen accounts for about 46% of the mass of Earth's crust. On an atom-for-atom basis the proportion is even more impressive: Of every 100 atoms of Earth's crust, 62 are oxygen. Most of this oxygen is not present as the gaseous element but is combined with other atoms into oxides and other compounds. Most of the familiar crustal rocks and minerals are oxides of aluminum, silicon, and iron (rust, for example, is iron oxide).

In Earth as a whole, iron is the most abundant element, making up 35% of the mass of the planet, while oxygen accounts for 30% overall. Remember, most of the mass of the universe is hydrogen gas. Oxygen and iron are abundant on Earth only because fusion reactions in stars can transform light elements like hydrogen into heavy ones.

4. How do geologists determine the location and magnitude of an earthquake?

Geologists use the time difference in the arrival of the seismic waves at their instruments to determine the distance to an earthquake. At least three seismographs in widely separated locations are needed to get a fix on the location.

The strength of the waves, adjusted for the distance, is used to calculate an earthquake's magnitude. Earthquake magnitude is often expressed on the **Richter scale.** Each full step on the Richter scale represents a ten-fold change in surface wave amplitude and a 32-fold change in energy release. Thus an earthquake with a Richter magnitude of 6.5 releases about 32 times more energy than an earthquake with a magnitude of 5.5, and about 1,000 times that of a 4.5-magnitude quake. People rarely notice an earthquake unless the Richter magnitude is 3.2 or higher, but the energy associated with a magnitude-6 quake may cause significant destruction.

The energy released by the 1964 Alaska earthquake was more than a billion times greater than the energy released by the smallest earthquakes felt by humans. The energy release was equal to about twice the energy content of world coal and oil production for an entire year. Very low or very high Richter magnitudes are not easy to measure accurately. The Alaska earthquake's magnitude was initially calculated as between 8.3 and 8.6 on the Richter scale, but recent reassessment has yielded an extraordinary magnitude of 9.2. Earth rang like a great silent bell for ten days after that earthquake.

5. What's the potential for serious loss of life and property damage due to tectonic plate movement?

Relatively great. About 40% of the world's largest cities lie within 160 kilometers (100 miles) of a plate boundary. In the year 2000, about 290 million people were living in high-risk areas. About 80% of those at risk live in developing nations, where seismic safety is not a high priority in building design. Another 210 million people are also threatened by active volcanoes; most live along the subduction zones of the Pacific Ring of Fire, where 75% of Earth's 850 active volcanoes are located. Since 1600 there have been approximately 262,000 deaths from volcanic eruptions, about 76,000 in this century. See **Figure 3.37** for a graphic assessment of the risk.

6. How common are large earthquakes?

About every two days, somewhere in the world, there's an earthquake of from 6 to 6.9 on the Richter scale—roughly equivalent to the quake that shook Northridge and the rest of southern California in January 1994, or Kobe, Japan, in January 1995. Once or twice a month, on average, there's a 7 to 7.9 quake somewhere. There is about one 8 to 8.9 earthquake—similar in magnitude to the 1964 earthquake in Alaska or the 1812 New Madrid quake—each year.

Northridge- and Kobe-sized quakes are moderate in size. Large losses of life and property can result when these earthquakes occur in populated areas, however. Damage estimates from the Northridge earthquake exceeded $40 billion. In Kobe more than 5,000 people died, and more than 26,000 were injured. Some 56,000 buildings were destroyed; estimates of the cost of reconstruction exceeded $400 billion.

Earthquake fatalities from A.D. 1000 to 1999

Fatalities exceeding:

300,000   90,000   10,000   • Cities with population of 2 million in the year 2000   — Convergent or transform plate boundaries

FIGURE 3.37 Population centers and earthquake areas. Large cities near plate boundaries in developing countries are likely to have the most fatalities in a massive earthquake. (Source: Reprinted by permission of Dr. Roger Billham, University of Colorado.)

7. Do other Earth-like planets or moons have hot internal layers and slowly drifting lithospheric plates?

The two bodies that planetary geologists have studied—the moon and Mars—appear to have much thicker crusts than Earth. The crusts of the moon and Mars make up about 10% of their masses, while the crust of Earth constitutes only about 0.4% of its mass. The crust of the moon appears to be contiguous; one researcher has called it a "single-plate" planet.

Although plate tectonics, at least in the terrestrial sense, almost certainly does not occur on the moon, there is some controversial evidence that the surface of Mars may at one time have consisted of two plates. Mars is pocked by giant volcanoes and marked by gigantic lava plains; it probably had an active interior at some time in the not-too-distant past. Martian volcanoes are so huge that some researchers believe hot spots may exist beneath a stationary lithosphere. The case for Martian plate tectonics was bolstered in May 1999 when *Mars Global Explorer* reported evidence of paleomagnetic stripes on Mars. The stripes are about ten times wider than those seen on Earth.

Chapter in Perspective

In this chapter you learned that Earth is composed of concentric spherical layers, with the least dense layer on the outside and the densest as the core. The layers may be classified by chemical composition into crust, mantle, and core; or by physical properties into lithosphere, asthenosphere, lower mantle, and core. Geologists have confirmed the existence and basic properties of the layers by analysis of seismic waves, which are generated by the forces that cause large earthquakes.

The theory of plate tectonics explains the nonrandom distribution of earthquake locations, the curious jigsaw-puzzle fit of the continents, and the patterns of magnetism in surface rocks. Plate tectonics theory suggests that Earth's surface is not a static arrangement of continents and ocean, but a dynamic mosaic of jostling lithospheric plates. The plates have converged, diverged, and slipped past one another since Earth's crust first solidified and cooled, driven by slow heat-generated currents rising and falling in the asthenosphere. Continental and oceanic crusts are generated by tectonic forces, and most major continental and seafloor features are shaped by plate movement. Plate tectonics explains why our ancient planet has surprisingly young seafloors; the oldest is only as old as the oldest dinosaurs—that is, about $\frac{1}{23}$ of the age of Earth.

In the next chapter you will learn about seabed features, which the slow process of plate movement has made and re-made on our planet. The continents are old; the ocean floors young. The seabed bears the marks of travels and forces only now being understood. Hidden from our view until recent times, the contours and materials of the seabed have their own stories to tell.

Key Ideas

✓ Some seismic waves—energy associated with earthquakes—can pass through Earth. Analysis of how these waves are changed, and the time required for their passage, has told researchers much about conditions inside Earth.
Review Figure 3.3, page 56

✓ Earth is composed of concentric spherical layers, with the least dense layer on the outside and the most dense as the core. The lithosphere, the outermost solid shell that includes the crust, floats on the hot, deformable asthenosphere. The mantle is the largest of the layers.
Review Figure 3.5, page 59

✓ Large regions of Earth's continents are held above sea level by isostatic equilibrium, a process analogous to a ship floating in water.
Review Figures 3.6 and 3.7, pages 60, 61

✓ Seismic events and volcanic activity do not occur at random on Earth's surface. Instead, they are concentrated in lines across Earth's surface.
Review Figure 3.11, page 64

✓ Plate tectonics theory suggests that Earth's surface is not a static arrangement of continents and ocean, but a dynamic mosaic of jostling segments called lithospheric plates. The plates have collided, moved apart, and slipped past one another since Earth's crust first solidified.
Review Figures 3.14–3.16, pages 67–69

✓ Plate motion is driven by slow convection (heat-generated) currents flowing in the mantle. Most of the heat that drives the plates is generated by the decay of radioactive elements within Earth.
Review Figures 3.13 and 3.14 and Box 3.2, pages 67, 66

✓ Most of the large-scale features seen at Earth's surface may be explained by the interactions of plate tectonics. Plate tectonics also explains why our ancient planet has surprisingly young seafloors, the oldest of which is only as old as the dinosaurs—that is, about $\frac{1}{23}$ of the age of Earth.
Review Figures 3.20, 3.22, and 3.23, pages 72–73

✓ The confirmation of plate tectonics rests on diverse scientific studies from many disciplines. Among the most convincing is the study of paleomagnetism, the orientation of Earth's magnetic field frozen into rock as it solidifies.
Review Figures 3.26–3.28, pages 76–77

Terms and Concepts to Remember

absolute dating, 66
asthenosphere, 58
basalt, 58
buoyancy, 60
catastrophism, 62
conduction, 61
continental crust, 58
continental drift, 63
convection, 61
convection current, 65
convergent plate boundary, 70
core, 58
crust, 58
Curie point, 76
density, 54
density stratification, 54
divergent plate boundary, 70
earthquake, 54
echo sounder, 65
fault, 60
granite, 58
half-life, 66
hot spot, 78
isostatic equilibrium, 60
lithosphere, 58
lower mantle, 58
magma, 70
magnetometer, 76
mantle, 58

mantle plume, 77
oceanic crust, 58
ophiolite, 80
P wave (primary wave) , 55
Pacific Ring of Fire, 64
paleomagnetism, 76
Pangaea, 63
Panthalassa, 63
plate, 67
plate tectonics, 65
radioactive decay, 61
radiometric dating, 64, 66
relative dating, 66
Richter scale, 84
S wave (secondary wave) , 55
seafloor spreading, 65
seismic waves, 54
seismograph, 55
shadow zone, 55
spreading center, 65
subduction, 65
subduction zone, 65
superplume, 77
terrane, 80
transform fault, 74
transform plate boundary, 74
uniformitarianism, 62
Wegener, Alfred, 63
Wilson, John Tuzo, 67

Study Questions

Reviewing What You've Learned

1. Name Earth's interior layers and their properties. Which classification method (chemical or physical) is more useful in explaining plate tectonics?
2. How is crust different from lithosphere?
3. Would the most violent earthquakes be associated with spreading centers or with subduction zones? Why?
4. Describe the mechanism that powers the movement of the lithospheric plates.
5. Why are the continents about 20 times older than the oldest ocean basins?
6. How does the principle of density stratification influence the plate tectonic process?

Thinking Critically

1. Seawater is more dense than fresh water. A ship moving from the Atlantic into the Great Lakes goes from seawater to fresh water. Will the ship sink farther into the water during the passage, stay at the same level, or rise slightly?

2. Some earthquakes are linked to adjustments of isostatic equilibrium. How can this occur? Where would you be likely to experience such an earthquake?
3. Where are the youngest rocks in the ocean crust? The oldest? Why?
4. Why did geologists have such strong objections to Wegener's ideas when he proposed them in 1912?
5. What biological evidence supports plate tectonics theory?
6. Imagine a tectonic plate moving westward. What geological effects would you expect to see on its northern edge? Western edge? Eastern edge?
7. What evidence can you cite to support the theory of plate tectonics? What questions remain unanswered? Which side would you take in a debate?
8. **InfoTrac College Edition Project** Continental drift illustrations sometimes give the impression that the continents have always looked the same. However, plate tectonics research has shown that the areas of land we now identify as continents may have had interesting transformations during the long periods of geological time. North and South America, for example, contain rocks and structures in common with regions now found at much greater distances. Do some research using InfoTrac College Edition to discover some of the connections between the American continents and other landmasses. Write an essay on your findings and speculate on how these connections can be explained—or not explained—by plate tectonics theory.

Thinking Analytically

1. How much farther would Columbus have to sail if he crossed the Atlantic today?
2. Look at Figure 3.28. Knowing the Atlantic's spreading rate, how wide would the orange reversed-polarity blocks be in the figure?

Oceanography Now™

Assess your understanding of this chapter's topics with additional quizzing and conceptual-based problems at http://oceanography.brookscole.com/garrison5e

Study and Research on the Internet

Assess your understanding of this chapter's topics with additional quizzing and conceptual-based problems at:

http://oceanography.brookscole.com/garrison5e

- **InfoTrac College Edition Research** leads you to critical thinking exercises that use InfoTrac College Edition® as a research tool.
- **Regional InfoTrac College Edition** articles are organized into East Coast, West Coast, and Gulf Coast regions, allowing you to study oceanography on a more local level.

 For more readings, go to InfoTrac College Edition, your online research library, at:

http://infotrac.thomsonlearning.com

a

(**a**) *Alvin*, the most famous and accomplished of the small research submarines. Divers prepare *Alvin* for a descent to the resting place of the RMS *Titanic*. (**b**) *Kaiko*, the deepest-diving vehicle presently in operation. The small ROV (remotely operated vehicle) sends information back to operators in the mothership by fiber-optic cable. *Kaiko* is operated by JAMSTEC, a Japanese marine science consortium.

b

CONTINENTAL MARGINS AND OCEAN BASINS

Going Deep

In this chapter you will read about ocean bottom features. The ideas in Chapter 3 are put to use here: Tectonic forces, erosion, and deposition have built and shaped the seabed.

First consider how difficult it is to see the signature of these forces. Light doesn't penetrate far into the ocean, so "seeing" must be by other means. Our knowledge of the seabed has been gained largely by people working at a distance from their goal. They use remote sensors to measure sound waves, radar beams, and differences in the pull of gravity; then they combine this information to draw details of the seafloor. Sometimes, though, there is no substitute for actually seeing—focusing a well-trained set of eyes on the ocean floor. Here's where research submarines come in handy.

Alvin, the best-known and oldest of the deep-diving manned research submarines now in operation, has made more than 4,200 dives (**Figure a**). This trusty little sub has explored the Mid-Atlantic Ridge near the Azores at 2,700 meters (9,000 feet), measured rock temperatures, collected water samples for chemical analysis, and taken photographs. *Alvin* carries three people in a sealed titanium sphere and is capable of diving to 4,000 meters (13,120 feet) and remaining on the bottom for as long as 5 hours. It's not very speedy—the optimum across-the-bottom speed is only about 2 knots (2.3 miles an hour). *Alvin* bristles with sensors, cameras, sample baskets and bottles, and remotely controlled arms and claws. The 6.7-meter (22-foot) sub was commissioned in 1964 and is operated jointly by the Woods Hole Oceanographic Institution, the National Science Foundation, the Office of Naval Research, and the National Oceanic and Atmospheric Administration.

Remotely operated robots are becoming more popular with researchers. Known as ROVs (remotely operated vehicles), they're cheaper to operate and safer than people-carrying vehi-

cles. The deepest-diving of these vehicles, *Kaiko*, descended to a measured depth of 10,914 meters (35,798 feet) near the bottom of the Challenger Deep on 24 March 1995 (**Figure b**). *Kaiko* took photographs, collected samples of sediments, and planted a small plaque to commemorate the event. *Kaiko* has recently been fitted with a small instrumented rover that can venture away from the ROV for close-up inspection of targets.

Bathymetry

In February 2000 the *Mars Global Surveyor* spacecraft completed a map of the surface of our planetary neighbor Mars. The photos from orbit were clear and detailed, and objects as small as about 2 meters (7 feet) across could be seen almost anywhere on the planet. There was no ocean and few storms to spoil the view.

Mapping Earth is much more difficult because water and clouds hide more than three-quarters of the surface. Until surprisingly recently we have known more about the global contours of the moon and the inner planets than we knew about our own home. Thanks to modern bathymetry, our view is clearing.

The discovery and study of ocean floor contours is called **bathymetry** (*bathy* = deep, *meter* = measure). The earliest known bathymetric studies were carried out in the Mediterranean by a Greek named Posidonius in 85 B.C. He and his crew let out nearly 2 kilometers (1.25 miles) of rope until a stone tied to the end of the line touched bottom. Bathymetric technology had not improved by the time Sir James Clark Ross obtained soundings of 4,893 meters (16,054 feet) in the South Atlantic in 1818. In the 1870s the researchers aboard HMS *Challenger* added the innovation of a steam-powered winch to raise the line and weight, but the method was the same (**Figure 4.1**). The *Challenger* crew made 492 bottom soundings and confirmed Matthew Maury's earlier discovery of the Mid-Atlantic Ridge.

Echo Sounding

The sinking of the RMS *Titanic* in 1912 stimulated research that finally ended slow, laborious weight-on-a-line efforts. By April 1914 Reginald A. Fessenden, a former employee of Thomas Edison, had developed an "Iceberg Detector and Echo Depth Sounder." The detector directed a powerful sound pulse ahead of a ship and then listened for an echo from the submerged portion of an iceberg. It was easy to direct the beam downward to sense the distance to the bottom. It might take most of a day to lower and raise a weighted line, but echo sounders could take many bottom recordings in a minute.

During World War I Fessenden's talents were turned to enemy submarine detection devices, but when peace returned he continued his echo sounder research. In June 1922 an echo sounder based on his designs made the first continuous profile

FIGURE 4.1 Seamen handling the steam winch aboard HMS *Challenger*. The winch was used to lower a weight on the end of a line to the seabed to find the ocean depth. This illustration is from the *Challenger Report* (1880).

across an ocean basin aboard the USS *Stewart*, a U.S. Navy vessel. Using an improved echo sounder based on Fessenden's design, the German research vessel *Meteor* made 14 profiles across the Atlantic from 1925 to 1927. The wandering path of the Mid-Atlantic Ridge was revealed, and its obvious coincidence with coastlines on both sides of the Atlantic stimulated

a

b

Robert Profeta

FIGURE 4.2 How an echo sounder operates. (**a**) The accuracy of an echo sounder can be affected by water conditions and bottom contours. The pulses of sound energy, or "pings," from the sounder spread out in a narrow cone as they travel from the ship. When depth is great, the sounds reflect from a large area of seabed. Because the first sound of the returning echo is used to sense depth, measurements over deep depressions are often inaccurate. (**b**) An echo sounder trace. A sound pulse from a ship is reflected off the seabed and returns to the ship. Transit time provides a measure of depth. For example, it takes about 2 seconds for a sound pulse to strike the bottom and return to the ship when the water is 1,500 meters (4,900 feet) deep. Bottom contours are revealed as the ship sails a steady course. In this trace the horizontal axis represents the course of the ship; the vertical axis represents the water depth. The ship has sailed over a small submarine canyon.

the discussions that culminated in our present understanding of plate tectonics.

Echo sounding wasn't perfect. The ship's exact position was sometimes uncertain. The speed of sound through seawater varies with temperature, pressure, and salinity, and those variations made depth readings slightly inaccurate. Simple depth sounder images (such as that shown in **Figure 4.2**) were also unable to resolve the fine detail that oceanographers needed to explore seabed features. Even so, researchers using depth

sounder tracks had painstakingly compiled the first comprehensive charts of the ocean floor by 1959 (**Figure 4.3**).

Since then, two new techniques—made possible by improved sensors and fast computers—have been perfected to minimize inaccuracies and speed the process of bathymetry. Multibeam echo sounder systems and satellite altimetry (as well as other systems) have been used to study the features discussed in this chapter. Any of them is surely an improvement over lowering rocks into the ocean!

FIGURE 4.3 (**a**) A detail from the first clear map of the seabed, hand-drawn from echo sounder data by Marie Tharp at Columbia University's Lamont–Doherty Earth Observatory. This detail shows the southern tip of South America and adjacent seafloor. Tharp's remarkable 1959 map was pivotal in debates over the theory of plate tectonics discussed in Chapter 3. Compare it to the satellite-derived, computer-processed image of the same area in Figure 4.5c.

a

Marie Tharp & Charles D. Hollister, WHOI

CONTINENTAL MARGINS AND OCEAN BASINS **89**

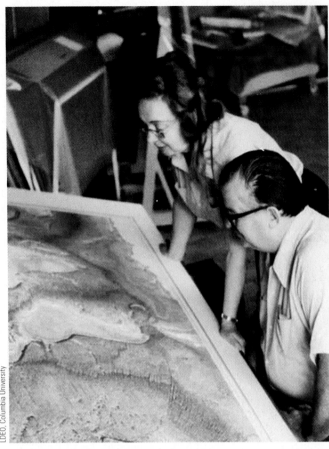

b

FIGURE 4.3 (**b**) Marie Tharp and geologist Bruce Heezen study the final version of Heinrich Berann's World Ocean Floor Panorama in Berann's studio, Austria, 1977. Berann was engaged by the U.S. Navy and the National Geographic Society to incorporate Tharp's work and data from other sources in a unified form. One of Berann's beautiful maps is included as Figure 4.22a.

Multibeam Systems

Like other echo sounders, a multibeam system bounces sound off the seafloor to measure ocean depth. Unlike a simple echo sounder, however, a multibeam system may have as many as 121 beams radiating from a ship's hull. Fanning out at right angles to the direction of travel, these beams can cover a 120° arc (**Figure 4.4a**). Typically a pulse of sound energy is sent toward the seabed every 10 seconds. Listening devices record sounds reflected from the bottom, but only from the narrow corridors corresponding to the outgoing pulse. Successive observations build a continuous swath of coverage beneath the ship. By "mowing the lawn"—moving the ship in a coverage pattern similar to one you would follow in cutting grass—researchers can build a complete map of an area (**Figure 4.4b**). Further processing can yield remarkably detailed images like those of Figures 4.12, 4.13, and 4.15. Fewer than 200 research vessels are equipped with multibeam systems. At the present rate, charting the entire seafloor in this way would require more than 125 years.

a

b

FIGURE 4.4 How a multibeam system collects data. (**a**) A multibeam echo sounder uses as many as 121 beams radiating from a ship's hull. Fanning out at right angles to the direction of travel, these beams can cover a 120° arc and measure a swath of bottom about 3.4 times as wide as the water is deep. Typically a "ping" is sent toward the seabed every 10 seconds. Listening devices record sounds reflected from the bottom, but only from the narrow corridors corresponding to the outgoing pulse. Thus a multibeam system is much less susceptible to contour error than the single-beam device shown in Figure 4.2. (**b**) A multibeam record of a fragment of seafloor near the East Pacific Rise south of the tip of Baja California, Mexico. The uneven coverage reflects the path of the ship across the surface. Detailed analysis requires sailing a careful back-and-forth pattern. Computer processing provides extraordinarily detailed images, such as those in Figures 4.12, 4.13, and 4.15.

Satellite Altimetry

Satellites cannot measure ocean depths directly, but they can measure small variations in the elevation of surface water. Using about a thousand radar pulses each second, the U.S. Navy's *Geosat* satellite (**Figure 4.5a**) measured its distance from the ocean surface to within 0.03 meter (1 inch)! Because the precise position of the satellite can be calculated, the

a

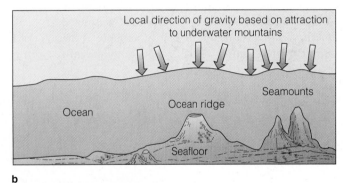

b

FIGURE 4.5 Satellite altimetry. (**a**) *Geosat*, a U.S. Navy satellite that operated from 1985 through 1990, provided measurements of sea surface height from orbit. Moving above the ocean surface at 7 kilometers (4 miles) a second, *Geosat* bounced a thousand pulses of radar energy off the ocean every second; its measurements were accurate to within 0.03 meter (1 inch). (**b**) Differences in sea surface height above a seabed feature are caused by the extra gravitational attraction of the feature, which "pulls" water toward it from the sides and forms a mound of water over itself. (**c**) This view of the complex system of ridges, trenches, and fracture zones on the South Atlantic seafloor east of the tips of South America and Antarctica's Palmer Peninsula was derived in large part from *Geosat* data declassified by the U.S. Navy in 1995. Compare this image to the hand-drawn map of the same area in Figure 4.3a.

c

average height of the ocean surface can be known with great accuracy.

Disregarding waves or tides or currents, researchers have found that the ocean surface can vary from the ideal smooth (ellipsoid) shape by as much as 200 meters (660 feet). This is because the pull of gravity varies across Earth's surface depending on the nearness (or distance away) of massive parts of Earth. An undersea mountain or ridge "pulls" water toward it from the sides, forming a mount of water over itself (**Figure 4.5b**). For example, a typical undersea volcano with a height of 2,000 meters (6,600 feet) above the seabed and a radius of 20 kilometers (32 miles) would produce a 2-meter (6.6-foot) rise in the ocean surface. (This mound cannot be seen with the unaided eye because the slope of the surface is very gradual.) The large features of the seabed are amazingly and accurately reproduced in the subtle standing irregularities of the sea surface (**Figure 4.5c**).

Geosat and its successors, *TOPEX/Poseidon* and *Jason-1*, have allowed the rapid mapping of the world ocean floor from space. Hundreds of previously unknown features have been discovered through the data they have provided.

The Topography of Ocean Floors

Most people think an ocean basin is shaped like a giant bathtub. They imagine that the continents drop off steeply just beyond the surf zone and that the ocean is deepest somewhere out in the middle. As is clear in **Figure 4.6,** bathymetric studies have shown that this picture is wrong.

Why? As you read in Chapter 3, plate tectonics theory suggests that Earth's surface is not a static arrangement of continents and ocean, but a dynamic mosaic of jostling lithospheric plates. The lighter continental lithosphere floats in isostatic equilibrium above the level of the heavier lithosphere of the ocean basins. The great density of the seabed partly explains why more than half of Earth's solid surface is at least 3,000 meters (10,000 feet) below sea level (**Figure 4.7**). Continental crust is thicker than oceanic crust, and continental lithosphere is less dense than oceanic lithosphere. Therefore the less dense lithosphere containing the continents floats in isostatic equilibrium above the level of the denser lithosphere containing ocean basins. The seabed topography we observe is the result of this dynamic balance *and* the jostling of tectonic plates.

Notice in **Figure 4.8** the transition between the thick (and less dense) granitic rock of the continents and the relatively thin (and denser) basalt of the deep seafloor. Near shore the features of the ocean floor are similar to those of the adjacent continents because they share the same granitic basement. The transition to basalt marks the *true* edge of the continent and divides ocean floors into two major provinces. The submerged outer edge of a continent is called the **continental margin.** The deep seafloor beyond the continental margin is properly called the **ocean basin. Figure 4.9** shows the relative amounts of Earth's surface composed of continental and oceanic structures and the important subdivisions of each.

FIGURE 4.6 Cross sections of the Atlantic Ocean basin and the continental United States, showing the range of elevations. The vertical exaggeration is 100:1. Although the ocean depth is clearly greater than the average height of the continent, the general range of contours is similar.

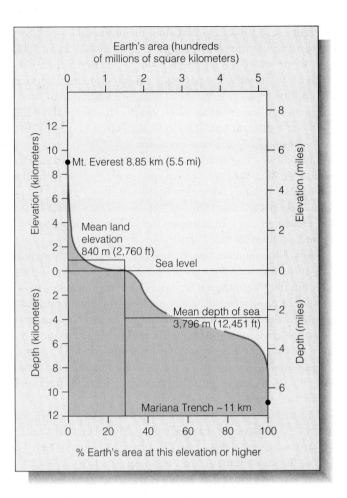

FIGURE 4.7 A graph showing the distribution of elevations and depths on Earth. There are two dominant levels on our planet: one exposed and the other submerged. This curve is not a land-to-sea profile of Earth but rather a plot of the area of Earth's surface above any given elevation or depth below sea level. Note that more than half of Earth's solid surface is at least 3,000 meters (10,000 feet) below sea level. The average depth of the world ocean (3,796 meters or 12,451 feet) is much greater than the average height of the continents (840 meters or 2,760 feet).

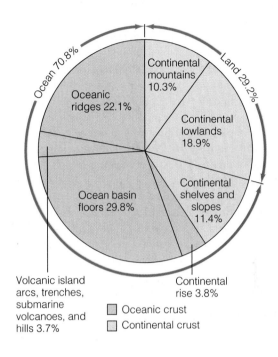

FIGURE 4.9 Features of Earth's solid surface shown as percentages of the planet's total surface.

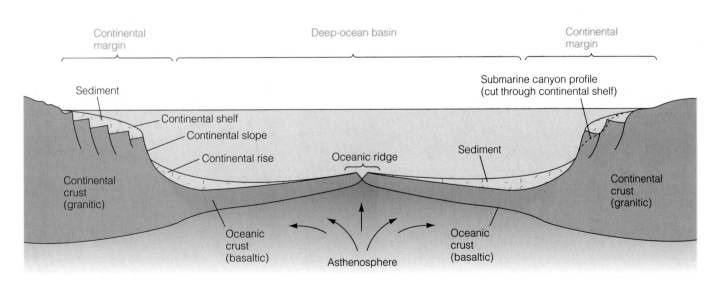

FIGURE 4.8 A cross section of a typical ocean basin flanked by passive continental margins. (The vertical scale has been greatly exaggerated to emphasize the basin contours.)

Continental Margins

You learned in Chapter 3 that lithospheric plates converge, diverge, or slip past one another. As you might expect, the submerged edges of continents—continental margins—are greatly influenced by this tectonic activity. Continental margins facing the edges of *diverging* plates are called **passive margins** because relatively little earthquake or volcanic activity is now associated with them. Because they surround the Atlantic, passive margins are sometimes referred to as *Atlantic-type* margins. Continental margins near the edges of *converging* plates (or near places where plates are slipping past one another) are called **active margins** because of their earthquake and volcanic activity. Because of their prevalence in the Pacific, active margins are sometimes referred to as *Pacific-type* margins.

Active and passive margins west and east of South America are shown in **Figure 4.10**. Note that active margins coincide with plate boundaries but passive margins do not. Passive margins are also found outside the Atlantic, but active margins are confined mostly to the Pacific.

Continental margins have three main divisions: a shallow, nearly flat continental *shelf* close to shore, a more steeply sloped continental *slope* to seaward, and an apron of sediment—the continental *rise*—that blends the continental margins into the deep-ocean basins.

Continental Shelves

The shallow submerged extension of a continent is called the **continental shelf.** Continental shelves, extensions of the adjacent continents, are underlain by granitic continental crust. They are much more like the continent than like the deep-ocean floor, and they may have hills, depressions, sedimentary rocks, and mineral and oil deposits similar to those on the dry land nearby. Taken together, the area of the continental shelves is 7.4% of Earth's ocean area.

Figure 4.11 shows a passive-margin continental shelf characteristic of Atlantic Ocean edges. The broad shelf extends far from shore in a gentle incline, typically 1.7 meters per kilometer (0.1°, or about 9 feet per mile), much more gradual than the slope of a well-drained parking lot. Shelves along the margin of the Atlantic Ocean often reach 350 kilometers (220 miles) in width and end at a depth of about 140 meters (460 feet), where a steeper drop-off begins.

The passive-margin shelves of the Atlantic Ocean formed as the fragments of Pangaea were carried away from each other by seafloor spreading. The continental lithosphere, thinned during initial rifting, cooled and contracted as it moved away from the spreading center, submerging the trailing edges of the continents and forming the shelves.

Most of the material that makes up a shelf comes from erosion of the adjacent continental mass. Rivers assist in passive shelf building by transporting huge amounts of sediments to the shore from far inland. In some places the sediments accumulate behind natural dams formed by ancient reefs or ridges of granitic crust (see again Figure 4.8). The weight of the sediment isostatically depresses the continental edges and allows the sediment load to grow even thicker. Sediment at the outer edge of a shelf can be up to 15 kilometers (9 miles) thick and 150 million years old.

The width of a shelf is usually determined by its proximity to a plate boundary. You can see in Figure 4.10 that the shelf at the *passive* margin (east of South America) is broad, but the shelf at the *active* margin (west of South America) is very narrow. The widest shelf, 1,280 kilometers (800 miles) across, lies north of Siberia in the tectonically quiet Arctic Sea. Shelf width depends not only on tectonics but also on marine processes: Fast-moving ocean currents can sometimes prevent sediments from accumulating. For example, the east coast of Florida has a very narrow shelf because no natural offshore dam is formed by ridges of granitic crust and because the swift current of the nearby Gulf Stream scours surface sediment

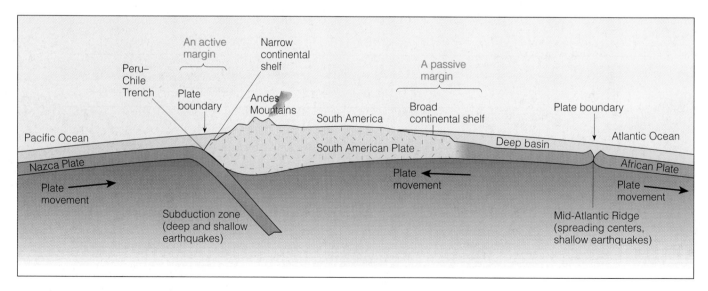

FIGURE 4.10 Typical continental margins bordering the tectonically active (Pacific-type) and tectonically passive (Atlantic-type) edges of a moving continent. (The vertical scale has been exaggerated.)

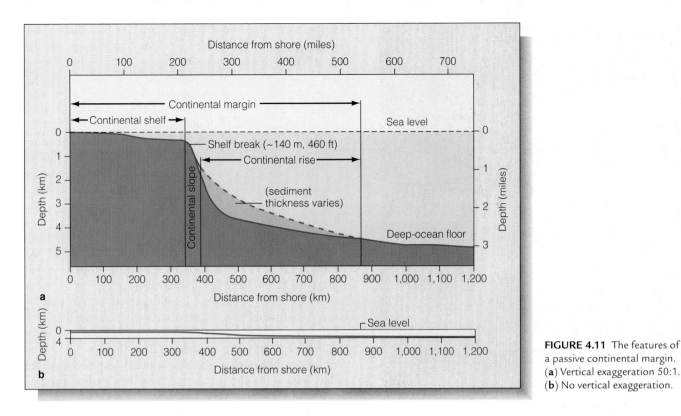

Distance from shore (miles)

Continental margin

Continental shelf

Sea level

Shelf break (~140 m, 460 ft)

Continental rise

Continental slope

(sediment thickness varies)

Deep-ocean floor

Depth (km)

Depth (miles)

Distance from shore (km)

a

Sea level

Depth (km)

Distance from shore (km)

b

FIGURE 4.11 The features of a passive continental margin. (**a**) Vertical exaggeration 50:1. (**b**) No vertical exaggeration.

away. Florida's west coast, however, has a broad shelf with a steep terminating slope (**Figure 4.12**).

A broad shelf has resulted from sediment accumulation along the sheltered edges of the Gulf of Mexico (**Figure 4.13**). Sediment eroded from the Rocky Mountains has been carried by the Mississippi River and deposited south of Texas and Louisiana. This sediment overlies salt deposited about 180 mil-lion years ago when much of the water in the gulf evaporated. The weight of the overlying sediment causes domes of salt to rise, spread out, and dissolve; collapsed salt domes create the pockmarked bottom characteristic of the area.

The shelves of the active Pacific margins are generally not as broad and flat as the Atlantic shelves. An example is the ab-breviated shelf off the west coast of South America, where the

Florida coast

FIGURE 4.12 A cliff more than 1.6 kilometers (1 mile) high marks the edge of the continental shelf west of central Florida. The very steep continental slope seen here is unusual. Perhaps fresh water seep-ing from the adjacent land under-mined the slope and caused it to collapse. Currents have removed much of the material that would otherwise be found at the base of the slope. Vertical exaggeration is about 50:1. Multibeam systems were used to obtain data for this image (and for Figures 4.13 and 4.15).

FIGURE 4.13 The broad continental shelf along the edge of the Gulf of Mexico south of Texas and Louisiana (looking east). Sediments carried by the Mississippi have overlain ancient salt deposits. The weight of the sediments causes salt domes to form and dissolve, leaving pockmarks in the continental shelf.

steep western slope of the Andes Mountains continues nearly uninterrupted beneath the sea into the depths of the Peru–Chile Trench (see again Figure 4.10). Active-margin shelves have more varied topography than passive-margin shelves; the character of continental shelves at an active margin may be determined more by faulting, volcanism, and tectonic deformation than by sedimentation.

A few Pacific shelves are broad, however. As in the Atlantic, natural offshore dams trap sediments and form shelves; but in the Pacific the sediment-trapping dams are more commonly offshore chains of volcanoes or lines of coral reefs. Volcanic activity east of China and Southeast Asia has formed a broad basin that is now filling with sediments and is one of the largest shelves in the Pacific.

Because of their gentle slope, continental shelves are greatly influenced by changes in sea level. Around 18,000 years ago—at the height of the last **ice age** (period of widespread glaciation)—massive ice caps covered huge regions of Earth's continents. The water that formed these thick ice sheets came from the ocean, and sea level fell about 125 meters (410 feet) below its present position (**Figure 4.14**). The continental shelves were almost completely exposed, and the surface area of the continents was about 18% greater than it is today. Rivers and waves cut into the sediments that had accumulated during periods of higher sea level, and they transported some coarse sediments to their present locations at the shelves' outer edges. Sea level began to rise again when the ice caps melted, and sediments again began to accumulate on the shelves. (More on the history and effects of sea-level change will be found in the discussion of coasts in Chapter 12.)

FIGURE 4.14 Changes in sea level over the last 250,000 years, as traced by data taken from ocean floor cores. The rise and fall of sea level are due largely to the coming and going of ice ages—periods of increased and decreased glaciation, respectively. Water that formed the ice-age glaciers came from the ocean, and this caused sea level to drop. Point **a** indicates a low stand of −125 meters (−410 feet) at the climax of the last ice age some 18,000 years ago. Point **b** indicates a high stand of +6 meters (+19.7 feet) during the last interglacial period about 120,000 years ago. Point **c** shows the present sea level. The rate of sea-level rise has nearly leveled off, but sea level is rising slowly as we continue to emerge from the last ice age.

Note that sea level has been considerably below its present position for nearly all of the last quarter-million years. During the low periods, there would have been fewer estuaries and much narrower continental shelves. Consider the implications for coastal civilizations. During the time between 18,000 and 8,000 years ago, sea level rose more than 100 meters (330 feet) at a rate of about 1 centimeter ($\frac{1}{2}$ inch) a year—more than 0.5 meter in a human lifetime. Could this have given rise to the flood legends common to many religions?

Oregon coast

William Haxby, Lamont-Doherty Earth Observatory of Columbia

FIGURE 4.15 Folded ridges of sediment cover the ocean floor west of Oregon (looking south). The crinkles result from the collision between the North American Plate and the Juan de Fuca Plate. Like a bulldozer, the North American Plate scrapes sediment from the subducting Juan de Fuca Plate and piles it into folds.

The continental shelves have been the focus of intense exploration for natural resources. Because shelves are the submerged margins of continents, any deposits of oil or minerals along a coast are likely to continue offshore. The water depth over shelves averages only about 75 meters (250 feet), so large areas of the shelves are accessible to mining and drilling activities. Many of the techniques used to find and exploit natural resources on land can also be used on the continental shelves. Resource development requires intensive scientific investigations, and our understanding of the geology of the shelves has benefited greatly from the search for offshore oil and natural gas. (The economic, environmental, and legal implications of the exploitation of these riches are touched on in Chapter 17's discussion of the Law of the Sea and the establishment of exclusive economic zones.)

Continental Slopes

The **continental slope** is the transition between the gently descending continental shelf and the deep-ocean floor. Continental slopes are formed of sediments that reach the built-out edge

of the shelf and are transported over the side. At active margins a slope may also include marine sediments scraped off a descending plate during subduction (**Figure 4.15**). The inclination of a typical continental slope is about 4° (70 meters per kilometer, 370 feet per mile), slightly steeper than the steepest road slope allowed on the interstate highway system. As Figure 4.11b implies, even the steepest of these slopes is not precipitous: A 25° slope is the greatest incline yet discovered. In general, continental slopes at active margins are steeper than those at passive margins. Continental slopes average about 20 kilometers (12 miles) wide and end at the continental rise, usually at a depth of about 3,700 meters (12,000 feet). The bottom of the continental slope is the solid edge of a continent.

The **shelf break** marks the abrupt transition from continental shelf to continental slope. The depth of water at the shelf break is surprisingly constant—about 140 meters (460 feet) worldwide—but there are exceptions. The great weight of ice on Antarctica, for example, has isostatically depressed that continent, and the depth of water at the shelf break is 300 to 400 meters (1,000 to 1,300 feet). The shelf break in Greenland is similarly depressed.

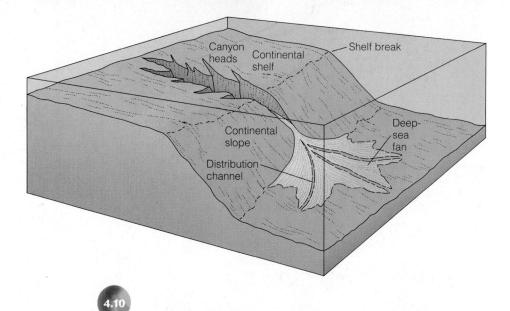

FIGURE 4.16 A submarine canyon.

Submarine Canyons

Submarine canyons cut into the continental shelf and slope, often terminating on the deep-sea floor in a fan-shaped wedge of sediment (**Figure 4.16**). More than a hundred submarine canyons nick the edge of nearly all of Earth's continental shelves. The canyons generally trend at right angles to the shoreline (and shelf edge), sometimes beginning very close to shore. Congo Canyon actually extends into the African continent as a deep estuary at the mouth of the Congo River. These enigmatic features can be quite large. In fact, submarine canyons similar in size and profile to Arizona's Grand Canyon have been discovered!

Hudson Canyon, a typical large canyon on a passive margin, is shown in **Figure 4.17**. Like many submarine canyons, Hudson Canyon is located just offshore of the mouth of a river or stream—in this case, New York's Hudson River. Because of their similarity to canyons on land, submarine canyons appear to have been created by erosion, so marine geologists initially thought the canyons were carved into the shelves by stream erosion at times of lower sea level. But most researchers agree that sea level has never fallen more than 200 meters (660 feet) below its present level in the last 600 million years. Stream erosion could account for the shape of the uppermost parts of the canyons. However, because submarine canyons can be traced to depths in excess of 3,000 meters (10,000 feet) below sea level, stream erosion could not have played a direct role in cutting their lower depths.

What, then, caused the submarine canyons to form? As is so often the case in the history of marine science, the answer to this riddle was associated with a peculiar event. A sharp earthquake struck the Grand Banks south of Newfoundland at 1531 (3:31 P.M.) Greenwich time on 18 November 1929. The quake's **epicenter,** the point on the surface of Earth directly above the focus of an earthquake, was located on the continental slope at a water depth of about 2,200 meters (7,200 feet). Ordinarily such an earthquake would not attract much attention, but a number of important transatlantic communication cables lay on the ocean floor south of the earthquake area. One

Elevation (in meters)					
−304.8	−152.4	0	152.4	304.8	457.2
−1000	−500	0	500	1,000	1,500

Elevation (in feet)

FIGURE 4.17 A side-scan sonar image of Hudson Canyon east of New Jersey. The shelf in this area is broad. The canyon can be seen nicking the shelf–slope junction and then continuing toward the abyssal plain to the southwest. The underwater topography has been exaggerated by a factor of 5 relative to the land topography. The fine black line marks sea level.

FIGURE 4.18 A turbidity current flowing down a submerged slope off the island of Jamaica. The propeller of a submarine caused the turbidity current by disturbing sediment along the slope.

FIGURE 4.19 A continuous cascade of sediment at the head of San Lucas submarine canyon (off the coast of Baja California, Mexico), which assists in eroding the narrow gorge in conjunction with occasional turbidity currents.

by one these cables failed. A segment of the Western Union cable nearest the epicenter broke almost immediately, and a French cable about 160 kilometers (100 miles) south of the Western Union cable parted at 1835 (6:35 P.M.). Another Western Union cable to the south snapped early the next morning, and cables even farther south were damaged later that day. Officials had expected an occasional random failure, but a *sequence* of cable failures, from north to south along 480 kilometers (300 miles) of seafloor in the mid-Atlantic, was baffling.

Local landslides or sediment liquefaction triggered by the earthquake probably caused the first cable breaks, and although no one thought of it right away, an abrasive underwater "avalanche" almost certainly caused the more distant breaks. These avalanche-like sediment movements, called **turbidity currents,** are caused when turbulence mixes sediments into water above a sloping bottom. The sediment-filled water is denser than the surrounding water, so the thick, muddy fluid runs down the slope at speeds up to 27 kilometers (17 miles) per hour. **Figure 4.18** is a rare photograph of a turbidity current.

What is the connection between turbidity currents and submarine canyons? Small amounts of debris may cascade continuously down the canyons (**Figure 4.19**), but earthquakes can shake loose huge masses of sediment that rush down the edge of the shelf, scouring the canyon deeper as they go. Most geologists believe that the canyons have been formed by abrasive turbidity currents plunging down the canyons. In this way the canyons can be cut to depths far below the reach of streams even during the low sea levels of the ice ages.

Continental Rises

Along passive margins, the oceanic crust at the base of the continental slope is covered by an apron of accumulated sediment called the **continental rise** (see again Figure 4.8). Sediments from the shelf slowly descend to the ocean floor along the whole continental slope, but most of the sediments that form the continental rise are transported to the area by turbidity cur-

rents. The width of the rise varies from 100 to 1,000 kilometers (63 to 630 miles), and its slope is gradual—about one-eighth that of the continental slope. One of the widest and thickest continental rises has formed in the Bay of Bengal at the mouths of the Ganges–Brahmaputra River, the most sediment-laden of the world's great rivers.

Deep-ocean currents are an important factor in shaping continental rises, especially along the western boundaries of most ocean basins. Deep boundary currents held against the continental slopes by the Coriolis effect (news of which awaits you in Chapter 8) pick up volcanic debris and sediments transported by turbidity currents and sweep them along the ocean floor. As the currents are forced around bends and across depressions, they slow and deposit the suspended material on the rises in the form of ridges and mud waves (**Figure 4.20**).

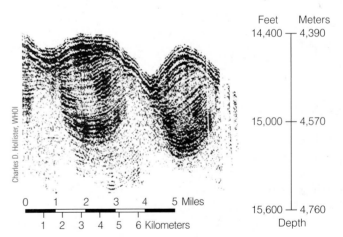

FIGURE 4.20 Large stationary waves of sediment deposited at the base of the continental rise in the western North Atlantic. Deep boundary currents pick up sediments and sweep them across the ocean floor. As the currents are forced around bends and across depressions, they slow and deposit the suspended material at the base of the rises in the form of ridges and mud waves.

Deep-Ocean Basins

Away from the margins of continents the structure of the ocean floor is quite different. Here the seafloor is a blanket of sediments up to 5 kilometers (3 miles) thick overlying basaltic rocks. Deep-ocean basins constitute more than half of Earth's surface.

The deep-ocean floor consists of mainly oceanic ridge systems and the adjacent sediment-covered plains. Deep basins may be rimmed by trenches or by masses of sediment. Flat expanses are interrupted by islands, hills, active and extinct volcanoes, and active zones of seafloor spreading. The sediments on the deep-ocean floor reflect the history of the surrounding continents, the biological productivity of the overlying water, and the ages of the basins themselves.

Oceanic Ridges

If the ocean evaporated, the oceanic ridges would be Earth's most remarkable and obvious feature. An **oceanic ridge** is a mountainous chain of young basaltic rock at the active spreading center of an ocean. Stretching 65,000 kilometers (40,000 miles), more than $1\frac{1}{2}$ times Earth's circumference, oceanic ridges girdle the globe like seams surrounding a softball (**Figure 4.21**). The rugged ridges, which often are devoid of sediment, rise about 2 kilometers (1.25 miles) above the seafloor. In places they project above the surface to form islands such as Iceland, the Azores, and Easter Island. Oceanic ridges and their associated structures account for 22% of the world's solid surface area (all the land above sea level accounts for 29%). Although these features are often called mid-ocean ridges, less than 60% of their length is actually along the centers of ocean basins.

As we saw in our discussion of plate tectonics, the rift zones associated with oceanic ridges are sources of new ocean floor where lithospheric plates diverge. The oceanic ridges are widest where they are most active. The youngest rock is located at the active ridge center, and rock becomes older with distance from the center. As the lithosphere cools, it shrinks and subsides. Slowly spreading ridges have a steeper profile than rapidly spreading ones because slowly diverging seafloor cools and shrinks closer to the spreading center. **Figure 4.22a**, a bathymetric map of the North Atlantic, clearly shows the great extent of the Mid-Atlantic Ridge, a typical oceanic ridge. **Figure 4.22b**, a multibeam image, provides a detailed look at the young central rift. A cross section is provided as **Figure 4.22c**.

As you may remember from Chapter 3, much has been learned about the ridge system on and near the country of Iceland. Iceland straddles the Mid-Atlantic Ridge *and* sits atop a vast mantle plume—geological excitement is guaranteed! The section of the Mid-Atlantic Ridge of which Iceland is a part spreads about 15 centimeters (6 inches) per year. Iceland is broken by parallel fissures stretching north and south along a central rift zone (**Figure 4.23**); they are similar to fissures on the adjacent submerged ridge. As along the rest of the ridge, tectonic forces are stretching Iceland in the east–west direction, and magma is rising from the plume to fill the fissures

FIGURE 4.21 The oceanic ridge system (in colors) stretches some 65,000 kilometers (40,000 miles) around Earth. If the ocean evaporated, the ridge system would be Earth's most remarkable and obvious feature. The thickness of the red lines indicate the rate of spreading for some of the most rapidly spreading sections, and the numbers give spreading rates in centimeters per year. The East Pacific Rise typically spreads about six times faster than the Mid-Atlantic Ridge.

and become new seafloor. The lava venting from Iceland's volcanoes is similar to lava rising into nearby submerged segments of the Mid-Atlantic Ridge, but the cooling lava doesn't assume the overstuffed pillow shape of submerged lava—for two reasons: First, lava in contact with cold seawater cools and solidifies more rapidly than lava in contact with air; second, the water pressure at the ocean bottom keeps most of the volatile gases in solution during the cooling process, thus maintaining the extruded bunlike shape. Instead, the Icelandic lava blasts from the crevices, explosively releases its gases, and spreads over the desolate countryside.

As can be seen in Figure 4.22a and b, the Mid-Atlantic Ridge does not run in a straight line. It is offset at more or less

regular intervals by **transform faults**. A fault, recall, is a fracture in the lithosphere along which movement has occurred, and transform faults are fractures along which lithospheric plates slide horizontally (**Figure 4.24**). As you saw in Figure 3.24, when segments of a ridge system are offset, the fault connecting the axis of the ridge is a transform fault. Shallow earthquakes are common along transform faults. Since the ocean floor cannot expand evenly on the surface of a sphere, plate divergence on the spherical Earth can be only irregular and asymmetrical, and transform faults and fracture zones result.

Transform faults are the active parts of **fracture zones.** Extending outward from the ridge axis, fracture zones are seismically inactive areas that show evidence of past transform fault

a

FIGURE 4.22 (a) Heinrich Berann's hand-drawn map of a portion of the Atlantic Ocean floor showing some major oceanic features: oceanic ridge, transform faults, fracture zones, submarine canyons, seamounts, continental rises, trenches, and abyssal plains. Depths are in feet. The map is vertically exaggerated.

FIGURE 4.22 (**b**) The fine structure of the central portion of the Mid-Atlantic Ridge between Florida and western Africa. The depressed central valley (the spreading center, shown in blue) is clearly visible in this computer-generated multibeam image. (**c**) As this seismic profile shows, the rugged relief of the North Atlantic's oceanic ridge is gradually being buried by slowly accumulating sediments.

28°N 50°W

3,940 meters; 12,000 feet

4,400 meters; 14,400 feet

5,860 meters; 19,200 feet

10 miles
16 kilometers

FIGURE 4.23 Thingvellir Graben, a rift valley in Iceland. The valley is actually the central part of the Mid-Atlantic Ridge, exposed above sea level. The arrows show the direction of spreading on each side of the rift.

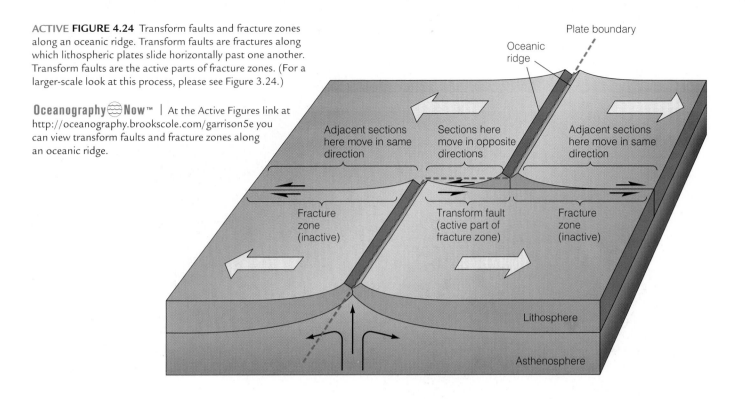

ACTIVE **FIGURE 4.24** Transform faults and fracture zones along an oceanic ridge. Transform faults are fractures along which lithospheric plates slide horizontally past one another. Transform faults are the active parts of fracture zones. (For a larger-scale look at this process, please see Figure 3.24.)

Oceanography≋Now™ | At the Active Figures link at http://oceanography.brookscole.com/garrison5e you can view transform faults and fracture zones along an oceanic ridge.

activity. While segments of a lithospheric plate on either side of a transform fault move in *opposite* directions from each other, the plate segments adjacent to the outward segments of a fracture zone move in the *same* direction, as Figure 4.24 shows.

Oceanography≋Now™ | Click Oceanography Interactive to work through an activity on plate boundaries through plate tectonics.

Hydrothermal Vents

Some of the most exciting features of the ocean basins are the **hydrothermal vents.** Hot springs were discovered on oceanic ridges in 1977 by Robert Ballard and J. F. Grassle of the Woods Hole Oceanographic Institution. Diving in *Alvin* at 3 kilometers (1.9 miles) near the Galápagos Islands along the East Pacific Rise (an oceanic ridge), they came across rocky chimneys up to 20 meters (66 feet) high, from which dark, mineral-laden water was blasting at 350°C (660°F) (**Figure 4.25**). Only the great pressure at this depth prevented the escaping water from flashing to steam. These *black smokers*, as they were quickly nicknamed, fascinate marine geologists. It is believed that water descends through fissures and cracks in the ridge floor until it comes into contact with very hot rocks associated with active seafloor spreading. There the superheated, chemically active water dissolves minerals and gases and escapes upward through the vents by convection (**Figure 4.26**).

Since that first discovery, vents have been found on the Mid-Atlantic Ridge east of Florida, in the Sea of Cortez east and south of Baja California, and on the Juan de Fuca Ridge off the coast of Washington and Oregon. Scientists now believe

FIGURE 4.25 A black smoker discovered at a depth of about 2,800 meters (9,200 feet) along the East Pacific Rise.

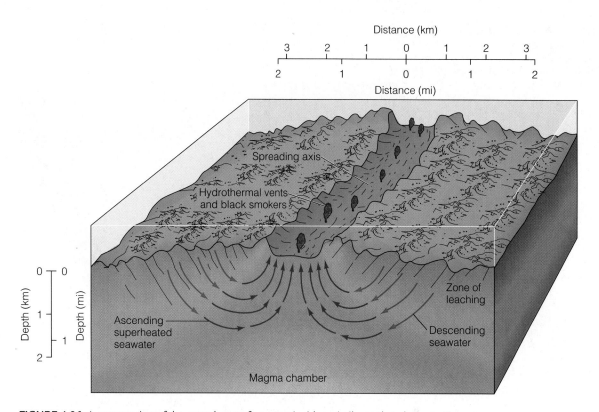

Distance (km)

Distance (mi)

Spreading axis

Hydrothermal vents
and black smokers

Zone of
leaching

Ascending
superheated
seawater

Descending
seawater

Depth (km)

Depth (mi)

Magma chamber

FIGURE 4.26 A cross section of the central part of an oceanic ridge—similar to that shown in Figure 4.22b—showing the origin of hydrothermal vents. Cool water (blue arrows) is heated as it descends toward the hot magma chamber, leaching sulfur, iron, copper, zinc, and other materials from the surrounding rocks. The heated water (red arrows) returning to the surface carries these elements upward, discharging them at hydrothermal springs on the seafloor. The areas around the vents support unique communities of organisms (see Chapter 16).

that hydrothermal vents may be very common on oceanic ridges, especially in zones of rapid seafloor spreading. In July 1990 vents were discovered in fresh water, at the bottom of Lake Baikal in southern Siberia. This discovery suggests that the world's oldest and deepest lake may someday become part of the ocean as Asia slowly breaks apart.

Not all vents form chimneys of mineral deposits; some are simply cracks in the seabed, or porous mounds, or broad segments of ocean ridge floor through which warm, mineral-laden water percolates upward. Cooler vents result when hot, rising water mixes with cold bottom water before reaching the surface. Water temperature in the vicinity of most hydrothermal vents averages 8–16°C (46–61°F), much warmer than usual for ocean bottom water, which has an average temperature of 3–4°C (37–39°F). We will study the unusual communities of animals that populate the vents in Chapter 16.

A volume of water equal to the volume of the world ocean is thought to circulate through the hot oceanic crust at spreading centers about every 10 million years! The water coming from the vents and seeping from the floor is more acidic, is enriched with metals, and has higher concentrations of dissolved gas than seawater does. The heat and chemicals issuing from these structures may play important roles in the chemical composi-

tion of seawater and the atmosphere, and in the formation of mineral deposits.

Abyssal Plains and Abyssal Hills

4.15

A quarter of Earth's surface consists of abyssal plains and abyssal hills. *Abyssal* is an adjective derived from a Greek word meaning "without bottom." Although this is obviously not literally true, you can appreciate how the term came into use following the *Challenger* expedition's laborious soundings of these extremely deep areas!

Abyssal plains are flat, featureless expanses of sediment-covered ocean floor found on the periphery of all oceans. They are most common in the Atlantic, less so in the Indian Ocean, and relatively rare in the active Pacific, where peripheral trenches trap most of the sediments flowing from the continents. Abyssal plains lie between the continental margins and the oceanic ridges about 3,700 to 5,500 meters (12,000 to 18,000 feet) below the surface (see again Figure 4.22a). The Canary Abyssal Plain, a huge plain west of the Canary Islands in the North Atlantic, has an area of about 900,000 square kilometers (350,000 square miles).

5,120 meters; 16,800 feet

6,585 meters; 21,600 feet

10 miles
16 kilometers

Charles D. Hollister

FIGURE 4.27 The deep, smooth sediments of the Atlantic's Northern Madeira Abyssal Plain bury 100-million-year-old mountains. The image was generated by a powerful echo sounder.

Abyssal plains are extraordinarily flat. A 1947 survey by the Woods Hole Oceanographic Institution ship *Atlantis* found that a large Atlantic abyssal plain varies no more than a few meters in depth over its entire area. Such flatness is caused by the smoothing effect of the layers of sediment, which are often more than 1,000 meters (3,300 feet) thick. Most of the sediment that forms the abyssal plains appears to be of terrestrial or shallow-water origin, not derived from biological activity in the ocean above. Some of it may have been transported to the plains by winds or turbidity currents. These deep sediment layers mask irregularities in the underlying ocean crust, but a powerful type of echo sounder can "see" through this sediment to reveal the complex topography of the basaltic basin floor below (**Figure 4.27**). The broad basaltic shoulders of the Mid-Atlantic Ridge extend beneath this cloak of sediment almost as far as the bordering continental slopes.

Abyssal plain sediments may not be thick enough to cover the underlying basaltic floor near the edges bordering the oceanic ridges. Here the plains are punctuated by **abyssal hills**—small, sediment-covered extinct volcanoes or intrusions of once-molten rock, usually less than 200 meters (650 feet) high (one is seen extending above the flat surface in Figure 4.27). These abundant features are associated with seafloor spreading; they form when newly formed crust moves away from the center of a ridge, stretches, and cracks. Some blocks of the crust drop to form valleys, and others remain higher as hills. Lava erupting from the ridge flows along the fractures, coating the hills. This helps explain why abyssal hills occur in lines parallel to the flanks of the nearby oceanic ridge, and why they occur most abundantly in places where the rate of seafloor spreading is fastest. Abyssal plains and hills account for nearly all the area of deep-ocean floor that is not part of the oceanic ridge system.

Seamounts and Guyots

The ocean floor is dotted with thousands of volcanic projections that do not rise above the surface of the sea. These projections are called **seamounts.** Seamounts are circular or elliptical, more than 1 kilometer (0.6 mile) in height, with relatively steep slopes of 20° to 25°. (Abyssal hills, in contrast, are much more abundant, less than a kilometer high, and not as steep.) Seamounts may be found alone or in groups of 10 to 100. Though many form at hot spots, most are thought to be submerged inactive volcanoes that formed at spreading centers (**Figure 4.28**). Movement of the lithosphere away from spreading centers has carried them outward and downward to their present positions. As many as 10,000 seamounts are thought to exist in the Pacific, about half the world total.

Guyots are flat-topped seamounts that once were tall enough to approach or penetrate the sea surface. Generally they are confined to the west-central Pacific. The flat top suggests that they were eroded by wave action when they were near sea level. Their plateaulike tops eventually sank too deep for wave erosion to continue wearing them down. Like the more abundant seamounts, most guyots were formed near spreading centers and transported outward and downward as the seafloor moves away from a spreading center and cools.

Trenches and Island Arcs

A **trench** is an arc-shaped depression in the deep-ocean floor. These creases in the seafloor occur where a converging oceanic plate is subducted. The water temperature at the bottom of a trench is slightly cooler than the near-freezing temperatures of the adjacent flat ocean floor, reflecting the fact that trenches are underlain by old, relatively cold ocean crust sinking into the upper mantle. Trenches (and their associated island arcs topped by erupting volcanoes) are among the most active geological features on Earth. Great earthquakes and tsunami (huge waves we will discuss in Chapter 10) often originate in them. **Figure 4.29** shows the distribution of the ocean's major trenches. It is not surprising that most are around the edges of the active Pacific.

FIGURE 4.28 (a) The process by which guyots (G) and seamounts (S) form. Guyots have flat tops because they were once tall enough to be eroded by waves at the ocean's surface. Seamounts have a similar origin but retain their more pointed volcano shape because they never reached the surface. (b) An undersea volcano east of the easternmost island of the Samoan chain in the Pacific. Vailulu'u, as it has been named, rises from an ocean depth of 4,800 meters (15,700 feet) to within 590 meters (1,900 feet) of the ocean surface.

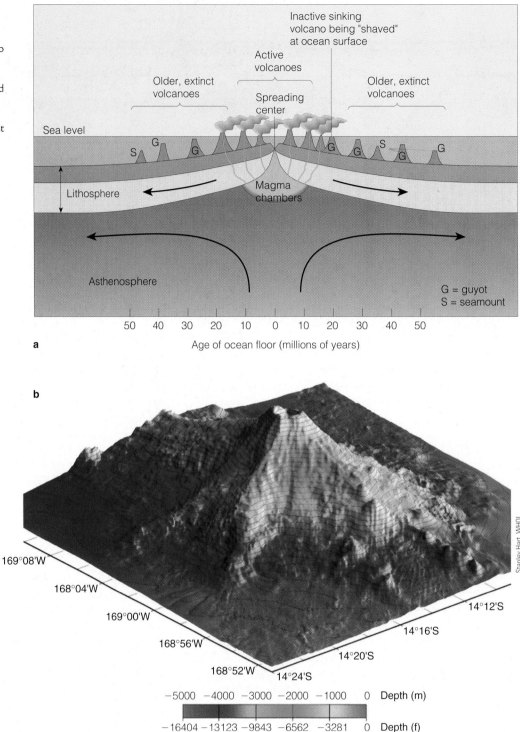

Trenches are the deepest places in Earth's crust, 3 to 6 kilometers (1.9 to 3.7 miles) deeper than the adjacent basin floor. The ocean's greatest depth is the Mariana Trench of the western Pacific, where *Trieste* set the world's deep-diving record described in **Box 4.1**. Near the place where *Trieste* came to rest, the ocean bottom is 11,022 meters (36,163 feet) below sea level, 20% deeper than Mt. Everest is high (**Figure 4.30**). The Mariana Trench is about 70 kilometers (44 miles) wide and 2,550 kilometers (1,600 miles) long, typical dimensions for these structures.

Trenches are curving chains of V-shaped indentations. The trenches are curved because of the geometry of plate interactions

FIGURE 4.29 Oceanic trenches of the world. Note their prevalence in the active Pacific.

on a sphere. The convex sides of these curves generally face the open ocean (see again Figure 4.29). The trench walls on the island side of the depressions are steeper than those on the seaward side, indicating the direction of plate subduction. The sides of trenches become steeper with depth, normally reaching angles of about 10° to 16° before flattening to a floor underlain by thick sediment. (Parts of the concave wall of the Kermadec–Tonga Trench are the world's steepest at 45°.) No continental rise occurs along coasts with trenches because the sediment that would form the rise ends up at the bottom of the trench.

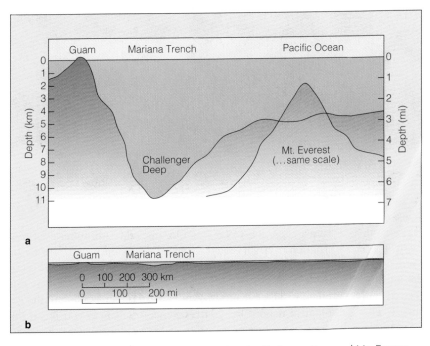

FIGURE 4.30 The Mariana Trench. (**a**) Comparing the Challenger Deep and Mt. Everest on the same scale shows that the deepest part of the Mariana Trench is about 20% deeper than the mountain is high. (**b**) The Mariana Trench shown without vertical exaggeration.

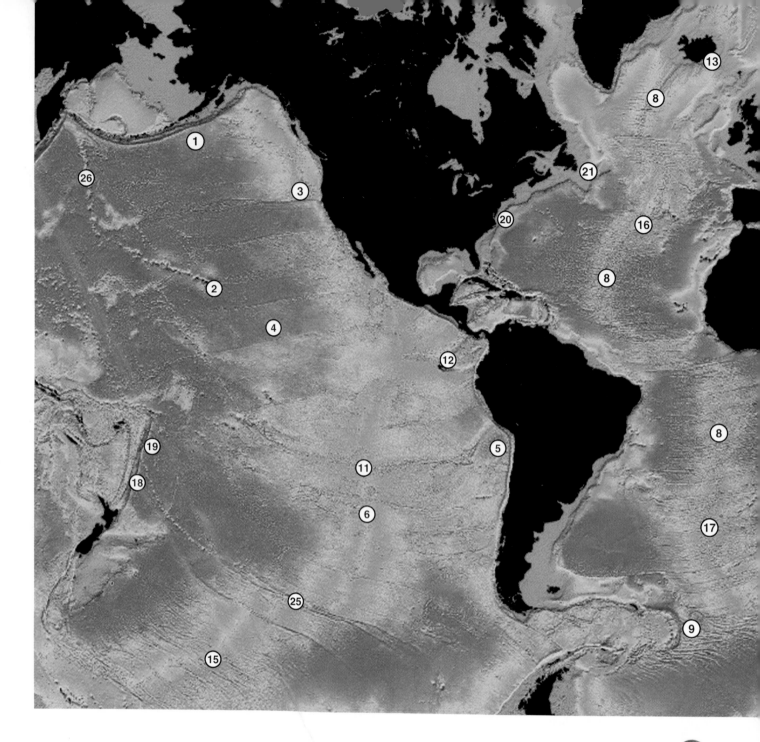

Island arcs, curving chains of volcanic islands and seamounts, are almost always found parallel to the concave edges of trenches. As you may remember from Chapter 3, trenches and island arcs are formed by tectonic and volcanic activity associated with subduction. The descending lithospheric plate contains some materials that melt as the plate sinks into the mantle. These materials rise to the surface as magmas and lavas that form the chain of islands behind the trench. The Aleutian Islands, most Caribbean islands, and the Mariana Islands are island arcs. (See Figure 3.22 for a review of the processes involved in their construction.)

The Grand Tour

Researchers at the National Oceanic and Atmospheric Administration have generated a map of the world ocean floor based on satellite observations of the shape of the sea surface (**Figure 4.31**). The graphic shows all the features discussed in this chapter. These features—and a basic understanding of the geological reasons for their existence—will help you recall the dramatic nature and history of the seafloor that we have discussed in the past two chapters.

FIGURE 4.31 A technological *tour de force:* a map that shows all the features discussed in this chapter, derived from data provided to the National Geophysical Data Center from satellites and shipborne sensors. These features—and a basic understanding of the geological reasons for their existence—will help you recall the dramatic nature and history of the seafloor that we have discussed in the past two chapters.

Key to features:
(1) Aleutian Trench
(2) Hawaiian Islands
(3) Juan de Fuca Ridge
(4) Clipperton Fracture Zone
(5) Peru–Chile Trench
(6) East Pacific Rise
(7) Mariana Trench, Challenger Deep
(8) Mid-Atlantic Ridge
(9) South Sandwich Trench
(10) Red Sea
(11) Easter Island
(12) Galápagos Rift
(13) Iceland
(14) Gulf of Aden
(15) Pacific–Antarctic Ridge

(16) Azores
(17) Tristan de Cunha
(18) Kermadec Trench
(19) Tonga Trench
(20) Hatteras Abyssal Plain
(21) Grand Banks
(22) Mid-Indian Ridge
(23) Atlantic–Indian Ridge
(24) Maldive Islands
(25) Eltanin Fracture Zone
(26) Emperor Seamounts
(27) Java Trench
(28) Great Barrier Reef
(29) Kuril Trench
(30) Japan Trench

BOX 4.1   *Trieste* at the Bottom

Don Walsh

(**a**) *Trieste* before launching. Note the crew sphere suspended below the gasoline-filled flotation hull. *Trieste* is 15.7 meters (59.5 feet) long.

Naval Historical Center

Studying deep-ocean basins has always presented daunting obstacles. The most difficult problem is to reach extreme depths, but amazingly scientists have visited the bottom of the deepest ocean basin. On 23 January 1960 U.S. Navy Lieutenant Don Walsh and Dr. Jacques Piccard descended to a depth of 11,022 meters (6.85 miles) into the Challenger Deep, an area of the Mariana Trench discovered in 1951 by the British oceanographic research vessel *Challenger II*.[1] The vehicle used in the descent was *Trieste* (**Figures a, b,** and **c**), a deep-diving submersible designed like a blimp with a very strong and thick (and cramped) steel crew sphere suspended below. A blimp uses helium gas for buoyancy, but a gas

(**b**) *Trieste*'s steel personnel sphere, which varies from 13 to 18 centimeters (5 to 7 inches) in thickness. The small, thick window is cone-shaped to withstand the pressure of the deepest spot in the world ocean—11 million kilograms per square meter (8 tons per square inch)!

would be compressed by water pressure; so gasoline, which is relatively incompressible, was used to provide lift.

---

[1] The position of the Challenger Deep is marked in Figure 4.29, and its depth in relation to Mt. Everest is shown in Figure 4.30a.

Questions from Students

1. Shouldn't the ocean be deeper in the middle? And why is Iceland one of only a few places in the world where oceanic crust is found above sea level?

Understanding *why* there are ocean basins explains their contours. A basin usually contains an expanding ridge that is higher than the surrounding bed. Oceanic crust is thin and dense. Be-

cause of isostasy, the ocean floor lies at a lower elevation than the thicker, less dense, higher continents. Water filled the lower elevations first, submerging nearly all the basaltic basement we now call ocean floor. In areas of rapid seafloor spreading, or at hot spots, peaks are occasionally pushed toward the ocean surface by the large volume of mantle material rising in plumes from below. Large quantities of erupted magma (lava) then build the crests above sea level to form islands like Iceland or the Azores.

(**c**) Andreas Rechnitzer and Don Walsh prepare *Trieste* for a deep dive. The submersible made 130 dives, including three world-record dives. *Trieste's* last voyage was in 1963.

First *Trieste* made a series of increasingly deep practice dives; then the big day arrived. Don Walsh wrote of the experience in 1979:

At about 600 feet we entered a zone of deepening twilight where colors faded into gray. By 1,000 feet the light had gone completely. We turned out the lights in the sphere to watch for the luminescent creatures that are sometimes visible at this level. We saw very few. Eventually we turned the cabin lights back on and briefly tested the forward lights that throw a beam in front of the observation window. Formless plankton streamed past, giving us a sensation of great speed.

There were minor incidents such as the small leak that always developed in one of the hull connectors—a place where wires from lights and instruments on the outside of the sphere pass through the hull. The leak started at about 10,000 feet. It was an old friend, a tiny drip, drip, drip. I timed the drips and found no change from before, which meant that it had not become more serious. We expected it to disappear at about 15,000 feet, when the water pressure packed the plastic sealer in more tightly—and it did.

At 27,000 feet we checked our rate of descent to 2 feet per second by dumping some shot ballast. We were not too sure of the underwater currents here, and we did not want to go crashing into a wall of the trench by mistake. As we neared 30,000 feet I started thinking about the changes we had planned to make when we got to within 1,000 feet of the bottom—which we now were expecting to find only another 3,500 feet below us. I was running through a mental checklist when we heard and felt a powerful crack. The sphere rocked as though we were on land and going through a mild earthquake.

We waited anxiously for what might happen next. Nothing did. We flipped off the instruments and the underwater telephone so that we could hear better. Still nothing happened. We switched the instruments back on and studied the dials that would tell us if something critical had occurred. No, we had our equilibrium and were descending exactly as before.

We checked our speed to half a foot a second and continued. At that rate, time and distance pass very slowly, and I think for the first time in the dive both of us had the feeling of awe that comes from exploring the totally unknown.

I did not take my eyes off the fathometer, and Jacques never stopped watching out of the tiny porthole with its weak probe of light. No bottom was in sight. . . . Soon Jacques could see a difference in the effect of our light in the water, as the rays reflected off the bottom. As we approached the floor I called the fathometer readings to Jacques in fathoms: "Thirty . . . twenty . . . ten . . ." At eight he called that he could see the bottom.

At 1:10 P.M. we sank gently onto the soft floor. A great cloud of silt rose around us. The fifteen-man, Navy civilian/military team had set a record that could not be broken.

The voyagers took temperature readings and spotted a pair of bottom fish and some crustaceans before ascending. With the possible exception of the moon, no place visited by humans has been more hazardous to explore.

4.19

2. Wouldn't wave action and tides hopelessly clutter the radar signals sent from satellites to determine sea surface height?

Satellite altimetry is one of the most sophisticated oceanographic uses of high-speed computer processing. Imagine the processing power needed to reduce the data generated by more than a thousand radar pulses from orbit each second! Programmers subtract predicted tidal height from the measurements and then use algorithms to average and cancel wave crests and troughs. The remaining sea surface height—determined to perhaps 2 centimeters (slightly less than 1 inch)—is due to gravitational variations from submerged features. Still more processing is needed to generate graphic images like Figure 4.5c. The procedures were pioneered by the U.S. Navy and the Office of Naval Research. Their original goal was to provide detailed seafloor maps for use in antisubmarine defense.

3. How far away is the horizon when I stand on the beach and look out to sea?

That depends on how tall you are. If you're about 6 feet tall, the distance to the horizon is about 3 miles (4.8 kilometers). For a more precise estimate, subtract 4 inches (a third of a foot) to find the height of your eyes (in feet) above the ground; then divide that number by 0.5736. Take the square root of the result and you have the number of miles to the horizon. Lifeguards standing on a 10-foot tower see a horizon roughly 8.5 kilometers (5.3 miles) distant.

4. Is *Alvin* the deepest-diving human-carrying research submersible?

No, that honor belongs to *Shinkai 6500* (**Figure 4.32**). Operated by a consortium of Japanese research institutions and industries, *Shinkai 6500* is the newest deep-diving vehicle capable of carrying a human crew. Now the deepest-diving submersible in operation, *Shinkai 6500* safely descended to a depth of 6,527 meters (21,409 feet) on 11 August 1989, and it can reach all but the deepest trench floors, or about 98% of the world ocean bottom.

5. You mentioned that Walsh and Piccard saw animals at a depth of 11,022 meters (6.85 miles). The pressure there is more than 8 tons per square inch. How could any living thing withstand such crushing pressure?

It seems impossible until you think about your own situation. Humans live at the bottom of a surprisingly heavy "ocean" of air: A column of air 1 inch square extending from sea level to the top of the atmosphere weighs about 6.7 kilograms (14.7 pounds). This pressure presents no problem to us because we have the same pressure inside of us pushing out. The animals at great depths also have the same pressure inside as outside—that is, they live in perfect hydrostatic balance. They notice the weight of water above them no more than you notice the weight of air above you.

6. What's next in deep-sea exploration?

Perhaps the most imaginative new technology is *telepresence*, the extension of a person's senses by remote manipulators. A scientist might wear a helmet containing small stereo television screens and earphones, and place his hands in special gloves equipped with tactile feedback units. His head and

FIGURE 4.32 Now the deepest-diving human-carrying submersible in operation, *Shinkai 6500* safely descended to a depth of 6,527 meters (21,409 feet) on 11 August 1989 and thus can reach all but the deepest trench floors, or about 98% of the world ocean bottom. *Shinkai* is operated by JAMSTEC, a Japanese oceanographic research consortium.

hand movements would be duplicated by a robot on the seafloor, and the sensations "felt" by the robot would be relayed back to the scientist through the TVs, earphones, and gloves. Scientists would thus have the sensation of being on the seafloor, and they could take samples, manipulate tools, or just look around. Other researchers could watch or participate at distant locations via the information superhighway. Oceanographers have set a course for interesting times!

Chapter in Perspective

In this chapter you learned how difficult it has been to discover the shape of the seabed. Even today, the surface contours of Mars are better known than those of our ocean floor.

We now know that seafloor features result from a combination of tectonic activity and the processes of erosion and deposition. The ocean floor can be divided into two regions: continental margins and deep-ocean basins. The continental margin, the relatively shallow ocean floor nearest the shore, consists of the continental shelf and the continental slope. The continental margin shares the structure of the adjacent continents, but the deep-ocean floor away from land has a much different origin and history. Prominent features of the deep-ocean basins include rugged oceanic ridges, flat abyssal plains, occasional deep trenches, and curving chains of volcanic islands. The processes of plate tectonics, erosion, and sediment deposition have shaped the continental margins and ocean basins.

In the next chapter you will learn that nearly all the ocean floor is blanketed with sediment. Except for the spreading centers themselves, the broad shoulders of the oceanic ridge systems are buried according to their age—the older the seabed, the greater the sediment burden. Some oceanic crust near the trailing edges of plates may be overlain by sediments more than 1,500 meters (5,000 feet) thick.

Sediments have been called the "memory of the ocean." The memory, however, is not a long one. Before continuing, can you imagine why that is so?

Key Ideas

✓ New bathymetric devices used to study seabed features include multibeam echo sounder systems and satellites that use sensitive radar for altimetry. Their use has revolutionized our understanding of ocean floor formation and topography.
 Review Figures 4.4 and 4.5, pages 90–91

✓ Seafloor features result from a combination of tectonic activity and the processes of erosion and deposition.
 Review Chapters 3 and 4 and Figure 4.8, page 93

✓ Near shore the features of the ocean floor are similar to those of the adjacent continents because they share the same granitic basement. The transition to basalt marks the true edge of the continent and divides ocean floors into two major provinces. The submerged outer edge of a continent is called the *continental margin*. The deep-sea floor beyond the continental margin is properly called the *ocean basin*.
 Review Figure 4.11, page 95

✓ Features of the continental margins include continental shelves, continental slopes, submarine canyons, and continental rises.
 Review Figures 4.8 and 4.10, pages 93, 94

✓ Features of the deep-ocean basins include oceanic ridges, hydrothermal vents, abyssal plains and hills, seamounts, guyots, trenches, and island arcs.
 Review Figure 4.22a, page 101

Terms and Concepts to Remember

abyssal hill, 105	ice age, 96
abyssal plain, 104	island arc, 108
active margin, 94	ocean basin, 92
bathymetry, 88	oceanic ridge, 100
continental margin, 92	passive margin, 94
continental rise, 99	seamount, 105
continental shelf, 94	shelf break, 97
continental slope, 97	submarine canyon, 98
epicenter, 98	transform fault, 101
fracture zone, 101	trench, 105
guyot, 105	turbidity current, 99
hydrothermal vent, 103	

Study Questions

Reviewing What You've Learned

1. How can satellites be used to measure ocean surface height? Why does the surface of the ocean "bunch up" over submerged mountains and ridges?
2. How does a multibeam sonar system work?
3. Imagine that you could walk from the shore to the deep-sea bed off the east coast of the United States. What features would you encounter?
4. What happens to continental shelves during ice ages?
5. Which is greater: the height of Mt. Everest or the depth of the Mariana Trench?
6. What do scientists think causes submarine canyons?
7. Why are abyssal plains relatively rare in the Pacific?

8. Why are trenches and island arcs curved? Is the descent to the bottom steeper on the convex side of the arc or on the concave side? Why the difference? Why do you think most trenches are in the western Pacific? (Hint: Check the position and action of the East Pacific Rise.)
9. Distinguish among abyssal hills, seamounts, guyots, and island arcs.
10. What is an oceanic ridge? Are they always literally in mid-ocean? How are oceanic ridges and trenches related?

Thinking Critically

1. Why did people think an ocean was deepest at its center? What changed their minds?
2. What do the facts that (a) granite underlies the edges of continents and (b) basalt underlies deep-ocean basins suggest? (Hint: Consider thicknesses and densities.)
3. The terms *leading* and *trailing* are also used to describe continental margins. How do you suppose these words relate to *active* and *passive*, or *Atlantic-type* and *Pacific-type*, used in the text?
4. What forces control the shape of a continental shelf? A continental slope? A continental rise?
5. Answer this question if you have already read Chapter 3: Your time machine has been programmed to deliver you to Frankfurt, Germany, on a chilly evening in January 1912 to hear Wegener's lectures on continental drift. What two illustrations from this chapter would you take with you to cheer him up after the lecture? Why did you select those particular illustrations?
6. **InfoTrac College Edition Project** In the twentieth century, ships carrying sensors of various kinds were the dominant means of exploring the oceans, including the ocean floor. In recent years, satellite data have also contributed much to our knowledge. What new opportunities for exploration of the ocean floor do recent technological improvements provide? What strategies are scientists proposing to observe undersea processes and collect data? Do some research using Infotrac College Edition to find out about proposals for future undersea investigations. Write a brief essay on your findings, and describe some types of observations that could not be made using traditional research tools.

Thinking Analytically

1. The speed of sound through seawater is about 1,500 meters per second. If a ship equipped with a multibeam mapping system is surveying a feature 3,500 meters below the surface, and if the researchers wish to obtain an image of the feature at a resolution of 10 meters, what is the maximum speed the ship can steam?
2. Review the speed of a turbidity current. In the unlikely event that a fast-running current formed near the shoreline of a trailing edge coast, how long would it take for the current to traverse a typical continental shelf and arrive at the shelf break? Would you expect the current to move at a constant speed during this traverse?
3. How much wider has Iceland become because of seafloor spreading since the last huge sequence of eruptions in the 1500s? [Hint: Use the data in Figure 4.21]

Oceanography🌐Now™
Assess your understanding of this chapter's topics with additional quizzing and conceptual-based problems at http://oceanography.brookscole.com/garrison5e

Study and Research on the Internet

Assess your understanding of this chapter's topics with additional quizzing and conceptual-based problems at:

 http://oceanography.brookscole.com/garrison5e

- **InfoTrac College Edition Research** leads you to critical thinking exercises that use InfoTrac College Edition® as a research tool.
- **Regional InfoTrac College Edition** articles are organized into East Coast, West Coast, and Gulf Coast regions, allowing you to study oceanography on a more local level.

For more readings, go to InfoTrac College Edition, your online research library, at:

http://infotrac.thomsonlearning.com

CHAPTER 5

Age (Ma)   cm   Events
64.9

Tertiary

50 —
                                        Post-extinction layer:
                                        Sediments containing
                                        microfossils from
                                        after the dinosaurs

65.0 —
                                        Fireball layer:
60 —   Dust and ash fallout from
                                        the asteroid impact

                                        Ejecta layer:
                                        Material blasted from the
                                        crater and deposited here
                                        within days to months

Cretaceous

70 —

Don Davies/NASA

Richard Norris/SIO

a

Portrait of a disaster: 65 million years ago on one very bad day! (**a**) An artist's conception of a catastrophic asteroid strike. (**b**) This cross section of a seabed core shows clear evidence of the impact and its aftermath.

b

SEDIMENTS

The Memory of the Ocean

A glance at the first few photographs in this chapter will show you the true face of the ocean floor. The basalt and lava we have been discussing are nearly always hidden—covered by dust and gravel, silt and mud. This sediment includes particles from land, from biological activity in the ocean, from chemical processes within water, and even from space. Analysis of this sedimentary material can tell us the recent history of an ocean basin and sometimes the recent history of the whole Earth. In a sense, seafloor sedimentary deposits are the memory of the ocean.

Sometimes that memory records catastrophic events. For example, many lines of evidence suggest that Earth was struck 65 million years ago by an asteroid about 10 kilometers (6 miles) across. The cataclysmic collision is thought to have propelled shock waves and huge clouds of seabed and crust all over Earth, producing a time of cold and dark that contributed to the extinction of many species, including the dinosaurs. The accompanying photograph of a deep-sea core shows evidence of this disaster, and you'll learn more about it in Chapter 13.

Because seafloors are recycled by tectonic forces, the ocean's sedimentary memory is not long. Records of events older than about 180 million years are recycled as oceanic crust and its overlying sediment reach subduction zones. The ocean has forgotten much in the last 4 billion years.

What Sediments Look Like

Sediment is particles of organic or inorganic matter that accumulate in a loose, unconsolidated form. The particles originate from the weathering and erosion of rocks, from the activity of living organisms, from volcanic eruptions, from chemical processes within the water itself, and even from space. Most of the ocean floor is being slowly dusted by a continuing rain of sediments. Accumulation rates on the deep-sea floor vary from a few centimeters per year to the thickness of a dime every thousand years.

Marine sediments occur in a broad range of sizes and types. Beach sand is sediment; so are the muds of a quiet bay and the mix of silt and tiny shells found on the continental margins. Less familiar sediments are the fine clays of the deep-ocean floor, the biologically derived oozes of abyssal plains, and the nodules and coatings that form around hard objects on the seafloor. The origin of these materials—and the distribution and sizes of the particles—depends on a combination of physical and biological processes.

What do sediments look like? That depends on where you look. **Figure 5.1** shows the Mid-Atlantic Ridge, 48°N, at a depth of 2,629 meters (8,626 feet). Sponges and some relatives of coral grow from young rocky outcrops only lightly powdered with sediment. Contrast that rough ridge with the smooth seafloor shown in **Figure 5.2**. The sediment there is about 35 meters (115 feet) thick and marked by the tracks of brittle stars. These widely distributed organisms feed on surface bacteria and fallen particles of organic sediment.

The surface of the sediment is not always smooth. Where bottom currents are swift and persistent, they can cause ripples like those on a streambed (**Figure 5.3**).

The extraordinary thickness of some layers of marine sediment can be seen in **Figure 5.4**, a seismic profile of the eastern edge of a seamount in the North Atlantic's Sohm Abyssal Plain south of Nova Scotia. The sediment at the eastern boundary of this profile covers the oceanic crust to a depth of more than 1.8 kilometers (1.1 miles).

The colors of marine sediments are often striking. Sediments of biological origin are white or cream-colored, with deposits high in silica tending toward gray. Some deep-sea clays—though traditionally termed "red clays" from the rusting (oxidation) of iron within the sediments to form iron oxide—can range from tan to chocolate brown. Other clays are shades of green or tan. Nodular sediments are a dark sooty brown or black. Some nearshore sediments contain decomposing organic material and smell of hydrogen sulfide, but most are odorless.

A very few areas of the seabed are altogether free of overlying sediments. The water over these areas is not completely sediment-free, but for some reason sediment does not collect on the bottom. Strong currents may scour the sediments away, the seafloor may be too young in these areas for sediments to have had time to accumulate, or hot water percolating upward through a porous seafloor may dissolve the material as fast as it settles.

Classifying Sediment by Particle Size

Particle size is frequently used to classify sediments. The scheme shown in **Table 5.1** was first devised in 1898 and with refinements has been used by geologists, soil scientists, and oceanographers ever since. In this classification the coarsest particles are boulders, which are more than 256 millimeters (about 10 inches) in diameter. Although boulders, cobbles, and pebbles occur in the ocean, most marine sediments are made of finer particles: **sand, silt,** and **clay.** The particles are defined by their size.

FIGURE 5.1 Sediment near the crest of the Mid-Atlantic Ridge. The rock outcrops are dusted only lightly with sediment; crinoids and gorgonians (relatives of coral) attach to the rocks.

FIGURE 5.2 Brittle stars and their tracks on the continental slope off New England. The depth is 1,476 meters (4,842 feet).

FIGURE 5.3 Ripples on the sediment beneath the swift Antarctic Circumpolar Current in the northern Drake Passage. The depth here is 4,010 meters (13,153 feet).

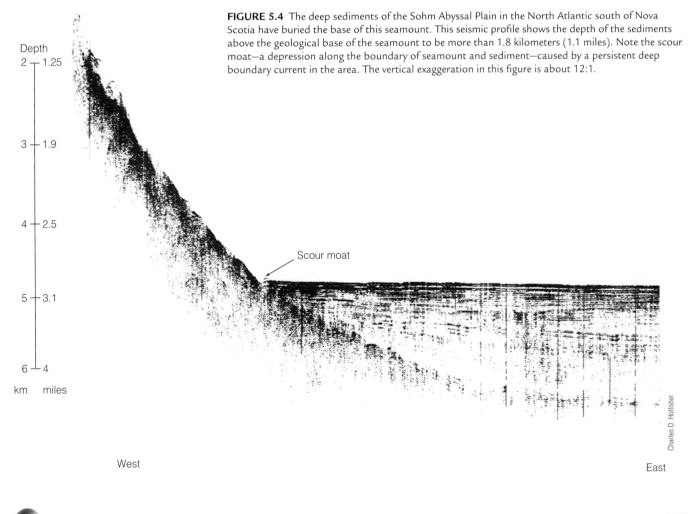

FIGURE 5.4 The deep sediments of the Sohm Abyssal Plain in the North Atlantic south of Nova Scotia have buried the base of this seamount. This seismic profile shows the depth of the sediments above the geological base of the seamount to be more than 1.8 kilometers (1.1 miles). Note the scour moat—a depression along the boundary of seamount and sediment—caused by a persistent deep boundary current in the area. The vertical exaggeration in this figure is about 12:1.

Depth

km	miles
2	1.25
3	1.9
4	2.5
5	3.1
6	4

Scour moat

West

East

Table 5.1	Particle Sizes and Settling Rate in Sediment		
Type of Particle	Diameter	Settling Velocity in Still Water	Time to Settle 4 km (2.5 mi)
Boulder	>256 mm (10 in.)	—	—
Cobble	64–256 mm ($>2\frac{1}{2}$ in.)	—	—
Pebble	4–64 mm ($\frac{1}{6}$–$2\frac{1}{2}$ in.)	—	—
Granule	2–4 mm ($\frac{1}{12}$–$\frac{1}{6}$ in.)	—	—
Sand	0.062–2 mm	2.5 cm/sec (1 in./sec)	1.8 days
Silt	0.004–0.062 mm	0.025 cm/sec ($\frac{1}{100}$ in./sec)	6 months
Clay	<0.004 mm	0.00025 cm/sec.	50 years[a]

[a] Though the theoretical settling time for individual clay particles is usually very long, under certain conditions clay particles in the ocean can interact chemically with seawater, clump together, and fall at a faster rate. Small biogenous particles are often compressed by organisms into fecal pellets that can fall more rapidly than would otherwise be possible. A fecal pellet is shown in Figure 5.16.

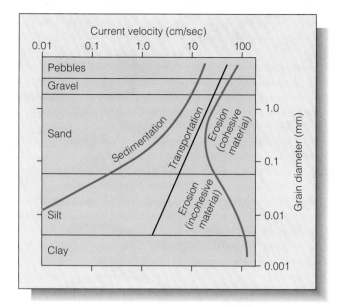

FIGURE 5.5 The velocities of currents required for erosion, transportation, and deposition (sedimentation) of sediment particles of different sizes. The green curve indicates that rapid currents are needed to erode clay, fine silt, and coarse sand. Weaker currents are sufficient to move fine or medium sand off the sea floor. Note that the *cohesiveness* of smaller particles can be more important than grain size in determining whether they will be eroded and transported. Mud, a mixture of clay and silt, is "stickier" than sand and resists erosion. As the black line indicates, less sticky materials would behave in flowing water in the near-linear manner we might expect.

Generally, the smaller the particle, the more easily it can be transported by streams, waves, and currents. As sediment is transported it tends to be sorted by size; coarser grains, which are moved only by turbulent flow, tend not to travel as far as finer grains, which are more readily moved. The clays, particles less than 0.004 millimeter in diameter, can remain suspended for very long periods and may be transported great distances by ocean currents before they are deposited. As indicated in **Figure 5.5,** the cohesiveness of smaller particles can be as important as grain size in determining whether they will be eroded and transported. Once in suspension, the finest clays may circulate in the ocean for decades.

A layer of sediment can contain particles of similar size, or it can be a mixture of different-sized particles. Sediments composed of particles of one size are said to be **well-sorted sediments.** Sediments with a mixture of sizes are **poorly sorted sediments.** Sorting is a function of the energy of the environment—the exposure of that area to the action of waves, tides, and currents. Well-sorted sediments occur in an environment where energy fluctuates within narrow limits. Sediments of the calm deep-ocean floor are typically well sorted (see again Figure 5.2). Poorly sorted sediments form in environments where energy fluctuates over a wide spectrum. The mix of rubble at the base of a rapidly eroding shore cliff is a good example of poorly sorted sediment. Sediments that have been transported by turbidity currents (which can transport a wide range of grain sizes) tend to be poorly sorted.

Classifying Sediment by Source

Another way to classify marine sediments is by their origin. Such a scheme was first proposed in 1891 by Sir John Murray and A. F. Renard after a thorough study of sediments collected during the *Challenger* expedition. A modern modification of their organization is shown in **Table 5.2.** This scheme separates sediments into four categories by source: terrigenous, biogenous, hydrogenous (or authigenic), and cosmogenous.

Terrigenous Sediments

Terrigenous sediments (*terra* = Earth, *generare* = to produce) are the most abundant; as the name implies, they originate on continents or islands from erosion, volcanic eruptions, and blown dust.

The rocks of Earth's crust are made up of **minerals,** inorganic crystalline materials with specific chemical compositions. The texture of igneous rocks—rocks that crystallize from molten material—is determined by how rapidly they cool. Igneous rocks that cool rapidly, such as the basalt that forms the ocean floor at spreading centers or pours from volcanic vents on land, solidify so quickly that obvious crystals do not have a chance to form. Slower cooling produces the most commonly encountered crystals, those about the size of a grain of rice or the head of a pin. Nearly all terrigenous sediments are derived directly or indirectly from these crystals. You have probably seen the crystals in granite, the most familiar continental igneous rock. Granite is the source of quartz and clay, the two most common components of terrigenous marine sediments.

Table 5.2   Classification of Marine Sediments by Source of Particles

Sediment Type	Source	Examples	Distribution	Percent of All Ocean Floor Area Covered
Terrigenous	Erosion of land, volcanic eruptions, blown dust	Quartz sand, clays, estuarine mud	Dominant on continental margins, abyssal plains, polar ocean floors	~45%
Biogenous	Organic; accumulation of hard parts of some marine organisms	Calcareous and siliceous oozes	Dominant on deep-ocean floor (siliceous ooze below about 5 km)	~55%
Hydrogenous (authigenic)	Precipitation of dissolved minerals from water, often by bacteria	Manganese nodules, phosphorite deposits	Present with other, more dominant sediments	<1%
Cosmogenous	Dust from space, meteorite debris	Tektite spheres, glassy nodules	Mixed in very small proportion with more dominant sediments	<<1%

Sources: Kennett, *Marine Geology,* 1982; Weihaupt, *Exploration of the Oceans,* 1979; Sverdrup, Johnson, and Fleming, *The Oceans: Their Physics, Chemistry, and General Biology,* 1942.

Terrigenous sediments are part of a slow, massive cycle (**Figure 5.6**). Over the great span of geological time, mountains rise as plates collide, fuse, and subduct. Then the mountains erode. The resulting sediments are transported to the sea by wind and water, where they collect on the seafloor. The sediments travel with the plate and are either uplifted or subducted. The material is made into mountains. The cycle begins anew.

Although estimates vary, it appears that about 15 billion metric tons (16.5 billion tons) of terrigenous sediments are transported in rivers to the sea each year, with an additional 100 million metric tons transported annually from land to ocean as fine airborne dust and volcanic ash (**Figure 5.7**).

Biogenous Sediments

Biogenous sediments (*bio* = life, *generare* = to produce) are the next most abundant marine sediment. The siliceous (silicon-containing) and calcareous (calcium carbonate–containing) compounds that make up these sediments of biological origin were originally brought to the ocean in solution by rivers or dissolved in the ocean at oceanic ridges. The siliceous and calcareous materials were then extracted from the seawater by the normal activity of tiny plants and animals to build protective shells and skeletons. Some of this sediment derives from larger mollusk shells or from stationary colonial animals such as corals, but most of the organisms that produce biogenous sediments drift free in the water as plankton. After the death of their owners, the hard structures fall to the bottom and accumulate in layers. Biogenous sediments are most abundant where ample nutrients encourage high biological productivity, usually near continental margins and areas of upwelling. Over millions of years, organic molecules within these sediments can form oil and natural gas (see Chapter 17 for details).

Note in Table 5.2 that biogenous sediments cover a larger percentage of the *area* of the ocean floor than terrigenous sediments do, but the terrigenous sediments dominate in total *volume.*

Hydrogenous Sediments

Hydrogenous sediments (*hydro* = water, *generare* = to produce) are minerals that have precipitated directly from seawater. The sources of the dissolved minerals include submerged rock and sediment, leaching of the fresh crust at oceanic ridges, material issuing from hydrothermal vents, and substances flowing to the ocean in river runoff. As we shall see, the most prominent hydrogenous sediments are manganese nodules, which litter some deep seabeds, and phosphorite nodules, seen along some continental margins. Hydrogenous sediments are also called **authigenic sediments** (*authis* = in place, on the spot) because they were formed in the place they now occupy.

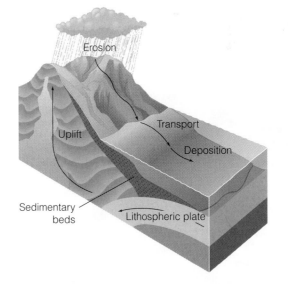

FIGURE 5.6 The sediment cycle. Over geological time mountains rise as lithospheric (crustal) plates collide, fuse, and subduct. Water and wind erode the mountains and transport the resulting sediment to the sea. The sediments are deposited on the seafloor, where they travel with the plate and are either uplifted or subducted. Thus the material is made into mountains again.

a

Liam Gumley, University of Wisconsin-Madison

b

Department of Geology, University of Delaware

NASA/GSFC, Orbimage, SeaWiFS

c

FIGURE 5.7 Sources of terrigenous sediments. (**a**) Rivers are the main source of terrigenous sediments. This photo, taken from space, shows sediment entering the Gulf of Mexico from the Mississippi River. (**b**) The wind may transport ash from a volcanic eruption for hundreds of kilometers and deposit it in the ocean. This ash cloud was caused by the summer 1991 eruption of Mount Pinatubo in the Philippines. (**c**) Dust from the Gobi Desert blows eastward across the Pacific on 18 March 2002. The particles will fall to the ocean surface and descend slowly to the bottom to end up as terrigenous sediments.

Though they usually accumulate very slowly, rapid deposition of hydrogenous sediments is possible—in a rapidly drying lake, for example.

Cosmogenous Sediments

5.8

Cosmogenous sediments (*cosmos* = universe, *generare* = to produce), which are of extraterrestrial origin, are the least abundant. These sediments are typically greatly diluted by other sediment components and rarely constitute more than a few parts per million of the total sediment in any layer. Scientists believe that cosmogenous sediments come from two major sources: interplanetary dust that falls constantly into the top of the atmosphere and rare impacts by large asteroids and comets.

Interplanetary dust consists of silt- and sand-sized micrometeoroids that come from asteroids and comets or from collisions between asteroids. The silt-sized particles settle gently to Earth's surface, but larger, faster moving dust is heated by friction with the atmosphere and melts, sometimes glowing as the meteors we see in a dark night sky. Though much of this material is vaporized, some may persist in the form of iron-rich cosmic spherules. Most of these dissolve in seawater before reaching the ocean floor. About 15,000 to 30,000 metric tons (16,500 to 33,000 tons) of interplanetary dust enters Earth's atmosphere every year.

The highest concentrations of cosmogenous sediments occur when large volumes of extraterrestrial matter arrive all at once. Fortunately, this happens only rarely, when Earth is hit by a large asteroid or comet. Very few examples of this are known, but most geologists believe that an impact like the one described at the beginning of this chapter would have blown vast quantities of debris into space around Earth. Much of it would fall back and be deposited in layers. Cosmogenous components may make up between 10% and 20% of these extraordinary sediments!

Occasionally cosmogenous sediment includes translucent oblong particles of glass known as **microtektites (Figure 5.8)**. Microtektites are thought to form from the violent impact of large meteors or small asteroids on the crust of Earth. The impact melts some of the crustal material and splashes it into space; the material melts again as it rushes through the atmosphere, producing the various raindrop shapes shown in the photo. Microtektites do not dissolve easily and usually reach the ocean floor. Most are smaller than 1.5 millimeters ($\frac{1}{16}$ inch) long.

Alan Hildebrand, University of Calgary

FIGURE 5.8 Microtektites, very rare particles that began a long journey when a large body impacted Earth and ejected material from Earth's crust. Some of this material traveled through space, reentered Earth's atmosphere, melted, and took on a rounded or teardrop shape. These specimens of sculptured glass range from 0.2 to 0.8 millimeter long. Glassy dust much finer in size, as well as nut-size chunks, has also fallen on Earth.

The Distribution of Marine Sediments

Sediments on the ocean floor only rarely come from a single source; most sediment deposits are a mixture of biogenous and terrigenous particles, with an occasional hydrogenous or cosmogenous supplement. The patterns and composition of sediment layers on the seabed are of great interest to researchers studying conditions in the overlying ocean. Different marine environments have characteristic sediments, and these sediments preserve a record of past and present conditions within those environments.

The sediments on the continental margins are generally different in quantity, character, and composition from those on the deeper basin floors. Continental shelf sediments—called **neritic sediments** (*neritos* = of the coast)—consist primarily of terrigenous material. Deep-ocean floors are covered by finer sediments than those of the continental margins, and a greater proportion of deep-sea sediment is of biogenous origin. Sediments of the slope, rise, and deep-ocean floor that originate in the ocean are called **pelagic sediments** (*pelagios* = of the sea).

The average thickness of the marine sediments in each oceanic region is shown in **Figure 5.9** and **Table 5.3**. Note that 72% of the total volume of all marine sediment is associated with continental slopes and rises, which constitute only about 12% of the ocean's area. **Figure 5.10** shows the worldwide distribution of marine sediment types. Put a bookmark in this page—you'll want to refer to these images as our discussion continues.

The Sediments of Continental Margins

The bulk of neritic sediments are terrigenous; they are eroded from the land and carried to streams, where they are transported to the ocean. Currents distribute sand and larger particles along the coast, while wave action carries the silts and clays to deeper water. When the water is too deep to be disturbed by wave action, the finest sediment may come to rest or continue to be transported by the turbulence of deep currents toward the deeper ocean floor. Ideally, these processes produce an orderly sorting of particles by size from relatively large grains near the coast to relatively small grains near the shelf break.

There are exceptions, however. Shelf deposits are subject to further modification and erosion as sea level fluctuates: Larger particles may be moved toward the shelf edge when sea level is low, as it was during periods of widespread glaciation—ice ages. Poorly sorted sediments are also found as glacial deposits. In polar regions, glaciers and ice shelves give rise to icebergs. These carry particles of all sizes, and when they melt they distribute their mixtures of rocks, gravel, sand, and silt onto high-latitude continental margins and deep-ocean floor (**Figure 5.11**). Turbidity currents also disrupt the orderly sorting of sediments on the continental margin by transporting coarse-grained particles away from coastal areas and onto the deep-ocean floor.

Thickness in Meters

0   500   1000   5000   10000   20000

NGDC/NOAA

FIGURE 5.9 Total sediment thickness of the ocean floor, with the thinnest deposits in dark blue and the thickest in red. Note the abundant deposits along the east and gulf coasts of North America, in the South China Sea, and in the Bay of Bengal east of India.

Ice ages have other effects on sediment deposition. Note in Figure 4.14 that continental shelves are almost completely exposed by the lowered sea level during times of widespread glaciation. Rivers carry their sediment right to the shelf edge, and it goes straight to the continental slope and deep seabed, mostly in turbidity currents.

Between ice ages, when the shelves are covered with water, the rate of sediment deposition on continental shelves is variable, but it is almost always greater than the rate of sediment deposition in the deep ocean. Near the mouths of large rivers, 1 meter (about 3 feet) of sediment may accumulate every thousand years. Along the east coast of the United States, however, many large rivers terminate in estuaries, which trap most of the sediment brought to them. The continental shelf of eastern North America is therefore covered mainly by sediments laid down during the last period of glaciation, when sea level was lower.

In addition to terrigenous material, neritic sediments almost always contain biological material. Biological productivity in coastal waters is often quite high, and biogenous sediments—the skeletal remains of creatures living on the bottom or in the water above—mix with the terrigenous sediments and dilute them.

Sediments can build to impressive thickness on continental shelves. In some cases, neritic sediments undergo **lithification:** They are converted into sedimentary rock by pressure-induced compaction or by cementation. If these lithified sediments are thrust above sea level by tectonic forces, they

Table 5.3   The Distribution and Average Thickness of Marine Sediments

Region	Percent of Ocean Area	Percent of Total Volume of Marine Sediments	Average Thickness
Continental shelves	9	15	2.5 km (1.6 mi)
Continental slopes	6	41	9 km (5.6 mi)
Continental rises	6	31	8 km (5 mi)
Deep-ocean floor	78	13	0.6 km (0.4 mi)

Sources: Emery in Kennett, *Marine Geology,* 1982 (Table 11-1); Weihaupt, *Exploration of the Oceans,* 1979; Sverdrup, Johnson, and Fleming, *The Oceans: Their Physics, Chemistry, and General Biology,* 1942

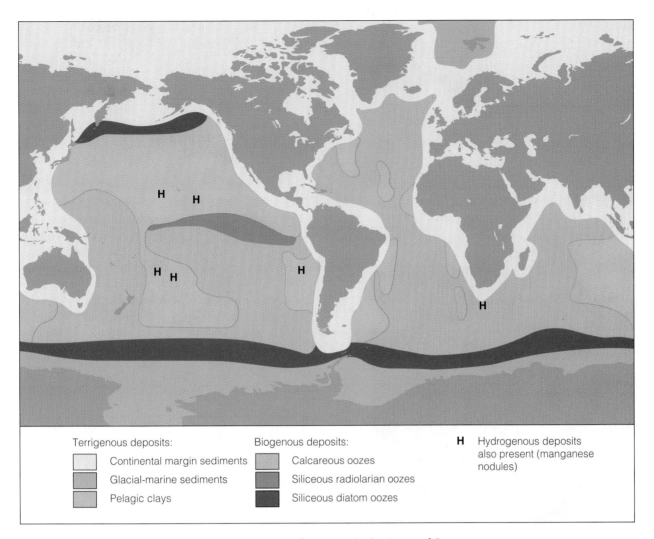

FIGURE 5.10 The general pattern of sediments on the ocean floor. Note the dominance of diatom oozes at high latitudes.

can form mountains or plateaus. The top of Mt. Everest, the world's tallest peak, is a shallow-water biogenic marine limestone (a calcareous rock). Much of the Colorado Plateau, with its many stacked layers, was formed by sedimentary deposition and lithification beneath a shallow continental sea beginning about 570 million years ago. The Colorado River has cut and exposed the uplifted beds to form the Grand Canyon. Hikers walking from the canyon rim down to the river pass through spectacular examples of continental-shelf sedimentary deposits. Their journey takes them deep into an old ocean floor!

The Sediments of Deep-Ocean Basins

The thickness of pelagic sediments is highly variable. When averaged, the Atlantic Ocean bottom is covered by sediments to a thickness of about 1 kilometer (3,300 feet), while the Pacific floor has an average sediment thickness of less than

0.5 kilometer (1,650 feet). There are two reasons for this difference. First, the Atlantic Ocean is fed by a greater number of rivers laden with sediment than the Pacific is, but the Atlantic is smaller in area; thus it gets more sediment for its size than the Pacific does. Second, in the Pacific Ocean many oceanic trenches trap sediments moving toward basin centers. Beyond this, the composition and thickness of pelagic sediments also vary with location, being thickest on the abyssal plains and thinnest (or absent) on the oceanic ridges.

Turbidites

Dilute mixtures of sediment and water periodically rush down the continental slope in turbidity currents, the erosive force of which is thought to help cut submarine canyons (see again Figures 4.16 and 4.17). These underwater avalanches of thick muddy fluid can reach the continental rise and often continue moving onto an adjacent abyssal plain before eventually coming to rest. The resulting deposits are called **turbidites,** graded layers of terrigenous sand interbedded with the finer pelagic

FIGURE 5.11 Researchers suspended over an iceberg take samples of the dust and gravel scraped off a nearby continent by the iceberg's parent glacier. When the iceberg melts, this sediment will fall to the seabed.

Mark Drinkwater

sediments typical of the deep-sea floor. This process and its results are shown in **Figure 5.12.**

Clays

5.13

About 38% of the deep seabed is covered by clays and other fine terrigenous particles. As we have seen, the finest terrigenous sediments are easily transported by wind and water currents. Microscopic waterborne particles and tiny bits of windborne dust and volcanic ash settle slowly to the deep-ocean floor, forming fine brown, olive-colored, or reddish clays. As Table 5.1 shows, the velocity of particle settling is related to particle size, and clay particles usually fall very slowly indeed. Terrigenous sediment accumulation on the deep-ocean floor is typically about 2 millimeters (1/8 inch) every thousand years.

Oozes

5.14

Seafloor samples taken farther from land usually contain a greater proportion of biogenous sediments than those obtained near the continental margins. This is not because biological productivity is higher farther from land (the opposite is usually true), but because there is less terrigenous material far from shore and thus pelagic deposits contain a greater proportion of biogenous material.

FIGURE 5.12 Turbidites are layers of terrigenous sand and gravel interspersed with finer pelagic sediments. They reach the continental rise in underwater avalanches. Sediments farther from the coast are more evenly bedded and have a smaller particle size. (Source: Based on K.O. Emery, et. al., Continental rise off eastern North America, American Association of Petroleum Geologists Bulletin 54 (1970): 44–108.)

Deep-ocean sediment containing at least 30% biogenous material is called an **ooze** (surely one of the most descriptive terms in the marine sciences). Oozes are named after the dominant remnant organism constituting them. The organisms that contribute their remains to deep-sea oozes are small, single-celled, drifting, plantlike organisms and the single-celled animals that feed on them. The hard shells and skeletal remains of these creatures are composed of relatively dense glasslike silica or calcium carbonate. When these organisms die, their shells settle slowly toward the bottom, mingle with fine-grained terrigenous silts and clays, and accumulate as ooze. The silica-rich residues give rise to **siliceous ooze**, the calcium-containing material to **calcareous ooze.**

Oozes accumulate slowly, at a rate of about 1 to 6 centimeters ($\frac{1}{2}$ to $2\frac{1}{2}$ inches) per thousand years. But they collect more than ten times more quickly than deep-ocean terrigenous clays. The accumulation of any ooze therefore depends on a delicate balance between the abundance of organisms at the surface, the rate at which they dissolve once they reach the bottom, and the rate of accumulation of terrigenous sediment.

Calcareous ooze forms mainly from shells of the amoeba-like **foraminifera** (**Figure 5.13a** and **b**), small drifting mollusks called **pteropods,** and tiny algae known as **coccolithophores** (**Figure 5.13c**). Although these creatures live in nearly all surface ocean water, calcareous ooze does not accumulate everywhere on the ocean floor. Shells are dissolved by seawater at great depths because it contains more carbon dioxide than seawater near the surface and thus becomes slightly acid. This acidity, combined with the increased solubility of calcium carbonate in cold water under pressure, dissolves the shells more rapidly, as you will see in Figure 7.10. At a certain depth, called the **calcium carbonate compensation depth** (**CCD**), the rate at which calcareous sediments are supplied to the seabed equals the rate at which those sediments dissolve. Below this depth, the tiny skeletons of calcium carbonate dissolve on the seafloor, so no calcareous oozes accumulate. Calcareous sediment dominates the deep-sea floor at depths of less than about 4,500 meters (14,800 feet), the usual calcium carbonate compensation depth. Sometimes a line analogous to a snow line on a terrestrial mountain can be seen on undersea peaks: Above the line the white sprinkling of calcareous ooze is visible; below it the "snow" is absent (**Figure 5.14**). About 48% of the surface of deep-ocean basins is covered by calcareous oozes.

a

b

c

FIGURE 5.13 Organisms that contribute to calcareous ooze. (**a**) A living foraminiferan, an amoeba-like organism. The shell of this beautiful foram, genus *Hastigerina,* is surrounded by a bubble-like capsule. It is one of the largest of the planktonic species with spines reaching nearly 5 centimeters (2 inches) in length. (**b**) The shell of a smaller foraminiferan—the snail-like, planktonic *Globigerina*—is visible in this scanning electron micrograph. (**c**) Coccoliths, individual plates of coccolithophores, a form of planktonic algae. Because of their tendency to dissolve, calcareous oozes very rarely occur at bottom depths below 4,500 meters (14,800 feet).

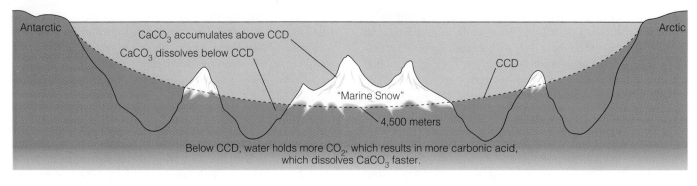

FIGURE 5.14 The dashed line shows the calcium carbonate ($CaCO_3$) compensation depth (CCD). At this depth, usually about 4,500 meters (14,800 feet), the rate at which calcareous sediments accumulate equals the rate at which those sediments dissolve.

Siliceous (silicon-containing) ooze predominates at greater depths and in colder polar regions. Siliceous ooze is formed from the hard parts of another amoeba-like animal, the beautiful glassy **radiolarian** (**Figure 5.15a**), and from single-celled algae called **diatoms** (**Figure 5.15b**). After a radiolarian or diatom dies, its shell will also dissolve back into the seawater, but this dissolution occurs much more slowly than the dissolution of calcium carbonate. Slow dissolution at all depths, combined with very high diatom productivity in some surface waters, leads to the buildup of siliceous ooze. Diatom ooze is most common in the deep-ocean basins surrounding Antarctica because strong ocean currents and seasonal upwelling in this area support large populations of diatoms. Radiolarian oozes occur in equatorial regions, most notably in the zone of equatorial upwelling west of South America (as was seen in Figure 5.10). About 14% of the surface of the deep-ocean floor is covered by siliceous oozes.

The very small particles that make up most of these pelagic sediments would need between 20 and 50 years to sink to the bottom. By that time they would have drifted a great lateral distance from their original surface position. But researchers have noted that the composition of pelagic sediments is usually similar to the particle composition in the water directly above. How could such tiny particles fall quickly enough to avoid great horizontal displacement? The answer appears to involve their compression into fecal pellets (**Figure 5.16**). Though still quite small, the fecal pellets of small animals are much larger than

a

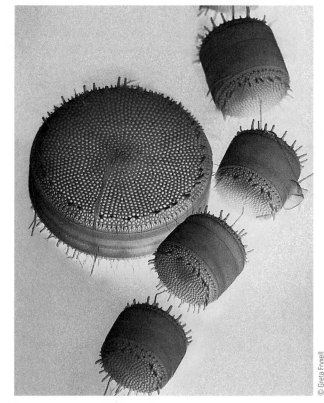

b

FIGURE 5.15 Scanning electron micrographs of siliceous oozes, which are most common at great depths. (**a**) Shells of radiolarians, amoeba-like organisms. Radiolarian oozes are found primarily in the equatorial regions. (**b**) Shells of diatoms, single-celled algae. Diatom oozes are most common at high latitudes.

a

b

Susumi Honjo

Susumi Honjo

FIGURE 5.16 A fecal pellet of a small planktonic animal. (**a**) The compressed pellet is about 80 micrometers long. (**b**) The pellet's surface magnified about 2,000 times. The pellet consists of the indigestible remains of small microscopic plantlike organisms, mostly coccolithophores. Unaided by pellet packing, these remains might take months to reach the seabed, but compressed in this way, they can be added to the ooze in perhaps two weeks.

the tiny individual skeletons of diatoms, foraminifera, and other plantlike organisms that they consumed, so they fall much faster, reaching the deep-ocean floor in about two weeks.

Some deep-sea oozes have been uplifted by geological processes and are now visible on land. The calcareous chalk White Cliffs of Dover in eastern England are partially lithified deposits composed largely of foraminifera and coccolithophores. Fine-grained siliceous deposits called *diatomaceous earth* are mined from other deposits. This fossil material is a valued component in flat paints, pool and spa filters, and mildly abrasive car and tooth polishes.

Hydrogenous Materials

Hydrogenous sediments also accumulate on deep-sea floors. They are associated with terrigenous or biogenous sediments and rarely form sediments by themselves. Most hydrogenous sediments originate from chemical reactions that occur on particles of the dominant sediment.

The most famous hydrogenous sediments are manganese **nodules,** which were discovered by the hardworking crew of HMS *Challenger*. The nodules consist primarily of manganese and iron oxides but also contain small amounts of cobalt, nickel, chromium, copper, molybdenum, and zinc. They form in ways not fully understood by marine chemists, "growing" at an average rate of 1 to 10 millimeters (0.04 to 0.4 inch) per *million* years, one of the slowest chemical reactions in nature. Though most are irregular lumps the size of a potato, some nodules exceed 1 meter (3.3 feet) in diameter. Manganese nodules often form around nuclei such as sharks' teeth, bits of bone, microscopic alga and animal skeletons, and tiny crystals—as the cross section of a manganese nodule in **Figure 5.17a** shows. Bacterial activity may play a role in the development of a nodule. Between 20% and 50% of the Pacific Ocean floor may be strewn with nodules (**Figure 5.17b**).

5.15

Unocal Research

a

Unocal Research

b

FIGURE 5.17 Manganese nodules. (**a**) A cross section cut through a manganese nodule, showing the concentric layers of manganese and iron oxides. This nodule is about 11 centimeters ($4\frac{1}{2}$ inches) long, a typical size. (**b**) Lemon-sized manganese nodules littering the abyssal Pacific.

Why don't these heavy lumps disappear beneath the constant rain of accumulating sediment? Possibly the continuous churning of the underlying sediment by creatures living there keeps the dense lumps on the surface, or perhaps slow currents in areas of nodule accumulation waft particulate sediments away.

Challenger scientists also discovered nodules of phosphorite. The first of these irregular brown lumps was taken from the continental rise off South Africa, and phosphorite nodule fields have since been found on shallow bank tops, outer continental shelves, and the upper parts of continental slope areas off California, Argentina, and Japan. Phosphorus is an important ingredient in fertilizer, and the nodules may someday be collected as a source of agricultural phosphates. Like manganese nodules, phosphorite nodules are found only in areas with low rates of sediment accumulation.

For now, the low market value of the minerals in manganese and phosphorite nodules makes them too expensive to recover. As techniques for deep-sea mining become more advanced and raw material prices increase, however, the nodules' concentration of valuable materials will almost certainly be exploited.

Powdery deposits of metal sulfides have been found in the vicinity of hydrothermal vents at oceanic ridges. Hot, metal-rich brines blasting from the vents meet cold water, cool rapidly, and lose the heavy metal sulfides by precipitation. Iron sulfides and manganese precipitates fall in thick blankets around the vents. The cobalt crusts of rift zones also seem to be associated with this phenomenon. These areas may one day be mined for their metal content.

Evaporites

Evaporites are an important group of hydrogenous deposits that include many salts important to humanity. These salts precipitate as water evaporates from isolated arms of the ocean or from landlocked seas or lakes. For thousands of years people have collected sea salts from evaporating pools or deposited beds. Evaporites are forming today in the Gulf of California, the Red Sea, and the Persian Gulf. The first evaporites to precipitate as water's salinity increases are the carbonates, such as calcium carbonate (from which limestone is formed). Calcium sulfate, which gives rise to gypsum, is next. (The M-reflector at the bottom of the Mediterranean consists of a layer of gypsum-rich evaporites.) Crystals of sodium chloride (table salt) will form if evaporation continues.

Oolite Sands

Not all hydrogenous calcium carbonate deposits are caused by evaporation, however. A small decrease in the acidity of seawater, or an increase in its temperature, can cause calcium carbonate to precipitate from water of normal salinity. In shallow areas of high biological productivity where sunlight heats the water, microscopic plants use up dissolved carbon dioxide, making seawater slightly less acidic (see Figure 7.9). Molecules of calcium carbonate then may precipitate around shell fragments or other particles. These white, rounded grains are called

ooliths (*oon* = egg) because they resemble fish eggs. **Oolite sands**—sands composed of ooliths—are abundant in many warm, shallow waters such as those of the Bahama Banks.

Sediments: An Overview

Look again at the types and distribution of marine sediments in Figures 5.9 and 5.10. Notice especially the lack of radiolarian deposits in much of the deep North Pacific; the strand of siliceous oozes extending west from equatorial South America; and the broad expanses of the Atlantic, South Pacific, and Indian Ocean floors covered by calcareous oozes. The broad, deep, relatively old Pacific contains extensive clay deposits, most delivered in the form of airborne dust. Why? Though some of the world's largest and muddiest rivers empty into the Pacific, most of their sediments are trapped in the peripheral trenches and cannot reach the mid-basins. And as you might expect, the poorly sorted glacial deposits are found only at high latitudes.

Figures 5.9 and 5.10 summarize more than a century of effort by marine scientists. Studies of sediments will continue because of their importance to natural resource development and because of the details of Earth's history that remain locked beneath their muddy surfaces.

Studying Sediments

Deep-water cameras have enabled researchers to photograph bottom sediments. The first of these cameras was simply lowered on a cable and triggered by a trip wire. Other more elaborate cameras have been taken to the seafloor on towed sleds or deep submersibles.

Actual samples usually provide more information than photographs do. *Challenger* scientists used weighted, wax-tipped poles and other tools attached to long lines to obtain samples, but today's oceanographers have more sophisticated equipment. Shallow samples may be taken using a **clamshell sampler** (named because of its method of operation, not its target; **Figure 5.18**). Deeper samples are taken by a **piston corer** (**Figure 5.19**), a device capable of punching through as much as 25 meters (82 feet) of sediment and returning an intact plug of material. Using a rotary drilling technique similar to that used to drill for oil, the drilling ship *JOIDES Resolution* (**Figure 5.20**) returned much longer core segments, some more than 1,100 meters (3,600 feet) long! These cores are stored in core libraries, a valuable scientific resource (**Figure 5.21**). Analysis of sediments and fossils from the Deep Sea Drilling Project cores helped verify the theory of plate tectonics. It has also shed light on the evolution of life-forms and helped researchers to decipher the history of changes in Earth's climate over the last 100,000 years.

Powerful new continuous seismic profilers have also been used to determine the thickness and structure of layers of sediment on the continental shelf and slope and to assist in the search for oil and natural gas (**Figure 5.22**). Typically, in seismic profiling a moving ship tows a sound transmitter and

a

b   c   d

FIGURE 5.18 (**a**) On board research vessel *Robert Gordon Sproul,* a scoop of muddy ocean-bottom sediments collected with a clamshell sampler is dumped onto the deck for study. (**b**) Before sampling, (**c**) during sampling, and (**d**) after the sample has been taken. Note that the sample is relatively undisturbed.

a

b   c   d   e

FIGURE 5.19 (**a**) A piston corer. (**b**) The corer is allowed to fall toward the bottom. (**c**) The corer reaches the bottom and continues, forcing a sample partway into the cylinder. (**d**) Tension on the cable draws a small piston within the corer toward the top of the cylinder, and the pressure of the surrounding water forces the corer deeper into the sediment. (**e**) The corer and sample are hauled in.

FIGURE 5.20 *JOIDES Resolution,* the deep-sea drilling ship operated by the Joint Oceanographic Institutions for Deep Earth Sampling. The vessel is 124 meters (407 feet) long, with a displacement of more than 16,000 tons. The rig can drill to a depth of 9,150 meters (30,000 feet) below sea level.

FIGURE 5.21 Sediment cores in storage. Cores are sectioned longitudinally, placed in trays, and stored in hermetically sealed cold rooms. The Gulf Coast Repository of the Ocean Drilling Program, located at Texas A&M University (pictured here), stores about 75,000 sections taken from more than 80 kilometers (50 miles) of cores recovered from the Pacific and Indian Oceans. Smaller core libraries are maintained at the Scripps Institution in California (Pacific and Indian Oceans) and at the Lamont–Doherty Earth Observatory in New York State (Atlantic Ocean).

FIGURE 5.22 A typical method of continuous seismic profiling. A moving ship trails a sound transmitter and receiver at a distance sufficient to minimize interference from the ship's noise. Bubbles and turbulence from a burst of compressed air act as a sound source. The sound is reflected from sediment layers beneath the surface and is detected by a sensitive hydrophone for analysis.

receiver behind it. Sounds from the transmitter reflect from the sediment layers beneath the bottom surface. Recent improvements in computerized image processing of the echoes returning from the seabed now permit detailed analysis of these deeper layers. Figure 5.4 was made in that way.

Sediments As Historical Records

In 1899 the British geologist W. J. Sollas theorized that deep-sea deposits could reveal much of the planet's history. In the era before plate tectonics theory this certainly seemed reasonable; the deep-ocean bottom was thought to be a calm, changeless place where an unbroken accumulation of sediment could be probed to discover the entire history of the ocean. Unfortunately for this promising idea, difficulties began to crop up almost immediately. For one thing, the sediments should have been much *thicker* than early probes indicated. If Earth's ocean is truly older than a few hundred thousand years, and if life has existed within it for most of that time, then the sediment layer should be thicker than had been observed. Another difficulty lay in the uneven distribution of sediments. Sollas thought that the center of an ocean basin should contain the thickest layers of sediment, yet the ridged mid-Atlantic bottom was nearly naked. There didn't seem to be any difference in the nature of the overlying seawater that could account for the variations in thickness and composition of the sediments across the bottom of the Atlantic. Oozes were especially puzzling: The organisms that form oozes grow well at the surface of the middle Atlantic, yet the mid-Atlantic floor seemed to bear little ooze.

Turn-of-the-century geologists were understandably confused, but today we know the tectonic reasons for these

BOX 5.1   Deep-Ocean Cores

The Ocean Drilling Program (ODP), currently the largest and most successful multinational earth science research project, is the direct successor to the Deep Sea Drilling Project (DSDP), which began in 1968. The two drilling ships commissioned for the projects pioneered drilling technology to retrieve cores, long cylinders of sediment and rock extracted from beneath the seafloor. The cores arrive on the deck of *JOIDES Resolution*—the ship currently in active service—in 9.5-meter (30-foot) sections encased in plastic sheaths. Immediately after retrieval, a core is marked to indicate its original location on the seafloor, coded to distinguish top from bottom, measured, and cut into sections for study and storage. Each section is slit lengthwise; one half is stored for the ODP archives, and the other is taken to the first of many shipboard laboratories for study.

Paleontologists then examine the sediments at the bottom of the core to determine the age of the oldest material. Other researchers measure its density, strength, molecular composition, radioactivity, and ability to conduct heat. Magnetic specialists read paleomagnetic data from the core to determine the ages of the rock fragments and the latitude at which they probably formed. Sensors are also lowered into the hole from which the core was removed, to gather additional information on the physical and chemical properties of the site.

ODP scientists have recently recovered fragments of the oldest remaining seafloor. The sample is about 175 million years old, a relic of the Middle Jurassic period, when the continents were massed in one huge cluster. From this and other cores they have learned about cycles of global climate change, information that will be useful in evaluating the present potential for global warming. They have also discovered how fluids move through the lithosphere, found evidence of an ice-free Antarctica, and noted the influence of plate tectonics on worldwide weather and current patterns. From trapped pollen grains, it is even possible to tell what land plants were thriving on Earth at the time the core sediments were laid down. As analysis technology evolves, stored cores are restudied and new information obtained.

The size and shape of a typical deep-core section may be seen in **Figure a**. This core was taken 250 meters (820 feet) beneath the seafloor off the northwestern coast of Australia by *JOIDES Resolution* on 22 July 1988. The ocean in the area was about 2,000 meters (6,500 feet) deep. The material in the core progresses from sandstone (near the bottom of the core, at the right side of the photo) to fine claystone (near the top). Foraminifera and nanofossils are abundant in the sandstone, and small squidlike fossils, easily visible to the unaided eye, are present in the claystone. The sediments date from the Late Jurassic period, about 140 million years ago. The progression of slowly deposited sediments and fossils suggests that the seabed in this area has slowly subsided, possibly because of tectonic effects.

Details in a core from DSDP Hole 480, leg 64, are shown in **Figure b.** This core was obtained by the now-retired drilling ship *Glomar Challenger* on 1 January 1979 near Guaymas in the Gulf of California. The gulf is an area of active seafloor spreading (see Figure 3.26), and researchers were interested in sampling the bottom in this place where basaltic magma is intruding into soft, wet, young sediments. The sediments here are very deep and have been accumulating at the extremely rapid rate of about 1,200 meters (4,000 feet) per million years.

(**a**) Researchers examine a deep core taken off the northwestern coast of Australia.

James Broda/Woods Hole Oceanographic Institution

(b) This core from the Gulf of California shows thin alternating bands of clay and ooze. Voids in the core show where samples were removed for study.

Glomar Challenger arrived in the area on 29 December and used a conventional core barrel to penetrate 444 meters (1,456 feet) into diatomaceous ooze and mudstones. Nearly 273 meters (900 feet) of core were recovered from the hole drilled before Hole 480. Because of the disturbance caused by the conventional coring process, researchers were unable to determine the details of fine structure tantalizingly seen in a few lengths of that core.

On 31 December *Glomar Challenger* moved to a new location 7 kilometers (4.4 miles) northwest. Hole 480, source of the core segment pictured here, was begun later that day. Unlike the previous hole, Hole 480 was drilled using a newly designed hydraulic corer developed by three DSDP engineers. The new device allowed 80% recovery of a core containing essentially undisturbed laminated diatomaceous ooze and muds. As can be seen in Figure b, the core consists of thin alternating brown and gray bands of sediments. These are believed to be annual couplets, formed about 2 million years ago in response to two seasonal events that still occur in the gulf: the winter rains that introduce terrigenous clays into the region, and diatom blooms produced by seasonal upwelling and northwest winds. Preservation of these fine details depends upon a low level of free oxygen at the seafloor. Burrowing animals would normally churn these fine layers into mush, but in a low-oxygen environment these animals are absent, so the thin alternating sheets of clay and diatom tests (shells) are preserved.

Note that parts of the core have already been removed for study. Trenches across the width of the core mark places where particular sets of layers have been removed for isotope analysis. A larger sample for paleomagnetic study was taken from the square depression in the sample. Considering the vast expense and skill necessary to retrieve and analyze them, these small, gray, gritty bits of sediment are probably more costly than their weight in fine diamonds!

discrepancies. Because the deep-sea sediment record is ultimately destroyed in the subduction process, the ocean's sedimentary "memory" does not start with the ocean's formation as originally reasoned by early marine scientists. Modern studies of deep-sea sediments using seafloor samples, cores obtained by deep drilling, and continuous seismic profiling have demonstrated that these deposits contain a remarkable record of relatively recent (that is, about the last 180 million years) ocean history. However, these same data have also shown that the record is not uninterrupted, as early workers had originally assumed. In fact, some of the gaps in the deep-sea deposits represent erosional events and constitute evidence of changes in deep-sea circulation, and hence are valuable in their own right. The analysis of layered sedimentary deposits, whether in the ocean or on land represents the discipline of **stratigraphy** (*stratum* = layer, *graph* = a drawing). Deep-sea stratigraphy utilizes variations in the composition of rocks, microfossils, depositional patterns, geochemical character, and physical character (density and such) to trace or correlate distinctive sedimentary layers from place to place, establish the age of the deposits, and interpret changes in ocean and atmospheric circulation, productivity, and other aspects of past ocean behavior. In turn, these sorts of studies and the advent of deep-sea drilling have given rise to the emerging science of **paleoceanography** (*palaios* = ancient), the study of the ocean's past.

Early attempts to interpret ocean and climate history from evidence in deep-sea sediments occurred in the 1930s through 1950s as cores became available. These initial studies relied primarily on identifying variations in the abundance and distribution of glacial marine sediments, carbonate and siliceous oozes, and temperature-sensitive microfossils in the cores. Modern paleoceanographic studies continue to utilize these same features, but researchers have much greater understanding of their significance and are aided by seismic imaging of the deposits over large areas. In addition, newer and more precise methods of dating deep-sea sediments have allowed events to be placed in a proper time context. Finally, scientists now have instruments capable of analyzing very small variations in the relative abundances of the stable isotopes of oxygen preserved within the carbonate shells of microfossils found in deep-sea sediments. These instruments allow them to interpret changes in the temperature of surface and deep water over time. These same data are also used to estimate variations in the volume of ice stored in continental ice sheets and thus to track the ice ages. Other geochemical evidence contained in the shells of marine microfossils, includ-

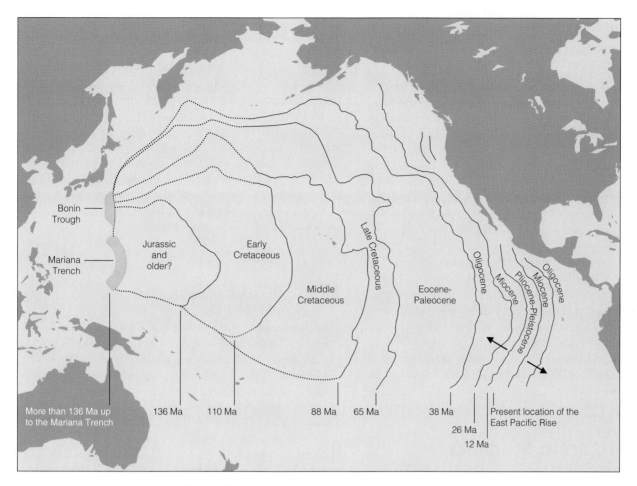

Bonin
Trough

Mariana
Trench

Jurassic
and
older?

Early
Cretaceous

Late Cretaceous

Middle
Cretaceous

Eocene-
Paleocene

Oligocene

Miocene

Pliocene-Pleistocene

Miocene

Oligocene

More than 136 Ma up
to the Mariana Trench

136 Ma

110 Ma

88 Ma

65 Ma

38 Ma

26 Ma

12 Ma

Present location of the
East Pacific Rise

FIGURE 5.23 The ages of portions of the Pacific Ocean floor, based on core samples of sediments just above the basalt seabed, in millions of years ago (Ma, mega-annum). The youngest sediments are found near the East Pacific Rise and the oldest close to the eastern side of the trenches. Contrast this figure with Figure 3.29.

ing variations in carbon isotopes and trace metals such as cadmium, provide insights into ancient patterns of ocean circulation, productivity of the marine biosphere, and upwelling. These sorts of data have already provided quantitative records of the glacial–interglacial climatic cycles of the past 2 million years. Future drilling and analysis of deep-sea sediments are poised to extend our paleoceanographic perspective much farther back in time.

Figure 5.23 shows the age of regions of the Pacific Ocean floor using data obtained largely from analyses of the overlying sediment. Note that sediments get older with increasing distance from the East Pacific Rise spreading center.

Earth might not be the only planet where marine sediments have left historical records. As you read in Chapter 2, Mars probably had an ocean between 3.2 and 1.2 billion years ago. In January 2004 the Mars rover *Opportunity* photographed bedrock inside the small crater where it landed (**Figure 5.24**). Layers in the rocks suggest it may have formed from layers of marine sediment. One can only wonder what stories their memories hold.

NASA

FIGURE 5.24 Lithified marine sediments on Mars? Many researchers believe Mars was once home to an ocean. On 27 January 2004 the newly landed Mars rover *Opportunity* photographed outcrops of layered bedrock inside the small crater where it landed. The layered appearance suggests the rocks may have been formed from marine sediments.

The Economic Importance of Marine Sediments

The study of sediments has brought practical benefits. You probably have more daily contact with marine sediment than you think. Components of the building materials for roads and structures, toothpaste, paint, and swimming pool filters come directly from sediments. In 2005 an estimated 36% of the world's crude oil and 28% of its natural gas will be extracted from the sedimentary deposits of continental shelves and continental rises. Offshore hydrocarbons currently generate annual revenues in excess of US$125 billion. Deposits within the sediments of continental margins account for about one-third of the world's estimated oil and gas reserves.

In addition to oil and gas, in 2000 sand and gravel valued at more than US$510 million were taken from the ocean. This is about 1% of world needs. Commercial mining of manganese nodules has also been considered. In addition to manganese, these nodules contain substantial amounts of iron and other industrially important chemical elements. The high iron content of these nodules has prompted a proposal to rename them *ferromanganese* nodules. We will investigate these resources in more detail in Chapter 17.

Questions from Students

1. The question of sediment age seems to occupy much of sedimentologists' time. Why?

The dating of sediments has been a central problem in marine science for many years. In 1957, the International Geophysical Year, sedimentologists designed a coordinated effort to determine sediment age, which included plans for the *Glomar Explorer* and *Glomar Challenger* drilling surveys. Their primary interest was to seek evidence of the hypothesis of the then-new idea of seafloor spreading. Cores returned by the Deep Sea Drilling Project in 1968 enabled researchers including J. Tuzo Wilson, Harry Hess, and Maurice Ewing to put the evidence together. Much of the proof for plate tectonics rests on the interpretation of sediment cores.

2. Where are sediments thickest?

Sediments are thickest close to eroding land and beneath biologically productive neritic waters, and thinnest over the fast-spreading oceanic ridges of the eastern South Pacific. The thickest accumulations of sediment may be found along and beneath the continental margins (especially on continental rises). Some are typically more than 1,500 meters (5,000 feet) thick. Remember, much of the rocky material of the Grand Canyon was once marine sediment atop an isostatically depressed ancient seabed. The Grand Canyon is nearly 2 kilometers ($1\frac{1}{4}$ miles) deep, and the uppermost layer of sedimentary rock has already been eroded completely away!

3. What's the relationship between deep-sea animals and the sediments on which they live?

Though microscopic bacteria and benthic foraminifera may be abundant on the seabed, visible life is not abundant on the bottom of the deep ocean. No plants live at great depths because there is no light, but animals do live there. Some, like the brittle stars in Figure 5.2, move slowly along the surface searching for bits of organic matter to eat. Others burrow through the muck in search of food particles. Worms eat quantities of sediment to extract any nutrients that may be present and then deposit strings of fecal material as they move forward. The deeps are uninviting places, but life is tenacious and survives even in this hostile environment.

Chapter in Perspective

In this chapter you learned that the sediments covering nearly all of the seafloor are parts of the great cycles of formation and destruction assured by Earth's hot interior. Marine sediment is composed of particles from land, from biological activity in the ocean, from chemical processes within water, and even from space. The blanket of seafloor sediment is thickest at the continental margins and thinnest over the active oceanic ridges.

Sediments may be classified by particle size, source, location, or color. Terrigenous sediments, the most abundant, originate on continents or islands. Biogenous sediments are composed of the remains of once-living organisms. Hydrogenous sediments are precipitated directly from seawater. Cosmogenous sediments, the ocean's rarest, come to the seabed from space.

The position and nature of sediments provide important clues to Earth's recent history, and valuable resources can sometimes be recovered from them.

In the next chapter you will learn about our story's main character—water itself. You know something about how water molecules were formed early in the history of the universe, something about the inner workings of our planet, and something about the nature of the ocean's "container." Now let's fill that container with water and see what happens.

Key Ideas

✓ Sediment is particles of organic or inorganic matter that accumulate in a loose, unconsolidated form. Sediment may be classified by grain size or by the origin of the majority of the particles.
Review page 116

✓ Most marine sediments are made of fine-grained particles: sand, silt, and clay.
Review Table 5.1, pages 118; 116–118

✓ Sediment is part of the great cycle of formation and destruction driven by plate tectonics.
Review Figure 5.6, page 119

✓ Sediment deposited on a quiet seabed can provide a sequential record of events in the water column above. Sediments can act only as the *recent* memory of the ocean; sediments are not of great age because they are recycled into Earth at subduction zones.
Review the chapter opener and Box 5.1, pages 131–132

✓ Marine sediments are broadly classified by origin into four categories. Terrigenous sediments are of geological origin and arise on the continents or islands near them; they are the most abundant. Biogenous sediments are of biological origin and may be calcareous or siliceous. Hydrogenous sediments are formed directly from seawater. Of less importance are cosmogenous sediments, which come from space.
Review Table 5.2, page 119

✓ Though there are exceptions, the sediments of continental margins tend to be mostly terrigenous, while the generally finer sediments of the deep-ocean floor contain a larger proportion of biogenous material.
Review Figures 5.9 and 5.10, pages 122, 123

✓ Deep-sea oozes—forms of biogenous sediment—contain the remains of some of the ocean's most abundant and important organisms.
Review Figures 5.13 and 5.15, pages 124–126

✓ Sediments are economically important. By 2005 an estimated 36% of the world's crude oil will be extracted from marine sedimentary deposits.
Review page 133

Terms and Concepts to Remember

Study Questions

Reviewing What You've Learned

1. In what ways are sediments classified?
2. List the four types of marine sediments. Explain the origin of each.
3. How are neritic sediments generally different from pelagic ones?
4. Where are sediments thickest? Are any areas of the ocean floor free of sediments?
5. Are sediments commercially important? In what ways?

Thinking Critically

1. Is the thickness of ooze always an accurate indication of the biological productivity of surface water in a given area? (Hint: See next question.)
2. What is the calcium carbonate compensation depth? Is there a compensation depth for the siliceous components of once-living things?
3. What sediments accumulate most rapidly? Least rapidly?
4. Can marine sediments tell us about the history of the ocean from the time of its origin? Why?
5. What problems might arise when working with deep-ocean cores? Imagine the process of taking a core sample, and think of what can go wrong!
6. **InfoTrac College Edition Project** The continental slopes and ocean bottom are thought to contain oil and gas reserves as well as other mineral reserves. But many oppose extraction of these minerals because of the dangers of contamination. Can these reserves be tapped safely? Would you support or oppose these efforts? Why? Use information from articles found on InfoTrac College Edition to support your answers.

Thinking Analytically

1. Given an average rate of accumulation, how much time would it take to build an average-size manganese nodule?
2. Assuming an average rate of sediment accumulation and seafloor spreading, how far from the Mid-Atlantic Ridge would one need to travel before encountering a layer of sediment 1,000 meters thick?
3. Microtektites are often found in "fields"—elongated zones of relative concentration a few hundred kilometers long. Why do you suppose that is?

http://oceanography.brookscole.com/garrison5e

4. How much faster do fragments of diatoms fall when they are compacted into fecal pellets than when they are not? (Hint: See Table 5.1.)

Oceanography⬡Now™

Assess your understanding of this chapter's topics with additional quizzing and conceptual-based problems at http://oceanography.brookscole.com/garrison5e

Study and Research on the Internet

Assess your understanding of this chapter's topics with additional quizzing and conceptual-based problems at:

http://oceanography.brookscole.com/garrison5e

■ **InfoTrac College Edition Research** leads you to critical thinking exercises that use InfoTrac College Edition® as a research tool.

■ **Regional InfoTrac College Edition** articles are organized into East Coast, West Coast, and Gulf Coast regions, allowing you to study oceanography on a more local level.

For more readings, go to InfoTrac College Edition, your online research library, at:

http://infotrac.thomsonlearning.com

CHAPTER 6

© Kaz Mori/The Image Bank/Getty Images

WATER AND OCEAN STRUCTURE

Familiar, Abundant, and Odd

Water is so familiar and abundant that we don't always appreciate its unusual characteristics. Here you'll meet the water you never knew. This chapter introduces the characteristics that make water unusual—the molecule's polarity and the bonds that hold it together, the large amount of heat needed to change its temperature, and the heat needed to change its physical state. And, no, heat and temperature are not the same thing.

Two big lessons follow. One lesson is the influence of water on global temperatures. Liquid water's thermal char-

acteristics prevent broad swings of temperature during day and night and, through a longer span, during winter and summer. Heat is stored in the ocean during the day and released at night. A much greater amount of heat is stored through the summer and given off during the winter. Liquid water has an important thermostatic balancing effect; an oceanless Earth would be much colder in winter and much hotter in summer than the moderate temperatures we experience.

Sea ice in the polar regions contributes to Earth's moderate surface temperatures in a different way. Because water expands and floats when it freezes, ice can absorb the morning warmth of the sun, melt, and then refreeze at night, giving back to the atmosphere the heat it stored through the daylight hours. The *heat content* of the water changes through the day; its *temperature* does not. Clearly, water is anything but ordinary!

The other lesson is the influence of density on ocean structure. You'll see that the ocean's structure and large-scale movement depend on changes in the density of seawater, with density dependent on temperature and salt content.

The chapter continues with an explanation of light and sound in the ocean. Why is the ocean blue? Why do noises sound different—and go farther—in water? Here are answers.

In Chapter 7 we'll extend our examination by looking at the many dissolved substances that produce the *salinity of* seawater, and we'll see how those substances alter seawater's physical properties.

The Water Molecule

Pure water is a **compound**—that is, a substance that contains two or more different elements in a fixed proportion. The familiar chemical formula for water, H_2O, shows that it is composed of the elements hydrogen (H) and oxygen (O) in a fixed proportion of two to one. An **element** is a substance composed of identical particles, called **atoms**, that cannot be broken into simpler substances by chemical means.

Water is a **molecule,** a group of atoms held together by chemical bonds. **Chemical bonds**, the energy relationships between atoms that hold them together, are formed when **electrons**—tiny negatively charged particles found toward the outside of an atom—are shared between atoms or moved from one atom to another. A water molecule forms when electrons are shared between two hydrogen atoms and one oxygen atom (**Figure 6.1**). The bonds formed by shared pairs of electrons are known as **covalent bonds**. Covalent bonds hold together many familiar molecules besides water, including CO_2 (carbon dioxide), CH_4 (methane gas), and O_2 (atmospheric oxygen).

Because of the way a water molecule's oxygen electrons are distributed, the overall geometry of the molecule is a bent or angular shape. The angle formed by the two hydrogen atoms and the central oxygen atom is about 105°. The angular shape of the water molecule makes it electrically asymmetrical. Each water molecule can be thought of as having a positive (+) end and a negative (−) end. This is because **protons** of the hydrogen atoms—the positively charged particles in the **nucleus** (center)—are left partially exposed when the negatively charged electrons bond more closely to oxygen. The water molecule behaves something like a magnet; its positive end attracts particles that have a negative charge, and its negative end (or pole) attracts particles that have a positive charge. For this reason water is called a **polar molecule**. When water comes into contact with compounds whose elements are held together by the attraction of opposite electrical charges (most salts, for example), the polar water molecule will separate that compound's component elements from each other. This explains why water can easily dissolve so many other compounds.

The polar nature of water also permits it to attract other water molecules. When a hydrogen atom (the positive end) in one water molecule is attracted to the oxygen atom (the negative end) of an adjacent water molecule, a **hydrogen bond**

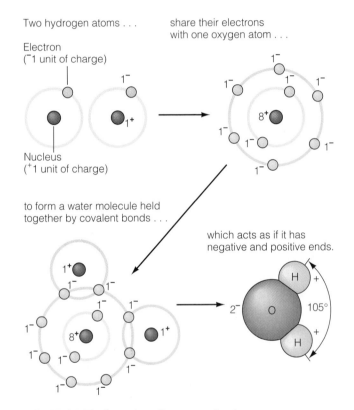

FIGURE 6.1 The formation of a water molecule.

forms. A hydrogen bond *between* molecules is about 5% to 10% as strong as a covalent bond *within* a molecule. Hydrogen bonds link water molecules together by electrostatic forces. The resulting loosely held webwork of water molecules is shown in **Figure 6.2**. Hydrogen bonds greatly influence the properties of water by allowing individual water molecules to stick to each other, a property called **cohesion**. Cohesion gives water an unusually high surface tension, which results in a surface "skin" capable of supporting needles, razor blades, and even walking insects. **Adhesion**, the tendency of water to stick to other materials, allows water to adhere to solids—that is, to make them wet. Cohesion and adhesion are the causes of capillary action, the tendency of water to spread through a towel when one corner is dipped in water.

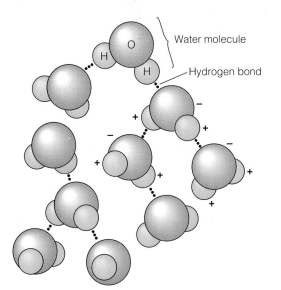

Water molecule

Hydrogen bond

FIGURE 6.2 Hydrogen bonds in liquid water. The attractions between adjacent polar water molecules form a webwork of hydrogen bonds. These bonds are responsible for cohesion and adhesion, the properties of water that cause surface tension and wetting. Hydrogen bonds among water molecules also make it difficult for individual molecules to escape from the surface.

Hydrogen bonds are also what give pure water its pale blue hue. When water molecules vibrate, adjacent molecules tug and push against their hydrogen-bonded neighbors. This action absorbs a small amount of red light, leaving proportionally more blue light to scatter back to our eyes. The same blue color is seen in ice formations.

If hydrogen bonds did not hold water molecules together, they would form a gas rather than a liquid. The compound hydrogen sulfide (H_2S) is chemically similar to water but lacks water's ability to form networks of hydrogen bonds. H_2S is therefore a gas (with individual molecules separated from each other) at normal temperatures and pressures.

Water and Heat

Perhaps the most important physical properties of water are related to its behavior as it absorbs or loses heat. Water's unusual thermal characteristics prevent wide temperature variation from day to night and from winter to summer, permit vast amounts of heat to flow from equatorial to polar regions, and power Earth's great storms, wind waves, and ocean currents. To see why, we first need to examine water's thermal characteristics in some detail.

Heat and Temperature

Heat and temperature are related concepts, but they are not the same. **Heat** is energy produced by the random vibration of atoms or molecules. On the average, water molecules in hot water vibrate more rapidly than water molecules in cold water. Heat is a measure of *how many* molecules are vibrating and *how rapidly* they are vibrating. Temperature records only *how rapidly* the molecules of a substance are vibrating. **Tempera-**

ture is an object's response to an input (or removal) of heat. The amount of heat required to bring a substance to a certain temperature varies with the nature of that substance.

An example will help. Which has a higher temperature: a candle flame or a bathtub of hot water? The flame. Which contains more heat? The tub. The molecules in the flame vibrate very rapidly, but there are relatively few of them. The molecules of water in the tub vibrate more slowly, but there are a great many of them, so the total amount of heat energy in the tub is greater.

Temperature is measured in **degrees**. One degree Celsius (°C) = 1.8 degrees Fahrenheit (°F). Though we are more familiar with the older Fahrenheit scale, Celsius degrees are more useful in science because they are based on two of pure water's most significant properties: its freezing point (0°C) and its boiling point (100°C).

Heat Capacity

Heat capacity is a measure of the heat required to raise the temperature of 1 gram (0.035 ounce) of a substance by 1°C (1.8°F). Different substances have different heat capacities. *Not all substances respond to identical inputs of heat by rising in temperature the same number of degrees* (**Table 6.1**). Heat capacity is measured in calories per gram. A **calorie** is the amount of heat required to raise the temperature of 1 gram of pure water by 1°C.[1]

Because of the great strength and large number of the hydrogen bonds between water molecules, more heat energy must be added to speed up molecular movement and raise water's temperature than would be necessary in a substance

---

[1] A nutritional Calorie, the unit we see on cereal boxes, equals 1,000 of these calories. A gram is about ten drops of seawater.

Table 6.1	Heat Capacity of Common Substances
Substance	**Heat Capacity**[a] **in calories/gram/°C**
Silver	0.06
Granite/sand	0.20
Aluminum	0.22
Alcohol (ethyl)	0.30
Gasoline	0.50
Acetone	0.51
Ice (not freezing or thawing)	0.51
Pure liquid water	**1.00**
Ammonia (liquid)	1.13

[a]Heat capacity is a measure of the heat required to raise the temperature of 1 gram (0.035 ounce) of a substance by 1°C (1.8°F). Different substances have different heat capacities. *Not all substances respond to identical inputs of heat by rising in temperature the same number of degrees.* Notice how little heat is required to raise the temperature of 1 gram of silver 1 degree. Of all common substances, only liquid ammonia has a higher heat capacity than liquid water.

held together by weaker bonds. Liquid water's heat capacity is therefore among the highest of all known substances. This means that *water can absorb (or release) large amounts of heat while changing relatively little in temperature.*

Anyone who waits by a stove for water to boil knows a lot about water's heat capacity—it seems to take a very long time to warm water for soup or coffee. Compared with water, ethyl alcohol has a much lower heat capacity. If both liquids absorb heat from identical stove burners at the same rate, pure ethyl alcohol (the active ingredient in alcoholic beverages) will rise in temperature about three times as fast as an equal mass of water. Beach sand has an even lower heat capacity. A gram of sand requires as little as 0.2 calorie to rise 1°C (1.8°F). So, on sunny days beaches can get too hot to stand on with bare feet, while the water remains pleasantly cool.

As we will soon see, the concept of heat capacity is very important in oceanography. But for now, remember this: Water has an extraordinarily high heat capacity; it resists changing temperature *when heat* is added or removed.

Water Temperature and Density

The uniqueness of water becomes even more apparent when we consider the effect of a temperature change on water's **density** (its mass per unit of volume). You may recall from Chapter 3 that the density of pure water is 1 gram per cubic centimeter (1 g/cm³). Granite rock is heavier, with a density of about 2.7 g/cm³, and air is lighter, with a density of about 0.0012 g/cm³.

Most substances become denser (weigh more per unit of volume) as they get colder. Pure water generally becomes denser as heat is removed and its temperature falls, but water's density behaves in an unexpected way as the temperature approaches the freezing point.

A **density curve** shows the relationship between the temperature (or salinity) of a substance and its density. Most substances become progressively denser as they cool; their temperature–density relationships are linear (that is, appear as a straight line on graphs). But **Figure 6.3** shows the unusual temperature–density relationship of pure water. Imagine heat being removed from some water placed in a freezer. Initially the water is at room temperature (20°C, 68°F), point A on the graph. As expected, the density of water increases as its temperature drops along the line from point A toward point B. As the temperature approaches point B, the density increase slows, reaching a maximum at point B of 1 g/cm³ at 3.98°C (39.16°F). As the water continues to cool, its framework of hydrogen bonds becomes more rigid, which causes the liquid to expand slightly because the molecules are held slightly farther apart. Water thus becomes slightly less dense as cooling continues, until point C (0°C, 32°F) is reached. At point C the water begins to freeze—to change state by crystallizing into ice.

State is an expression of the internal form of a substance (**Figure 6.4**). Changes in state are accompanied by either an input or an output of energy. Water exists on Earth in three physical states: liquid, gas (water vapor), and solid (ice). If the freezer continues to remove heat from the water at point C in Figure

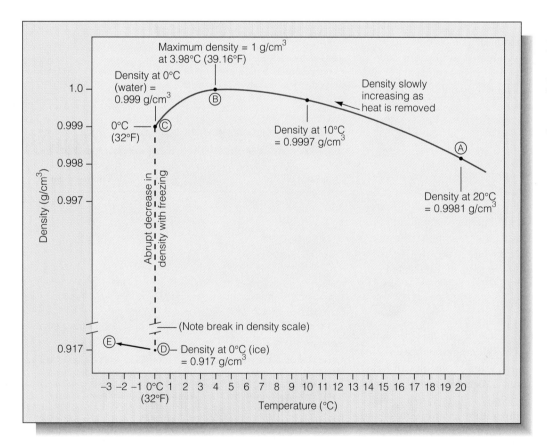

FIGURE 6.3 The relationship of density and temperature for pure water. Note that points C and D both represent 0°C (32°F) but different densities and thus different states of water. Ice floats because the density of ice is lower than the density of liquid water.

GAS

Fills closed container uniformly

Molecules in high-speed motion

Collisions and rebounds occur

Density very low

Piston | Readily compressed into smaller volume

More frequent collisions

LIQUID

Free upper surface

Flows freely to lower level

Molecules close together, glide easily past one another

SOLID
(crystalline)

Strong, rigid

Fractures when sudden, strong stress is applied

Density high

Stress

Rupture

Molecules locked into strict geometrical order

FIGURE 6.4 The three common states of matter: solid, liquid, and gas. A *gas* is a substance that can expand to fill any empty container. Atoms or molecules of a gas are in high-speed motion and move in random directions. A *liquid* is a substance that flows freely in response to unbalanced forces but has a free upper surface in a container it does not fill. Atoms or molecules of a liquid move freely past one another as individuals or small groups. Liquids compress only slightly under pressure. Gases and liquids are classed as *fluids* because both substances flow easily. A *solid* is a substance that resists changes of shape or volume. A solid can typically withstand stresses without yielding permanently. A solid usually breaks suddenly. On Earth, water can occur in all three states: gas, liquid, and solid. (Source: Fig. 2.5, p. 29 from *Physical Geology* by Arthur N. Strahler. Copyright © 1981 by Arthur N. Strahler. Reprinted by permission of Pearson Education, Inc.)

6.3, the water will change from liquid to solid state. Through this transition from water to ice—from point C to point D—the density of the water *decreases* abruptly. Ice is therefore lighter than an equal volume of water. Ice increases in density as it gets colder than 0°C. No matter how cold it gets, however, ice never reaches the density of liquid water. Being less dense than water, ice "freezes over" as a floating layer instead of "freezing under" like the solid forms of virtually all other liquids.

As we'll see in a moment, the implications of water's high heat capacity and the ability of ice to float are vital in maintaining Earth's moderate surface temperature. First, however, we look at the transition from point C to point D in Figure 6.3.

Freezing Water

During the transition from liquid to solid state at the **freezing point**, the bond angle between the oxygen and hydrogen atoms in water widens from about 105° to slightly more than 109°. This change allows the hydrogen bonds in ice to form a crystal lattice (**Figure 6.5**). The space taken by 27 water molecules in the liquid state will be occupied by only 24 water mol-

ecules in the solid lattice, however; so water expands about 9% as the crystal forms. Ice is less dense than liquid water—and thus floats—because the molecules are packed less efficiently. A cubic centimeter of ice at 0°C (32°F) has a mass of only 0.917 gram, but a cubic centimeter of liquid water at 0°C has a mass of 0.999 gram.

The ice lattice is not a flat sheet but a three-dimensional network. The water molecules in the ice lattice are still linked by hydrogen bonds, but they now form regular hexagons. This hexagonal pattern explains the lovely six-sided symmetry of snowflakes and other ice crystals.

The transition from liquid water to ice crystal (point C to point D in Figure 6.3) requires continued removal of heat energy; the change in state does not occur instantly throughout the mass when the cooling water reaches 0°C (32°F). Again, consider water in a freezer. **Figure 6.6**, a plot of heat versus temperature, illustrates the water's progress to ice. As in Figure 6.3, point A represents 20°C (68°F) water just placed in the freezer. The removal of heat does not stop when the water reaches point C, *but the decline in temperature stops*. Even though heat continues to be removed, the water will not get

http://oceanography.brookscole.com/garrison5e

WATER AND OCEAN STRUCTURE **141**

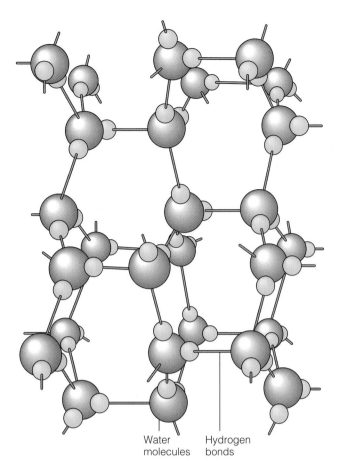

Water   Hydrogen
molecules   bonds

ACTIVE FIGURE 6.5 The lattice structure of an ice crystal, showing its hexagonal arrangement at the molecular level. The space taken by 24 water molecules in the solid lattice could be occupied by 27 water molecules in the liquid state, so water expands about 9% as the crystal forms. Because molecules of liquid water are packed less efficiently, ice is less dense than liquid water and floats.

Oceanography⌇Now™ | At the Active Figures link at http://oceanography.brookscole.com/garrison5e you can view the lattice structure of an ice crystal.

colder until all of it has changed state from liquid (water) to solid (ice). Heat may therefore be removed from water when it is changing state (that is, when it is freezing) without the water dropping in temperature. Indeed the continued removal of heat is what makes the change in state possible. Heat is released as hydrogen bonds form to make ice, and that heat must be removed to allow more ice to form.

The removal of heat from point A to point C in Figures 6.3 and 6.6 produces a *measurable* lowering of temperature detectable by a thermometer. Removing just 1 calorie of heat from a gram of liquid water causes its temperature to drop 1°C. This detectable decrease in heat is called **sensible heat** loss. But the loss of heat as water freezes between points C and D is not measurable (that is, not sensible) by a thermometer. Removing a calorie of heat from freezing water at 0°C (32°F) won't change its temperature at all; 80 calories of heat energy must be removed per gram of pure water at 0°C (32°F) to form ice. This heat is called the **latent heat of fusion** (*latere* = to be hidden). The straight line between points C and D in Figure

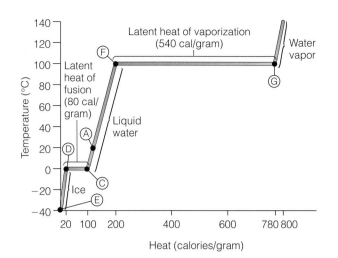

FIGURE 6.6 A graph of temperature versus heat as water freezes, melts, and vaporizes. The horizontal line between points C and D represents the latent heat of fusion, when heat is being added or removed but the temperature is not changing. The horizontal line between points G and F represents the latent heat of vaporization, when heat is being added or removed but the temperature is not changing. (Note that points A–E on this graph are the same as those in Figure 6.3.)

6.6 represents water's latent heat of fusion. No more ice crystals can form when all the water in the freezer has turned to ice. If the removal of heat continues, the ice will get colder and will soon reach the temperature inside the freezer, point E in Figures 6.3 and 6.6.

Latent heat of fusion is also a factor during thawing. When ice melts, it *absorbs* large quantities of heat (the same 80 calories per gram) but it does not change in temperature until all the ice has turned to liquid. This is why ice is so effective in cooling drinks.

Evaporating Water

Let's reverse the process now and warm the ice. Imagine the water resting at −40°C (−40°F), point E at the lower left of Figure 6.6. Add heat and the ice warms toward point D. It begins to melt. The horizontal line between points D and C represents the latent heat of fusion: Heat is absorbed but temperature does not change as the ice melts. All liquid now at point C, the water warms past our original point A and arrives at point F. It begins to boil—it *vaporizes*.

When water vaporizes (or evaporates), individual water molecules diffuse into the air (**Figure 6.7**). Because each water molecule is hydrogen-bonded to adjacent molecules, heat energy is required to break those bonds and allow the molecule to fly away from the surface. Evaporation cools a moist surface because departing molecules of water vapor carry this energy away with them. (This is how perspiring cools us when we're hot. The heat energy required to evaporate water from our skin is taken away from our bodies, thus cooling us.)

Hydrogen bonds are quite strong, and the amount of energy required to break them—known as the **latent heat of vaporization**—is very high. The long horizontal line between points

FIGURE 6.7 Water vapor is invisible, but as it evaporates from the sea surface and rises into cool air it can condense into tiny droplets that form clouds and fog. The low-lying fog bank that often obscures the Golden Gate Bridge at the entrance to San Francisco Bay forms when warm, moist air passes over the cold waters of the California Current. The air is cooled to the saturation point, and some of its water vapor condenses into fog droplets.

F and G in Figure 6.6 represents the latent heat of vaporization. As before, the term *latent* applies to heat input that does not cause a temperature change but does produce a change of state—in this case from liquid to gas. Even though more heat is applied, the water cannot get warmer until all of it has vaporized. At 540 calories per gram at 20°C (68°F), water has the highest latent heat of vaporization of any known substance.

About 1 meter (3.3 feet) of water evaporates each year from the surface of the ocean, a volume of water equivalent to 334,000 cubic kilometers (80,000 cubic miles). The great quantities of solar energy that cause this evaporation are carried from the ocean by the escaping water vapor. When a gram of water vapor condenses back into liquid water, the same 540 calories is again available to do work. As we shall see, winds, storms, ocean currents, and wind waves are all powered by that heat.

Why the big difference between water's latent heat of *fusion* (80 calories per gram) and its latent heat of *vaporization* (540 calories per gram)? Only a small percentage of hydrogen bonds are broken when ice melts, but *all* of them must be broken during evaporation. Breaking these bonds requires additional energy in proportion to their number.

A summary of water's unusual properties is provided in **Figure 6.8** and **Table 6.2**.

Seawater and Pure Water

Seawater is about 96.5% pure water and 3.5% dissolved solids and gases. The solids dissolved in seawater change its thermal characteristics, lowering its latent heat by about 4%. Only 0.96 calorie of heat energy is needed to raise the temperature of 1 gram of seawater by 1°C.

The dissolved solids also interfere with the formation of the ice lattice, acting as "antifreeze" to lower the freezing point. The saltier the water, the lower the freezing point. The temperature of maximum density moves toward the freezing point as salinity (salt content) increases (**Figure 6.9**), finally coinciding with the freezing point at a salinity of 24.7‰ (−1.33°C, 29.61°F).[2] Seawater at 35‰, typical ocean salinity, freezes at −1.91°C (28.6°F). Seawater's density simply increases smoothly with decreasing temperature until it freezes. The crystals that form are pure water ice, with the seawater salts excluded. The leftover cold, salty water is very dense. Some of this water may be trapped among the ice crystals, but most is free to fall toward the seabed, pulled rapidly downward by its great density.

Seawater evaporates more slowly than fresh water under identical circumstances because the dissolved salts tend to attract and hold water molecules. The latent heat of evaporation, however, is essentially the same for both fresh water and seawater. Salts are left behind as seawater evaporates. The remaining cool, salty water is also very dense, and it too sinks toward the ocean floor.

---

[2] The symbol % (percent) represents parts-per-hundred. Oceanographers use the symbol ‰ to represent parts-per-thousand. So 2.47% = 24.7‰, and saying that water has a salinity of 24.7‰, for example, means that there are 24.7 grams of salts dissolved in 1,000 grams of water.

Solid   Liquid   Gas

Add 80 calories   Add 100 calories   Add 540 calories

Melting   0°C   Warming   100°C   Evaporating

Water vapor

Water

Ice

Freezing   Cooling   0°C   100°C   Condensing

Remove 80 calories   Remove 540 calories

Latent heat of fusion—80 calories   Remove 100 calories   Latent heat of vaporization—540 calories

FIGURE 6.8 We must add 80 calories of heat energy to change a gram of ice to liquid water. After the ice is melted, about 1 calorie of heat is needed to raise each gram of water by 1°C. But 540 calories must be added to each gram of water to vaporize it—to boil it away. The process is reversed for condensation and freezing.

Table 6.2   Properties of Water

Property	Remarks	Importance to the Ocean Environment
Physical state	Only substance occurring naturally in all three phases as solid, liquid, and gas (vapor) on Earth's surface	Transfer of heat between ocean and atmosphere by phase change
Dissolving ability	Dissolves more substances in greater quantities than any other common liquid	Important in chemical, physical, and biological processes
Density: mass per unit volume	Density determined by (1) temperature, (2) salinity, and (3) pressure, in that order of importance. The temperature of maximum density for pure water is 4°C. For seawater, the freezing point decreases with increasing salinity.	Controls oceanic vertical circulation, aids in heat distribution, and allows seasonal stratification
Surface tension	Highest of all common liquids	Controls drop formation in rain and clouds; important in cell physiology
Conduction of heat	Highest of all common liquids	Important on the small scale, especially on cellular level
Heat capacity: quantity of heat required to raise the temperature of 1 g of a substance 1°C	Highest of all common solids and liquids	Prevents extreme range in Earth's temperatures; great heat moderator
Latent heat of fusion: quantity of heat gained or lost per unit mass by a substance changing from a solid to a liquid or a liquid to a solid phase without an accompanying rise in temperature	Highest of all common liquids and most solids (80 cal/g)	Thermostatic heat-regulating effect due to the release of heat on freezing and absorption on melting
Latent heat of vaporization: quantity of heat gained or lost per unit mass by a substance changing from a liquid to a gas or a gas to a liquid phase without an increase in temperature	Highest of all common substances (540 cal/g)	Immense importance: a major factor in the transfer of heat in and between ocean and atmosphere, driving weather and climate
Refractive index	Increases with increasing salinity and decreases with increasing temperature	Objects appear closer than in air
Transparency	Relatively great for visible light; absorption high for infrared and ultraviolet	Important in photosynthesis
Sound transmission	Good compared with other fluids	Allows sonar and precision depth recorders to rapidly determine water depth, detect subsurface features and animals; sounds can be heard great distances underwater
Compressibility	Only slight	Density changes only slightly with pressure/depth
Boiling and melting points	Unusually high	Allows water to exist as a liquid on most of Earth

Sources: Sverdrup et al., 1942; Ingmanson and Wallace, 1995.

Every semester my students ask the same question: "Does hot water freeze faster than cold?" Every semester I patiently answer "No." Why should it? If two equal amounts of water are placed in a freezer and one is warmer than the other, then the cooler one will always freeze first. After all, the hot water must lose heat to come down to the starting temperature of the cool water, and by that time the water that started out cool might be near the freezing point.

Right?

Well, yes and no.

Yes, because the logic above makes complete thermodynamic sense!

No, because other factors can intervene!

Curious? So was G. S. Kell, who wrote a paper on the subject for the *American Journal of Physics* (May 1969). He lived in Canada where many people apparently believe that, if left outside on a cold night, a bucket of hot water will freeze more quickly than a bucket of cold water. Kell tested this belief using covered buckets. The freezing went exactly as predicted above: The warmer water must first fall to the starting temperature of the cold water, and then its cooling curve (a graph of the fall in temperature with time) will follow the cooling curve already taken by the cooler bucket. The bucket containing cooler water always froze first. Good!

So why do many Canadians (not to mention American students) persist in their belief? Could there be a grain of truth here? Kell tried the experiment using buckets without lids. Now the water can evaporate, and hot water evaporates much more rapidly than cold water, especially in cold, dry air. He demonstrated that a bucket of water cooling from the boiling point to the freezing point would lose about 16% of its mass. Now there's less water to freeze. A smaller volume of water initially at 32°F will freeze faster than a larger volume *if* the same surface area is exposed to the cold.

And there's more. As you know, the latent heat of vaporization for water is very high (540 calories per gram). Water molecules escaping the surface of the hot water take away a very large amount of heat. The hot water's cooling curve plummets. All of these conditions together could make the hot water seem to freeze faster than the cold.

Even so, if the buckets are made of metal, much heat would be lost through the container walls to the cold air. It is unlikely that hot water in a metal bucket would freeze first. But what if the buckets are made of wood, an excellent insulator? Then most energy loss would occur through evaporation, and the initially-hot water might actually freeze faster. (But this is not a variable that Kell tested for his paper.)

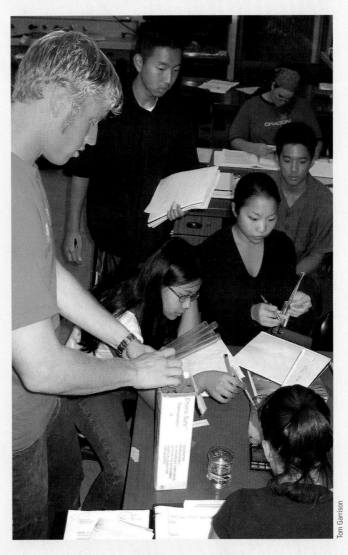

Lab students compare the freezing points of fresh water and seawater.

So should you make ice cubes with warm water? No. First you pay the electric (or gas) company to warm the water. Then you pay the electric company to run the refrigerator to bring the water down to tap temperature. Now you freeze the water. If you started with warm water, you'd have less ice. Would it really freeze faster? Re-read the discussion of the scientific method in Chapter 1, and then check it out!

6.10

Global Thermostatic Effects

The **thermostatic properties** (*therme* = heat, *stasis* = standing still) of water are those properties that act to moderate changes in temperature. Water temperature rises as the sun's energy is absorbed and changed to heat, but as we've seen, water has a very high heat capacity, so its temperature will not rise very much even if a large quantity of heat is added. This tendency of a substance to resist a change in temperature with the gain or loss of heat energy is called **thermal inertia**.

Remember the hot sand and cool water on a hot summer afternoon? Think about Earth as a whole. The highest temperatures on land, in the North African desert, exceed 50°C (122°F); the lowest, on the Antarctic continent, drop below −90°C (−129°F). That's a difference of 140°C (or 250°F). On the ocean surface, however, the range is from −2°C (29°F) where sea ice is forming to about 32°C (90°F) in the tropics—a difference of only 34°C (61°C). Consisting of water, the ocean rises very little in temperature as it absorbs heat. The ocean's thermal inertia is much greater than the land's.

A practical example of thermal inertia can be seen in **Figure 6.10**. San Francisco, California, and Norfolk, Virginia, are on the same line of latitude—each is the same distance from the equator. Wind tends to flow from west to east at this latitude (for reasons discussed in Chapter 8). Compared to Norfolk, San Francisco is warmer in the winter and cooler in the

FIGURE 6.9 The dependence of freezing temperature and temperature of maximum density on salinity (salt content). As we saw in Figure 6.3, pure water is densest at 3.98°C (39.2°F) (point B) and its freezing point is 0°C (32°F). Seawater with 15‰ salinity is densest at 0.73°C (33.31°F) and its freezing point is −0.80°C (30.56°F). The temperature of maximum density and the freezing point coincide at −1.33°C (29.61°F) in seawater with a salinity of 24.7‰. At salinities greater than 24.7‰ the density of water always decreases as temperature increases. Note that the symbol ‰ represents parts-per-thousand, so 2.47% = 24.7‰.

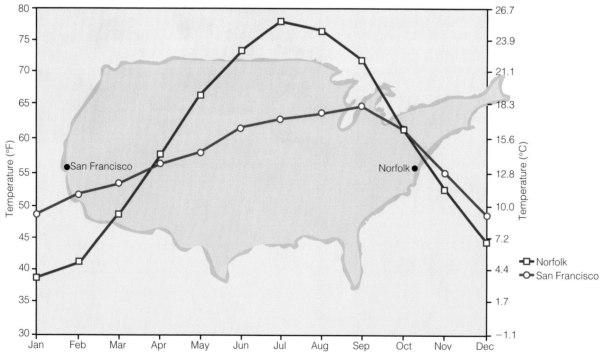

FIGURE 6.10 San Francisco, California, and Norfolk, Virginia, are on the same line of latitude, yet San Francisco is warmer in the winter and cooler in the summer than Norfolk. Part of the reason is that wind tends to flow from west to east at this latitude. Thus air in San Francisco has moved over the ocean while air in Norfolk has approached over land. Water doesn't warm as much as land in the summer, nor cool as much in winter—a demonstration of thermal inertia.

FIGURE 6.11 About 20 million square kilometers (7.7 million square miles) of ocean surface thaws and refreezes in the Southern Hemisphere each year—an area of ocean larger than South America! The autumn cooling of the atmosphere is delayed because heat energy is released as masses of water turn to ice. Heat is absorbed during the ice melt in the spring. Seasonal extremes are moderated by the absorption and release of heat energy as ice thaws and refreezes. (Remember, the seasons are reversed in the Southern Hemisphere.) The scale shows the percentage of ocean surface completely covered by ice.

summer, in part because air in San Francisco has moved over the ocean while air in Norfolk has approached over land.

Thermostatic Effects of Ice

As you know, removing a calorie of heat from freezing pure water at 0°C (32°F) won't change its temperature at all; 80 calories of heat energy must be removed per gram of liquid water to form ice. More than 18,000 cubic kilometers (4,300 cubic miles) of polar ice, covering as many as 20 million square kilometers (7.7 million square miles) of surface, thaws and refreezes in the Southern Hemisphere each year—an area of ocean larger than South America (**Figure 6.11**)! The annual change in sea ice cover is less in the Arctic, averaging about 5 million square kilometers (2 million square miles).

Ice provides a moderating thermostatic effect even if it doesn't get warm enough to melt. The heat capacity of solid ice is about half that of liquid water (0.51 calorie per gram). Although ice warms about twice as fast as liquid water with the same input of heat, it is more effective at moderating temperature than, say, granite rock (with a heat capacity of about 0.20 calorie per gram).

Thermostatic Effects of Water and Air Movement

Earth's North and South Poles have a marked deficiency of heat, and the equator has a pronounced surplus. Why don't the polar oceans freeze solid and the equatorial ocean boil away? The rea-

son is that currents in the atmosphere and ocean are moving huge amounts of heat from the tropics toward the poles. Water's high heat capacity makes it an ideal fluid to equalize the polar–tropical heat imbalance. Ocean currents and atmospheric weather result from the response of water and air to unequal solar heating. Although weather and currents are discussed in more detail in the next two chapters, here's a brief overview.

Ocean currents carry heat from the tropics (where incoming energy exceeds outgoing) to the polar regions (where outgoing energy exceeds incoming). The amount of heat transferred in this way is astonishing (**Figure 6.12**). For example, "outbound" water in the warm Gulf Stream (a large northward-flowing ocean current just offshore of the eastern United States) is about 10°C (18°F) warmer than "inbound" water coming from the central eastern Atlantic to replace it, meaning that about 10 million calories are transported per cubic meter. Since the flow rate of the Gulf Stream is about 55 million cubic meters per second, some 550 trillion calories are being transported northward in the western North Atlantic each *second*! Nearly half of these calories reach the high latitudes above 40°N. This warmth has a dramatic moderating influence on the winter climate of northwestern Europe.

As impressive as the figures are for ocean currents, the amount of heat transported by water vapor in the atmosphere is even greater. About half of the solar energy entering water results in evaporation. The solar energy required for this evaporation is later surrendered during condensation and cloud formation (and rain), but usually at a distance from where the initial evaporation occurred. So the ocean surface near Cuba may be

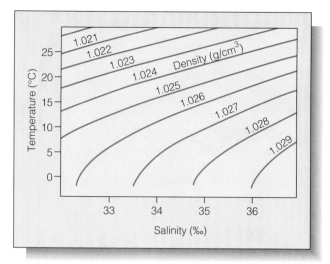

FIGURE 6.13 The complex relationship among the temperature, salinity, and density of seawater. Note that two samples of water can have the *same* density at *different* combinations of temperature and salinity. (Source: From G. P. Kuiper, ed., *The Earth as a Planet,* © 1954. The University of Chicago Press. Reprinted by permission.)

FIGURE 6.12 A satellite image of sea-surface temperature averaged through the period 1–8 January 2001 shows the tropical ocean brimming with heat. Blue, purple, red, yellow, and white represent progressively warmer water. Warm water is shown streaming northward in the Gulf Stream along the United States' east coast. In a week's time this water will have a moderating effect on the European winter climate.

cooled by evaporation today, the water vapor may then be moved north by winds, and eastern Canada may be warmed by condensation of the same water in a rainstorm later in the week.

Both atmosphere and ocean transfer heat by movement, but water's exceptionally high latent heat of vaporization means that water vapor transfers much more heat (per unit of mass) than liquid water. Masses of moving air account for about two-thirds of the poleward transfer of heat; ocean currents move the other third.

The Density Structure of the Ocean

As we've said, the density of water is mainly a function of its temperature and salinity. We will present more details about salinity in Chapter 7, but for now you need to know that a liter of seawater weighs between 2% and 3% more than a liter of pure water because of the solids (often called salts) dissolved in seawater. The density of seawater is thus between 1.020 and 1.030 g/cm³ compared with 1.000 g/cm³ for pure water at the same temperature. *Cold, salty water is more dense than warm, less salty water.* Seawater's density increases with increasing salinity, increasing pressure, and decreasing temperature. Figure 6.13 shows the relationship among temperature, salinity, and density. Notice that two samples of water can have the *same* density at *different combinations* of temperature and salinity.

The Density Zones

Much of the ocean is divided into three density zones: the surface zone, the pyconocline, and the deep zone. The **surface zone**, or **mixed layer**, is the upper layer of ocean (**Figure 6.14a**). Temperature and salinity are relatively constant with depth in the surface zone because of the action of waves and currents. The surface zone consists of water in contact with the atmosphere and exposed to sunlight; it contains the ocean's least dense water and accounts for only about 2% of total ocean volume. The surface zone (or mixed layer) typically extends to a depth of about 150 meters (500 feet), but depending on local conditions, it may reach a depth of 1,000 meters (3,300 feet) or be absent entirely.

The **pycnocline** (*pyknos* = strong, *clinare* = slope, to lean) is a zone in which density increases with increasing depth. This zone isolates surface water from the denser layer below. The pycnocline contains about 18% of all ocean water.

The **deep zone** lies below the pycnocline at depths greater than about 1,000 meters (3,300 feet) in mid-latitudes (40°S to 40°N). There is little additional change in water density with increasing depth through this zone. This deep zone contains about 80% of all ocean water.

Temperature, Salinity, and Stratification

The pycnocline's rapid density increase with depth is due mainly to a decrease in water temperature. **Figure 6.14b** shows the general relationship of temperature with depth in the open sea. The surface zone is well mixed, with little decrease in temperature with depth. In the next layer temperature drops rapidly with depth. Beneath it lies the deep zone of cold, stable water. The middle layer, the zone in which temperature changes rapidly with depth, is called the **thermocline** (*therm* = heat), and

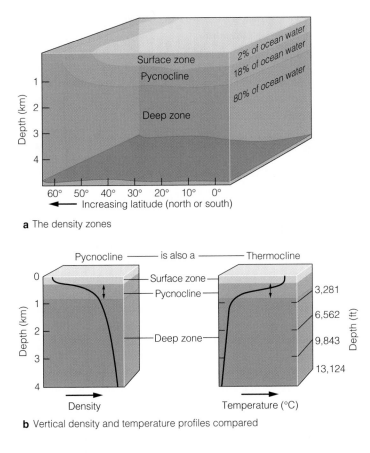

a The density zones

b Vertical density and temperature profiles compared

c Vertical density and salinity profiles compared

FIGURE 6.14 Density stratification in the ocean. (**a**) In most of the ocean, a surface zone (or mixed layer) of relatively warm, low-density water overlies a layer called the *pycnocline*. Density increases rapidly with depth in the pycnocline. Below the pycnocline lies the deep zone of cold, dense water—about 80% of total ocean volume. (**b**) The rapid density increase in the pycnocline is mainly due to a decrease in temperature with depth in this area—the *thermocline*. (**c**) In some regions, especially in shallow water near rivers, a pycnocline may develop in which the density increase with depth is due to vertical variations in salinity. In this case the pycnocline is a *halocline*.

falling temperature is the major contributor to the formation of the pycnocline.

Thermoclines are not identical in form in all areas or latitudes. Surface temperature is proportional to available sunlight. More solar energy is available in the tropics than in the polar regions, so the water there is warmer. The ocean's sunlit upper layer is thicker in the tropics, both because the solar

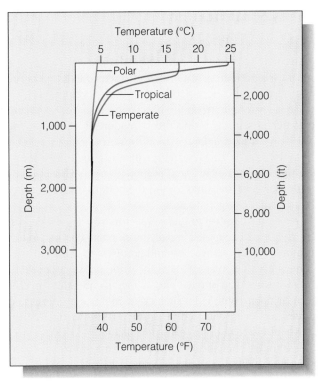

FIGURE 6.15 Typical temperature profiles at polar, tropical, and middle (temperate) latitudes. Note that polar waters lack a thermocline.

angle there is more nearly vertical and because water in the open tropical ocean contains fewer suspended particles (and is therefore clearer than water in open temperate or polar regions). Because the ocean is heated to a greater depth, the tropical thermocline is deeper than thermoclines at higher latitudes. It is also much more pronounced: The transition to the colder, denser water below is more abrupt in the tropics than at high latitudes.

Polar waters, which receive relatively little solar warmth, are not stratified by temperature and generally lack a thermocline because surface water in the polar regions is nearly as cold as water at great depths.

Figure 6.15 contrasts polar, tropical, and temperate thermal profiles, showing that the thermocline is primarily a mid- and low-latitude phenomenon. Thermocline depth and intensity also vary with season, local conditions (storms, for example), currents, and many other factors.

Below the thermocline, water is very cold, ranging from −1° to 3°C (30.5° to 37.5°F). Because this deep and cold layer contains the bulk of ocean water, the average temperature of the world ocean is a chilly 3.9°C (39°F).

Low salinity can also contribute to the pycnocline, especially in cool regions where precipitation is high or along coasts where freshwater runoff mixes with surface water. Wherever precipitation exceeds evaporation, salinity will be low. These differences produce the **halocline** (*halos* = salt), a zone of rapid salinity increase with depth (**Figure 6.14c**). The halocline often coincides with the thermocline, and the combination produces a pronounced pycnocline.

Density Stratification and Water Masses

The layers we have been discussing are distinct water masses. A **water mass** is a body of water with characteristic temperature and salinity and therefore density. Curiously, even the deepest of these layers originates at the ocean's surface. Very cold and salty water produced during the formation of sea ice at the polar ocean surface is denser than the surrounding water and sinks to the seabed. In a few marginal basins, the most notable being the Mediterranean Sea, evaporation can also produce salty, dense water that sinks toward the bottom until it reaches a layer of water of equal density.

Layering by density traps dense water masses at great depths, where they are not exposed to daily heating and cooling, to surface circulation driven by winds and storms, or to light. The pycnocline effectively isolates 80% of the world ocean's water from the 20% involved in surface circulation. Dense water masses form near polar continental shelves (as cold water freezes and excludes salt) or in enclosed areas such as the Mediterranean Sea (where evaporation exceeds precipitation and river input, raising salinity). These heavy water masses sink, sometimes overlapping one another and often retaining their identity for long periods. Separate water masses below the pycnocline tend not to merge because little energy is available for mixing in these quiet depths. (Density differences in water masses power the deep-ocean currents, as will be discussed in Chapter 9.)

Density Stratification and Vertical Water Movement

The vertical movement of large volumes of water from the surface to great depths (and vice versa) is possible only where surface-water density is similar to deep-water density. The great difference in temperature, and therefore density, between surface water and deep water in the tropics makes the water column very stable and prevents an exchange of surface and deep water. This stability is maintained despite the fact that the surface of the tropical ocean is in constant horizontal motion, churned by tropical cyclones and stirred by currents.

Vertical movement of water in the northern polar ocean is also limited. There, however, the stratification is caused largely by a salinity difference between surface water and water at great depths. The surface of the Arctic Ocean receives a large volume of freshwater runoff from Siberian and Canadian rivers. Continental masses block the formation of large currents, and the landlocked northern ocean communicates sluggishly with other ocean areas, so the surface water tends not to mix with deeper water or to flow to lower latitudes.

In contrast, the southern polar ocean is only weakly stratified. The cold temperature of southern ocean surface water closely matches that of deep water, so no thermo-cline divides surface water from deep water (see again Figure 6.15). The absence of confining continental margins and mixing at the boundaries of the Antarctic Circumpolar Current minimize salinity differences. Turbulence and weak stratification encourage a huge volume of deep-water upwelling, which contributes to high surface nutrient levels and high biological productivity.

An Overview of Ocean Surface Conditions

Table 6.3 summarizes a few important properties of seawater at polar, temperate, and tropical latitudes. The temperature data show the effects of solar radiation at various latitudes, along with the ratio of evaporation to precipitation. Notice that the temperature of ocean surface water is more variable through the year in the temperate zone than in either the polar or tropical areas, but that temperate zone salinity stays relatively constant. Notice also that evaporation generally exceeds precipitation in the tropics, but that precipitation dominates in temperate and polar zones.

Ocean surface temperature is highest in the Pacific, north and east of Borneo, where westward-flowing currents funnel seawater warmed by long exposure to tropical sunlight. The highest open-ocean surface temperatures range to 32°C (90°F). Surface salinity is especially high in the warm central North and South Atlantic, where evaporation rates are high and surface water is isolated by the currents flowing around the ocean's periphery. Some of this information is summarized graphically in **Figures 6.16** and **6.17**, which show typical ocean surface temperatures and salinities worldwide.

Remember, though, that variations in temperature, salinity, and density between high and low latitudes are usually confined to the uppermost 2,000 meters (6,500 feet) of water. Below this depth conditions are similar at all latitudes. The deep ocean is nearly as cold, salty, and dense in the equatorial Pacific as it is off the Siberian coast.

Table 6.3 Some Characteristics of the World Ocean Surface, by Latitude			
Characteristic	Tropical Oceanic Waters	Temperate Oceanic Waters	Polar Oceanic Waters
Winter temperature	20–25°C (68–77°F)	5–20°C (41–68°F)	About −2°C (28°F)
Annual variation of temperature	Less than 5°C (9°F)	About 10°C (18°F)	Less than 5°C (9°F)
Average salinity	35‰–37‰	About 35‰	28‰–32‰[a]
Annual variation of air temperature	Less than 5°C (9°F)	About 10°C (18°F)	Up to 40°C (72°F)
Precipitation–evaporation balance	E exceeds P	P exceeds E	P exceeds E

Source: H. Charnock, 1971.

[a]Polar surface water is a bit less saline because the very dense, cold, salty water left behind after the formation of sea ice sinks rapidly toward the seabed.

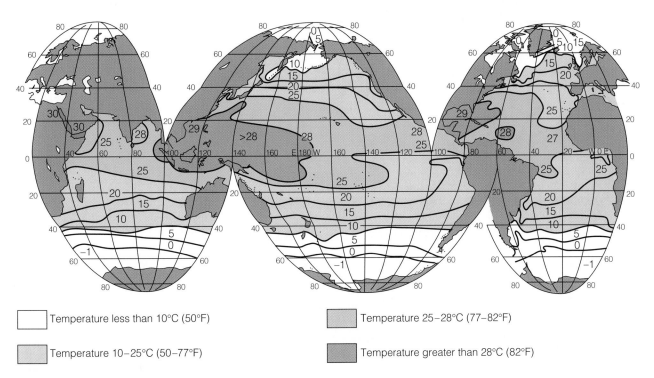

☐ Temperature less than 10°C (50°F)	☐ Temperature 25–28°C (77–82°F)
☐ Temperature 10–25°C (50–77°F)	☐ Temperature greater than 28°C (82°F)

FIGURE 6.16 Sea surface temperatures in degrees Celsius during Northern Hemisphere summer.

☐ Salinity greater than 36 parts per thousand	☐ Salinity 34–36 parts per thousand	☐ Salinity less than 34 parts per thousand

FIGURE 6.17 Sea surface salinities in parts per thousand (‰) during Northern Hemisphere summer.

Refraction, Light, and Sound

The ring of light sometimes seen around the moon and the safe concealment of a submarine may not seem related, but both events depend on **refraction**, the bending of waves. Light and sound are both wave phenomena. When a light wave or a sound wave leaves a medium of one density—such as air—and enters a medium of a different density—such as water—at an angle other than 90°, it is bent from its original path. The reason for this bending is that light or sound waves travel at different speeds in the different media.

The situation is analogous to a line of people marching along a desert highway with their arms linked over each other's shoulders. The marchers can walk faster if they stay on the pavement

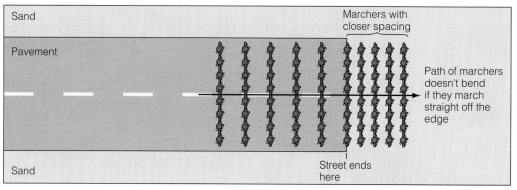

FIGURE 6.18 An analogy for refraction. The ranks of marchers represent light or sound waves; the pavement and sand represent different media. The marchers can walk faster if they stay on the pavement than if they walk in the sand next to the highway. (**a**) If the marchers head off the pavement at an angle other than 90°, their path will bend (refract) as they hit the sand because some will be walking more slowly than others. (**b**) If they march straight off the pavement, the ranks will slow down but not bend as they hit the sand.

than if they walk in the sand next to the highway. Their speed on the pavement, then, is greater than their speed in the sand. As long as they stay on the pavement, they won't change direction. But if their marching angle gradually takes them off the edge (into a medium in which their speed is *lower*), the people who reach the sand first will suddenly slow down, and the line will pivot quickly off the highway. They have been refracted. Their progress is depicted in **Figure 6.18a**. Note that the transition from one medium to another must occur at an angle other than 90° for refraction to occur; our marchers will not change direction if they march straight off the asphalt into the sand. They will still slow down, however (**Figure 6.18b**).

The refraction of light waves by water happens in much the same way. The speed of light in water is only about three-quarters its speed in air, so water effectively refracts light. (Glass bends light even more.) The degree to which light is refracted from one medium to another is expressed as a ratio called the **refractive index**. The higher the refractive index, the greater the bending of waves between media. The refractive index of water increases with increasing salinity. A beautiful example of the difference between the refractive index of fresh water and seawater is seen in **Figure 6.19**.

Examples of the refraction of light by water are all around you. A pencil sticking out of a glass of water looks bent because of refraction, the submerged steps of a swimming pool ladder look closer than they are because of refraction, and refraction

magnifies objects and causes fishermen to exaggerate the size of the fish that got away.

Light in the Ocean

6.21

Light is a form of electromagnetic radiation, or radiant energy, that travels as waves through space, air, and water. The visible spectrum—the wavelengths of light that human eyes can detect—is only a small part of the electromagnetic spectrum, which also includes radio waves, infrared, ultraviolet, and X-rays, for example. The wavelength of light determines its color: Shorter wavelengths are bluer, longer wavelengths are redder. Except for very long radio waves, water rapidly absorbs nearly all electromagnetic radiation. Only blue and green wavelengths pass through water in any appreciable quantity or distance.

Scattering and Absorption

6.22

Sunlight has a difficult time reaching and penetrating the ocean. Clouds and the sea surface reflect light, while atmospheric gases and particles scatter and absorb it. Once past the sea surface, light is rapidly attenuated by scattering and absorption. **Scattering** occurs as light is bounced between air or water molecules, dust particles, water droplets, or other objects before being absorbed. The greater density of water (along with the

FIGURE 6.19 A broken rainbow, a beautiful demonstration of the difference between the refractive index of fresh water and that of seawater. Broken rainbows occur when an atmospheric rainbow (caused by raindrops of pure water) meets a surf droplet rainbow. The rainbow angle for rainwater is 0.8° larger than for seawater.

Gunther Können, by permission Bob Greenler

greater number of suspended and dissolved particles) makes scattering more prevalent in water than in air. The **absorption** of light is governed by the structure of the water molecules it happens to strike. When light is absorbed, molecules vibrate and the light's electromagnetic energy is converted to heat.

Even the clearest seawater is not perfectly transparent. If it were, the sun's rays would illuminate the greatest depths of the ocean and seaweed forests would fill its warmed basins. The thin film of lighted water at the top of the surface zone is called the **photic zone** (*photo* = light). In the very clearest tropical waters the photic zone may extend to a depth of 600 meters (2,000 feet), but a more typical value for the open ocean is 100 meters (330 feet). In contrast, light typically penetrates the coastal waters in which we swim to only about 40 meters (130 feet). All the production of food by photosynthetic marine organisms takes place in this thin, warm surface layer. Here, water is heated by the sun, heat is transferred from the ocean into the atmosphere and space, and gases are exchanged with the atmosphere. The thermostatic effects we've discussed function largely within this zone. Most of the ocean's life is found here. The photic zone may be extraordinarily thin, but it is also extraordinarily important.

The ocean below the photic zone lies in blackness. Except for light generated by living organisms, the region is perpetually dark. This dark water beneath the photic zone is called the **aphotic zone** (*a* = without, *photo* = light).

Water Colors

6.23

The energy of some colors of light is converted into heat—that is, its wavelengths are absorbed—nearer to the surface than the energy of other colors. **Figure 6.20** shows this differential absorption by color. The top meter (3.3 feet) of the ocean absorbs nearly all the infrared radiation that reaches the ocean surface, significantly contributing to surface warming. The top meter also absorbs 71% of red light. The dimming light becomes bluer with depth because the red, yellow, and orange wavelengths are being absorbed. By 300 meters (1,000 feet) even the blue light has been converted into heat.

From above, clear ocean water looks blue because blue light can travel through water far enough to be scattered back through the surface to our eyes. Divers near the surface see an even brighter blue color. Because nearly all red light is converted to heat in the first few meters of ocean water, red objects a short distance beneath the surface look gray. If you were a diver working at a depth of 10 meters (33 feet) and you cut your hand, you would see gray blood rather than red because there is not enough red light at that depth to reflect from blood's red pigment and stimulate your eye. The underwater pictures of red organisms you've seen are possible only because the photographer has brought along a source of white light (which contains all colors). This is shown in **Figure 6.20c** and **d**.

Suspended particles scatter some colors of light and absorb others. Some sediments reflect yellow light, giving the ocean a yellow cast. The Red Sea gets its name from its abundance of cyanobacteria, small plantlike organisms that contain a reddish pigment.

Sound in the Ocean

6.24

Sound is a form of energy transmitted by rapid pressure changes in an elastic medium. Sound intensity decreases as it travels through seawater because of spreading, scattering, and absorption. Intensity loss due to spreading is proportional to the square of the distance from the source. Scattering occurs as sound bounces off bubbles, suspended particles, organisms, the

Absorption of light of different wavelengths (colors) by seawater

Color	Wavelength (nm)	% Absorbed in 1 m of Water	Depth in Which 99% Is Absorbed (m)
Ultraviolet (UV)	310	14.0	31
Violet (V)	400	4.2	107
Blue (B)	475	1.8	254
Green (G)	525	4.0	113
Yellow (Y)	575	8.7	51
Orange (O)	600	16.7	25
Red (R)	725	71.0	4
Infrared (IR)	800	82.0	3

a

b

c

d

FIGURE 6.20 Only a thin film of seawater is illuminated by the sun. Except for light generated by living organisms, most of the ocean lies in complete blackness. (**a**) The table shows the percentage of light absorbed in the uppermost meter of the ocean, and the depths at which only 1% of the light of each wavelength remains. (**b**) The bars show the depths of penetration of 1% of the light of each wavelength (as in the last column of the table). (**c**) A fish photographed in normal oceanic light; the color blue predominates. (**d**) The same fish photographed with a strobe light. The flash contains all colors, and the distance from the strobe to the fish and back to the camera is not far enough to absorb all the red light. The fish shows bright warm colors otherwise invisible.

surface, the bottom, or other objects. Eventually sound is absorbed and converted by molecules into a very small amount of heat. The absorption of sound is proportional to the square of the frequency of the sound: Higher frequencies are absorbed sooner. Sound waves can travel for much greater distances through water than light waves can before being absorbed. Because sound travels through water so efficiently, many marine animals use sound rather than light to "see" in the ocean.

The speed of sound in seawater of average salinity is about 1,500 meters per second (3,345 miles per hour) at the surface, almost five times the speed of sound in air. The speed of sound in seawater increases as temperature and pressure increase. Sound travels faster at the warm ocean surface than it does in deeper, cooler water. Its speed decreases with depth, eventually reaching a minimum at about 1,000 meters (3,300 feet). Below that depth, however, the effect of increasing pressure offsets the effect of decreasing temperature, so speed increases again. Near the bottom of an ocean basin the speed of sound may actually be higher than at the surface. Though these variations are important in the behavior of oceanic sound, they amount to only 2% or 3% of the average speed of sound in seawater. The relationship between depth and sound speed is shown in **Figure 6.21**.

Sofar Layers and Shadow Zones

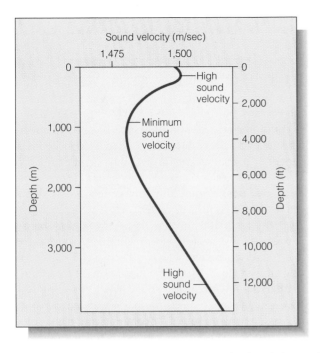

FIGURE 6.21 The relationship between water depth and the speed of sound.

The depth at which the speed of sound reaches its minimum varies with conditions, but it is usually near 1,200 meters (3,900 feet) in the North Atlantic or about 600 meters (2,000 feet) in the North Pacific. Although its speed is relatively slow, the transmission of sound in this minimum-velocity layer is very efficient because refraction tends to cause sound energy to remain within the layer. The outer edges of sound waves escaping from this layer will enter water in which the speed of sound is higher. This will cause the wave to speed up but then pivot back into the minimum-velocity layer, as shown in **Figure 6.22.** Upward-traveling sound waves that are generated within the minimum-velocity layer will tend to be refracted downward, and downward-traveling sound waves will tend to be refracted upward. In short, sound waves bend *toward* layers of lower sound velocity and so tend to stay within the zone. Therefore loud noises made at this depth can be heard for thousands of kilometers.

Navy depth charges detonated in the minimum-velocity layer in the Pacific have been heard 3,680 kilometers (2,290 miles) away from the explosion. In a recent test, sound generated by a U.S. Navy ship in the Indian Ocean was heard at the Oregon Coast! (See **Box 6.2.**) In the early 1960s the U.S. Navy

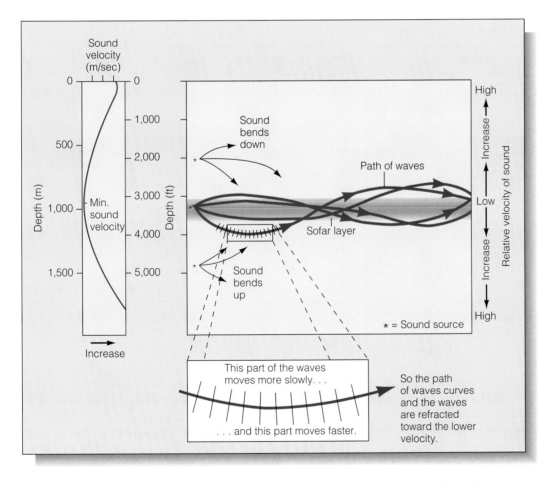

FIGURE 6.22 The sofar layer, in which sound waves travel at minimum speed. Sound transmission is particularly efficient—that is, sounds can be heard for great distances—because refraction tends to keep sound waves within the layer.

BOX 6.2   Is Climate Change Sofar Away?

Early in 1991 underwater sounds generated by a ship near Heard Island in the southern Indian Ocean were detected off the west coast of the United States, 17,960 kilometers (11,160 miles) away. The foghorn-like sounds were produced by powerful underwater loudspeakers suspended 200 meters (660 feet) beneath a research vessel. The ship's five transmitters emitted a booming, low-frequency hum in all directions for an hour and then were silent for two hours. The pattern was repeated through most of five days.

Some of the sound was trapped in the Indian Ocean's minimum-velocity (sofar) layer, in which it traveled both into the Atlantic and around Australia into the Pacific. The sound was detected by a ship equipped with sensitive hydrophones lowered to the sofar depth off the California coast, near Point Conception. It took about $3\frac{1}{2}$ hours for the sound to travel that far, nearly halfway around the world. Similarly equipped ships and offshore stations detected the sound at other locations.

The experiment was part of a plan to measure ocean temperature within a few thousandths of a degree. Because sound travels more rapidly through warmer water, water-temperature differences can be calculated by recording the time taken for sound to reach receivers around the world. The temperature of the ocean is an important measure of potential global warming, and the sound-transmission experiment collected baseline data for a study designed to help settle the argument over whether carbon dioxide and other greenhouse gases are warming the planet artificially. Computer models of Earth's climate suggest that greenhouse gases could cause

the ocean to warm by about 0.005°C (0.009°F) per year at a depth of 1,000 meters (3,300 feet). Such a subtle temperature change would be very difficult to detect by conventional monitoring, but over a decade this warming would cut a few seconds from the travel time of a sound signal. (More information on the possibilities and consequences of global warming may be found in Chapter 18.)

In early 1993 nearly 100 researchers at 13 institutions began a larger five-year, $40-million experiment. They constructed and tested sound sources capable of projecting a 75-cycle-per-second signal with 260 watts of acoustic power. To address concerns that the low-frequency sound would in some way harm marine organisms, especially whales and other marine mammals that depend on sound for communication and feeding, the scientists carefully analyzed mammal distribution and behavior around transmitters being tested. No significant evidence of disruption was found. Environmental impact statements were prepared for two transmitter sites: off the coast of Kauai, Hawaii, and on the Pioneer Seamount about 80 kilometers (50 miles) off the central California coast.

The Acoustic Thermography of Ocean Climate (ATOC) program began transmitting acoustic signals in late 1995 from the Pioneer Seamount. The signals were recorded on horizontal receiving arrays at 11 U.S. Navy SOSUS stations in the northeast Pacific and on two vertical receiving arrays, one near the island of Hawaii and the other near Kiritimati (Christmas) Island, just north of the equator in the central Pacific. The signals were also recorded by a single receiver off New Zealand.

experimented with the use of sound transmission in the minimum-velocity layer as a lifesaving tool. Survivors in a life raft would drop a small charge into the water that was set to explode at the proper depth. A number of widely spaced listening stations ashore would compare the differences in the arrival times of the signal and then compute the position of the raft. The project—which has since been abandoned in favor of radio beacons—was called **sofar** (for *sound fixing and ranging*). The minimum-velocity layer has come to be known as the **sofar layer**.

Sound travels slowly in the sofar layer, but it moves rapidly near the bottom of the well-mixed surface layer. Temperature and salinity conditions are homogeneous there, so they do not produce any refraction. Pressure still increases with depth, however, causing a thin high-velocity layer at around 80 meters (260 feet), just above the pycnocline.

Some ships project pulses of sound into the water to search for animals, submarines, or hazards to navigation. Depending on the angle at which the sound waves arrive at the high-velocity layer, they will sometimes split and refract to the surface or bend into

the depths. An object beyond the area of divergence may be undetectable; it would be within a **shadow zone**, a region into which very little sound energy penetrates. Shadow zones are of particular interest to submariners and to the ship captains who use sound to hunt them. As depicted in **Figure 6.23**, a smart submarine captain can use the shadow zone to hide from a pursuer.

Sonar

Crews aboard surface ships and submarines employ **active sonar** (*sound navigation and ranging*), the projection and return through water of short pulses (*pings*) of high-frequency sound, to search for objects in the ocean (**Figure 6.24**). In a modern system, electrical current is passed through crystals to produce powerful sound pulses pitched above the limit of human hearing. (Though this high-frequency sound is absorbed rapidly in the ocean, its use greatly improves the resolution of images.) Some of the sound from the transmitter bounces off any object larger than the wavelength of sound employed and returns to a

By the end of 1998, 27 months of data had been collected by some of these receivers (**Figures a** and **b**). There are three key results. First, acoustic travel times can be measured with a precision of about 20–30 milliseconds at 3,000–5,000 kilometer (1,860–3,100 mile) ranges. (For comparison the total travel time at 5,000 kilometers or 3,100 miles is nearly an hour.) This measurement is more precise than originally thought possible.

Second, the observed travel-time changes can be clearly related to known ocean processes. One of the significant sources of lower frequency variability in ocean temperature is the seasonal change, with the upper ocean warming during summer and cooling during winter. The ATOC data show corresponding seasonal changes in travel times, as expected, particularly for acoustic paths that travel north of the Subarctic Front, where the seasonal temperature changes extend to significant depths rather than being confined to a shallow seasonal thermocline.

Third, the range and depth-averaged temperatures derived from ATOC are consistent with and complementary to related estimates derived from measurements of sea-surface height made by the *TOPEX/Poseidon* satellite altimeter. Sea-surface height is related to ocean temperature because of thermal expansion. One of the primary goals of the ATOC program is to use the acoustic data to refine models of the ocean circulation. Consistent results for the seasonal heat storage are found when the acoustic and altimetric data are combined with a computer model of the ocean general circulation. The two data types are both found to be important in fine-tuning the model, with the combination providing more information than either data type alone.

As experiments proceed, the extent of ocean warming—and thus global warming—may become clear.

(**a**) A map of sound-speed anomalies in the North Pacific Ocean. This map is based on ATOC data measuring the speed of sound through water to a depth of 1,000 meters (3,300 feet). Variations from normal speed range from −0.5 meter per second (dark blue) to +0.5 meter per second (dark red). (**b**) Times of transmissions from ATOC sources used to compile data for Figure a.

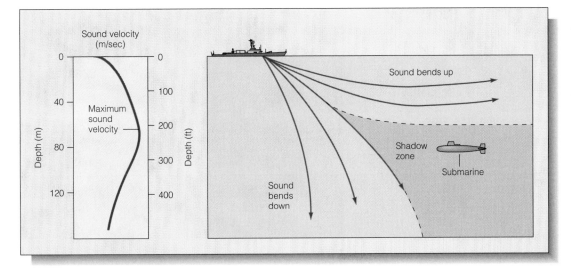

FIGURE 6.23 The shadow zone. The thin, high-sound-velocity layer, which forms at a depth of about 80 meters (260 feet), deflects sound. The shadow zone is a good place for submarines to hide from sonar. (Note the difference in depth scales in Figures 6.21 and 6.22.)

FIGURE 6.24 The principle of active sonar. Pulses of high-frequency sound are radiated from the sonar array of the sending vessel. Some of the energy of this ping reflects from the submerged submarine and returns to the sending vessel. The echo is analyzed to plot the position of the submarine.

microphone-like sensor. Signal processors then amplify the echo and reduce the frequency of the sound to within the range of human hearing. An experienced sonar operator can tell the direction of the contact, its size and heading, and even something about its composition (whale or submarine or school of fish) by analyzing the characteristics of the returned ping.

Side-scan sonar is a type of active sonar. Operating with as many as 60 transmitter/receivers tuned to high sound frequencies, side-scan systems towed in the quiet water beneath a ship are sometimes capable of near-photographic resolution (**Figure 6.25**). Side-scan systems are used for geological investigations, archaeological studies, and locating downed ships and airplanes. The multibeam system you read about in Chapter 4 is a form of side-scan sonar (see again Figure 4.4)

For deeper soundings, or to "see" into sediment layers below the surface, geologists use *seismic reflection profilers* employing powerful electrical sparks, explosives, or compressed air to generate a very energetic low-frequency sound pulse. Again, the round-trip travel time of the sound waves is crucial. The low-frequency sound cannot resolve great detail, but the echo can usually provide an image of the sedimentary layers beneath the surface (see Figures 4.27 and 5.4). Low-frequency sound also has the advantage of efficient travel with less absorption.

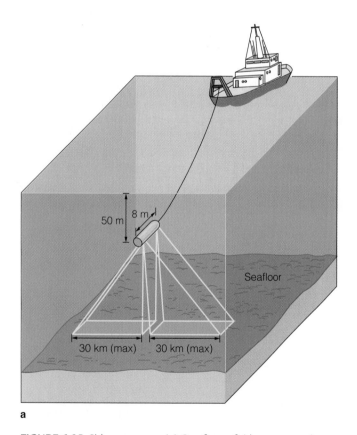

a

FIGURE 6.25 Side-scan sonar. (**a**) One form of side-scan sonar in action. Sound pulses leave the submerged towed array (**b**), bounce off the bottom, and return to the device. One of the two slits through which sound pulses depart and return is visible on its side. Computers process the impulses into images: (**c**) A side-scan sonar record showing a downed World War II–era PB4Y-2 Privateer at a depth of 50 meters (165 feet). The artificial colors enhance target detail and clarity.

b

c

The first human use of sea sound was passive: Mariners listened through their hulls to the whistles and clicks of whales and other animals. When submarine warfare emerged during World War I, the British invented a simple underwater listening device to detect noises made by enemy submarines. Operators would listen through a sensitive directional microphone for the telltale sounds of a propeller, a torpedo, or even a dropped wrench or slammed hatch cover. The first systems were primitive, but **passive sonar**, as these listening-only devices are now called, is currently undergoing a renaissance. Modern passive devices are much more sophisticated and sensitive than their World War I predecessors. Passive sonar confers the benefit of surprise—unlike active sonar, in which a listener can hear the loud ping long before an operator aboard the sending vessel can hear the faint echo. Usually it's safer just to listen. The combination of computerized signal processing, microphones towed at a distance from the listener's noisy ship, and better knowledge of ocean physics will improve the usefulness of both kinds of sonar.

Humans are not the only organisms to use sonar. Whales and other marine mammals use clicks and whistles to find food and avoid obstacles. We'll discuss this form of active sonar in Chapter 15.

Questions from Students

1. If water has the highest latent heat of vaporization, why does a drop of alcohol make my hand feel colder when it evaporates than a drop of water does?

Heat versus temperature again. The alcohol makes your hand *colder*, but it evaporates much *more quickly*. The water stays around longer (takes longer to evaporate), and although it may not cause your skin to become as *cold*, it will remove more *heat*.

2. I learned in physics that water's latent heat of evaporation is 585 calories per gram. In this chapter you say it's 540 calories per gram. What's up?

Actually, I wrote that water's latent heat of *vaporization* is 540 calories per gram. The latent heat of vaporization is the amount of heat that must be added to 1 gram of a substance, at its boiling point, to change it completely from a liquid to a gas. But water doesn't have to be at its boiling point to vaporize. When water molecules evaporate at temperatures below the boiling point (as from your skin during a workout), more heat must be supplied to evaporate each molecule than is needed to vaporize it at the boiling point. The latent heat of vaporization increases with decreasing temperature; at 20°C (68°F) the latent heat of vaporization is 585 calories per gram. This is sometimes called the latent of heat of *evaporation*.

3. Liquid methane appears to be the only other substance found in quantity in a liquid state in the solar system. If our ocean were of liquid methane rather than liquid water, how would conditions here be different?

Liquid methane freezes at −183°C (−272°F) and boils at −162°C (−234°F). Methane has a much lower heat capacity than liquid water because it does not form a lattice of hydrogen bonds as water does and consequently does not require a large energy input to release molecules. Because of methane's low heat capacity, the difference between Earth's polar and tropical daytime temperatures would be drastically greater unless the circulation rate of liquid and vapor currents accelerated to keep pace. Computer modeling yields a very unattractive vision of a planet with a boiling equatorial ocean, crushing atmospheric pressures from the greater amount of vapor in the air, torrential polar methane rainfall, cataclysmic cyclonic storms, and average wind speeds at mid-latitudes of hundreds of kilometers per hour. We miss all this excitement because our ocean is made of water.

4. Input of solar radiation in the Northern Hemisphere peaks on 22 June and reaches a minimum on 22 December because of Earth's orbital tilt (more about that in Chapter 8). Why, then, do our warmest days occur in August or September and our coldest days in January or February?

Because of thermal inertia. There is a lag between maximum sunlight and maximum warmth because of water's great heat capacity. The sun must shine on this watery planet for many weeks to raise the summer hemisphere's temperature. Of course, water also retains heat well, so the coldest days in the winter hemisphere come well after the darkest ones.

5. What about pressure? Is seawater compressible? Does compression have anything to do with water density?

Seawater is nearly incompressible—it doesn't squeeze very much as pressure builds. Still, there is some effect. Pressure in the deepest ocean trench is equal to the weight of 1,100 atmospheres (about $8\frac{1}{2}$ tons per square inch), and a cubic centimeter of water lowered to that level will lose about 2% of its volume. The average pressure acting on the world ocean's volume compresses water enough to lower sea level by about 37 meters (121 feet). If gravity (and therefore pressure) were released, sea level would rise that much. Still, the effect of pressure on density is very small in comparison to the effects of temperature and salinity.

6. Why don't sounds seem to travel easily from air to ocean, or vice versa?

Sound waves can make the transition from one medium to another with little energy loss only when the speeds of sound waves in the two different media are similar. The speeds of sound in water and air, however, are too different for an efficient transition to be made. Too great a contrast in speed produces reflection (not refraction) of the sound waves at the junction. This is why you can't hear people shouting from the edge of the pool while you're underwater, even though the weak sound of a submerged pebble clicking against the side is very clear and sharp.

If you place a solid medium in which the speed of sound is intermediate between air and water, the sound can move across one junction and then the other, for a more efficient total transition. Wood works well for this, which explains why ocean noises are easy to hear in wooden boats. Even if the speed of sound in the intermediate medium is higher than in water (as in steel, for instance), some sound will be audible simply because the hard surface provides a good radiating surface for noises coming from the water.

7. I can't tell where sound is coming from when I'm underwater. It seems to be coming from inside my head. Why?

Our normal sensation of stereo hearing depends in part on the difference in the arrival times of sound from one ear to the other. The speed of sound in water, however, is more than four times greater than its speed in air, so our brains are unable to sense arrival-time differences from sounds originating nearby. You'd need to be more than four times as sensitive—or have a head more than four times as large—to hear stereophonically underwater.

Chapter in Perspective

In this chapter you learned about the polar nature of the water molecule. This polarity and the hydrogen bonds that form between water molecules result in water's unexpected thermal properties. You found that liquid water is remarkably resistant to temperature change with the addition or removal of heat, and ice, with its large latent heat of fusion and low density, melts and refreezes over large areas of the ocean to absorb or release heat with no change in temperature. These thermostatic effects, combined with the mass movement of water and water vapor, prevent large swings in Earth's surface temperature. And, of course, you also learned that the words *heat* and *temperature* are not interchangeable.

Changes in temperature and salinity greatly influence water density. Ocean water is usually layered by density, with the densest water on or near the bottom.

The physical characteristics of the world ocean are largely determined by the physical properties of seawater. These properties include water's heat capacity, density, salinity, and ability to transmit light and sound.

In the next chapter you will learn what happens when solids and gases dissolve in seawater. Most of the properties of seawater are different from those of pure water because of the substances dissolved in the seawater.

Key Ideas

✓ Water is a polar chemical compound composed of two hydrogen atoms and one oxygen atom. Its remarkable thermal properties result from the large number and relatively great strength of hydrogen bonds between water molecules.
Review Figures 6.1 and 6.2, pages 138, 139

✓ Heat and temperature are not the same thing. Heat is energy produced by the random vibration of atoms or molecules. Heat is a measure of how many molecules are vibrating and how rapidly they are vibrating. Temperature records only how rapidly the molecules of a substance are vibrating. Temperature is an object's response to an input (or removal) of heat.
Review Figure 6.6, page 142

✓ Without water's unique thermal properties, temperatures on Earth's surface would change dramatically with only minor changes in atmospheric transparency or solar output. Water acts as a "global thermostat."
Review Figures 6.6 and 6.8 and Table 6.2, pages 142, 144

✓ Water density is greatly influenced by changes in temperature and salinity. Water masses are usually layered by density, with the densest (coldest and saltiest) water on or near the ocean floor. Differences in the density of water masses power deep-ocean circulation.
Review Figures 6.14 and 6.15, page 149

✓ Light and sound are affected by the physical properties of water, with refraction and absorption effects playing important roles.
Review Figures 6.18, 6.22, and 6.23, pages 152, 155–157

Terms and Concepts to Remember

absorption, 153
active sonar, 156
adhesion, 138
aphotic zone, 153
atom, 138
bond, 138
calorie, 139
chemical bond, 138
cohesion, 138
compound, 138
covalent bond, 138
deep zone, 148
degree, 139
density, 140
density curve, 140
electron, 138
element, 138
freezing point, 141
halocline, 149
heat, 139
heat capacity, 139
hydrogen bond, 138
latent heat of fusion, 142
latent heat
  of vaporization, 142
light, 152

mixed layer, 148
molecule, 138
nucleus, 138
passive sonar, 159
photic zone, 153
polar molecule, 138
proton, 138
pycnocline, 148
refraction, 151
refractive index, 152
scattering, 152
sensible heat, 142
shadow zone, 156
side-scan sonar, 158
sofar, 156
sofar layer, 156
sonar, 156
sound, 153
state, 140
surface zone, 148
temperature, 139
thermal inertia, 146
thermocline, 148
thermostatic property, 146
water mass, 150

Study Questions

Reviewing What You've Learned

1. Why is water a polar molecule? What properties of water derive from its polar nature?
2. How is heat different from temperature?
3. What is the difference between sensible and nonsensible heat? What is meant by *latent* heat?
4. Why does ice float? Why is this fact important to thermal conditions on Earth?
5. How is heat transported from tropical regions to polar regions?
6. What factors affect the density of water? Why does cold air or water tend to sink? What role does salinity play?
7. How is the ocean stratified by density? What physical factors are involved? What names are given to the ocean's density zones?
8. What factors influence the intensity and color of light in the sea? What factors affect the depth of the photic zone? Could there be a photocline in the ocean?
9. Which moves faster through the ocean: light or sound?
10. How is the pycnocline related to the thermocline and halocline?

Thinking Critically

1. What is the latent heat of vaporization? Of fusion? Which requires more heat to transform water's physical state? Why?
2. How does water's high latent heat influence the ocean? Leaving aside its effect on beach parties, how do you think conditions on Earth would differ if our ocean consisted of ethyl alcohol?
3. How do the seasonal freezing and thawing of the polar ocean areas influence global temperatures?
4. How does refraction permit sound to be transmitted in the ocean for thousands of miles?
5. What factors influence the density of seawater?
6. **InfoTrac College Edition Project** Sound travels in water both faster and farther than it does in air. Light, by contrast, does not travel as far, and shorter wavelengths, such as red, do not penetrate water more than a few meters. What do these differences suggest for the use of sound rather than light for underwater life and exploration? Use InfoTrac College Edition to back up your answers.

Thinking Analytically

1. How many Calories (note capital C) are required to raise the temperature of a can of diet cola (12 fluid ounces) from refrigerator temperature to body temperature? What percentage of the energy (Calories) in the can of diet cola is required to do this? Assume, for the moment, that the diet cola has the same thermal properties as pure water. (Hint: See footnote 1 on page 139.)
2. How much heat is required to melt a quart of ice resting at 0°C?
3. How much heat is required to vaporize a quart of pure water that has just come to a boil?

Oceanography Now ™

Assess your understanding of this chapter's topics with additional quizzing and conceptual-based problems at http://oceanography.brookscole.com/garrison5e

Study and Research on the Internet

Assess your understanding of this chapter's topics with additional quizzing and conceptual-based problems at:

 http://oceanography.brookscole.com/garrison5e

■ **InfoTrac College Edition Research** leads you to critical thinking exercises that use InfoTrac College Edition® as a research tool.
■ **Regional InfoTrac College Edition** articles are organized into East Coast, West Coast, and Gulf Coast regions, allowing you to study oceanography on a more local level.

For more readings, go to InfoTrac College Edition, your online research library, at:

 http://infotrac.thomsonlearning.com

CHAPTER 7

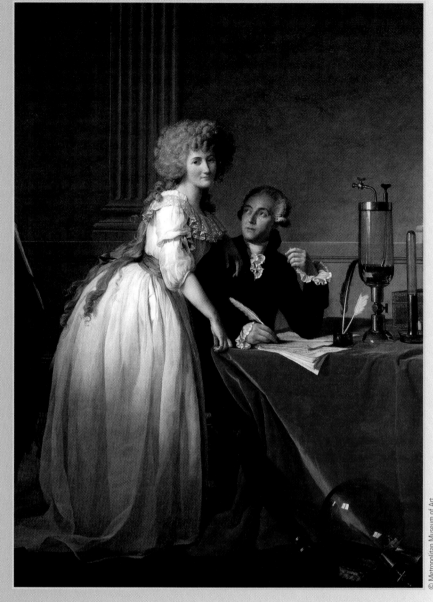

Monsieur and Madame Antoine-Laurent Lavoisier, in a painting by Jacques-Louis David, 1788. M. Lavoisier is credited with the "discovery" of water in 1783. Though a pioneering chemist and one of the founders of the metric system of measurements, Lavoisier could not escape the terrors of the French Revolution. He was sentenced to death by a revolutionary tribunal and guillotined on 8 May 1794.

© Metropolitan Museum of Art

SEAWATER CHEMISTRY

Cooking Up the Recipe for Water

In the fifth century B.C. the Greek philosopher Empedocles suggested that all matter was composed of four primary substances in various combinations: earth, air, fire, and water. Divide anything into small enough bits, he wrote, and you would end up with one or more of these four elements, none of which could be further divided. A century later this concept was endorsed by Aristotle, a philosopher considered by many to be the father of modern science. Aristotle's stature lent credence to the idea.

By the 1700s the four-element theory was being seriously challenged. Chemists were making many discoveries that could not adequately be explained using Empedocles's theory. A test that would unequivocally disprove the four-element theory would be to separate one of the four elements into smaller components, and water became the target. On 24 June 1783, the French chemist Antoine-Laurent Lavoisier discovered that water could be subdivided, a landmark event in the history of chemistry.

Lavoisier burned two gases in a special vessel. One gas was produced by dripping acid onto iron or copper. This clear, odorless gas was called "inflammable air" because it burned when ignited. The other gas—also colorless and odorless—was produced by heating a red precipitate of mercury. This second gas was known as the "acid maker" because of its tendency to combine with other substances to make mild acids. During the combustion of these gases, droplets of liquid condensed on the cool interior of an attached flask. Tests showed this liquid to be "as pure as distilled water." Lavoisier argued that in chemistry, as in mathematics, the whole is equal to the sum of its parts. "Since only water and nothing else was formed . . . the water was the sum of the [masses] of the two gases from which it was produced."

In short, water was not the irreducible element it had always been thought to be. It was a combination—a compound—of the acid maker (*oxys* = acid, *gen* = making) and the newly named "begetter of water" (*hydro* = water, *gen* = making). Oxygen and hydrogen were found to be simpler substances than water. Elements were clearly more numerous than Aristotle had thought, and water more complex.

We know today that water *is,* in a way, more than the sum of its parts. The parts that Lavoisier described are simple enough: hydrogen, the most abundant atom in the universe, and oxygen, an element that accounts for nearly half the mass of Earth's crust. But as you read in Chapter 6, their combination—water itself—behaves in complex and remarkable ways. Here you will see how easily water can dissolve substances—a legacy of its polar construction. Some of these substances we informally call "salts," but salts don't actually exist in seawater. Ions do—watch for them. Gases also dissolve in seawater. Marine organisms "breathe" dissolved oxygen, and the ocean is an important reservoir of carbon dioxide, a gas critical to photosynthesis and climate control.

When you've finished reading, consider Chapters 6 and 7 together. Your appreciation of the uniqueness of water will grow as your oceanographic perspective expands.

The Dissolving Power of Water

In Chapter 2 you saw that most of Earth's surface waters are thought to have escaped from the crust and mantle through the process of outgassing. Outgassing of other substances, and water's ability to dissolve crustal material as it cycles from the ocean to the atmosphere and the land and back again (**Figure 7.1**), have added solids and gases to the ocean. Water can dissolve more substances than almost any other liquid. Why?

As you have learned, water is a polar molecule. It can be thought of as having a positive (+) end and a negative (−) end (look again at Figures 6.1 and 6.2). In the polar water molecule, opposites attract: Its positive end attracts particles having a negative charge, and its negative end attracts particles having a positive charge. When water comes into contact with

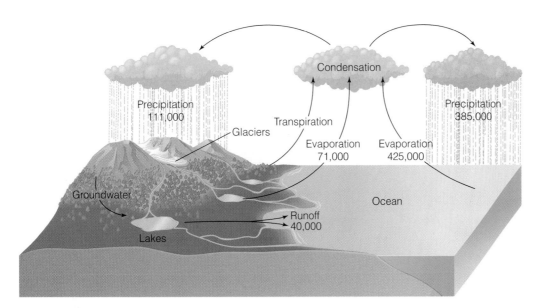

FIGURE 7.1 A simplified hydrologic cycle. Water moves from ocean to air, onto land, to lakes and streams and groundwater, back to the sky and ocean, in a continuous cycle. The numbers indicate the volumes of water in cubic kilometers per year (km³/yr). Water is also stored in the ocean, ice, groundwater, lakes and streams, and the atmosphere.

compounds whose elements are held together by the attraction of opposite electrical charges (most salts, for example), the polar water molecule will separate that compound's component elements from each other.

No wonder, then, that seawater and most other liquids in nature are water solutions. A **solution** is made of two components: The **solvent**, usually a liquid, is always the more abundant constituent; the **solute**, often a dissolved solid or gas, is the less abundant. In a true solution (sugar in well-stirred coffee, for example), the molecules of the solute are homogeneously dispersed among the molecules of solvent—that is, the solution has uniform properties throughout. In a **mixture**, different substances are closely intermingled but retain separate identities. The properties of a mixture are heterogeneous; they may vary from place to place within the mixture. Think of noodle soup as a mixture of noodles and liquid.

Water's dissolving power results from the polar nature of the water molecule. Consider how water dissolves sodium chloride (or NaCl), the most common salt, into its constituents, sodium ions (Na^+) and chloride ions (Cl^-).[1] An **ion** is an atom (or a small group of atoms) that has an unbalanced electrical charge because it has gained or lost one or more electrons. Unlike the electron *sharing* found in covalently bonded molecules such as water, the sodium atoms in NaCl have *lost* electrons and the chlorine atoms have *gained* them. The resulting ions are linked in a salt by the mutual attraction of their opposite electrical charges. The ions of sodium and chloride in NaCl are said to be held together by **ionic bonds** (**Figure 7.2**), electrostatic attraction

---

[1] Na, the chemical symbol for sodium, is derived from *natrium*, the Latin name for sodium.

that exists between ions that have opposite charge. When NaCl dissolves in water, the polarity of water reduces the electrostatic attraction (ionic bonding) between the Na^+ and Cl^- ions (**Figure 7.3**). This causes the two ions to separate. They move away from the salt crystal, permitting water to attack the next layer of NaCl. *Note that NaCl does not exist as "salt" in seawater; its components are separated when salt crystals dissolve in water, but they are joined when crystals re-form as water evaporates.*

Some other salts form ions in which groups of atoms bond together and carry an electrical charge. The sulfate ion (SO_4^{2-}), for example, carries two extra electrons. The charge on the overall ion is therefore -2.

By contrast, oil doesn't dissolve in water even if the two are thoroughly shaken together. When oil is dispersed in water, it forms a mixture because molecules of oil are nonpolar in character. This means that oil has no positive or negative charges to attract the polar water molecule. In a way this is fortunate: Living tissues would readily dissolve in water if the oils within their membranes didn't blunt water's powerful attack.

As we'll examine in more detail in a moment, seawater is a complex solution of ions and nonionic solutes. These particles can move through still water by **diffusion**, the random movement of particles through a solution. If you were to drop a large crystal of NaCl into a container of pure water, for example, the water would quickly begin to dissolve the salt. At first the concentration of Na^+ and Cl^- ions would be greater next to the crystal than a short distance away from it, but eventually the crystal would dissolve completely and the ions would diffuse evenly through the entire water volume, forming a homogeneous solution. Atmospheric gases dissolve in the ocean surface and diffuse easily through water. Dissolved substances

FIGURE 7.2 Salt in the solid state. In the compound sodium chloride (NaCl), ionic bonds hold the sodium ions (Na^+) and chloride ions (Cl^-) together.

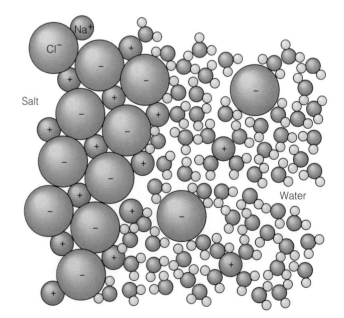

FIGURE 7.3 Salt in solution. When a salt such as NaCl is placed in water (left side of figure), the positively charged hydrogen end of the polar water molecule is attracted to the negatively charged Cl^-, and the negatively charged oxygen end is attracted to the positively charged Na^+. The ions are surrounded by water molecules that are attracted to them, and they become solute ions in the solvent (right side of figure).

tend to diffuse from regions of high concentration of those substances to regions of low concentration.

When no more of a substance will dissolve in water, the water is said to be *saturated* with that substance. At **saturation** the rate at which molecules of the solute are being dissolved equals the rate at which they are **precipitating** (re-forming into crystals) at another location in the solution.

Seawater

About 97.2% of the 1,370 million cubic kilometers (329 million cubic miles) of Earth's surface water is marine. By mass, seawater is about 96.5% water and 3.5% dissolved substances, most of which are salts of various kinds. The world ocean contains some 5,000 trillion kilograms (5.5 trillion tons) of salts. If the ocean's water evaporated completely, leaving its salts behind, the dried residue could cover the entire planet with an even layer 45 meters (150 feet) thick.

Salinity: Dissolved Solids and Water Together

The total quantity (or concentration) of dissolved inorganic solids in water is its **salinity**. The ocean's salinity varies from about 3.3% to 3.7% by mass, depending on such factors as evaporation, precipitation, and freshwater runoff from the continents, but the average salinity is usually given as 3.5%. Most of the dissolved solids in seawater are salts that have been separated into ions. Sodium and chloride are the most abundant of these.

The many ions present in seawater react with each other (and with water molecules) in complex ways to modify the physical properties of pure water. Consider the following:

■ The heat capacity of water decreases with increasing salinity—that is, less heat is necessary to raise the tempera-

ture of seawater by 1° than is required to raise the temperature of fresh water by the same amount.

■ Dissolved salts disrupt the webwork of hydrogen bonding in water. As salinity increases, the freezing point of water becomes lower; the salts act as a sort of antifreeze. Sea ice therefore forms at a lower temperature than ice in freshwater lakes.

■ Because dissolved salts tend to attract water molecules, seawater evaporates more slowly than fresh water does. Swimmers usually notice that fresh water evaporates quickly and completely from their skin, but seawater lingers.

■ Osmotic pressure, the pressure exerted on a biological membrane when the salinity of the environment is different from that within the cells, rises with increasing salinity. This is a key factor in transmitting water into and out of cells.

These four properties, which vary with the quantity of solutes dissolved in the water, are called water's **colligative properties** (*colligatus* = to bind together). Because colligative properties are the properties of *solutions*, the more concentrated (saline) water is, the more important these properties become. (Because it is not a solution, pure water has no colligative properties.)

The Components of Salinity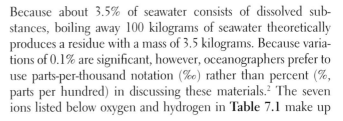

Because about 3.5% of seawater consists of dissolved substances, boiling away 100 kilograms of seawater theoretically produces a residue with a mass of 3.5 kilograms. Because variations of 0.1% are significant, however, oceanographers prefer to use parts-per-thousand notation (‰) rather than percent (%, parts per hundred) in discussing these materials.[2] The seven ions listed below oxygen and hydrogen in **Table 7.1** make up

---

[2] Note that 3.5% = 35‰. If you begin with 1,000 kilograms of seawater, you would expect 35 kilograms of residue.

Table 7.1	Major Constituents of Seawater at 34.4‰ Salinity		
Constituent	**Concentration in Parts per Thousand (‰), or Grams per Kilogram (g/kg)**	**Percent by Mass**	**Total Percent**
Water Itself			
Oxygen	857.8	85.8%	96.5%
Hydrogen	107.2	10.7%	
The Most Abundant Ions			
Chloride (Cl^-)	18.980	1.9%	
Sodium (Na^+)	10.556	1.1%	
Sulfate (SO_4^{2-})	2.649	0.3%	
Magnesium (Mg^{2+})	1.272	0.1%	3.4%
Calcium (Ca^{2+})	0.400	0.04%	
Potassium (K^+)	0.380	0.04%	
Bicarbonate (HCO_3^-)	0.140	0.01%	
Total	**999.377 g/kg**	**99.99%**	

Source: Adapted from Walton-Smith, 1994.

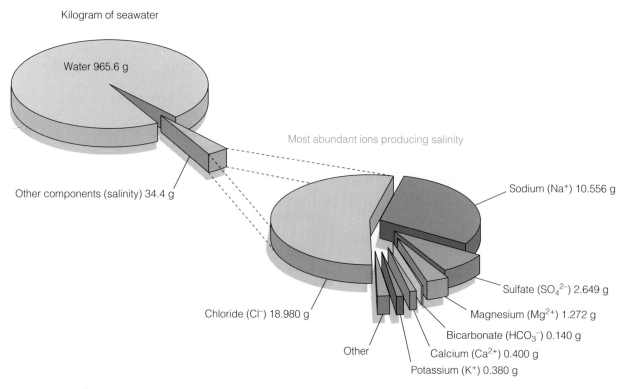

Kilogram of seawater

Water 965.6 g

Other components (salinity) 34.4 g

Most abundant ions producing salinity

Sodium (Na$^+$) 10.556 g

Sulfate (SO$_4{}^{2-}$) 2.649 g

Magnesium (Mg^{2+}) 1.272 g

Bicarbonate (HCO$_3{}^-$) 0.140 g

Calcium (Ca^{2+}) 0.400 g

Potassium (K$^+$) 0.380 g

Other

Chloride (Cl$^-$) 18.980 g

FIGURE 7.4 A diagrammatic representation of the most abundant components of a kilogram of seawater at 35‰ salinity. Note that the specific ions are represented in grams per kilogram, equivalent to parts-per-thousand (‰).

more than 99% of this residual material. When seawater evaporates, its ionic components combine in many different ways to form table salt (NaCl), epsom salts (MgSO$_4$), and other mineral salts. **Figure 7.4** shows the proportions of ions in seawater.

Seawater also contains minor constituents. The ocean is sort of an "Earth tea": Nearly every element present in the crust and atmosphere is also present in the oceans, though sometimes in extremely small amounts. Only 14 elements have concentrations in seawater greater than one part per million (ppm). Elements present in amounts less than 0.001‰ (1 ppm) are known as **trace elements.** These are more easily given in parts per billion (ppb). **Table 7.2** lists some of these minor and trace elements, many of which are crucial to life processes.[3]

The Source of the Ocean's Salts

Remembering the effectiveness of water as a solvent, you might think that the ocean's saltiness has resulted from the ability of rain, groundwater, or crashing surf to dissolve crustal rock. Much of the sea's dissolved material originated in that way, but is crustal rock the source of all the ocean's solutes? An easy way to find out would be to investigate the composition of salts in river water and compare these figures to those of the ocean as a whole. If crustal rock is the only source, then the salts in the ocean should be like those of concentrated river water. But they are not. River water is usually a dilute solution of calcium and

bicarbonate ions, while the principal ions in seawater are chloride and sodium. The magnesium content of seawater would also be higher if seawater were simply concentrated river water. The proportions of salts in isolated salty inland lakes, such as Utah's Great Salt Lake or the Dead Sea, are much different from the proportions of salts in the ocean. So, weathering and erosion of crustal rocks cannot be the only source of sea salts.

The components of ocean water whose proportions are *not* accounted for by the weathering of surface rocks are called **excess volatiles.** To find the source of these excess volatiles, we must look to Earth's deeper layers. The upper mantle appears to contain more of the substances found in seawater (including the water itself) than are found in surface rocks, and their proportions are about the same as found in the ocean. As you read in Chapter 3, convection currents slowly churn Earth's mantle, causing the movement of tectonic plates. Because of this activity, some deeply trapped volatile substances escape to the exterior, outgassing through volcanoes and rift vents. These excess volatiles include carbon dioxide, chlorine, sulfur, hydrogen, fluorine, nitrogen, and, of course, water vapor. This material, along with residue from surface weathering, accounts for the chemical constituents of today's ocean.

Some of the ocean's solutes are hybrids of the two processes of weathering and outgassing (**Figure 7.5**). Table salt, or sodium chloride, is an example. The sodium ions come from the weathering of crustal rocks, while the chloride ions come from the mantle by way of volcanic vents and outgassing from mid-ocean rifts. As for the lower-than-expected quantity of magnesium and sulfate ions in the ocean, recent research at a spreading center

---

[3] The periodic table of the elements in Appendix VIII provides more information on individual elements.

Table 7.2	Minor and Trace Elements in Seawater at 35‰ Salinity				
Constituent	Concentration in Parts per Thousand (‰), or g/kg		Concentration in Parts per Million (ppm), or mg/kg (mg/L)		Concentration in Parts per Billion (ppb), or mg/1,000 kg
Minor Elements					
Bromine (Br)	0.065	=	65		
Strontium (Sr)			8		
Boron (B)			4		
Silicon (Si)			3		
Fluorine (F)			1		
Important Trace Elements					
Nitrogen (N)[a]			0.28	=	280
Lithium (Li)					125
Iodine (I)					60
Phosphorus (P)					30
Zinc (Zn)					10
Iron (Fe)					6
Aluminum (Al)					2
Manganese (Mn)					2
Lead (Pb)					0.04
Mercury (Hg)					0.03
Gold (Au)					0.000004

Source: Adapted from Kennish, 1994.

[a] Refers to nitrogen available as a nutrient, not as dissolved gas.

east of the Galápagos Islands suggests that mid-ocean rifts may play a role in reducing the magnesium and sulfate content and increasing the calcium and potassium content of seawater. The water that circulates through new ocean floor at these sites is apparently stripped of magnesium and a few other elements. The magnesium seems to be incorporated into mineral deposits, but calcium is added as hot water dissolves adjacent rocks. All the water in the ocean is thought to cycle through the seabed at rift zones every 1 to 10 million years.

The Principle of Constant Proportions

In 1865 the chemist Georg Forchhammer noted that although the total *amount* of dissolved solids (salinity) might vary among samples, the *ratio* of major salts was constant in samples of seawater from many locations. In other words, the percentage of various salts in seawater is the same in samples from many places, regardless of how salty the water is. For example, when the solids are isolated from any seawater sample, whether from the high-salinity North Atlantic or the low-salinity Arctic oceans, 55.04% of those solids will be chloride ions. This constant ratio is known as **Forchhammer's principle,** or the **principle of constant proportions.** Forchhammer was also the first to observe that seawater contains fewer silica and calcium ions than concentrated river water, and the first to realize that removal of these compounds by marine animals and plants, to form shells and other hard parts, might account for part of the difference. The English chemist William Dittmar, working with HMS *Challenger* samples ten years after they had been collected, con-

firmed Forchhammer's principle of constant proportions. Building on Forchhammer's and Dittmar's work and taking advantage of improved analytical techniques, researchers have established a reliable way to determine salinity.

Determining Salinity

Water's salinity by weight would seem an easy property to measure. Why not simply evaporate a known weight of seawater and weigh the residue? This simple method yields imprecise results because some salts will not release all the molecules of water associated with them. If these salts are heated to drive off the water, then other salts (carbonates, for example) will decompose to form gases and solid compounds not originally present in the water sample.

Because of the principle of constant proportions we need to measure the concentration of only one major constituent to find the salinity of a water sample. The chloride ion is abundant and comparatively easy to measure, and it always accounts for the same proportion of dissolved solids (55.04%), so marine chemists devised the concept of chlorinity to simplify the measurement of salinity. **Chlorinity** is a measure of the total mass of halogen ions (the halogens are a chemical family that includes fluorine, chlorine, bromine, and iodine) in seawater. By international agreement, the following formula is used to determine salinity:

$$\text{Salinity in ‰} = 1.80655 \times \text{Chlorinity in ‰}$$

Average chlorinity is about 19.2‰, so average salinity is around 34.7‰.

Sulfur

SO₂ (Sulfur dioxide)
CO₂ (Carbon dioxide)
H₂S (Hydrogen sulfide)
HCl (Hydrochloric acid)

Cl⁻ (chloride)
SO₄²⁻ (sulfate)

Volcano

Sea spray (ions)

Adsorption
Biological and
processes precipitation

Ca²⁺ (calcium)
K⁺ (potassium)

Dissolution

River discharge (ions)

HCO₃⁻ (bicarbonate)
Ca²⁺ (calcium)
SO₄²⁻ (sulfate)
Na⁺ (sodium)

Uplift

Mg²⁺ (magnesium)
SO₄²⁻ (sulfate)

Chemical reactions
Bottom sediments

Ridge system

a

Woods Hole Oceanographic Institution

b

FIGURE 7.5 Processes that regulate the major constituents in seawater. (**a**) Ions are *added* to seawater by rivers running off crustal rocks, volcanic activity, groundwater, hydrothermal vents and cold springs, and the decay of once-living organisms. Ions are *removed* from the ocean by chemical entrapment as water percolates through the mid-ocean ridge systems, sea spray, uptake by living organisms, incorporation into sediments, and ultimately by subduction. (**b**) An active hydrothermal vent spouts superheated seawater. Recent research at a spreading center east of the Galápagos Islands suggests that mid-ocean rifts strip the water that circulates through new ocean floor of magnesium and a few other elements, while calcium is added as hot water dissolves adjacent rocks.

Seawater samples can be obtained by methods ranging from tossing a clean bucket over the side of the ship to sophisticated tube-and-pump systems. Typically water samples are collected using a group of sampling bottles (**Figure 7.6**). The bottles are lowered from a ship and triggered to close at specific depths by an electronic signal. The bottles are then hauled to the surface and their contents analyzed.

Until recently, marine chemists used a sensitive chemical procedure involving a silver nitrate solution to measure the chlorinity of seawater. Conversion to salinity was made using a set of mathematical tables. The procedure was calibrated against a standard sample of seawater of precisely known chlorinity. Today's marine scientists usually use an electronic device called a **salinometer** that measures the electrical conductivity of seawater (**Figure 7.7**). Conductivity varies with the concentration and mobility of ions present and with the water temperature. Circuits in the salinometer adjust for water temperature, convert conductivity to salinity, and then display salinity. Salinometers are calibrated against a reference sample of known conductivity and salinity. The best salinometers can determine salinity to an accuracy of 0.001%. Some salinometers are designed for remote sensing—the electronics stay aboard ship while the sensor coil is lowered over the side.[4]

---

[4] For a look at ocean surface salinity, review Figure 6.17.

Chemical Equilibrium and Residence Times

7.9

If outgassing and the chemical weathering of rock are continuing processes, shouldn't the ocean become progressively saltier with age? Landlocked seas and some lakes usually become saltier as they grow older, but the ocean does not. The ocean appears to be in **chemical equilibrium**—that is, the proportion *and amounts* of dissolved salts per unit volume of ocean are nearly constant. Evidently whatever goes in must come out somewhere else.

Geologists in the 1950s developed the concept of a *steady-state ocean*. The idea suggests that ions are added to the ocean at the same rate as they are being removed. This theory helps explain why the ocean is not growing saltier. The idea was quantified by T.F.W. Barth, who in 1952 devised the concept of **residence time,** the average length of time an atom of an element spends in the ocean. Residence time for a particular element may be calculated by this equation:

$$\text{Residence time} = \frac{\text{Amount of element in the ocean}}{\begin{array}{c}\text{Rate at which the element is added}\\ \text{to (or removed from) the ocean}\end{array}}$$

FIGURE 7.6 A "rosette" of sampling bottles. Each of these 10-liter bottles may be separately sealed by a signal from the research ship when the array reaches predetermined depths. The bottles are hauled to the surface and their contents analyzed.

Table 7.3 Approximate Residence Times for Constituents of Seawater

Constituent	Residence Time (years)
Chloride (Cl^-)	100,000,000
Sodium (Na^+)	68,000,000
Magnesium (Mg^{2+})	13,000,000
Potassium (K^+)	12,000,000
Sulfate (SO_4^{2-})	11,000,000
Calcium (Ca^{2+})	1,000,000
Carbonate (CO_3^{2-})	110,000
Silicon (Si)	20,000
Water (H_2O)	4,100
Manganese (Mn)	1,300
Aluminum (Al)	600
Iron (Fe)	200

Sources: Data from Broecker and Peng, 1982; Bruland, 1983; Riley and Skirrow, 1975.

Additions of salts from the mantle or from the weathering of rock are balanced by subtractions of minerals being bound into sediments. Dissolved salts precipitate out of the water, and the hard parts of living organisms containing silicon and calcium carbonate drift slowly down to the seabed. Some of these sediments are removed from the ocean and drawn into the mantle at subduction zones by the cycling of crustal plates. Input (from runoff and outgassing) equals outfall (binding into sediments) for each dissolved component.

The residence time of an element depends on its chemical activity. Atoms (or ions) of some elements, such as aluminum and iron, remain in seawater for a relatively short time before becoming incorporated in sediments; others, such as chloride, sodium, and magnesium, remain in water for millions of years. The approximate residence times for the major constituents of seawater are shown in **Table 7.3**. Even seawater itself has a residence time (**Box 7.1**).

Mixing Time

If constituent minerals remain in ocean water longer than the ocean's mixing time, they will become evenly distributed throughout the ocean. Because of the vigorous activity of currents, the **mixing time** of the ocean is thought to be on the order of 1,600 years, so the ocean has been mixed hundreds of thousands of times during its long history. The relatively long residence times of seawater's major constituents assure thorough mixing, the foundation of Forchhammer's principle of constant proportions.

Conservative and Nonconservative Constituents

As we have seen, the major constituent salts maintain constant ratios in seawater, but the quantity and proportions of minor and

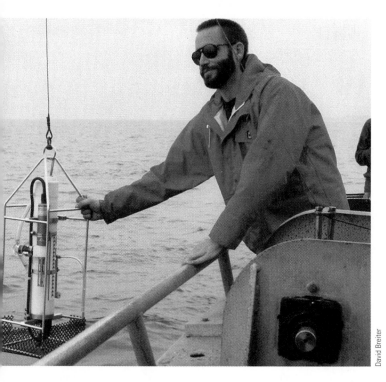

FIGURE 7.7 This portable salinometer reads temperature, pH, and dissolved oxygen as well as conductivity. Designed to be lowered over the side of a research vessel at the end of a line, the self-powered device contains a small pump that passes water over sensors. This model takes two readings every second and operates down to about 200 meters (660 feet). Data may be transmitted over a cable to the research vessel or stored in the salinometer's memory to be retrieved when it is brought back aboard ship. A microprocessor inside the white tube converts conductivity to salinity and, when connected to a portable computer, displays the results of all readings as graphs.

Ocean water has a residence time. Scientists estimate that about 334,000 cubic kilometers (80,000 cubic miles) of pure water evaporates from the ocean each year. Assuming a total ocean volume of 1,370 million cubic kilometers (329 million cubic miles), we can calculate that about 0.024% of the water in the ocean vaporizes each year. Thus, if it were not replenished by precipitation and runoff from land, the ocean would evaporate completely in about 4,100 years.

For a complete picture, more than oceanic water should be considered. Earth's total surface water volume—including the ocean, lakes, rivers, groundwater, ice caps, and so on—is about 1,399 million cubic kilometers (336 million cubic miles). The total yearly evaporation/precipitation estimate is 396,000 cubic kilometers (95,000 cubic miles) of pure water. From these figures we again calculate a recycling time of about 4,100 years. With the exception of water subducted into the mantle along with sediments at subduction zones, then, water molecules have a residence time in the ocean of about 4,100 years.

Earth's ocean is more than 4 billion years old. Thus, on average, individual water molecules have evaporated from and returned to the ocean more than a million times since the world ocean formed in its basins!

© Rommel/Masterfile

trace elements can change because living things take them up or excrete them or local geological events add or remove them.

Those constituents of seawater that occur in constant proportion or change very slowly through time are **conservative constituents.** Conservative elements have long residence times. Not surprisingly, these are the most abundant dissolved constituents of water—the ones that constitute the bulk of the ocean's dissolved material, seen at the top of Table 7.3. The inert gases dissolved in the ocean (and the water of the ocean itself) are also conservative constituents.

Nonconservative constituents are those substances dissolved in seawater that are tied to biological or seasonal cycles or to very short geological cycles. They have short residence times. Biologically important nonconservative constituents include dissolved oxygen produced by plants, carbon dioxide produced by animals, silica and calcium compounds needed for plant and animal shells, and the nitrates and phosphates needed for production of protein and other biochemicals. Aluminum, with a residence time of only 600 years, is rapidly removed by adsorption onto clay sediment particles and other objects, so it is a nonconservative element. Aluminum is very rare in seawater (10 parts per billion). Many trace elements are in great biological demand or tend to form insoluble compounds in water.

Ocean residence times calculated to be less than 1,600 years—the ocean's mixing time—are mainly a function of strong local demand variations. The definition of residence time assumes the existence of steady-state conditions and a well-mixed ocean, so a residence time less than the ocean's mixing time is mainly an indication of an element's reactivity.

Dissolved Gases

Most gases in the air dissolve readily in seawater at the ocean's surface. Plants and animals living in the ocean require these dissolved gases to survive. No marine animal has the ability to break down water molecules to obtain oxygen directly, and no marine plant can manufacture enough carbon dioxide to support its own metabolism. In order of their relative abundance, the major gases found in seawater are nitrogen, oxygen, and carbon dioxide (**Table 7.4**). The proportions of dissolved gases in the ocean are very different from the proportions of the same gases in the atmosphere because of differences in their solubility in water and air.

Unlike solids, gases dissolve most readily in *cold* water. A cubic meter of chilly polar water usually contains a greater volume of dissolved gases than a cubic meter of warm tropical water. Dissolved nitrogen gas is a conservative component of seawater, while biologically active oxygen and carbon dioxide are nonconservative components.

Nitrogen

About 48% of the dissolved gas in seawater is nitrogen. (In contrast, the atmosphere is slightly more than 78% nitrogen by volume.) The upper layers of ocean water are usually saturated with nitrogen gas—that is, additional nitrogen gas will not dissolve. Living organisms require nitrogen to build proteins and other important biochemicals, but they cannot use the nitrogen gas in the atmosphere and ocean directly. It must first be *fixed* into usable chemical forms by specialized organisms. Though some species of bottom-dwelling bacteria can manufacture usable nitrates from the nitrogen gas dissolved in seawater, most of the nitrogen compounds needed by living organisms must be recycled among the organisms themselves.

Oxygen

About 36% of the gas dissolved in the ocean is oxygen, but there is about a hundred times more gaseous oxygen in Earth's atmosphere than is dissolved in the whole ocean. An average of 6 milligrams of oxygen is dissolved in each liter of seawater (that is, 6 parts per million parts of gaseous oxygen per liter of seawater, by mass). Yet this small amount of oxygen is a vital resource for animals that extract oxygen with gills. The sources of the ocean's dissolved oxygen are the photosynthetic activity of plants and plantlike organisms and the diffusion of oxygen from the atmosphere.

Carbon Dioxide

The amount of carbon dioxide (CO_2) in the atmosphere is very small (0.03%) because CO_2 is in great demand by photosynthetic plants as a source of carbon for growth. Carbon dioxide is very soluble in water, though; the proportion of dissolved CO_2 in water is about 15% of all dissolved gases. Because CO_2 combines chemically with water to form a weak acid (carbonic acid, H_2CO_3), water can hold perhaps a thousand times more carbon dioxide than either nitrogen or oxygen at saturation. Carbon dioxide is quickly used for marine photosynthesis, however; so dissolved quantities of CO_2 are almost always much less than this theoretical maximum. Even so, about 60 times as much CO_2 is dissolved in the ocean as in the atmosphere.

CO_2 moves quickly from atmosphere to ocean, but more slowly from ocean to atmosphere. This is because some dissolved CO_2 forms carbonate ions, which combine with calcium ions in seawater to form the calcium carbonate ($CaCO_3$) used by many marine organisms to build shells and skeletons. When these organisms die, their coverings and bones sink to form sediments that may, in time, become limestone rock. Most of Earth's surface carbon—10,000 times that in the total mass of all life on Earth—is stored in sediments. This carbon slowly reenters the cycle as sediments dissolve and re-form CO_2 that can enter the atmosphere. Geological uplift can also bring sediments to the surface, exposing the carbonate rock to chemical attack by oxygen and conversion to CO_2. Acidic rain falling on exposed limestone rock also releases some CO_2 back into the atmosphere. (This cycle can be seen in Figure 7.10.)

Figure 7.8 illustrates how carbon dioxide and oxygen concentrations vary with depth. Carbon dioxide concentrations increase with depth, but oxygen concentrations usually decrease through the middle depths and then rise again toward

Gas	Percent of Gas in Atmosphere, by Volume	Percent of Dissolved Gas in Seawater, by Volume	Concentration in Seawater in Parts per Million, by Mass
Nitrogen (N_2)	78.08%	48%	10–18 ppm
Oxygen (O_2)	20.95%	36%	0–13 ppm
Carbon dioxide (CO_2)[a]	0.035%	15%	64–107 ppm

Table 7.4 Major Gases in the Atmosphere and Ocean

Sources: Data from Weihaupt, 1979; Hill, 1963.

[a] Also present in seawater as carbonic acid, carbonate ions, and bicarbonate ions.

FIGURE 7.8 How concentrations of oxygen and carbon dioxide vary with depth. Oxygen is abundant near the surface because of the activity of marine photosynthesizers (plants and plantlike organisms). The oxygen concentration decreases below the sunlit layer because of the respiration of marine animals and bacteria, and because of the oxygen consumed by the decay of tiny dead organisms slowly sinking through the area. In contrast, plants and plantlike organisms use carbon dioxide during photosynthesis, so surface levels of CO_2 are low. Because photosynthesis cannot take place in the dark, CO_2 given off by animals and bacteria tends to build up at depths below the sunlit layer. Levels of CO_2 also increase with depth because its solubility increases as pressure increases and temperature decreases.

the bottom. High concentrations of oxygen at the surface are usually by-products of photosynthesis in the ocean's brightly lit upper layer. Because plants and plantlike organisms require carbon dioxide for photosynthesis, surface CO_2 concentrations tend to be low. A decrease in oxygen below the sunlit upper layer usually results from the respiration of bacteria and marine animals, activity that tips the balance in favor of carbon dioxide. Oxygen levels are slightly higher in deeper water because fewer animals are present to take up the oxygen that reaches these depths and because oxygen-rich polar water that sinks from the surface is the greatest source of deep water.

Acid–Base Balance   7.17

Water can separate to form hydrogen ions (H^+) and hydroxide ions (OH^-). In pure water these two ions are present in equal concentrations. An imbalance in the ions produces either an acidic or a basic solution. An **acid** is a substance that *releases* a hydrogen ion in solution. A **base** is a substance that *combines* *with* a hydrogen ion in solution. A solution containing a base is also called an **alkaline** solution.

The acidity or alkalinity of a solution is measured in terms of the **pH scale,** which measures the concentration of hydrogen ions in a solution. An excess of hydrogen ions (H^+) in a solution makes that solution acidic. An excess of hydroxide ions (OH^-) makes a solution alkaline. **Figure 7.9** shows a pH scale and the pH values of a few familiar solutions. The scale is logarithmic, which means that a change of one pH unit represents a ten-fold change in the hydrogen ion concentration. A modern nonphosphate detergent is a thousand times more alkaline than seawater, and black coffee is a hundred times more acidic than pure water. Pure water, which is neutral (neither acidic nor basic), has a pH of 7; lower numbers indicate greater acidity (more H^+), and higher numbers indicate greater alkalinity (fewer H^+).

Seawater is slightly alkaline; its average pH is about 8.0. This seems odd because of the large amount of CO_2 dissolved in the ocean. If dissolved CO_2 combines with water to form carbonic acid, why is the ocean mildly alkaline and not slightly acidic? When dissolved in water, CO_2 is actually present in several different forms. Carbonic acid (H_2CO_3) is only one of these. In water solutions some carbonic acid breaks down to

FIGURE 7.9 The pH scale.

produce the hydrogen ion (H^+), the bicarbonate ion (HCO_3^-), and the carbonate ion (CO_3^{2-}). It works like this:

Step One:	$CO_2 + H_2O \Leftrightarrow H_2CO_3$
Step Two:	$H_2CO_3 \Leftrightarrow HCO_3^- + H^+$
Step Three:	$HCO_3^- + H^+ \Leftrightarrow CO_3^{2-} + 2H^+$

(The double-headed arrows indicate that the reactions can proceed in either direction.)

In Step One carbon dioxide and water combine to form carbonic acid. Carbonic acid rapidly dissociates into a bicarbonate ion (HCO_3^-) and a hydrogen ion (H^+) in Step Two. Most of the CO_2 dissolved in seawater ends up as HCO_3^-. Some of the bicarbonate and hydrogen ions will then combine to form carbonate (CO_3^{2-}) ions and two hydrogen ions ($2H^+$) in Step Three. At any given pH the CO_2, H_2CO_3, HCO_3^-, CO_3^{2-}, and H^+ exist in equilibrium. But what if the pH changes?

If acid is added to seawater (at a sewage outfall or volcanic eruption) and the pH of the ocean drops (becomes more acidic), the reaction proceeds toward Step One (to the left), raising the pH by removing the excess H^+. If an alkaline solution is added and the pH of the ocean rises (becomes more alkaline), the reaction proceeds toward Step Three (to the right), lowering the pH by releasing H^+. This behavior acts to **buffer** the water, preventing broad swings of pH when acids or bases are introduced. Rivers and lakes lack buffering, making them more susceptible to pH swings when acids or bases are added at waste outfalls.

Figure 7.10 shows the progress of these reactions in the ocean. At the normal pH of seawater (slightly above 8.0), about 80% of the carbon compounds are in the form of HCO_3^-. Carbon dioxide is used in photosynthesis by plantlike organisms for the production of sugars. A decrease in dissolved CO_2 will cause bicarbonate to dissociate and form more CO_2. Carbonate ions combine with hydrogen ions to form more bicarbonate.

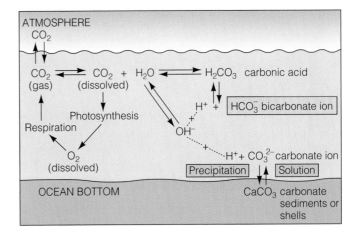

FIGURE 7.10 Carbon dioxide (CO_2) combines readily with seawater to form carbonic acid (H_2CO_3). Carbonic acid can then lose a H^+ to become a bicarbonate ion (HCO_3^-), or it can lose two H^+ to become a carbonate ion (CO_3^{2-}). Some bicarbonate ions dissociate to form carbonate ions, which combine with calcium ions in seawater to form calcium carbonate ($CaCO_3$), used by some organisms to form hard shells and skeletons. When their builders die, these structures may fall to the seabed as carbonate sediments, eventually to be redissolved. As the double-headed arrows indicate, all these reactions may move in either direction.

Calcium carbonate ($CaCO_3$) can then dissolve from the carbonate shells of marine organisms or from seabed sediments.

Though seawater remains slightly alkaline, it is subject to some variation. In areas of rapid photosynthesis, for example, pH will rise because CO_2 is used by plants and plantlike organisms. The reactions move toward Step One, removing free H^+. Because temperatures are generally warmer at the surface, less CO_2 can dissolve in the first place. So, the surface pH in warm productive water is usually around 8.5. **Figure 7.11** shows the relative abundance of carbonic acid, bicarbonate ions, and

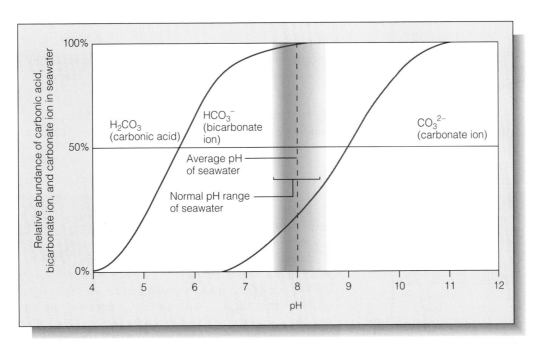

FIGURE 7.11 The relative abundance of carbonic acid, bicarbonate ions, and carbonate ions in seawater is a function of pH. Seawater has an average pH of about 8, at which bicarbonate ions are most prevalent. As the graph shows, carbonic acid dominates at low pH and carbonate ions at higher pH.

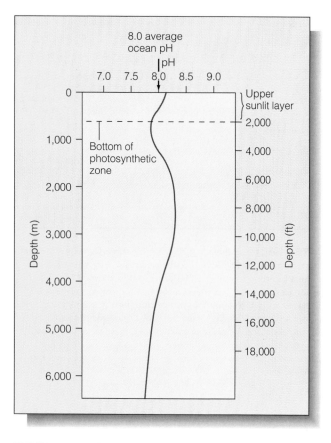

FIGURE 7.12 Variation in pH with depth.

carbonate ions in seawater as a function of pH. Seawater has an average pH of about 8, and bicarbonate ions are most prevalent. As the graph shows, carbonic acid dominates at low pH and carbonate ions at higher pH.

At middle depths and in deep water, more CO_2 may be present than at the surface. Its source is the respiration of animals and bacteria and the decay of the remains of tiny organisms falling from the sunlit layer above. With cold temperatures, high pressure, and no photosynthetic plants to remove it, this CO_2 will lower the pH of water, making it more acid with depth (**Figure 7.12**). Deep, cold seawater below 4,500 meters (15,000 feet) has a pH of around 7.5. This lower pH can dissolve calcium-containing marine sediments. A drop to pH 7 can occur at the deep-ocean floor when bottom bacteria consume oxygen and produce hydrogen sulfide. You may recall from Chapter 5 that sediments containing calcium carbonate are rarely found in deep water. Now does Figure 5.14—the calcium carbonate compensation depth (CCD)—make more sense?

Questions from Students

1. If outgassing continues today, why aren't the oceans getting bigger? By now shouldn't they have covered the continents?

Well, yes, outgassing does continue, but the water is being drawn from the mantle. Some marine geologists believe that the mantle shrinks a little, making the ocean basins a bit deeper, which accommodates the ocean's greater volume. Also, geological processes have increased the amount of continental crust, and water is drawn back into the mantle at subduction zones. Though some of this water eventually reappears at the surface in volcanic vapor, some is incorporated back into mantle material. It probably just about balances out.

2. I see that there is a hundred times more oxygen in the atmosphere than in the ocean. How can that be? Isn't water 86% oxygen by weight?

Yes, but the oxygen of water (H_2O) is bonded tightly to hydrogen atoms. Unlike atmospheric oxygen, the oxygen in water molecules cannot be used by organisms for respiration. Fish can't extract this oxygen with their gills. Marine animals must depend on dissolved oxygen, paired molecules of oxygen (O_2) present in the water and free to move through the gill membranes. Compared to the atmosphere, very little of this free oxygen is present in the ocean.

3. There is more than $4.1 million worth of gold (at 2003 prices) in each cubic kilometer of seawater. Why don't we remove some of it and pay off the national debt?

Easy to say, difficult to do. For one thing, extracting the gold would require sophisticated chemical treatment. But for the sake of argument, let's assume you perfect a magic method of precipitating gold from seawater with a simple wave of your hand. All you have to do is pump water out of the ocean into a large holding tank to do your trick. Unfortunately the energy cost of lifting the water just a few millimeters into the tank would eat up all the profits, and then some. German economists made a study after World War I to see if this method could be used to retire their war debt. It was shown to be fruitless. So much for get-rich-quick schemes.

4. Some essential resources are relatively abundant in ocean water. Are there any prospects for chemical "mining" of seawater?

Yes. Commercial extraction of magnesium from seawater began in 1940. Seawater is treated with calcium hydroxide and hydrochloric acid to form magnesium chloride. This substance is dried and electrolytically separated into chlorine gas (which can be used to make more hydrochloric acid) and magnesium metal. Magnesium is an essential component of the aluminum alloys from which aircraft, beverage cans, and automobile parts are manufactured.

Many nonmetal resources are also obtained from seawater. For example, bromine—an element important in the production of motor fuels, photographic film, dyes, and insecticides—is extracted from concentrated brines or from crystallized sea salts. And, as we'll see in Chapter 17, the salts themselves are among the most important of the ocean's physical resources.

Chapter in Perspective

In this chapter you learned that water has the remarkable ability to dissolve more substances than any other natural solvent. Though most solids and gases are soluble in water, the ocean is in chemical equilibrium and neither the proportion nor the amount of most dissolved substances changes significantly through time. Most of the properties of seawater are different from those of pure water because of the substances dissolved in the seawater.

About 3.5% (35‰) of seawater consists of dissolved substances. These almost always exist as ions—"salts" don't exist in the ocean. The most abundant ions dissolved in seawater are chloride, sodium, and sulfate. Seawater is not concentrated river water or rainwater; its chemical composition has been altered by circulation through the crust at oceanic spreading centers and by other chemical and biological processes.

Most gases in the air dissolve readily in seawater at the ocean's surface. Plants and animals living in the ocean require these dissolved gases to survive. In order of their relative abundance, the major gases in seawater are nitrogen, oxygen, and carbon dioxide. The proportions of dissolved gases in the ocean are very different from the proportions of the same gases in the atmosphere because of differences in their solubility in water and air.

In the next chapter you will learn how atmosphere and ocean work together to shape Earth's weather, climate, and natural history. Our past discussions of heat, temperature, convection, dissolved gases, and geological cycles will be put to use. You'll find surprising oceanic connections among storms, tropical organisms, deserts, and balloon races!

Key Ideas

✓ The polar nature of the water molecule produces some unexpected chemical properties. One of the most important is water's remarkable ability to dissolve more substances than any other natural solvent.
Review Figures 6.2, 7.2, and 7.3; Tables 7.1 and 7.2; pages 139, 164, 165, 167

✓ The quantity of dissolved inorganic solids in water is its salinity. The proportion of ions in seawater is not the same as the proportion in concentrated river water, which indicates that ongoing geological and chemical processes affect the ocean's salinity.
Review Figures 7.4, and 7.5, pages 166–167

✓ The most abundant ions dissolved in seawater are chloride, sodium, and sulfate.
Review Figure 7.4, page 166

✓ Though most solids and gases are soluble in water, the ocean is in chemical equilibrium, and neither the proportion nor the amount of most dissolved substances changes significantly through time.
Review page 168

✓ Gases dissolve in water in proportions that vary with their physical properties. Nitrogen is the most abundant dissolved gas in seawater; oxygen is the second most abundant. Carbon dioxide is the most soluble gas and is one of many substances that affect the ocean's pH balance.
Review Table 7.4 and Figure 7.8, pages 171, 172

✓ Seawater acts as a buffer to prevent broad swings of pH when acids or bases are introduced.
Review Figure 7.9, page 172

Terms and Concepts to Remember

acid, 172	mixture, 164
alkaline, 172	nonconservative
base, 172	constituent, 170
buffer, 173	pH scale, 172
chemical equilibrium, 168	precipitate, 165
chlorinity, 167	principle of constant propor-
colligative properties, 165	tions, 167
conservative constituent, 170	residence time, 168
diffusion, 164	salinity, 165
excess volatiles, 166	salinometer, 168
Forchhammer's	saturation, 165
principle, 167	solute, 164
ion, 164	solution, 164
ionic bond, 164	solvent, 164
mixing time, 169	trace element, 166

Study Questions

Reviewing What You've Learned

1. Why is water a polar molecule? What properties of water derive from its polar nature?
2. How does a crystal of common salt dissolve in water?
3. How does a solution differ from a mixture?
4. What are some of seawater's colligative properties? Does pure water have colligative properties?
5. Other than hydrogen and oxygen, what are the most abundant elements in seawater?
6. How is salinity determined?
7. What factors affect seawater's pH? How does the pH of seawater change with depth? Why?

Thinking Critically

1. Where did the water of the ocean come from?
2. How are modern methods of determining salinity dependent on the principle of constant proportions?

flow caused by Earth's uneven solar heating and daily rotation. This circulation of air is important in creating climate and weather, driving ocean currents, and steering storms. The trip took 14 days.

Fossett is no stranger to the power of global air circulation. He made ocean sailing history in October 2001 by setting a trans-Atlantic record of 4 days 17 hours aboard a huge twin-hulled sailboat with a mast 45 meters (148 feet) tall. Again powerful winds propelled him. Fossett and his team owe their many successes to thorough preparation, superior materials, excellent design, and (most important) the uneven heating of Earth by the sun!

The sun's rays make the equatorial regions of Earth warmer than the poles. The atmosphere responds to this uneven solar heating by flowing in three great circulating cells over each hemisphere. The flow of air within these cells is influenced by the rotation of Earth, which causes moving air (or any moving mass) in the Northern Hemisphere to curve to the right of its initial path, and in the Southern Hemisphere to curve to the left. The apparent curvature of path is known as the Coriolis effect. The Coriolis effect is the mysterious "force" that people believe causes toilets to flush clockwise in one hemisphere, counterclockwise in the other. Does that really happen? Read on.

Weather and storms are part of atmospheric circulation. Large storms are spinning areas of unstable air between or within air masses. Look for the similarities and differences in the two kinds of large storms—extratropical cyclones (frontal storms) and tropical cyclones (hurricanes, typhoons).

The large-scale movement of the atmosphere greatly influences the movement of ocean water. In this chapter we investigate the movement of air; in the next chapter, the movement of water.

Atmosphere and Ocean

The **atmosphere** is the volume of gases, water vapor, and airborne particles enveloping Earth. Earth's atmosphere and ocean are intertwined, their gases and waters freely exchanged. Gases entering the atmosphere from the ocean have important effects on climate, and gases entering the ocean from the atmosphere can influence sediment deposition, the distribution of life, and some of the physical characteristics of the seawater itself. Water evaporated from the ocean surface and moved by **wind**, the mass movement of air, helps minimize worldwide extremes of surface temperature; through rain, it provides moisture for agriculture. Thus the weather that so profoundly affects our daily lives is shaped at the junction of wind and water, and the flow of air greatly influences the movement of water in the ocean. **Weather** is the state of the atmosphere at a specific time and place, while **climate** is the long-term average of weather in an area. To understand the interactions of atmosphere and ocean that produce weather and climate, we begin by examining the composition and properties of the atmosphere.

Composition and Properties of the Atmosphere

The lower atmosphere is a nearly homogeneous mixture of gases, most plentifully nitrogen (78.1%) and oxygen (20.9%). Other elements and compounds, as shown in **Figure 8.1**, make up less than 1% of its composition.

Air is never completely dry; **water vapor**, the gaseous form of water, can occupy as much as 4% of its volume. Sometimes liquid droplets of water are visible as clouds or fog, but more often the water is simply there—invisible in vapor form, having entered the atmosphere from the ground, plants, and sea surface. The residence time of water vapor in the lower atmosphere is about ten days. Water leaves the atmosphere by condensing into dew, rain, or snow.

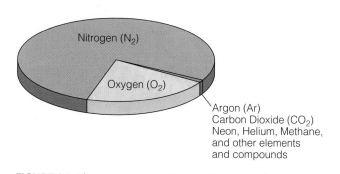

FIGURE 8.1 The average composition of dry air, by volume.

Air has mass. A 1-square-centimeter (0.16 square inch) column of dry air, extending from sea level to the top of the atmosphere, weighs about 1.04 kilograms (2.3 pounds). A 1-square-foot column of air the same height weighs more than a ton.

The temperature and water content of air greatly influence its density. Because the molecular movement associated with heat causes a mass of warm air to occupy more space than an equal mass of cold air, warm air is less dense than cold air. But contrary to what we might guess, humid air is *less* dense than dry air at the same temperature—because molecules of water vapor weigh less than the nitrogen and oxygen molecules that the water vapor displaces.

Near Earth's surface, air is packed densely by its own weight. Air lifted from near sea level to a higher altitude is subjected to less pressure and will expand. As anyone knows who has felt the cool air rushing from an open tire valve, air becomes *cooler* when it *expands*. The opposite effect is also familiar: Air *compressed* in a tire pump becomes *warmer*. Air descending from high altitude toward sea level warms as it is compressed by the higher atmospheric pressure near Earth's surface.

Warm air can hold more water vapor than cold air. Water vapor in rising, expanding, cooling air will often condense into

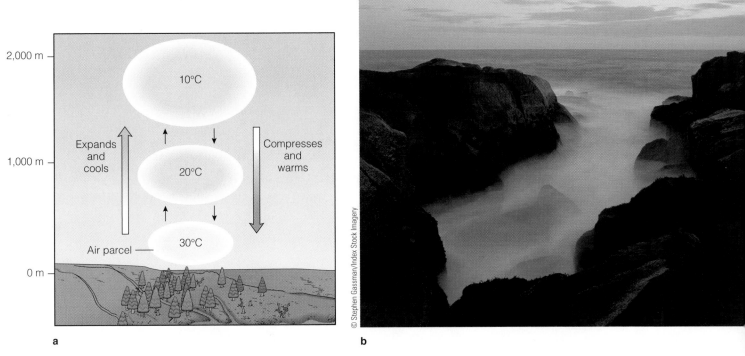

2,000 m

10°C

Expands
and
cools

1,000 m

20°C

Compresses
and
warms

Air parcel ——

30°C

0 m

© Stephen Gassman/Index Stock Imagery

a

b

ACTIVE FIGURE 8.2 (**a**) Ascending air cools as it expands. Cooler air can hold less water in solution, so water vapor condenses into tiny droplets—clouds. Descending air warms as it compresses, so the droplets (clouds) evaporate. (**b**) Steam fog over the ocean indicates rapid evaporation. Water vapor is invisible, but as water vapor rises into cool air it can condense into visible droplets.

Oceanography Now™ | At the Active Figures link at http://oceanography.brookscole.com/garrison5e you can view the adiabatic process.

clouds (aggregates of tiny droplets) because the cooler air can no longer hold as much water vapor. If rising and cooling continue, the droplets may coalesce into raindrops or snowflakes. The atmosphere will then lose water as **precipitation,** liquid or solid water that falls from the air to Earth's surface. These rising-expanding-cooling and falling-compressing-heating relationships are important in understanding atmospheric circulation, weather, and climate (**Figure 8.2**).

Atmospheric Circulation

Atmospheric circulation is powered by sunlight. Only about 1 part in 2.2 billion of the sun's radiant energy is intercepted by Earth, but that amount averages 7 million calories per square meter per day at the top of the atmosphere—or, for Earth as a whole, an impressive 17 trillion kilowatts (23 trillion horsepower)! As may be seen in **Figure 8.3**, on a global basis about 51% of the incoming energy is absorbed by Earth's land and water surface. How much light penetrates the ocean depends greatly on several factors: the angle at which it approaches, the sea state (surface turbulence), the presence of an ice covering or light-colored foam, and others.

The 51% of short-wave solar energy (light) striking land and sea is converted to longer-wave infrared radiation (heat) and then transferred into the atmosphere by conduction, radiation, and evaporation. The atmosphere, like the land and ocean, eventually radiates this heat back into space in the form of long-wave infrared radiation. The heat input and outflow

"account" for Earth is its **heat budget.** As in your personal financial budget, income must eventually equal outgo. Over long periods of time the total incoming heat (plus that from earthly sources) equals the total heat radiating into the cold of space, so Earth is in **thermal equilibrium:** It is growing neither significantly warmer nor colder.[1]

Uneven Solar Heating and Latitude

Although the heat budget for Earth *as a whole* is in balance, the heat budget for its *different latitudes* is not. As can be seen in **Figure 8.4**, sunlight striking polar latitudes spreads over a greater area (that is, polar areas receive less radiation per unit area) than sunlight at tropical latitudes. Near the poles, light also filters through more atmosphere and approaches the surface at a low angle, favoring reflection. Polar regions receive no sunlight at all during the depths of local winter. By contrast, at tropical latitudes sunlight strikes at a more nearly vertical angle, which distributes the same amount of sunlight over a much smaller area. The light passes through less atmosphere and minimizes reflection. Tropical latitudes thus receive significantly more solar energy than the polar regions, and mid-latitude areas receive more heat in summer than in winter.

---

[1] Changes in heat balance do occur over short periods of time. Increasing amounts of carbon dioxide and methane in Earth's atmosphere may contribute to an increase in surface temperature called the *greenhouse effect.* More on this subject may be found in Chapter 18.

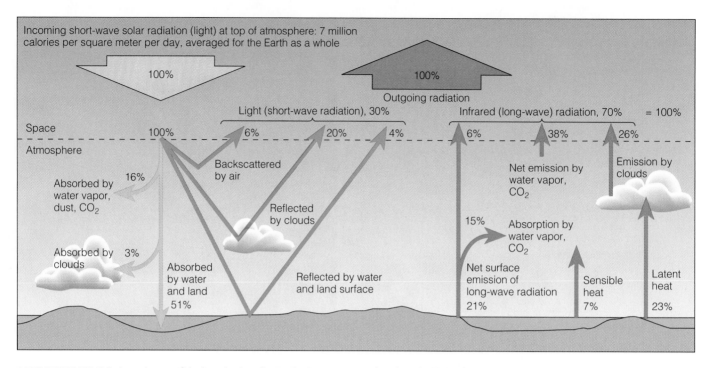

Incoming short-wave solar radiation (light) at top of atmosphere: 7 million calories per square meter per day, averaged for the Earth as a whole

100%

100%

Outgoing radiation

Light (short-wave radiation), 30%   Infrared (long-wave) radiation, 70%   = 100%

Space   100%   6%   20%   4%   6%   38%   26%

Atmosphere

Absorbed by water vapor, dust, CO_2 16%

Backscattered by air

Net emission by water vapor, CO_2

Emission by clouds

Reflected by clouds

15%   Absorption by water vapor, CO_2

Absorbed by clouds   3%

Absorbed by water and land   51%

Reflected by water and land surface

Net surface emission of long-wave radiation   21%

Sensible heat   7%

Latent heat   23%

ACTIVE FIGURE 8.3 An estimate of the heat budget for Earth. On an average day about half of the solar energy arriving at the upper atmosphere is absorbed at Earth's surface. Light (short-wave) energy absorbed at the surface is converted into heat. Heat leaves Earth as infrared (long-wave) radiation. Since input equals output over long periods of time, the heat budget is balanced.

Oceanography Now™ | At the Active Figures link at http://oceanography.brookscole.com/garrison5e you can view the energy budget for Earth.

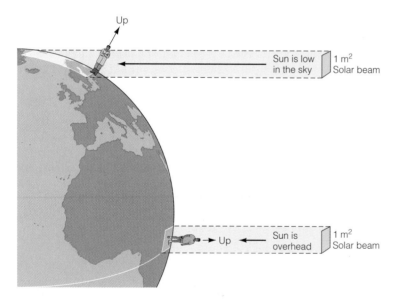

Up

Sun is low in the sky   1 m² Solar beam

Up   Sun is overhead   1 m² Solar beam

FIGURE 8.4 How solar energy input varies with latitude. Equal amounts of sunlight are spread over a greater surface area near the poles than in the tropics. Ice near the poles reflects much of the energy that reaches the surface there. (Source: From Levin Danielson, *Meteorology*, Fig. 3.4, p. 73. © 1998 McGraw-Hill Companies. Reprinted with permission.)

Figure 8.5 shows heat received versus heat reradiated at different latitudes. Near the equator the amount of solar energy received by Earth greatly exceeds the amount of heat radiated into space. In the polar regions the opposite is true.

So why doesn't the polar ocean freeze solid and the equatorial ocean boil away? The reason is that water itself is moving huge amounts of heat between tropics and poles. Water's thermal properties make it an ideal fluid to equalize the polar–tropical heat imbalance. Water's heat capacity moves heat poleward in ocean currents, but water's exceptionally high latent heat of vaporization (540 calories per gram) means that water vapor transfers much more heat (per unit of mass) than liquid water. Masses of moving air account for about two-thirds of the poleward transfer of heat; ocean currents move the other third.

Uneven Solar Heating and the Seasons

At mid-latitudes the Northern Hemisphere receives about three times as much solar energy per day in June as it does in December. This difference is due to the $23\frac{1}{2}°$ tilt of Earth's rotational axis relative to the plane of its orbit around the sun (**Figure 8.6**). As Earth revolves around the sun, the constant tilt of its rotational axis causes the Northern Hemisphere to lean *toward* the sun in June but *away* from it in December. The sun therefore appears higher in the sky in summer but lower in winter. The inclination of Earth's axis also causes days to become longer as summer approaches but to become shorter

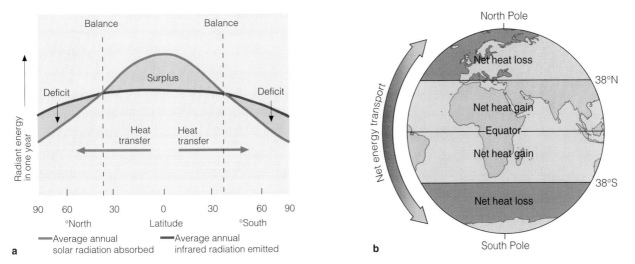

a

b

FIGURE 8.5 Areas of heat gain and loss on Earth's surface. (**a**) The average annual incoming solar radiation *absorbed* by Earth (red line) is shown along with the average annual infrared radiation *emitted* by Earth (blue line). Note that polar latitudes lose more heat to space than they gain, and tropical latitudes gain more heat than they lose. Only at about 38°N and 38°S latitudes does the amount of radiation received equal the amount lost. Because the area of heat gained (orange area) equals the area of heat lost (blue areas), Earth's total heat budget is balanced. (**b**) The ocean does not boil away near the equator or freeze solid near the poles because heat is transferred by winds and ocean currents from equatorial to polar regions.

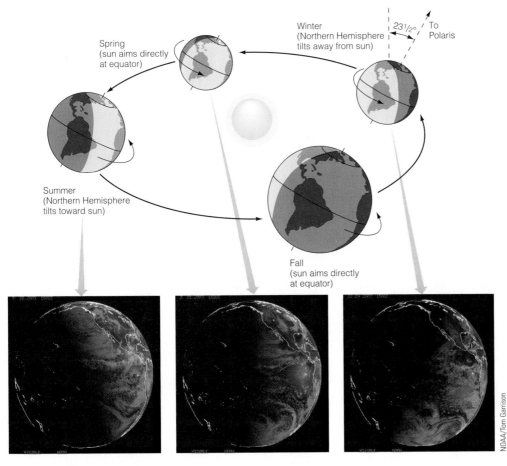

ACTIVE FIGURE 8.6 The seasons (shown for the Northern Hemisphere) are caused by variations in the amount of incoming solar energy as Earth makes its annual rotation around the sun on an axis tilted by $23\frac{1}{2}°$. During the Northern Hemisphere winter, the Southern Hemisphere is tilted toward the sun and the Northern Hemisphere receives less light and heat. During the Northern Hemisphere summer, the situation is reversed. The satellite images clearly show the significant difference in illumination angles in December, September, and June.

Oceanography⟳Now™ | At the Active Figures link at http://oceanography.brookscole.com/garrison5e you can view the reasons for seasons.

FIGURE 8.7 A convection current forms in a room when air flows from a hot radiator to a cold, closed window and back. (For a practical oceanic application of this principle, look ahead to Figure 8.17.)

with the coming of winter. Longer days mean more time for the sun to warm Earth's surface. The tilt causes the seasons.

Mid-latitude heating is strongly affected by season: The mid-latitude regions of the Northern Hemisphere receive about three times more light energy in June than in December.

Uneven Solar Heating and Atmospheric Circulation

The concentration of solar energy near the equator affects the atmosphere. We know that warm air rises and cool air sinks.

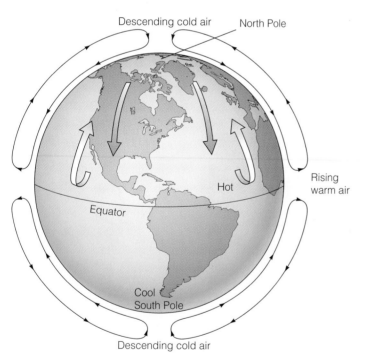

FIGURE 8.8 A hypothetical model of Earth's air circulation if uneven solar heating were the only factor to be considered. (The thickness of the atmosphere is greatly exaggerated.)

Think of air circulation in a room with a hot radiator opposite a cold, closed window (**Figure 8.7**). Air warms, expands, becomes less dense, and *rises* over the radiator. Air cools, contracts, becomes more dense, and *falls* near the cold window. The circular current of air in the room, a **convection current,** is caused by the difference in air density resulting from the temperature difference between the ends of the room.

A similar process occurs over the surface of Earth. As we have seen, surface temperatures are higher at the equator than at the poles, and air can gain heat from warm surroundings. Because air is free to move over Earth's surface, it would be reasonable to assume that an air circulation pattern like the one shown in **Figure 8.8** would develop over Earth. In this ideal model, air heated in the tropics would expand and become less dense, rise to high altitude, turn poleward, and "pile up" as it converged near the poles. The air would then cool by radiating heat into space, become more dense, sink to the surface, and turn equatorward, flowing along the surface back to the tropics to complete the circuit.

But this is *not* what happens. Global circulation of air is governed by two factors: uneven solar heating *and* Earth's rotation. The eastward rotation of Earth on its axis deflects the moving air or water (or any moving object that has mass) away from its initial course. This deflection is called the **Coriolis effect** in honor of **Gaspard Gustave de Coriolis,** the French scientist who worked out its mathematics in 1835.

An understanding of the Coriolis effect is important to an understanding of atmospheric and oceanic circulation.

The Coriolis Effect

To an earthbound observer, any object moving freely across the globe appears to curve slightly from its initial path. In the Northern Hemisphere this curve is to the right, or *clockwise,* from the expected path; in the Southern Hemisphere it is to

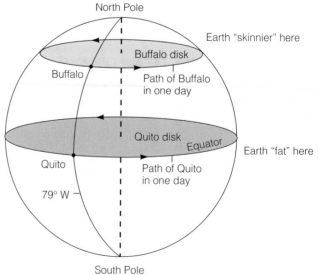

FIGURE 8.9 Sketch of the thought experiment in the text, showing that Buffalo travels a shorter path on the rotating Earth than Quito does.

FIGURE 8.10 Same idea, different approach. (Calvin and Hobbes. © 1990 Universal Press Syndicate. Reprinted by permission of Universal Press Syndicate.)

the left, or *counterclockwise*. To earthbound observers the deflection is very real; it isn't caused by some mysterious force, and it isn't an optical illusion or some other trick caused by the shape of the globe itself. *The observed deflection is caused by the observer's moving frame of reference on the spinning Earth.*

The influence of this deflection can be illustrated by performing a mental experiment involving concrete objects—in this case, cities and cannonballs—and then applying the principle to atmospheric circulation. Let's pick as examples for our experiment the equatorial city of Quito, the capital of Ecuador, and Buffalo, New York. Both cities are on almost the same line of longitude (79°W), so Buffalo is almost exactly north of Quito (as **Figure 8.9** shows). Like everything else attached to the rotating Earth, both cities make one trip around the world each 24 hours. Through one day, the north–south relationship of the two cities never changes: Quito is *always* due south of Buffalo.

A complete trip around the world is 360°, so each city moves eastward at an angular rate of 15° per hour (360°/24 hours = 15°/hour). Yet, even though their angular rates are the same, the two cities move eastward at different speeds. Quito is on the equator, the "fattest" part of Earth. Buffalo is farther north at a "skinnier" part. Imagine both cities isolated on flat disks, and imagine Earth's sphere being made of a great number of these disks strung together on a rod connecting the North Pole to the South Pole. From Figure 8.9 you can see that Buffalo's disk has a smaller circumference than Quito's. Buffalo doesn't have as far to go in one day as Quito because its disk is not as large. That means that Buffalo must move eastward *more slowly* than Quito to maintain its position due north of Quito. (**Figure 8.10** puts another spin on this idea.)

Look at Earth from above the North Pole in **Figure 8.11**. The Quito disk and the Buffalo disk must turn through 15° of longitude each hour (or Earth would rip itself apart), but the city on the equator must move faster to the east to turn its 15° each hour because its "slice of the pie" is larger. Buffalo must move at 1,260 kilometers (783 miles) per hour to go around the world in one day, while Quito must move at 1,668 kilometers (1,036 miles) per hour to do the same.

Now imagine a massive object moving between the two cities. A cannonball shot north from Quito toward Buffalo would carry Quito's eastward component as it goes; that is, regardless of its northward speed, the cannonball is also moving *east* at 1,668 kilometers (1,036 miles) per hour. The fact of being fired northward by the cannon does not change its eastward movement in the least. As the cannonball streaks north, an odd thing happens: The cannonball veers from its northward path, angling slightly to the right (east) (**Figure 8.12**). Actually this first cannonball is moving just as an observer from space would expect it to, but to those of us on the ground the cannonball "gets ahead of Earth." As cannonball 1 moves north, the ground beneath it is no longer moving eastward at 1,668 kilometers per hour. During the ball's time of flight, Buffalo (on its smaller disk) *has not moved eastward enough to be where the ball will hit.* If the time of flight for the cannonball is

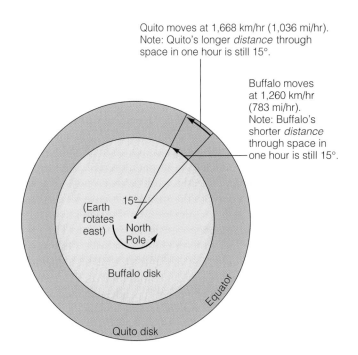

Quito moves at 1,668 km/hr (1,036 mi/hr). Note: Quito's longer *distance* through space in one hour is still 15°.

Buffalo moves at 1,260 km/hr (783 mi/hr). Note: Buffalo's shorter *distance* through space in one hour is still 15°.

FIGURE 8.11 A continuation of the thought experiment. A look at Earth from above the North Pole shows that Buffalo and Quito move at different speeds.

ACTIVE FIGURE 8.12 The final step in the thought experiment. As observed from space, cannonball 1 (shot northward) and cannonball 2 (shot southward) move as we might expect; that is, they travel straight away from the cannons and fall to Earth. Observed from the ground, however, cannonball 1 veers slightly east and cannonball 2 veers slightly west of their intended targets. The effect depends on the observer's frame of reference.

Oceanography☼Now™ | At the Active Figures link at http://oceanography.brookscole.com/garrison5e you can view the Coriolis force.

one hour, then a city 408 kilometers east of Buffalo (1,668 [Quito's speed] − 1,260 [Buffalo's speed] = 408) will have an unexpected surprise. Albany may be in for some excitement!

If a cannonball were fired south from Buffalo toward Quito, the situation would be reversed. This second cannonball has an eastward component of 1,260 kilometers (783 miles) per hour even while it sits in the muzzle. Once fired and moving southward, cannonball 2 travels over portions of Earth that are moving ever faster in an eastward direction. The ball again appears to veer off course to the right (see Figure 8.12 again), falling into the Pacific to the west of Ecuador. Don't be deceived by the word *appears* in the last sentence. The cannonballs really do veer to the right, or *clockwise*. Only to an observer in space would they appear to go straight, and points on Earth would appear to move out from underneath them.

The Coriolis effect is a real effect dependent on our rotating frame of reference. Part of that frame of reference involves the direction from which you view the problem. Thus, in Figure 8.12, cannonball 2 looks to you as if it is veering left, but to the citizens of Buffalo facing south to watch the cannonball disappear, it moves to the right (west). Coriolis deflection works *counterclockwise* in the Southern Hemisphere because the frame of

reference there is reversed. Also, except at the equator (where the Coriolis effect is nonexistent), the Coriolis effect influences the path of objects moving from east to west, or west to east.

An easy way to remember the Coriolis effect: In the Northern Hemisphere moving objects veer off course clockwise, to the right; in the Southern Hemisphere moving objects veer off course counterclockwise, to the left.

Because the Coriolis effect influences any object with mass—*as long as that object is moving*—it plays a large role in the movements of air and water on Earth. The Coriolis effect is most apparent in mid-latitude situations involving the almost frictionless flow of fluids: between layers of water in the ocean and in the circuits of winds. Does the Coriolis effect influence the directions of cars and airplanes? Yes, but in these cases friction (of tires on pavement, of wings on air) is much greater than the influence of the Coriolis effect, so the deflection is not observable.

The Coriolis Effect and Atmospheric Circulation Cells

We can now modify our original model of atmospheric circulation (Figure 8.8) into the more correct representation provided in **Figure 8.13.** Yes, air does warm, expand, and rise at the equator; and air does cool, contract, and fall at the poles. But instead of continuing all the way from equator to pole in a continuous loop in each hemisphere, air rising from the equatorial region moves poleward and is gradually deflected eastward; that is, it turns to the *right* in the Northern Hemisphere and to the *left* in the Southern Hemisphere. This eastward deflection is caused by the Coriolis effect. (Note that the Coriolis effect does not *cause* the wind; it only *influences* the wind's direction.)

As air rises at the equator it loses moisture by precipitation (rainfall) caused by expansion and cooling. This drier air now grows denser in the upper atmosphere as it radiates heat to space and cools. When it has traveled about a third of the way from the equator to the pole—that is, to about 30°N and 30°S latitude—the air becomes dense enough to fall back toward the surface. Most of the descending air turns back toward the equator when it reaches the surface. In the Northern Hemisphere the Coriolis effect again deflects this surface air to the right, and the air blows across the ocean or land from the northeast. (This air, the northeasterly trades, is represented by arrows in Figure 8.13.) Though it has been heated by compression during its descent, this air is generally still colder than the surface over which it flows. The air soon warms as it moves equatorward, however, evaporating surface water and becoming humid. The warm, moist, less dense air then begins to rise as it approaches the equator and completes the circuit.

Such a large circuit of air is called an **atmospheric circulation cell,** and there are six of these cells in our model. A pair of cells exists in the tropics, one on each side of the equator. They are known as **Hadley cells** in honor of George Hadley, the London lawyer and philosopher who worked out an overall scheme of wind circulation in 1735. Look for them in Figure 8.13.

A more complex pair of circulation cells operates at mid-latitudes in each hemisphere. Some of the air descending at

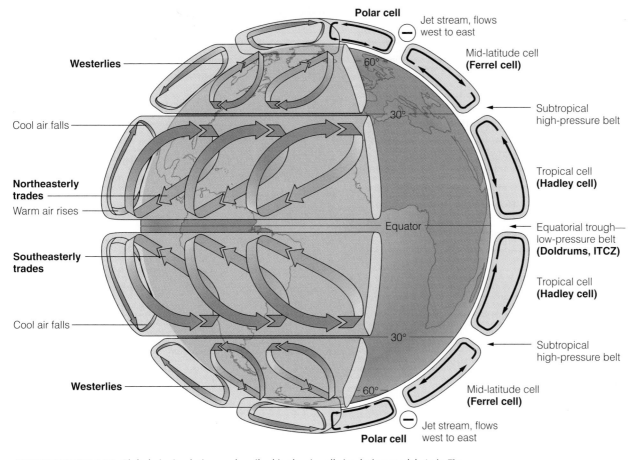

Polar cell

Westerlies

Jet stream, flows west to east

Mid-latitude cell **(Ferrel cell)**

60°

Cool air falls

30°

Subtropical high-pressure belt

Tropical cell **(Hadley cell)**

Northeasterly trades

Warm air rises

Equator

Equatorial trough— low-pressure belt **(Doldrums, ITCZ)**

Southeasterly trades

Tropical cell **(Hadley cell)**

Cool air falls

30°

Subtropical high-pressure belt

Westerlies

60°

Mid-latitude cell **(Ferrel cell)**

Polar cell

Jet stream, flows west to east

ACTIVE FIGURE 8.13 Global air circulation as described in the six-cell circulation model. As in Figure 8.8, air rises at the equator and falls at the poles. But instead of one great circuit in each hemisphere from equator to pole, there are three in each hemisphere. Note the influence of the Coriolis effect on wind direction. The circulation shown here is ideal—that is, a *long-term average* of wind flow. Contrast this view with Figure 8.15, a snapshot of a moment in 1996.

Oceanography Now™ | At the Active Figures link at http://oceanography.brookscole.com/ garrison5e you can view global air circulation.

30° latitude turns poleward rather than equatorward. Before this air descends to the surface it is joined at high altitude by air returning from the north. As can be seen in Figure 8.13, a loop of air forms a temperate cell between 30° and about 50°–60° latitude. As before, the air is driven by uneven heating and influenced by the Coriolis effect. Surface wind in this circuit is again deflected to the right, this time flowing from the west to complete the circuit (the westerlies in Figure 8.13). The mid-latitude circulation cells of each hemisphere are named **Ferrel cells** after William Ferrel, the American who discovered their inner workings in the mid-nineteenth century. They too can be seen in Figure 8.13.

Meanwhile, air that has grown cold over the poles begins blowing toward the equator at the surface, turning to the west as it does so. Between 50° and 60° latitude in each hemisphere, this air has taken up enough heat and moisture to ascend. However, this polar air is denser than the air in the adjacent Ferrel cell and does not mix easily with it. The unstable zone between these two cells generates most mid-latitude weather. At

high altitude the ascending air from 50°–60° latitude turns poleward to complete a third circuit. These are the **polar cells.**

Three large atmospheric circulation cells—a Hadley cell, a Ferrel cell, and a polar cell—exist in each hemisphere. Air circulation within and between cells, powered by uneven solar heating and influenced by the Coriolis effect, generates predictable long-term patterns of winds.

Wind Patterns

8.10

The model of atmospheric circulation described above has many interesting features. Look once more at Figure 8.13. At the boundaries between circulation cells, the air is moving *vertically* and surface winds are weak and erratic. Such conditions exist at the equator (where air rises and atmospheric pressure is generally low) and at 30° latitude in each hemisphere (where air falls and atmospheric pressure is generally high). Places within these circulation cells where air moves rapidly *horizontally* across the surface from zones of high pressure to zones of low pressure are characterized by strong, dependable winds.

Sailors have a special term for the calm equatorial areas where the surface winds of the two Hadley cells converge: the equatorial low called the **doldrums.** The word has come to be associated with a gloomy, listless mood, perhaps reflecting the sultry air and variable breezes found there. Scientists who study the atmosphere call this area the **intertropical convergence zone,** or **ITCZ,** to reflect the influence of wind convergence on conditions near the equator. Strong heating in the ITCZ causes surface air to expand and rise. The humid, rising, expanding air loses moisture as rain, some of which contributes to the success of tropical rain forests.

Sinking air, in contrast, is generally arid. The great deserts of both hemispheres, dry bands centered around 30° latitude, mark the intersection of Hadley and Ferrel cells. Because evaporation is higher than precipitation in these areas, ocean surface salinity tends to be highest at these latitudes (as can be seen in Figure 6.17). At sea, these areas of high atmospheric pressure and little surface wind are called the subtropical high, or **horse latitudes.** Spanish ships laden with supplies for the New World were often becalmed there, sometimes for weeks on end. When the mariners ran out of water and feed for their livestock, they were forced to eat the animals or throw them over the side.

Of much more interest to sailing masters were the bands of dependable surface winds *between* the zones of ascending and descending air. The most constant of these are the persistent **trade winds,** or easterlies, centered at about 15°N and 15°S latitude. The trade winds are the surface winds of the Hadley cells as they move from the horse latitudes to the doldrums. In the Northern Hemisphere they are the northeast trade winds; the southeast trade winds are the Southern Hemisphere counterpart.[2] The **westerlies,** surface winds of the Ferrel cells centered at about 45°N and 45°S latitude, flow between the horse latitudes and the boundaries of the polar cells in each hemisphere. The westerlies, then, approach from the southwest in the

Northern Hemisphere and from the northwest in the Southern Hemisphere. Sailors outbound from Europe to the New World learned to drop south to catch the trade winds and to return home by a more northerly route to take advantage of the westerlies. Trade winds and westerlies are shown in Figure 8.13.

Cell Circulation: Ideal Versus Actual

8.11

The six-cell model of atmospheric circulation discussed here represents an average air flow through many years over the planet as a whole. Though the model is accurate in a general sense, local details of cell circulation vary because surface conditions are different at different longitudes. The ocean's thermostatic effect is the major factor reducing irregularities in cell circulation over water.

For example, the equator-to-pole patterns of air flow within circulation cells along the 20°E line of longitude, a meridian crossing Africa and Europe, are much more complex than the patterns of flow along 170°W longitude, which is almost exclusively over the ocean. The ITCZ is also much narrower and more consistent over the ocean than over land. Because the Northern Hemisphere contains much less ocean surface than the Southern Hemisphere, and because landmasses have a lower heat capacity than the ocean, seasonal differences in temperature and cell circulation are more extreme in the north. Cell circulation is also much more symmetrical in the Southern Hemisphere.

Another consequence of the markedly different proportions of land to ocean surface in the two hemispheres is the position of the ITCZ. The convergence zone does not coincide with the **geographical equator** (0° latitude). Instead it lies at the **meteorological equator** (or *thermal equator*), an irregular imaginary line of thermal equilibrium between the hemispheres, situated about 5° north of the geographical equator. The positions of the meteorological equator and the ITCZ generally coincide and change with the seasons, moving slightly farther north in the northern summer and returning toward the equator in the northern winter (**Figure 8.14**). Atmospheric and oceanic circulation in the two hemispheres

---

[2] Winds are named by the direction *from* which they blow. A west wind blows *from* the west toward the east; a northeast wind blows *from* the northeast toward the southwest.

FIGURE 8.14 Seasonal changes in the position of the intertropical convergence zone (ITCZ). The zone reaches its most northerly location in July and its most southerly location in January. Because of the thermostatic effect of water, the seasonal north-south movement is generally less over the ocean than over land. (Source: Meteorology, 5/e by Moran and Morgan. © 1997 Prentice-Hall, Inc. Reprinted with permission.)

NASA

FIGURE 8.15 Winds over the Pacific Ocean on 20 and 21 September 1996. Wind speed increases as colors change from blue-purple to yellow-orange, with the strongest winds at 20 meters per second (45 miles per hour). Wind direction is shown by the small white arrows. The measurements were made with a NASA radar scatterometer aboard Japan's Advanced Earth Orbiting Satellite, launched 16 August 1996. Note the Hawaiian Islands in the midst of the persistent northeast trade winds, the vigorous westerlies driving toward western Canada, a large extratropical cyclone east of New Zealand, and the last remnants of a tropical cyclone off the coast of Japan. Although *instantaneous* views such as this one depart substantially from wind flow predicted in the six-cell model developed in Figure 8.13, the *average* wind flow over many years looks remarkably like what we would expect.

is approximately symmetrical about the meteorological equator, not the geographical equator. Thus the doldrums, trade winds, horse latitudes, and westerlies shift north in the northern summer, and south in the northern winter.

There are also east-west variations in the expected patterns of the circulation cells. In the northern winter, air above the chilled continental masses of North America and Siberia becomes very cold and dense. This air sinks and forms zones of high atmospheric pressure over the continents. Air over the relatively warmer waters near the Aleutians and Iceland rises and forms zones of low atmospheric pressure. Air flows from the high-pressure zones toward the low-pressure zones, modifying the flow of air within the cells. In summer the situation is reversed: Low pressure forms over the heated landmasses, and higher pressure forms over the cooler ocean. These effects are most pronounced in the middle latitudes of the Northern Hemisphere, where land and water are present in near-equal amounts.

Figure 8.15 is a depiction of winds over the Pacific on two days in September 1996. As you can see, the patterns can depart from what we would expect based on the six-cell model of Figure 8.13. Most of the difference is caused by the geographical distribution of landmasses, the different responses of

land and ocean to solar heating, and chaotic flow. But, as noted earlier, over long periods of time (many years) *average* flow looks remarkably like what we would expect.

The major surface wind and pressure systems of the world, and their prevailing weather conditions, are summarized in **Table 8.1**. These great wind patterns are responsible for about two-thirds of the heat transfer from the tropics to the polar regions. (Ocean currents account for the other third.)

Monsoons

A **monsoon** is a pattern of wind circulation that changes with the season. (The word *monsoon* is derived from *mausim*, Arabic for "season.") Areas subject to monsoons generally have wet summers and dry winters.

Monsoons are linked to the different heat capacities of land and water and to the annual north-south movement of the ITCZ. In the spring, land heats more rapidly than the adjacent ocean. Air above the land becomes warmer and so rises. Relatively cool air flows from over the ocean to the land to take its place. Continued heating causes this humid air to rise, condense, and form clouds and rain. In autumn, the land cools more rapidly than the adjacent ocean. Air cools and sinks over

Table 8.1 Major Wind and Pressure Systems and Related Weather

Region	Name	Pressure	Surface Winds	Weather
Equator (0°)	Doldrums (ITCZ) (equatorial low)	Low	Light, variable winds	Cloudiness, abundant precipitation in all seasons; breeding ground for hurricanes. Relatively low sea surface salinity because of rainfall (see Figure 6.17)
0°–30°N and S	Trade winds (easterlies)	–	Northeast in Northern Hemisphere, southeast in Southern Hemisphere	Summer wet, winter dry; pathway for tropical disturbances
30°N and S	Horse latitudes (subtropical high)	High	Light, variable winds	Little cloudiness; dry in all seasons. Relatively high sea surface salinity because of evaporation
30°–60°N and S	Prevailing westerlies	–	Southwest in Northern Hemisphere, northwest in Southern Hemisphere	Winter wet, summer dry; pathway for subtropical high and low pressure
60°N and S	Polar front	Low	Variable	Stormy, cloudy weather zone; ample precipitation in all seasons
60°–90°N and S	Polar easterlies	–	Northeast in Northern Hemisphere, southeast in Southern Hemisphere	Cold polar air with very low temperatures
90°N and S	Poles	High	Southerly in Northern Hemisphere, northerly in Southern Hemisphere	Cold, dry air; sparse precipitation in all seasons

Note: Compare to Figure 8.13. (*Source:* From *Earth in Crisis: An Introduction to Earth Sciences,* 2/e, Thomas L. Burrus, Herbert J. Spiegel, 1980. C.V. Mosby Co. Reprinted by permission of Thomas L Burrus.)

the land, and dry surface winds move seaward. The intensity and location of monsoon activity depend on the position of the ITCZ. Note that the monsoons follow the ITCZ south in the Northern Hemisphere's winter (**Figure 8.16a**) and north in its summer (**Figure 8.16b**).

In Africa and Asia more than 2 billion people depend on summer monsoon rains for drinking water and agriculture. The most intense summer monsoons occur in Asia. Heating of the great landmass of Asia draws vast quantities of warm, moist air from the Indian Ocean. Winds from the south drive this moisture toward Asia, where it rises and condenses to produce a months-long deluge. The volume of water vapor carried inland is astonishing; the town of Cherrapunji, on the slopes of the Khasi Hills in northeastern India, receives about 10 *meters* (425 inches) of rain each year, most of it between April and October (**Figure 8.16c**)! Much smaller monsoons occur in North America as warming and rising air over the South and West draws humid air and thunderstorms from the Gulf of Mexico.

Sea Breezes and Land Breezes

Land breezes and sea breezes are small, daily mini-monsoons. Morning sunlight falls on land and the adjacent sea, warming both. The temperature of the water doesn't rise as much as the temperature of the land, however. The warmer inland rocks transfer heat to the air, which expands and rises, creating a zone of low atmospheric pressure over the land. Cooler air from over the sea then moves toward land; this is the **sea breeze** (**Figure 8.17a**). The situation reverses after sunset, with land losing heat to space and falling rapidly in temperature.

After a while the air over the still-warm ocean will be warmer than the air over the cooling land. This air will then rise and the breeze direction will reverse, becoming a **land breeze** (**Figure 8.17b**). Land breezes and sea breezes are common and welcome occurrences in coastal areas.

Storms

Storms are regional atmospheric disturbances characterized by strong winds often accompanied by precipitation. Few natural events underscore human insignificance like a great storm. When powered by stored sunlight, the combination of atmosphere and ocean can do fearful damage.

In Bangladesh on 13 November 1970, a tropical cyclone (a hurricane) with wind speeds of more than 200 kilometers (125 miles) per hour roared up the mouths of the Ganges River, carrying with it masses of seawater up to 12 meters (40 feet) high. Water and wind clawed at the aggregation of small islands, most just above sea level, that makes up this impoverished country. In only 20 minutes at least 300,000 lives were lost, and estimates ranged up to a million dead! Property damage was essentially absolute. Photographs taken soon after the storm showed a horizon-to-horizon morass of flooded, deep-gashed ground tortured by furious winds. There was almost no trace of human inhabitants, farms, domestic animals, or villages. Another great storm that struck in May 1991 killed another 200,000 people. The economy of the shattered country may not recover for decades.

A much different type of storm hammered the U.S. East Coast in March 1993. Mountainous snows from New York to North Carolina (1.3 meters, or 50 inches, at Mount Mitchell),

a January

b July

c

FIGURE 8.16 Monsoon patterns. During the monsoon circulations of January (**a**) and July (**b**), surface winds are deflected to the right in the Northern Hemisphere and to the left in the Southern Hemisphere. (**c**) Detail of summer Asian monsoon, showing the location of Cherrapunji, India, one of the world's wettest places. Rainfall amounts there can exceed 10 meters (425 inches) per year! (Source for 8.16a, b: Reprinted with permission of Alan D. Iselino.)

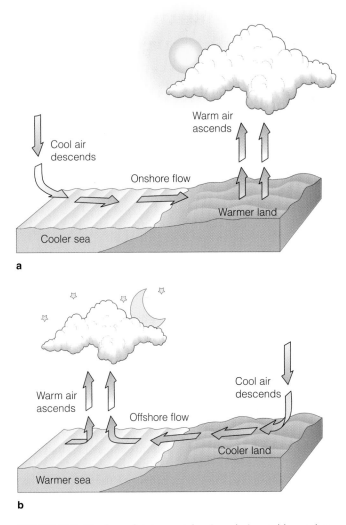

a

b

FIGURE 8.17 The flow of air in coastal regions during stable weather conditions. (**a**) In the afternoon, the land is warmer than the ocean surface, and warm air rising from the land is replaced by an onshore sea breeze. (**b**) At night, as the land cools, the air over the ocean is now warmer than the air over the land. The ocean air rises. Air flows offshore to replace it, generating an offshore flow (a land breeze).

winds of 175 kilometers (109 miles) per hour in Florida, and record cold in Alabama (-17°C, or 2°F, in Birmingham) were elements of a four-day storm that spread chaos from Canada to Cuba. At least 238 people died on land; another 48 were lost at sea. At one point more than 100,000 people were trapped in offices, factories, vehicles, and homes; 1.5 million were without electricity. Damage exceeded $1 billion.

These two great storms are examples of *tropical cyclones* and *extratropical cyclones* at their worst. As the name implies, tropical cyclones like the hurricane that struck Bangladesh are primarily a tropical phenomenon. Extratropical cyclones—the winter weather disturbances with which residents of the U.S. Eastern Seaboard and other mid-latitude dwellers are most familiar—are found mainly in the Ferrel cells of each hemisphere. (Note that the prefix *extra-* means "outside" or "beyond," so *extratropical* refers to the location of the storm, not its intensity.)

FIGURE 8.18 Vilhelm Bjerknes, the Norwegian meteorologist who, with his son Jacob, formulated the air-mass theories on which our understanding of weather is based.

Both kinds of storms are **cyclones,** huge rotating masses of low-pressure air in which winds converge and ascend. The word *cyclone,* derived from the Greek noun *kyklon* (meaning an object moving in a circle), underscores the spinning nature of these disturbances. (Don't confuse a cyclone with a **tornado,** a much smaller funnel of fast-spinning wind associated with severe thunderstorms.)

Air Masses

Cyclonic storms form between or within air masses. An **air mass** is a large body of air with nearly uniform temperature, humidity, and therefore density throughout. Air pausing over water or land will tend to take on the characteristics of the surface below. Cold, dry land causes the mass of air above to become chilly and dry. Air above a warm ocean surface will become hot and humid. Cold, dry air masses are dense and form zones of high atmospheric pressure. Warm, humid air masses are less dense and form zones of lower atmospheric pressure.

Air masses can move within or between circulation cells. Density differences, however, will prevent the air masses from mixing when they approach one another. Energy is required to mix air masses. Because that energy is not always available, a dense air mass may slide beneath a lighter air mass, lifting the lighter one and causing its air to expand and cool. Water vapor in the rising air may condense. All of these effects contribute to turbulence at the boundaries of the air masses.

The boundary between air masses of different density is called a **front.** The term was coined by **Vilhelm Bjerknes (Figure 8.18),** a pioneering Norwegian meteorologist who saw a similarity between the zone where air masses meet and the violent battle fronts of World War I.

Extratropical cyclones form at a front between *two* air masses. Tropical cyclones form from disturbances within *one* warm and humid air mass.

Extratropical Cyclones

Extratropical cyclones form at the boundary between each hemisphere's polar cell and its Ferrel cell—the **polar front.**

These great storms occur mainly in the winter hemisphere when temperature and density differences across the polar front are most pronounced. Remember, the cold wind poleward of the front is generally moving from the east; the warmer air equatorward of the front is generally moving from the west (see again Figure 8.13). The smooth flow of winds past each other at the front may be interrupted by zones of alternating high and low atmospheric pressure that bend the front into a series of waves. Because of the difference in wind direction in the air masses north and south of the polar front, the wave shape will enlarge and a twist will form along the front. The different densities of the air masses prevent easy mixing, so the cold, dense air mass will slide beneath the warmer, lighter one. Formation of this twist in the Northern Hemisphere, as seen from above, is shown in **Figure 8.19.** The twisting mass of air becomes an extratropical cyclone.

The twist that generates an extratropical cyclone circulates counterclockwise in the Northern Hemisphere, seemingly in opposition to the Coriolis effect. The reasons for this paradox become clear, however, when we consider the wind directions and the nature of interruption of the air flow between the cells. (In fact, the counterclockwise motion of the cyclone *is* Coriolis-driven because the large-scale air flow pattern at the edges of the cells is generated in part by the Coriolis effect.) Wind speed increases as the storm "wraps up" in much the same way that a spinning skater increases rotation speed by pulling in his or her arms close to the body. Air rushing toward the center of the spinning storm rises to form a low-pressure zone at the center. Extratropical cyclones are embedded in the westerly winds and thus move eastward. They are typically 1,000 to 2,500 kilometers (620 to 1,600 miles) in diameter and last from two to five days. **Figure 8.20** provides a beautiful example.

Precipitation can begin as the circular flow develops. **Figure 8.19b** shows why. Precipitation is caused by the lifting and consequent expansion and cooling of the mass of mid-latitude air involved in the twist. As it rises and cools, this air can no longer hold all of its water vapor, so clouds and rain result. When *cold* air advances and does the lifting (as on the left side of Figure 8.19b), a *cold front* occurs. A *warm front* happens when *warm* air is blown on top of the retreating edge of cold air (as on the right side of Figure 8.19b). The wind and precipitation associated with these fronts are sometimes referred to as **frontal storms.** They are the principal cause of weather in the mid-latitude regions, where most of the world's people live.

North America's most violent extratropical cyclones are the **nor'easters** (northeasters) that sweep the Eastern Seaboard in winter. The name indicates the direction from which the storm's most powerful winds approach. About 30 times a year, nor'easters moving along the mid-Atlantic and New England coasts generate wind and waves with enough force to erode beaches and offshore barrier islands, disrupt communication and shipping schedules, damage shore and harbor installations, and break power lines. About every hundred years a nor'easter devastates coastal settlements. In spite of a long history of destruction, people continue to build on unstable exposed coasts (**Figure 8.21**).

North

Cold air

Atmospheric pressure increases

Front

Atmospheric pressure increases

Warm air

South

Stage 1

Low

Stage 2

Cold air

Low

Warm front

Cold front

Warm air

A

B

a

Winds aloft

26,000 ft

Thunderstorms

Widespread precipitation

0°C (32°F)

Warm air

0 ft

0°C (32°F)
A

Cold air

5°C (41°F)

11°C (52°F)

Cold air receding

B

4°C (39°F)

50 km

Cold front

L

9°C (48°F)

Warm front

600 km

0°C (32°F)

−6°C (22°F)

b

ACTIVE **FIGURE 8.19** (**a**) The genesis and early development of an extratropical cyclone in the Northern Hemisphere. (The arrows depict air flow.) (**b**) How precipitation develops in an extratropical cyclone. These relationships between two contrasting air masses are responsible for nearly all the storms generated in the polar frontal zone, and thus for the high rainfall within these belts and the decreased salinities of surface waters below.

Oceanography Now™ | At the Active Figures link at http://oceanography.brookscole.com/garrison5e you can view the genesis and early development of an extratropical cyclone.

FIGURE 8.20 A well-developed extratropical cyclone swirls over the northeastern Pacific on 27 October 2000. Looking like a huge comma, the cloud-dense cold front extends southward and westward from the storm's center. Spotty cumulus clouds and thunderstorms have formed in the cold, unstable air behind the front. This picture was taken by *GOES-10* in visible light.

Tropical Cyclones

Tropical cyclones are great masses of warm, humid, rotating air. They occur in all tropical oceans except the equatorial South Atlantic. Large tropical cyclones are called **hurricanes** (*Huracan* is the god of the wind of the Caribbean Taino people) in the North Atlantic and eastern Pacific, *typhoons* (*Tai-*

fung is Chinese for "great wind") in the western Pacific, *tropical cyclones* in the Indian Ocean, and *willi-willis* in the waters near Australia. To qualify formally as a hurricane or typhoon, the tropical cyclone must have winds of at least 119 kilometers (74 miles) per hour. About 100 tropical cyclones grow to hurricane status each year. A very few of these develop into superstorms, with winds near the core that exceed 250 kilometers

FIGURE 8.21 Storm damage from a particularly severe nor'easter, the Ash Wednesday (7 March) storm of 1962. Massive damage such as this, on Fire Island, New York, is most common for buildings near the shoreline.

(155 miles) per hour! (To imagine what winds in such a storm might feel like, picture yourself clinging to the wing of a twin-engine private airplane in flight!) Tropical cyclones containing winds less than hurricane force are called *tropical storms* and *tropical depressions*.

From above, tropical cyclones appear as circular spirals (**Figure 8.22**). They may be 1,000 kilometers (620 miles) in diameter and 15 kilometers (9.3 miles, 50,000 feet) high. The

calm center, or *eye* of the storm—a zone some 13 to 16 kilometers (8 to 10 miles) in diameter—is sometimes surrounded by clouds so high and dense that the daytime sky above looks dark. Farther out, churned by furious winds, the rainband clouds condense huge amounts of water vapor into rain. A tropical cyclone is diagrammed in **Figure 8.23.**

Unlike extratropical cyclones, these greatest of storms are generated within *one* warm, humid air mass that forms between 10° and 25° latitude in either hemisphere (**Figure 8.24**). (Though air conditions would be favorable, the Coriolis effect closer to the equator is too weak to initiate rotary motion.)

The origins of tropical cyclones are not well understood. A tropical cyclone usually develops from a small tropical depression. Tropical depressions form in easterly waves, areas of lower pressure within the easterly trade winds that are thought to originate over a large, warm landmass. When air containing the disturbance is heated over tropical water with a temperature of about 26°C (79°F) or higher, circular winds begin to blow in the vicinity of the wave, and some of the warm, humid air is forced upward. Condensation begins, and the storm takes shape.

Under ideal conditions the embryo storm reaches hurricane status—that is, with wind speeds in excess of 119 kilometers (74 miles) per hour—in two to three days. The centers of most tropical cyclones move westward and poleward at 5 to 40 kilometers (3 to 25 miles) per hour. Typical tracks of these storms are shown in **Figure 8.25.**

Although its birth process is somewhat mysterious, the source of the storm's power is well understood. Its strength comes from the same seemingly innocuous process that warms a chilled soft-drink can when water from the atmosphere condenses on its surface. As you may recall from Chapter 6, it takes quite a bit of energy to break the bonds that hold water molecules together and evaporate water into the atmosphere; water's latent heat of vaporization is very high. That heat energy is released when the water vapor recondenses as liquid. It tends to warm your drink very quickly, and the more humid the air, the faster the condensing and warming. Think of the situation with a hair dryer. The heat energy generated by the dryer causes water to evaporate rapidly from your hair. When

FIGURE 8.22 Hurricane Alberto spins in the North Atlantic east of Bermuda. Note the thinness of Earth's atmosphere in this oblique view.

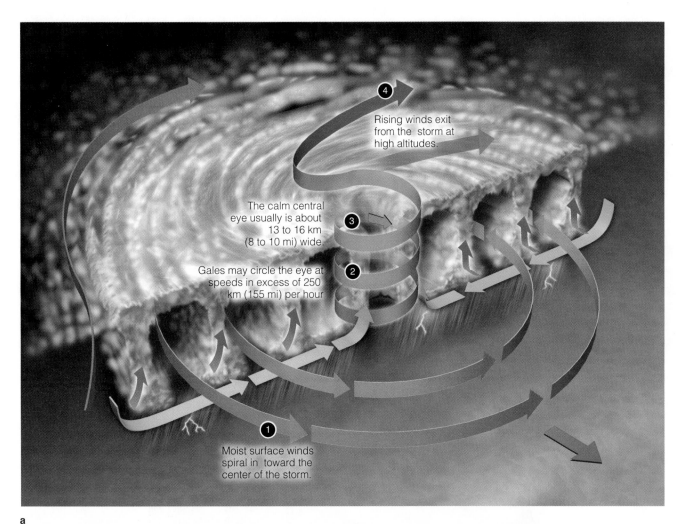

④ Rising winds exit from the storm at high altitudes.

The calm central eye usually is about 13 to 16 km (8 to 10 mi) wide.

③

② Gales may circle the eye at speeds in excess of 250 km (155 mi) per hour

① Moist surface winds spiral in toward the center of the storm.

a

© M. Laca/Weatherstock

b

FIGURE 8.23 (**a**) The internal structure of a mature tropical cyclone, or hurricane. (The vertical dimension is greatly exaggerated.) (**b**) High winds and strong rain can inflict great damage when a large tropical storm comes ashore. Here, Hurricane Kate slams into Key West, Florida, on 20 November 1985. This hurricane was responsible for seven deaths in the United States and more than $1 billion in damage.

JPL/NASA

FIGURE 8.24 Tropical cyclones can develop in zones of high humidity and warm air over a sea surface exceeding 26°C (79°F), the areas shown in red on the map. The base map is derived from satellite data showing water vapor in the atmosphere in October 1992.

a

b

NOAA/National Weather Service

FIGURE 8.25 The tracks of tropical cyclones. (**a**) The breeding grounds of tropical cyclones are shown as orange shaded areas. The storms follow curving paths: First they move westward with the trade winds. Then they either die over land or turn eastward until they lose power over the cooler ocean of the mid-latitudes. Cyclones are not spawned over the South Atlantic or the southeast Pacific because their waters are too chilly, nor in the still air—the doldrums—within a few degrees of the equator. (**b**) A composite of infrared satellite images of Hurricane Georges from 18 to 28 September 1998. Its westward and poleward trek across the Caribbean and into the United States is clearly shown.

BOX 8.1   The Galveston Disaster, 1900

The greatest natural disaster to strike the United States was the tropical cyclone that hit Galveston, Texas, on the night of 8 September 1900. The combination of high wind, great waves, and storm surge killed about 8,000 people—more than the Johnstown Flood, the San Francisco Earthquake, the 1938 New England Hurricane, and the Great Chicago Fire combined. Indeed this one event accounts for more than a third of all tropical storm- or hurricane-related fatalities ever recorded in the United States.

In 1900 Galveston, home to about 37,000 people, was one of the most important cotton markets in America. The island city was (and is) located at the eastern end of Galveston Island, a low sand barrier island off the Texas coast about 48 kilometers (30 miles) long and 3.2 kilometers (2 miles) wide. When the hurricane struck, the highest point in Galveston was only 2.7 meters (8.7 feet) above sea level.

Unlike today, there were no geosynchronous satellites, ship-to-shore radios, or networks of weather forecasters to warn of the coming storm. Forecasting was done by experience and hunch, and few were better guessers than Dr. Isaac Cline, chief of the U.S. Weather Bureau's Galveston station. On the evening of 7 September, as many of Galveston's residents were settling down to dinner, Cline became increasingly concerned about the 26-kilometer- (16-mile-) per-hour northerly wind that had blown steadily all day. Something was amiss—the wind was from the wrong direction, and high clouds at sunset were moving in nearly the opposite way, from the southeast. By midnight the wind had shifted toward the northeast and had grown to about 60 kilometers (50 miles) per hour. At first light residents flocked to the shore to gawk at the monstrous waves crashing on the beaches. Then the water began to rise. Cline sensed a hurricane was imminent and believed he knew what was going to happen next, but as he spread emergency warnings to evacuate the island, a steamship was torn from its moorings and smashed through the three bridges connecting the island to the mainland. There would be no escape.

The atmospheric pressure plummeted and the wind speed increased as the storm approached and intensified. The tropical cyclone's low pressure drew the ocean into a broad mound, and the winds drove this mound ashore—a phenomenon known as a storm surge. As misfortune would have it, the surge arrived at a time of high tide. Waters from the Gulf of Mexico and Galveston Bay rose to meet each other. Residents scrambled to the second, third, or fourth stories of buildings to avoid the rising water. Winds that reached 200 kilometers (125 miles) per hour collapsed the structures, freeing masses of flotsam that hammered anyone outside. By the afternoon of 8 September, buildings crumbled and people were battered by debris and drowned. At 8:30 that evening, the water stood 3.4 meters (11 feet) above Galveston Island's highest point. People died in the thousands, clinging to heaving rafts of wreckage and each other. Property damage was extraordinary (**figure**).

Galveston was rebuilt. In 1902 residents began to construct a 5-meter- (16-foot-) thick, 5.2-meter- (17-foot-) high sea wall covering 3 miles of oceanfront. (The seawall today extends for 16 kilometers, or 10 miles.) They also dredged enough sediment from Galveston Bay to raise the island 2.5 meters (8 feet).

Between 80% and 90% of the residents of hurricane-prone areas have not experienced a major hurricane. The smaller storms they have seen often leave them with the false impression of a hurricane's true potential for damage. Galveston proves otherwise.

8.18

that water recondenses to liquid (on a nearby can of soda, for instance), the original heat used to evaporate the water is released. The cycle of evaporation and condensation has carried heat from the hair dryer to your soda. In tropical cyclones the condensation energy generates air movement (wind), not more heat. Fortunately only 2% to 4% of this energy of condensation is converted into motional energy!

A tropical cyclone is an ideal machine for "cashing in" water vapor's latent heat of vaporization. Warm, humid air forms in great quantity only over a warm ocean. As already noted, tropical cyclones originate in ocean areas that have surface temperatures in excess of 26°C (79°F) (see Figure 8.24). When hot, humid tropical air rises and expands, it cools and is unable to retain the moisture it held when warm. Rainfall begins. The rainfall rate in some parts of the storm routinely exceeds 2.5 centimeters (1 inch) per hour, and 20 billion metric tons of water can fall from a large tropical cyclone in a day! Tremendous energy is released as this moisture changes from water vapor to liquid. In one day, a large tropical cyclone generates about 2.4 trillion kilowatt-hours of power, equivalent to the electrical energy needs of the entire United States for a year! So, solar energy ultimately powers the storm in a cycle of heat absorption, evaporation, condensation, and conversion of heat energy to kinetic energy. This energy is available as long as the storm stays over warm water and has a ready source of hot, humid air.

Three aspects of a tropical cyclone can cause property damage and loss of life: wind, rain, and storm surge. The destructive force of winds of 250 kilometers (155 miles) per hour or more is self-evident (**Box 8.1**). Rapid rainfall can cause severe flooding when the storm moves onto land. But the most danger lies in a **storm surge,** a mass of water driven by the storm. The low atmospheric pressure at the storm's center produces a dome of seawater that can reach a height of 1 meter (3.3 feet) in the open sea. The water height increases when waves and strong hurricane winds ramp the water mass ashore. If a high tide coincides with the arrival of all this water at a coast or if the coastline converges (as is the case at the mouths of the Ganges in Bangladesh), rapid and catastrophic flooding will occur. Storm surges of up to 12 meters (40 feet) were reported at Bangladesh in 1970. Much of the $3 billion in damage done by Hurricane Hugo to Charleston, South Carolina, in September 1989 was caused by a 5-meter (16-foot) storm surge arriving at high tide. You'll learn more about storm surges in Chapter 10's discussion of large waves.

You may have noticed that tropical cyclones turn counterclockwise in the Northern Hemisphere and clockwise in the Southern Hemisphere. Does this mean that the Coriolis effect does not apply to tropical cyclones? No. This apparent anomaly is caused by the Coriolis deflection of winds approaching the center of a low-pressure area from great distances. In the Northern Hemisphere there is rightward deflection of the *approaching air.* The edge spin given by this approaching air causes the storm to spin counterclockwise in the Northern Hemisphere.

Tropical cyclones last from 3 hours to 3 weeks; most have lives of 5 to 10 days. They eventually run down when they move over land or over water too cool to supply the humid air that sustains them. The friction of a land encounter rapidly drains a tropical cyclone of its energy, and a position above ocean water cooler than 24°C (75°F) is a sure harbinger of the storm's demise. When deprived of energy, the storm "unwinds" and becomes a mass of unstable humid air pouring rain, lightning, and even tornadoes from its clouds. Tropical cyclones can be dangerous to the end: Torrential rain streaming from the remnants of Hurricane Agnes in 1972 caused more than $2 billion in damage, mostly to Pennsylvania. Chesapeake and Delaware Bays were flooded with fresh water and sediments, destroying much of the shellfish industry there.

Tropical cyclones are nature's escape valves, flinging solar energy poleward from the tropics. They are beautiful, dangerous examples of the energy represented by water's latent heat of vaporization.

Questions from Students

1. Earth's orbit brings it closer to the sun in the Northern Hemisphere's winter than in its summer. Yet it's warmer in summer. Why?

Earth's orbit around the sun is elliptical, not circular. The whole Earth receives about 7% more solar energy through the half of the orbit during which we are closer to the sun than through the other half. The time of greater energy input comes during our winter, but the entire Northern Hemisphere is tilted toward the sun during the summer, which results in much more light reaching it in the summer. Three times more energy enters the Northern Hemisphere each day at midsummer than at midwinter.

2. How did the trade winds get their name? Is their name a reminder of the assistance they provided to shipboard traders eager to sell their wares in distant corners of the world?

The trade winds are not named after their contribution to commerce in the days of sail. This use of the word *trade* derives from an earlier English meaning equivalent to our adverbs *steadily* or *constantly.* These persistent winds were said to "blow trade."

3. If the Coriolis effect acts on all moving objects, why doesn't it pull my car to the right? And does the Coriolis effect really make explorers wander to the left in the snows of Antarctica, or tree trunks grow in rightward spirals in Canadian forests, or water swirl clockwise down a toilet in Springfield?

The Coriolis effect depends on the speed, mass, and latitude of the moving object. Your car's motion is affected. When you drive along the road at 70 miles per hour in an average-sized car, Coriolis "force" would pull your car to the right about 460 meters (1,500 feet) for every 160 kilometers (100 miles) you travel *if* it were not for the friction between your tires and the road surface.

Small, lightweight objects moving slowly are subject to many forces and conditions (such as wind currents, natural variations in basin shape, and friction) that overwhelm Coriolis acceleration. For example, think of how *very* small the difference in eastward speed of the northern edge of a toilet is in comparison to the southern edge. Any small irregularity in the toilet's shape will be

FIGURE 8.26 A vast extratropical cyclone on Mars. The storm is more than 1,600 kilometers (1,000 miles) across.

hundreds of times more important than the Coriolis effect in determining whether water will exit in a rightward spin or a leftward spin! Explorers and trees aren't massive enough and don't move quickly enough to be affected by the Coriolis effect.

But if the moving object is at mid-latitude, is heavy, and is moving quickly, it *will* be deflected to the right of its intended path. The computers aboard jetliners subtly nudge the flight path in the appropriate direction, but unguided devices (artillery shells, etc.) move noticeably.

4. Does the ocean affect weather at the centers of continents?

Absolutely. In a sense, *all* large-scale weather on Earth is oceanically controlled. The ocean acts as a solar collector and heat sink, storing and releasing heat. Most great storms (tropical and extratropical cyclones alike) form over the ocean and then sweep over land.

5. On average, the meteorological equator lies about 5° north of the geographical equator. Why the displacement?

The meteorological equator lies in the Northern Hemisphere because, overall, that hemisphere is slightly warmer than the Southern Hemisphere. At least three factors are responsible for this. First, the unbroken ice covering of Antarctica reflects more light back into space than the surface of the Arctic Ocean, which contains occasional light-absorbing patches of open water. Second, there is more land surface in the tropical latitudes of the Northern Hemisphere than in the Southern Hemisphere. Because land warms more than water with the same input of solar energy, tropical latitudes are warmer in the Northern Hemisphere. And third, ocean currents transport more warm water toward the north than toward the south.

6. Are any of the results of large-scale atmospheric circulation apparent to the casual observer?

Yes. The view of atmospheric circulation developed in this chapter explains some phenomena you may have experienced.

For instance, flying from Los Angeles to New York takes about 40 minutes less than flying from New York to Los Angeles because of westerly headwinds (indicated in Figure 8.13).

Because of these same prevailing westerlies, most storms travel over the United States from west to east. Weather prediction is based on observations and samples taken from the air masses as they move. Forecasting is often easier in the East and Midwest than in the West because more data are available from an air mass when it's over land.

7. Has anything similar to Earth's weather patterns been seen on other planets?

Yes, indeed. Saturn, Jupiter, and Neptune have tremendous cyclonic storms. A few on Jupiter are large enough to be seen from Earth through small telescopes. Tracks of tornadoes have recently been identified on Mars, and a huge cyclonic storm was photographed there in 1999 (**Figure 8.26**). Venus has huge cloud banks suggestive of polar fronts and extratropical cyclones. The Coriolis effect and uneven solar heating are common to all planets with atmospheres in this solar system, so we shouldn't be surprised by similarities.

Oceanography⊜Now™ | Click Oceanography Interactive to work through an activity on weather forecasting through weather and climate.

Chapter in Perspective

In this chapter you learned that Earth and ocean are in continuous contact, and conditions in one are certain to influence conditions in the other. The interaction of ocean and atmosphere moderates surface temperatures, shapes Earth's weather and climate, and creates most of the sea's waves and currents.

The atmosphere responds to uneven solar heating by flowing in three great circulating cells over each hemisphere. This circulation of air is responsible for about two-thirds of the heat transfer from tropical to polar regions. The flow of air within these cells is influenced by Earth's rotation. To observers on the surface, Earth's rotation causes moving air (or any moving mass) in the Northern Hemisphere to curve to the right of its initial path, and in the Southern Hemisphere to curve to the left. The apparent curvature of path is known as the Coriolis effect.

Uneven flow of air within cells is one cause of the atmospheric changes we call *weather*. Large storms are spinning areas of unstable air that occur between or within air masses. Extratropical cyclones originate at the boundary between air masses; tropical cyclones, the most powerful of Earth's atmospheric storms, occur within a single humid air mass. The immense energy of tropical cyclones is derived from water's latent heat of vaporization.

In the next chapter you will learn how movement of the atmosphere can cause movement of ocean water. Wind blowing over the ocean creates surface currents, and deep currents form when the ocean surface is warmed or cooled as the seasons change. Currents join with the atmosphere to form a giant heat engine that moves energy from regions of excess (tropics) to regions of scarcity (poles). This energy keeps the tropical seas from boiling away and the polar ocean from freezing solid in its basins.

The ocean's surface currents are governed by some of the principles you've learned here; the Coriolis effect and uneven solar heating continue to be important concepts in our discussion. Taken together, an understanding of air and water circulation is at the heart of physical oceanography.

Key Ideas

✓ Different amounts of solar energy are absorbed at different latitudes. The tropics are warmer than the polar regions because of this difference.
   Review Figures 8.4, 8.5, and 8.6, pages 180–181

✓ The interaction of ocean and atmosphere moderates surface temperatures, shapes Earth's weather and climate, and creates most of the sea's waves and currents.
   Review pages 179–182

✓ To observers on the surface, Earth's rotation causes moving air (or any moving mass) in the Northern Hemisphere to curve to the right of its initial path, and in the Southern Hemisphere to curve to the left. The apparent curvature of path is known as the Coriolis effect.
   Review Figures 8.9, 8.10, and 8.11, pages 182–184

✓ The Coriolis effect does not initiate the movement of air or water, but once moving, it influences the direction of their movement.
   Review Figures 8.12, and 8.13, pages 184–185

✓ The atmosphere responds to uneven solar heating by flowing in three great circulating cells over each hemi-

sphere: a Hadley cell, a Ferrel cell, and a polar cell. Air circulation within these cells is powered by uneven solar heating and influenced by the Coriolis effect.
   Review Figure 8.13, page 185

✓ Circulation of air is responsible for about two-thirds of the heat transfer from tropical to polar regions. (Ocean currents account for the other third.)
   Review page 180

✓ Large storms are spinning areas of unstable air that develop between or within air masses. Extratropical cyclones originate at the boundary between air masses. Tropical cyclones, the most powerful of Earth's atmospheric storms, occur within a single humid air mass.
   Review Figures 8.19 and 8.23, pages 191, 193, and Box 8.1, page 195

Terms and Concepts to Remember

air mass, 190
atmosphere, 178
atmospheric circulation cell, 184
Bjerknes, Vilhelm, 190
climate, 178
convection current, 182
Coriolis, Gaspard Gustave de, 182
Coriolis effect, 182
cyclone, 190
doldrums, 186
extratropical cyclone, 190
Ferrel cell, 185
front, 190
frontal storm, 190
geographical equator, 186
Hadley cell, 184
heat budget, 179
horse latitudes, 186
hurricane, 191

intertropical convergence zone (ITCZ) , 186
land breeze, 188
meteorological equator, 186
monsoon, 187
nor'easter (northeaster) , 190
polar cell, 185
polar front, 190
precipitation, 179
sea breeze, 188
storm, 188
storm surge, 196
thermal equilibrium, 179
tornado, 190
trade winds, 186
tropical cyclone, 191
water vapor, 178
weather, 178
westerlies, 186
wind, 178

Study Questions

Reviewing What You've Learned

1. What is the composition of air?
2. Can more water vapor be held in warm air or cool air?
3. Which is denser at the same temperature and pressure: humid air or dry air?
4. What happens when air containing water vapor rises?
5. What causes precipitation (rain and snow)?
6. How many atmospheric circulation cells exist in each hemisphere?

7. Where does air rise, and fall, in each hemisphere. How does this movement affect weather?

8. What are the two kinds of large storms? How do they differ? How are they similar?

9. What causes an extratropical cyclone? What happens in one?

10. What triggers a tropical cyclone? From what is its great power derived? What causes the greatest loss of life and property when a tropical cyclone reaches land?

Thinking Critically

1. What factors contribute to the uneven heating of Earth by the sun?

2. How does the atmosphere respond to uneven solar heating? How does the rotation of Earth affect the resultant circulation?

3. Why doesn't the ocean boil away at the equator and freeze solid near the poles?

4. Describe the atmospheric circulation cells in the Northern Hemisphere. At what latitudes does air move vertically? Horizontally? What are the trade winds? The westerlies? Where are deserts located? Why? What is ocean surface salinity like in these desert bands?

5. How are the geographical equator, meteorological equator, and ITCZ related? What happens at the ITCZ?

6. What is a monsoon? How is monsoon circulation affected by the position of the ITCZ?

7. If the Coriolis effect causes the clockwise deflection of moving objects in the Northern Hemisphere, why does air rotate counterclockwise around zones of low pressure in that hemisphere?

8. **InfoTrac College Edition Project** People have always been attracted to the seashore as a place to live. Suppose a friend of yours has just told you of a great opportunity to purchase beachfront property in an area on the Atlantic coast that is known to have hurricane warnings every year. Your friend asks what you think of the idea. Do some research using InfoTrac College Edition and compose an answer for your friend.

Thinking Analytically

1. There is no such thing as "suck." Imagine the palm of your hand covering an open, empty peanut butter jar. Now imagine the air being pumped out of the jar. Your hand is not being "sucked" into the jar. Your hand is forced tightly onto the jar by differential pressure—the air pressure outside the jar is greater than the air pressure inside. Knowing the mass of a 1-square-centimeter column of air, calculate the total inward force exerted on the palm of your hand if it were covering a jar with a diameter of 7.5 centimeters and all the air could be removed from inside the jar.

2. Why does water drip from the bottom of your car when the air conditioner is being used on an especially humid day?

3. Why is the longest day of the year almost never the hottest day of the year?

 Oceanography Now™

Assess your understanding of this chapter's topics with additional quizzing and conceptual-based problems at http://oceanography.brookscole.com/garrison5e

Study and Research on the Internet

Assess your understanding of this chapter's topics with additional quizzing and conceptual-based problems at:

http://oceanography.brookscole.com/garrison5e

■ **InfoTrac College Edition Research** leads you to critical thinking exercises that use InfoTrac College Edition® as a research tool.

■ **Regional InfoTrac College Edition** articles are organized into East Coast, West Coast, and Gulf Coast regions, allowing you to study oceanography on a more local level.

 For more readings, go to InfoTrac College Edition, your online research library, at:

http://infotrac.thomsonlearning.com

"Tropical" gardens on Britain's Scilly Isles. Only 48 kilometers (30 miles) off the coast of Cornwall at 50°N, these scenic islands lie in the path of the warm waters of the Gulf Stream. (Source: John Brunton, *Sky*, Delta Airlines' in-flight magazine, April 2003.)

Collections/Mike Pooler

CIRCULATION OF THE OCEAN

Palm Trees in Britain?

Commercial airliners flying to Europe from the west coast of the United States pass over central Ontario. In winter and spring months the ground is hidden under ice and snow, but passengers can often make out the frozen surface of James Bay. When those passengers land in London or Belfast, they find a much milder climate than the barren whiteness of central Canada. Yet London's latitude of 51°N is the same as that of the southern tip of James Bay, and Belfast lies at 54°N, the same latitude as Ontario's Polar Bear Provincial Park, a sanctuary for migrating polar bears. The gardens of the Scilly Isles at 50°N off the west coast of England feature a surprising variety of tropical plant species (**figure**).

Why the great difference in climate? The predominant direction of air flow is eastward at these latitudes. The air flowing toward James Bay loses heat as it passes over the frozen landmass of Canada, but the air moving over the ocean toward the British Isles is heated by contact with the warm Gulf Stream and North Atlantic Current. Ireland and England therefore have a mild maritime climate, and polar bears are found only in zoos. The cities of western Europe and Scandinavia are warmed by the energy of tropical sunlight transported to their northern latitudes by winds and by moving masses of water called **currents.** Currents also influence weather and climate, distribute nutrients, and scatter organisms.

The waters of the world ocean are layered. Surface currents driven by the wind affect the upper layer in and above the pycnocline, the zone in which density increases with depth. Some of these surface currents are rapid and riverlike with well-defined boundaries; others are slow and diffuse. Circulation of the deeper water beneath the pycnocline is driven by the force of gravity, as denser water sinks and less dense water rises. Currents near the seafloor move as slow streams in a few places, but the greatest volumes of deep water flow through the ocean at an almost imperceptible pace. The whole ocean slowly falls, rises, creeps from place to place.

As you read, notice the interplay between wind and water, heat and cold, salinity and density. Although El Niño and La Niña grab all the headlines, you'll find that ocean currents can influence your daily life (and the lives of polar bears) in surprising ways.

Surface Currents

About 10% of the water in the world ocean is involved in **surface currents,** water flowing horizontally in the uppermost 400 meters (1,300 feet) of the ocean's surface, driven mainly by wind friction. Most surface currents move water above the pycnocline, the zone of rapid density change with depth described in detail in Chapter 6.

The primary force responsible for surface currents is wind. As you read in Chapter 8, surface winds form global patterns within latitude bands (see Figure 8.13). Most of Earth's surface wind energy is concentrated in each hemisphere's trade winds (easterlies) and westerlies (**Figure 9.1**). Waves on the sea surface transfer some of the energy from the moving air to the water by friction. This tug of wind on the ocean surface begins a mass flow of water. The water flowing beneath the wind forms a surface current.

The moving water "piles up" in the direction the wind is blowing. Water pressure is higher on the "piled up" side, and the force of gravity pulls the water down the slope—against the *pressure gradient*—in the direction from which it came. But the Coriolis effect intervenes. Because of the Coriolis effect (discussed in Chapter 8), Northern Hemisphere surface currents flow to the *right* of the wind direction. Southern Hemisphere currents flow to the *left*. Continents and basin topography often block continuous flow and help to deflect the moving water into a circular pattern. This flow around the periphery of an ocean basin is called a **gyre** (*gyros* = a circle). Two gyres are shown in **Figure 9.2**.

Oceanography ⊜Now™ | Click Oceanography Interactive to work through an activity on world currents through waves, tides, and currents.

Flow Within a Gyre

Figure 9.3 shows the North Atlantic gyre in more detail. Though it flows continuously without obvious places where one current ceases and another begins, oceanographers subdi-

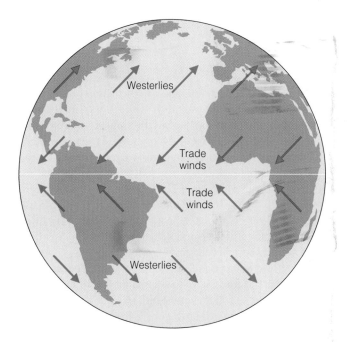

FIGURE 9.1 Winds, driven by uneven solar heating and Earth's spin, drive the movement of the ocean's surface currents. The prime movers are the powerful westerlies and the persistent trade winds (easterlies).

vide the North Atlantic gyre into four interconnected currents because each has distinct flow characteristics and temperatures. (Gyres in other ocean basins are similarly divided.) Notice that the east-west currents in the North Atlantic gyre flow to the right of the driving winds; once initiated, water flow in these currents continues in a roughly east-west direction. Where their flow is blocked by continents, the currents turn clockwise to complete the circuit.

Why does water flow around the *periphery* of the ocean basin instead of spiraling to the center? After all, the Coriolis effect influences any moving mass *as long as it moves,* so water

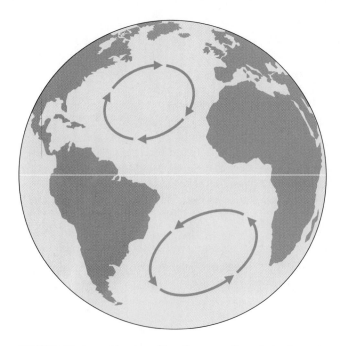

FIGURE 9.2 A combination of four forces—surface winds, the sun's heat, the Coriolis effect, and gravity—circulates the ocean surface clockwise in the Northern Hemisphere and counterclockwise in the Southern Hemisphere, forming gyres.

FIGURE 9.3 The North Atlantic gyre, a series of four interconnecting currents with different flow characteristics and temperatures.

in a gyre might be expected to curve to the center of the North Atlantic and stop. To understand this aspect of current movement, imagine the forces acting on the surface water at 45°N latitude (point A in **Figure 9.4**). Here the westerlies blow from the southwest, so initially the water will move toward the northeast. The rightward Coriolis deflection then causes the water to flow almost due east. A particle at 15°N latitude (point B) responds to the push of the trade winds from the northeast, however, and with Coriolis deflection it will flow almost due west. Water at the surface can flow at a velocity no greater than about 3% of the speed of the driving wind.

When driven by the wind, the topmost layer of ocean water in the Northern Hemisphere flows at about 45° to the right of the wind direction, a flow consistent with the arrows leading away from points A and B in Figure 9.4. But what about the water in the next layer down? It can't "feel" the wind at the surface; it "feels" only the movement of the water immediately above. This deeper layer of water moves *at an angle to the right* of the overlying water. The same thing happens in the layer below that, and the next layer, and so on, to a depth of about 100 meters (330 feet) at mid-latitudes. Each layer slides horizontally over the one beneath it like cards in a deck, with each lower card moving at an angle slightly to the right of the one above. Because of frictional losses, each lower layer also moves more slowly than the layer above. The resulting situation, portrayed in **Figure 9.5**, is known as **Ekman spiral** after the Swedish oceanographer who worked out the mathematics involved.

The term *spiral* is somewhat misleading; the water itself does not spiral downward in a whirlpool motion. Rather, the spiral is a way of conceptualizing the horizontal movements in a layered water column, with each layer moving in a slightly different hor-

izontal direction. An unexpected result of the Ekman spiral is that at some depth (known as the friction depth), water will be flowing in the opposite direction from the surface current!

The *net* motion of the water down to about 100 meters (330 feet), after allowance for the summed effects of the Ekman spiral (the sum of all the arrows indicating water direction in the affected layers), is known as **Ekman transport** (or *Ekman*

FIGURE 9.4 Surface water blown by the winds at point A will veer to the right of its initial path and continue eastward. Water at point B veers to the right and continues westward.

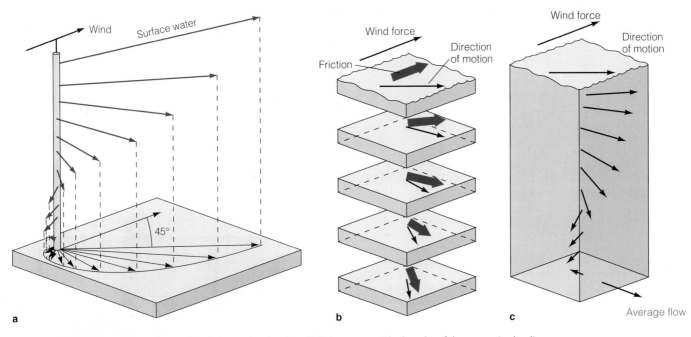

a   b   c

ACTIVE FIGURE 9.5 The Ekman spiral and the mechanism by which it operates. The lengths of the arrows in the diagrams are proportional to the speeds of the current in each layer. (**a**) The Ekman spiral model. (**b**) A body of water can be thought of as a set of layers. The top layer is driven forward by the wind, and each layer below is moved by friction. Each succeeding layer moves with a slower speed and at an angle to the layer immediately above it—to the right in the Northern Hemisphere, to the left in the Southern Hemisphere—until water motion becomes negligible. (**c**) Though the direction of movement varies for each layer in the stack, the theoretical net flow of water in the Northern Hemisphere is 90° to the right of the prevailing wind force. (Source for 9.5a: From *Laboratory Exercises in Oceanography* by Pipkin, Gorslin, Casey, and Hammond. © 1987 by W. H. Freemman and Co. Reprinted by permission. Source for 9.5b: NASA.)

Oceanography☉Now™ | At the Active Figures link at http://oceanography.brookscole.com/garrison5e you can view the mechanism of the Ekman spiral.

flow). In theory, the direction of Ekman transport is 90° to the *right* of the wind direction in the Northern Hemisphere and 90° to the *left* in the Southern Hemisphere.

Armed with this information, we can look in more detail at the area around point B in Figure 9.4, which is enlarged in **Figure 9.6**. In nature Ekman transport in gyres is less than 90°; in most cases the deflection barely reaches 45°. This deviation from theory occurs because of an interaction between the Coriolis effect and the pressure gradient. Some flowing Atlantic water has turned to the right and forms a hill of water; it followed the rightward dotted-line arrow in Figure 9.6.

Why does the water now go straight west from point B without turning? Because, as **Figure 9.7a** shows, to turn farther *right* the water would have to move uphill against the pressure gradient (and in defiance of gravity), but to turn *left* in response to the pressure gradient would defy the Coriolis effect. So the water continues westward and then clockwise around the whole North Atlantic gyre, dynamically balanced between the downhill urge of the pressure gradient and the uphill tendency of Coriolis deflection. The hill has consequences for deeper water as well. As **Figure 9.7b** shows, water flowing inward to form the hill sinks and depresses the thermocline.

Yes, there really *is* a hill near the middle of the North Atlantic, centered in the area of the Sargasso Sea; satellite images provide the evidence (**Figure 9.7c**). This hill is formed of surface water gathered at the ocean's center of circulation. It is not a steep mountain of water—its maximum height is an unspectacular

2 meters (6.5 feet)—but rather a gradual rise and fall from coastline to open ocean and back to opposite coastline. Its slope is so gradual you wouldn't notice it on a transatlantic crossing.

The hill is maintained by wind energy. If the winds did not continuously inject new energy into currents, then friction

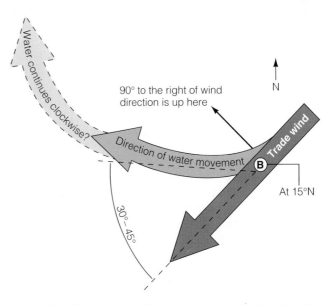

FIGURE 9.6 The movement of water away from point B in Figure 9.4 is influenced by the rightward tendency of the Coriolis effect and the gravity-powered movement of water down the pressure gradient.

a

b

Center of hill

NASA/JPL

c

FIGURE 9.7 The hill of water in the North Atlantic. (**a**) The surface of the North Atlantic is raised through wind motion and Ekman transport to form a low hill. Water from point B (see also Figures 9.4 and 9.6) turns westward and flows along the side of this hill. The westward-moving water is balanced between the Coriolis effect (which would turn the water to the right) and flow down the pressure gradient driven by gravity (which would turn it to the left). Thus water in a gyre moves along the outside edge of an ocean basin. (**b**) The hill is formed by Ekman transport. Water turns clockwise (inward) to form the dome and then descends, depressing the thermocline. (**c**) The average height of the surface of the North Atlantic is shown in color in this image derived from data taken in 1992 by the *TOPEX/Poseidon* satellite. Red indicates the highest surface; green and blue the lowest. Note that the measured position of the hill is offset to the west, as seen in (**a**). (The westward offset is explained in Figure 9.13.) The gradually sloping hill is only 2 meters (6.5 feet) high and would not be apparent to anyone traveling from coast to coast.

within the fluid mass and with the surrounding ocean basins would slow the flowing water, gradually converting its motion into heat. The balance of wind energy and friction, and of the Coriolis effect and the pressure gradient (through the effect of gravity), propels the currents of the gyre and holds them along the outside edges of the ocean basin.

Oceanography⊛Now™ | Click Oceanography Interactive to work through an activity to dissect a current through waves, tides, and currents.

Geostrophic Gyres

Gyres in balance between the pressure gradient and the Coriolis effect are called **geostrophic gyres** (*Geos* = Earth, *strophe* = turning), and their currents are *geostrophic currents*. Because of the patterns of driving winds and the present positions of continents, the geostrophic gyres are largely independent of one another in each hemisphere.

There are six great current circuits in the world ocean, two in the Northern Hemisphere and four in the Southern Hemisphere (**Figure 9.8**). Five are geostrophic gyres: the North Atlantic gyre, the South Atlantic gyre, the North Pacific gyre, the South Pacific gyre, and the Indian Ocean gyre. Though it is a closed circuit, the sixth and largest current is technically not a

gyre because it does not flow around the periphery of an ocean basin. The **West Wind Drift,** or **Antarctic Circumpolar Current,** as this exception is called, flows endlessly eastward around Antarctica, driven by powerful, nearly ceaseless westerly winds. This greatest of all the surface ocean currents is never deflected by a continent.

We might expect the two gyres in the North and South Pacific (and the two gyres in the North and South Atlantic) to converge exactly at the geographical equator. However, as Figure 9.8b shows, the junction of equatorial currents lies a few degrees north of the geographical equator, at the meteorological equator. As noted in Chapter 8, the meteorological equator and the intertropical convergence zone (the band at which the trade winds converge) are displaced 5° to 8° northward primarily because of the heat accumulated in the Northern Hemisphere's greater tropical land surface area. Ocean circulation, like atmospheric circulation, is balanced around the meteorological equator.

Currents Within Gyres

Because of the different factors that drive and shape them, the currents that form geostrophic gyres have different characteristics. Geostrophic currents may be classified by their position within the gyre as western boundary currents, eastern boundary currents, or transverse currents.

a

→ Warm-water current   → Cold-water current

b

ACTIVE FIGURE 9.8 Two ways of viewing the major surface currents of the world ocean. (**a**) An illustration of sea surface temperature showing the general direction and pattern of surface current flow. Sea surface temperatures were measured by a radiometer aboard *NOAA-7* in July 1984. The warmest temperatures, shown in red, are 25° to 28°C (77° to 82°F). Yellow represents 20° to 25°C (68° to 77°F); green 15° to 20°C (59° to 68°F); blue 15° to 0°C (59° to 32°F). The purple color around Antarctica and west of Greenland indicates water below 0°C (32°F), the freezing point of fresh water. Note the distortion of the temperature patterns we might expect from the effects of solar heating alone; the patterns twist clockwise in the Northern Hemisphere, counterclockwise in the Southern Hemisphere. (**b**) A chart showing the names and usual directions of the world ocean's major surface currents. The powerful western boundary currents flow along the western boundaries of ocean basins in *both* hemispheres.

Oceanography◯Now™ | At the Active Figures link at http://oceanography.brookscole.com/garrison5e you can view the major surface currents of the world ocean.

Western Boundary Currents The fastest and deepest geostrophic currents are found at the western boundaries of ocean basins (that is, off the *east* coast of continents). These narrow, fast, deep currents move warm water poleward in each of the gyres. As you can see in Figure 9.8b, there are five large **western boundary currents:** the Gulf Stream (in the North Atlantic), the Japan or Kuroshio Current (in the North Pacific), the Brazil Current (in the South Atlantic), the Agulhas Current (in the Indian Ocean), and the East Australian Current (in the South Pacific).

The **Gulf Stream** is the largest of the western boundary currents. Studies of the Gulf Stream, such as those undertaken by the drift submarine *Ben Franklin* (**Box 9.1**), have revealed that off Miami the Gulf Stream moves at an average speed of 2 meters per second (5 miles per hour) to a depth of more than 450 meters (1,500 feet). Water in the Gulf Stream can move more than 160 kilometers (100 miles) in a day. Its average width is about 70 kilometers (43 miles).

The volume of water transported in western boundary currents is extraordinary. The unit used to express volume transport in ocean currents is the **sverdrup (sv)**, named in honor of Harald Sverdrup, one of this century's pioneering oceanographers. A sverdrup equals 1 million cubic meters per second.[1] The Gulf Stream flow is at least 55 sv (55 million cubic meters per second), about 300 times the usual flow of the Amazon, the greatest of rivers. In **Figure 9.9** the surface currents of the North Atlantic gyre are shown with their volume transport (in sverdrups) indicated.

Water in a current, especially a western boundary current, can move for surprisingly long distances within well-defined boundaries, almost as if it were a river. In the Gulf Stream the current-as-river analogy can be startlingly apt: The western edge of the current is often clearly visible. Water within the current is usually warm, clear, and blue, often depleted of nutrients and incapable of supporting much life. By contrast, water over the continental slope adjacent to the current is often cold, green, and teeming with life. **Figure 9.10** shows its distinct appearance.

Long, straight edges are the exception rather than the rule in western boundary currents, however. Unlike rivers, ocean currents lack well-defined banks, and friction with adjacent water can cause a current to form waves along its edges. Western boundary currents meander as they flow poleward. The looping meanders sometimes connect to form turbulent rings, or **eddies**, that trap cold or warm water in their centers and then separate from the main flow. For example, *cold-core eddies* form in the Gulf Stream as it meanders eastward upon leaving the coast of North America off Cape Hatteras (**Figure 9.11**). *Warm-core eddies* can form north of the Gulf Stream when the warm current loops into the cold water lying to the north. When the loops are cut off, they become freestanding spinning masses of water. Warm-core eddies rotate clockwise, and cold-core eddies rotate counterclockwise.

The slowly rotating eddies move away from the current and are distributed across the North Atlantic. Some may be 1,000 kilometers (620 miles) in diameter and retain their identity for more than three years. In mid-latitudes as much as one-fourth of the surface of the North Atlantic may consist of old, slow-moving, cold-core eddy remnants! Both cold and warm eddies are visible in the satellite image in **Figure 9.12a**. Recent research suggests that their influence reaches to the seafloor. Warm- and cold-core eddies may be responsible for slowly

---

[1] One million cubic meters is about half the volume of the Louisiana Superdome.

FIGURE 9.9 The general surface circulation of the North Atlantic. The numbers indicate flow rates in sverdrups (1 sv = 1 million cubic meters of water per second).

FIGURE 9.10 Moving at a speed of about 10 kilometers (6 miles) an hour, the Gulf Stream departs the coast at Cape Hatteras. Its warm, clear, blue water contrasts with the cooler, darker, more productive water to the north and west. Clouds form over the warm current as water vapor evaporates from the ocean surface. Look for this area in Figure 9.12.

BOX 9.1

Streaming Along

In July 1969 six researchers sealed themselves into a specially designed submarine for a 2,640-kilometer (1,650-mile) drift within the Gulf Stream, the great ocean current that flows northward off the east coast of the United States. Their vehicle, named the *Ben Franklin* in honor of the man who first took a scientific interest in the current, was designed by Dr. Jacques Piccard, the builder of the bathyscaphe *Trieste* and father of balloonist Bertrand Piccard.

The 15-meter (50-foot) *Ben Franklin* was towed into the Gulf Stream off West Palm Beach, Florida, and positioned at a depth of 150 meters (500 feet). The month-long program of observations was designed to track the Gulf Stream and measure physical conditions within it. The scientists aboard were in regular contact with two surface ships that followed the drifting submarine and monitored its exact position. They kept a nearly continuous record of water temperature, salinity, and density.

The team made surprising discoveries right from the start. First, the submarine drifted more rapidly than expected. The speed of the surface current often exceeds 8 kilometers (about 5 miles) per hour, but oceanographers believed the current would move more slowly at greater depths. It didn't; the surface ships drifted at about the same rate as the submarine. And the drift was not always smooth. Underwater hills looming from the seabed, many of them previously uncharted, disrupted the flow of the submerged current and caused the submarine to rise and fall alarmingly. Gulf Stream eddies also caused problems. Thirteen days into the mission, a huge eddy spun the submarine 80 kilometers (50 miles) out of the Gulf Stream core, forcing it to surface in order to be towed back into the center of the stream. These great eddies were found to reach much deeper than previously believed (we now know that they can reach all the way to the ocean bottom).

Another surprise was the general lack of marine life. The crew did sight several species of sharks, a huge jellyfish with tentacles 10 centimeters (4 inches) thick, and an aggressive broadbilled swordfish that jousted with their vessel before swimming away unharmed; but this was less life than they had expected. The biology program did make some strides, however. Tape recordings of underwater sounds of biological origin were studied on board, and some of these recordings were used in the first detailed analyses of whale vocalizations. Dr. Piccard photographed some unusual bioluminescent organisms, a few of which were new to science. The crew also completed medical and psychological investigations on each other and analyzed the performance of the vehicle's life-support systems. These studies were used in later submarine and spacecraft design.

When the *Ben Franklin* finally bobbed to the surface 510 kilometers (317 miles) south of Halifax, Nova Scotia, researchers were confident of their biggest finding: They hadn't known as much about the Gulf Stream as they thought they did.

The submarine *Ben Franklin* enters New York harbor in August 1969 after drifting for 31 days within the Gulf Stream.

The Bettman Archive

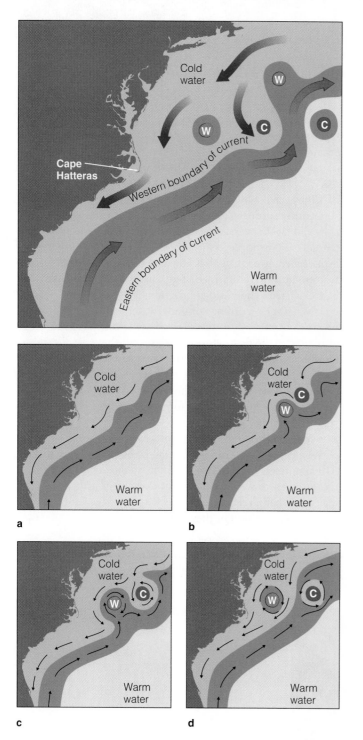

FIGURE 9.11 Eddy formation. The western boundary of the Gulf Stream is usually distinct, marked by abrupt changes in water temperature, speed, and direction. Meanders (eddies) form at this boundary as the Gulf Stream leaves the U.S. coast at Cape Hatteras (**a**). The meanders can pinch off (**b**), and eventually become isolated cells of warm water between the Gulf Stream and the coast (**c**). Likewise, cold cells can pinch off and become entrained in the Gulf Stream itself (**d**). (C = cold water, W = warm water; blue = cold, red = warm). Figure 9.12 shows the Gulf Stream from space with meanders and eddies clearly visible.

moving *abyssal storms*, which leave ripple marks that have been observed in deep sediments. Nutrients brought toward the surface by turbulence in eddies sometimes stimulate the growth of tiny marine plantlike organisms (**Figure 9.12b**).

Eastern Boundary Currents

As shown in Figure 9.8b, there are five **eastern boundary currents** at the eastern edge of ocean basins (that is, off the *west* coast of continents): the Canary Current (in the North Atlantic), the Benguela Current (in the South Atlantic), the California Current (in the North Pacific), the West Australian Current (in the Indian Ocean), and the Peru or Humboldt Current (in the South Pacific).

Eastern boundary currents are the opposite of their western boundary counterparts in nearly every way: They carry cold water equatorward; they are shallow and broad, sometimes more than 1,000 kilometers (620 miles) across; their boundaries are not well defined; and eddies tend not to form. Their total flow is less than that of their western counterparts. The Canary Current in the North Atlantic carries only 16 sv of water at about 2 kilometers (1.2 miles) per hour. The current is so shallow and broad that sailors may not notice it. Contrast the flow rates of the North Atlantic's western and eastern boundary currents in Figure 9.9. **Table 9.1** summarizes the major differences between boundary currents in the Northern Hemisphere.

Transverse Currents

As we have seen, most of the power for ocean currents is derived from the trade winds at the fringes of the tropics and from the mid-latitude westerlies. The stress of winds on the ocean in these bands gives rise to the **transverse currents**—currents that flow from east to west and west to east, linking the eastern and western boundary currents.

The tradewind-driven North and South Equatorial Currents in the Atlantic and Pacific are moderately shallow and broad, but each transports about 30 sv westward. Because of the thrust of the trades, Atlantic water at Panama is usually 20 centimeters (8 inches) higher, on average, than water across the isthmus in the Pacific. The Pacific's greater expanse of water at the equator and stronger trade winds develop more powerful westward-flowing equatorial currents, and the height differential between the western and eastern Pacific is thought to approach 1 meter (3.3 feet)!

Westerly winds drive the eastward-flowing transverse currents of the mid-latitudes. Because they are not shepherded by the trade winds, eastward-flowing currents are wider and flow more slowly than their equatorial counterparts. The North Pacific and North Atlantic Currents are Northern Hemisphere examples.

As can be seen in Figure 9.8, the westward flow of the transverse currents near the equator proceeds unimpeded for great distances, but the eastward flow of transverse currents at middle and high latitudes in the northern ocean basins is interrupted by continents and island arcs. In the far south, however, eastward flow is almost completely free. Intense westerly winds over the southern ocean drive the greatest of all ocean currents, the unobstructed West Wind Drift (or Antarctic Circumpolar Current). This current carries more water than any other—at least 100 sv west-to-east in the Drake Passage between the tip of South America and the adjacent Palmer Peninsula of Antarctica.

a

b

ACTIVE FIGURE 9.12 Eddies in western boundary currents. (**a**) The Gulf Stream viewed from space. This image is a composite of temperature data returned from NOAA polar orbiting meteorological satellites during the first week of April 1984. The composite image is printed with an artificial color scale: Reds and oranges are a warm 24° to 28°C (76° to 84°F); yellows and greens are 17° to 23°C (63° to 74°F); blues are 10° to 16°C (50° to 61°F); and purples are a cold 2° to 9°C (36° to 48°F). The Gulf Stream appears like a red (warm) river as it moves from the southern tip of Florida ① north along the east coast. Moving offshore at Cape Hatteras ② , it begins to meander, with some meanders pinching off to form warm-core ③ and cold-core ④ eddies. As it moves northeastward, the water cools dramatically, releasing heat to the atmosphere and mixing with the cooler surrounding waters. By the time it reaches the middle of the North Atlantic, it has cooled so much that its surface temperature can no longer be distinguished from that of the surrounding waters. (**b**) Eddies in another western boundary current—the Kuroshio, off Japan's east coast. The green color in this natural-color photograph indicates areas in which the growth of small plantlike organisms has been stimulated by nutrients brought to the surface by turbulence.

Oceanography ≋ Now™ | At the Active Figures link at http://oceanography.brookscole.com/garrison5e you can view the Gulf Stream from space.

O. Brown, R. Evans, M. Carle/University of Miami Rosenstiel School of Marine and Atmospheric Science

Table 9.1   Boundary Currents in the Northern Hemisphere

Type of Current (example)	General Features	Speed	Transport (millions of cubic meters per second)	Special Features
Western boundary currents	**Warm**			
Gulf Stream, Kuroshio (Japan) Current	Narrow, <100 km. Deep—substantial transport to depths of 2 km.	Swift, hundreds of kilometers per day.	Large, usually 50 sv or greater.	Sharp boundary with coastal circulation system; little or no coastal upwelling; waters tend to be depleted in nutrients, unproductive; waters derived from trade wind belts.
Eastern boundary currents	**Cold**			
California Current, Canary Current	Broad, ~1,000 km. Shallow, <500 m.	Slow, tens of kilometers per day.	Small, typically 10–15 sv.	Diffuse boundaries separating from coastal currents; coastal upwelling common; waters derived from mid-latitudes.

Source: From M. Grant Gross, *Oceanography: A View of the Earth,* 5/e, © 1990, p. 173. Prentice-Hall. Reprinted by permission.

Without the Coriolis effect, ocean gyres would look like this:

With the Coriolis effect, they look like this:

a

Center of geostrophic "hill" is offset to the west.

Gulf Stream
Steep slope
Top of hill
Sargasso Sea
Water surface
Gentle slope
Canary Current

Narrow, deep, warm, strong currents

Broad, shallow, cold, weak currents

b

FIGURE 9.13 (**a**) The influence of the Coriolis effect on westward intensification. Without the Coriolis effect water currents would form a regular and symmetrical gyre. However, because the Coriolis effect is strongest near the poles, water flowing eastward at high latitudes (red arrows) turns sooner to the right (clockwise), "short circuiting" the gyre. And because the Coriolis effect is nonexistent at the equator, water flowing westward near the equator (green arrows) tends not to turn clockwise until it encounters a blocking continent. Western boundary currents are therefore faster and deeper than eastern boundary currents, and the geostrophic hill is offset to the west. (**b**) A cross section of geostrophic flow in the North Atlantic. The Gulf Stream, a western boundary current, is narrow and deep and carries warm water rapidly northward. The Canary Current, an eastern boundary current, is shallow and wide and carries cold water at a much more leisurely pace. Although the gradually sloping hill is only 2 meters (6.5 feet) high and would not be apparent to anyone traveling from coast to coast, it is large enough to steer currents in the North Atlantic gyre. (This figure is vertically exaggerated.)

Westward Intensification Why should western boundary currents be concentrated and eastern boundary currents be diffuse? The reasons are complex, but as you might expect, the Coriolis effect is involved. Due to the Coriolis effect—which increases as water moves farther from the equator—eastward-moving water on the north side of the North Atlantic gyre is turned sooner and more strongly toward the equator than westward-flowing water at the equator is turned toward the pole. So the peak of the hill described in Figure 9.7 is not in the center of the ocean basin, but closer to its western edge. Its slope is steeper on the western side. If an equal volume of water flows around the gyre, this means the current on the eastern boundary (off the coast of Europe) is spread out and slow, and the current on the western boundary (off the U.S. coast) is concentrated and rapid. Western boundary currents are faster (up to ten times), deeper, and narrower (up to 20 times) than eastern boundary currents. The effect on current flow is known as **westward intensification** (**Figure 9.13**), a phenomenon clearly visible in Figures 9.7c and 9.9. Westward intensification doesn't happen just in the North Atlantic. The western boundary currents in the gyres of both hemispheres are more intense than their eastern counterparts.

Countercurrents and Undercurrents

Equatorial currents are typically accompanied by **countercurrents**, which flow on the surface in the opposite direction

from the main current. As you may remember, at the meteorological equator air is rising and the trade winds do not blow across the boundary. Without persistent winds to drive water to the west, some backward flow of water occurs at the meteorological equator. Some water flows away from the main equatorial currents (which flow westward a bit north and south of the meteorological equator) to return on the surface to the east. Look for this in the Pacific in Figure 9.8b.

Countercurrents can also exist *beneath* surface currents. These are sometimes referred to as **undercurrents**. The first undercurrent was discovered in 1951 in the central Pacific by Townsend Cromwell, a researcher employed by the U.S. Fish and Wildlife Service to investigate deep long-line fishing techniques pioneered by the Japanese. The Pacific Equatorial Undercurrent, or Cromwell Current, flows eastward beneath the North Equatorial Current with an average velocity of 5 kilometers (3 miles) per hour at a depth of 100 to 200 meters (330 to 660 feet). It is about 300 kilometers (190 miles) wide and carries a volume equivalent to about half the Gulf Stream. It has been traced for more than 14,000 kilometers (8,700 miles), from New Guinea to Ecuador. Undercurrents have since been found under most major currents. They can be very large—their volumes sometimes approach the volume of the current above.

Undercurrents can influence conditions at the ocean surface. The ocean around the Galápagos Islands is cold at the surface, even though the islands straddle the equator. For many years the low temperature was blamed on Antarctic water driven north in the Peru (Humboldt) Current, but recent research has shown that this surface water turns west before reaching the islands. The chill is actually caused by the upwelling of the Pacific Equatorial Undercurrent (the Cromwell Current) near the eastern end of its travel. Once at the surface, the cool water moves westward as part of the South Equatorial Current. The Galápagos Islands lie in its path.

Oceanographers are uncertain about what causes undercurrents, but near the equator (where the Coriolis effect is negligible) some water is not deflected north or south when it reaches the west side of the basin. It stays on the equator and underflows the surface current back to the east. This eastward-flowing current is kept on course by the Coriolis effect. If the current veers north from the equatorial region, the Coriolis effect bends it to the right, or south, back to its original course. Should the current wander too far south, the Coriolis effect in the Southern Hemisphere deflects the water to the left (north).

Exceptional Surface Currents

Some currents don't behave in ways you might first expect. Local or seasonal variations in wind and geography sometimes intrude on the flow patterns we have discussed. Two of these exceptions are noted here.

Monsoon Currents From November to March the northeast trade winds propel water in the usual way, producing normal surface circulation. In the summer months, however, the intertropical convergence zone moves north. Winds over the Indian Ocean change direction to blow from the south and southwest (see again Figure 8.16). Water in the relatively small northern Indian Ocean begins to flow clockwise, and the normally westward-flowing North Equatorial Current south of India reverses flow to become the eastward-moving Southwest Monsoon Current. (Due to its temporary nature, this current is not shown in Figure 9.8b.)

High-Latitude Currents Another exception to expected surface flow is seen in the many lesser current circuits at high northern latitudes in Figure 9.8b. Some of these—such as the Greenland and Labrador Currents in the North Atlantic and the Kamchatka and Alaska Currents in the North Pacific—move in apparent contradiction to the influence of the Coriolis effect, gravity, and friction. These currents form when larger currents are split and deflected by collision with a continent. The polar easterly winds at these latitudes keep the water moving westward. The size of these high-latitude currents on the map is exaggerated by the distortion inherent in the Mercator projection (see Appendix IV). The direction and rate of flow of these smaller currents are controlled primarily by the tendency of water to flow around obstacles rather than by the direct action of winds or Coriolis deflection. Technically these are not geostrophic currents, but their importance is often great. The Greenland and Labrador Currents, for example, transport icebergs to the Atlantic shipping lanes and deliver nutrients to the rich fishing grounds off Newfoundland.

Gyres: A Final Word

Although I have stressed individual currents in our discussion, remember that gyres consist of currents that blend into one another. Flow is continuous without obvious places where one current ceases and another begins. The *balance* of wind energy, friction, the Coriolis effect, and the pressure gradient propels gyres and holds them along the outside of ocean basins.

Effects of Surface Currents on Climate

Along with the winds, surface currents distribute tropical heat worldwide. Warm water flows to higher latitudes, transfers heat to the air and cools, moves back to low latitudes, absorbs heat again, and repeats the cycle. The greatest amount of heat transfer occurs at mid-latitudes, where about 10 million billion calories of heat are transferred each second—more than a million times the power consumed by all the world's human population in the same length of time! This combination of water flow and heat transfer from and to water influences climate and weather in several ways.

In winter, for example, Edinburgh, Dublin, and London are bathed in eastward-moving air only recently in contact with the relatively warm North Atlantic Current. Scotland, Ireland, and England have a maritime climate. As you read in this chapter's opener, they are warmed in part by the energy of tropical sunlight transported to high latitudes by the Gulf Stream (see once again Figure 9.8b).

At lower latitudes on an ocean's eastern boundary the situation is often reversed. Mark Twain is supposed to have said that the coldest winter he ever spent was a summer in San Francisco. Summer months in that West Coast city are cool, foggy, and mild, while Washington, D.C., on nearly the same line of latitude (but on the western boundary of an ocean basin), is known for its August heat and humidity. Why the difference? Look at Figure 9.8b and follow the currents responsible. The California Current, carrying cold water from the north, comes close to the coast at San Francisco. As shown in **Figure 9.14,** air normally

FIGURE 9.14 General summer air circulation patterns of the east and west coasts of the United States. Warm ocean currents are shown in red; cold currents in blue. Air is chilled as it approaches the west coast and warmed as it approaches the east coast.

a

FIGURE 9.15 Equatorial upwelling. (**a**) The South Equatorial Current, especially in the Pacific, straddles the geographic equator (see again Figure 9.8b). Water north of the equator veers to the right (northward), and water to the south veers to the left (southward). Surface water therefore diverges, causing upwelling. Most of the upwelled water comes from the area above the equatorial undercurrent, at depths of 100 meters (330 feet) or less. (**b**) The phenomenon of equatorial upwelling is worldwide but most pronounced in the Pacific. Red, orange, and yellow mark the areas of greatest upwelling as determined by biological productivity.

May Global Wind-induced Upwelling (cm/day)

b

flows clockwise in summer around an offshore zone of high atmospheric pressure. Wind approaching the California coast loses heat to the cold sea and comes ashore to chill San Francisco. Summer air often flows around a similar high off the east coast (the Bermuda High). Winds approaching Washington, D.C., therefore blow from the south and east. Heat and moisture from the Gulf Stream contribute to the capital's oppressive summers. (In winter, on the other hand, Washington, D.C., is colder than San Francisco because westerly winds approaching Washington are chilled by the cold continent they cross.)

Upwelling and Downwelling

9.11

The wind-driven *horizontal* movement of water can sometimes induce *vertical* movement in the surface water. This movement is called **wind-induced vertical circulation.** Upward movement of water is known as **upwelling;** the process brings deep, cold, usually nutrient-laden water toward the surface. Downward movement is called **downwelling.**

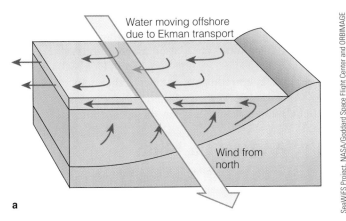

Water moving offshore due to Ekman transport

Wind from north

a

Chlorophyll Concentration (mg / m³)
<0.04 .05   0.1   0.2 0.3 0.5   1   2   3 4 >5

SeaWiFS Project, NASA/Goddard Space Flight Center and ORBIMAGE

FIGURE 9.16 Coastal upwelling. (**a**) In the Northern Hemisphere coastal upwelling can be caused by winds from the north blowing along the west coast of a continent. Water moved offshore by Ekman transport is replaced by cold, deep, nutrient-laden water. (**b**) A satellite view of the U.S. west coast shows (in artificial color) the growth of small plantlike organisms stimulated by upwelled nutrients. The color bar indicates the concentration of chlorophyll in milligrams per cubic meter of seawater. Notice that chlorophyll concentration—and biological productivity—is highest near the coast.

b

Equatorial Upwelling

The South Equatorial Currents of the Atlantic and Pacific straddle the equator. Though the Coriolis effect is weak near the equator (and absent *at* the equator), water moving in the currents on either side of the equator is deflected slightly poleward and replaced by deeper water (**Figure 9.15**). Thus **equatorial upwelling** occurs in these westward-flowing equatorial surface currents. Upwelling is an important process because this water from within and below the pycnocline is often rich in the nutrients needed by marine organisms for growth. The long, thin band of upwelling and biological productivity extending along the equator westward from South America is clearly visible in Figures 9.19b and 9.22. The layers of ooze on the equatorial Pacific seabed (see Figure 5.10) are testimony to the biological productivity of surface water there. By contrast, generally poor conditions for growth prevail in most of the open tropical ocean—because strong layering isolates deep, nutrient-rich water from the sunlit ocean surface.

Coastal Upwelling

Wind blowing parallel to shore or offshore can cause **coastal upwelling**. The friction of wind blowing along the ocean sur-

face causes the water to begin moving, the Coriolis effect deflects it to the right (in the Northern Hemisphere), and the resultant Ekman transport moves it offshore. As shown in **Figure 9.16a**, coastal upwelling occurs when this surface water is replaced by water rising along the shore. Again, because the new surface water is often rich in nutrients, prolonged wind can result in increased biological productivity. Coastal upwelling along the coast of California is visible in **Figure 9.16b**.

Upwelling can also influence weather. Wind blowing from the north along the California coast causes offshore movement of surface water and subsequent coastal upwelling. The overlying air becomes chilled, contributing to San Francisco's famous fog banks and cool summers. Wind-induced upwelling is also common in the Peru Current, along the west coast of Antarctica's Palmer Peninsula, in parts of the Mediterranean, and near some large Pacific islands.

Downwelling

Water driven toward a coastline will be forced downward, returning seaward along the continental shelf. This downwelling (**Figure 9.17**) helps supply the deeper ocean with dissolved gases and nutrients, and it assists in the distribution of living

 http://oceanography.brookscole.com/garrison5e

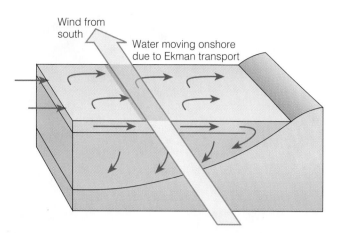

FIGURE 9.17 Wind blowing from the south along a Northern Hemisphere west coast for a prolonged period can result in downwelling. Areas of downwelling are often low in nutrients and therefore relatively low in biological productivity.

organisms. Unlike upwelling, downwelling has no direct effect on the climate or productivity of the adjacent coast.

Langmuir Circulation

Winds that blow steadily across the ocean, and the small waves that such winds generate, can induce long sets of counterrotating vortices (or cells) in the surface water. These slowly twisting vortices align in the direction of the wind (**Figure 9.18**). It takes about an hour for a particle in a vortex to complete one revolution. Streaks of foam or seaweed or debris, known as *windrows*, collect in areas where adjacent vortices converge, while regions of divergence remain relatively clear. Observed by mariners for hundreds of years, these windrow lines (and the underlying vortices) were first explained in 1938 by Irving

Langmuir, who observed them while crossing the calm Sargasso Sea in the center of the North Atlantic gyre. Named in his honor, **Langmuir circulation** rarely disturbs the ocean below a depth of about 20 meters (66 feet). Unlike the equatorial and coastal upwellings mentioned earlier, Langmuir circulation operates within the surface layer and thus does not lift nutrients trapped within or below the pycnocline.

El Niño and La Niña

Surface winds across most of the tropical Pacific normally move from east to west (review Figure 8.13). The trade winds blow from the normally high-pressure area over the eastern Pacific (near Central and South America) to the normally stable low-pressure area over the western Pacific (north of Australia). However, for reasons that are still unclear, these pressure areas change places at irregular intervals of roughly three to eight years: High pressure builds in the western Pacific, and low pressure dominates the eastern Pacific. Winds across the tropical Pacific then reverse direction and blow from west to east—the trade winds weaken or reverse. This change in atmospheric pressure (and thus in wind direction) is called the **Southern Oscillation**. Ten of these attention-getting oscillations have occurred since 1950.

The trade winds normally drag huge quantities of water westward along the ocean's surface on each side of the equator, but as the winds weaken these equatorial currents crawl to a stop. Warm water that has accumulated at the western side of the Pacific—the warmest water in the world ocean—can then build to the east along the equator toward the coast of Central and South America. The eastward-moving warm water usually arrives near the South American coast around Christmastime. In the 1890s it was reported that Peruvian fishermen were using the expression *Corriente del Niño* ("current of the Christ Child") to describe the flow—thus the current's name,

FIGURE 9.18 Winds that blow steadily across the ocean, and the small waves that such winds generate, can induce long sets of counterrotating vortices in the surface water. These shallow, slowly twisting cells of surface water are known as Langmuir circulation in honor of the researcher who explained their motion.

Windrows (convergences where seaweed, debris, and foam accumulate)

Divergence

Sea surface

Wind

Helical vortices

~50 m
~165 ft

~6 m
~20 ft

Level of no motion

Downwelling (2–6 cm/sec)

Upwelling (1–2 cm/sec)

Mobil Oil Corporation

a

b

NASA

FIGURE 9.19 A non–El Niño year. (**a**) Normally the air and surface water flow westward, the thermocline rises, and upwelling of cold water occurs along the west coast of Central and South America. (**b**) This map from satellite data shows the temperature of the equatorial Pacific on 31 May 1988. The warmest water is indicated by dark red, and progressively cooler upwelled water by yellow and green. Note the coastal upwelling along the coast at the lower right of the map, and the tongue of recently upwelled water extending westward along the equator from the South American coast.

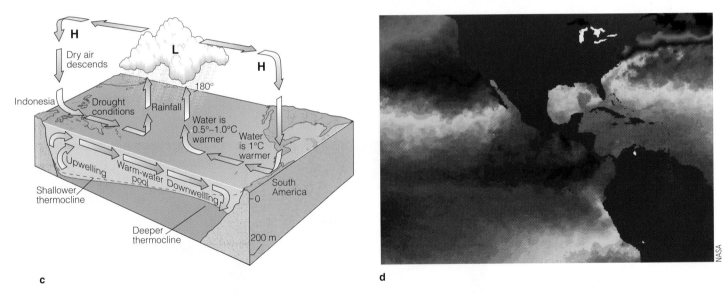

c

d

NASA

An El Niño year. (**c**) When the Southern Oscillation develops, the trade winds diminish and then reverse, leading to an eastward movement of warm water along the equator. The surface waters of the central and eastern Pacific become warmer, and storms over land may increase. (**d**) Sea surface temperatures on 13 May 1992, a time of El Niño conditions. The thermocline was deeper than normal, and equatorial upwelling was suppressed. Note the absence of coastal upwelling along the coast and the lack of the tongue of recently upwelled water extending westward along the equator.

El Niño. The phenomena of the Southern Oscillation and El Niño are coupled, so the terms are often combined to form the acronym **ENSO**, for El Niño/Southern Oscillation. An ENSO event typically lasts about a year, but some have persisted longer than three years. The effects are felt not only in the Pacific; all ocean areas at trade-wind latitudes in both hemispheres can be affected.

Normally, a current of cold water, rich in upwelled nutrients, flows north and west away from the South American continent (**Figure 9.19a** and **b**). When the propelling trade winds falter during an ENSO event, warm equatorial water that would normally flow westward in the equatorial Pacific backs up to flow east (**Figure 9.19c** and **d**). The normal northward flow of the cold Peru Current is interrupted or overridden by the warm water. Upwelling within the nutrient-laden Peru Current is responsible for the great biological productivity of the ocean off the coasts of Peru and Chile. Although upwelling may continue during an ENSO event, the source of the upwelled water is nutrient-depleted water in the thickened surface layer approaching from the west. When the Peru Current slows and its upwelled water lacks nutrients, the fish and seabirds dependent on the abundant life it contains die or migrate elsewhere. Peruvian fishermen are never cheered by this Christmas gift!

During major ENSO events, sea level rises in the eastern Pacific, sometimes by as much as 20 centimeters (8 inches) in the Galápagos. Water temperature also increases by up to 7°C (13°F). The warmer water causes more evaporation, and the area of low atmospheric pressure over the eastern Pacific intensifies. Humid air rising in this zone, centered some 2,000 kilometers (1,200 miles) west of Peru, causes high precipitation in normally dry areas. The increased evaporation intensifies coastal storms, and rainfall inland may be much higher than normal. Marine and terrestrial habitats and organisms can be affected by these changes.

The two most severe ENSO events of this century occurred in 1982–83 and 1997–98 (**Figure 9.20**). In both cases, effects associated with El Niño were spectacular over much of the Pacific and some parts of the Atlantic and Indian Oceans. In February 1998, 40 people were killed and 10,000 buildings damaged by a "wall" of tornadoes advancing over the southeastern United States. This record-breaking tornado event was spawned by the collision of warm, moist air that had lingered over the warm Pacific and a polar front that dropped from the north. In the eastern Pacific, heavy rains throughout the 1997–98 winter in Peru left at least 250,000 people homeless, destroyed 16,000 dwellings, and

closed every port in the country for at least one month. Hawaii, however, experienced record drought, and some parts of southwestern Africa and Papua New Guinea received so little rain that crops failed completely and whole villages were abandoned due to starvation. Most of the United States escaped serious consequences; indeed, the midwestern states, Pacific Northwest, and Eastern Seaboard enjoyed a relatively mild fall, winter, and spring. But California's trials were widely reported: Rainfall in most of the state exceeded twice normal amounts, and landslides, avalanches, and other weather-related disasters crowded the evening news. Conditions did not return to near normal until the late spring of 1998. Estimates of worldwide 1997–98 ENSO-related damage exceed 23,000 deaths and $33 billion.

Normal circulation sometimes returns with surprising vigor, producing strong currents, powerful upwelling, and chilly and stormy conditions along the South American coast. These contrasting colder-than-normal events are given a contrasting name: **La Niña** ("the girl"). As conditions to the east cool off, the ocean to the west (north of Australia) warms rapidly. The renewed thrust of the trade winds piles this water upon itself, depressing the upper curve of the thermocline to more than 100 meters (328 feet). In contrast, the thermocline during a La Niña event in the eastern equatorial Pacific rests at about 25 meters (82 feet). A vigorous La Niña followed the 1997–98 El Niño and persisted for nearly a year (**Figure 9.20e**).

Studies of the ocean and atmosphere in 1982–83 and 1997–98 have given researchers new insight into the behavior and effects of the Southern Oscillation. Some researchers believe that the 1982–83 event was triggered by the violent 1982 eruption of El Chichón, a Mexican volcano, which injected huge quantities of sun-obscuring dust and sulfur-rich gases into

FIGURE 9.20 Development of the 1997–98 El Niño, observed by the *TOPEX/Poseidon* satellite.

(**a**) March 1997. The slackening of the trade winds and westerly wind bursts allow warm water to move away from its usual location in the western Pacific Ocean. Red and white indicate sea level above average height.

(**b**) April 1997. About a month after it began to move, the leading edge of the warm water reaches South America.

the atmosphere. No similar trigger occurred before the 1997–98 ENSO, however. Though the exact cause or causes of the Southern Oscillation are not yet understood, subtle changes in the atmosphere permit meteorologists to predict a severe El Niño nearly a year in advance of its most serious effects.

Oceanography◎Now™ | Click Oceanography Interactive to work through an activity about El Niño through weather and climate.

Thermohaline Circulation

The surface currents we have discussed affect the uppermost layer of the world ocean (about 10% of its volume), but horizontal and vertical currents also exist below the pycnocline in the ocean's deeper waters. The slow circulation of water at great depths is driven by density differences rather than by wind energy. Because density is largely a function of water temperature and salinity, the movement of water due to differences in density is called **thermohaline circulation** (*therme* = heat, *halos* = salt). Virtually the entire ocean is involved in slow thermohaline circulation, a process responsible for most of the vertical movement of ocean water and the circulation of the global ocean as a whole.

Water Masses

As you may recall from Chapter 6, the ocean is density stratified, with the most dense water near the seafloor and the least dense near the surface. Each water mass has specific tempera-ture and salinity characteristics. Density stratification is most pronounced at temperate and tropical latitudes because the temperature difference between surface water and deep water is greater there than near the poles.

The water masses possess distinct, identifiable properties. Like air masses, water masses don't often mix easily when they meet due to their differing densities; instead, they usually flow above or beneath each other. Water masses can be remarkably persistent and will retain their identity for great distances and long periods of time. Oceanographers name water masses according to their relative position. In temperate and tropical latitudes, there are five common water masses:

- *Surface water*, to a depth of about 200 meters (660 feet)
- *Central water*, to the bottom of the main thermocline (which varies with latitude)
- *Intermediate water*, to about 1,500 meters (5,000 feet)
- *Deep water*, water below intermediate water but not in contact with the bottom, to a depth of about 4,000 meters (13,000 feet)
- *Bottom water*, water in contact with the seafloor

Surface currents move in the relatively warm upper environment of surface and central water. The boundary between central water and intermediate water is the most abrupt and pronounced.

No matter at what depth they are located, each water mass has characteristics that are usually determined by the conditions of heating, cooling, evaporation, and dilution that occurred at the ocean surface when the mass was formed. The densest (and deepest) masses were formed by surface

(**c**) May 1997. Warm water piles up against the South American continent. The white area of sea level is 13 to 30 centimeters (5 to 12 inches) above normal height, and 1.6° to 3°C (3° to 5°F) warmer.

(**d**) October 1997. By October sea level is as much as 30 centimeters (12 inches) lower than normal near Australia. The bulge of warm water has spread northward along the coast of North America from the equator to Alaska. Fisheries in Peru are severely affected: The warm water prevents upwelling of cold, nutrient-rich water necessary for the support of large fish populations.

(**e**) Normal circulation sometimes returns with surprising vigor after an El Niño event, producing strong currents, powerful up-welling, and chilly and stormy conditions along the South American coast. This image was prepared from data for 17 February 1999. Note the mass of cold surface water and relatively low sea level (purple). Such cold water tends to deflect winds around it, changing the course of weather systems locally and the nature of weather patterns globally.

conditions that caused the water to become very cold and salty. Water masses near the surface can be warmer and less saline; they may have formed in warm areas where precipitation exceeded evaporation. Water masses at intermediate depths are intermediate in density.

In spite of this differentiation, the relatively cold water masses lying beneath the thermocline exhibit smaller variations in salinity and temperature than the water in the currents that move across the ocean's surface.

The Temperature–Salinity Diagram

Perhaps the best way to visualize ocean layering is with a **temperature–salinity (T-S) diagram** like the one in **Figure 9.21**. The S-shaped curve through the center of the figure shows the temperature and salinity of water at each depth indicated in this area of ocean. Note that many *combinations* of temperature and salinity can yield the *same* density, and that the density of the water tends to increase with depth. The shape of the S-curve is governed by the position and nature of water masses.

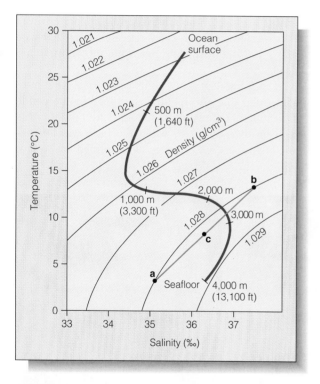

FIGURE 9.21 A general temperature–salinity (T-S) diagram, with density curves added. The S curve tracks the combination of temperature and salinity (and therefore density) with increasing depth. Note that the top of the S curve represents the temperature and salinity (and therefore density) of water at the ocean surface; the bottom of the curve represents the density of water at the ocean floor. Intermediate depths are shown on the curve.

Points a and b are on the same gently curving *isopycnal*—line of equal density. Note that seawater at points a and b have *different* temperatures and salinities but the *same* density. If these two masses of water were to merge, their combined density would be greater than either of their separate densities—a situation represented at point c. This process is known as *caballing* and can lead to the formation of deep water (see Figures 9.22 and 9.25).

Formation and Downwelling of Deep Water

Antarctic Bottom Water, the most distinctive of all deep-water masses, is characterized by a salinity of 34.65‰, a temperature of −0.5°C (30°F), and a density of 1.0279 grams per cubic centimeter. This water is noted for its extreme density (the densest in the world ocean), for the great amount of it produced near Antarctic coasts, and for its ability to migrate north along the seafloor.

Most Antarctic Bottom Water forms in the Weddell Sea during winter. Salt is concentrated in pockets between crystals of pure water and then squeezed out of the freezing mass to form a frigid brine. Between 20 and 50 million cubic meters of this brine form every second! The water's great density causes it to sink toward the continental shelf, where it mixes with nearly equal parts of water from the southern Antarctic Circumpolar Current.

The mixture settles along the edge of Antarctica's continental shelf, descends along the slope, and spreads along the deep-sea bed, creeping north in slow sheets. Antarctic Bottom Water flows many times more slowly than the water in surface currents: In the Pacific it may take a thousand years to reach the equator. Six hundred years later it may be as far away as the Aleutian Islands at 50°N! Antarctic Bottom Water also flows into the Atlantic Ocean basin, where it flows north at a faster rate than in the Pacific. Antarctic Bottom Water has been identified as high as 40° *north* latitude on the Atlantic floor, a journey that has taken some 750 years.

Some dense bottom water also forms in the northern polar ocean, but the topography of the Arctic Ocean basin prevents most of it from escaping, except in the deep channels formed in the submarine ridges separating Scotland, Iceland, and Greenland. These channels allow the cold, dense water formed in the Arctic to flow into the North Atlantic to form **North Atlantic Deep Water** (look ahead to Figure 9.25). Very little deep Arctic Ocean water enters the Pacific.

North Atlantic Deep Water forms when the relatively warm and salty North Atlantic Ocean cools as cold winds from northern Canada sweep over it. Exposed to the chilled air, water at the latitude of Iceland releases heat, cools from 10° to 2°C (50° to 36°F), and sinks. (Transferred to the air, this bonus heat helps to moderate European winters.)

Other distinct deep-water masses exist. Their positions in relation to each other are always determined by their relative densities. As can be seen in **Figure 9.22,** similar water forms at the Antarctic Circumpolar Current in the South Atlantic. Pacific Deep Water also forms in the Antarctic Circumpolar Current and along the east coast of the Kamchatka Peninsula. Atlantic or Pacific Deep Water is less dense than Antarctic Bottom Water and so floats above it, out of contact with the ocean floor.

A deep-water mass also forms in the enclosed Mediterranean Sea, where surface water is made more saline by the excess of evaporation over freshwater input. About 300,000 cubic kilometers (72,000 cubic miles) more water evaporates annually from the Mediterranean than is replaced by river runoff or precipitation. In the cool winter months, Mediterranean water with a salinity of about 38‰ flows past the lip of

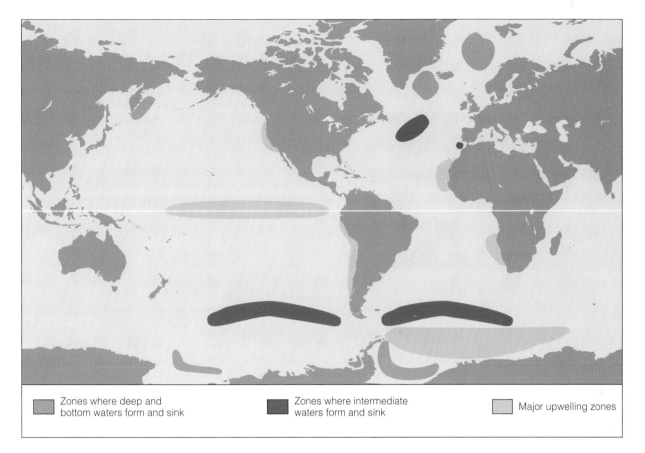

FIGURE 9.22 Places where most surface water sinks and where most deep or intermediate water rises.

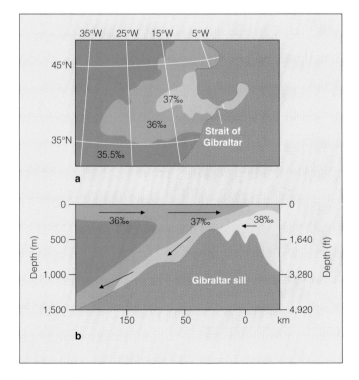

FIGURE 9.23 Saline flow over the sill of the Mediterranean Sea at Gibraltar. (**a**) A map showing the Strait of Gibraltar; salinity was measured at 1,000 meters. (**b**) A vertical section at the strait, showing the movement of water masses of different salinity.

Gibraltar and spreads into the Atlantic as Mediterranean Deep Water (**Figure 9.23**). Mediterranean Deep Water underlies much of the central water mass in the Atlantic, and some of this water can be traced as far south as the basins of the Antarctic. Though saltier than Antarctic Bottom Water or Atlantic Deep Water, Mediterranean Deep Water is considerably warmer and therefore not as dense. It will lie atop the layers of denser water at high southern latitudes.

Researchers can determine the age of deep water by analyzing its dissolved oxygen content. Ocean water picks up free oxygen only at or near the surface by contact with the atmosphere or through the action of photosynthetic organisms. After leaving the surface, the water gradually loses its oxygen through the respiration of organisms or by chemical reactions with sediments, rocks, or dissolved components. The dissolved oxygen content of a water mass is therefore a rough index of its age—the length of time since the water left the surface. Researchers have also found that water masses slowly mix with the surrounding water as they flow away from their sources, losing their individual identity as they grow older.

Thermohaline Circulation Patterns and Water Masses

9.21

The great quantities of dense water sinking at ocean basin edges must be offset by equal quantities of water rising

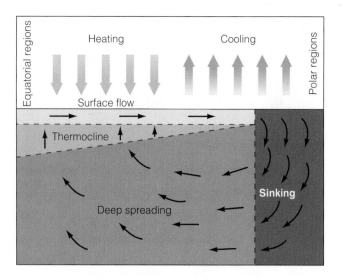

FIGURE 9.24 The classic model of a pure thermohaline circulation, caused by heating in lower latitudes and cooling in higher latitudes. Compare this figure to Figure 8.7.

found everywhere at low and mid-latitudes. This slow upward movement is estimated to be about 1 centimeter ($\frac{1}{2}$ inch) per day over most of the ocean. If this rise were to stop, downward movement of heat would cause the thermocline to descend and would reduce its steepness. In a sense, the thermocline is "held up" by the continual slow upward movement of water.

Most features of this ideal circulation pattern exist in nature. **Figure 9.25** shows deep circulation in the Atlantic. The water masses, each of distinct density and sandwiched in layers, are slowly propelled by gravity. The water masses butt against one another in **convergence zones,** and the heavier water can slide beneath the lighter water (**Figure 9.26**).

In some cases, however, two distinct water masses with the same *density* but with different *temperatures and salinities* combine at a convergence to produce a new water mass of greater density. This mixing-and-sinking process is called **caballing.** To understand how caballing works, imagine the two water masses indicated by points a and b in Figure 9.21. Though each has a different temperature and salinity, they lie on the same density line—they are equally dense. Now imagine the waters at these points mixing together. The density of the new water mass could be represented by a point halfway along a straight line connecting them—point c. The combined water mass is more dense than 1.0280—perhaps 1.0283—and will tend to sink. North Atlantic Intermediate Water, Antarctic Intermediate Water, and some Antarctic Bottom Water are produced by caballing.

Hundreds of years may pass before water masses complete a circuit or blend to lose their identities. Remember, Antarctic Bottom Water in the Pacific retains its character for up to

elsewhere. **Figure 9.24** shows an idealized model of thermohaline flow. Note that water sinks relatively rapidly in a small area where the ocean is very cold, but it rises much more gradually across a very large area in the warmer temperate and tropical zones. It then slowly returns poleward near the surface to repeat the cycle. The continual diffuse upwelling of deep water maintains the existence of the permanent thermocline

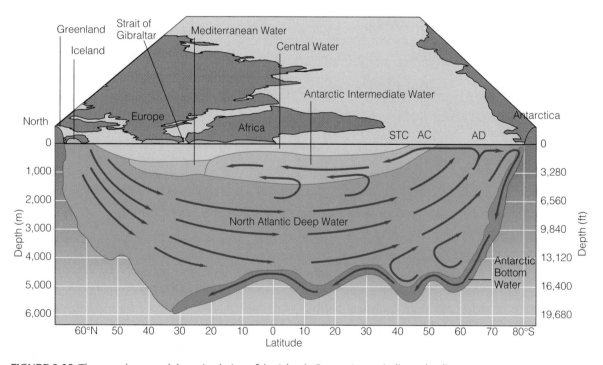

FIGURE 9.25 The water layers and deep circulation of the Atlantic Ocean. Arrows indicate the direction of water movement. STC indicates the position of the Subtropical Convergence, AC the Antarctic Convergence, and AD the Antarctic Divergence. The surface layer is too thin to show clearly in this scale. The vertical scale is greatly exaggerated. (For a view of the convergence zones from above, please see Figure 9.26.)

FIGURE 9.26 The major convergence zones, areas where surface water masses meet.

1,600 years! The residence time of most deep water is less, however; it takes 200 to 300 years to rise to the surface. (By contrast, a bit of surface water in the North Atlantic gyre may take only a little more than a year to complete a circuit.)

Not all thermohaline circulation is so sedate. Ripple marks in sediments, scour lines, and the erosion of rocky outcrops on deep-ocean floors are evidence that relatively strong, localized bottom currents exist (see Figure 5.3). Some of these currents may move as rapidly as 60 centimeters (24 inches) per second. These relatively fast currents are strongly influenced by bottom topography, and they are sometimes called **contour currents** because their dense water flows around (rather than over) seafloor projections. Bottom currents generally move equatorward at or near the western boundaries of ocean basins (below the western boundary surface currents). The subtle density differences between deep-water masses are not capable of moving water at the speed of the wind-driven surface currents. Water in some of these currents may move only 1 to 2 meters (3 to 7 feet) per day. Even at that slow speed the Coriolis effect modifies their pattern of flow.

Convergence Zones and the Surface and Central Water Masses

The most pronounced convergence zone is the *Antarctic Convergence*, which encircles the continent of Antarctica between about 50°S and 60°S. South of the Antarctic Convergence is cold *Antarctic Circumpolar Water*, in which flows the Antarc-

tic Circumpolar Current, the greatest of all ocean currents. The Southern Hemisphere's *Subtropical Convergence*, between about 40°S and 50°S, marks the northern boundary of *Subantarctic Surface Water*. The position of the Subtropical Convergence varies with the time of year, the position of currents, and other factors.

The surface water equatorward of the Subtropical Convergence—appropriately known as *Central Water*—is the most abundant, warmest, and saltiest of the surface water masses. It extends to the Northern Hemisphere's Subtropical Convergence, which occurs at about 45°N in the Pacific and about 60°N in the Atlantic.

The *Subarctic Surface Water* stretches north from the Northern Hemisphere's Subtropical Convergence. The transition to *Arctic Water* is a subtle one, and the *Arctic Convergence* is often poorly defined. The organization of surface water masses in the Northern Hemisphere is less orderly than in the Southern Hemisphere because of the different configuration of continents there. A circumpolar current like that of the southern polar ocean does not form in the north because continents interrupt its flow.

Thermohaline Flow and Surface Flow: The Global Heat Connection

As we have seen, swift and narrow currents along the western margins of ocean basins carry warm tropical surface waters toward the poles. In a few places, as shown in Figure 9.22, the

FIGURE 9.27
The global pattern of deep circulation resembles a vast "conveyor belt" that carries surface water to the depths and back again. Begin with the formation of North Atlantic Deep Water north of Iceland. This water mass flows south through the Atlantic and then flows over (and mixes with) deep water formed near Antarctica. The combined mass circumnavigates Antarctica and then moves north into the Indian and Pacific Ocean basins. Diffuse upwelling in all of the ocean returns some of this water to the surface. Water in the conveyor gradually warms and mixes upward to be returned to the North Atlantic by surface circulation. The whole slow-moving system is important in transporting water and heat.

→ Warm, shallow currents

→ Cold and salty deep currents

● Some areas of deep-water formation

water loses heat to the atmosphere and sinks to become deep water and bottom water. This sinking is most pronounced in the North Atlantic. The cold, dense water moves at great depths toward the Southern Hemisphere and eventually wells up into the surface layers of the Indian and Pacific Oceans. Almost a thousand years are required for this water to make a complete circuit.

The transport of tropical water to the polar regions is part of a global conveyor belt for heat. A simplified outline of the global circuit, the result of three decades of concentrated effort to understand deep circulation, is shown in **Figure 9.27**. This slow circulation straddles the hemispheres and is superimposed upon the more rapid flow of water in surface gyres. Recent analysis of this global circuit suggests that some of the heat warming the coasts of Europe enters the ocean in the vicinity of Indonesia and Australia, travels to the Indian Ocean, and enters the Gulf Stream by way of the Agulhas Current rounding the southern tip of Africa. The surface water that leaves the Pacific is driven in part by excess rainfall and river runoff throughout the Pacific basin. The slow, steady, three-dimensional flow of water in the conveyor belt distributes dissolved gases and solids, mixes nutrients, and transports the juvenile stages of organisms among ocean basins.

Studying Currents

Traditional methods of measuring currents divide into two broad categories: the **float method** and the **flow method.** The float method depends on the movement of a drift bottle or other free-floating object. In the flow method the current is measured as it flows past a fixed object.

Surface currents can be traced with drift bottles or drift cards. These tools are especially useful in determining coastal circulation, but they provide no information on the path the drift bottle or card may have taken between its release and collection points. Researchers who want to know the precise track taken by a drifting object can deploy more elaborate drift devices, such as the drogue arrangement shown in **Figure 9.28a**. These drogues can be tracked continuously by radio direction finders or radar. Surface currents can also be tracked by noting the difference between the daily expected and observed positions of ships at sea.

Bottle, card, or drogue studies are almost always carefully planned and executed, but not all surface drift releases have been intentional. In May 1990 a violent storm struck the containership *Hansa Carrier* enroute between Korea and Seattle. Twenty-one boxcar-sized cargo containers were lost overboard,

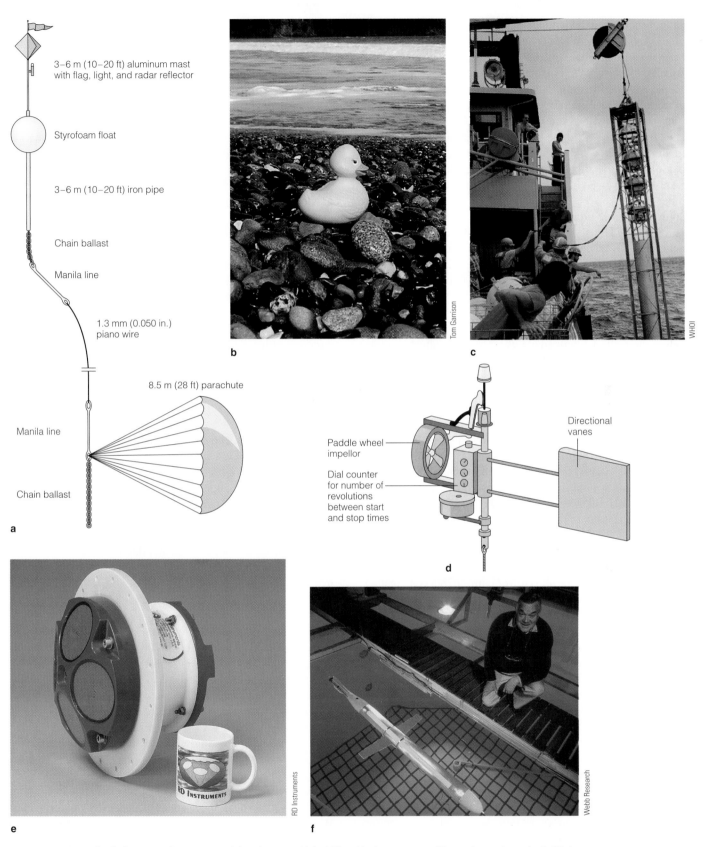

a

b — Tom Garrison

c — WHOI

3–6 m (10–20 ft) aluminum mast with flag, light, and radar reflector

Styrofoam float

3–6 m (10–20 ft) iron pipe

Chain ballast

Manila line

1.3 mm (0.050 in.) piano wire

Manila line

Chain ballast

8.5 m (28 ft) parachute

Paddle wheel impellor

Dial counter for number of revolutions between start and stop times

Directional vanes

d

e — RD Instruments

f — Webb Research

FIGURE 9.28 Methods for measuring currents. (**a**) A drogue, which drifts with the current and is continuously tracked. (**b**) A rubber duck of the kind lost overboard in large numbers in 1992 in an unintentional drift experiment (see text). (**c**) A sofar float being launched from the Woods Hole Oceanographic Institution's research ship *Oceanus*. The probe will drop to a depth of 3,500 meters (11,500 feet) and produce a low-frequency tone once each day for tracking. (**d**) An Ekman flow meter, which measures current speed and direction at a fixed position. (**e**) A through-hull transducer for an ADCP system. (**f**) A Slocum glider, a probe that uses energy from gravity, buoyancy, heat, and batteries to power long-range exploration of water masses.

among them containers holding 30,910 pairs of Nike athletic shoes. About six months later shoes from the broken containers began to wash up on beaches from British Columbia to Oregon. Because the shoes were not tied together in pairs, beachcombers placed advertisements in local newspapers and held swap meets to exchange the shoes (which were in excellent condition despite having been exposed to the ocean). Oceanographers noticed the ads and asked the media to request that individuals let them know where and when they found shoes. By knowing the place where the shoes were lost and the places where they were found, researchers have been able to refine their computer models of the North Pacific gyre. Some of the shoes completed a full circuit of the North Pacific. A similar spill occurred on 10 January 1992, when a storm-beset freighter lost a container filled with 29,000 rubber ducks, turtles, and other bathtub toys in the North Pacific (**Figure 9.28b**). More than 400 of the toys were recovered from 500 miles of Alaskan shoreline.

Deeper currents can also be surveyed by free-floating devices. The Swallow float (named after its developer) is used to detect the drift of intermediate water masses. Adjusted to descend to a specific density, the Swallow float emits sonar "pings" as it drifts so that a tracking vessel can follow the movement of the water mass in which it is embedded. More sophisticated devices developed in the 1970s and 1980s depend on sofar channels (see Chapter 6) to transmit sound. Low-frequency tones are broadcast from submerged probes to moored listening stations (**Figure 9.28c**). These sofar probes work autonomously—that is, on their own without human guidance. They have accumulated more than 240 float-years of data in the North Atlantic, at depths from 700 to 2,000 meters (2,300 to 6,600 feet). One of these drifting probes sent data for nine years!

Current meters, or flow meters, measure the speed and direction of a current from a fixed position. Most flow meters, such as the Ekman type shown in **Figure 9.28d**, use rotating vanes to measure current speed and a recording compass to measure direction. Bottom-water movements are usually too slow to be measured by flow meters.

Advances in electronics and computer design have made possible several new methods of study. One new device pioneered by the U.S. Office of Naval Research measures current speed by sensing the electromagnetic force generated by seawater as it moves in Earth's magnetic field. Buoys equipped with these sensors can record current speed and direction without dependence on delicate moving parts. Research vessels are often equipped with a different kind of current sensor that uses reflected sound to sense current strength and direction. Some ships carry through-hull mounted acoustic Doppler current profilers (ADCPs) (**Figure 9.28e**). Their transducers project three or four beams of sonic pulses ("pings") into the water each second. Unlike the echo sounders and acoustic profilers we discussed in Chapter 4, which listen for sounds reflected from the seafloor, these sounds are tailored to reflect from suspended particles, tiny organisms, or bubbles in the water itself. The echoes arrive back at the instrument at different times, with echoes from shallow depths arriving sooner than ones from farther away. Computers analyze the frequency shift and other characteristics of the returned signal, and determine the direction of the currents beneath the ship relative to the ship's motion. ADCP systems can measure currents to a depth of about 200 meters (660 feet).

A promising new research tool that works autonomously is the Slocums, named after Joshua Slocum, the first person to circumnavigate the globe alone. These little gliders "fly" smoothly up and down through the water column powered by gravity and buoyancy (**Figure 9.28f**). Energy to pump ballast overboard, allowing the glider to rise, is provided by a simple heat engine powered by the difference in temperature between ocean surface and great depths. To fall again, seawater is pumped aboard. Fleets of gliders could map the ocean's thermohaline depth profiles for years, transmitting data to satellites on their occasional visits to the surface.

Yet another method, developed for studying thermohaline circulation, senses the presence in seawater of chemical tracers—artificial substances with known histories of production or release. Because they dissolve easily in seawater, chlorofluorocarbons (CFCs) can be used as such tracers. A totally artificial chemical first produced in the 1930s for use as refrigerants, aerosol propellants, and blowing agents for foam, CFCs spread through the ocean like a dye, following oceanic circulation. The speed of deep currents has been measured by careful analysis of their CFC content.

As you saw in Figure 9.16b, satellites can sense sea surface temperature and topography, ocean color, and even chlorophyll content (an indication of biological productivity), giving fresh insight into ocean circulation. So far, satellites can observe only the ocean surface, but satellite-borne lasers may someday probe the ocean to greater depths.

Acoustical tomography (*tomos* = to cut or slice, *graph* = to write) uses pulses of low-frequency sound to sense variations in water temperature, salinity, and movement beneath the surface. Experiments begun in 1981 have used multiple transmitters and receivers to investigate the formation and movement of Gulf Stream eddies and deep-ocean circulation at a depth of 760 meters (2,500 feet). Preliminary results suggest that a great deal of the ocean's energy of motion is involved in relatively small fluctuating flows analogous to atmospheric weather systems. Unlike weather systems—which generally extend across 1,000 kilometers (620 miles) and last less than a week—this midscale oceanic turbulence typically extends only 100 kilometers (62 miles) but may persist for more than 100 days.

Researchers are also using acoustic tomography to conduct a long-term study of the formation of North Atlantic Deep Water south of Greenland. The technique may revolutionize our understanding of deep-sea circulation.

Currents are the very heart of physical oceanography. Their global effects, great masses of water, complex flow, and possible influence on human migrations make their study of particular importance.

Questions from Students

1. If the Gulf Stream warms Britain during the winter and keeps Baltic ports free of ice, why doesn't it moderate New England winters? After all, Boston is much closer to the warm core of the Gulf Stream than London is.

Yes, but remember the direction of prevailing winds in winter. Winter winds at Boston's latitude are generally from the west, so any warmth is simply blown out to sea. On the other side of the Atlantic the same winds blow toward London. It does get cold in London, but generally winters in London are much milder than those in Boston.

2. Could currents be used as a source of electrical power? With the Gulf Stream so close to Florida, it seems that some way could be devised to take advantage of all that water flow to turn a turbine.

It's been considered. The total energy of the Gulf Stream flowing off Miami has been estimated at 25,000 megawatts! A Woods Hole Oceanographic Institution team has proposed a honeycomb-like array of turbines for the layer between 30 and 130 meters (100 and 430 feet) across 20 kilometers (13 miles) of the current. They estimate a power output of around 1,000 megawatts, equal to the generation potential of two large nuclear power plants. Engineering difficulties would be prohibitive, however. Refer to Figure 17.7 for a practical application of this principle.

3. Are there any *nongeostrophic* currents? Are any currents not noticeably influenced by gravity, the Coriolis effect, uneven solar heating, planetary winds, and so forth?

Yes, there are small-scale currents that are not noticeably affected. Currents of fresh water from river mouths, rip currents in surf, and tidal currents in small harbors are much more affected by basin and bottom topography than by the Coriolis effect and gravity.

4. Why are western boundary currents strong in *both* hemispheres? I thought things went the other way (counterclockwise) in the Southern Hemisphere. Shouldn't the *eastern* boundary currents be stronger down there?

Western boundary currents are strong in part because the "Coriolis hill" is offset to the west, forcing water to move in a relatively narrow path along the ocean's western boundary. Truly, the Coriolis effect works in a clockwise direction in the Northern Hemisphere and counterclockwise in the Southern Hemisphere, so the "hill" is offset to the west in both hemispheres. Thus western boundary currents are strong in both hemispheres.

5. Which takes the lead in producing the hill in the middle of a geostrophic gyre: the pressure gradient or the Coriolis effect?

The question reminds me of the "chicken and egg" question—which came first? In ocean currents, both act together, in bal-

ance, to form both the hill and the circular flow around its crest. Imagine the situation some 150 million years ago when the Atlantic was first forming—Pangaea was splitting and the rift began to fill with water. Driven by winds, a small amount of water would have turned right to begin forming the hill. A pressure gradient was immediately formed, and water would have been forced back downhill by gravity. On its way back down, that water would achieve a balance with the Coriolis effect and come to a "compromise" position of clockwise flow around the apex.

6. A north wind comes from the north, but a north current is going north. Why the difference?

Traditions die hard, it seems. For thousands of years winds have been named by where they come *from*. A north wind comes from the north, and a west wind comes from the west. Currents, though, are named by where they are *going*. A southern current is headed south; a western current is moving west. An exception is the Antarctic Circumpolar Current, or West Wind Drift, which moves eastward. This current, however, is named after the wind that drives it, the powerful polar westerlies.

It may also be a matter of perspective. Ancient peoples took shelter *from* winds (and referred to the wind's place of origin). But early oceanic travelers were aware of where the currents were carrying them *to*.

Chapter in Perspective

In this chapter you learned that ocean water circulates in currents. Surface currents affect the uppermost 10% of the world ocean. The movement of surface currents is powered by the warmth of the sun and by winds. Water in surface currents tends to flow horizontally, but it can also flow vertically in response to wind blowing near coasts or along the equator. Surface currents transfer heat from tropical to polar regions, influence weather and climate, distribute nutrients, and scatter organisms. They have contributed to the spread of humanity to remote islands, and they are important factors in maritime commerce.

Circulation of the 90% of ocean water beneath the surface zone is driven by the force of gravity, as dense water sinks and less dense water rises. Because density is largely a function of temperature and salinity, the movement of deep water due to density differences is called *thermohaline circulation*. Currents near the seafloor flow as slow, riverlike masses in a few places, but the greatest volumes of deep water creep through the ocean at an almost imperceptible pace. The Coriolis effect, gravity, and friction shape the direction and volume of surface currents and thermohaline circulation.

In the next chapter you will learn about ocean waves. The traveling crests produce the appearance of movement we see in a wave. In an ocean wave, a ribbon of *energy* is moving at the speed of the wave, but *water* is not. In a sense, an ocean wave is an illusion. How can you be knocked off your surfboard by an illusion? Well, there's much to learn!

Key Ideas

✓ Ocean water circulates in currents caused mainly by wind friction at the surface and by differences in water mass density beneath the surface zone.
Review page 201

✓ Surface currents affect the uppermost 10% of the world ocean, mostly above the pycnocline. Some surface currents are rapid and riverlike, with well-defined boundaries; others are slow and diffuse. The largest surface currents are organized into huge circuits known as gyres.
Review Figure 9.8, page 205

✓ Water near the ocean surface moves to the right of the wind direction in the Northern Hemisphere and to the left in the Southern Hemisphere.
Review Figure 9.5 and 9.6, page 203

✓ The Coriolis effect modifies the courses of currents, with currents turning clockwise in the Northern Hemisphere and counterclockwise in the Southern Hemisphere. The Coriolis effect is largely responsible for the phenomenon of westward intensification in both hemispheres.
Review Figures 9.7, 9.8, and 9.13, pages 204, 205, 210

✓ Upwelling and downwelling describe the vertical movements of water masses. Upwelling is often due to the divergence of surface currents; downwelling is often caused by surface current convergence or an increase in the density of surface water.
Review Figures 9.15, 9.16, and 9.17, pages 212–214

✓ El Niño, an anomaly in surface circulation, occurs when the trade winds falter, allowing warm water to build eastward across the Pacific at the equator.
Review Figures 9.19 and 9.20, pages 215–217

✓ Circulation of the 90% of ocean water beneath the surface zone is driven by gravity, as dense water sinks and less dense water rises. Because density is largely a function of temperature and salinity, the movement of deep water due to density differences is called thermohaline circulation.
Review Figures 9.21–9.24, pages 218–220

✓ Water masses almost always form at the ocean surface. The densest (and deepest) masses were formed by surface conditions that caused water to become very cold and salty.
Review Figures 9.25 and 9.27, pages 220, 222

✓ Because they transfer huge quantities of heat, ocean currents greatly affect world weather and climate.
Review chapter opener, pages 201, 214–217

✓ Currents are the very heart of physical oceanography. Their global effects, vast masses of water, complex flow, and possible influence on human migrations make their study of particular importance.
Review pages 222–224

Terms and Concepts to Remember

acoustical tomography, 224
Antarctic Bottom Water, 218
Antarctic Circumpolar
 Current (West Wind
 Drift), 204
caballing, 220
coastal upwelling, 213
contour current, 221
convergence zone, 220
countercurrent, 210
current, 201
downwelling, 212
eastern boundary
 current, 208
eddy, 206
Ekman spiral, 202
Ekman transport, 202
El Niño, 215
ENSO, 215
equatorial upwelling, 213
float method, 222
flow method, 222
geostrophic gyre, 204
Gulf Stream, 206

gyre, 201
La Niña, 216
Langmuir circulation, 214
North Atlantic Deep
 Water, 218
Southern Oscillation, 214
surface current, 201
sverdrup (sv), 206
temperature–salinity (T-S)
 diagram, 218
thermohaline
 circulation, 217
transverse current, 208
undercurrent, 210
upwelling, 212
western boundary
 current, 206
westward
 intensification, 210
West Wind Drift (Antarctic
 Circumpolar
 Current), 204
wind-induced vertical
 circulation, 212

Study Questions

Reviewing What You've Learned

1. What forces are responsible for the *movement* of ocean water in currents?
2. What forces and factors influence the *direction* and *nature* of ocean currents?
3. What is a gyre? How many large gyres exist in the world ocean? Where are they located?
4. What are water masses? Where are distinct water masses formed? What determines their relative positions in the ocean?
5. What drives the vertical movement of ocean water? What is the general pattern of thermohaline circulation?
6. What methods are used to study ocean currents?

Thinking Critically

1. Why does water tend to flow around the periphery of an ocean basin? Why are western boundary currents the fastest ocean currents? How do they differ from eastern boundary currents?
2. What causes El Niño? How does an El Niño situation differ from normal current flow? What are the usual consequences of El Niño?
3. What are countercurrents? Undercurrents? How might El Niño be related to these currents?

4. What is the role of ocean currents in the transport of heat? How can ocean currents affect climate? Contrast the climate of a mid-latitude coastal city at a western ocean boundary with a mid-latitude coastal city at an eastern ocean boundary.
5. Can you think of ways ocean currents have (or *might* have) influenced history?
6. What holds up the thermocline? Wouldn't water slowly warm in an even gradient all the way to the bottom? (Hint: See Figure 9.24.)
7. **InfoTrac College Edition Project** The El Niño Southern Oscillation has been going through its cycles for hundreds of thousands of years. Is El Niño getting worse? Do some research using InfoTrac College Edition to gather information about El Niño, and use your findings to back up your view about El Niño's severity and predictability.

Thinking Analytically

1. Look again at Figure 9.7b and consider the descending water. Why does it stop descending? Which way is it most likely to go when it stops descending? (Hint: The Coriolis effect is greatest near the poles and nonexistent at the equator.)
2. Calculate the time it would take for an ideally situated rubber duck to make a loop of the North Atlantic. Now calculate the time it would take for an ideally situated rubber duck to make a loop of the North Pacific. Why the difference?

Oceanography☰Now™

Assess your understanding of this chapter's topics with additional quizzing and conceptual-based problems at http://oceanography.brookscole.com/garrison5e

Study and Research on the Internet

Assess your understanding of this chapter's topics with additional quizzing and conceptual-based problems at:

 http://oceanography.brookscole.com/garrison5e

- **InfoTrac College Edition Research** leads you to critical thinking exercises that use InfoTrac College Edition® as a research tool.
- **Regional InfoTrac College Edition** articles are organized into East Coast, West Coast, and Gulf Coast regions, allowing you to study oceanography on a more local level.

For more readings, go to InfoTrac College Edition, your online research library, at:

http://infotrac.thomsonlearning.com

CHAPTER 10

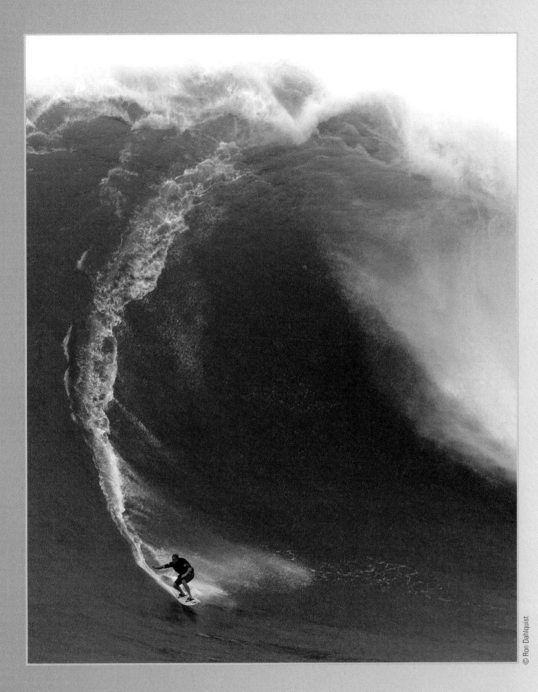

Makua Rothman rides his winning wave on the north shore of the Hawaiian island of Maui.

© Ron Dahlquist

WAVES

Surfing

Waves are not what they seem. Waves transmit energy, not water mass, across the ocean's surface. Beware of the wave illusion: *Water* isn't moving at the speed of the wave, but *energy* is.

More than 2 million Americans use this energy regularly: They surf. The thrill of rushing down the face of a growing, breaking wave is exhilarating! People willingly endure cold, boredom, and some danger for a ride lasting only a few seconds. If you're a good swimmer, give it a try. You will need to paddle your board or swim vigorously to match your speed to that of the advancing wave crest. As the wave rises to

228

break, your forward speed (and sense of timing) will place you on the leading edge of the crest, accelerating downward and forward. The technique takes time to master, but the feeling is worth the effort!

Extreme-wave surfing is a new twist on this ancient sport, which was invented by Hawaiian royalty and well developed by the time of Captain Cook's visits in the late eighteenth century. Unlike the Hawaiians of old (who surfed large waves on huge, heavy wooden boards), modern surfers are towed into monstrous waves behind jet skis and use long but relatively lightweight fiberglass boards. Their goal: to ride the ocean's largest waves—waves long thought impossible to catch or surf.

A contest in 2003 offered US$60,000 to the rider of the surfing season's highest wave. Everybody knew that Hawaii would be the place to win. Was everybody wrong? In March of that year a French surfer and his fellow enthusiasts tracked a massive storm system as it entered the North Atlantic east of Newfoundland. The storm sent 8-meter (26-foot) waves to-

ward the southeast. When the waves struck the coast of southwestern France on 10 March, the surfers were ready. They rode breaking waves towering to 18 meters (60 feet) and—as required by contest organizers—captured their feat on film.

The Hawaiian contingent was not about to be outdone. On 18 April 18-year-old Makua Rothman overcame the unexpected French challenge to win the prize (**figure**). Although the French wave was almost as high as the Hawaiian wave, contest judges voted for Mr. Rothman because he began his ride at a higher point on the wave and the face of the wave was steeper.

What generates ocean waves? From personal experience we think first of the wind, but can you think of other forces that could be responsible? In this chapter you will read about the basic properties of all ocean waves, the occasionally enormous "small" waves generated by wind and storms, the rocking waves of seiches, and the often-destructive tsunami. The next chapter will extend this information to the greatest of all waves, the tides.

10.1

Ocean Waves

 10.2

To most people an ocean wave in deep water appears to be a massive moving object—a ridge of water traveling across the sea surface. An ocean wave is one of several kinds of **waves,** all of which are disturbances caused by the movement of energy from a source through some medium (solid, liquid, or gas). As the energy of the disturbance travels, the medium through which it passes moves in specific ways. Sometimes this movement is visible to us as crests or ridges in the medium. The traveling crests produce the appearance of movement we see in a wave. In an ocean wave a ribbon of *energy* is moving at the speed of the wave, but *water* is not. In a sense, the wave is an illusion.

Picture a resting sea gull as it bobs on the wavy ocean surface far from shore. The gull moves in *circles*—up and forward as the tops of the waves move to its position, down and backward as the tops move past. Each circle's diameter is equal to the wave's height. As may be seen in **Figure 10.1,** energy in waves flows past the resting bird, but the gull and its patch of water move only a very short distance forward in each up-and-forward, down-and-back wave cycle. The water on which the bird rests does not move continuously across the sea surface as the wave illusion suggests.[1]

The transfer of energy from water particle to water particle in these circular paths, or **orbits,** transmits wave energy across the ocean surface and causes the wave form to move. This kind of wave is known as an **orbital wave**—a wave in which parti-

cles of the medium (water) move in closed circles as the wave passes. Orbital ocean waves occur at the boundary between two fluid media (between air and water) and between layers of water of different densities. Because the wave form moves forward, these waves are a type of **progressive wave.**

The progressive wave that moved the gull was probably caused by wind. Other forces can generate much larger progressive waves in which water molecules move through larger circular or elliptical orbits. Some of these waves are so large that they do not appear to us as waves at all, but rather as the slow sloshing of water in a harbor or bay, as dangerous flooding surges of water, or as rhythmic and predictable ocean tides.

Ocean waves have distinct parts. The **wave crest** is the highest part of the wave above average water level; the **wave trough** is the valley between wave crests below average water level. **Wave height** is the vertical distance between a wave crest and the adjacent trough; **wavelength** is the horizontal distance between two successive crests (or troughs). The relationship between these parts is shown in **Figure 10.2.** The time it takes for a wave to move a distance of one wavelength is known as the **wave period. Wave frequency** is the number of waves that pass a fixed point per second.

The circular motion of water particles at the surface of a wave continues underwater. As **Figure 10.3** shows, the diameter of the orbits through which water particles move decreases rapidly with depth. For all practical purposes wave motion is negligible below a depth of one-half the wavelength, where the circles are only $\frac{1}{23}$ the diameter of those at the surface. This means divers in 20 meters of water would not notice the passage of a wind wave of 30-meter wavelength and might barely notice the wave if they were at a depth of 15 meters. Because most ocean waves have moderate wavelengths, the circular disturbance of the ocean that propagates these waves affects only the uppermost layer of water. *Note that the movement of water*

---

[1] To clarify the important idea of wave-as-illusion, imagine yourself at a sports stadium where spectators are doing "the wave." Your role in wave propagation is simple: You stand up and sit down in precise synchronization with your neighbors. Though you move only a few feet vertically, the wave of which you were a part circles the arena at high speed. You and all the other participants stay in place, but the wave moves faster than anyone can run.

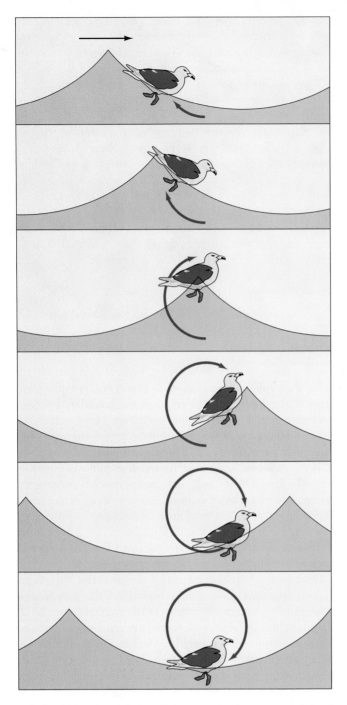

FIGURE 10.1 A floating sea gull demonstrates that *waves* travel ahead but the *water* itself does not. In this sequence a wave moves from left to right as the gull (and the water in which it is resting) revolves in an imaginary circle, moving slightly to the left up the front of an approaching wave, then to the crest, and finally sliding to the right down the back of the wave.

in circles doesn't resemble interlocking mechanical gears. Instead, there is a coordinated, uniform circular movement of water molecules in one direction as the waves pass.

Figure 10.3 indicates that water molecules in the crest of a passing wave move in the same direction as the wave, but molecules in the trough move in the opposite direction. If particle speed decreases with depth, then molecules in the top half of

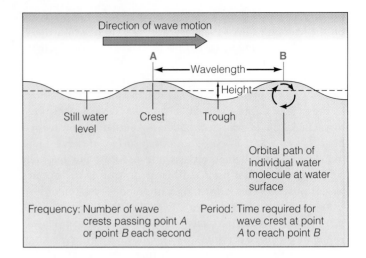

FIGURE 10.2 The anatomy of a progressive wave.

ACTIVE FIGURE 10.3 The orbital motion of water particles in a wave, which extends to a depth of about half of the wavelength.

Oceanography Now™ | At the Active Figures link at http://oceanography.brookscole.com/garrison5e you can view orbital motion of water in waves.

the orbit will move farther forward in the direction the wave is moving than molecules in the bottom half of the orbit will move backward (**Figure 10.4**). This flow, known as **Stokes drift**, is important in driving the ocean surface currents you studied in Chapter 9.

Oceanography Now™ | Click Oceanography Interactive to work through an activity on wave properties through waves, tides, and currents.

Classifying Waves

Ocean waves are classified by the *disturbing force* that creates them, the extent to which the disturbing force *continues* to influence the waves once they are formed, the *restoring force* that tries to flatten them, and their *wavelength*. (Wave height is not

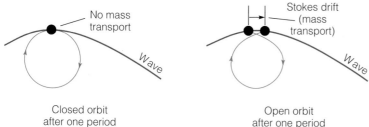

FIGURE 10.4 Stokes drift, the small net amount of mass transport of water in the direction of a wave. Although the orbits in Figures 10.1 and 10.3 are drawn as shown in the left figure, actually they are not completely closed. Instead, each particle moves forward a very small distance (exaggerated in the figure on the right) during each orbit. Stokes drift contributes to the movement of the wind-driven surface currents discussed in Chapter 9.

often used for classification because it varies greatly depending on water depth, interference between waves, and other factors.)

Disturbing Force

Energy that causes ocean waves to form is called a **disturbing force.** Wind blowing across the ocean surface provides the disturbing force for *wind waves.* The arrival of a storm surge or seismic sea wave in an enclosed harbor or bay, or a sudden change in atmospheric pressure, is the disturbing force for the resonant rocking of water known as a *seiche.* Landslides, volcanic eruptions, and the faulting of the seafloor associated with earthquakes are the disturbing forces for seismic sea waves (also known as *tsunami*). The disturbing forces for *tides* are changes in the magnitude and direction of gravitational forces among Earth, moon, and sun, combined with Earth's rotation.

Free Waves, Forced Waves

A wave that is formed and then propagates across the sea surface without any further influence of the force that formed it is known as a **free wave.** When wind waves move away from the storm that created them, or when the storm ceases, they continue without the injection of additional wind energy. Likewise, tsunami—waves caused by submerged landslides or earthquakes—continue to move across the ocean surface long after the landslide or earthquake has stopped moving.

In contrast, a **forced wave** is maintained by its disturbing force. Tides are forced waves dependent on the gravitational attraction of the moon and the sun.

Restoring Force

A **restoring force** is the dominant force that returns the water surface to flatness after a wave has formed in it. If the restoring force of a wave were quickly and fully successful, a disturbed sea surface would immediately become smooth and the energy of the embryo wave would be dissipated as heat. But that isn't what happens. Waves continue after they form because the restoring force overcompensates and causes oscillation. The situation is analogous to a weight bobbing at the bottom of a very flexible spring, constantly moving up and down past its normal resting point.

The dominant restoring force for very small water waves—those with wavelengths less than 1.73 centimeters (0.68 inch)—is cohesion, the property that enables individual water molecules to stick to each other by means of hydrogen bonds (see Figure 6.2). These **capillary waves** are transmitted across a puddle because cohesion, the force that makes the tea creep up on the sides of a teacup, tugs the tiny wave troughs and crests toward flatness.

Capillary waves are the first waves to form when the wind blows. These small ripples are important in transferring energy from air to water to drive ocean currents, but they are of little consequence in the overall picture of ocean waves because they are tiny and carry very little energy.

All waves with wavelengths greater than 1.73 centimeters depend mostly on gravity to provide the restoring force. Gravity pulls the crests downward, but the inertia of the water causes the crests to overshoot and become troughs. The repetitive nature of this movement, like the spring weight moving up and down, gives rise to the circular orbits of individual water molecules in an ocean wave. These larger waves are called **gravity waves.** Because the circular motion of water molecules in a wave is nearly friction-free, gravity waves can travel across thousands of miles of ocean surface without disappearing, eventually to break on a distant shore.

Wavelength

Wavelength is an important measure of wave size. **Table 10.1** lists the typical wavelengths and the disturbing and restoring forces for capillary waves, wind waves, seiches, seismic sea waves, and tides. **Figure 10.5** shows the relationships between disturbing and restoring forces, period, and relative amounts of energy present in the ocean's surface for each wave type. Note

Table 10.1	Wavelength, Disturbing Forces, and Restoring Forces for Ocean Waves		
Wave Type	**Typical Wavelength**	**Disturbing Force**	**Restoring Force**
Capillary wave	Up to 1.73 cm (0.68 in.)	Usually wind	Cohesion of water molecules
Wind wave	60–150 m (200–500 ft)	Wind over ocean	Gravity
Seiche	Large, variable; a function of ocean basin size	Change in atmospheric pressure, storm surge, tsunami	Gravity
Seismic sea wave (tsunami)	200 km (125 mi)	Faulting of seafloor, volcanic eruption, landslide	Gravity
Tide	Half Earth's circumference	Gravitational attraction, rotation of Earth	Gravity

FIGURE 10.5 Wave energy in the ocean as a function of the wave period. As the graph shows, most wave energy is typically concentrated in wind waves. However, large tsunami, rare events in the ocean, can transmit more energy than all wind waves for a brief time. Tides are waves—their energy is concentrated at periods of 12 and 24 hours.

that more energy is stored in wind waves than in any of the other wave types.

Deep-Water Waves, Shallow-Water Waves

Most characteristics of ocean waves depend on the relationship between their wavelength and water depth. Wavelength determines the *size* of the orbits of water molecules within a wave, but water depth determines the *shape* of the orbits. The paths of water molecules in a wind wave are circular only when the wave is traveling in deep water. A wave cannot "feel" the bottom when it moves through water deeper than half its wavelength because too little wave energy is contained in the small circles below that depth. Waves moving through water deeper than half their wavelength are known as **deep-water waves**. A wave has no way of "knowing" how deep the water is, only that it is in water deeper than about half its wavelength. For example, a wind wave with a 20-meter wavelength will act as a deep-water wave if it is passing through water more than 10 meters deep (**Figure 10.6**).

The situation is different for wind-generated waves close to shore. The orbits of water molecules in waves moving through shallow water are flattened by the proximity of the bottom. Water just above the seafloor cannot move in a circular path, only forward and backward. Waves in water shallower than $\frac{1}{20}$ their original wavelength are known as **shallow-water waves**. A

wave with a 20-meter wavelength will act as a shallow-water wave if the water is less than 1 meter deep.

Transitional waves travel through water deeper than $\frac{1}{20}$ their original wavelength but shallower than half their original wavelength. In our example, this would be water between 1 and 10 meters deep. The flattened orbital motion of water molecules in a transitional wave is shown in Figure 10.6.

Of the wave types listed in Table 10.1, only capillary waves and wind waves can be deep-water waves. To understand why, remember that most of the ocean floor is deeper than 125 meters (400 feet), half the wavelength of very large wind waves. The wavelengths of the larger waves are *much* longer: The wavelength of seismic sea waves usually exceeds 100 kilometers (62 miles). No ocean is 50 kilometers (31 miles) deep, so seiches, seismic sea waves, and tides are always in water that to them is shallow or transitional in depth. Their huge orbit circles flatten against a distant bottom that is always less than half a wavelength away.

In general, the longer the wavelength, the faster the wave energy will move through the water. For *deep-water* waves this relationship is shown in the formula

$$C = \frac{L}{T}$$

in which C represents speed (celerity), L is wavelength, and T is time or period (in seconds).

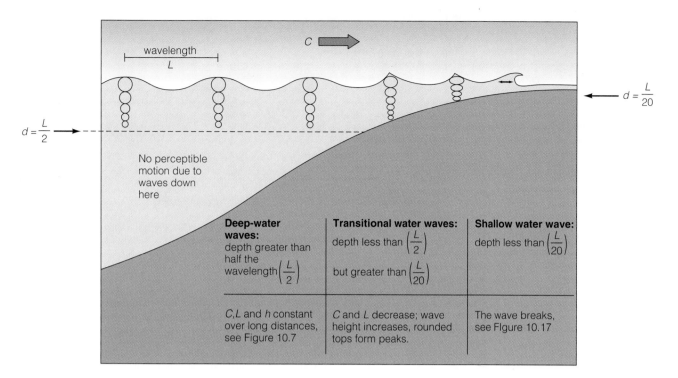

FIGURE 10.6 Progressive waves. Classification depends on their wavelength relative to the depth of water through which they are passing. The diagram is not to scale.

The speed of all ocean waves is controlled by gravity, wavelength, and water depth. The speed of a deep-water wave may also be approximated by

$$C = \sqrt{\frac{gL}{2\pi}}$$

where g is acceleration due to gravity, or 9.8 meters per second per second. Because g and π (3.14) are constants, we have

$$C = 1.25\sqrt{L}$$

when C is measured in meters per second and L in meters. Note in both instances that wave speed is proportional to wavelength.

Now look at **Figure 10.7,** a diagram showing the relationship between wavelengths of *deep-water waves* and their speed and period. In the example shown by the red lines, the speed of a wave with a wavelength of 233 meters and a period of 12 seconds is

$$C = \frac{L}{T} = \frac{233 \text{ meters}}{12 \text{ seconds}} = \frac{19.4 \text{ meters}}{\text{second}}$$

Wavelength is difficult to determine at sea, but period is comparatively easy to find; for example, an observer simply times the movement of waves past the bow of a stopped ship. If the period (T) is known, then the speed (S) can be calculated from the relationship

$$C \text{ (in meters per second)} = \frac{gT}{2\pi}$$

$$= \frac{9.8 \text{ meters/sec}^2 \times T \text{ (in seconds)}}{2 \times (3.14)} = 1.56T$$

where g is acceleration due to gravity.

FIGURE 10.7 The theoretical relationship among speed, wavelength, and period in deep-water waves. Speed is equal to wavelength divided by period. If one characteristic of a wave can be measured, then the other two can be calculated. The easiest characteristic to measure exactly is period. In the example shown in red, the speed of a wave with a wavelength of 233 meters and a period of 12 seconds is 19.4 meters per second.

The speed of *shallow-water* waves is described by a different equation,

$$C = \sqrt{gd} \quad \text{or} \quad C = 3.1\sqrt{d}$$

where C is speed (in meters per second), g is acceleration due to gravity (9.8 meters per second per second), and d is depth of the water (in meters). The *period* of a wave remains unchanged regardless of the depth of water through which it is moving. As deep-water waves enter the shallows and feel bottom, however, their speed is reduced and their crests "bunch up," so their wavelength shortens.

Comparing deep-water wind waves and shallow-water seismic sea waves is like comparing apples and oranges, but we can state the *general relationship* between wavelength and wave speed: The longer the wavelength, the greater the speed.

Wind Waves (Deep-Water Waves)

- Period to about 20 *seconds*
- Wavelength to perhaps 600 meters (2,000 feet) in extreme cases
- Speed to perhaps 112 kilometers (70 miles) per hour in extreme cases

Seismic Sea Waves (Shallow-Water Waves)

- Period to perhaps 20 *minutes*
- Wavelength typically 200 kilometers (125 miles)
- Speeds of 760 kilometers (470 miles) per hour

Remember that it is *energy*—not the water mass itself—that is moving through the water at the astonishing speed of 760 kilometers (470 miles) per hour—the speed of a jet airliner!—in seismic sea waves.

Wind Waves

Wind waves are gravity waves formed by the transfer of wind energy into water. Most wind waves are less than 3 meters (10 feet) high. Wavelengths from 60 to 150 meters (200 to 500 feet) are most common in the open ocean.

Wind waves grow from capillary waves. Capillary waves form as wind friction stretches the water surface and as surface tension tries to restore it to smoothness. These waves are nearly always present on the ocean. A capillary wave interrupts the smooth sea surface, deflecting the surface wind upward, slowing it, and causing some of the wind's energy to be transferred into the water to drive the capillary wave crest forward. The wind may eddy briefly downwind of the tiny crest, creating a slight partial vacuum there. Atmospheric pressure pushes the trailing crest forward (downwind) toward the trough, adding still more energy to the water surface. The process is shown in **Figure 10.8.** The increasing energy in the water surface expands the circular orbits of water particles in the direction of the wind, enlarging the small wave's size. The capillary wave becomes a wind wave when its wavelength exceeds 1.73 cen-

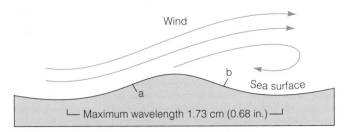

FIGURE 10.8 Wind forces acting on a capillary wave. A capillary wave interrupts the smooth sea surface, deflecting surface wind upward, slowing it, and causing some of the wind's energy to be transferred into the water to drive the capillary wave crest forward (point a). The wind may eddy briefly downwind of the tiny crest, creating a slight partial vacuum there. Atmospheric pressure pushes the trailing crest forward (downwind) toward the trough (point b), adding still more energy to the water surface.

timeters (0.68 inch), the wavelength at which gravity supersedes capillary action as the dominant restoring force.

If the wind wave remains in water deeper than half its wavelength and the wind continues to blow, the wave becomes larger. Its crest is thrust higher into faster wind, extracting even more energy from the moving air. The circular orbits of water particles within the wave grow larger with more energy input; height, wavelength, and period increase proportionally. The irregular peaked waves in the area of wind wave formation are called **sea**; the chaotic surface is formed by simultaneous wind waves of many wavelengths, periods, and heights. When the wind slows or ceases, as it does away from a storm, the wave crests become rounded and regular.

Oceanography⬡Now™ | Click Oceanography Interactive to work through an activity on wave progression through waves, tides, and currents.

Swell Formation and Dispersion

Because they move faster, waves with longer wavelengths leave the area of wave formation sooner.[2] They outrun their smaller relatives. Mature waves from a storm sort themselves into groups with similar wavelengths and speeds. The process of wave separation, or **dispersion,** produces the familiar smooth undulation of the ocean surface called **swell** (**Figure 10.9**). Swell often move thousands of kilometers from a storm to a shore, announcing the storm's impending arrival. Contrary to what you might expect, observers far from the storm would first encounter large, quick-moving waves of long wavelength, then middle-sized waves, and then slow, small ones. Because the circular movement of water particles in deep-water waves is virtually friction-free, the waves continue until they break upon a shore. There they release their absorbed wind energy as random movement, heat, and sound.

---

[2] Remember, speed is directly proportional to wavelength in deep-water waves: $C = 1.25\sqrt{L}$.

FIGURE 10.9 Swell—mature, regular wind waves sorted by dispersion—off the Oregon coast. Small waves superimposed on the large swell are the result of local wind conditions.

Tom Garrison

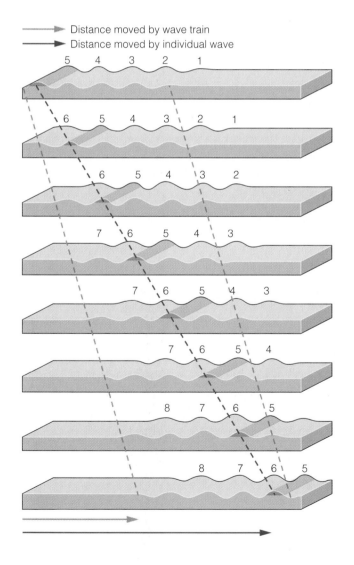

Progressing groups of swell with the same origin and wavelength are called **wave trains.** The leading waves in the wave train are drained of energy because they must begin the circular movement of the undisturbed water into which they are intruding. These leading waves gradually disappear, but after the wave train has passed some energy remains behind in the circles to form new waves. New waves thus form behind as the leading waves disappear at the front of the wave train. This process is shown in **Figure 10.10.**

The implications of this detail are surprising. Though each *individual* wave moves forward with a speed proportional to its wavelength in deep water (C), the *wave train* itself moves forward at only half that speed. Groups of waves therefore move ahead at half the speed of individual waves within the group. The half-speed advance of the wave train is called **group velocity** and is the speed with which wave energy advances. Group velocity is often represented by V in wave equations.

Note that individual waves do not persist in the ocean. Individual waves last only as long as they take to pass through the group. Only deep-water waves are subject to dispersion. As deep-water

ACTIVE FIGURE 10.10 Waves travel in groups called wave trains. As the leading wave of the group travels forward, it transfers half of its energy forward to initiate motion in the undisturbed surface ahead. The other half is transferred to the wave behind to maintain wave motion. The leading wave in the wave train continuously disappears, while a new wave is continuously formed at the back of the train. Follow wave number 5 in this diagram. The wave train travels at *half* the speed of any individual wave, a speed known as group velocity (*V*).

Oceanography�================Now™ | At the Active Figures link at http://oceanography.brookscole.com/garrison5e you can view the mechanics of wave train.

FIGURE 10.11 The fetch, the uninterrupted distance over which the wind blows without significant change in direction. Wave size increases with increased wind speed, duration, and fetch. A strong wind must blow continuously in one direction for nearly three days for the largest waves to develop fully.

waves move into shallow water, the speed of individual waves within the group slows until wave speed equals group velocity.

Factors That Affect Wind Wave Development

Three factors affect the growth of wind waves. First, the wind must be moving faster than the wave crests for energy transfer from air to sea to continue, so the mean speed, or **wind strength,** of the wind is clearly important to wind-wave development. A second factor is the length of time the wind blows, or **wind duration;** high winds that blow only a short time will not generate large waves. The third factor is the uninterrupted distance over which the wind blows without significant change in direction, the **fetch** (**Figure 10.11**).

A strong wind must blow continuously in one direction for nearly three days for the largest waves to develop fully. A **fully developed sea** is the maximum wave size theoretically possible for a wind of a specific strength, duration, and fetch. Longer ex-

posure to wind at that speed will not increase the size of the waves because energy is lost due to the breaking of wave tops and the formation of whitecaps. The first three columns of **Table 10.2** give examples of the combinations of wind speed, fetch, and duration required to form waves of the fully developed sea described in the remaining columns. Note that waves in a fully developed sea are small if the wind speed is low. However, if the wind speed is 74 kilometers (46 miles) through 1,313 kilometers (816 miles) per hour for 42 hours—conditions not wholly unrealistic in the Pacific—waves that average 8.5 meters (28 feet) high can result. The largest 10% of the waves in this sea will exceed 17.2 meters (56.6 feet) in height!

The combination of factors required to produce truly great seas doesn't occur very often. In the example given at the bottom of Table 10.2, it seems unlikely that a 92-kilometer- (58-mile-) per-hour wind would blow steadily in one direction for 69 hours over 2,627 kilometers (1,633 miles). Of course, as Lieutenant Marggraff's experience in 1933 suggests (**Box 10.1**), the ocean is full of surprises!

Table 10.2 Conditions Necessary for a Fully Developed Sea at Given Wind Speeds, and the Parameters of the Resulting Waves					
Wind Conditions			**Wave Size**		
Wind Speed in One Direction	Fetch	Wind Duration	Average Height	Average Wavelength	Average Period
19 km/hr (12 mi/hr)	19 km (12 mi)	2 hr	0.27 m (0.9 ft)	8.5 m (28 ft)	3.0 sec
37 km/hr (23 mi/hr)	139 km (86 mi)	10 hr	1.5 m (4.9 ft)	33.8 m (111 ft)	5.7 sec
56 km/hr (35 mi/hr)	518 km (322 mi)	23 hr	4.1 m (13.6 ft)	76.5 m (251 ft)	8.6 sec
74 km/hr (46 mi/hr)	1,313 km (816 mi)	42 hr	8.5 m (27.9 ft)	136 m (446 ft)	11.4 sec
92 km/hr (58 mi/hr)	2,627 km (1,633 mi)	69 hr	14.8 m (48.7 ft)	212.2 m (696 ft)	14.3 sec

Source: Data from U.S. Army Corps of Engineers Coastal Research Center, Richmond, Virginia.

Think of the ocean and you think of waves. On the night of 7 February 1933, Lieutenant Frederick Marggraff's world consisted of little *but* waves. Mr. Marggraff was standing a predawn watch on the bridge of the USS *Ramapo,* a large U.S. Navy tanker steaming from Manila to San Diego. February 7 was the third full day of a furious storm, and the tanker's crew had become professionally and emotionally interested in the intense wind and mountainous seas sweeping past their ship. A few of the officers may have thought about the so-called 60-foot law, a paragraph in a Navy Hydrographic Office bulletin implying that all reported sightings of waves over 60 feet (about 18 meters) high were exaggerations. Indeed, despite persistent reports to the contrary, few theorists believed that wind-driven ocean waves exceeded 18 meters in height. Lieutenant Marggraff was about to prove them very wrong.

February is the stormiest month in the North Pacific, and *Ramapo* had encountered the largest and most severe storm of the year, a storm made more intense by the coalescence of three low-pressure centers. For days a steady wind had blown at 107 kilometers (67 miles) per hour, and gusts to 126 kilometers (78 miles) per hour often lashed the decks. But the wind blew persistently from one direction, and though the monstrous waves it generated dwarfed the tanker, they were surprisingly orderly in form.

"The conditions for observing the seas from the ship were ideal," wrote *Ramapo's* executive officer. "We were running directly down the wind with the sea. There were no cross seas and therefore no peaks along wave crests. There was practically no rolling, and the pitching motion was easy because of the fact that the sides of the waves were much longer than the ship. The moon was out astern and facilitated observations during the night. The sky was partly cloudy." At about three in the morning, Mr. Marggraff observed a train of giant waves looming in the moonlight. As the trough of the first wave approached the ship, he noted that its distant crest was on a level with the crow's nest on the mainmast. At that instant the ship's stern sank into the bottom of the onrushing trough. The next two waves were about the same size. Not surprisingly, such immense waves made an indelible impression on all who witnessed them.

How big were these waves? When the executive officer had some time to spare, he did some calculations. The **figure** illustrates the attitude of the ship when the largest waves were measured. The height of the waves was determined by using a set of the ship's plans, a calculation of the height of the observer above the sea surface, the draft of the ship, and a sight to the horizon. The largest wave for which a dependable observation had been made was 34 meters (112 feet) high, still a record! After the executive officer published his findings, the paragraph on the 60-foot law was quietly dropped from the next edition of the Hydrographic Office bulletin.

10.12

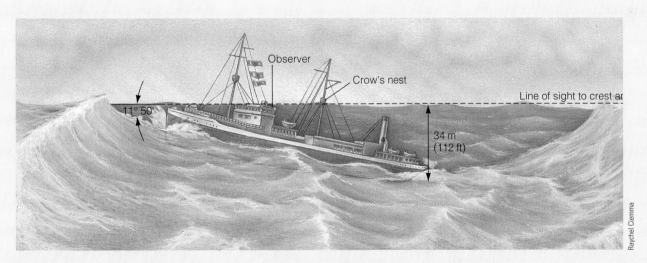

How the great wave observed from the USS *Ramapo* was measured. An officer on the bridge was looking toward the stern and saw the crow's nest in his line of sight to the crest of the wave, which had just come in line with the horizon. Wave height was calculated based on the geometry of the situation.

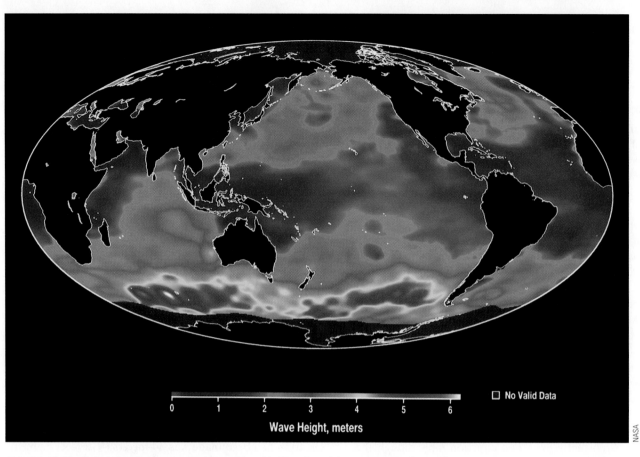

FIGURE 10.12 Global wave height acquired by a radar altimeter aboard the *TOPEX/Poseidon* satellite in October 1992. In this image the highest waves occur in the southern ocean, where waves more than 6 meters (19.8 feet) high (represented in white) were recorded. The lowest waves (indicated by dark blue) are found in the tropical and subtropical ocean, where wind speed is lowest.

The greatest potential for large waves occurs beneath the strong and nearly continuous winds of the West Wind Drift surrounding Antarctica. The early nineteenth-century French explorer of the South Seas Jules Dumont d'Urville encountered a wave train with heights estimated "in excess" of 30 meters (100 feet) in Antarctic waters. In 1916 Ernest Shackleton contended with occasional waves of similar size in the West Wind Drift during a heroic voyage to remote South Georgia Island in an open boat. Satellite observations (like those of **Figure 10.12**) have shown that wave heights to 11 meters (36 feet) are fairly common in the West Wind Drift.

In zones of high winds a less than fully developed sea can also attract attention. Though wind speed within cyclonic storms is often very strong, the circular motion of air doesn't allow long fetches, and fully developed seas rarely occur beneath such storms. Officers standing deck watches during storms rarely quibble with theoretical maximum height versus observed height, however. Wind waves can be overwhelming even if they are not fully developed (**Figure 10.13**).

Wind Wave Height

Wave height is not represented directly in the deep-water wave formula $C = L/T$. During their formation, moderate-sized wind waves in the open ocean exhibit a maximum 1:7 ratio of wave height to wavelength (**Figure 10.14**); this ratio is the **wave steepness.** Waves 7 meters long will not be more than 1 meter high, and waves with a 70-meter wavelength will not exceed 10 meters in height. The angle at their crest will not exceed 120°. A peaked appearance usually indicates the continuing injection of wind energy. If a wave gets any higher than the 1:7 ratio for its wavelength, it will break and excess energy from the wind will be dissipated as turbulence: Think of the *whitecaps* or *combers* associated with a fully developed sea.

As wind waves mature into swell, their wavelengths increase somewhat and the ratio of wave height to wavelength decreases: The waves become less steep. Swell continue in this rounded-top shape until they break on shore.

Wavelength Records

How big can wind waves get? This is an interesting question, and one not dispassionately answered. Calculating the wavelength from the wave period (T in $C = L/T$), Bigelow and Edmondson (1953) reported that mariners have sighted waves with wavelengths of 451 meters (1,480 feet) off the west coast of Ireland, 583 meters (1,913 feet) off the Cape of Good Hope, and 829 meters (2,720 feet) in the equatorial Atlantic. (That

FIGURE 10.13 A U.S. Navy destroyer battles large waves generated by a storm in the Pacific. Despite the battering, the ship is holding a steady enough course to take on fuel and stores from the aircraft carrier barely visible on the right.

last monster would have had a period of 25 seconds!) The nineteenth-century French admiral J. Mottez reported a wave with a wavelength of 790 meters (2,600 feet), a period of 23 seconds, and a speed of 123 kilometers (76 miles) per hour in the equatorial Atlantic west of Africa. The wavelength of the *Ramapo* wave discussed in Box 10.1 was calculated at 360 meters (1,180 feet), so these sightings are perhaps exaggerated. I can attest to the fact that smaller waves can also add *much* excitement to a deck officer's day (as you saw in Figure 10.13)!

Interferences and Rogue Waves

The real situation of wind waves at sea is not as simple as has been suggested here. The ideal vision of one set of waves moving in one direction at one speed across an otherwise smooth surface is almost never seen in the ocean.

Independent wave trains exist simultaneously in the ocean most of the time. Because long waves outrun shorter ones,

wind waves from different storm systems can overtake and interfere with one another. One wave doesn't crawl over the others when they meet; instead, they add to or subtract from one another. Such interaction is known as **interference**. In **Figure 10.15a** one wave is represented as a blue line and a second wave with a slightly longer wavelength as a green line. In the sea surface where these waves coincide (shown in **Figure 10.15b**), you can see the alternation between addition (large crests and troughs) and subtraction (almost no waves at all). The cancellation effect of subtraction is termed **destructive interference**—not because of harm to lives or property but

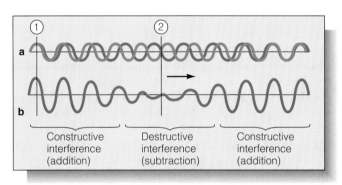

FIGURE 10.15 Constructive and destructive interference. (**a**) Two overlapping waves of different wavelength are shown, one in blue and one in green. Note that the wave in blue has a slightly longer wavelength. (**b**) If both are present in the ocean at the same time, they will interfere with each other to form a composite wave At the position of line 1, the two waves in **a** will constructively interfere to form very large crests and troughs, as shown in **b**. At the position of line 2, the two waves will destructively interfere and the crests and troughs will be very small (again shown in **b**).

FIGURE 10.14 A wind wave of moderate size shown during its formation. The ratio of height to wavelength, called wave steepness, is 1:7; the crest angle does not exceed 120°.

FIGURE 10.16 A rare photo of a rogue wave, taken in 1935 from the liner SS *Washington*, eastbound across the North Atlantic. Rogue waves occur when two or more wave crests add together to form a short-lived high wave.

W.C. Stillwell

because wave interference destroys or cancels waves. **Constructive interference** is the additive formation of large crests or deep troughs, the size of which exceeds the size of each participating wave. The huge waves observed by officers aboard *Ramapo* were surely the result of constructive interference.

You have probably noticed that the surf along a coast seems to rise to a few big waves, diminish, and then build again. Surfers wait for the big "sets" to arrive, ride toward the shore, and then use the relatively calm interval to swim out into position for the next big set of waves. Constructive and destructive interferences explain this behavior, called **surf beat**. Constructive interference between waves of different wavelengths creates the sought-after big waves; destructive interference diminishes the waves and makes it easier to swim back out. The characteristics of surf beat explain why in some instances every ninth wave might be quite large. As wavelengths and interference change, though, every seventh wave might be large, or every fifth, or twelfth. Contrary to folklore, there is no set ratio.

Interference can have sudden unpleasant consequences on the open sea. In or near a large storm, wind waves at many wavelengths and heights may approach a single spot from different directions. If such a rare confluence of crests occurred at your position, a huge wave crest would suddenly erupt from a moderate sea to threaten your ship. The freak wave—called a **rogue wave**—would be much larger than any noticed before or after, and it would be higher than the theoretical maximum wave capable of being sustained in a fully developed sea. In such conditions one wave in about 1,175 is more than three times average height, and one in every 300,000 is more than four times average height! **Figure 10.16** is a very rare photograph of a rogue wave.

Currents can contribute to the formation of a kind of rogue wave that does not depend on interference. These waves are formed by the interaction of a wind wave and a swift surface current. The southeastern tip of Africa seems to breed a dispro-

portionately large number of these giant waves. The Agulhas Current, a strong western boundary current, flows close to shore there, moving in the opposite direction to wind waves generated by high-latitude southwesterly gales. When large waves suddenly hit the 5-kilometer- (3-mile-) per-hour current, they can "trip" on it, rise suddenly to double their original height, and break precipitously. Mathematical modeling suggests that a swift current or large fields of random eddies can act like a lens to refract and focus wave trains at a point. Waves at that location would have a steep forward face preceded by a deep trough. Mariners who have experienced these kinds of rogue waves describe them as "holes in the sea." Such waves have broken tankers in half; many lives have been lost.

Wind Waves Approaching Shore

Most wind waves eventually find their way to a shore and break, dissipating all their order and energy. The process begins with the transition of our now-familiar deep-water wave to a transitional wave in water less than half a wavelength deep. **Figure 10.17** outlines the events that lead to the break:

1. The wave train moves toward shore. When the depth of the water is less than half the wavelength, the wave "feels" bottom.

2. The circular motion of water molecules in the wave is interrupted. Circles near the bottom flatten to ellipses. The wave's energy must now be packed into less water depth, so the wave crests become peaked rather than rounded.

3. Interaction with the bottom slows the wave. The waves behind it continue toward shore at the original rate. Wavelength therefore decreases, but period remains unchanged.

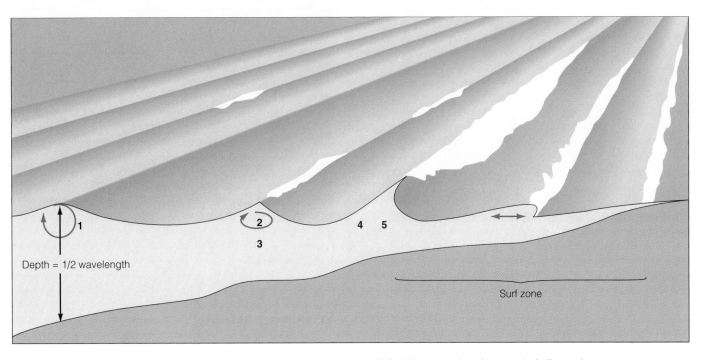

ACTIVE FIGURE 10.17 How a wave train breaks against the shore. (1) The swell "feels" bottom when the water is shallower than half the wavelength. (2) The wave crests become peaked because the wave's energy is packed into less water depth. (3) Constraint of circular wave motion by interaction with the ocean floor slows the wave, while waves behind it maintain their original speed. Therefore the wavelength shortens, but the period remains unchanged. (4) The wave approaches the critical 1:7 ratio of wave height to wavelength. (5) The wave breaks when the ratio of wave height to water depth is about 3:4. The movement of water particles is shown in red. Note the change from a deep-water wave—through the transitional wave stage—to a shallow-water wave.

Oceanography Now™ | At the Active Figures link at http://oceanography.brookscole.com/ garrison5e you can view how a wave train breaks against the shore.

4. The wave becomes too high for its wavelength, approaching the critical 1:7 ratio.

5. As the water becomes even shallower, the part of the wave below average sea level slows because of the restricting effect of the ocean floor on wave motion. When the wave was in deep water, molecules at the top of the crest were supported by the molecules ahead (thus transferring energy forward). This is now impossible because the *water* is moving faster than the *wave*. As the crest moves ahead of its supporting base, the wave breaks. The break occurs at about a 3:4 ratio of wave height to water depth (that is, a 3-meter wave will break in 4 meters of water). The turbulent mass of agitated water rushing shoreward during and after the break is known as **surf.** The **surf zone** is the region between the breaking waves and the shore.

Waves break against the shore in different ways, depending in part on the slope of the bottom. The break can be violent and toppling, leaving an air-filled channel (or *tube*) between the falling crest and the foot of the wave. These **plunging waves** form when waves approach a steeply sloping bottom (as in this chapter's opener) (**Figure 10.18a**) A more gradually sloping bottom generates a milder **spilling wave** as the crest slides down the face of the wave (**Figure 10.18b**).

Slope alone does not determine the position and nature of the breaking wave. The contour and composition of the bottom can also be important. Gradually shoaling bottoms can sap waves of their strength because of prolonged interaction against the bottom of the lowest elliptical water orbits. Energy may be lost even more rapidly if the bottom is covered with loose gravel or irregular growths of coral. Masses of moving seaweed or jostling chunks of sea ice can also extract energy from a wave. In rare cases the shore is configured in such a way that waves don't break at all: The waves have lost virtually all their energy by the time their remnants arrive at the beach.

Wave Refraction

What happens when a wave line approaches the shore at an angle, as it almost always does (**Figure 10.20**)? The line does not break simultaneously because different parts of it are in different depths of water. The part of the wave line in shallow water slows down, but the attached segment still in deeper water continues at its original speed, so the wave line bends, or refracts. The bend can be as much as 90° from the original direction of the wave train. This slowing and bending of waves in shallow water is called **wave refraction.**[3] The refracted waves break in a line almost parallel to the shore.

Wave refraction can produce some odd surf patterns. A swell with a wavelength of 600 meters "feels" bottom at 300 meters (1,000 feet), while a swell with a wavelength of

---

[3] To review the principle of refraction, please see Chapter 6, page 151.

a&b: Dennis Junor, Creation Captured

a

b

FIGURE 10.18 Types of breaking waves. (**a**) Plunging waves form when the bottom slopes steeply toward shore. (**b**) Spilling waves form when the bottom slopes gradually.

200 meters "feels" bottom at 100 meters (330 feet). You may remember that the average depth at the outer continental shelves is only about 150 meters (500 feet). As these two wave trains approach shore, the longer swell will be slowed by the shelf for perhaps 60 kilometers (37 miles) before the shorter swell is slowed by the bottom. With one wave bending and the other heading in its original direction, the potential for complex interference is great. Waves may break for a time at one spot, cease, and then begin breaking at another spot a few hundred meters down the beach. The ocean surface might have a checkerboard appearance from the constructive and destructive interference of these crests (**Figure 10.21**).

Wave Diffraction

10.18

Wave diffraction is the propagation of a wave around an obstacle. Unlike wave refraction (which depends on a wave's response to a change in speed), wave diffraction depends on the interruption of the wave trains by an obstacle to provide a new point of departure for the waves. This is illustrated by the gap in the breakwater in **Figure 10.22**. Wave crests excite the water in the gap. Water moving in the gap generates smaller waves in the harbor, which radiate from the gap and disturb the quiet water. The diffracted waves are much smaller than those in the open ocean, but boats in the harbor might still be jostled.

FIGURE 10.19 *Neptune's Horses* by Walter Crane, a graphic representation of the power of breaking waves. The energy of a wave is proportional to the square of its height. Researchers report that each linear meter of a wave 2 meters (6.6 feet) above average sea level represents an energy flow of about 25 kilowatts (34 horsepower), enough to light 250 100-watt light bulbs; a wave twice as high would contain four times as much energy. A single wave 1.2 meters (4 feet) high striking the west coast of the United States may release as much as 50 million horsepower.

Walter Crane/Erich Lessing, Art Resource, NY

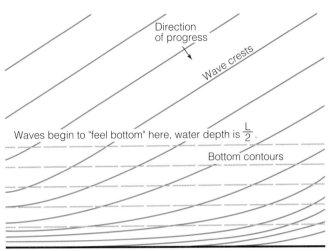

Direction
of progress

Wave crests

Waves begin to "feel bottom" here, water depth is $\frac{L}{2}$.

Bottom contours

Shoreline

a

b

FIGURE 10.20 Wave refraction. (**a**) Diagram showing the elements that produce refraction. (**b**) Wave refraction around Maili Point, Oahu, Hawaii. Note how the wave crests bend almost 90° as they move around the point.

A more complex case of diffraction is shown in **Figure 10.23**, in which waves are interrupted by a chain of islands. Some of the wave energy "squeezes" through the spaces between the islands. These spaces act as if they were new sources of waves. The waves spread into the space past the island, creating areas of reinforcement and cancellation. Polynesian navigators learned early in their training to sense the disturbances in wave patterns caused by diffraction. The best among them could feel the presence of islands beyond the horizon by subtle irregularities in the rocking of their canoe.

Wave Reflection

10.19

All the waves discussed so far in this chapter—the familiar ocean waves in which the disturbance travels in one direction along the surface of the transmission medium—are progressive waves. A vertical barrier such as a seawall, large ship hull, or smooth jetty will reflect progressive waves with little loss of energy. If the waves approach the obstruction straight on, the reflected waves will move away from the obstruction in the di-

FIGURE 10.21 A "checkerboard" sea surface beyond the surf zone off La Jolla, California—the result of wave interference.

FIGURE 10.22 Diffraction of waves by a breakwater gap at Morro Bay, California.

rection from which they came. This **wave reflection** will cause interference in the form of vertical oscillations called **standing waves**. As their name suggests, standing waves do not progress but appear as alternative crests and troughs at a fixed position. **Figure 10.24** shows how a standing wave oscillates in a motion that resembles water sloshing back and forth in a half-filled bathtub. Because of constructive interference between crests (and troughs), these waves can be dangerous to boats or swimmers near the obstruction. As you will soon see, standing waves are important in the physics of tsunami and tides.

Waves approaching the obstruction at an angle can also reflect; the sea surface near the reflector will not form standing waves, but complicated sea-surface motions develop. Watch for these effects the next time you visit a solid breakwater or a steep beach.

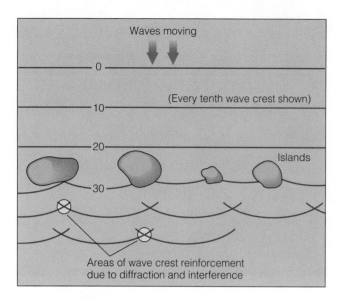

FIGURE 10.23 Wave diffraction past an island chain. Polynesian navigators used diffraction patterns to sense the presence of islands out of sight over the horizon (see Figure 1.10).

Internal Waves

Progressive waves can occur at the junction between air and water, as we have seen, or they can form at the boundary between water layers of different densities. These subsurface waves are called **internal waves.** As is the case with ocean waves at the air–ocean interface, internal waves possess troughs, crests, wavelength, and period. "Desktop-ocean" devices, which are sold at some gift shops, demonstrate internal waves. These sealed bottles contain nonmixing liquids of contrasting colors and slightly different densities. When you tilt the bottle, a very slow internal wave forms at the junction of the liquids, moves slowly to the low end, and breaks.

Normal ocean waves move rapidly because the difference in density between air and ocean is relatively great. Internal waves usually move very slowly because the density difference between the joined media is very small. Internal waves occur

in the ocean at the base of the pycnocline, especially at the bottom edge of a steep thermocline (**Figure 10.25a**). The wave height of internal waves may be greater than 30 meters (100 feet), causing the pycnocline to undulate slowly through a considerable depth (**Figure 10.25b**). Their wavelength often exceeds 0.8 kilometer (0.5 mile); periods are typically 5 to 8 minutes. Internal waves are generated by wind energy, tidal energy, and ocean currents. Surface manifestations of internal waves have been photographed from space (**Figure 10.25c**).

Are internal waves important? They may mix nutrients into surface water and trigger plankton blooms. They can also affect submarines and oil platforms. In 1963 the nuclear-powered USS *Thresher*, a fast attack submarine, was lost off the coast of Massachusetts with all hands. Running at high speed, *Thresher* may have encountered an internal wave and been forced beyond her test depth. In 1980 a production oil platform was slowly rotated nearly 90° from its original orientation by a series of internal waves. Also, the slow-motion breaking of internal waves against a shore may occasionally exaggerate tidal height.

A Word About "Tidal Waves"

Now that we've described the formation and characteristics of deep-water waves, we turn to the topic of very large, shallow-water waves: storm surges, seiches, and tsunami (including seismic sea waves). The popular media and general public tend to label *any* unusually large wave a "tidal wave" regardless of its origin, and the waves described next are prime candidates for this error. Press accounts of storms at sea usually list a rogue wave as a tidal wave, and very large sets of wind waves are called tidal waves by some yacht owners or surfers. The sea waves associated with earthquakes are almost always called tidal waves in media damage reports. The waves caused by the approach of a tropical cyclone to land may also incorrectly be termed tidal waves. The term *tidal wave* is *not* synonymous with *large wave*, however. As we will see in Chapter 11, the only true tidal waves are relatively harmless waves associated with the tides themselves.

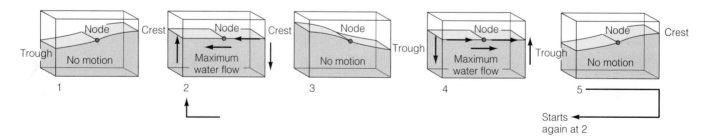

FIGURE 10.24 A standing wave in a basin. The wave oscillates about a node, a location at which there is no vertical movement. The rocking movement of water generates alternative crests and troughs at fixed positions a distance from the node. As their name suggests, standing waves do not progress, and there is no net movement of water in them. (See also Figure 10.28).

FIGURE 10.25 Internal waves can form between masses of water with different densities, especially at the base of the pycnocline. (**a**) The crest of an internal wave. (**b**) Movement of internal waves through the Straits of Messina (between Italy and the island of Sicily). The colors indicate differences in water density (blue represents less dense water; red represents more dense water). The arrows indicate the speed and direction of water flow. (**c**) Internal waves diffracting around the Seychelles Islands in the Indian Ocean, photographed by the crew of the space shuttle *Columbia* in January 1990. The waves are visible because their crests have altered the reflectivity of the ocean surface. Patterns of constructive and destructive interference can be seen.

Storm Surges

The abrupt bulge of water driven ashore by a tropical cyclone (hurricane) or frontal storm is called a **storm surge.** The low atmospheric pressure associated with a great storm will draw the ocean surface into a broad dome as much as 1 meter (more than 3 feet) higher than average sea level. This dome of water accompanies the storm to shore, becoming much higher as the water gets shallower at the coast. There the water ramps ashore, driven forward by large, storm-generated wind waves (**Figure 10.26**). The crest of a storm surge can temporarily add up to 9 meters (30 feet) to coastal sea level. Water can reach even greater heights when the surge is funneled into a confined bay or estuary.

A storm surge is a short-lived phenomenon. Technically it is not a wave because it is only a crest, so wavelength and period cannot be assigned to it. Water in a storm surge does not come ashore as a single breaking wave but rushes inland in what

looks like a sudden, very high, windblown tide. Indeed storm surges are sometimes called *storm tides.* The wrong combination of low atmospheric pressure, strong onshore winds, high tide, and bottom contour can be very dangerous—especially if estuaries in the area have been swollen by heavy rainfall preceding the storm.

Storm surges have had catastrophic consequences. The frightful tropical storm of November 1970 in Bangladesh (described in Chapter 8) generated a storm surge up to 12 meters (40 feet) high that caused the death of more than 300,000 people. Storm surges associated with extratropical cyclones (frontal storms) can also do tremendous damage. On 1 February 1953 a storm surge and high tide arrived simultaneously against the Dutch coast, a coincidence estimated to occur there only once in 400 years. Wind waves breached the dikes and flooded the low country, covering more than 3,200 square kilometers (800,000 acres) and drowning 1,783 people.

a

b

AP/Wide World Photos

FIGURE 10.26 A storm surge. (**a**) The low pressure and high winds generated within a hurricane can produce a storm surge up to 9 meters (30 feet) high. (**b**) In 1992 storm surge waves up to 4.5 meters (15 feet) pounded homes on Peggotty Beach, Scituate, Massachusetts.

Rebuilding the dikes higher and reclaiming the land from the North Sea cost billions of dollars. On the opposite side of the North Sea, Londoners have spent $1.3 billion on a flood defense system at the mouth of the Thames River. The centerpiece of the project is an immense barrier against storm surges (**Figure 10.27**). Experts expect the barrier to prevent a devastating flood on an average of once every 50 years.

The coast of the United States is also at risk. In 1900 a storm surge topped the Galveston, Texas, seawall, swept into the city, and killed more than 6,000 residents (see again the figure in Box 8.1). In contrast, no lives were lost during a similar Texas storm in 1961 because coastal barriers had been constructed and advance warning permitted preparation and evacuation. In August 1954 the combined effects of a high tide and Hurri-

a

Tom Garrison

b

Thames Barrier Visitors Centre

FIGURE 10.27 The Thames tidal barrier in London. The natural funnel shape of the Thames estuary concentrates surges as they near London. The barrier sections (**a**) can be raised to prevent flooding upstream. Each barrier section is controlled by hydraulic arms within towers. The great size of the $1.3-billion project can be seen in (**b**).

cane Carol caused water to rise almost 5 meters (16 feet) above normal to flood much of Providence, Rhode Island. Damage exceeded $41 million. City officials have since built a protective surge barrier across upper Narragansett Bay. Anyone who lives in a low-lying coastal area frequented by violent storms should be aware of the potential danger of storm surge and should heed calls for evacuation if conditions warrant.

Seiches

When disturbed, water confined to a small space (a bucket, a bathtub, or a bay) will slosh back and forth at a specific resonant frequency. The frequency changes with different amounts of water or with different sizes or shapes of containers. If you carry a shallow tray of water from one place to another, you must be careful not to move the tray at its resonant frequency to avoid a spill. Most of the water's random motion quickly settles down after you place the tray on a table top, but the water in the tray may rock gently for some seconds at this one resonant frequency. That rocking is a shallow-water wave phenomenon called a **seiche** (pronounced *saysh*).

The seiche phenomenon was first studied in Switzerland's Lake Geneva by eighteenth-century researchers curious about why the water level at each end of the long, narrow lake rises and falls at regular intervals after wind storms. They found that a constant, steady wind tends to push water into the downwind end of the lake. When the wind stops, the water is released to rock slowly back and forth at the lake's resonant frequency, completing a crest-trough-crest cycle in a little more than an hour. At the ends of the lake the water rises and falls nearly a meter; at the center it moves back and forth without changing height. Thus a seiche is a form of standing wave that oscillates vertically with no forward movement (see again Figure 10.24). The point (or line) of no vertical wave action in a standing wave—the place in the lake where the water moves only back and forth—is called a **node**. In Lake Geneva the wavelength of the seiche is twice the length of the lake itself; the node lies at the center (**Figure 10.28**). The lake acts like a large version of the tray of water in our example.

Energy can be added rhythmically to a seiche. The process is analogous to adding energy to a child in a swing. Insert a little energy at the right time in each cycle, and the child will swing higher and higher. (As our daughter discovered at an early age, the same thing applies to water in a bathtub. By moving her body in rhythm with the growing wave, she quickly learned how to slop most of the bathwater onto the floor.)

This kind of activity can also occur in confined areas of ocean ranging in size from bays and harbors to entire ocean basins. As we have seen, a drop in atmospheric pressure beneath a storm can draw water to a slightly higher level. When the storm moves inland, the water is released and the oscillation begins. The wavelength of the oscillation is a multiple of the length of the ocean basin, often thousands of miles. Tides and surf beat can also cause seiches in bays or harbors, as can internal waves moving toward shore. Seiche periods may range from a few minutes to more than a day. In large areas the wavelength of these standing waves may rival that of tsunami.

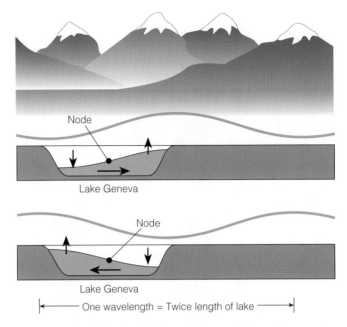

FIGURE 10.28 Seiche in Lake Geneva. After a windstorm the water in the lake sloshes back and forth, forming a standing wave (seiche) with a wavelength of 220 kilometers (132 miles)—twice the length of the lake. The period of the seiche is about 72 minutes, and its height is about 0.5 meter (1.5 feet). At the node, water moves sideways and does not rise or fall (see also Figure 10.24). The blue line represents the hypothetical whole wave of which the seiche is a part.

Damage from seiches along most ocean coasts is rare. The wavelength may be tremendous, but seiche wave height in the open ocean rarely exceeds a few inches. Larger seiches can occur in harbors. Coastal seiches at Nagasaki, on the southern coast of Japan, occasionally reach 3 meters (10 feet). Seiches may disturb shipping schedules by interfering with the predicted arrival times of tides, or they may cause currents in harbors, which could snap mooring lines. In rare instances seiches contribute to the peril of people on shore, but usually because the seiche is coincidentally accompanied by wind waves of considerable height or by a tsunami.

Tsunami and Seismic Sea Waves

Long-wavelength, shallow-water progressive waves caused by the rapid displacement of ocean water are called **tsunami**, a descriptive Japanese term combining *tsu* ("harbor") with *nami* ("wave"). The word is both singular and plural. Tsunami caused by the sudden vertical movement of Earth along faults (the same forces that cause earthquakes) are properly called **seismic sea waves**. Tsunami can also be caused by landslides, icebergs falling from glaciers, volcanic eruptions, and other direct displacements of the water surface. Note that all seismic sea waves are tsunami, but not all tsunami are seismic sea waves.

Oceanography⬡Now™ | Click Oceanography Interactive to work through an activity on tsunami through earthquakes and tsunami.

FIGURE 10.29 Failure of the edge of the continental shelf off the Oregon coast produced an underwater landslide that probably caused a large tsunami sometime around the end of the last ice age. The pocket from which the debris originated is about 6 kilometers (3.7 miles) wide. (See Figure 4.15 for a larger view of this area.)

William F. Haxby & Lincoln. F. Praston

Origins

"Small" tsunami are caused by the displacement of surface water. Although less energy is released by landslides than by most seismic fractures, the resulting sea waves are still very destructive for people or structures near their point of origin. This is especially true if the wave is formed within a confined area.

Tsunami originate on the ocean floor when underwater landslides, Earth movement along faults, or other activities displace ocean water. **Figure 10.29** shows a debris trail off the Oregon coast, the remains of an underwater landslide some 6 kilometers (3.7 miles) across. Probably triggered by an earthquake, the slump displaced water and probably caused a tsunami. (Faults recently discovered in the Hawaiian Islands and off the coasts of southern California, southernVirginia, and North Carolina suggest that similar events could happen in those areas.)

As an example of tsunami formation, **Figure 10.30** shows the birth of a seismic sea wave in the Aleutians. Rupture along a submerged fault lifts the sea surface above. Gravity pulls the crest downward, but the momentum of the water causes the crest to overshoot and become a trough. The oscillating ocean surface generates progressive waves that radiate from the epicenter in all directions (**Figure 10.31**). Waves would also form if the fault movement were downward. In that case a depression in the water surface would propagate outward as a trough.

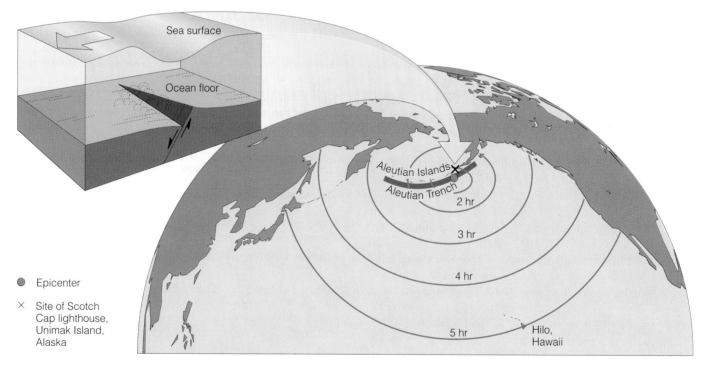

FIGURE 10.30 A tsunami occurred on 1 April 1946 when a rupture along the Aleutian Trench, a submerged fault in Alaska's Aleutian Islands, lifted the sea surface above. The resulting seismic sea wave quickly engulfed the Scotch Cap lighthouse, located on a point overlooking the sea on Unimak Island (see also Figure 10.33). The wave moved outward at a speed of about 212 meters per second (472 miles per hour) and took only about five hours to travel to the Hawaiian Islands (see Figure 10.34).

a

b

c

d

e

ACTIVE **FIGURE 10.31** A computer simulation of the movement of a 1960 tsunami that originated in western Chile and sent destructive waves to Japan. The images represent the successive positions of the waves (**a**) 2.5 hours after the earthquake, (**b**) 5 hours after, (**c**) 12.5 hours after, (**d**) 17.5 hours after, and (**e**) reaching Japan after 22.5 hours of travel. (Source: Research by Philip L.F. Liv, Seung Nam Seo, and Sung Bum Yoon, and Civil and Environmental Engineers, Cornell University. Visualization by Catherine Devine, Cornell Theory Center. Used by permission.)

Oceanography≋Now™ | At the Active Figures link at http://oceanography.brookscole.com/garrison5e you can view a computer simulation of a tsunami.

The trough would be followed by smaller crests and troughs caused by surface oscillation.

It seems strange to refer to tsunami—waves with wavelengths of up to 200 kilometers (125 miles)—as shallow-water waves. Yet half their wavelength would be 100 kilometers (62 miles), and even the deepest ocean trenches do not exceed 11 kilometers (7 miles) in depth. These immense waves are therefore never in water deeper than half their wavelength. Like any transitional or shallow-water wave, seismic sea waves are affected by the contour of the bottom and are commonly refracted, sometimes in unexpected ways.

Speed

The speed (C) of a tsunami is given by the formula for the speed of a shallow-water wave:

$$C = \sqrt{gd}$$

Because the acceleration due to gravity (g) is 9.8 meters (32.2 feet) per second per second, and a typical Pacific abyssal depth (d) is 4,600 meters (15,000 feet), solving for C shows that the wave would move at 212 meters per second (470 miles per

a

b

c

FIGURE 10.32 Sequential photos showing the arrival of a tsunami at Laie Point, Oahu, Hawaii, on 9 March 1957. An 8.3 earthquake had occurred in the Aleutian Islands about five hours earlier. Note how the wave resembles a fast, onrushing flood of water, not the huge, plunging breaker of movies and folklore.

hour). At this speed a seismic sea wave will take only about five hours to travel from Alaska's seismically active Aleutians to the Hawaiian Islands!

Encountering Tsunami

We are familiar with the steepness of a wind wave and the short period of a few seconds between its crests. Tsunami are much different. Once a tsunami is generated, its steepness (ratio of height to wavelength) is extremely low. This lack of steepness, com-

bined with the wave's very long period (5 to 20 minutes), enables it to pass unnoticed beneath ships at sea. A ship on the open ocean that encounters a tsunami with a 16-minute period would rise slowly and imperceptibly for about 8 minutes, to a crest only 0.3 to 0.6 meter (1 or 2 feet) above average sea level. It would then ease into the following trough 8 minutes later. With all the wind waves around, such a movement would not be noticed.

As the tsunami crest approaches shore, however, the situation changes rapidly and often dramatically. The period of the wave remains constant, velocity drops, and wave height increases greatly. As the crest arrives at the coast, observers would see water surge ashore in the manner of a very high, very fast tide (**Figure 10.32**). In confined coastal waters relatively close to their point of origin, tsunami can reach a height of perhaps 30 meters (100 feet). The wave is a fast, onrushing flood of water, not the huge, plunging breaker of popular folklore.

As a tsunami expands from its point of origin, the wave energy spreads through an enlarging circumference. People on shore near the generating shock have reason to be concerned because the energy will not have dissipated very much. On 1 April 1946 a fracture along the Aleutian Trench generated a seismic sea wave that quickly engulfed the Scotch Cap lighthouse on Unimak Island in the Aleutians. The lighthouse was completely destroyed and the five coast guardsmen operating the lighthouse died (**Figure 10.33**).

By the time the same seismic sea wave reached the Hawaiian Islands about five hours later, the wave circumference was enormous and its energy more dispersed. Even so, successive waves surged onto Hawaii's beaches at 15-minute intervals for more than two hours. One 9-meter (30-foot) wave struck the town of Hilo, and water rose to a height of 17 meters (56 feet) in an exposed valley near Polulu! At least 150 people were killed in Hawaii that morning, and property damage exceeded $25 million. **Figure 10.34** is a photograph taken in Hilo.

Note that the destruction in Hawaii was not caused by one wave (as at Scotch Cap) but by a series of waves following one another at regular intervals. Some energy from the main tsunami wave was distributed into smaller waves ahead of or behind the main wave as it moved. If the source of energy responsible for a tsunami is very far away, sea level at shore will rise and fall as these waves arrive. The interval between crests (the wave period) is usually about 15 minutes. Coastal residents far from a tsunami's origin can be lulled into thinking the waves are over; they return to the coastline only to be injured or killed by the next crest. This behavior has contributed to loss of life in Hawaii.

Tsunami in History

Destructive tsunami strike somewhere in the world an average of once each year. An earthquake along the Peru–Chile Trench on 22 May 1960 killed more than 4,000 people. The resulting tsunami reached Japan, 14,500 kilometers (9,000 miles) away, killing 180 people and causing $50 million in structural damage. Los Angeles and San Diego harbors were badly disrupted by seiches excited by the tsunami. In 1992 a tsunami struck the coast of Nicaragua and killed 170 people; 13,000 Nicaraguans were left homeless. In 1993 an earthquake in the Sea of Japan

D. Millsap

FIGURE 10.33 Artist's conception of the Scotch Cap lighthouse moments before its destruction on 1 April 1946 by a tsunami. The wreckage of the lighthouse, located at Unimak Pass in Alaska's Aleutian Islands, is shown in the inset. The remnants of the lighthouse's foundations are 15 meters (49 feet) above sea level, and rubble has been carried to the upper plateau 36 meters (118 feet) above sea level.

USGS

a   b

generated a tsunami that washed over areas 31 meters (101 feet) above sea level and killed 239 people (**Figure 10.35**). In 1998 an underwater landslide at the edge of the local continental shelf triggered a wave 15 meters (50 feet) high that destroyed remote villages in Papua New Guinea, killing more than 2,200 people. Sometimes even a relatively minor wave can cause considerable loss of life and property. On 17 February 1996, 53 people died at Biak, Indonesia, in a wave that advanced less than 9 meters (28 feet) past the normal high-tide line. Some recent lethal tsunami are shown in **Figure 10.36.**

Half the time a tsunami's trough will arrive first. This occurred with catastrophic results in the Tagus estuary at Lisbon, Portugal, on 1 November 1755. Many curious people were attracted to the shore when sea level dropped several meters during about 15 minutes. People ventured out to pick up fish, inspect grounded vessels, and look around. Thousands were killed when the first large crest arrived, and would-be rescuers were drowned by later crests, some of which exceeded 10 meters (33 feet) in height! In all, more than 50,000 people died from the effects of the earthquake and tsunami in and around Lisbon.

There have been even greater tsunami disasters. In 1703 at Awa, Japan, 100,000 people died. On 27 August 1883 the enormous volcanic explosion of Krakatoa in Indonesia generated 35-meter (115-foot) waves that destroyed 163 villages and killed more than 36,000 people. Evidence of even larger waves has been discovered. Researchers have found signs of a huge

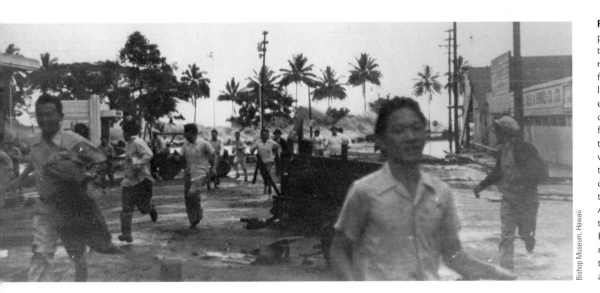

FIGURE 10.34 A tsunami in progress. The largest wave of the 1 April 1946 tsunami rushes ashore in Hilo. Terrified residents run for their lives. V-shaped Hilo Bay is an especially dangerous place during a tsunami because its funnel-like outline concentrates the energy of the waves. Fourteen years after the 1946 disaster, a seismic disturbance in the subduction zone off western South America generated a tsunami that drowned 61 people in Hilo. Wave-cut scars on cliffs north of the town suggest that visits by large tsunami are not rare occurrences.

Bishop Museum, Hawaii

FIGURE 10.35 Massive property damage in the fishing village of Aonae on the island of Okushiri, Japan, was caused by a tsunami generated by the Hokkaido Nansei-Oki earthquake in 1993. Maximum wave height was 31 meters (101 feet); 239 people died. This event has become the best-documented tsunami disaster in history.

Kyodo Photo Service

wave, perhaps as high as 91 meters (300 feet), that crashed against the Texas coast 66 million years ago. It may have been caused by a comet or asteroid striking the Gulf of Mexico near Yucatan (see Box 13.1). The wave scoured the floor of the gulf; picked up sand, gravel, and shark's teeth; and deposited the material in what is now central Texas.

The Tsunami Warning Network

Since 1948, an international tsunami warning network has been in operation around the seismically active Pacific to alert coastal residents of possible danger. Data obtained from bottom-

mounted pressure sensors (able to detect very long-wavelength waves), seismographs, and subtle changes in the expected arrival times of tide waves are analyzed by computer models. If circumstances warrant, warnings must be issued rapidly because of the speed of these waves through the water. Telephone books in coastal Hawaiian towns contain maps and evacuation instructions for use when the warning siren sounds.

The tsunami warning system was responsible for averting the loss of many lives after the great 27 March 1964 earthquake in Alaska (discussed in Chapter 3). A 3.7-meter (12-foot) wave, probably the fourth crest to reach the coast, swept into Crescent City, California, about six hours after the quake. Though

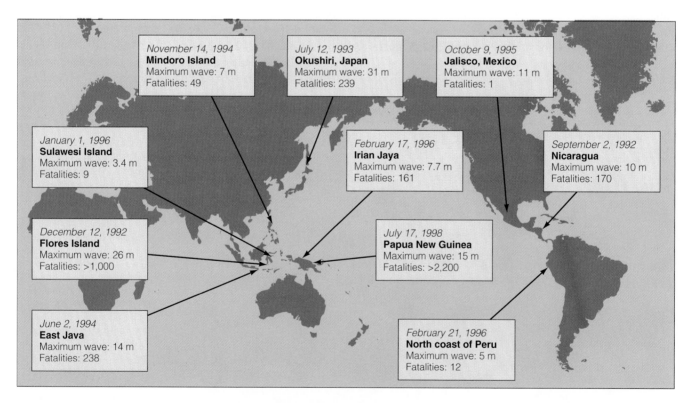

November 14, 1994
Mindoro Island
Maximum wave: 7 m
Fatalities: 49

July 12, 1993
Okushiri, Japan
Maximum wave: 31 m
Fatalities: 239

October 9, 1995
Jalisco, Mexico
Maximum wave: 11 m
Fatalities: 1

January 1, 1996
Sulawesi Island
Maximum wave: 3.4 m
Fatalities: 9

February 17, 1996
Irian Jaya
Maximum wave: 7.7 m
Fatalities: 161

September 2, 1992
Nicaragua
Maximum wave: 10 m
Fatalities: 170

December 12, 1992
Flores Island
Maximum wave: 26 m
Fatalities: >1,000

July 17, 1998
Papua New Guinea
Maximum wave: 15 m
Fatalities: >2,200

June 2, 1994
East Java
Maximum wave: 14 m
Fatalities: 238

February 21, 1996
North coast of Peru
Maximum wave: 5 m
Fatalities: 12

FIGURE 10.36 Ten destructive tsunami have claimed more than 4,000 lives since 1990.

b

FIGURE 10.37 (**a**) A tsunami hazard warning sign in a town in central coastal Oregon. The configuration of the coast and the slope and instability of the bottom in this area make tsunami particularly dangerous (see Figure 10.29). (**b**) An unfortunate effect of the tsunami warning network. Sightseers flocked to Oahu's Makapu'u Beach in Hawaii to await a tsunami generated by a 6.5 earthquake centered in the Aleutian Trench on 7 May 1986.

a

more than 300 buildings were destroyed or damaged, five gasoline storage tanks set ablaze, and 27 blocks of the city demolished, there were relatively few casualties. Unfortunately the models are not always accurate: Of 20 warnings issued by the system since its inception, 15 have been false alarms.

It has been more than 30 years since a tsunami caused substantial damage in the United States. Some public safety experts suggest we have become complacent about the risks associated with these destructive waves. As can be seen in **Figure 10.37**, though, some coastal communities remind residents and visitors of the danger.

Questions from Students

1. If so much energy is expended when wind waves break, why doesn't water in the surf zone get hot?

The amount of energy moving through the ocean in progressive waves is impressive. The energy of a wave is proportional to the square of its height. Each linear meter of a wave 2 meters above average sea level represents an energy flow of about 25 kilowatts (34 horsepower), enough to light 250 100-watt light bulbs. A wave twice as high contains four times as much energy. A single wave 1.2 meters high striking the west coast of the United States may release as much as 50 million horsepower!

The energy of wind waves is dissipated mostly as heat in the surf, but because water has such a high heat capacity (discussed in Chapter 7), the injection of heat into the surf zone doesn't significantly increase water temperature. The surf zone is also an area of vigorous mixing, so any heat released is quickly distributed through a large volume of water.

A number of methods have been proposed to take advantage of this energy before it is dissipated. An installation by Kvaerner Industrier A/S, a Norwegian energy company, is now producing power on the coast of Scotland. Its method of operation is shown in Figure 17.7.

2. What about wind waves and the Coriolis effect? Do wind waves turn right in the Northern Hemisphere as they move across the ocean surface?

The Coriolis effect has no influence on waves that have periods shorter than about 5 minutes. Large shallow-water waves such as tsunami, seiches, and tides involve the mass movement of water and are influenced by Earth's rotation. Wind waves, however, with a period rarely exceeding 20 seconds, are not.

3. Could a method be invented to surf internal waves?

Think of the possibilities: a wave 30 meters high moving in slow motion with a period of 5 to 8 minutes. It would be a calm and stately ride, and the density differences would require a very delicate balancing act at the interface between the two water densities. Still, with the right equipment to adjust buoyancy and provide air, it might be done.

Indeed, surfing an internal wave may already have been inadvertently and disastrously accomplished. As noted in the text, the slow undulation of an internal wave may have led to the loss in 1963 of USS *Thresher*. An equipment failure at high speed coupled with a surprise wave encounter may have driven the new submarine past its designed pressure limit, causing the violent implosion that cost 129 lives.

4. Will people in California be in great danger from a seismic sea wave when a major earthquake occurs on the San Andreas Fault?

Probably not, for two reasons. First, San Andreas quakes are usually lateral-displacement quakes: The ground moves suddenly sideways rather than up or down. Tsunami are usually generated by vertical movement. Second, earthquake motion along the San Andreas Fault is parallel to the coast, not at right angles to it. If the ground movement were toward the coast (as could happen north of San Francisco), the "shove" might result in a wave. There may be some slopping about at the coastline during a large earthquake, but probably not a classic tsunami. One caveat: If massive offshore slumping occurs (see Figure 10.29), all bets are off.

5. One of my friends has discovered I'm studying waves. He demands to know about the world's largest wave. What's the record, anyway?

If your friend classifies the largest wave by wavelength, the answer is easy: Tides take the prize, with a wavelength that is ideally half Earth's circumference. Tides also win the speed race: Under ideal circumstances their crests can move at speeds up to 1,600 kilometers (1,000 miles) per hour.

If your friend classifies the largest wave by height, things become more complicated. As described in Box 10.1, the crew of *Ramapo* measured a wave 34 meters (112 feet) high in 1933. In general, the highest wind waves probably develop in the West Wind Drift, the nasty area of windy ocean ringing Antarctica. Multiple storms there, some with huge fetch distances, can cause very large waves and opportunity for constructive interference between them. In 1916 during an agonizing small-boat trip from an island on which their expedition was marooned, Frederic Worsley, captain of Ernest Shackleton's ship

Endurance, encountered tremendous waves greater than any he had encountered in a lifetime at sea. He estimated a few to be more than 30 meters (100 feet) high!

Some tsunami have been higher, however. The greatest height recently recorded for a tsunami in the ocean was 84 meters (278 feet) in 1971 off Ishigaki, Japan. But don't forget the 91-meter (300-foot) tsunami thought to have occurred 66 million years ago on the coast of what is now Texas.

6. Is there a reliable way to standardize the measurement of wave size or surface conditions at sea?

Yes, and it is a venerable method indeed. In the summer of 1805, when Rear Admiral Sir Francis Beaufort was captain of HMS *Woolwich*, a British warship, he invented a numbered scale to describe sea conditions. The lowest number, Beaufort 0, describes a "sea like a mirror." Beaufort 1 clearly describes capillary waves; 5 describes moderate waves and whitecaps; 9 describes dense streaks of foam and huge waves. Beaufort 12 (the highest number) conjures ". . . exceptionally high waves; sea covered with white foam; visibility greatly reduced." This method of describing **sea state** by number quickly spread through the fleet.

In 1912 the International Commission for Weather Telegraphy sought to correlate the Beaufort scale with specific wind velocities. A uniform set of equivalents was accepted in 1926, with revisions in 1946 and 1955. With new extension above Beaufort 12, the scale is still used today. A modification of the U.S. Navy's version is provided as **Table 10.3.**

Table 10.3	The Beaufort Scale				

Beaufort Number	Wind Speed		Effects Observed at Sea	Effects Observed on Land
	Kilometers Per Hour	Miles Per Hour		
0	<1	<1	Sea like a mirror	Calm; smoke rises vertically
1	1–5	1–3	Ripples with appearance of scales (capillary waves), no foam	Smoke drifts to indicate wind direction; weather vanes don't move
2	6–11	4–7	Small wavelets, crest glassy and not breaking	Wind felt on face; leaves rustle; weather vanes begin to move
3	12–19	8–12	Wave crests begin to break; scattered whitecaps	Leaves in constant motion; lightweight flags extended
4	20–28	13–18	Small waves of longer wavelengths; numerous whitecaps	Dust, leaves, loose paper raised up; small branches move
5	29–38	19–24	Moderate waves, many whitecaps, some spray	Small trees begin to sway
6	39–49	25–31	Larger waves forming whitecaps everywhere, more spray	Large trees in motion; whistling heard in wires
7	50–61	32–38	Sea heaps up; white foam from breaking waves begins to be blown in streaks	Large trees in motion; resistance felt when walking against wind
8	62–74	39–46	Moderately high waves of long wavelength; foam is blown from crests in well-marked streaks	Small branches break off trees, walking progress impeded
9	75–88	47–54	High waves, dense streaks of foam, spray may reduce visibilty	Slight structural damage; shingles are blown from roofs
10	89–102	55–63	High waves with overhanging crests; sea takes on white appearance from foam; visibility reduced	Trees broken or uprooted; considerable structural damage
11	103–117	64–72	Exceptionally high waves; sea covered with white foam; visibility sporadic	Very rarely experienced on land; usually accompanied by widespread damage
12	>118	>73	Hurricane conditions; air filled with foam; sea white with driven spray; visibilty greatly reduced	Catastrophic damage to structures

Source: U.S. Navy, with modifications.

Chapter in Perspective

In this chapter you learned that waves transmit energy, not water mass, across the ocean's surface. The speed of ocean waves usually depends on their wavelength, with long waves moving fastest. Arranged from short to long wavelengths (and therefore from slowest to fastest), ocean waves are generated by very small disturbances (capillary waves), wind (wind waves), rocking of water in enclosed spaces (seiches), seismic and volcanic activity or other sudden displacements (tsunami), and gravitational attraction (tides). The behavior of waves depends largely on the relationship between a wave's size and the depth of water through which it is moving. Waves can refract and reflect, break, and interfere with one another.

Wind waves can be deep-water waves if the water is more than half their wavelength deep. The waves of very long wavelengths are always in "shallow water" (water less than half their wavelength deep). These long waves travel at high speeds, and some may have great destructive power.

In the next chapter you will learn that even greater waves exist: the tides. You may be interested to know that two "tidal waves" move along most coasts each day. You don't read of daily mayhem and destruction from these passages because the crests are the day's high tides and the troughs the day's low tides. Tides are shallow-water waves no matter how deep the ocean they're moving through. Tides can be destructive, but their ability to cause damage is fortunately not proportional to their wavelengths.

Key Ideas

✓ Waves transmit energy, not water mass, across the ocean's surface.
   Review Figures 10.1, 10.2, and 10.3, pages 229–230

✓ Waves are classified by the disturbing force that creates them, the extent to which the disturbing force continues to influence the waves once they are formed, the restoring force that tries to flatten them, and their wavelength.
   Review pages 230–231

✓ The speed of ocean waves usually depends on their wavelength, with long waves moving fastest.
   Review Figure 10.7, page 233

✓ Arranged from short to long wavelengths (and therefore from slowest to fastest), ocean waves are generated by very small disturbances (capillary waves), wind (wind waves), rocking of water in enclosed spaces (seiches), seismic and volcanic activity or other sudden displacements (tsunami), and gravitational attraction (tides).
   Review pages 231–234

✓ The behavior of a wave depends largely on the relationship between the wave's size and the depth of water through which it is moving. Deep-water waves move through water deeper than half their wavelength. Only wind waves can be deep-water waves.
   Review Figure 10.6, page 233

✓ Wind waves form when energy is transferred from wind into water. Wave size depends on wind strength and duration.
   Review Figures 10.8 and 10.11, pages 234 and 236

✓ Most waves change shape and speed as they approach shore. They may plunge or spill at the surf zone and bend to break nearly parallel to shore.
   Review Figure 10.6, page 233

✓ Waves can change direction by refraction and diffraction, interfere with one another, and reflect from solid objects.
   Review Figures 10.15 and 10.17–10.23, pages 239, 241–244

✓ The great waves described in this chapter have such long wavelengths that they are always in "shallow water" (water that is shallower than half their wavelength). In wavelength order, they are seiches, tsunami, and tides. (Technically a storm surge is not a wave because it has only a crest.)
   Review Figures 10.6 and 10.17–10.19, pages 233, 241–242

✓ A seiche is a pendulum-like rocking of water in a basin. Its resonant frequency depends on the dimensions of the basin. Seiches take the form of standing waves.
   Review Figure 10.28, page 247

✓ Because wave speed is proportional to wavelength, great waves move rapidly through the ocean. Tides move fastest of all—up to 1,600 kilometers (1,000 miles) per hour.
   Review page 232

✓ Tsunami are caused when water is displaced by the forces that cause earthquakes or by landslides, eruptions, or asteroid impacts. Typically tsunami resemble a fast-onrushing tide, not a breaking wave. They are not dangerous in the open sea.
   Review Figures 10.29–10.37, pages 248–253

Terms and Concepts to Remember

$C = \sqrt{gd}$, 234	gravity waves, 231
$C = L/T$, 232	group velocity, 235
capillary waves, 231	interference, 239
constructive interference, 240	internal wave, 244
deep-water wave, 232	node, 247
destructive interference, 239	orbit, 229
dispersion, 234	orbital wave, 229
disturbing force, 231	plunging wave, 241
fetch, 236	progressive wave, 229
forced waves, 231	restoring force, 231
free waves, 231	rogue wave, 240
fully developed sea, 236	sea, 234

Tides and the Forces That Generate Them

Tides are periodic, short-term changes in the height of the ocean surface at a particular place caused by a combination of the gravitational force of the moon and the sun and the motion of Earth (**Figure 11.1**). With a wavelength that can equal half of Earth's circumference, tides are the longest of all waves. Unlike the other waves we have met, these huge shallow-water waves are never free of the forces that cause them and thus are called *forced* waves. (After they are formed, wind waves, seiches, and tsunami are *free* waves; that is, they are no longer being acted upon by the force that created them and they do not require a maintaining force to keep them in motion.)

The Greek navigator and explorer Pytheas first wrote of the connection between the position of the moon and the height of a tide around 300 B.C., but a full understanding of tides had to await Newton's analysis of gravitation.

Among many other things, Isaac Newton's brilliant 1687 book *Philosophiae Naturalis Principia Mathematica* (*Mathematical Principles of Natural Philosophy*) described the motions of planets, moons, and all other bodies in gravitational fields. A central finding: The pull of gravity between two bodies is proportional to the masses of the bodies but inversely proportional to the square of the distance between them. This means that heavy bodies attract each other more strongly than light bodies do, and that gravitational attraction quickly weakens as the distance grows larger. This relationship may be expressed mathematically as

$$F = G\left(\frac{m_1 m_2}{r^2}\right)$$

where F is the gravitational attraction, G is the universal gravitational constant, m_1 and m_2 are the masses of the two bodies,

and r is the distance between their centers. We can use this equation to calculate the gravitational attraction between the sun and Earth or between the moon and Earth.

Although the main cause of tides is the combined gravitational attraction of the moon and the sun acting on the ocean, the forces that actually generate the tides vary inversely with the *cube* of the distance from Earth's center to the center of the tide-generating object (the moon or the sun). Distance is thus even more important in this relationship, which may be expressed as

$$T = G\left(\frac{m_1 m_2}{r^3}\right)$$

where T is the tide-generating force, G is the universal gravitational constant, m_1 and m_2 are the masses of the two bodies, and r is the distance between their centers.[1] The sun is about 27 million times more massive than the moon, but the sun is about 387 times farther away than the moon, so the sun's influence on the tides is only 46% that of the moon's.

As we will see, Newton's gravitational model of tides—the *equilibrium theory*—deals primarily with the position and attraction of Earth, the moon, and the sun and does not factor in the influence on tides of ocean depth or the positions of continental landmasses. The equilibrium theory would accurately describe tides on a planet uniformly covered by water. A modification proposed by Pierre-Simon Laplace about a century later—the *dynamic theory*—takes into account the speed of the long-wavelength tide wave in relatively shallow water, the presence of interfering continents, and the circular movement or rhythmic back-and-forth rocking of water in ocean basins. We will explore the idealized situation of the equilibrium theory before moving to the real-world dynamic view.

---

[1] For a mathematical treatment of the tide-generating force, please see Appendix VII.

Scott Walking Adventures, Halifax, Nova Scotia

Scott Walking Adventures, Halifax, Nova Scotia

FIGURE 11.1 Tides in the eastern Bay of Fundy on the Atlantic coast of Canada. The difference between high and low water (called tidal range) is almost 15 meters (50 feet). At the peak of the incoming tide, water rises 1 meter (3.3 feet) in 23 minutes. These photos were taken near St. John, New Brunswick (see Figure 11.18). Other places in the world that have large tides include the Port of Bristol in England; the Sea of Okhotsk, northeast of Japan; Turnagain Arm in Alaska; the Bay of St. Malo in France; and the Feuilles River in Ungava Bay, Quebec. All have tidal ranges of about 10 meters (33 feet). However, not as much water is moved at these locations as in the Bay of Fundy, where the tidal contribution equals the total daily discharge of all the world's rivers—about 100 cubic kilometers (24 cubic miles) of water.

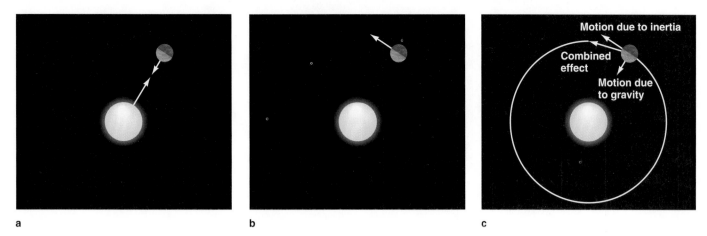

a   b   c

FIGURE 11.2 A planet orbits the sun in a balance between gravity and inertia. (**a**) If the planet is not moving, gravity will pull it into the sun. (**b**) If the planet is moving, the inertia of the planet will keep it moving in a straight line. (**c**) In a stable orbit, gravity and inertia together cause the planet to travel in a fixed path around the sun.

The Equilibrium Theory of Tides

The **equilibrium theory of tides** explains many characteristics of ocean tides by examining the balance and effects of the forces that allow a planet to stay in a stable orbit around the sun, or the moon to orbit Earth. The equilibrium theory assumes that the seafloor does not influence the tides and that the ocean conforms instantly to the forces that affect the position of its surface. The ocean surface is presumed always to be in equilibrium (balance) with the forces acting on it.

Oceanography ⟨≋⟩Now™ | Click Oceanography Interactive to work through an activity on tidal forces through waves, tides, and currents.

The Moon and Tractive Forces

We begin our examination of these forces by looking at the moon's effect on the ocean surface. Gravity tends to pull Earth and the moon toward each other, but inertia—the tendency of

moving objects to continue in a straight line—keeps them apart. (In this context, we sometimes call inertia *centrifugal force*, the "force" that keeps water against the bottom of a bucket when you swing it overhead in a circle.) Earth and the moon don't smash into each other (or fly apart) because they are in a stable orbit; their mutual gravitational attraction is exactly offset by their inertia (**Figure 11.2**).

Contrary to what you might think, the moon does not revolve around the center of Earth. Rather, the Earth–moon *system* revolves once a month (27.3 days) around the system's center of mass. Because Earth's mass is 81 times that of the moon, this common center of mass is located not in space but 1,650 kilometers (1,023 miles) *inside* the Earth. This center of mass is shown in **Figure 11.3**.

The moon's gravity attracts the ocean surface toward the moon. Earth's motion around the center of mass of the Earth–moon system throws up a bulge on the opposite side of Earth. Two tidal bulges result (**Figure 11.4**).

Let's look more closely at the two bulges in Figure 11.4c. **Figure 11.5** shows four places on Earth's surface marked with numbers 1 through 4. Each place has three arrows drawn to represent forces: The outward-flinging force of inertia is shown in

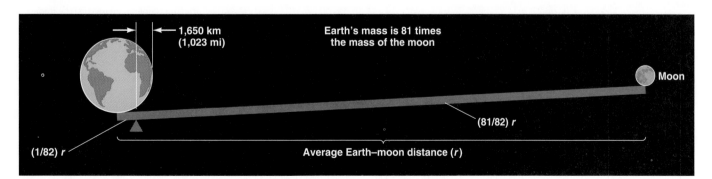

FIGURE 11.3 The moon does not rotate around the center of Earth. Earth and the moon together—the Earth-moon system—rotate around a common center of mass about 1,650 kilometers (1,023 miles) beneath Earth's surface.

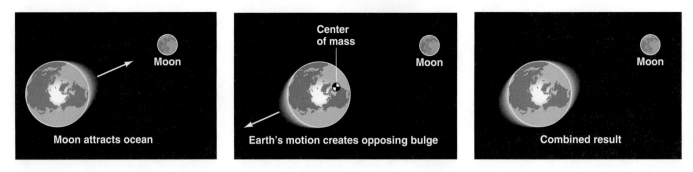

FIGURE 11.4 The moon's gravity attracts the ocean toward it. The motion of Earth around the center of mass of the Earth–moon system throws up a bulge on the side of Earth opposite the moon. The combination of the two effects creates two tidal bulges.

blue, and the inward-pulling force of gravity is shown in red. Collectively, they are called *tractive forces.* Note that the inward pull of gravity and the outward-moving tendency of inertia don't always act in exactly the same balanced way on each particle of Earth and moon. The net strength and direction that result when the two forces are combined are shown as brown arrows.

Points one and two are closer to the moon, so gravitational attraction at those points slightly exceeds the outward-moving tendency of inertia. Water there tends to be attracted toward the moon and so is pulled along the ocean surface toward a spot beneath the moon. At points three and four, slightly farther from the moon, inertia exceeds gravitational attraction. Water at those points tends to be flung away from the moon and so moves along the ocean surface toward a spot opposite the moon. Together, the tractive forces cause the two small bulges in the ocean: one in the direction of the moon, the other in the opposite direction.

Note that there is no point on Earth's surface where the force of the moon's gravity exactly equals the outward-moving tendency of inertia. Only at point **CE**—the center of Earth—are the inward pull of gravity and the outward-moving tendency of inertia exactly equal and opposite. The solid Earth cannot move

much in response to these forces, but the fluid atmosphere and ocean can. We don't notice the changes in the height of the atmosphere, but changes in the water level are visible to coastal observers. In **Figure 11.6** tractive forces pull water toward a point beneath the moon and toward a point opposite the moon.

How do these bulges cause the rhythmic rise and fall of the tides? In the idealized equilibrium model we are discussing, the bulges tend to stay aligned with the moon as Earth spins around its axis. **Figure 11.7** shows the situation in Figure 11.6 as it would look from above the North Pole. As Earth turns eastward, an island on the equator is seen to move in and out of these bulges through one rotation (one day). The bulges are the crests of the planet-sized waves that cause **high tides. Low tides** correspond to the troughs, the areas between bulges. Starting at 0000 (midnight) we see the island in shallow water at low tide. Around six hours later, at 0613 (6:13 A.M.), the island is submerged in the lunar bulge at high tide. At 1226 (about noon) the island is within the tide wave trough at low tide. At 1838 (6:38 P.M.) the island is again submerged, this time in the opposite crest caused by inertia. About an hour after midnight (0050) on the next day, the island is back in shallow water, where it began.

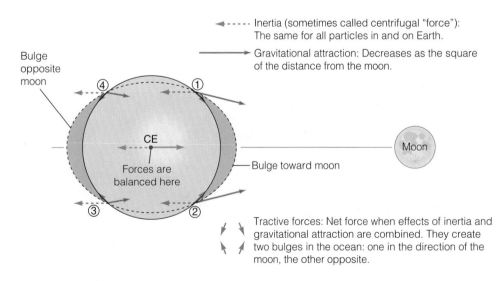

Inertia (sometimes called centrifugal "force"): The same for all particles in and on Earth.

Gravitational attraction: Decreases as the square of the distance from the moon.

Bulge opposite moon

CE
Forces are balanced here

Bulge toward moon

Moon

Tractive forces: Net force when effects of inertia and gravitational attraction are combined. They create two bulges in the ocean: one in the direction of the moon, the other opposite.

The two forces that can move the ocean—inertia and gravitational attraction—are precisely equal in strength but opposite in direction, and thus balanced, only at the center of Earth (point CE).

ACTIVE FIGURE 11.5 The actions of gravity and inertia on particles at five different locations on Earth. At points 1 and 2, the gravitational attraction of the moon slightly exceeds the outward-moving tendency of inertia; the imbalance of forces causes water to move along Earth's surface, converging at a point toward the moon. At points 3 and 4, inertia exceeds gravitational force, so water moves along Earth's surface to converge at a point opposite the moon. Forces are balanced only at the center of Earth (point CE).

Oceanography ⬡Now™ |
At the Active Figures link at http://oceanography. brookscole.com/garrison5e you can view the Earth–Moon forces that cause tides.

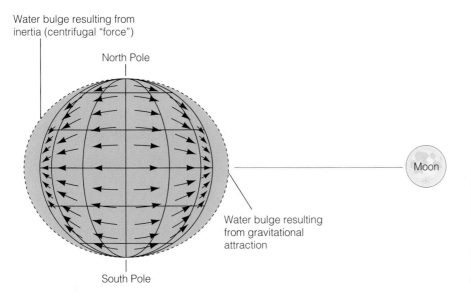

Water bulge resulting from inertia (centrifugal "force")

North Pole

South Pole

Water bulge resulting from gravitational attraction

Moon

ACTIVE FIGURE 11.6 The formation of tidal bulges at points toward and away from the moon. (This and the similar diagrams that follow are not drawn to scale.)

Oceanography☰Now™ | At the Active Figures link at http://oceanography. brookscole.com/garrison5e you can view the formation of tidal bulges.

The wave crests and troughs that cause high and low tides are actually very small: A 2-meter (7-foot) rise or fall in sea level is insignificant in comparison to the ocean's great size. Earth rotates beneath the bulges (tide wave crests) at about 1,600 kilometers (1,000 miles) per hour at the equator. The bulges appear to move across the ocean surface at this speed in an attempt to keep up with the moon. Theoretically the wavelength of these tide waves is as long as 20,000 kilometers (12,500 miles)! The bulges tend to stay aligned with the moon as Earth spins around its axis. *The key to understanding the equilibrium theory of tides is to see Earth turning beneath these bulges.*

There are complications, of course. For example, **lunar tides,** tides caused by gravitational and inertial interaction of the moon and Earth, complete their cycle in a tidal day (also called a lunar day). A complete tidal day is 24 hours 50 minutes long because the moon, which exerts the greatest tidal in-

fluence, rises 50 minutes later each day (**Figure 11.8**). Thus the highest tide also arrives 50 minutes later each day.

Another complication arises from the fact that the moon does not stay directly over the equator; each month it moves from a position as high as $28\frac{1}{2}°$ above Earth's equator to $28\frac{1}{2}°$ below.[2] When the moon is above the equator, the bulges are offset accordingly (**Figure 11.9**). When the moon is $28\frac{1}{2}°$ north of the equator, an island north of the equator will pass through the bulge on one side of Earth but miss the bulge on the other side. During one day the island passes through a very high tide, a low tide, a lower high tide, and another low tide. This is shown in **Figure 11.10**.

---

[2] If Earth, the moon, and the sun were all moving in the same plane, lunar and solar eclipses would happen every two weeks.

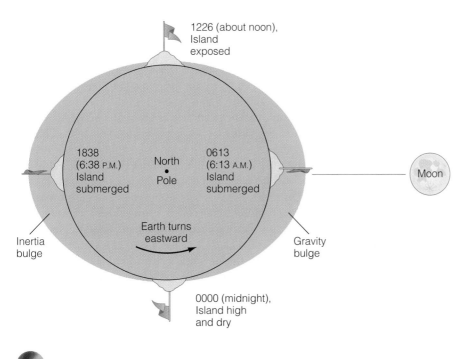

1226 (about noon), Island exposed

1838 (6:38 P.M.) Island submerged

North Pole

0613 (6:13 A.M.) Island submerged

Earth turns eastward

Inertia bulge

Gravity bulge

0000 (midnight), Island high and dry

Moon

ACTIVE FIGURE 11.7 How Earth's rotation beneath the tidal bulges produces high and low tides. Notice that the tidal cycle is 24 hours 50 minutes long because the moon rises 50 minutes later each day.

Oceanography☰Now™ | At the Active Figures link at http://oceanography. brookscole.com/garrison5e you can view how Earth's rotation produces high and low tides.

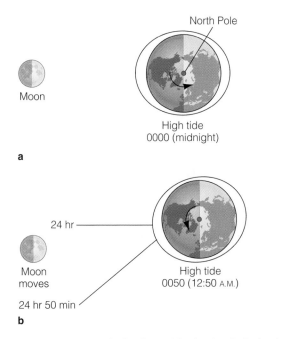

FIGURE 11.8 (a) A high tide at midnight. (b) The high tide one day later. Earth rotates once a day, and the Earth–moon system revolves around its center of mass once a month. Both turn in the same direction (eastward). During one complete Earth rotation (one day), the moon moves eastward 12.2°. So Earth must rotate for an additional 50 minutes to place the moon in the same position in the sky on consecutive days. Because Earth must turn an additional 50 minutes for the same tidal alignment, lunar tides usually arrive 50 minutes later each day.

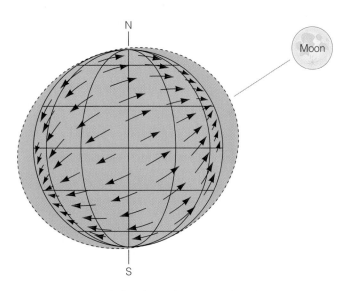

FIGURE 11.9 Tidal bulges follow the moon. When the moon's position is north of the equator, the gravitational bulge toward the moon is also located north of the equator, and the opposite inertial bulge is below the equator. (Compare with Figure 11.6.)

The Sun's Role

The sun's gravity also attracts particles on Earth. Remember, closeness counts for much in determining the strength of gravitational attraction. As we saw earlier, the sun is about 27 million times more massive than the moon but about 387 times farther

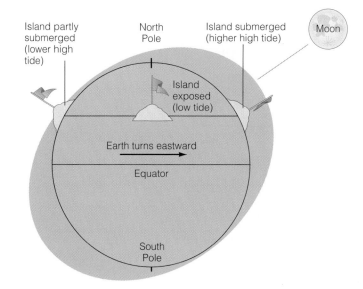

FIGURE 11.10 How the changing position of the moon relative to the Earth's equator produces higher and lower high tides. Sometimes the moon is below the equator; sometimes it is above.

from Earth than the moon, so the sun's influence on the tides is only 46% that of the moon's. The sun's tractive forces develop in the same way as the moon's, and the smaller solar bulges tend to follow the sun through the day. These are the **solar tides,** caused by the gravitational and inertial interaction of the sun and Earth.

Like the moon, the sun also appears to move above and below the equator ($23\frac{1}{2}°$ north to $23\frac{1}{2}°$ south, as you may recall from Chapter 8), so the position of the solar bulges varies like that of the lunar bulges. Earth revolves around the sun only once a year, however, so the position of the solar bulges above or below the equator changes much more slowly than the position of the lunar bulges. (Figure 8.6, used to explain the cause of the seasons, shows this well.)

Sun and Moon Together

The ocean responds simultaneously to inertia and to the gravitational force of both the sun and the moon. If Earth, the moon, and the sun are all in a line (as shown in **Figure 11.11a**), then the lunar and solar tides will be additive, resulting in higher high tides and lower low tides. But if the moon, Earth, and the sun form a right angle (as shown in **Figure 11.11b**), the solar tide will tend to diminish the lunar tide. Because the moon's contribution is more than twice that of the sun, the solar tide will not completely cancel the lunar tide.

The large tides caused by the linear alignment of the sun, Earth, and the moon are called **spring tides** (*springen* = to move quickly). During spring tides, high tides are very high and low tides very low. These tides occur at two-week intervals corresponding to the new and full moons. (Please note that spring tides don't happen only in the spring of the year.) **Neap** (*naepa* = hardly disturbed) **tides** occur when the moon, Earth, and the sun form a right angle. During neap tides, high tides are not very high and low tides not very low. Neap tides also

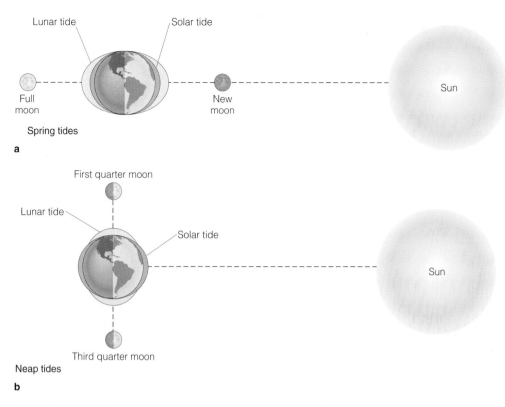

Lunar tide Solar tide

Full moon

New moon

Spring tides

a

First quarter moon

Lunar tide

Solar tide

Sun

Third quarter moon

Neap tides

b

ACTIVE **FIGURE 11.11** Relative positions of the sun, the moon, and Earth during spring and neap tides. (**a**) At the new and full moons, the solar and lunar tides reinforce each other, making spring tides, the highest high and lowest low tides. (**b**) At the first and third quarter moons, the sun, Earth, and the moon form a right angle, creating neap tides, the lowest high and the highest low tides.

Oceanography Now™ | At the Active Figures link at http://oceanography.brookscole.com/ garrison5e you can view the difference between spring tides and neap tides.

occur at two-week intervals, with the neap tide arriving a week after the spring tide. **Figure 11.12** plots tides at two coastal sites through spring and neap cycles.

Because their orbits are ellipses, not perfect circles, the moon and the sun are closer to Earth at some times than at others. The difference between **apogee** (the moon's greatest distance from Earth) and **perigee** (its closest approach) is 30,600 kilometers (19,015 miles). Because the tidal force is inversely proportional to the cube of the distance between the bodies, the closer moon raises a noticeably higher tidal crest. The difference between **aphelion** (Earth's greatest distance from the sun) and **perihelion** (its closest approach) is 3.7 million kilometers (2.3 million miles). If the moon and the sun are over nearly the same latitude and if Earth is also close to the sun, extreme spring tides will result. Interestingly, spring tides will have greater ranges in the Northern Hemisphere winter than in the Northern Hemisphere summer because Earth is closest to the sun during the northern winter.

The Dynamic Theory of Tides

11.7

Newton knew his explanation was incomplete. For one thing, the maximum theoretical range of a lunar tidal bulge is only

55 centimeters (about 22 inches) and of a solar tide only 24 centimeters (about 10 inches), both considerably less than the 2-meter (7-foot) average tidal range we observe in the world ocean. This is because the ocean surface never comes completely to the equilibrium position at any instant. The moon and the sun change their positions so rapidly that the water cannot keep up.

The **dynamic theory of tides,** first proposed in 1775 by Laplace, added a fundamental understanding of the problems of fluid motion to Newton's breakthrough in celestial mechanics. The dynamic theory explains the differences between predictions based on Newton's model and the observed behaviors of tides.

Remember that tides are a form of *wave.* The crests of these waves—the tidal bulges—are separated by a distance of half of Earth's circumference (see again Figure 11.7). In the equilibrium model, the crests would remain stationary, pointing steadily toward (or away from) the moon (or the sun) as Earth turned beneath them. They would appear to move across the idealized water-covered Earth at a speed of about 1,600 kilometers (1,000 miles) per hour. But how deep would the ocean have to be to allow these waves to move freely? For a tidal crest to move at 1,600 kilometers per hour, the ocean would have to be 22 kilometers (13.7 miles) deep. As you may recall, the average depth of the ocean is only 3.8 kilometers (2.4 miles). So, tidal crests (tidal bulges) move as forced waves, and their velocity is determined by ocean depth.

http://oceanography.brookscole.com/garrison5e

FIGURE 11.12 Tidal records for a typical month at (**a**) New York and (**b**) Port Adelaide, Australia. Note the relationship of spring and neap tides to the phases of the moon.

Oceanography⊜Now™ | Click Oceanography Interactive to work through an activity on high/low tides and tidal shifts through waves, tides, and currents.

Tidal Patterns and Amphidromic Points

This behavior of tides as *shallow-water waves* is only one variation from the ideal that the dynamic theory explains. The continents also get in the way. As Earth turns, landmasses obstruct the tidal crests, diverting, slowing, and otherwise complicating their movements. This interference produces different patterns in the arrival of tidal crests at different places. Imagine, for example, a continent directly facing the moon. There would be no oceanic bulge, and the shores of the continent would experience high tide. A few hours later the moon would be over the ocean. When the continent was not aligned with the moon but the ocean was, the tidal bulge would re-form and the continent's edges experience low tide.

The shape of the basin itself has a strong influence on the patterns and heights of tides. As we have seen, water in large basins can rock rhythmically back and forth in seiches. Though they are small, tidal crests can stimulate this resonant oscillation, and the configuration of coasts around a basin can alter its rhythm.

For these and other reasons some coastlines experience **semidiurnal** (twice daily) **tides**: two high tides and two low tides of nearly equal level each lunar day. Others have **diurnal** (daily) **tides**: one high and one low. The tidal pattern is called **mixed** if successive high tides or low tides have significantly different heights throughout the cycle. **Figure 11.13** shows an example of each tidal pattern. At Cape Cod two tidal crests arrive per lunar day, a semidiurnal pattern (Figure 11.13a). The natural tendency of water in an enclosed ocean basin to rock at a specific frequency modifies the pattern in the Gulf of Mexico, so Mobile sees one crest per lunar day, a diurnal pattern (Figure 11.13b). The Pacific coast of the United States has a mixed tidal pattern: often a *higher* high tide, followed by a *lower* low tide, a *lower* high tide, and a *higher* low tide each lunar day. (That's not as confusing as it seems—see Figure 11.13c.)

The Pacific Ocean has a unique pattern of diurnal, semidiurnal, and mixed tides. As **Figure 11.14** shows, it has the most complex of all tidal patterns. The east coast of Australia, all of New Zealand, and much of the west coasts of Central and South America have a semidiurnal tidal pattern. The Aleutians have diurnal tides. The Pacific coasts of North America and some of South America have mixed tides. *Why the differences?*

Remember the surface of Lake Geneva in our discussion of seiches? The water level at the center of the lake remains at the same height while water at the ends rises and falls (see again Figure 10.28). The long axis of the lake stretches east and west. Because of the Coriolis effect, water moving east at the center of the lake is deflected slightly to the right (to the south). If the lake were larger and the flow of water greater (and thus the Coriolis effect stronger), water would hug the southern shore as it

FIGURE 11.13 Tide curves for the three common types of tides. (**a**) A semidiurnal tide pattern at Cape Cod, Massachusetts. (**b**) A diurnal tide pattern at Mobile, Alabama. (**c**) A mixed tide pattern at Los Angeles, California.

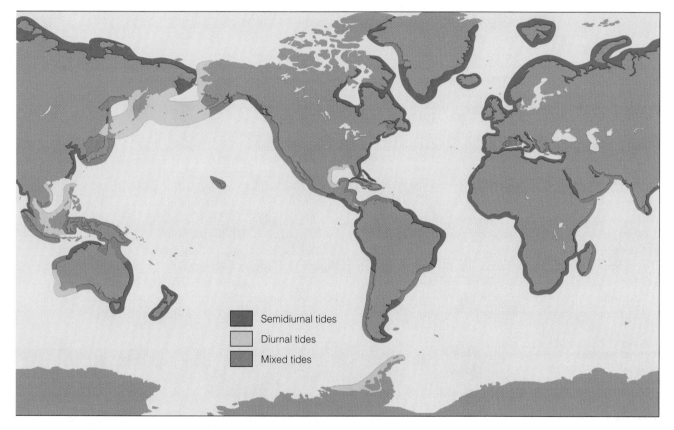

FIGURE 11.14 The worldwide geographic distribution of the three tidal patterns. Most of the world's ocean coasts have semidiurnal tides.

traveled eastward. When the water began to rock the other way—back to the west—the Coriolis effect would move the water to the right toward (and along) the northern shore. Note that the overall movement of water would be counterclockwise.

Water moving in a tide wave tends to stay to the right of an ocean basin for the same reason. As water moves north in a Northern Hemisphere ocean, it moves toward the eastern boundary of the basin; as it moves south, it moves toward the western boundary. A wave crest moving counterclockwise will develop around a node if this motion continues to be stimulated by tidal forces. This rotary motion is shown in **Figure 11.15.**

The node (or nodes) near the center of an ocean basin is called an **amphidromic point** (*amphi* = around, *dromas* = running). An amphidromic point is a no-tide point in the

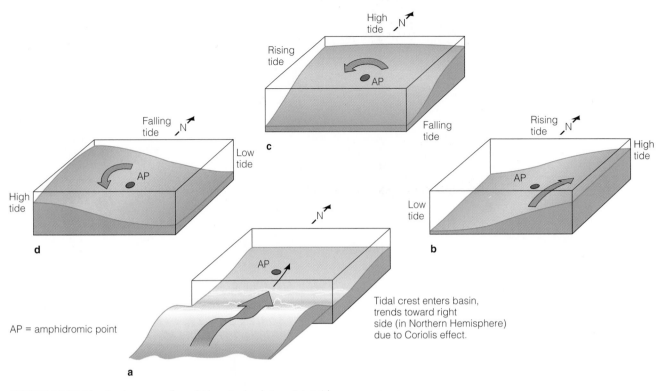

FIGURE 11.15 The development of amphidromic circulation. (**a**) A tide wave crest enters an ocean basin in the Northern Hemisphere. (**b**) The wave trends to the right because of the Coriolis effect, causing a high tide on the basin's eastern shore. (**c**) Unable to continue turning to the right because of the interference of the shore, the crest moves northward, following the shoreline and causing a high tide on the basin's northern shore. (**d**) The wave continues its progress around the basin in a counterclockwise direction, forming a high tide on the western shore and completing the circuit. The point around which the crest moves is an amphidromic point (AP).

ocean around which the tidal crest rotates through one tidal cycle. Because of the shape and placement of landmasses around ocean basins, the tidal crests and troughs cancel each other at these points. The crests sweep around amphidromic points like wheel spokes from a rotating hub, radiating crests toward distant shores. Tide waves are influenced by the Coriolis effect because a large volume of water moves with the waves. They move counterclockwise around the amphidromic point in the Northern Hemisphere and clockwise in the Southern Hemisphere. The height of the tides increases with distance from an amphidromic point.

About a dozen amphidromic points exist in the world ocean; **Figure 11.16** shows their location. Notice the complexity of the Pacific, which contains five. It's no wonder that the arrival of tide wave crests at the Pacific's edges produces such a complex mixture of tide patterns, depending on the shoreline location.

FIGURE 11.16 Amphidromic points in the world ocean. Tidal ranges generally increase with increasing distance from amphidromic points. The colors indicate where tides are most extreme (highest highs, lowest lows), with blues being least extreme. White lines radiating from the points indicate tide waves moving around these points. In almost a dozen places on this map the lines converge. Notice how at each of these places the surrounding color—the tidal force for that region—is blue, indicating little or no apparent tide. These convergent areas are called amphidromic points. Tide waves move around these points, counterclockwise in the Northern Hemisphere and clockwise in the Southern Hemisphere.

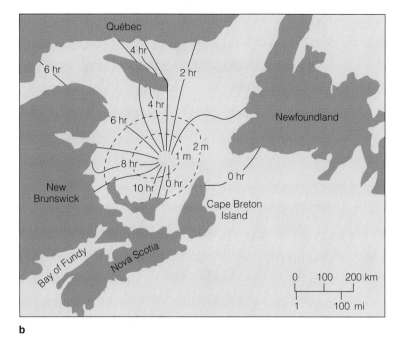

a

b

ACTIVE **FIGURE 11.17** Tides in broad, confined basins. (**a**) An imaginary amphidromic system in a broad, shallow basin. The numbers indicate the hourly positions of tide crests as a cycle progresses. (**b**) The amphidromic system for the Gulf of St. Lawrence between New Brunswick and Newfoundland, southeastern Canada. Dashed lines show the tide heights when the tide crest is passing.

Oceanography☰Now™ | At the Active Figures link at http://oceanography.brookscole.com/garrison5e you can view tides in confined basins.

Tidal Datum

The reference level to which tidal height is compared is called the **tidal datum.** The tidal datum is the zero point (0.0) seen in tide graphs such as Figures 11.12 and 11.13. This reference plane is not always set at **mean sea level,** which is the height of the ocean surface averaged over a few years' time. On coasts with mixed tides, the zero tide level is the average level of the lower of the two daily low tides (*mean lower low water,* or MLLW). On coasts with diurnal and semidiurnal tides, the zero tide level is the average level of all low tides (*mean low water,* or MLW).

Tides in Confined Basins

The **tidal range** (high-water to low-water height difference) varies with basin configuration. In small areas such as lakes, the tidal range is small. In larger enclosed areas such as the Baltic and Mediterranean Seas, the tidal range is also moderate. The tidal range is not the same over a whole ocean basin; it varies from the coasts to the centers of oceans. The largest tidal ranges occur at the edges of the largest ocean basins, especially in bays or inlets that concentrate tidal energy because of their shape.

If a basin is wide and symmetrical, like the Gulf of St. Lawrence in eastern Canada, a miniature amphidromic system develops that resembles the large systems of the open ocean (**Figure 11.17**). If the basin is narrow and restricted, the tide wave crest cannot rotate around an amphidromic point and simply moves into and out of the bay (**Figure 11.18**). Extreme

tides occur in places where arriving tide crests stimulate natural oscillation periods of around 12 or 24 hours. In rare cases, water in the bay naturally resonates (seiches) at the same frequency as the lunar tide (12 hours, 25 minutes). This rhythmic sloshing results in extreme tides. In the eastern reaches of the Bay of Fundy near Moncton, New Brunswick (Canada), the tidal range is especially wide: up to 15 meters (50 feet) from highs to lows (**Figure 11.19**). The northern reaches of the Sea of Cortez east of Baja California have a tidal range of about 9 meters (30 feet). Tide waves sweeping toward the narrow southern end of the North Sea can build to great heights along the southeast coast of England and the north coast of France.

As you saw in this chapter's opener, a **tidal bore** (*bara* = wave) will form in some inlets (and their associated rivers) exposed to great tidal fluctuation. Here, at last, is a true **tidal wave**—a steep wave moving upstream generated by the action of the tide crest in the enclosed area of a river mouth. The confining river mouth forces the tide wave to move toward land at a speed that exceeds the theoretical shallow-water wave speed for that depth. The forced wave then breaks, forming a spilling wave front that moves upriver. Though most are less than 1 meter (3 feet) high, some bores may be up to 8 meters (26 feet) high and move 11 meters per second (25 miles per hour). Their potential danger is lessened by their predictability. Accurately predicting the arrival of tidal bores is essential to safe navigation. In addition to southwestern China and the Bay of Fundy, tidal bores are common in the Amazon, the Ganges Delta, and England's Severn River.

http://oceanography.brookscole.com/garrison5e

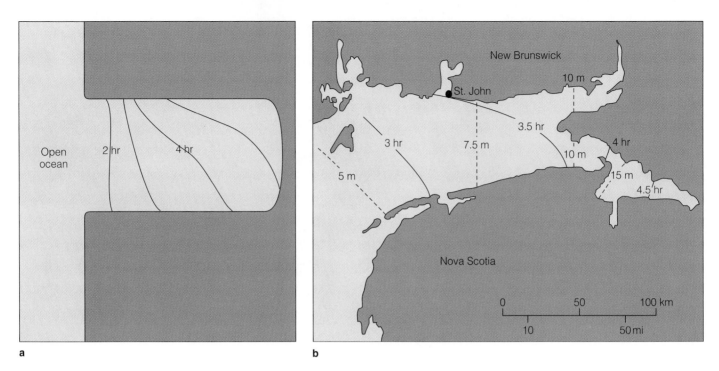

a   b

FIGURE 11.18 Tides in narrow, restricted basins. (**a**) True amphidromic systems do not develop in narrow basins because there is no space for rotation. (**b**) Tides in the Bay of Fundy, Nova Scotia, are extreme because water in the bay naturally resonates (seiches) at the same frequency as the lunar tide (see Figure 11.1).

Tidal Currents

The rise or fall in sea level as a tide crest approaches and passes will cause a **tidal current** of water to flow into or out of bays and harbors. Water rushing into an enclosed area because of the rise in sea level as a tide crest approaches is called a **flood current**. Water rushing out because of the fall in sea level as the tide trough approaches is called an **ebb current**. (The terms *ebb tide* and *flood tide* have no technical meaning.) Tidal currents reach maximum velocity midway between high tide and low tide. **Slack water,** a time of no currents, occurs at high and low tides when the current changes direction.

Anyone who has stood at the narrow mouth of a large bay or harbor cannot help but be impressed with the speed and volume

of the tidal current that occurs between tidal extremes. Midway between high and low spring tides, the ebb current rushing from San Francisco Bay strikes the base of the south tower of the Golden Gate Bridge with such force that a bow wave is formed, giving the convincing illusion that the bridge itself is moving rapidly. Tidal currents at the Golden Gate can reach 3 meters per second (about 7 miles per hour) because of the volume of enclosed water and the narrowness of the channel through which it must escape. Navigators must know the times of tidal currents to safely negotiate any harbor entrance or other narrow strait; in some places, this knowledge may save their lives (**Box 11.1**).

Tidal currents become more complex in the open sea. One's position relative to an amphidromic point, the shape of the basin, and the magnitude of gravitational forces and inertia

FIGURE 11.19 Tides in the eastern Bay of Fundy on the Atlantic coast of Canada. The tidal range is almost 15 meters (50 feet). At the peak of the flood, water rises 1 meter (3.3 feet) in 23 minutes.

BOX 11.1
Maelstrom!

According to Homer's account in The Odyssey, across Ulysses's homeward path lay a lethal whirlpool named Charybdis:

> We sailed up the narrow strait. On one side was shining Charybdis, who made her terrible ebb and flow of the sea's water. When she vomited it up, like a caldron over a strong fire, the whole sea would boil up in turbulence, and the foam flying spattered the pinnacles of the rocks in either direction; but when in turn again she sucked down the sea's salt water, the turbulence showed all the inner sea, and the rock around it groaned terribly, and the ground showed at the sea's bottom, black with sand. My crew turned sallow with fright, staring into this abyss from which we expected our immediate death.

For millennia, the prospect of navigating past Charybdis terrified sailors in the Strait of Messina, which separates Italy and Sicily. Its activity was greatly reduced in 1908 when the great Calabria–Messina earthquake and tsunami altered the geology of the area.

Edgar Allan Poe wrote about a whirlpool called the Maelstrom, which lies among the southern Lofoten Islands off Norway's west coast. The rapid spinning of water in the Maelstrom raises its outer edge and depresses the central core. Here is Poe's description, from his 1841 story "A Descent into the Maelstrom":

> The edge of the whirl was represented by a broad belt of gleaming spray; but no particle of this slipped into the mouth of the terrific funnel, whose interior, as far as the eye could fathom it, was a smooth, shining, and jet-black wall of water, inclining to the horizontal at an angle of some forty-five degrees, speeding dizzily round and round with swaying and sweltering motion, and sending forth to the winds an appalling voice, half shriek, half roar, such as not even the mighty cataract of Niagara ever lifts up in its agony to Heaven.

Do these whirlpools genuinely deserve the dread they inspired, or is their reputation largely a figment of artistic imagination (**Figure a**)?

Whirlpools arise in shallow, restricted straits connecting two large bodies of deep water. They are caused by the tides. If the large bodies of water move to different tidal cycles, high tide in one will not occur at the same time as high tide in the other. Turbulence develops when one mass of water tries to pass the other during changes in tide. Under ideal conditions, bottom contours and the rush of opposing tidal currents can cause seawater in the confined area to spin vigorously. The larger the opposing bodies of water and the smaller the passage through which the confined water must pass, the greater the vortex caused by the tidal currents.

Arthur Rackham

(**a**) Maelstrom: fiction. A representation by the early twentieth-century English illustrator Arthur Rackham.

(**b**) Maelstrom: fact. The maelstrom is located in the strait between the Norwegian islands of Moskenes and Moskenesøya.

BØDØ Arrangement, Norway

Maelstrøm is a Norwegian word derived from the Dutch *malen* ("to grind in a circle," as a millstone grinds grain) and *strøm* ("stream"). Today's maelstrom is a heaving mass of active tidal water connecting Vest Fjord and the Norwegian Sea between the islands of Moskenesøya and Moskenes. Look for the area on a chart at 67°48′N, 12°50′E, about 165 kilometers (100 miles) north of the Arctic Circle. Currents in the passage can reach a speed of 13 kilometers (8 miles) per hour when the tides change, and strong local winds make the passage even more dangerous. The shallow "saddle" separating the large bodies of water to the east and west is only about 20 meters (70 feet) deep, and water rushing over its rocky bottom is driven to spin.

But is the chaos as horrifying as Homer, Poe, Jules Verne, and others have suggested? Well, no. Fishing boats in this rich fishing region carefully avoid the area during times of maximum tidal difference, and no sailor likes to risk his or her life in turbulent confined waters during times of high winds. But as can be seen in **Figure b,** the sucking, roaring, cavernous maw of the maelstrom is more fiction than fact.

must all be considered to calculate the speed and direction of tidal currents over a deep bottom. Tidal currents are slower in the open sea because the water is not confined, as it is in a harbor. The speed of open-sea tidal currents has been measured at a few centimeters per second, and their velocity tends to decrease with depth.

Tidal Friction

The daily rise and fall of the tides consumes a very large amount of energy, and this energy is ultimately dissipated as heat. Most of this energy comes directly from the rotation of Earth itself, and tidal friction is gradually slowing Earth's rotation by a few hundredths of a second per century. Even such a small change has long-term planetary effects, however. Geologists studying the daily growth rings of fossil corals and clams estimate that the days have grown longer, so the number of days in a year has decreased as planetary rotation has slowed. Evidence suggests that 350 million years ago a year contained between 400 and 410 days, with each day being about 22 hours long; 280 million years ago there were about 390 $22\frac{1}{2}$-hour days in a year.

Tidal friction affects other bodies. Tidal forces have locked the rotation of the moon to that of Earth. As a result, the same side of the moon is always facing Earth, and a day on the moon is a month long.

Predicting and Studying Tides

There are at least 140 tide-generating and tide-altering forces and factors in addition to the ones we have discussed. About seven important factors must be considered if we wish to predict tides mathematically. The interactions of all the forces and factors are so complex that if a previously unknown continent were discovered on Earth, the coastal tide times and ranges on its shores could not be predicted accurately because it is the study of past records that allows tide tables to be projected into the future. Experience permits prediction of tidal height to an accuracy of about 3 centimeters (1.2 inches) for years in advance.

Even so, extraneous factors can affect the estimates. For example, the arrival of a storm surge will greatly affect the height or timing of a tide—as will gentle, atmospherically induced seiching of the basin or excitement of large-scale resonances by a tsunami. Even a strong, steady wind onshore or offshore will affect tidal height and the arrival time of the crest. Weather-related alterations are sometimes called **meteorological tides** after their origin.

Systems for studying the tides range from the simple (teams of students chasing oranges dropped into a harbor) to the complex (lines of pressure-sensing transducers arrayed on the ocean floor). As our discussion has shown, tidal behavior is so complex—the number of variables is so large—that any new method of tracking and predicting tidal movements would be welcome. Researchers in Germany and the United States have recently found a way to study tides through extremely subtle variations in Earth's main magnetic field sensed by orbiting satellites. How can this be done? The ocean is an electrically conducting fluid. As it flows through Earth's magnetic field, it generates a small but distinct secondary magnetic field that interferes with the main field. The trace of the lunar semidiurnal tide has been identified from satellite observations, and efforts are under way to increase the sensitivity of the readings.

Tides and Marine Organisms

Not surprisingly, as pervasive a phenomenon as the tides has a significant influence on coastal marine life. Organisms that live between the high-tide and low-tide marks experience very different conditions from those that reside below the low-tide line. Within the intertidal zone itself, organisms are exposed to varying amounts of emergence and submergence. Because some organisms can tolerate many hours of exposure while others are able to tolerate only a very few hours per week or month, the animals and plants sort themselves into three or more horizontal bands, or subzones, within the intertidal zone. Each distinct zone is an aggregation of animals and plants best adapted to the conditions within that particular narrow habitat. The zones are often strikingly different in appearance, even to a person unfamiliar with shoreline characteristics. (This zonation is clearly evident in the rocky shore of Figure 16.7b.)

Less obvious are periodic visitors that time their arrivals and departures to the rise and fall of the tide At low tide the tiny diatom (one-celled alga) *Hantzschia* migrates upward through wet sand to photosynthesize at the sunlit surface. At the first hint of the returning tide, these plantlike organisms descend to a relatively safe depth in the sand, protected from wave action. Fiddler crabs (genus *Uca*) return to their burrows at high tide to avoid marine predators but emerge at low tide to look for any bits of food the ocean might have deposited at their doorsteps. Filter-feeding animals such as sand crabs (*Emerita*) and bean clams (*Donax*) migrate up and down the beach to stay in the surf zone where the chaos can provide both food and protection (see Figure 16.8).

Among the most famous tide-driven visitors are the grunion (*Leuresthes*), a small fish named after the Spanish word *gruñon*, "grunter," a reference to the squeaking noise they sometimes make during spawning. From late February through early September, these small fish (which reach a length of 15 centimeters, or 6 inches) swim ashore at night in large numbers just after the highest spring tides, deposit and fertilize their eggs below the sand surface, and return to the sea (**Figure 11.20**). Safe from marine predators, the eggs develop and are ready to hatch nine days after spawning. After a few more days, the spring tides return, soak and erode the beach, and stimulate the eggs to hatch. No one is certain how grunion time their reproductive behavior so precisely to the tidal cycle, but some research suggests that they sense very small changes in hydrostatic pressure caused by tidal change, or that they visually time their spawning three or four nights following each full and new moon. These accommodating little fish were an important food for Native Americans.

Power from the Tides

Humans have found ways to use the tides. Ships sail to sea and return to port with the tides. Intentional grounding of a ship with the fall of a tide can provide a convenient, if temporary, drydock. To these traditional uses has been added a potential alternative to our growing dependence on fossil fuels: taking advantage of trapped high-tide water to generate electricity.

Tidal power is the only marine energy source that has been successfully exploited on a large scale. The first major tidal power station was opened in 1966 in France on the estuary of the river Rance (**Figure 11.21**), where tidal range reaches a maximum of 13.4 meters (44 feet). Built at a cost of $75 million, this 850-meter- (2,800-foot-) long dam contains 24 turbo-alternators capable of generating 544 million kilowatt-hours of electricity annually. At high tide seawater flows from the ocean through the generators into the estuary. At low tide the seawater and river water from the estuary flow out through the same generators. Power is generated in both directions. A smaller but similar installation is generating power on the Annapolis River in Nova Scotia. A much larger generating facility has been proposed for Passamaquoddy Bay, part of the Bay of Fundy between Maine and New Brunswick, Canada.

FIGURE 11.20 Students enjoying a grunion run. During spring and summer months these small fish (genus *Leuresthes*) swim ashore at night in large numbers just after the highest spring tides, deposit and fertilize their eggs below the sand surface, and return to the sea. Nearly two weeks later, when spring tides return, the eggs hatch. No one is certain how grunion time their reproductive behavior so precisely to the tidal cycle. The grunion are found only along the Pacific coast of North America and in the Gulf of California. Unlike the Pacific coast species, Gulf of California grunion spawn during daylight.

Cabrillo Marine Museum

FIGURE 11.21 A tidal power installation at the Rance estuary in western France.

Tidal power has many advantages: Operating costs are low, the source of power is free, and no carbon dioxide or other pollutants are added to the atmosphere. But even if tidal power stations were built at every appropriate site worldwide, the power generated would amount to less than 1% of current world needs. And, of course, this method of power generation is not free of trade-offs. The dam and electrical generators can be damaged by storms, and the large, finely made metal valves and vanes at the heart of the plant are easily corroded by seawater. Computer simulations have suggested that installing a dam would change the resonance modes of a bay or estuary—and therefore the height of the tide wave. Studies also suggest that sensitive planktonic and benthic marine life would be disrupted and even that increased tidal friction would cause a tiny decrease in the rate of Earth's rotation.

Questions from Students

1. Are there tides in the solid Earth?

Yes. Even Earth isn't stiff enough to resist the tidal pulls of the moon and the sun. Bulges occur in the solid Earth just as they appear in the ocean or the atmosphere. The crests (bulges) of Earth are, of course, much smaller; 25 to 30 centimeters (10 to 12 inches) is about average. They pass unnoticed beneath us twice a day. In contrast, tidal variability in the height of the atmosphere has been measured in miles.

2. In my newspaper one of today's low tides is listed as −1.0, 1445. What does that mean?

In the United States it means that the water will be 1.0 feet below tidal datum (that is, below mean lower low water

[MLLW]—the long-term average position of the lower of the daily low tides) at 2:45 P.M. local time. This might be a good afternoon to spend at the shore digging for clams because a tide that low would expose intertidal organisms only rarely seen above water. See Figure 16.7 for more information on the relationship between tidal height and exposure.

3. Are there tides in a glass of water?

Yes. They're too small to detect, but equilibrium tides do exist. Each molecule of water in the glass responds to the same planetary forces that affect molecules in the ocean.

4. You mentioned that the process of extracting power from the tides slows Earth's rotation. What would happen if all the world's electrical power needs could somehow be met by tidal generating plants?

If we assume it could actually be done, and if we assume a constant energy demand of today's requirement of 2×10^{20} joules per year (somewhat unrealistic in light of the recent strong increase in electrical power demand), then the length of a day would increase by an additional second in roughly a million and a half years. Clearly the "cashing in" of Earth's rotational kinetic energy to supply energy needs is not of great concern. Much more serious hazards would be disruptions of ocean currents, biological cycles, and aesthetic values.

5. The TV news reporter said to expect "astronomical high tides" tonight. Should we pack our stuff and head for the hills?

Not necessarily. The reporter is calling attention to an alignment of the sun and the moon that produces a high spring tide. "Astronomical" here refers to an alignment of heavenly bodies; it is not synonymous with "gigantic" or "spectacular."

Chapter in Perspective

In this chapter you learned that tides have the longest wavelengths of the ocean's waves. They are caused by a combination of the gravitational force of the moon and the sun, the motion of Earth, and the tendency of water in enclosed ocean basins to rock at a specific frequency. Unlike other waves, these huge shallow-water waves are never free of the forces that cause them and so act in unusual but generally predictable ways. Basin resonances and other factors combine to cause different tidal patterns on different coasts. The rise and fall of the tides can be used to generate electrical power, and tides are important in many physical and biological coastal processes.

In the next chapter you will learn how the interaction of wind, waves, and weather affects the edges of the land—the coasts. Coasts are complex, dynamic places where the only constant is change.

Key Ideas

✓ Tides are huge shallow-water waves—the largest waves in the ocean. Tides are caused by a combination of the gravitational force of the moon and the sun and the motion of Earth.
Review Figures 11.2, 11.4, and 11.5, pages 259–260

✓ The moon's influence on tides is about twice that of the sun's.
Review Figure 11.11, page 263

✓ Gravitational attraction and inertia cause the ocean surface to bulge. Tides occur as Earth rotates beneath the bulges.
Review Figures 11.5–11.10, pages 260–262

✓ Because wave speed is proportional to wavelength, tides move rapidly through the ocean—up to 1,600 kilometers (1,000 miles) per hour.
Review page 261

✓ Tides have such long wavelengths that they are always in "shallow water" (in water that is shallower than half their wavelength).
Review pages 232–234, 258

✓ The *equilibrium theory* of tides deals primarily with the position and attraction of the Earth, the moon, and the sun. It assumes that the ocean conforms instantly to the forces that affect the position of its surface, and only approximately predicts the behavior of the tides.
Review Figure 11.11, pages 263, 259–263

✓ The *dynamic theory* of tides takes into account the speed of the long-wavelength tide wave in water of varying depth, the presence of interfering continents, and the circular movement or rhythmic back-and-forth rock-ing of water in ocean basins. It predicts the behavior of the tides more accurately than the equilibrium theory.
Review Figures 11.15–11.18, pages 263–268

✓ Tidal patterns—the heights and arrival times of tide crests—at a continent's edge may be semidiurnal (twice daily), diurnal (daily), or mixed depending on local combinations of gravity, inertia, and basin shape.
Review Figures 11.12–11.14, pages 264–265

✓ Tides caused by the interaction of the gravity of the sun, the moon, and Earth are known as astronomical tides. Meteorological tides, caused by weather, can add to or detract from the height of tide crests.
Review page 270

✓ The rise and fall of the tides can be used to generate electrical power, and tides are important in many physical and biological coastal processes
Review Figure 11.21, pages 270–272

Terms and Concepts to Remember

amphidromic point, 265
aphelion, 263
apogee, 263
diurnal tide, 264
dynamic theory of tides, 263
ebb current, 268
equilibrium theory of tides, 259
flood current, 268
high tide, 260
low tide, 260
lunar tide, 261
mean sea level, 267
meteorological tide, 270

mixed tide, 264
neap tide, 262
perigee, 263
perihelion, 263
semidiurnal tide, 264
slack water, 268
solar tide, 262
spring tide, 262
tidal bore, 267
tidal current, 268
tidal datum, 267
tidal range, 267
tidal wave, 267
tide, 258

Study Questions

Reviewing What You've Learned

1. What causes the rise and fall of the tides? What celestial bodies are most important in determining tides? Are there such things as "tidal waves"?
2. What is a high tide and a low tide? A spring tide and a neap tide? A tidal bore?
3. How does the equilibrium theory of tides differ from the dynamic theory?
4. How is an astronomical tide different from a meteorological tide? Are the tides separate and independent of each other?

Thinking Critically

1. Though they move through all the ocean, tides are referred to as shallow-water waves. How can that be?

2. What are the most important factors influencing the heights and times of tides? What tidal patterns are observed? Are there tides in the open ocean? If so, how do they behave?

3. How does the latitude of a coastal city affect the tides there—or does it?

4. From what you learned about tides in this chapter, where would you locate a plant that generates electricity from tidal power? What would be some advantages and disadvantages of using tides as an energy source?

5. **InfoTrac College Edition Project** High tides inundate land at the shore, often with heavy wave action, and low tides expose land. The intertidal zone, between high tide and low tide, is an area of rich biodiversity. What adaptations would you expect plants and animals living in the intertidal zone to have? Do some research using InfoTrac College Edition to find examples for your answer.

Thinking Analytically

1. If you live near a coast, find a source of local tidal data (newspaper, Internet site, pamphlet from bait shop, etc.). Plot the rise and fall of the tides for two weeks. Note the cycle, and point out spring and neap tides. How does the tide correlate with the position of the moon and the sun?

2. You know that tides always act like shallow-water waves. What is the speed of a tide wave in the open ocean? Assume the average depth of the ocean is 4,000 meters.

Oceanography⊜Now™

Assess your understanding of this chapter's topics with additional quizzing and conceptual-based problems at http://oceanography.brookscole.com/garrison5e

Study and Research on the Internet

Assess your understanding of this chapter's topics with additional quizzing and conceptual-based problems at:

⬤ http://oceanography.brookscole.com/garrison5e

- ■ **InfoTrac College Edition Research** leads you to critical thinking exercises that use InfoTrac College Edition® as a research tool.
- ■ **Regional InfoTrac College Edition** articles are organized into East Coast, West Coast, and Gulf Coast regions, allowing you to study oceanography on a more local level.

For more readings, go to InfoTrac College Edition, your online research library, at:

⬤ http://infotrac.thomsonlearning.com

CHAPTER 12

Sydney Harbour, Australia, a drowned river valley.

Tom Garrison

COASTS

". . . the finest harbour in the world."

On 20 January 1788 Captain Arthur Phillip led a fleet of ships into a shallow bay on the east coast of Australia. Discovered and explored by Captain James Cook eight years before, Botany Bay (as it was called) was insufficiently protected from the elements and offered little fresh water or fertile soil. Captain Phillip's ships were carrying more than a thousand people, most of them convicts and their jailers. His mission: to establish a prison colony as far from England as possible.

Phillip knew immediately that Botany Bay would be inadequate to the task. A day after the landing he sent expeditions to look for a better location for the colony. Cook had mentioned a gap in the sandstone headlands to the north, and two small boats headed in that direction. Within two days the scouts returned with news of ". . . the finest harbour in the world." Less than a week later the whole colony moved to Sydney Cove (named after Lord Sydney, the British home secretary whose idea it was to transport criminals to this newly discovered continent).

Sydney Harbour seems an ideal coastal haven. But it was not always so. About 200 million years ago, layers of sediments were deposited in the huge delta of the Parramatta River system. The sands hardened to form the Hawkesbury Sandstones, which now form the local headlands and base rock. These were uplifted to near their present height by tectonic forces and eroded by the river. At the end of the last ice age, water released to the ocean from the melting ice caps caused a general rise in sea level. The river valleys were flooded to form a complex system of bays and promontories (**figure**). Phillip's sailing master wrote, "Here you are locked and it is impossible for the wind to do you the least damage."

Classifying Coasts

Coastal areas join land and sea. Our personal experience with the ocean usually begins at the coast. Have you ever wondered why a coast is in a particular location, or why it is shaped as you see it? These temporary, often beautiful junctions of land and sea are subject to rearrangement by waves and tides, by gradual changes in sea level, by biological processes, and by tectonic activity.

The place where ocean meets land is usually called the **shore**, while the term **coast** refers to the larger zone affected by the processes that occur at this boundary. A sandy beach might form the shore in an area, but the coast (or coastal zone) includes the marshes, sand dunes, and cliffs just inland of the beach as well as the sandbars and troughs immediately offshore. The world ocean is bounded by about 440,000 kilometers (273,000 miles) of shore.

Because of its proximity to both ocean and land, a coast is subject to natural events and processes common to both realms. A coast is an active place. Here is the battleground on which wind waves break and expend their energy. Tides sweep water on and off the rim of land, rivers drop most of their sediments at the coasts, and ocean storms pound the continents. The *location* of a coast depends primarily on global tectonic activity and the volume of water in the ocean. The *shape* of a coast is a product of many processes: uplift and subsidence, the wearing down of land by **erosion,** and the redistribution of material by sediment transport and deposition.

As we saw in Chapter 3, no area of geology has been left undisturbed by the revelations of plate tectonics. In the 1960s geologists began to classify coasts according to their tectonic position. *Active* coasts, near the leading edge of moving continental plates, were found to be fundamentally different from the more *passive* coasts near trailing edges. The shapes, compositions, and ages of coasts are better understood by taking plate movements into account. But, as we'll see, the slow forces of plate movement are frequently obscured by the more rapid action of waves, by the erosion of land, and by the transport of sediments.

Another important consideration in the classification of coasts is long-term change in sea level. Five factors can cause sea level to change. Three of these factors are responsible for **eustatic change**—variations in sea level that can be measured all over the world ocean:

- The amount of water in the world ocean varies. Sea level is lower during periods of global glaciation (ice ages) because there is less water in the ocean. It is higher during warm periods, when the glaciers are smaller. Periods of abundant volcanic outgassing can also add water to the ocean and raise sea level.

- The volume of the ocean's "container" may vary. High rates of seafloor spreading are associated with the expansion in volume of the oceanic ridges. This displaces the ocean's water, which climbs higher on the edges of the continents. Sediments shed by the continents during periods of rapid erosion can also decrease the volume of ocean basins and raise sea level.

- The water itself may occupy more or less volume as its temperature varies. During times of global warming, seawater expands and occupies more volume, raising sea level.

Of course the continents rarely stay still as sea level rises and falls. Local changes are bound to occur, and two other factors produce variations in *local* sea level:

- Tectonic motions and isostatic adjustment can change the height and shape of a coast. Coasts can experience uplift as lithospheric plates converge, or they can be weighted down by masses of ice during a period of widespread glaciation. The continents slowly rise when the ice melts. (See Figure 3.6 to review the principle of isostatic equilibrium.)

- Wind and currents, seiches, storm surges, an El Niño or La Niña event, and other effects of water in motion can force water against the shore or draw it away.

Sea level has been at its current elevation (give or take 0.5 meter, 1.5 feet) for only about 2,500 years. Over the past 2 million years, worldwide sea level has varied from about 6 meters (20 feet) above to about 125 meters (410 feet) below its present position. The recent low point occurred about 18,000 years ago at the height of the most recent glaciation (see Figure 4.14). Indeed sea level has been at the modern "high" only rarely in the past 2 million years; the dominant state for Earth is a much lower eustatic sea level position. It is important to realize that coastlines have not yet come into equilibrium with modern sea level.

Changes in sea level produce major differences in the position and nature of coastlines, especially in areas where the edge of the continent slopes gradually or where the coast is rising or sinking. **Figure 12.1** shows an estimate of shore positions along the southern coast of the United States in the geologically recent past, and a prediction for the distant future should the present warming trend cause more of the polar ice to melt.

Because coasts are influenced by so many factors, perhaps the most useful scheme for classifying a coast is based on the pre-

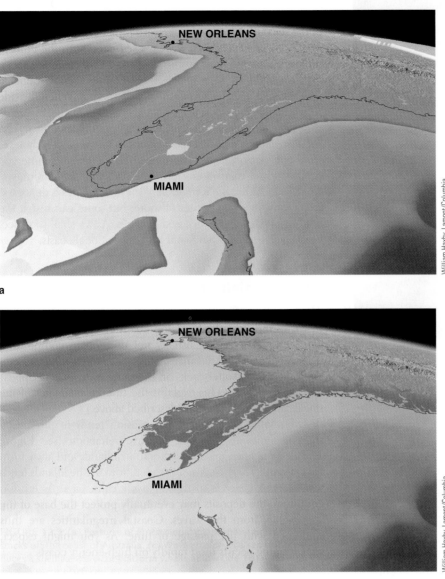

a

b

William Haxby, Lamont/Columbia

William Haxby, Lamont/Columbia

FIGURE 12.1 The southeastern coast of the United States, past and future. (**a**) About 18,000 years ago, during the last ice age, sea level was much lower. The position of the gently sloping eastern coast was as much as 200 kilometers (125 miles) seaward from the present shoreline, leaving much of the continental shelf exposed. (**b**) In the distant future, if the ocean were to expand and the polar ice caps were to melt because of global warming, sea level could rise perhaps 60 meters (200 feet), driving the coast inland as much as 250 kilometers (160 miles).

dominant events that occur there: erosion and deposition. **Erosional coasts** are new coasts in which the dominant processes are those that *remove* coastal material. **Depositional coasts** are *steady or growing* because of their rate of sediment accumulation or the action of living organisms (such as corals). The rocky shores of Maine are erosional because erosion exceeds deposition there; the sandy coastline from New Jersey to Florida is typically depositional because deposits of sediment tend to protect the shore from new erosion. The rocky central California coast is erosional, while the broad beaches of southern California are depositional. About 30% of the U.S. coastline is depositional and 70% is erosional. We will use the erosional–depositional classification scheme in the rest of this chapter.

Erosional Coasts

Land erosion and marine erosion both work to modify the nature of a rocky coast. Erosional coasts are shaped and attacked from the land by stream erosion, the abrasion of wind-driven grit, the alternating freezing and thawing of water in rock cracks, the probing of plant roots, glacial activity, rainfall, dissolution by acids from soil, and slumping.

From the sea, large storm surf routinely generates tremendous pressures. The crashing waves push air and water into tiny rock crevices. The repeated buildup and release of pressure within these crevices can weaken and fracture the rock. But it is not the hydraulic pressure of moving water alone that abrades the coasts. Tiny pieces of sand, bits of gravel, or stones hurled by the waves are even more effective at eroding the shore. Some indication of the violence of this activity may be inferred from **Figure 12.2**. Water dissolves minerals in the rocks, contributing to the erosion of easily soluble coastal rocks such as limestone. Even the digging and scraping of marine organisms have an effect.

The rate at which a shore erodes depends on the hardness and resistance of the rock, the violence of the wave shock to which it is exposed, and the local range of tides. Hard rock resists wear. Coasts made of granite or basalt may retreat an insignificant amount over a human lifetime; the granite coast of

FIGURE 12.10 A calm depositional shore—a beach scene on Florida's southwestern coast.

onshore, increasing the beach's slope. If water returning to the ocean—the **backwash**—carries back the same amount of material as it delivered, the beach slope will be in equilibrium; that is, the beach will not become larger or steeper.

On fine-grain beaches, the ability of small, sharp-edged particles to interlock discourages water from percolating down into the beach itself, so water from waves runs quickly back down the beach, carrying surface particles toward the ocean. This process results in a very gradual slope. Broad, flat beaches also have a large area on which to dissipate wave energy, and they can provide a calm environment for the settling of fine sediment particles. In contrast, coarse particles (gravel, pebbles) do not fit together well and readily allow water to drain between them. Onrushing water disappears *into* a beach made of coarse particles, so little water is left to rush down the slope, thereby minimizing the transport of sediments back to the ocean. Larger particles tend to build up at the back of the beach where they are thrown by large waves, increasing the steepness of the beach.

Wave action and beach permeability combine to build beaches. Small and moderate waves tend to predominate in places where there are beaches. These waves keep sand on or near shore because the orbital motion of water within them before they break moves sand more easily *toward* shore than *away* (**Figure 12.11**). After the waves break, a certain amount of the water that rushes onto the beach soaks into the sand. That much less water is therefore available to push sand grains downhill and out to sea. Coarse-grain beaches are proportionally steeper because a higher percentage of the uprushing water percolates into the beach, and so less is left to rush down the slope and "unpack" the beach.

Beach Shape

12.11

Figure 12.12 shows a profile, or cross section, of a beach affected by small to moderate wave and tidal action. The key feature of any beach is the **berm** (or berms), an accumulation of

Table 12.1	The Relationship Between the Particle Size of Beach Material and the Average Slope of the Beach	
Type of Beach Material	**Size (mm)**	**Average Slope of Beach**
Very fine sand	0.0625–0.125	1°
Fine sand	0.125–0.25	3°
Medium sand	0.25–0.50	5°
Coarse sand	0.50–1.0	7°
Very coarse sand	1–2	9°
Granules	2–4	11°
Pebbles	4–64	17°
Cobbles	64–256	24°

Source: Shepard, 1973.

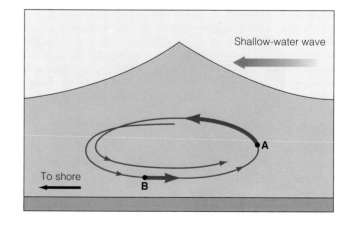

FIGURE 12.11 In small or moderate waves, a sand grain at point A moves easily *toward* shore, free of most bottom friction. Movement *off-shore* by the sand grain at point B is impeded by friction with the bottom. New flow may be reversed from onshore to offshore by high waves because of strong backwash.

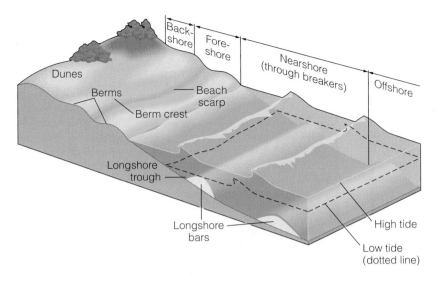

FIGURE 12.12 A typical beach profile. The scale is exaggerated vertically to show detail.

sediment that runs parallel to shore and marks the normal limit of sand deposition by wave action. The peaked top of the highest berm, called the **berm crest,** is usually the highest point on a beach. It corresponds to the shoreward limit of wave action during the most recent high tides. Inland of the berm crest, extending to the farthest point where beach sand has been deposited, is the **backshore.** The backshore is the relatively inactive portion of the beach, which may include wind-blown dunes and grasses. The **foreshore,** seaward of the berm crest, is the active zone of the beach, washed by waves during the daily rise and fall of the tides. It extends from the base of the berm—where a **beach scarp** (a vertical wall of variable height) is often carved by wave action at high tide—to the low-

tide mark where the offshore zone begins. Below the low-tide mark, wave action, turbulent backwash, and longshore currents excavate a **longshore trough** parallel to shore. Irregular **longshore bars** (submerged or exposed accumulations of sand) complete the seaward profile.

This beach profile is only temporary, generated by the interplay of sediments, waves, and tides. Great storm waves can rearrange a beach in a day, transporting thousands of tons of sediment from the beach to hidden **sandbars** offshore. Most temperate-climate beaches undergo a seasonal transformation. Beaches are cut to a lower level in winter than in summer because higher waves accompany winter storms. Changes from summer to winter on a beach are shown in **Figure 12.13.**

a

b

FIGURE 12.13 As seasons change sand moves on and off Boomer Beach near La Jolla, California. Gentle summer waves move sand onshore (**a**), but larger winter waves remove the sand to offshore bars, exposing the basement rock (**b**).

http://oceanography.brookscole.com/garrison5e

a

b

c

FIGURE 12.14 Minor features of beach sand. (a) Ripple marks. (b) The streamlike pattern of rill marks. (c) Backwash marks. Each V-shaped feature is caused by a small irregularity in the sand; each is about 15 centimeters (6 inches) long. (d) Layers of sediments are clearly seen in this freshly eroded beach face. The alternating layers were deposited by small waves during many ebb and flood tide cycles. Larger storm-driven waves have exposed the layers to view. The keys provide scale.

d

Minor Beach Features

Small features are among the things that make beaches such interesting places. *Ripples* in the sand are caused by rushing currents (**Figure 12.14a**); similar structures can be seen in dry river bottoms or stream beds. *Rills* are small, branching surface depressions that channel water back to the ocean from a saturated beach during a falling tide (**Figure 12.14b**). Diamond-shaped *backwash marks* form when projecting shells, pebbles, or animals interrupt the backwash along the **low-tide terrace** and cause uneven deposition or erosion of very fine sediment of a contrasting color (**Figure 12.14c**). Another result of uneven deposition is beach layering. A vertical slice into a calm beach will often reveal layer upon layer of sediments. The sediments are often separated by their density and may have different colors and textures (**Figure 12.14d**).

Longshore Transport

If the submerged slope of the seafloor is steep, eroded sediments will soon drain to deeper waters. If the slope is not too steep, sediments will be transported along the coast by wave and current action. The movement of sediment (usually sand) along the coast, driven by wave action, is referred to as **longshore drift.** Longshore drift occurs in two ways: the wave-driven movement of sand along the exposed beach, and the current-driven movement of sand in the surf zone just offshore.

Most wind waves approach at an angle and then refract in shallow water to break almost parallel to shore. Refraction is usually incomplete, however, and some angle remains when the waves break. If sediments have accumulated to form a beach, water from the breaking wave will rush up the beach at a slight angle but return to the ocean by running straight

FIGURE 12.15 (**a**) A longshore current moves sediment along the shoreline between the surf zone and the upper limit of wave action. (**b**) Groins built at right angles to the shore at Cape May, New Jersey, to slow the migration of sand. The groins interrupt the flow of longshore currents, so sand is trapped on their upcurrent sides. This view is toward the south, and south of the groins, on the downcurrent sides, sand is eroded.

a

b

downhill under the influence of gravity. The sand grains disturbed by the wave will follow the water's path, moving up the beach at an angle but retreating down the beach straight down the slope (**Figure 12.15**). The net transport of the grains is longshore, parallel to the coast, away from the direction of the approaching waves.

Sediments are also transported in the surf zone in a **longshore current.** The waves breaking at a slight angle distribute a portion of their energy away from their direction of approach. This energy propels a narrow current in which sediment already suspended by wave action can be transported downcoast. The speed of the longshore current can approach 4 kilometers (about $2\frac{1}{2}$ miles) per hour.

Sand moving in the wash of waves along the beach and sediments propelled in the longshore current just offshore are often joined by much greater loads of sediment brought to the coast by rivers. The net southward transport of all this material along the central California coast exceeds 230,000 cubic meters (300,000 cubic yards) per year. Typical figures for the Atlantic coast of the United States are about two-thirds of this value. Net sand flow along both the Pacific and Atlantic coasts of the United States is usually to the south because the waves that drive the transport system usually approach from the north, where storms most commonly occur.

Rip Currents

You may have noticed currents moving seaward through the surf on high-energy beaches (**Figure 12.16**). These small cur-

rents are caused by the local offshore movement of a large amount of water at one time. Although sometimes improperly termed *rip tides*, they're not caused by large-scale gravitational and inertial forces and thus are properly called **rip currents.**

Rip currents form when a group of incoming waves pile an excess of water on the landward side of the surf zone faster than the longshore current can carry it. The water breaks through the wave line in a few places and flows rapidly through the surf

FIGURE 12.16 A rip current is clearly visible in this aerial photograph of Black's Beach north of Scripps Canyon in southern California. The canyon influences wave action and circulation, and strong rip currents sometimes form here.

Despite these dangers, about 70 barrier islands off the U.S. coast have been commercially developed, and millions of people live on them. The most famous barrier islands include Atlantic City, New Jersey; Ocean City, Maryland; Miami Beach and Palm Beach, Florida; and Galveston, Texas. Roughly once every hundred years a winter storm has catastrophic effects on populated areas of Atlantic barrier islands, and the southeastern Atlantic and Gulf coasts must contend with occasional large hurricanes. The continuing subsidence of these passive coasts (combined with changes caused by commercial development and the ongoing rise in sea level) will undoubtedly cost lives and destroy property. **Figure 12.22** suggests the extent of the threat.

Unlike barrier islands, **sea islands** contain a firm central core that was part of the mainland when sea level was lower. The rising ocean separated these high points from land, and sedimentary processes surrounded them with beaches. Hilton Head, South Carolina, and Cumberland Island, Georgia, are sea islands. If the island is close to shore, a bridge of sediments called a **tombolo** may accumulate to connect the island to the mainland. Tombolos can also connect offshore rocky outcrops or volcanoes to the mainland. A sea island and a tombolo are shown in Figure 12.18.

Deltas

12.20

In a few places, sediments washing off the land have built out the coasts extensively. The shoreline in such places is much different from its configuration at the end of the last ice age. The most important of these coastal features are **deltas**.[1]

Deltas do not form at the mouth of every sediment-laden river. A broad continental shelf must be present to provide a platform on which sediment can accumulate. The tidal range is usually low, and waves and currents generally mild. There are no large deltas along the Atlantic coast of the United States

---

[1] The term is derived from the triangular shape of the capital Greek letter delta, Δ.

a

b

c

FIGURE 12.22 Barrier island modification: actual and potential. (**a**) The beach extending along the Matagorda Peninsula (Texas) barrier in September 1960. (**b**) The same area six days after the passage of Hurricane Carla in September 1961. The beach and island have been breached, and washover deltas are clearly seen. (**c**) Ocean City, Maryland, a developed barrier island. Host to 8 million visitors a year, this city (and others similarly situated) has no effective protection against flooding and damage from severe storms. The northernmost of the jetties shown in Figure 12.21 is just out of view at the lower right of the photograph.

a

b

FIGURE 12.23 River deltas form at places where sediment-laden rivers enter enclosed or semi-enclosed seas, where wave energy is limited. (**a**) The bird's-foot shape of the Mississippi Delta is seen clearly in this photograph. Lobed and bird's-foot deltas form where deposition overwhelms the processes of coastal erosion and sediment transportation. The sediment-laden water looks brown or tan in this photograph taken from low orbit. (**b**) The mouths of the Ganges–Bramaputra river system on 28 February 2000. This tide-dominated delta, home to about 120 million people, is routinely flooded during cyclones and monsoon rains. Note the sediment (milky blue color) flowing from the delta into the Bay of Bengal.

because sediments that arrive at the coast are deposited in the sunken river mouths or dispersed by tides and currents. Also, there are no large deltas along the western margins of North and South America because these coasts are converging margins, where an oceanic plate is being subducted and the continental shelf is very narrow; sediment that would form a delta is swept down the continental slope or dispersed along the coast by waves. Deltas are most common on the low-energy shores of enclosed seas (where the tidal range is not extreme) and along the tectonically stable trailing edges of some continents. The largest deltas are those of the Gulf of Mexico (the Mississippi, **Figure 12.23a**), the Mediterranean Sea (the Nile), the Ganges–Brahmaputra river system in the Bay of Bengal (**Figure 12.23b**), and the huge deltas formed by the rivers of China that empty into the South China Sea.

The shape of a delta represents a balance between the accumulation of sediments and their removal by the ocean. For a delta to maintain its size or grow, the river must carry enough sediment to keep marine processes in check. The combined effects of waves, tides, and river flow determine the shape of a delta. In 1975 William Galloway, a geologist at the University of Texas, classified deltas by the relative influence of those three factors (**Figure 12.24**). Any delta that is strongly affected by only one of the three factors is placed at an apex of the triangle. A delta with a balance of forces is placed near the middle of the triangle. *River-dominated deltas* are fed by a strong flow of fresh water and continental sediments, and they form in protected marginal seas. They terminate in a well-developed set of *distributaries*—the split ends of the river—in a characteristic bird's-foot shape (as shown in the Mississippi, Figure 12.23a). In *tide-dominated deltas*, freshwater discharge is overpowered by tidal currents that mold sediments into long islands parallel to the river flow and perpendicular to the trend of the coast. The largest tide-dominated delta has formed at the mouths of the Ganges–Brahmaputra river system on the Bay of Bengal (see Figure 12.23b). *Wave-dominated deltas* are generally smaller than either tide- or river-dominated deltas and have a smooth shoreline punctuated by beaches and sand dunes. Instead of a bird's-foot pattern of distributaries, a wave-dominated delta has one primary exit channel.

Deltas are not the only types of coasts built out by the land. The glaciers that covered the poleward parts of the continents during the last ice age deposited great quantities of sediments and rocks near their outer margins. When the glaciers retreated, they left streamlined hills known as **drumlins** and hills and ridges of sediments called **moraines**—some of which still stand above sea level. Part of Long Island, New York, is a glacial moraine, and the oval-shaped hills of Boston and sections of the Puget Sound coast at Seattle were shaped in part by glaciers. Perhaps the most famous glacial moraine in the United States is the area around Cape Cod, Massachusetts. The cores of the islands of Martha's Vineyard and Nantucket represent the farthest advance of the glaciers around 18,000 years ago, and the spine of the cape itself is a remnant of material dropped by the glacier as it stabilized and then retreated northward when the climate warmed (**Figure 12.25**).

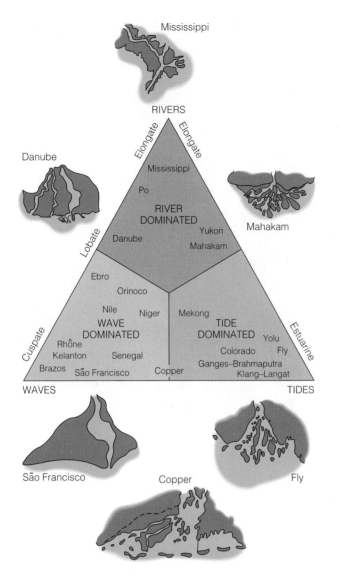

RIVERS

Mississippi

Danube

Mahakam

Mississippi
Po
RIVER DOMINATED
Yukon
Danube
Mahakam

Ebro
Orinoco
Nile   Niger   Mekong
WAVE DOMINATED   TIDE DOMINATED   Yolu
Rhône
Kelanton   Senegal   Colorado   Fly
Brazos   São Francisco   Copper   Ganges–Brahmaputra
Klang–Langat

Elongate   Elongate
Lobate
Cuspate   Estuarine

WAVES   TIDES

São Francisco   Copper   Fly

FIGURE 12.24 The Galloway classification of deltas. Each delta reflects the balance between the rate of sediment delivery at the river mouth and the mechanisms of its dispersal. This triangular diagram classifies river deltas according to the influence of the three major factors in their development: the river, waves, and tides. (Source: Used by permission of Dr. William Galloway, University of Texas, Austin, and the Houston Geological Survey.)

Coasts Formed by Biological Activity

Coasts can be extensively modified by the activities of animals and plants. Some kinds of marine algae and plants can build coasts, but the most dramatic biological modifications occur in the tropics, where coral organisms form reefs around volcanic islands or along the margin of a continent.

Coral Reefs

A **coral reef** is a linear mass of calcium carbonate assembled from and by multitudes of coral animals. Coral animals (which

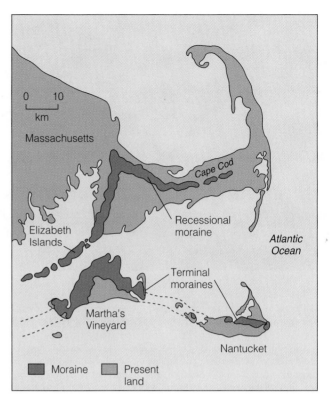

FIGURE 12.25 Most of the islands of Martha's Vineyard and Nantucket were formed from debris deposited or shaped by an advancing glacier. The spine of Cape Cod is built of material dropped as the glaciers stabilized and then retreated.

you will learn more about in Chapter 15) are related to the familiar sea anemones found along American coasts. Individual reef-building coral animals secrete a cup-shaped calcium carbonate skeleton, which remains behind after the animal dies; the accumulation of skeletons gradually forms the reef. These corals grow best in brightly lit water about 5 to 10 meters (16 to 33 feet) deep, and in ideal conditions they grow at a rate of about 1 centimeter ($\frac{1}{2}$ inch) per year.

The greatest of all reefs is the Australian Great Barrier Reef (**Figure 12.26**), which begins in the Torres Strait separating New Guinea and Australia and runs down the northeastern coast of Australia for 2,500 kilometers (1,500 miles).

Coral Reef Types

In 1842 Charles Darwin classified tropical coral reef structures into three types: Fringing reefs, barrier reefs, and atolls (**Figure 12.27**). We still use his classification today.

Fringing Reefs As their name implies, **fringing reefs** cling to the margin of land. As can be seen in Figure 12.27a, a fringing reef connects to shore near the water surface. Fringing reefs form in areas of low rainfall runoff primarily on the leeward (downwind side) of tropical islands. The greatest concentration of living material will be at the reef's seaward edge, where plankton and clear water of normal salinity are dependably available. Most new islands anywhere in the tropics have

Great Barrier Reef Marine Park Authority, Queensland Australia

FIGURE 12.26 A small section of the Great Barrier Reef, Queensland, Australia. This coast has been extensively modified by biological activity.

Fringing reef   Barrier reef   Atoll

a   b   c

© Douglas Faulkner, Science Source/Photo Researchers, Inc.

d

FIGURE 12.27 The development of an atoll. (**a**) A fringing reef forms around an island in the tropics. (**b**) The island sinks as the oceanic plate on which it rides moves away from a spreading center. In this case the island does not sink at a rate faster than coral organisms can build upward. (**c**) The island eventually disappears beneath the surface, but the coral remains at the surface as an atoll. (**d**) The typical ring shape of an atoll.

Oceanography ⬲ Now™ |
At the Active Figures link at http://oceanography.brookscole.com/garrison5e you can view the development of an atoll.

FIGURE 12.30 Types of estuaries in vertical cross sections. The salinity values show the amount of mixing between fresh water (0‰) and seawater (34‰) in the various types. (**a**) Salt wedge estuary. (**b**) Well-mixed estuary. (**c**) Partially mixed estuary. (**d**) A fjord estuary. (**e**) Reverse estuary, in which evaporation plays a major role.

FIGURE 12.31 Range of salinity in Cheasapeake Bay, an example of a partially mixed estuary. The colors indicate salinity in parts per thousand. The typical distribution of surface salinity in the estuary ranges from 28‰ at the mouth to 1‰ near the upper reaches. The Coriolis effect forces the inflowing salt water against the right (eastern) bank. (Notice how the 20‰ contour lines trend toward the right bank.) Compare this diagram to the photograph in Figure 12.5. (Source: Used by permission of Glen Wheless.)

Deeper estuaries exposed to similar tidal conditions but greater river flow become **partially mixed estuaries.** Partially mixed estuaries share some of the properties of salt wedge and well-mixed estuaries. Note in **Figure 12.30c** the influx of seawater beneath a surface layer of fresh water flowing seaward; mixing occurs along the junction. Energy for mixing comes from both tidal turbulence and river flow. England's Thames River, San Francisco Bay, and Chesapeake Bay are examples.

Fjord estuaries form where glaciers have gouged steep U-shaped valleys below sea level. Typically fjord estuaries have small surface areas, high river input, and little tidal mixing. River water tends to flow seaward at the surface with little contact with the seawater below (**Figure 12.30d**). In fjord estuaries with steep sills, a layer of stagnant water—cold water containing little oxygen and few nutrients—can form above the floor.

Reverse estuaries can form along arid coasts when rivers cease to flow. The evaporation of seawater in the uppermost reaches of these estuaries will cause water to flow from the ocean into the estuary, producing a gradient of *increasing* salinity from the ocean to the estuary's upper reaches (**Figure 12.30e**). Reverse estuaries are common on the Pacific coast of Mexico's Baja Peninsula and along the U.S. Gulf coast.

In well-mixed and partially mixed estuaries in the Northern Hemisphere, the incoming seawater presses against the right side of the estuary because of the Coriolis effect. Outflowing river water also trends to the right of its direction of travel. This rightward drift can be seen in the contour of lines representing surface salinity in Chesapeake Bay (**Figure 12.31**).

The Value of Estuaries

Some of the oldest continuous civilizations have flourished in estuarine environments. The lower regions of the Tigris and Euphrates Rivers, the Po River Delta region of Italy, the Nile Delta, the mouths of the Ganges, and the lower Hwang Ho Valley have supported dense human habitation for thousands of years. Estuaries continue to be irresistibly attractive to developers. In areas of high population density, estuaries are routinely dredged to provide harbors, marinas, and recreational resources and filled to make space for homes and agricultural land.

As we will see in Chapter 16's discussion of marine communities, estuaries often support a tremendous number of living organisms. The easy availability of nutrients and sunlight, protection from wave shock, and the presence of many habitats permit the growth of many species and individuals. Estuaries are frequently nurseries for marine animals; several species of perch, anchovy, and Pacific herring take advantage of the abundant food in estuaries during their first weeks of life. Unfortunately for their inhabitants, the high demand for development is incompatible with a healthy estuarine ecosystem.

Estuaries have also become the most polluted of all marine environments. Some of the plants growing in shallow temperate estuaries have the ability to "scrub" polluted water—to remove inorganic nitrogen compounds and metals from seawater polluted by sources on land. These plants use electrostatic attraction and sticky surface layers to flocculate clay-sized particles from the water and deposit them on their surfaces. When the tide falls, this material is deposited at the base of the plant, helping to protect it from erosion. Bacteria in the mud can decompose the nitrogen compounds and bind the metals, making the water cleaner. Ironically, development is destroying the very ecosystems that are capable of helping to clean the water. More than half the nation's estuaries and other wetlands have been lost. Of the original 870,000 square kilometers (215 million acres) of wetlands that once existed in the lower 48 states, only about 360,000 square kilometers (90 million acres) remain.

Lagoons and Wetlands

The term *lagoon* is used for the region between a barrier island or spit and the mainland and also for any shallow estuary that is essentially isolated from the ocean. Laguna Madre, one of the best known, extends along the Texas coast from Corpus Christi to the mouth of the Rio Grande River (see Figure 12.20). Most of the lagoon is less than 1 meter (3.3 feet) deep. It is protected from the ocean by Padre Island, a barrier island about 160 kilometers (100 miles) long. Laguna Madre is hypersaline; high evaporation keeps the salinity above 50‰ for most of the year. Because tidal range in the Gulf of Mexico is so small and the inlets at the ends of the barrier island are constricted, there is very little exchange of water between the lagoon and the ocean.

Characteristics of U.S. Coasts

Plate tectonic forces have had immense influence on the margins of continents, and the edges of the United States are no exception. The results of plate movement on the Pacific coast differ greatly from those on the Atlantic and Gulf coasts, primarily because the Pacific coast is near an active plate margin and the Atlantic and Gulf coasts are not.

The Pacific Coast

The Pacific coast is an actively rising margin on which volcanoes, earthquakes, and other indications of recent tectonic activity are easily observed. Pacific coast beaches are typically interrupted by jagged rocky headlands, volcanic intrusions, and the effects of submarine canyons. Wave-cut terraces are found as much as 400 meters (1,300 feet) above sea level in a number of places—evidence that tectonic uplift has exceeded the general rise in sea level over the past million years (**Figure 12.32**).

Most of the sediments on the Pacific coast originated from erosion of relatively young granitic or volcanic rocks of nearby mountains. The particles of quartz and feldspar that constitute

FIGURE 12.32 Wave-cut terraces on San Clemente Island off the coast of southern California. Tectonic uplift and the erosive forces shown in Figure 12.3c explain their origin.

© John S. Shelton

most of the sand were transported to the shore by flowing rivers. The volume of sedimentary material transported to Pacific coast beaches from inland areas greatly exceeds the amount originating at the coastal cliffs. Deltas tend not to form at Pacific coast river mouths because the continental shelf is narrow, river flow is generally low (except for the Columbia River), and beaches are usually high in wave energy. The predominant direction of longshore drift is to the south because northern storms provide most of the wave energy.

The Atlantic Coast

The Atlantic coast is a passive margin, tectonically calm and subsiding because of its trailing position on the North American Plate. Subsidence along the coast has been considerable—3,000 meters (10,000 feet) over the last 150 million years. A deep layer of sediment has built up offshore, material that helped produce today's barrier islands. Relatively recent subsidence has been more important in shaping the present coast, however. Except for the coast of Maine (which is still in isostatic rebound after the recent departure of the glaciers), coastal sinking and rising sea level have combined to submerge some parts of the Atlantic coast at a rate of about 0.3 meter (1 foot) per century. This process has formed the huge flooded valleys of Chesapeake and Delaware Bays, the landward-migrating barrier islands, and the shrinking lowlands of Florida and Georgia.

Rocks to the north (in Maine, for example) are among the hardest and most resistant to erosion of any on the continent, so beaches are uncommon in Maine. But from New Jersey southward the rocks are more easily fragmented and weathered, and beaches are much more common. As on the Pacific coast, sediments are transported coastward by rivers from eroding inland mountains, but the transported material is trapped in estuaries and therefore plays a less important role on beaches. Eastern beaches are typically formed of sediments from shores eroding nearby, or from the shoreward movement of offshore deposits laid down when the sea level was lower. The amount of sand in an area thus depends in part on the resistance or susceptibility of nearby shores to erosion. Sand moves generally south on these beaches just as it does on the Pacific coast, but the volume of moving sand is less in the East.

As we have seen, glaciers have also contributed to the shaping of the northern part of the Atlantic coast: Large portions of Long Island and all of Cape Cod are remnants of debris deposited by glaciers.

The Gulf Coast

The Gulf coast experiences a smaller tidal range and—hurricanes excepted—a smaller average wave size than either the Pacific or Atlantic coasts. Reduced longshore drift and an absence of interrupting submarine canyons allow the great volume of accumulated sediments from the Mississippi and other rivers to form large deltas, barrier islands, and a long raised "super berm" that prevents the ocean from inundating much of this sinking coast.

These are fortunate conditions because the rate of subsidence in the Gulf coast is greater than that for most of the Atlantic coast. Subsidence here is not the result of tectonic activity but rather is due to sediment compaction, dewatering, and the removal of oil and natural gas. Sediment starvation and dredging have made the situation worse around some large cities. At Galveston, Texas, for example, sea level appears nearly 64 centimeters (25 inches) higher than it was a century ago, and parts of New Orleans are now 2 meters (6.6 feet) below sea level. As we have seen, the results of hurricanes at such places can be tragic. The protective natural berm can easily be breached, and floodwaters can surge far inland.

Human Interference in Coastal Processes

Coasts are active areas where marine, terrestrial, atmospheric, and human factors converge. No single one of these factors dominates for long. We enjoy visiting and living near coasts, but human interference in coastal processes does not always produce the desired result. Steps taken to preserve or "improve" a stretch of rocky coast or a beach may have the opposite effect, and coastal residents do not always learn by example.

Beaches exist in a tenuous balance between accumulation and destruction. Human activity can tip the balance one way or the other. For example, consider the rocky **breakwater** shown in **Figure 12.33.** The breakwater interrupts the progress of waves to the beach, weakening the longshore current and allowing sand to accumulate there. Without dredging, the beach will eventually reach the breakwater and fill the small boat anchorage the breakwater was built to provide. This is a minor example of human alteration of a beach, yet it serves to introduce the growing problem of human influences on coastal processes.

We often divert or dam rivers, build harbors, and develop property with surprisingly little understanding of the impact our actions will have on the adjacent coast. Our role then becomes that of powerless observers. Residents of eroding coasts can only accept the inevitable loss of their property to the attack of natural forces, but residents of coasts in which deposition exceeds erosion are sometimes presented with alternatives. The choices are almost never simple. For example, should rivers be dammed to control devastating floods? If the dams are built, they will trap sediments on their way from mountains to coast. Beaches within the coastal cell fed by the dammed river will shrink because the sand on which they depend (to replenish losses at the shore) is blocked. Alarmed coastal residents will then take steps to hang onto whatever sand remains. They may try to trap "their" beaches by erecting **groins**: short extensions of rock or other material placed at right angles to longshore drift, to stop the longshore transport of sediments. This temporary expedient usually accelerates erosion downcoast (**Figure 12.34a** and Figure 12.15b). Diminished beaches then expose shore cliffs to accelerated erosion. Wind-wave energy that would have harmlessly churned sand grains now speeds the destruction of natural and artificial structures. Seawalls don't help either. They increase beach a   b

FIGURE 12.33 Growth of a beach protected by a breakwater: Santa Monica, California. (**a**) The shoreline as it appeared in 1931. (**b**) The same shoreline in 1949 after the breakwater was built. The boat anchorage formed by the breakwater is filling with sand deposited by disruption of the longshore current.

Groin
Groins are structures that extend from the beach into the water. They help counter erosion by trapping sand from the current. Groins accumulate sand on their updrift side, but erosion is worse on the downdrift side, which is deprived of sand.

Current

Seawall
Seawalls protect property temporarily, but they also increase beach erosion by deflecting wave energy onto the sand in front of and beside them. High waves can wash over seawalls and destroy them and property.

Importing sand
Importing sand to a beach is considered the best response to erosion. The new sand often is dredged from offshore, can cost tens of millions of dollars, and can disturb aquatic biodiversity. Because it is often finer than beach sand, dredged sand erodes more quickly.

FIGURE 12.34 Some measures taken to slow beach erosion.

 http://oceanography.brookscole.com/garrison5e

FIGURE 12.35 A resident of Rodanthe, North Carolina, rides his bicycle past a house undercut by Hurricane Dennis in September 1999. The extent of beach erosion is dramatically clear.

erosion by deflecting wave energy onto the sand. Churning by this increased energy eventually undermines the seawall, causing it to collapse (**Figure 12.34b**). The importation of sand trapped behind dams (or from other sources) is also only a temporary—and very expensive—expedient (**Figure 12.34c**). Scenes like the one in **Figure 12.35** will be more common.

What are the implications of these unlooked-for sand movements? Douglas Inman, director of the Center for Coastal Studies at the Scripps Institution of Oceanography, believes

that *at least 20% of the beach-bounded coastline of the United States is in danger of serious or catastrophic alteration.* On the Pacific coast a 30-year period of relatively mild weather may be ending. During this time people felt it was safe to build close to the shore. Increased dam building and breakwater, jetty, and groin construction have made southern California's beaches more vulnerable. In the 1997–98 El Niño, coastal California alone suffered losses exceeding $750 million. The barrier islands of the Atlantic and Gulf coasts are at least as vulnerable.

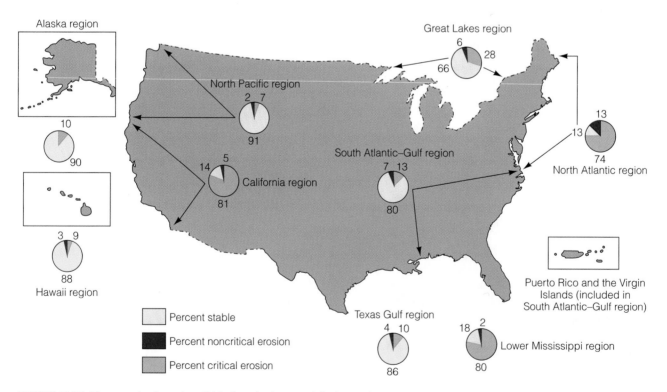

FIGURE 12.36 Shore erosion by region. Critical erosion is potentially threatening to property.

Figure 12.36 summarizes the progress of shore erosion by region.

Shores that look permanent through the short perspective of a human lifetime are in fact among the most temporary of all marine structures. Let's enjoy them in their present stages.

Questions from Students

1. The erosional and depositional categories seem limiting. I can imagine a situation where a shore would be erosional in the morning and depositional in the afternoon. Are there any other ways to classify shores?

Yes, and no classification system—including erosional–depositional—is completely satisfactory. One of the most used systems was proposed by marine geologist Francis Shepard. He separated coasts into two categories: *primary* and *secondary*. If a coast is essentially in almost the same condition as it was when sea level stabilized after the last ice age, Shepard called it a *primary coast*. Primary coasts are young coasts in which terrestrial influences (that is, processes that occur at the boundary between land and air) dominate. If the coast has been significantly changed by wave action and other marine processes since sea level stabilized, Shepard termed it a *secondary coast*. Secondary coasts are usually older than primary coasts; they have been exposed to marine action for a longer time. Secondary coasts retain little (if any) evidence of the nonmarine processes that produced them. Some coasts show both characteristics. A single long shoreline can consist of both primary and secondary coasts. In this chapter erosional is roughly analogous to primary, and depositional to secondary.

A better way (or so it seems to me) is a threefold classification proposed by Anthony Orme. His scheme is based on time scales. Coastal origins are related to the tectonics of the coastal zone acting over millions of years. More precise limits of coasts are dependent on sea-level change acting on shorter time scales. A shore's immediate condition is a product of contemporary processes—the processes that one can measure within a few hours or a lifetime. Orme's hierarchy of scales begins with the core tectonic system and progresses through the intermediate sea-level change scale, to the localized and specific effects of waves and tidal currents.

2. My foot tends to sink whenever I stand on the beach and let water from a wave run over it. The sand moves away from the edges of my foot, and I sink in. Why?

This is a good example of water's ability to carry more sediment as its speed of flow increases. Your foot interrupts the flow of water up or down the beach after a wave breaks, and the water must speed up to get around your foot. Fast-running water moves sand more effectively than slow-running water, so the sand immediately next to your foot is removed. This process, termed *scouring*, becomes a problem when structures are placed in shallow water.

3. What are those little white pellets I find on the beach along the high-tide line? Surfers call them "nurdles."

Those ubiquitous, insidious particles are the raw material for molded plastic goods. They are transported from producers to fabricators in containers loaded onto container ships. The pellets escape if a container is mishandled, breaks open during a storm, or is lost overboard. Virtually indestructible, "nurdles" float with the winds and currents until they encounter a shore. One researcher has calculated that just 25 containers would carry enough plastic pellets to spread 100,000 "nurdles" per mile along all the seashores of the world!

4. I hope someday to live near the ocean. What should I look for in buying property there?

Firm ground! Coastal Maine would be an ideal bet. The dense metamorphic rock of much of the Maine shore is stable (within the human time frame) and hard—ideal footings for a house. Make sure the site is far enough inland to avoid high surf, storm surge, and tides. If the winters in Maine don't appeal to you, coastal Florida might make a good choice—if your children aren't hoping to inherit the property. If you insist on building on a barrier island, make sure your home is on the mainland side of the southern end! Parts of the Pacific coast are all right, but local variability on that active margin makes some knowledge of the geological history of the area valuable. For example, some parts of the San Diego shoreline are eroding about 3 meters (10 feet) per year—hardly a solid investment.

Chapter in Perspective

In this chapter you learned that the *location* of a coast depends primarily on global tectonic activity and the ocean's water volume, while the *shape* of a coast is a product of many processes: uplift and subsidence, the wearing down of land by erosion, and the redistribution of material by sediment transport and deposition. Coasts are classified as erosional coasts (on which erosion dominates) or depositional coasts (on which deposition dominates). Natural rock bridges, tall stacks, and sea caves are found on erosional coasts. Depositional coasts often support beaches, accumulations of loose particles. Generally the finer the particles on the beach, the flatter its slope. Beaches change shape and volume as a function of wave energy and the balance of sediment input and removal. Coral reefs and estuaries are among the most complex and biologically productive coasts. Human interference with coastal processes has generally accelerated the erosion of coasts near inhabited areas.

In the next chapter you will learn that the study of oceanography includes a marvelous variety of living things. The next chapter's discussion of the general nature and characteristics of marine life will launch us into the biological part of our journey.

 http://oceanography.brookscole.com/garrison5e

Key Ideas

✓ Coasts are temporary structures, often subject to rapid change.
Review pages 267–268

✓ The *location* of a coast depends primarily on global tectonic activity and the volume of water in the ocean.
    The *shape* of a coast is a product of many processes: uplift and subsidence, the wearing down of land by erosion, and the redistribution of material by sediment transport and deposition.
Review pages 267–268

✓ Change in sea level greatly influences coastal processes. For most of Earth's recent history, sea level has been lower than we find it today.
Review Figures 4.14 and 12.1, pages 96 and 277

✓ Erosional coasts are new coasts in which the dominant processes are those removing coastal material. Depositional coasts are those coasts that are steady or growing because of their rate of sediment accumulation or the action of living organisms.
Review Figures 12.2–12.4 and 12.10–12.13, pages 276, 278–279, 282–283

✓ Erosional coasts are often punctuated by sharp cliffs, sea caves, rock stacks, and occasional natural bridges.
Review Figure 12.3, page 279

✓ Depositional coasts often support beaches, accumulations of loose particles. The shape and volume of beaches change as a function of wave energy and the balance of sediment input and removal.
Review Figures 12.13–12.16, pages 283–285

✓ Depositional coasts, especially along subsiding continental margins, often exhibit characteristic large-scale features such as barrier islands.
Review Figures 12.17–12.20, pages 286–288

✓ Deltas form on broad continental shelves where rivers deposit sediment, tidal range is low, and wave and current action is generally mild.
Review Figures 12.23–12.24, pages 290–292

✓ Coasts can be extensively modified by the actions of living organisms. Coral reefs and mangrove coasts are examples.
Review Figures 12.26–12.28, pages 292–294

✓ Estuaries can form at river mouths where fresh water mixes with seawater. They are among the most complex and biologically productive coasts.
Review Figures 12.29–12.31, pages 295–296

✓ Human interference in coastal processes rarely increases the long-term stability of a coast.
Review Figures 12.33 and 12.34, pages 299–300

Terms and Concepts to Remember

atoll, 294
backshore, 283
backwash, 282
barrier island, 288
barrier reef, 294
bay mouth bar, 287
beach, 281
beach scarp, 283
berm, 282
berm crest, 283
breakwater, 298
coast, 276
coastal cell, 287
coral reef, 292
delta, 290
depositional coast, 277
drumlin, 291
erosion, 276
erosional coast, 277
estuary, 295
eustatic change, 276
fjord, 278
fjord estuary, 296
foreshore, 283
fringing reef, 292

groin, 298
high-energy coast, 278
inlet, 287
lagoon, 288
longshore bar, 283
longshore current, 285
longshore drift, 284
longshore trough, 283
low-energy coast, 278
low-tide terrace, 284
moraine, 291
partially mixed estuary, 296
reverse estuary, 296
rip current, 285
salt wedge estuary, 295
sandbar, 283
sand spit, 287
sea cave, 278
sea cliff, 278
sea island, 290
shore, 276
swash, 281
tombolo, 290
wave-cut platform, 278
well-mixed estuary, 295

Study Questions

Reviewing What You've Learned

1. How is an erosional coast different from a depositional coast?
2. What features would you expect to see along an erosional coast? A depositional coast? What determines how long the features will last?
3. What are some of the features of a sandy beach? Are they temporary or permanent? Is there a relationship between wave energy on a coast and the size (or slope, or grain size) of beaches found there?
4. How are deltas classified? Why are there deltas at the mouths of the Mississippi and Nile Rivers but not at the mouth of the Columbia River?
5. What is a coastal cell? Where does sand in a coastal cell come from? Where does it go?
6. How are estuaries classified? Upon what does the classification depend? Why are estuaries important?
7. Compare and contrast the Pacific, Atlantic, and Gulf coasts of the United States.

Thinking Critically

1. What other ways can you think of to classify coasts? What are the advantages of your method of classification? Are there disadvantages?

2. What two processes contribute to longshore drift? What powers longshore drift? What is the predominant direction of drift on U.S. coasts? Why?

3. What physical conditions might limit or encourage the development of a coral reef? An atoll?

4. How do human activities interfere with coastal processes? What steps can be taken to minimize loss of life and property along U.S. coasts?

5. **InfoTrac College Edition Project** Do increased human use and occupation of coastal areas always result in erosion and pollution? Why or why not? What steps could be taken to prevent degradation of coastal areas? Use Info-Trac College Edition to answer this question.

Thinking Analytically

1. Hurricane Isabel struck the east coast of the United States in September 2003. Consult contemporaneous accounts and estimate the relative importance of high wind, heavy rain (and attendant flooding), and storm surge in the final damage toll.

2. Figure 12.1 shows the hypothetical result of a rise in sea level of 60 meters (200 feet) in Florida. Obtain an elevation map of your favorite coastal town and see where the coastline would be if sea level rose 60 meters (200 feet) there.

3. About 5,000 cubic meters of sand is eroding every year from the beach shown in Figure 12.17c. Find the weight limit for dump trucks in your state. Now find the weight of a cubic meter of sand. How many dump truck loads would be required to make up this deficit?

Study and Research on the Internet

CHAPTER 13

Sunlight and fishes in one of the ocean's most productive habitats.

LIFE IN THE OCEAN

The Ideal Place for Life

To the weekend sailor, the ocean may seem interesting mainly because of its winds, currents, and waves, but the seemingly empty seawater next to a small sailboat may support millions of invisible plantlike organisms. The waters beneath the hull can conceal wonderful worms and colorful crustaceans, gently swaying forests of seaweeds, whole communities of microscopic creatures drifting with the currents, schools of fishes, maybe even whales. On the dark, distant bottom animals locate one another with glowing lures or jostle for food near volcanic vents. The great wide sea is the ideal habitat for life.

Life on Earth is notable for both unity and diversity: *diversity* because Earth may house as many as 100 million different species (kinds) of living organisms; *unity* because all species share the same underlying mechanisms for capturing and storing energy, manufacturing proteins, and transmitting information between generations. That idea is especially important: In a sense, all life on Earth is fundamentally the same; it's just packaged in different ways. All Earth's life-forms are related; all have apparently evolved from a single instance of origin. Earth and life have grown old together, generation by generation, over some 4 billion years.

How does the ocean fit into this? In a sense all life on Earth is marine because life almost certainly began in the ocean. All life on our planet is water-based and shares the same basic underlying life processes. The ocean can support such a bewildering array of life-forms because of water's unique physical properties—its density, its dissolving power, its ability to absorb large quantities of heat yet rise very little in temperature. In part at least, life in the ocean is so successful because of the ocean's benign physical characteristics; the oceanic environment is a relatively easy place for cells to live.

13.1

Being Alive

13.2

It would seem easy to differentiate between the living and nonliving components of the marine environment, but we cannot always see the difference, as **Figure 13.1** points out. The same atoms move continuously into and out of living and nonliving systems. With your last breath you exhaled millions of carbon atoms that entered your body as food in your last meal. Before the day is done, some of those atoms may be incorporated into a nearby house plant. Nothing is special about the atoms or energy of life, nothing to distinguish them from their nonliving counterparts. The free exchange of identical components between life and nonlife complicates any attempts at formal definition. Yet intuitively we all make this distinction when we observe the organization of matter in living things, and especially when we consider the ways living things manipulate energy.

Life, Energy, and Food

13.3

Living matter cannot function without **energy,** the capacity to do work. Living organisms cannot create new energy, but they can transform one kind of energy to a different kind. A plant can transform light energy into chemical energy; an animal can transform chemical energy into energy of movement by the muscles, can transform energy of movement into heat, and so on. All life activity is involved, directly or indirectly, in energy transformation and transfer. Energy must therefore be central to anyone's definition of life.

Living organisms are complex assemblies; they need energy to build that complexity from large numbers of simple molecules. The **second law of thermodynamics** shows, however, that disorder inevitably tends to increase in the universe as time passes. Things run down, become disorganized, break. **Entropy** is a measure of this disorder.

Living things are not exempt from the second law of thermodynamics, but they can temporarily and locally delay that inevitable descent into disorganization through the transformation of energy. Living organisms are places where the flow of energy results in *increased* order—that is, areas of great complexity and

© Linda E. Tway

FIGURE 13.1 Living or nonliving? The brightly colored "rock" in this picture is a colony of small marine plants. The distinction between life and nonlife does not lie in composition or outward appearance, but in the ability to manipulate energy.

low entropy. Car engines use the energy in gasoline only to move. Organisms use the energy in food to move, to maintain their highly complex organization, and to grow. My car and my computer don't fix themselves; fish and people do. This sophisticated use of energy is a basic attribute of living things.

The main source of energy for living things on Earth is the sun. Life prospers, becomes ever more complex, and evolves into millions of forms by accepting sunlight and radiating waste heat to the cold of space. As we will see, with a few exceptions, organisms get their power directly or indirectly through the capture, storage, and transmission of energy from sunlight. Light energy is transformed into chemical energy and finally into heat as organisms temporarily forestall the disorderly fate decreed by the second law.

How does this work?

Converting Energy to Food

The sun produces enormous quantities of energy—some of it in the form of visible light, a tiny portion of which strikes Earth. Only about one part in 2,000 of the light that reaches Earth's surface is captured by organisms, but that "small" input of energy powers nearly all the growth and activity of living things. Light energy from the sun is trapped by **chlorophyll** in organisms called *producers* (certain bacteria, algae, and green plants) and changed into chemical energy. The chemical energy is used to build simple carbohydrates and other organic molecules—**food**—which are then used by the producer or eaten by animals (or other organisms) called *consumers*. Because light energy is used to synthesize molecules rich in stored energy, the process is called **photosynthesis** (*photos* = light, *syn* + *tithenai* = to place together). Here is a general formula for photosynthesis:

$$6\,CO_2 \;+\; 6\,H_2O \;\xrightarrow{\text{sunlight}}\; C_6H_{12}O_6 \;+\; 6\,O_2$$

6 molecules of carbon dioxide + 6 molecules of water (yields by photosynthesis) a molecule of glucose (food, a carbohydrate) + 6 molecules of oxygen

The stored energy is released when food such as glucose is used for growth, repair, movement, reproduction, and other functions of organisms. The breakdown of food eventually produces waste heat, which flows away from Earth into the coldness of space. This one-way flow of energy is shown in **Figure 13.2**.

Photosynthesis is the dominant method of binding energy into carbohydrates, but there is also another method. **Chemosynthesis,** used by a few relatively simple forms of life, is the production of usable energy directly from energy-rich inorganic molecules available in the environment rather than from the sun. Some unusual forms of marine life depend on chemosynthesis, as we will see in Chapter 16. Overall, chemosynthetic production of food in the ocean is very small in comparison to photosynthetic production.

Primary Productivity

The synthesis of organic materials from inorganic substances by photosynthesis or chemosynthesis is called **primary productivity** (**Figure 13.3**). Primary productivity is expressed in *grams of carbon bound into organic material per square meter of ocean surface area per year* (gC/m²/yr). The organic material produced is usually glucose, a carbohydrate. The source of carbon for glucose is dissolved carbon dioxide (CO_2). *Phytoplankton*—minute, drifting photosynthetic organisms that you will meet in Chapter 14—produce between 90% and 96% of oceanic carbohydrates. Seaweeds—larger marine photosynthesizers—contribute from 2% to 5% of the ocean's primary productivity. Chemosynthetic

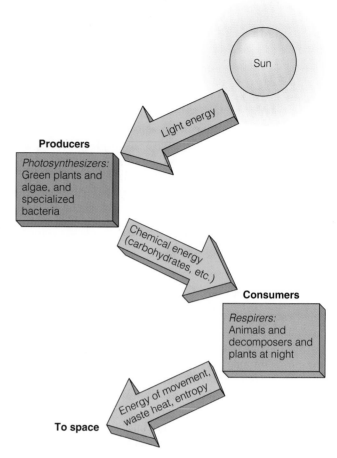

FIGURE 13.2 The flow of energy through living systems. At each step energy is degraded (transformed into a less useful form).

organisms probably account for between 2% and 5% of the total. Though estimates vary widely, recent studies suggest that total ocean productivity ranges from 75 to 150 grams of carbon bound into carbohydrates per square meter of ocean surface per year (75 to 150 gC/m²/yr). (For comparison, a well-tended alfalfa field produces about 1,600 gC/m²/yr.)

How does marine net productivity compare with terrestrial net productivity? Recent research suggests that global net productivity in *marine* ecosystems is 35 to 50 billion metric tons of carbon bound into carbohydrates per year; global net *terrestrial* productivity is roughly similar at 50 to 70 billion metric tons per year.[1] However, the total producer **biomass** (the mass of living tissue) in the ocean is only 1 to 2 billion metric tons, compared with 600 to 1,000 billion metric tons of living biomass on land! As you can see in **Table 13.1,** marine producers are *much* more efficient in their production of food than their land-based counterparts are. As the rapid turnover time indicates, nutrients cycle from producer to consumer and back much more quickly in marine ecosystems.

---

[1] 1 billion metric tons = 1.1 billion tons.

1 meter

1 meter

CO_2 → Carbon dioxide

Diatom

→ $C_6H_{12}O_6$ Glucose

Typically, oceanic primary productivity in this water column will bind ~120 grams of carbon into molecules of glucose each year.

to bottom of ocean

a

b

Wim von Egmond

COMMUNITY	NET PRIMARY PRODUCTIVITY (gC/m²/yr)
OCEAN COMMUNITIES	
Coral reef	880–2,200
Kelp bed	400–1,900
Shelf plankton	90–270
Open ocean	1–180
LAND COMMUNITIES	
Rain forest	460–1,600
Temperate forest	270–1,140
Freshwater swamp	360–1,820
Cropland	45–1,820

0   500 1,000 1,500 2,000 2,500

ALL OCEANS = 120 (average)   ALL LANDS = 150 (average)

c

FIGURE 13.3 (**a**) Oceanic productivity—the incorporation of carbon atoms into carbohydrates—is measured in grams of carbon bound into carbohydrates per square meter of ocean surface area per year (gC/m²/yr). (**b**) The diatom *Coscinodiscus*, an important marine primary producer. This diatom is about the size of the period at the end of this sentence. (**c**) The net annual primary productivity in some marine and terrestrial communities.

Table 13.1	Comparison of Global Net Productivity and Living Biomass in Marine and Terrestrial Ecosystems		
Ecosystem	Net Primary Productivity (10¹⁵ grams/year)[a]	Total Plant Biomass (10¹⁵ grams)	Turnover Time (years)
Marine	35–50	1–2	0.02–0.06
Terrestrial	50–70	600–1,000	9–20

Source: Falkowski and Raven, 1997.
[a] 10¹⁵ grams is equivalent to 1 billion metric tons.

Chlorophyll a Concentration mg/m3

.01   .02 .03   .05   .1   .2   .3   .5   1   2   3   5   10   15 20 30   50

FIGURE 13.4 Oceanic productivity can be observed from space. NASA's *SeaWiFS* satellite, launched in 1997, can detect the amount of chlorophyll in ocean surface water. Chlorophyll content provides an estimate of productivity. Red, yellow, and green areas indicate high primary productivity; blue areas indicate low. This image was derived from measurements made between September 1997 and August 1998.

The total mass of a **primary producer** is assumed to be about ten times the mass of the carbon it has bound into carbohydrates. Thus a primary productivity of 100 $gC/m^2/yr$ represents the yearly growth of about 1,000 grams of primary producers for each square meter of ocean surface. Because 35 to 50 billion metric tons of carbon is bound into carbohydrates in the ocean each year, between 350 and 500 billion metric tons of marine plants and plantlike organisms are produced annually (**Figure 13.4**). Each year this vast bulk is consumed by the metabolic activity of the producers themselves and by the consumers that graze on them. The component atoms are then reassembled by photosynthesis into carbohydrates in a continuous solar-powered cycle.

Feeding (Trophic) Relationships

Photosynthetic and chemosynthetic organisms can be called either primary producers or **autotrophs** (*auto* = self, *trophe* = nourishment) because they make their own food. The bodies of autotrophs are rich sources of chemical energy for any organisms capable of consuming them. **Heterotrophs** (*hetero* = other, different) are organisms, such as animals, that must consume other organisms because they are unable to synthesize their own food molecules. Some heterotrophs consume autotrophs, and some consume other heterotrophs.

We can label organisms by their positions in a "who eats whom" feeding hierarchy called a **trophic pyramid** (*trophos* = one who feeds). The primary producers shown at the bottom of the pyramid in **Figure 13.5** are mostly chlorophyll-containing photosynthesizers. The animal heterotrophs that eat primary producers are called **primary consumers** (or herbivores), the animals that eat primary consumers are called secondary consumers, and so on to the **top consumer** (or top carnivore).

Note that the mass of consumers becomes smaller as energy flows toward the top of the pyramid. There are many small primary producers at the base, and only a few large top consumers at the apex. Only about 10% of the energy from the organisms consumed is stored in the consumers as flesh, so each level is about one-tenth the mass of the level directly below. The rest of the energy is lost as waste heat as organisms live and work to maintain themselves.

Pyramids such as the one in Figure 13.5 can lead to the misconception that one kind of fish eats only one other kind of fish, and so on. Real communities are more accurately described as food webs; an example is included as **Figure 13.6**. A **food web** is a group of organisms linked by complex feeding relationships in which the flow of energy can be followed from primary producers through consumers. Organisms in a food web almost always have some choices of food species.

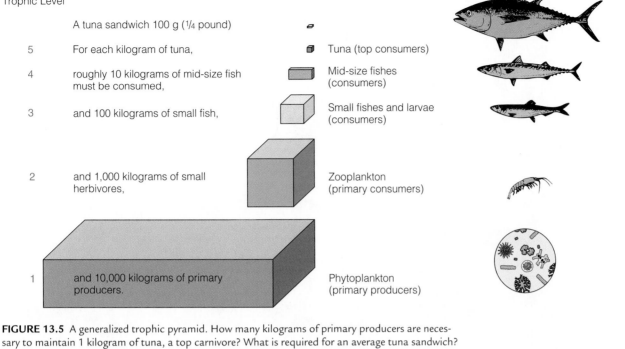

Trophic Level

5	For each kilogram of tuna,	Tuna (top consumers)
4	roughly 10 kilograms of mid-size fish must be consumed,	Mid-size fishes (consumers)
3	and 100 kilograms of small fish,	Small fishes and larvae (consumers)
2	and 1,000 kilograms of small herbivores,	Zooplankton (primary consumers)
1	and 10,000 kilograms of primary producers.	Phytoplankton (primary producers)

A tuna sandwich 100 g (¼ pound)

FIGURE 13.5 A generalized trophic pyramid. How many kilograms of primary producers are necessary to maintain 1 kilogram of tuna, a top carnivore? What is required for an average tuna sandwich? Using this trophic pyramid model, you can see that 1 kilogram of tuna at the fifth trophic level (the fifth feeding step of the pyramid) is supported by 10 kilograms of mid-size fish at the fourth, which in turn is supported by 100 kilograms of small fish at the third, which have fed on 1,000 kilograms of zooplankton (primary consumers) at the second, which have eaten 10,000 kilograms of phytoplankton (small autotrophs, primary producers) at the first. The quarter-pound tuna sandwich has a long and energetic history. (These figures have been rounded off to illustrate the general principle. The actual measurements are difficult to make and quite variable.)

So organisms interact with each other, feed on one another, and transfer energy as food from producing autotrophs through a web of consuming heterotrophs. Nearly all are ultimately dependent on sunlight and photosynthesis. What is the physical role of the ocean in this?

The Elements of Life

We've been discussing energy and life. Now let's turn to life's material substance. All Earth's organisms are composed of about 23 of the 107 known chemical elements.[2] Four of these elements—carbon, hydrogen, oxygen, and nitrogen—make up 99% of the mass of all living things, with nine additional elements comprising nearly all of the remainder (**Table 13.2**). Atoms of these elements combine to form the classes of biological chemicals common to all life. The main categories of these chemicals are probably familiar to you: carbohydrates, lipids (fats, waxes, oils), proteins, and nucleic acids (such as DNA, the primary molecule of heredity).

Despite the astonishing variety of life-forms on Earth, biologists have come to appreciate that *the central message of biology is unity,* not diversity (**Figure 13.7**). That is, the extraordinary sameness of biological chemicals as they are found in the structure of all living things suggests that all life is related through a common evolutionary history.

Elements in Cycles

The atoms and small molecules that make up the biochemicals, and thus the bodies, of organisms move between the living and nonliving realms in **biogeochemical cycles.** Living organisms are supported and sustained by huge nonliving chemical reserves, and there is a large-scale transport of elements between the reserves and the organisms themselves. Sometimes the environment contains enough of a required element to sustain life; sometimes the element is in short supply.

You learned in Chapter 6 about the ocean's density stratification. The tropical and temperate ocean is usually highly stratified, with a warm, less dense layer of water—the *mixed zone*—separated from the cold, dense *deep zone* by a strong pycnocline (see Figure 6.14). In the surface mixed layer, the atoms and small molecules that make up the bodies of organisms may cycle rapidly for a time among predators, prey, scavengers, and decomposers. When these organisms die, their bodies sink below the sunlit upper sea, beneath the pycnocline, where they are isolated from the rapid biological activity

---

[2] The periodic table of the elements—an arrangement of the elements by their characteristics—is included as Appendix VIII.

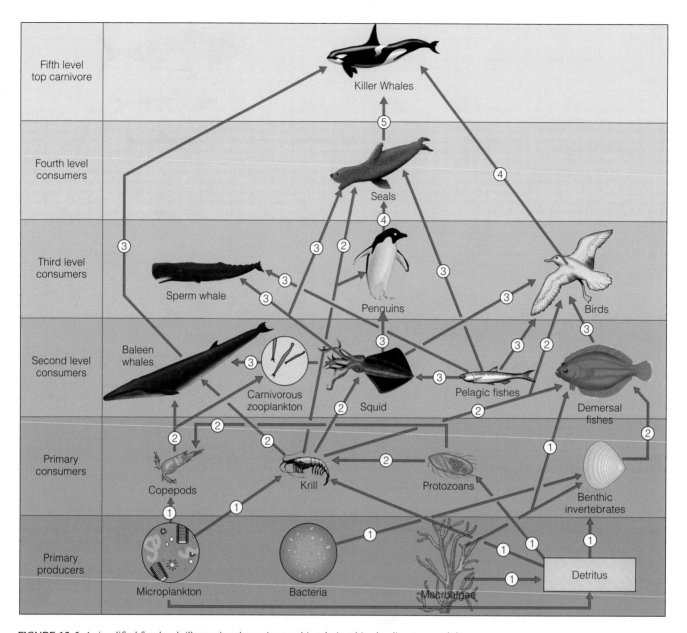

FIGURE 13.6 A simplified food web illustrating the major trophic relationships leading to an adult killer whale. The arrows show the direction of energy flow; the number on each arrow represents the trophic level at which the organism is feeding. Note that feeding relationships are not as simple as one might assume from Figure 13.5.

of the surface. We can appreciate the critical role of upwelling in returning these substances to the surface, if only for a short reprieve before their eventual incorporation into deep sediments from which only the very slow progress of tectonic cycles will liberate them.

As you read about the biogeochemical cycles described next, remember that the elements and small molecules that form the tissues of an organism *are always on the move.* They may cycle rapidly into and out of living things or they may be trapped in Earth for great spans of time, but the nature of the cycles dictates what will live where, which creatures will be successful, and ultimately the composition of the ocean and atmosphere itself.

Carbon

The largest of all biogeochemical cycles is the global **carbon cycle** (**Figure 13.8**). Because of its ability to form long chains to which other atoms can attach, carbon is considered the basic building block of all life on Earth. Carbon enters the atmosphere by the **respiration** of living organisms (such as carbon dioxide), volcanic eruptions that release carbon from rocks deep in Earth's crust, the burning of fossil fuels, and other processes.

As we have seen, large and small plants (and plantlike organisms) capture sunlight and use this energy to incorporate, or *fix,* CO_2 into organic molecules. Some of these molecules

Table 13.2 The Chemical Elements Used by Organisms in Their Metabolism

Element	Symbol	Representative Use
Major Components[a]		
Carbon	C	All organic molecules
Oxygen	O	Almost all organic molecules
Hydrogen	H	All organic molecules
Nitrogen	N	Proteins, nucleic acids
Macronutrients[b]		
Sodium	Na	Body fluid, osmotic regulation
Magnesium	Mg	Osmotic balance; in chlorophyll
Phosphorus	P	Nucleic acids, teeth/bone/shell
Sulfur	S	Proteins, cell division
Chlorine	Cl	Nerve discharge, osmotic balance, ATP formation
Potassium	K	Nerve discharge, osmotic balance, enzyme activation
Calcium	Ca	Shell, bone, coral, teeth
Iodine	I	Thyroid hormone
Silicon	Si	Rigid parts
Micronutrients[c]		
Iron	Fe	Electron transport, N assimilation
Copper	Cu	Electron transport
Zinc	Zn	Nucleic acid replication and Transcription
Manganese	Mn	Enzyme activity

Source: Adapted from Milne, 1995, p. 281.
[a]Major components are elements that make up 100,000 or more of every million atoms in an organism.
[b]Macronutrients constitute between 1,000 and 100,000 of every million atoms.
[c]Micronutrients constitute fewer than 1,000 of every million atoms.

a

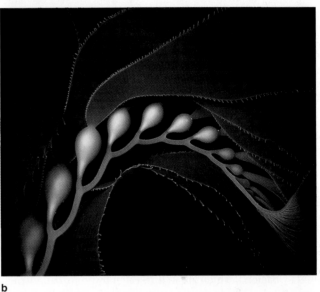

b

are used as food and some as structural components. When an animal eats a plant (or plantlike organism), three things can happen to the carbon: (1) it can be incorporated into the animal's body for growth, (2) it can be respired by the animal (taken apart to harvest the energy), or (3) it can be wasted, excreted back into the seawater as what is called **dissolved organic carbon (DOC)**. Typically about 45% of the carbon from an ingested plant is used for growth, 45% for respiration, and about 10% is lost as DOC. The end product of respiration is CO_2, a gas eventually lost to the atmosphere. The DOC is

FIGURE 13.7 Although this bryozoan (**a**), kelp (**b**), and person (**c**) are superficially quite different, their underlying biochemistry is startlingly similar. Each runs on glucose, each uses the same kinds of chemicals to transmit genetic information between generations, and each builds its proteins using the same coded set of instructions. The central message of all biology on Earth is unity, not diversity.

c

FIGURE 13.8 A simplified model of the global carbon cycle. The left portion shows the movement of carbon through marine systems, and the right shows its movement through terrestrial systems. Carbon reservoirs are shown as boxes; processes that change one form of carbon to another are shown in unboxed print. (From Cecie Starr, *Biology: Concepts and Applications*, 4th ed., Brooks/Cole ©2000)

used by bacteria, which are in turn eaten by protozoans, which are eaten by zooplankton, which are then eaten by fish, in what is termed the *microbial loop*. Eventually the organisms—or at least their hard parts containing calcium carbonate, $CaCO_3$—sink below the mixed layer and begin the long fall toward the seabed. Most of the carbon in this $CaCO_3$ is respired (turned into CO_2) by bacteria long before it hits the bottom, but a small percentage (<1%) reaches the sediments and is buried. The carbonate sediments can be uplifted over geological time and weathered, so that the carbon is eventually returned to the biologically active upper sea.

Because of the large amount of carbon dioxide available in the ocean and because CO_2 from the atmosphere dissolves readily in seawater, marine organisms almost never suffer from a deficit of available carbon. For life in the sea, the critical bottlenecks lie elsewhere—mainly in the nitrogen, phosphorus, and iron cycles.

Nitrogen

Nitrogen is a critical component of proteins, chlorophyll, and nucleic acids. Like carbon, nitrogen may be found in the bodies of organisms, as a dissolved gas (N_2), and as dissolved organic matter (**dissolved organic nitrogen, DON**). It also exists as the dissolved inorganic ions nitrate (NO_3^-), nitrite (NO_2^-), and ammonium (NH_4^+).

One might think nitrogen would be abundantly available in the ocean: Table 7.4 indicates that nitrogen accounts for 48% of the dissolved gas in seawater, by volume. But most organisms cannot use the free nitrogen in the atmosphere and ocean directly. Nitrogen must first be bound with oxygen or hydrogen, or *fixed*, into usable chemical forms by specialized organisms, usually bacteria or cyanobacteria. Thus oceanic regions are frequently nitrogen-limited; the growth of plants and plantlike organisms is often held back by a lack of available nitrogen.

Readily usable forms of nitrogen enter the ocean primarily as nitrate in rivers and precipitation or are created by nitrogen-fixing organisms in the ocean. After being assimilated by small plants and plantlike organisms, nitrogen is recycled as animals consume them and then excrete primarily ammonium and urea. These reduced forms of nitrogen are then oxidized back into nitrate, via nitrite, by **nitrifying bacteria**. In the deep ocean, most of the nitrogen is in the form of nitrate. In anoxic sediments and certain low-oxygen regions of the ocean, **denitrifying bacteria** use nitrate in respiration and convert nitrate back to nitrite and nitrogen gas, which is lost to the atmosphere. The other major loss occurs when nitrogen-containing

Atmosphere
(mainly carbon dioxide)

volcanic action

Combustion
of fossil
fuels

Terrestrial
rocks

photosynthesis

aerobic
respiration

combustion
of wood (for clearing
land; or for fuel)

weathering

Land food webs
Producers, consumers,
decomposers, detritivores

deforestation

Soil water
(dissolved carbon)

Peat,
fossil fuels

death, burial, compaction over geologic time

leaching,
runoff

organisms and debris are buried in ocean sediments. The **nitrogen cycle** is depicted in **Figure 13.9**.

Phosphorus and Silicon

All organisms use phosphorus to link the parts of nucleic acids, and it is used in molecules that carry energy within the cell. Calcium phosphate is used in the formation of bones, teeth, and some shells. Several groups of marine organisms, including diatoms and radiolarians, make their skeletons from silica (silicon dioxide).

The phosphorus and silicon cycles in the ocean are relatively simple. Both elements enter the ocean in rivers and precipitation, and both are taken up by organisms near the ocean's surface. Phosphorus is generally released in organic form when organisms decompose, but at the pH of seawater this organic phosphorus is rapidly converted back to phosphate, which is then available to be reused by phytoplankton and bacteria. Silica in shells is released back to the water when organisms die and is then available to be reused as the ionized, dissolved form, silicate.

Both phosphorus and silicon cycle in three loops. The most rapid recycling occurs in the daily feeding, death, and decay of surface organisms. A slower loop occurs as the bodies of organ-

isms fall below the pycnocline and phosphorus escapes downward into deep-ocean circulation. A few hundred years may pass before the phosphorus or silicon is again available in the sunlit surface waters where plants can take it up again. The longest loop, which may take millions of years, begins with the phosphorus or silicon locked into rocks or shells that become marine sediments. The sediment is subducted at converging plate margins, and the phosphorus- and silicon-containing compounds reemerge into the ocean through volcanoes. The phorphorus cycle is shown in **Figure 13.10**.

Iron and Other Trace Metals

Iron is used in minute quantities in the reactions of photosynthesis, in certain enzymes crucial to nitrogen fixation, and in the structure of proteins. Other essential trace metals such as zinc, copper, and manganese are also used by organisms in small quantities, primarily in enzymes. Although in absolute terms organisms require only very small quantities of iron, iron's concentration in seawater relative to the concentration of nutrients such as nitrogen and phosphorus can sometimes be so low that phytoplankton growth is limited by the availability of iron. This may seem strange—after all, iron is one of the most abundant elements in Earth's crust. But iron is nearly

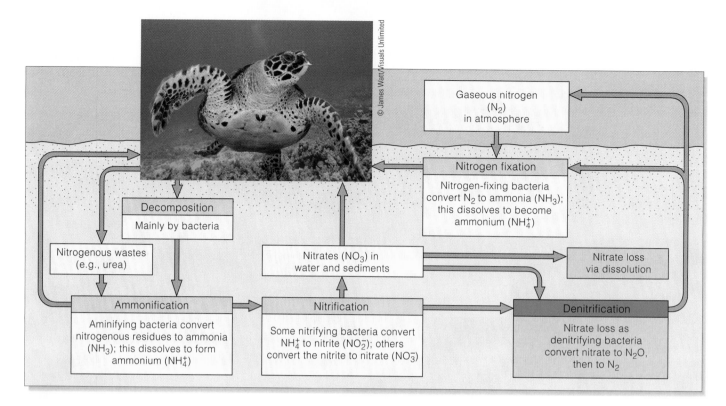

FIGURE 13.9 The nitrogen cycle. The atmosphere's vast reserve of nitrogen cannot be assimilated by living organisms until it is "fixed" by bacteria and cyanobacteria, usually in the form of ammonium and nitrite ions. Nitrogen is an essential element in the construction of proteins, nucleic acids, and a few other critical biochemicals.

FIGURE 13.10 The phosphorus cycle. Phosphorus is an essential part of the energy-transporting compounds used by all Earth's life-forms. Much of the phosphorus-containing material in the ocean falls to the seabed, is covered with sediment, is subducted, and millions of years later returns to the surface through volcanic eruptions.

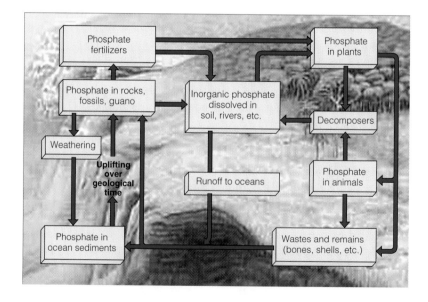

insoluble in oxygenated seawater, and the little dissolved iron that is present is highly reactive, sticks to falling particles, and sinks to the bottom of the water column.

In general, the biogeochemical cycles of the trace metals follow the pattern we have seen earlier: uptake and recycling in the surface ocean, and regeneration—sometimes over long periods of time—at depth. However, we still have much to learn about the interactions between living organisms and trace metals. Iron and other trace metals exist in many chemical forms in seawater. Discovering what these forms are, how transformation between forms occurs, and how available the different forms are to marine organisms is a major focus of current research in trace metal biochemistry.

Physical and Biological Factors That Affect Marine Life

Marine organisms depend on the ocean's chemical composition and physical characteristics for life support. Any aspect of the physical environment that affects living organisms is called a **physical factor.** Living in the ocean often has advantages over living on land: Physical conditions in the sea are usually milder and less variable than physical conditions on land. The most important physical factors for marine organisms are light, temperature, dissolved nutrients, salinity, dissolved gases, acid–base balance, and hydrostatic pressure. These physical factors work in concert to provide the physical environment for oceanic life. But some more **biological factors**—biologically generated aspects of the environment—also affect living organisms. These factors include diffusion, osmosis, active transport, and surface-to-volume ratio. Let's examine all of these factors in more detail.

Light

On land, most photosynthesis proceeds at or just above ground level. But seawater, unlike soil, is relatively transparent, which allows photosynthesis to proceed for some distance below the ocean surface. Incoming sunlight must run a gauntlet of difficulties, however, before it can be absorbed by the chlorophyll in marine autotrophs.

Most sunlight approaching at a low angle (near sunrise or sunset, or in the polar regions) reflects off the water surface and doesn't enter the ocean. Light that does penetrate the surface is selectively absorbed; water is more transparent to some colors of light than to others. In clear water, blue light penetrates to the greatest depth, while red light is absorbed near the surface. Light energy absorbed by water turns to heat.

The depth to which light penetrates is also limited by the number and characteristics of particles in the water. These particles, which may include suspended sediments, dustlike bits of once-living tissue, or the organisms themselves, scatter and absorb light. High concentrations of particles quickly absorb most blue and ultraviolet light. This absorption, combined with the reflection of green light by chlorophyll within the producers, changes the color of productive coastal waters to green.

How far down does light penetrate? **Figure 13.11** shows the depths reached by light of various wavelengths (colors) in the ocean. The **photic zone** (*photos* = light) is the uppermost layer of seawater lit by the sun. Because of the abundant small organisms and light-scattering particles, the photic zone near the coasts usually extends to about 100 meters (330 feet), and in mid-latitude waters it reaches down to about 150 meters (500 feet). In clear tropical waters in the open ocean, instruments much more sensitive than human eyes have detected light at greater depths: The present record is 590 meters (1,935 feet) in the tropical Pacific! The **aphotic zone** (*a* = without)—the permanently dark layer of seawater beneath the photic zone—extends below the sunlit surface to the seabed. The vast bulk of the ocean is never brightened by sunlight.

Photosynthesis proceeds slowly at low light levels. Most of the biological productivity of the ocean occurs in the upper part of the photic zone called the **euphotic zone** (*eu* = good), shown in **Figure 13.12**. This is where marine autotrophs can capture enough sunlight energy for plant primary production by photosynthesis to exceed the loss of carbohydrates through respiration. Though it is difficult to generalize for the ocean as a whole, the euphotic zone typically extends to a depth of approximately 70 meters (230 feet) in mid-latitudes, averaged over the whole year. The upper productive layer of ocean is a very thin skin indeed—the water within this zone amounts to less than 1% of world ocean volume—and yet nearly all marine life depends on this fine illuminated band.

Below the euphotic zone (and still in the photic zone) lies the **disphotic zone** (*dys* = difficult), also seen in Figure 13.12. Though light is present in this zone, it is not bright enough to allow photosynthesis to generate as much carbohydrate as would be used by an autotroph through a day. (Another view of these zones is provided in Figure 13.18.)

Temperature

Ocean temperature varies with depth and latitude. The average temperature of the world ocean is only a few degrees above freezing. Warmer water is found only in the lighted surface zones of the temperate and tropical ocean and in rare, deep, warm chemosynthetic communities. Though ocean temperatures have a wide range (**Figure 13.13**), they are much narrower than comparable ranges on land.

What are the implications of the ocean's temperature for living things? The rate at which chemical reactions occur in an organism is largely dependent on the molecular vibration we call heat. Because agitation brings reactants together, warmer temperatures increase the rate at which chemical reactions occur. Thus an organism's **metabolic rate,** the rate at which energy-releasing reactions proceed within an organism, increases with temperature. The metabolic rate approximately doubles with a 10°C (18°F) temperature rise. The interior temperature of an organism is directly related to the rate at which it moves, reacts, and lives.

The great majority of marine organisms are "cold blooded," or **ectothermic,** having an internal temperature that stays very close to the temperature of their surroundings. A few complex

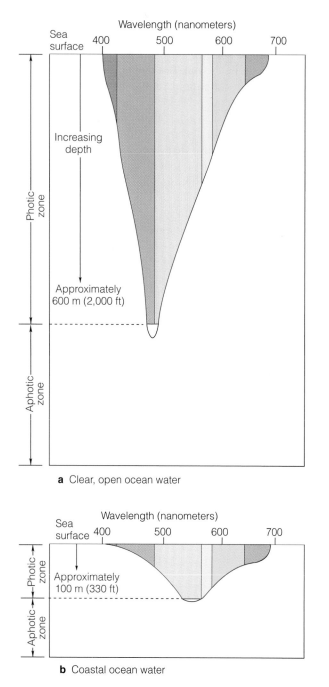

a Clear, open ocean water

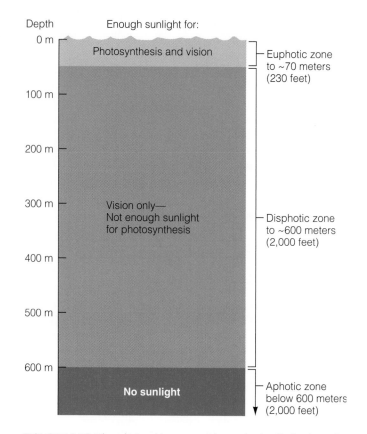

FIGURE 13.12 The relationships among the euphotic, disphotic, and aphotic zones. (The euphotic zone statistic is for mid-latitude waters averaged through a year.)

b Coastal ocean water

FIGURE 13.11 Penetration of light into the ocean. (**a**) In clear open ocean water, sensitive instruments can detect light to a depth of 600 meters (2,000 feet). (**b**) Because of the suspended particles often present in coastal waters, light cannot penetrate so far—about 100 meters (330 feet) is typical. The sunlit upper zone is called the *photic zone*. The dark ocean beneath is called the *aphotic zone*.

animals—mammals and birds and some of the larger, faster fishes—are "warm blooded," or **endothermic**, meaning that they have a stable, high internal temperature.

In general, the warmer the environment of an ectotherm within its tolerance range, the more rapidly its metabolic processes will proceed. Tropical fish in a heated aquarium will therefore eat more food and require more oxygen than goldfish of the same size living in an unheated but otherwise identical aquarium. The tropical fish will generally grow faster, have a faster heartbeat, reproduce more rapidly, swim more swiftly, and live shorter lives. But you can't just crank the heater up another notch for even faster fish—eventually the little fellows will cook. The upper limit of temperature that an ectotherm can tolerate is often not much higher than its optimum temperature. The lower limit is usually more forgiving because molecules are merely slowed.

Do endotherms have narrow temperature requirements? Yes and no. Endotherms can tolerate a tremendous range of *external* temperature compared to ectotherms; think of a whale migrating from polar waters to the tropics, or an Emperor penguin incubating an egg at −51°C (−60°F). Their *internal* temperatures vary only slightly. In our own case, consider the temperatures of places inhabited by humans in contrast to the narrow internal temperature range physicians consider normal. Sophisticated thermal regulation mechanisms make it possible for endotherms to live in a variety of habitats, but they pay a price. Their high metabolic rates make proportionally high demands on food supply and gas transport, but the bene-

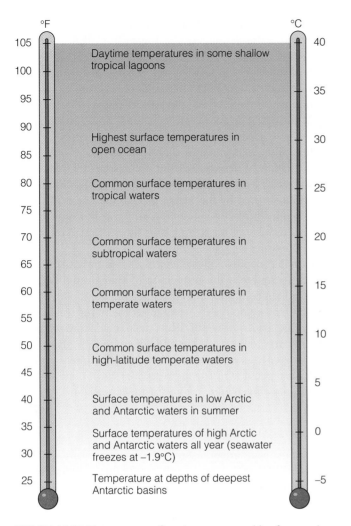

°F   °C

Daytime temperatures in some shallow tropical lagoons

Highest surface temperatures in open ocean

Common surface temperatures in tropical waters

Common surface temperatures in subtropical waters

Common surface temperatures in temperate waters

Common surface temperatures in high-latitude temperate waters

Surface temperatures in low Arctic and Antarctic waters in summer

Surface temperatures of high Arctic and Antarctic waters all year (seawater freezes at −1.9°C)

Temperature at depths of deepest Antarctic basins

FIGURE 13.13 Temperatures of marine waters capable of supporting life. Some isolated areas of the ocean, notably within and beneath hydrothermal vents, may support specialized living organisms at temperatures to 113°C (238°F).

fit of having a biochemistry finely tuned to a single efficient temperature is worth the regulatory difficulties involved.

Dissolved Nutrients

Another of the physical factors that are important to marine organisms is the level of dissolved nutrients. A **nutrient** is a compound required for the production of organic matter. Some nutrients help form the structural parts of organisms, some make up the chemicals that directly manipulate energy, and some have other functions. A few of these necessary nutrients are always present in seawater, but most are not readily available.

The main inorganic nutrients required in primary productivity are nitrogen (such as nitrate, NO_3^-) and phosphorus (as phosphate, PO_4^{3-}). As any gardener knows, plants require fertilizer—mainly nitrates and phosphates—for success. Ocean gardeners would have more trouble raising crops than their terrestrial counterparts, though, because the most fertile

ocean water contains only about $\frac{1}{10,000}$ the available nitrogen of topsoil. Phosphorus is even more scarce in the ocean, but fortunately less of it is required by living things, which have about one atom of phosphorus for every 16 atoms of nitrogen.

Nitrogen and phosphorus are often depleted by autotrophs during times of high productivity and rapid reproduction. Also in short supply during rapid growth are dissolved silicates and calcium compounds (used for shells and other hard parts) and trace elements such as iron, copper, and magnesium (used in enzymes, vitamins, and other large molecules). Marine plants have no choice but to recycle these nutrients.

Salinity

All the cells of every organism are enclosed by membranes, complex films through which a few selected substances can move. A cell's membranes are greatly affected by the salinity of the surrounding water.

The salinity of seawater (see Chapter 7) can vary in places because of rainfall, evaporation, runoff of water and salts from land, and other factors. Surface salinity varies most, with lows of 6‰ or less along the coast of the outer Baltic Sea in early summer to year-round highs exceeding 40‰ in the Red Sea. Salinity is less variable with increasing depth, with the ocean typically becoming slightly saltier with depth.

A change in salinity can physically damage cell membranes, and concentrated salts can alter protein structure. Salinity can affect the specific gravity and density of seawater and therefore the buoyancy of an organism. Salinity is also important because it can cause water to enter or leave a cell through the membrane, changing the cell's overall water balance. Seawater is nearly identical in salinity to the interior of all but the most advanced forms of marine life; this means that maintaining salt balance, and therefore water balance, is easy for most marine species.

Dissolved Gases

Nearly all marine organisms require dissolved gases—in particular, carbon dioxide and oxygen—to stay alive. Oxygen does not dissolve easily in water, and as a result there is about a hundred times more gaseous oxygen in the atmosphere than in the ocean. But CO_2, essential to primary productivity, is much more soluble and reactive in seawater than oxygen is (**Table 13.3**). Although as much as a thousand times more carbon dioxide than oxygen can dissolve in water, normal values at the ocean surface average around 50 milliliters per liter for CO_2 and around 6 milliliters per liter for oxygen. At present the ocean holds about 60 times as much carbon dioxide as the atmosphere does. Because of this abundance marine plants almost never run out of CO_2.

Deep water tends to contain more carbon dioxide than surface water. Why should this be? Table 13.3 also shows the relationship between water temperature and its ability to dissolve gases. Note that colder water contains more gas at saturation. You may recall that the deepest and densest seawater masses

Table 13.3	The Solubility of Gases in Seawater Decreases as Temperature Rises		
	Solubility (ml/l at atmospheric pressure and salinity of 33‰)[a]		
Temperature	**N_2**	**O_2**	**CO_2**
0°C (32°F)	14.47	8.14	8,700.0
10°C (50°F)	11.59	6.42	8,030.0
20°C (68°F)	9.65	5.26	7,350.0
30°C (86°F)	8.26	4.41	6,600.0

[a] Figures are given at *saturation*, the maximum amount of gas held in solution before bubbling begins.
Source: F.G. Walton-Smith, *CRC Handbook of Marine Science* (Cleveland, OH: CRC Press, 1974).

are formed at the surface in the cold polar regions, and, as we have seen, more CO_2 can dissolve in that low-temperature environment. The dense water sinks, taking its large load of CO_2 to the bottom, and the pressure at depth helps to keep it in solution. CO_2 also builds in deep water because only heterotrophs (animals) live and metabolize there and because CO_2 is produced as decomposers consume falling organic matter. No photosynthetic primary producers are present in the dark depths to use this excess CO_2 because there is not enough sunlight for photosynthesis to occur.

Rapid photosynthesis at the surface lowers CO_2 concentrations and increases the quantity of dissolved oxygen. Oxygen is least plentiful just below the limit of photosynthesis because of respiration by many small animals at middle depths. (These relationships were shown in Figure 7.8).

Low oxygen levels can sometimes be a problem at the ocean surface. Plants produce more oxygen than they use, but they produce it only during daylight hours. The continuing respiration of plants at night will sometimes remove much of the oxygen from the surrounding water. In extreme cases this oxygen depletion may lead to the death of the plants and animals in the area, a phenomenon most noticeable in enclosed coastal waters during spring and fall plankton blooms.

The greatest variability in levels of dissolved gas is found at the surface near shore. Less dramatic changes occur in the open sea.

Acid–Base Balance

Another of the physical conditions that affects life in the ocean is the acid–base balance of seawater. The complex chemistry of Earth's life-forms depends on precisely shaped enzymes, large protein molecules that speed up the rate of chemical reactions. When strong acids or bases distort the shapes of these vital proteins, they lose their ability to function normally.

The acidity or alkalinity of a solution is expressed in terms of a *pH scale*, a logarithmic measure of the concentration of hydrogen ions in a solution. Recall (from Figure 7.9) that 7 on the pH scale is neutral, with lower numbers indicating greater acidity and higher numbers indicating greater alkalinity.

Seawater is slightly alkaline; its average pH is about 8. The dissolved substances in seawater act to *buffer* pH changes, preventing broad swings of pH when acids or bases are introduced. The normal pH range of seawater is much less variable than that of soil. Terrestrial organisms are sometimes limited by the presence of harsh alkali soils that damage cell components.

Though seawater remains slightly alkaline, it is subject to some variation. When dissolved in water, some CO_2 becomes carbonic acid. In areas of rapid plant growth, pH will rise because CO_2 is used by the plants for photosynthesis. And because temperatures are generally warmer at the surface, less CO_2 can dissolve in the first place. Thus surface pH in warm productive water is usually around 8.5.

At middle depths and in deep water, more CO_2 may be present. Its source is the respiration of animals and bacteria. With cold temperatures, high pressure, and no photosynthetic plants to remove it, this CO_2 will lower the pH of water, making it more acid with depth. Thus deep, cold seawater below 4,500 meters (15,000 feet) has a pH of around 7.5. This lower pH can dissolve calcium-containing marine sediments. A drop to pH 7 can occur at the deep-ocean floor when bottom bacteria consume oxygen and produce hydrogen sulfide.

Hydrostatic Pressure

Marine organisms are often subject to great pressure from the constant weight of water above them, but this so-called **hydrostatic pressure** presents very little difficulty to them. In fact, the situation in the ocean is parallel to that on land. Land animals live in air pressurized by the weight of the atmosphere above them (1 kilogram per square centimeter, or 14.7 pounds per square inch, at sea level) without experiencing any problems. Indeed atmospheric pressure is necessary for breathing, flight, and some other physical necessities of life.

Pressures inside and outside an organism are virtually the same, both in the ocean and at the bottom of the atmosphere. Thus marine organisms do not need heavy shells to keep from being crushed by hydrostatic pressure. Great pressure does have some chemical effects: At high pressure gases become more soluble, some enzymes are inactivated, and metabolic rates for a given temperature tend to be slightly higher. These effects are felt only at great depth, however. Unless marine organisms have gas-filled spaces in their bodies, a moderate change in pressure has little effect.

Diffusion, Osmosis, and Active Transport

If a cube of dye is left undisturbed in a container of still water, the dye molecules will eventually diffuse evenly throughout the water. This process, known as **diffusion,** might take weeks to complete (**Figure 13.14**). The energy to distribute the dye comes from heat, the random vibration of molecules: The warmer the water, the faster the diffusion. The net transfer of

Day 1   Day 2   Day 5   Day 20

FIGURE 13.14 An example of diffusion. Molecules of dye gradually diffuse through the surrounding water as a dye cube dissolves. Random molecular movement spreads the dye away from the region of high concentration at the cube's surface. At the same time water moves from its own region of high concentration (next to the dye cube) to areas of lower concentration (within the disintegrating dye cube itself). After many days dye will be distributed evenly throughout the water. Stirring would greatly accelerate the mixing process, but this input of mechanical energy is not required if you don't mind waiting for diffusion alone to accomplish the task.

material in diffusion occurs from a region of high concentration to regions of lower concentration.

Diffusion, an important marine process, can be a physical factor. For example, minerals dissolve and their components tend to diffuse randomly throughout a liquid environment. Liquids and gases can also diffuse through water from zones of high concentration to zones of low concentration. But mass transport (the movement of substances in currents, for example) is more important than diffusion in moving dissolved substances over long distances. Diffusion is most important over short distances, particularly the distances within and between living cells. Selected substances can diffuse across cell membranes. They tend to diffuse from areas where they are highly concentrated to areas where they are less concentrated. For example, oxygen resulting from photosynthesis will diffuse through a membrane from inside a plant cell (a region of high oxygen concentration) to the outside (a region of lower oxygen concentration). Because biological membranes allow only certain kinds of small molecules to pass, they are considered *selectively permeable.*

Diffusion of water through a membrane is called **osmosis** (*osmos* = a thrusting). In osmosis water moves between two solutions of different water concentrations through a membrane permeable to water but not to salts. If the water outside a cell membrane contains less dissolved salt than the water inside, it will diffuse from the region of higher water concentration (outside the cell) to the region of lower water concentration (inside the cell), causing the cell to swell. The nature of the membrane prevents salt from moving outside to balance the situation. This is why human swimmers avoid getting fresh water in their nasal membranes: Our body fluids are more saline than fresh water, so cells in our sinuses take up the water and swell painfully.

Most simple marine organisms have nearly the same concentration of dissolved substances in their body fluids as seawater does. They are almost **isotonic** (*isos* = equal, *tonos* = strength) to their fluid environment and so experience little net flow of water through their outer membranes. In fresh water a marine animal would be **hypertonic** (*hyper* = over) to its surroundings; water would move *into* the animal through its cell

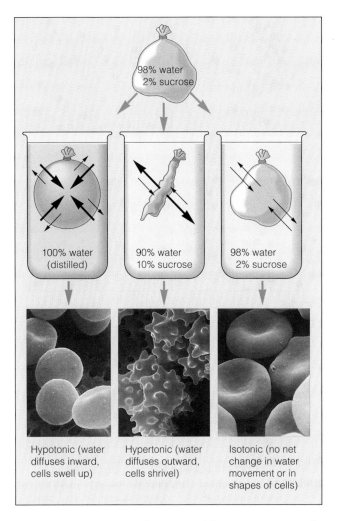

Hypotonic (water diffuses inward, cells swell up)

Hypertonic (water diffuses outward, cells shrivel)

Isotonic (no net change in water movement or in shapes of cells)

FIGURE 13.15 The effects of osmosis in different environments. The selectively permeable bag at the top of the figure represents the concentration of water and dissolved solids typical in the cells of higher forms of life. If the bag is placed into the solutions indicated, it will swell, shrink, or stay the same size. The cells pictured here are human red blood cells immersed in distilled water, concentrated seawater, and water of the same tonicity as human blood.

membranes. If the animal had no way to eliminate the water, its cells would burst. The same kind of marine animal moved to Utah's highly saline Great Salt Lake would be **hypotonic** (*hypo* = under), and water would flow *out of* its cells, causing it to dehydrate and collapse. **Figure 13.15** summarizes these relationships. Because they are nearly isotonic to their surroundings, simple marine organisms have a big advantage over their freshwater counterparts. Freshwater animals must expend large amounts of energy in excretory systems designed to transport water from their tissues back to the outside. These specialized excretory mechanisms are the reason very few freshwater and marine organisms can successfully change places.

Active transport is the reverse of passive diffusion. In active transport dissolved substances are "pumped" through a membrane "uphill" from a region of low concentration to a region

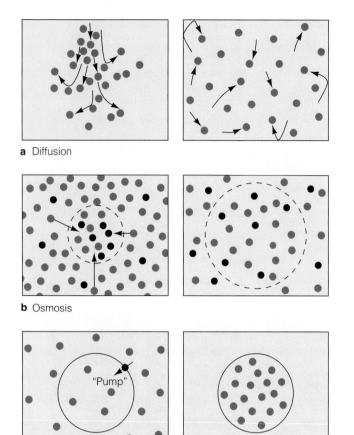

a Diffusion

b Osmosis

c Active transport

FIGURE 13.16 A comparison of the three main ways by which substances move into and out of cells. (**a**) In diffusion molecules introduced into a container (left) become evenly distributed after a period of time (right). (**b**) In osmosis the diffusion of water may cause a cell to swell. The blue dots represent water molecules, the black dots represent dissolved particles, and the arrows indicate the direction of water movement into the original cell. Under different conditions water may move out of the cell, causing it to shrink. (**c**) Active transport enables a cell to accumulate molecules even when there are more inside the cell than outside. Cells may expel other molecules by the same process.

Diameter (cm)	0.5	1.0	1.5
Surface area (cm^2)	0.79	3.14	7.07
Volume (cm^3)	0.06	0.52	1.77
Surface-to-volume ratio	13.17:1	6.04:1	3.99:1

FIGURE 13.17 Surface-to-volume ratio: the relationship between surface area and volume when a sphere is enlarged. Notice that as the diameter increases, the volume is increasing more quickly than the surface area. The surface-to-volume ratio thus decreases as the cell increases in size.

of high concentration. Active transport requires energy because this "uphill" movement defies the normal "downhill" functioning of the second law of thermodynamics. When a cell exports a finished product (the sugars made in photosynthesis, for example) through a membrane to a storage area in which this product is concentrated, it uses active transport. Active transport is a common process in living things, and much of any organism's energy is expended on facilitated movement against the normal concentration gradient.

Diffusion, osmosis, and active transport are all temperature-dependent, proceeding at a more rapid rate as temperature rises. These three processes are compared in **Figure 13.16**.

Surface-to-Volume Ratio

Though individual organisms are often large, the cells they contain are always small. The smaller the cell, the more effi-

ciently materials can cross the outer cell membrane to be distributed through the interior. A small cell has enough surface area to permit the passage of oxygen (or carbon dioxide, or glucose, or wastes) at a rate sufficient to supply the metabolic needs of its internal machinery.

If a round cell grew only by expanding its volume, the surface area of its outer membrane would not increase at an adequate rate. The cell would soon be unable to shuttle materials through the membrane in large enough quantities to keep itself alive. Note in **Figure 13.17** that the volume of a spherical cell increases with the *cube* of its diameter, while its surface area increases with only the *square* of its diameter. If the cell grew to 4 times its original diameter, its volume would increase 64 times but its surface area would increase only 16 times. Each square unit of outer membrane would have to serve four times as much interior volume as before. Past a certain point, the inward flow of nutrients and gases (and the outward flow of wastes) would not be fast enough to supply the cell's metabolic needs and the cell would die. Large cells have an inefficient ratio of surface area to internal volume.

Instead, cells grow by dividing, not by limitless expansion. Diffusion, osmosis, and active transport can be accomplished more efficiently by cells with a high **surface-to-volume ratio**— that is, by small cells with a relatively large surface area and a relatively small volume.

Limiting Factors

Often too much or too little of a single physical factor can adversely affect the function of an organism. We call that factor a **limiting factor,** a physical or biological necessity whose presence in inappropriate amounts limits the normal action of the organism. Imagine, for example, an ocean area in which everything is perfect for photosynthesis—warmth, nutrients, adequate CO_2—everything *except light*. In that circumstance no photosynthesis would occur; light is the limiting factor. If light were present but nitrates were absent, nitrate nutrients would be the limiting factor. Sometimes *too much* of something—heat, for instance—can be limiting.

Classifications of the Marine Environment

Scientists have found it useful to divide the marine environment into **zones,** areas with homogeneous physical features. Divisions can be made on the basis of light, temperature, salinity, depth, latitude, water density, or almost any of the other physical dimensions we have discussed. Some classifications, such as by light and location, are particularly useful, however. These classifications are shown in **Figure 13.18.**

The classification by light level, in which the primary division is between the aphotic and photic zones, has already been described. It is particularly valuable in the study of marine life because light powers photosynthesis and thus primary productivity.

In classification by location, the primary division is between water and ocean bottom. Open water is called the **pelagic zone** (*pelagius* = of the sea) and is divided into two subsections: the **neritic zone** (*neritos* = shallow), near shore over the continental shelf, and the deep-water **oceanic zone,** beyond the continental shelf. The oceanic zone is further divided by depth into zones. The *epipelagic zone* (*epi* = atop) corresponds to the lighted photic zone. In the aphotic depths are layered the *mesopelagic* (*mesos* = in the middle), *bathypelagic* (*bathos* = depth), and *abyssopelagic* (*a* = without, *byssos* = bottom) zones. Abyssopelagic water is the water in the deep trenches.

The divisions of the bottom are labeled **benthic** (*benthos* = bottom) and begin with the intertidal **littoral zone** (*litoral* = of shore), the band of coast alternately covered and uncovered by tidal action. (The *supralittoral zone*, the splash zone *above* the high intertidal, is not technically part of the ocean bottom.) Past the littoral is the **sublittoral zone** (*sub* = below), which is further divided into inner and outer segments: The *inner sublittoral* is ocean bottom near shore, and the *outer sublittoral* is

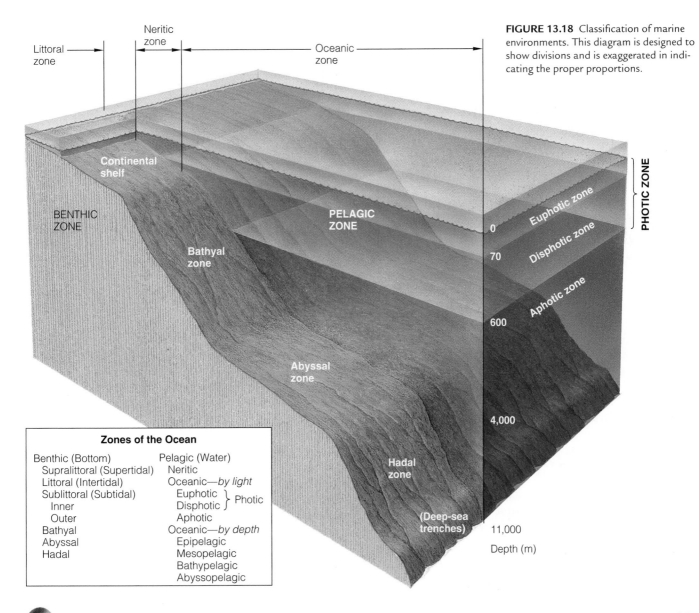

FIGURE 13.18 Classification of marine environments. This diagram is designed to show divisions and is exaggerated in indicating the proper proportions.

Zones of the Ocean

Benthic (Bottom)	Pelagic (Water)
Supralittoral (Supertidal)	Neritic
Littoral (Intertidal)	Oceanic—*by light*
Sublittoral (Subtidal)	Euphotic ⎱ Photic
Inner	Disphotic ⎰
Outer	Aphotic
Bathyal	Oceanic—*by depth*
Abyssal	Epipelagic
Hadal	Mesopelagic
	Bathypelagic
	Abyssopelagic

ocean floor out to the edge of the continental shelf.[3] The **bathyal zone** covers seabed on the slopes and down to great depths, where the **abyssal zone** begins. The **hadal zone** (*Hades* = underworld) is the deepest seabed of all, the trench walls and floors.

Classification by location is the basic, standard system most often used by oceanographers to describe everything from the position of their physical and chemical measurements to the realms where specific organisms are found.

Evolution and Life in the Ocean

Marine life has not accidentally capitalized on its rich fluid home. Earth's organisms did not arise in just a few thousand years. They have changed, generation by generation, over almost 4 billion years. The ability of living things to change through time to fit the physical and chemical environment, to become ever more efficient at extracting energy from their surroundings, to colonize virtually every location capable of sustaining them, and finally to investigate themselves by scientific logic has come about through **evolution.**

Evolution means change. Clothing styles evolve, systems of government evolve, our perceptions of the world evolve; *things change.* That animals and plants might be capable of change with the passage of time was not a popular idea in the nineteenth century. Indeed the proposal that organisms do change with time began what has since been called a revolution in biology. The unlikely revolutionaries were Charles Darwin (**Figure 13.19**), a quiet and thoughtful English naturalist, and Alfred Wallace, an English biologist working in the Malay Archipelago.

---

[3] Many workers use the words *supertidal*, *intertidal*, and *subtidal* instead of the *littoral* terms.

FIGURE 13.19 Charles Darwin, codiscoverer of the principle of natural selection, in about 1880. He is pictured between a Galápagos marine iguana (on the left) and the mainland iguana from which it presumably evolved (on the right). Darwin visited the Galápagos Islands off the Pacific coast of South America during his voyage aboard HMS *Beagle* in the early 1830s; his landmark book *On the Origin of Species* was published in 1859.

The Theory of Evolution by Natural Selection

Why was Darwin and Wallace's theory of evolution controversial then, and why is it capable of stirring passions even today? By the mid-1850s the two men had independently discovered a mechanism—now called **natural selection**—for *how* living things might evolve (change) with the passage of time. Here are Darwin's main points:

1. In any group of organisms, more offspring are produced than can survive to reproductive age.

2. Random variations occur in all organisms. Some of these variations are inheritable; that is, they can be passed on to the offspring.

3. Some inheritable traits increase the probability that the organisms possessing them will survive. These are favorable traits.

4. Because bearers of favorable traits are more likely to survive, they are also more likely to reproduce successfully than bearers of unfavorable traits. Thus favorable traits tend to accumulate in the population; they are *selected.*

5. The physical and biological (*natural*) environment itself does the selection. Favorable traits are retained because they contribute to the organism's success in its environment. These traits show up more often in succeeding generations if the environment stays the same. If the environment changes, other traits become favorable—and the organisms with those traits live most effectively in the environment.

It is easy to see how random variations are selected for or selected against by environmental pressures, but how do entirely new traits arise? They come about by spontaneous **mutation,** an inheritable change in an organism's genes (the structures that contain its assembly instructions). The vast majority of mutations are unfavorable, and the organisms that possess them are eliminated by other organisms or by the physical environment. For example, a tuna born with no eyes could not see to feed, so it would not live to reproductive age. But a tuna born with extraordinarily good eyesight might have more to eat than its cohorts, and being better nourished it might be especially effective in its reproductive efforts, spreading its genes far and wide. It is the accumulation of these favorable traits—either variations or mutations—and the elimination of unfavorable traits that make life possible in changing conditions on Earth (**Box 13.1**).

Note that although mutations occur randomly, evolution by natural selection is anything but random. The natural environment winnows favorable mutations from unfavorable ones— thus the term *natural selection.* The process takes a great deal of time, but time is in abundant supply now that geologists have shown Earth to be about 4.6 billion years old.

The evolutionary viewpoint has provided a new way of looking at life: An organism is a vessel that holds a particular combination of traits, *testing* that combination. If the organism is successful, it will reproduce and the genes responsible for its

BOX 13.1   Mass Extinctions

The environment for life changes as time passes, and the changes are not always gradual. The biological history of Earth has been interrupted—catastrophically—at least six times in the last 450 million years. In these events, known as **mass extinctions,** a great many species died off simultaneously (in geological terms). Scientists are not certain of the causes of the mass extinctions, but leading candidates for a couple of these events include the collision of Earth with an asteroid or comet.

Flocks of asteroids orbit the sun along with Earth and the other planets (**Figure a**). Some of these asteroids have orbits that cross our own, and meetings are inevitable. The Earth and its neighbors are pocked with impact craters as evidence of these meetings (**Figure b**). The consequences for Earth of a collision with even a small asteroid are all but unimaginable. An asteroid only 10 kilometers (6 miles) in diameter would strike with an energy equivalent to the explosion of 500 billion tons of TNT (**Figures c** and **d**). More than 100 million metric tons of Earth's crust and mantle would be thrown into the atmosphere, obscuring the sun for decades and causing sulfur-rich acid rain that would poison the planet's surface. Concussive shock waves would shatter structures, crush large organisms, and trigger earthquakes for a radius of hundreds of kilometers. If the impact occurred in the Atlantic Ocean— say, 1,600 kilometers (1,000 miles) east of Bermuda, the resulting tsunami would wash away the resort islands of the Caribbean and swamp most of Florida. Boston would be struck by a 100-meter (300-foot) wall of water. A hole more than 25 kilometers (16 miles) wide and perhaps 10 kilometers (6 miles) deep would mark the point of impact. Clouds of fine particles, accelerated to escape velocity, would travel around the sun in orbits that would intersect Earth's; a steady rain of fine debris might fall for tens of thousands of years.

These kinds of cataclysmic events have been disquietingly common in Earth's past. Where are the craters? As you may recall from the discussion of plate tectonics in Chapter 3, much of the ocean floor has been recycled by the movement of lithospheric plates, so any undersea impact craters more than about 100 million years old have disappeared. The distortion and erosion of continents have obscured the outlines of ancient craters on land, although they are more readily visible from space than from the surface—especially if we know what to look for (as in Figure b). Evidence for one massive impact has been bolstered by the discovery of a thin, worldwide layer of iridium-rich continental rock dated at the boundary between the Cretaceous and Tertiary periods (see Appendix II). Iridium is rare on Earth but common in asteroids. The thin, iridium-rich layer may have formed from the dust that settled after a collision some 65 million years ago.

As can be seen in **Table a,** vast numbers of marine organisms have perished in mass extinctions. (Extinctions of land families and genera are thought to have been roughly

NASA

(**a**) What one asteroid looks like, courtesy of the *Galileo* spacecraft en-route to study Jupiter. This asteroid would extend from Washington, D.C., halfway to Baltimore.

NASA

(**b**) What an impact crater looks like. Aerial view of Manicougan Crater, Quebec, where an asteroid struck Earth some 210 million years ago. The crater is about the size of Rhode Island. Asteroids are rocky, metallic bodies with diameters ranging from a few meters to 1,000 kilometers (600 miles). Most were swept into the planets during their formation, but about 6,000 large asteroids are still orbiting the sun in a belt between Mars and Jupiter. Unfortunately, the orbits of many dozens of others take them across Earth's orbit.

(c) An artist's conception of a large cometary nucleus, 13 to 16 kilometers (8 to 10 miles) across, striking Earth 65 million years ago. The cataclysmic release of energy is thought to have propelled shock waves and huge clouds of seabed, crust, and even mantle material across Earth, producing a time of cold and dark that contributed to the extinction of many species, including the dinosaurs. The resulting changes in the composition of the atmosphere may also have been toxic to animals.

comparable.) At the end of the Cretaceous period, about 65 million years ago, almost one of every five families, half the genera, and three-quarters of the species disappeared. Many scientists believe an asteroid impact triggered the mass extinction. Clearly these tumultuous interruptions do not represent biological business as usual. In a few instances rocky visitors from space appear to have massively disrupted the environment for life on this planet. The animals, plants, bacteria, and other single-celled organisms we see on Earth are descendants of the survivors.

13.27

(d) A gravity map of the Chicxulub crater off Mexico's Yucatan coast, the largest impact crater found on Earth. The crater is at least 180 kilometers (113 miles) in diameter. This impact has been blamed for the mass extinctions at the end of the Cretaceous period some 65 million years ago.

Table a The Six Great Mass Extinctions			
Geological Period in Which Extinction Occurred	Millions of Years Ago	Percent of Marine Extinctions for:[a]	
		Families	Genera
Late Ordovician	435	27	57
Late Devonian	365	19	50
Late Permian	245	57	83
Late Triassic[b]	220	23	48
Late Cretaceous[b]	65	17	50
Late Eocene	35	2	16

[a]Rough estimate of percentage of all families and genera of marine animals with hard parts (so we have fossil evidence of their existence) rendered extinct; numbers are given to the nearest 5 million years (see column 2).
[b]Mass extinctions for which asteroid or comet strikes may be responsible.
Source: C. Sagan and A. Druyan. Comet (New York: Random House, 1985).

good traits will continue in the population. If the organism is not successful, that combination will be eliminated.

There is no biological predeterminism. Organisms don't *want* to evolve, and individual organisms *don't* evolve. Rather, generation after generation, groups of individuals respond to environmental pressures by change. The changes can be in shape, size, color, biochemistry, behavior, or any other aspect of the organism. Evolution by natural selection is the accumulation of these beneficial inheritable structural or behavioral traits, known as favorable **adaptations.** Organisms with favorable adaptations have more reproductive success than less well adapted organisms.

A **species** is a group of actually (or potentially) interbreeding organisms that is reproductively isolated from all other forms of living things. How do new species arise? One way is by physical isolation, such as that created when land animals or birds are rafted or blown from a mainland shore to an isolated oceanic island. Because the number of breeding animals within a species on an island may be small, evolutionary change may be rapid; that is, favorable traits may accumulate quickly in the population, and generation after generation the species will change relatively rapidly to suit its new habitat. In general the smaller the reproducing population, the more rapid the rate of evolutionary change.

For example, on the Galápagos Islands off the coast of Ecuador, Darwin observed finches and marine lizards, most of which closely resembled their mainland South American ancestors (see again Figure 13.19). They were not the same species of birds and reptiles that occurred in mainland South America, however. This suggested to Darwin that isolation was a driving force of evolution.

Evolution, then, is the maintenance of life under changing conditions by the continuous adaptation of successive generations of a species to its environment. It is a remarkable, beautiful, productive theory.

Evolution in the Marine Environment

Evolutionary theory has important implications for marine science. The ocean has a much larger inhabitable volume than dry land, and as you saw in our discussion of physical factors, it is often an easier place to live than either the terrestrial or freshwater realm. The ocean contains only a fifth of the species known to science, but some scientists suggest that counting species is not a valid way of assessing the success of an environment. All major animal groups (known as *phyla*) are found in the sea, and one-third of them are exclusively marine. If plants and single-celled organisms are included, at least 80% of all phyla include marine species. Perhaps an examination of lifestyles (what an organism *does*) is more important in judging the biological diversity of an environment than what an organism *is*. There are more ways of "making a living" in the ocean than on land. Indeed some important oceanic lifestyles have no terrestrial counterparts. For example, filter feeding—practiced by clams, barnacles, and some whales—relies on a relatively dense fluid matrix and is therefore not seen on land. Also, marine food chains tend to be more complex than terrestrial ones and contain more trophic levels. Marine life has

evolved in countless ways to take advantage of nearly every scrap of energy, nearly every nook and cranny of space.

Since physical conditions in the open ocean are relatively uniform, large marine animals with similar lifestyles but different heritages eventually tend to look much the same. That is, similar conditions may result in coincidentally similar organisms. The shark (a fish), the ichthyosaur (an extinct marine reptile), the penguin (a bird), and the porpoise (a mammal) are only remotely related, yet they resemble each other in shape, because the physics of rapid movement through water requires a similar streamlined shape (**Figure 13.20**). Traits leading to this shape were independently selected by environmental conditions. These accumulated adaptations resulted in superficially similar animals, each derived from different and diverse stock. The process is known as **convergent evolution.**

Through these processes life and Earth change together. Life is tenacious; it has survived catastrophe and calm. In every instance, however, *the environment isn't right for the organisms; rather, the organisms are right for the environment.* One is reminded of author Pär Lagerkvist's observation, "Things need be as they are." Living things have adapted to the physical conditions of their environment, and they continue to increase in numbers, complexity, and efficiency.

Classification of Oceanic Life

Just as oceanographers found it necessary to develop a standard classification and naming system for position in the oceanic realm, biologists centuries earlier had realized the value of being able to classify living things into categories and give them universally understood names.

Systems of Classification

The study of biological classification is called **taxonomy.** Classification schemes have been around for as long as people have looked at living things. Putting animals in one category and plants in another is an ancient distinction, for example.

The Greek philosopher Aristotle proposed a system of classifying animals based on their exterior similarities, but his results were not very useful. With his system we would place airline pilots, gliding squirrels, flying fish, and grasshoppers into the same group because each can fly! Such a system is an **artificial system of classification.** (Another artificial system of classification is the arrangement of books by jacket color, or page size, or typeface.) In contrast, the **natural system of classification** for living organisms that biologists use today relies on organisms' evolutionary history and developmental characteristics. We place all insects together regardless of their flying ability just as we place all books by Melville together, all compositions of Brahms together, and all sea stars together because each group has a *common underlying natural origin.* The groups are arranged systematically—that is, in some order that makes structural and evolutionary sense.

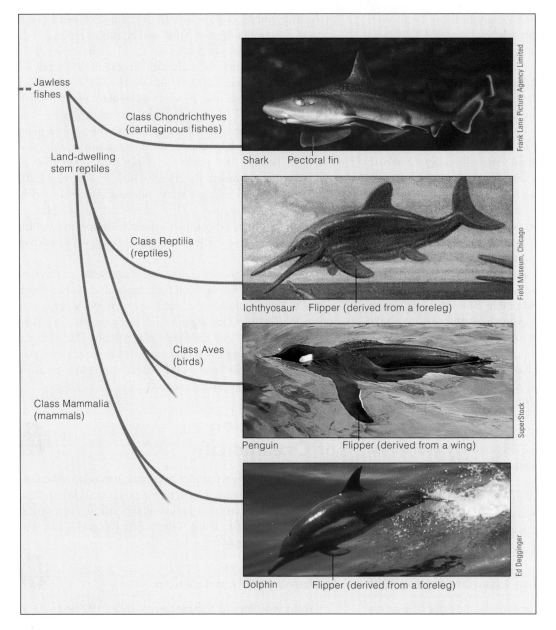

FIGURE 13.20 Convergent evolution in sharks, ichthyosaurs, penguins, and dolphins. Selection for adaptations that permitted rapid swimming resulted in superficially similar shapes among these four kinds of vertebrates, even though they are only remotely related.

One of the first to classify groups of organisms into natural categories was the eighteenth-century Swedish naturalist Carl von Linné or, as he called himself, **Linnaeus** (**Figure 13.21**). In his zeal to classify every aspect of the natural world, Linnaeus invented three supreme categories, or **kingdoms:** animal, vegetable, and mineral. Today's biologists leave the mineral kingdom to the geologists and have expanded Linnaeus's two living kingdoms to six. The names and characteristics of these six kingdoms are listed in **Figure 13.22** and **Table 13.4.**

Linnaeus's great contribution was a system of classification based on **hierarchy,** a grouping of objects by degrees of complexity, grade, or class. In this boxes-within-boxes approach, sets of small categories are nested within larger categories. Linnaeus devised names for the categories, starting with kingdom (the largest category) and passing down through phylum, class, order, family, and genus, to species (the smallest category). In 1758 he published a catalog of all animals then known, his monumental *Systema Naturæ (The System of Nature)*. **Figure 13.23** shows the classification of a popular fish, Rex sole, using the Linnaean method. Note the nested arrangement of category-within-category, each category becoming more specific with every downward step.

FIGURE 13.21 Carolus Linnaeus—the father of modern taxonomy—in Laplander costume. (He went on a scientific expedition to Lapland in 1732.)

Names

Linnaeus also perfected the technique of naming animals. The *genus* and *species* names—the names of the last two nested categories—constitute an organism's **scientific name,** or binomen. *Octopus bimaculatus* is the scientific name of a common west coast octopus: *Octopus* is the generic name, *bimaculatus* the specific name. A closely related species, *Octopus dofleini*, is a larger animal that ranges to Alaska. *Octopus bimaculatus* and *Octopus dofleini* are not interfertile (they're not the same species) but, as their shared generic name suggests, they *are* closely related.

The advantage of a scientific name over a common name is immediately apparent to anyone trying to identify a shell on the

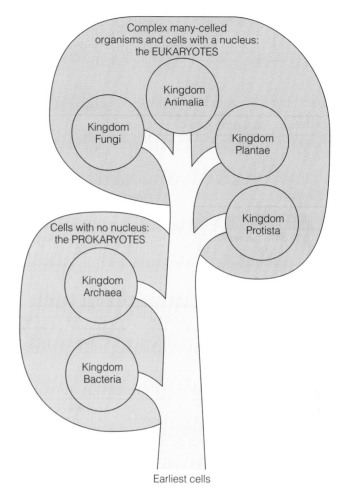

FIGURE 13.22 A family tree showing the relationship of kingdoms presumably evolved from a distant common ancestor. The Bacteria and Archaea contain single-celled organisms without nuclei or organelles; collectively, they are called prokaryotes. The fungi, protists, animals, and plants contain organisms with cells having nuclei and organelles; collectively, they are called eukaryotes.

Table 13.4	Classification of Organisms into Six Kingdoms		
Group	**Kingdom**	**Characteristics**	**Examples**
Prokaryotes: Single-celled organisms lacking a nucleus and other internal structural subdivisions; feed by absorption, photosynthesis, chemosynthesis.	Bacteria	Single chromosome, asexual reproduction, extreme metabolic diversity, no nucleus or cytoskeleton.	Bacteria, cyanobacteria ("blue-green algae").
	Archaea	Superficially similar to bacteria, but with many different genes capable of producing different kinds of enzymes; often live in extreme environments.	*Methanococcus, Pyrolobus,* "extremophiles."
Eukaryotes: Single- or multicelled organisms possessing a nucleus and other internal structural subdivisions; feed by absorption, photosynthesis, or ingestion of particles.	Protista	Usually unicellular, sexual or asexual reproduction, great genetic diversity.	Diatoms and dinoflagellates, radiolarians and foraminifera, single- and multicellular marine algae (seaweeds).
	Fungi	Usually multicellular, sexual or asexual reproduction; release enzymes that break down organic material for absorption.	Molds, mushrooms, symbionts within lichens.
	Plantae	Multicellular photosynthetic autotrophs, sexual or asexual reproduction.	Mosses, ferns, flowering plants.
	Animalia	Multicellular heterotrophs, sexual or asexual reproduction.	Invertebrates, vertebrates.

FIGURE 13.23 The modern system of biological classification, using the Rex sole, a type of flatfish (*Glyptocephalus zachirus*), as an example. Note the boxes-within-boxes approach, a hierarchy. Ellipses (· · ·) indicate groups not shown for clarity.

beach. The same shell may have many different common names in many different languages, but it has *only one scientific name*. When you discover that name in a good key to shells, you can use it to find references that will tell you what is known about the animal, its lifestyle, its range, and its evolutionary history.

Questions from Students

1. **Total terrestrial primary productivity appears to be roughly the same as total marine primary productivity. But the total *biomass* of producers in the ocean is**

at least 300 times smaller (and maybe 1,000 times smaller) than the total *biomass* of producers on land! How can that be? How can terrestrial and marine productivity be so similar?

Because of the astonishing efficiency of small marine autotrophs (phytoplankton), nutrient and carbohydrate molecules are cycled with great speed and efficiency. There may not be nearly as large a biomass of producers in the ocean, but they appear to be *very* busy indeed!

2. What would terrestrial productivity and food chains be like if 99% of the *land* environment were too dark for successful photosynthesis? In other words, what if land plants had to contend with an environment as dark as the ocean?

Productivity would plummet, of course. If the land were lighted as the ocean is, Milne (1995) estimated that all land animals would be dependent on the plant growth in a lighted area the size of the United States east of the Mississippi River. Life on land would be much less abundant than at present, and animals would be concentrated in or around the lighted region. Because land plants are less efficient than aquatic ones (in part because of the infrastructure needed for support, to pump juices around their bodies, and hold leaves to the light and air), total land productivity would be reduced to less than 1% of present values.

3. What proportion of total world productivity is achieved by chemosynthesis?

An interesting and controversial question! Chemosynthesis occurs in the marine environment primarily at hydrothermal vents associated with the 65,000-kilometer (41,000-mile) network of oceanic ridges. Specialized communities of organisms have evolved in the zones of warm or hot mineral-rich water percolating through the seafloor in these places. Once thought to be rare and isolated, vent communities have now been discovered in widely separated locations and may be quite common to oceanic ridges. The trophic pyramid in these deep communities is based on chemosynthetic bacteria. Some biological oceanographers have suggested that if vent communities are abundant on ridges, or if bacteria can grow at warmer temperatures than previously thought, then chemosynthesis there may account for a much larger proportion of total oceanic productivity than was previously thought. And recent discoveries have shown chemosynthetic communities deep *within* the seabed itself!

4. What's the difference between the photic zone and the euphotic zone?

The photic zone is the sunlit uppermost layer of the ocean. The euphotic zone is part of the photic zone. Within the euphotic zone, autotrophic organisms receive enough sunlight to make enough food (by photosynthesis) for their lives to continue. During daylight hours light is available below the euphotic zone, but it is not bright enough to allow the photosynthetic machinery of autotrophs to produce enough food to sustain them indefinitely. Unless they rise into the euphotic zone, they will eventually die.

5. Why is nitrogen often a limiting factor in marine organisms? If air is 78% nitrogen, there should be lots of nitrogen around even if the solubility of nitrogen in the ocean is low.

Yes, but marine plants cannot use dissolved atmospheric nitrogen (N_2) directly. The preferred forms of nitrogen for uptake are ammonium (NH_4^+) and nitrate (NO_3^-), an ion formed by the oxidation of ammonium and nitrite (NO_2^-). Nitrate runoff from soil is an especially rich source of this often limiting nutrient, which is why coastal water tends to support larger plankton populations than oceanic water. There is an important exception, however. Some kinds of bacteria and cyanobacteria can use dissolved atmospheric nitrogen, so they take N_2 directly from seawater (see again Figure 13.9). Their contribution to total oceanic productivity is generally relatively low, however. In certain areas of the world (for example, the subtropical north Pacific Ocean), evidence suggests that nitrogen fixation by very small organisms is a major source of nitrogen for primary productivity.

6. If humans have a fluid much like seawater bathing their cells, why can't we drink seawater and survive?

Human cells function in an environment hypotonic to seawater; that is, blood plasma is less saline than seawater. Drinking seawater therefore causes water to leave the intestinal walls, flood the intestine, and leave the body. There is a net loss via intestines or kidneys even if the seawater is diluted with fresh water before drinking. Moral: Never drink seawater in a survival situation at sea, and never dilute fresh water with seawater (in any proportion) to stretch your supply.

7. Other than listing it as a kingdom, you didn't mention Fungi. Are there any marine fungi?

Though of great terrestrial importance, few fungi exist in the ocean. Most of those that do are confined to the intertidal zone, where they live in close association with marine algae. On land we would call these symbioses "lichens," but botanists are hesitant to categorize the marine equivalent with that word. A few types of fungi have been found in subtidal sediments, where they play the same role as on land: decomposers of organic matter. Fungi are incapable of photosynthesis, and DNA studies have shown that they are more closely related to animals than to plants.

Chapter in Perspective

In this chapter you learned that the atoms in living things are no different from the atoms in nonliving things; in fact, they move between the living and nonliving realms in large biogeochemical cycles. Also, the energy that powers living things is the same energy found in inanimate objects. So how is it possible to distinguish between life and nonlife? The discussion of life in this chapter highlights the highly organized nature of living material and the complex ways that living things manipulate matter and energy.

Life runs on food, and some of the world's most important producers of food are found in the ocean. Understanding the term *primary producers* is central to this chapter: *Primary* indicates that food webs start with those organisms; *producers* emphasizes that the organisms make glucose, an all-important food molecule. Primary producers are organisms that synthesize energy-rich organic compounds (food) from inorganic substances. If someone asks you: "What is produced in primary productivity?" a safe answer is: "the carbohydrate glucose."

Marine life depends on the ocean's chemical composition and physical characteristics for life support. The aspects of the physical environment that affect living organisms are called physical factors; examples are water's transparency, temperature, dissolved nutrients, salinity, dissolved gases, hydrostatic pressure, and acid–base balance. Life and the ocean have evolved together—change in one is met by change in the other. The evolution of life on Earth and the scheme of natural classification we use to organize it are closely related. Our categorizations of living things are based on their presumed ancestry and history.

In the next chapter you will learn about the primary producers themselves and meet the ocean's largest community, the drifters of the plankton. Primary productivity will be revisited; you'll discover how researchers measure productivity and learn about some of the physical and biological factors that limit it. You will see how advanced marine plants fit into the overall productivity picture and begin thinking about animals.

Key Ideas

✓ Life on Earth is notable for both unity and diversity: *diversity* because there are perhaps 100 million different species (kinds) of living things on Earth; *unity* because each species shares the same underlying mechanisms for basic life processes.
   Review pages 305; Figure 13.7, page 311

✓ Living things are not exempt from the second law of thermodynamics, but they can temporarily and locally delay that inevitable descent into disorganization through the transformation of energy.
   Review pages 305–306

✓ Producers assemble food molecules using energy from the sun (photosynthesis) or from energy-rich inorganic molecules (chemosynthesis).
   Review Figure 13.2, page 306

✓ Productivity is expressed in grams of carbon bound into carbohydrates (food) per square meter of ocean surface per year (gC/m²/yr).
   Review Figure 13.3, page 307

✓ Primary producers—autotrophs—are organisms that synthesize food from inorganic substances by photosynthesis and chemosynthesis.
   Review Figure 13.3 and Figure 14.3, pages 307–308, 336

✓ Producers and heterotrophs (consumers) interact in often-complex energy relationships called food webs.
   Review Figure 13.6, page 310

✓ All of Earth's organisms are composed of about 23 of the 107 known chemical elements. Four elements—carbon, hydrogen, oxygen, and nitrogen—make up 99% of the mass of all living things.
   Review page 311

✓ Living organisms are supported and sustained by huge nonliving reserves, and there is a large-scale transport of elements between the reserves and the organisms themselves.
   Review Figures 13.8–13.10, pages 309–315

✓ The great bulk of the ocean lies in perpetual darkness. The upper part of the photic (lighted) zone sustains photosynthetic producers.
   Review Figures 13.11, 13.12, and 13.18, pages 316, 321

✓ Cold water can hold more gas in solution than warm water can.
   Review page 318

✓ Seawater tends to buffer solutions, preventing wide swings in pH (acid–base balance).
   Review Figure 7.9, pages 172, 318

✓ Diffusion, the movement of substances from regions of high concentration to low concentration, and osmosis, the movement of water through membranes, are important processes in moving substances within and between living cells.
   Review Figures 13.14–13.16, pages 319–320

✓ The environments populated by marine life may be classified by their physical characteristics.
   Review Figure 13.18, page 321

✓ Living things adapt to their environment by natural selection—the retention of favorable traits because they contribute to the organism's success in its environment.
   Review Figures 13.19 and 13.20, pages 322–326

✓ Marine organisms are naturally classified by their physical characteristics and by the degree to which they resemble other organisms.
   Review Figures 13.21–13.23, pages 327–328

Terms and Concepts to Remember

abyssal zone, 322
active transport, 319
adaptation, 325

aphotic zone, 315
artificial system
   of classification, 325

Study Questions

Reviewing What You've Learned

1. What is the ultimate source of the energy used by most living things?
2. What do primary producers produce? How is productivity expressed?
3. What is an autotroph? A heterotroph? How are they similar? How are they different?
4. What is a trophic pyramid? What is the relationship of organisms in a trophic pyramid? Does this have anything to do with food webs?
5. Briefly describe the operation of the carbon and nitrogen cycles.
6. Name and briefly discuss five physical factors of the marine environment that affect living organisms. How is each different in the ocean from on the land?
7. How is the marine environment classified? Which scheme is most useful? Justify your answer.
8. Would you support expenditure of governmental funds to search for asteroids or other bodies on a collision course

with Earth? What would the public's response be to discovery of a serious threat?

Thinking Critically

1. The second law of thermodynamics states that entropy (disorganization) tends to increase with time. But living things tend to become *more* complex with time (embryos grow to adults, populations evolve). How can that be?
2. Can you suggest any ways humans might be altering biogeochemical cycles?
3. What is a limiting factor? Can you think of some examples not given in the text?
4. How is evolution by natural selection thought to work?
5. How would you define biological success? Does success depend on the size of an organism? Its beauty? The amount of space it controls? Its numbers?
6. How does a natural system of classification differ from an artificial system? Can you give an example of each? Was the hierarchy-based system invented by Linnaeus natural or artificial? What *is* a hierarchy-based system?
7. **Infotrac College Edition Project** Explain how marine life can exist near undersea hydrothermal vents. What would such organisms lack, and what would they need and not need in order to exist in such an environment? Use Info-Trac College Edition to gather information for your answer.

Thinking Analytically

1. An apple contains about 50 grams of carbon. How many "apple equivalents" would typically be produced each year in a square meter of open-ocean surface area? In a kelp bed? (Hint: See Figure 13.3.)
2. Researchers believe there may be as many as 100 million species of living things on and in Earth. We know of about a million species so far. Where do you think the rest of the species are hiding out?

Oceanography Now™

Assess your understanding of this chapter's topics with additional quizzing and conceptual-based problems at http://oceanography.brookscole.com/garrison5e

Study and Research on the Internet

Assess your understanding of this chapter's topics with additional quizzing and conceptual-based problems at:

http://oceanography.brookscole.com/garrison5e

- **InfoTrac College Edition Research** leads you to critical thinking exercises that use InfoTrac College Edition® as a research tool.
- **Regional InfoTrac College Edition** articles are organized into East Coast, West Coast, and Gulf Coast regions, allowing you to study oceanography on a more local level.

 For more readings, go to InfoTrac College Edition, your online research library, at:

http://infotrac.thomsonlearning.com

A woman gently waves her arms in the shining waters of a sheltered bay on the island of Vieques, Puerto Rico. The disturbance causes the bioluminescent dinoflagellate Pyrodinium to emit a bright blue light.

PLANKTON, ALGAE, AND PLANTS

Glowing Steps

As a child I remember walking along a dark California beach where the sand in my footprints glowed after each step. Handfuls of sand thrown out over the wet beach surface burst into fans of blue light strong enough to cast a shadow, and swimmers emerged from the warm water with thousands of glowing points of light clinging to their bodies. The light I saw was caused by dinoflagellates that had multiplied rapidly in the nearshore waters off southern California. These dinoflagellates were bioluminescent—they glowed when disturbed. In the early 1960s wastewater treatment was not as thorough as it is today, and the effluent was not pumped as far out to sea. The combination of warm, calm water and abundant nutrients often triggered such "blooms." These displays are less common today, but on warm September or October nights after the northeast winds have blown the surface water out to sea and upwelling has brought nutrients to the shore, the sand sometimes glows and waves flash blue as they break.

Entranced by the light from the tiny organisms, I wanted to learn more. I later found that this beautiful display had been made by members of the ocean's largest and most important biological community—the plankton. Marine scientists have been inspired by the beauty and variety of plankton since first observing them under microscopes in the nineteenth century.

14.1

Plankton: Ocean Wanderers

The organisms identified as **plankton** are as important as they are inconspicuous. The word is derived from the Greek word *planktos*, meaning "wandering." Plankton drift or swim weakly, going where the ocean goes, unable to move consistently against waves or current flow. (In contrast, the word **nekton** describes organisms that actively swim.)

The diversity of planktonic organisms is astonishing. There are giant, drifting jellyfish with tentacles 8 meters (25 feet) long; small but voracious arrowworms; many single-celled creatures that glow brightly when disturbed; mollusks with slowly beating flaps that resemble butterfly wings; crustaceans that look like microscopic shrimp; miniature, jet-propelled animals that live in jellylike houses and filter food from water; and shimmering, crystal-shelled algae. The only feature common to all plankton is their inability to move consistently laterally through the ocean. However, many can and do move vertically in the water column.

Plankton include many different photosynthetic and chemosynthetic species and virtually every major group of animals. The terms *plankton* and *nekton* are not collective natural categories like mollusks or algae, which imply an ancestral relationship among the organisms; instead they describe a common ecological connection—a lifestyle. Members of the plankton community, also referred to as **plankters,** can and do interact with one another. Grazing, predation, parasitism, and competition occur among members of this dynamic group. **Figure 14.1** suggests the rich diversity of plankton and shows them in the context of the larger open-ocean community.

Collecting and Studying Plankton

The first large-scale, systematic study of plankton was carried out by biologists aboard the research vessel *Meteor* during the German Atlantic Oceanographic Expedition of 1925–26. Many of the tools and techniques they pioneered are still in use today. **Plankton nets** of the type perfected for the *Meteor* expedition are essential to plankton studies (**Figure 14.2**). These conical nets are customarily made of nylon or Dacron threads woven in a fine interlocking pattern to assure consistent openings. The net is hauled slowly for a known distance behind a ship, or it is cast to a set depth, and then reeled in. Trapped organisms are flushed to the pointed end of the net and carefully removed for analysis. Quantitative analysis of plankton requires both a count of the organisms and an estimate of the sampled volume of water.

Because very small plankton can slip through a plankton net, their capture and study require either concentration of water samples by centrifugation or entrapment by a filter. Filtration of water samples is currently the most common method used by biological oceanographers to collect plankton and bacteria for study and experimentation. Filters may be made of glass fibers, polycarbonate membranes, or even matrices of aluminum or silver. Some recent research suggests that *very*

small producers, some of which are archaeans, may be responsible for at least as much primary productivity as their larger and better-known relatives!

Measurements of physical ocean conditions, such as dissolved carbon dioxide and oxygen content, pH, temperature, and light intensity at the time and place of sampling, are of special importance in interpreting the samples. Simultaneous sampling at many locations can be useful in pinpointing the often subtle interplay between species and physical conditions that complicates our understanding of plankton biology.

Phytoplankton: The Autotrophs

Autotrophic plankton that generate glucose by photosynthesis—the primary producers—are generally called **phytoplankton,** a term derived from the Greek word *phyton*, meaning "plant." A huge, nearly invisible mass of phytoplankton drifts within the euphotic zone, the sunlit surface layer of the world ocean (see Figure 13.18). This upper productive layer of ocean is a very thin skin indeed. The water within the euphotic zone amounts to less than 2% of world ocean volume, but nearly all marine life depends on this fine illuminated band.

Phytoplankton are critical to marine life—and to all life on Earth—because of their great contribution to food webs and their generation of large amounts of atmospheric oxygen through photosynthesis. Planktonic autotrophs are thought to bind *at least* 35 billion metric tons of carbon into carbohydrates each year, at least 40% of the food made by photosynthesis on Earth! These easily overlooked, mostly single-celled, drifting photosynthesizers are much more important to marine productivity than are the larger and more conspicuous seaweeds.

Types of Phytoplankton

There are at least eight major types of phytoplankton, of which the most prominent are the diatoms and dinoflagellates. In the past decade, however, the small coccolithophores and silicoflagellates, and the extremely small autotrophic picoplankton, have been found to be more important in world productivity than most researchers could have imagined.

Diatoms The dominant and most productive photosynthetic organisms in the plankton—and in the world—are the **diatoms.** Diatoms evolved comparatively recently and began to dominate phytoplanktonic productivity in the Cretaceous period about 100 million years ago. Their abundance and photosynthetic efficiency increased the proportion of free oxygen in Earth's atmosphere. More than 5,600 species of diatoms are known to exist. The larger species are barely visible to the unaided eye. Most are round, but some are elongated or branched or triangular.

Typical diatoms are shown in **Figures 14.3**. *Diatom* means "to cut through" (*dia* = through, *tomos* = to cut), a reference to the patterns of perforations through the diatom's rigid cell wall, or **frustule.** As much as 95% of the mass of the frustule

a

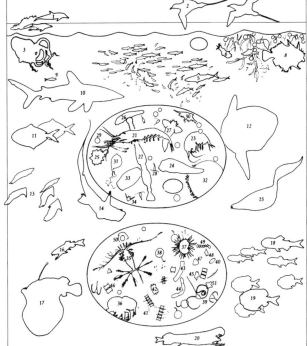

1 dolphins, *Delphinus*	26 mullet larva, *Mullus*
2 tropic birds, *Phaëthon*	27 sea butterfuly, *Clione*
3 paper nautilus, *Argonauta*	28 copepods, *Calanus*
4 anchovies, *Engraulis*	29 assorted fish eggs
5 mackerel, *Pneumatophorus*, and sardines, *Sardinops*	30 stomatopod larva
	31 hydromedusa, *Hybocodon*
6 squid, *Onykia*	32 hydromedusa, *Bougainvilli*
7 *Sargassum*	33 salp (pelagic tunicate), *Doliolum*
8 sargassum fish (*Histro*)	34 brittle star larva
9 pilot fish (*Naucrates*)	35 copepod, *Calocalaus*
10 white-tipped shark *Carcharhinus*	36 cladoceran, *Podon*
11 pompano, *Palometa*	37 foraminifer, *Hastigerina*
12 ocean sunfish, *Mola mola*	38 luminescent dinoflagellates, *Noctiluca*
13 squids, *Loligo*	39 dinoflagellates, *Ceratium*
14 rabbitfish, *Chimaera*	40 diatom, *Coscinodiscus*
15 eel larva, *Leptocephalus*	41 diatoms, *Chaetoceras*
16 deep sea fish	42 diatoms, *Cerautulus*
17 deep sea angler, *Melanocetus*	43 diatom, *Fragilaria*
18 lantern fish, *Diaphus*	44 diatom, *Melosira*
19 hatchetfish, *Polyipnus*	45 dinoflagellate, *Dinophysis*
20 "widemouth," *Malacosteus*	46 diatoms, *Biddulphia regia*
21 euphausid shrimp, *Nematoscelis*	47 diatoms, *B. arctica*
	48 dinoflagellate, *Gonyaulax*
22 arrowworm, *Sagitta*	49 diatom, *Thalassiosira*
23 amphipod, *Hyperoche*	50 diatom, *Eucampia*
24 sole larva, *Solea*	51 diatom, *B. vesiculosa*
25 sunfish larva, *Mola mola*	

b

FIGURE 14.1 (**a**) Representative plankton and nekton of the pelagic zone in the subtropical Atlantic Ocean. Note the relative magnification of organisms in the plankton community. (**b**) Key. This stylized representation shows organisms much more crowded than they are in real life.

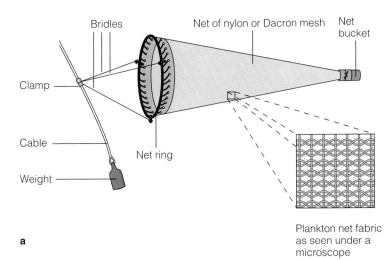

Bridles
Net of nylon or Dacron mesh
Net bucket
Clamp
Cable
Weight
Net ring
Plankton net fabric as seen under a microscope

a

FIGURE 14.2 Plankton nets. (**a**) The standard conical net is made of fine mesh and has a mouth up to 1 meter (3.3 feet) in diameter. The net is towed behind a ship for a set distance. The number of organisms present in the water can be estimated if the trapped organisms are counted and the volume of sampled water is known. (**b**) The net shown here has a somewhat coarser mesh because its target organisms, small shrimplike crustaceans known as *krill*, are relatively large.

Dennis Kelly, Orange Coast College

b

a

b

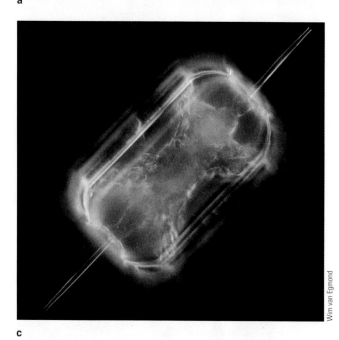

c

FIGURE 14.3 Diatoms. (**a**) The transparent frustule of the diatom *Coscinodiscus* as shown with a light micrograph. The many small perforations that give diatoms their name are clearly visible. Note also the many green chloroplasts—cell organelles responsible for photosynthesis. (**b**) A closer view using an electron microscope shows the perforations to be groups of still smaller holes, each small enough to exclude bacteria and some marine viruses. The holes allow the diatom to pass gases, nutrients, and waste products through the otherwise impermeable silica covering. (**c**) *Ditylum,* a diatom, photographed in visible light. Note the junction between the valves. The cells in this figure are about the size of the period at the end of this sentence.

consists of silica (SiO_2), giving this heavy but beautiful covering the optical, physical, and chemical characteristics of glass—an ideal protective window for a photosynthesizer. Magnification reveals that the frustule consists of two closely fitting halves, or **valves,** that fit together like a well-made gift box, with the top valve adhering tightly over the lip of the bottom one. The pattern of perforations, slits, striations, dots, and lines on the surface of the valves is different for each diatom species.

Inside the diatom's tailored valves lies an extraordinary photosynthetic machine. Fully 55% of the energy of sunlight absorbed by a diatom can be converted into the energy of carbohydrate chemical bonds, one of the most efficient energy conversion rates known. Excess oxygen not needed in the cell's respiration is released through the perforations in the frustule

into the water. Some oxygen is absorbed by marine animals, some is incorporated into bottom sediments, and some diffuses into the atmosphere. Most of the oxygen we breathe has moved recently through the glistening pores of diatoms.

For more effective light absorption, chlorophyll, the main photosynthetic pigment, is accompanied in diatoms by accessory pigments called **xanthophylls.** These yellow or brown pigments give most diatoms a yellow-green or tan appearance. Diatoms store energy as fatty acids and oils, compounds that are lighter than their equivalent volume of water and assist in flotation. As you might guess, flotation is a potential problem for diatoms because the weight of their heavy silica frustule seems at odds with their need to stay near the sunlit ocean surface. Oil floats, glass sinks, and a balanced amount of both reduces cell density and

a

Shih and Kessel

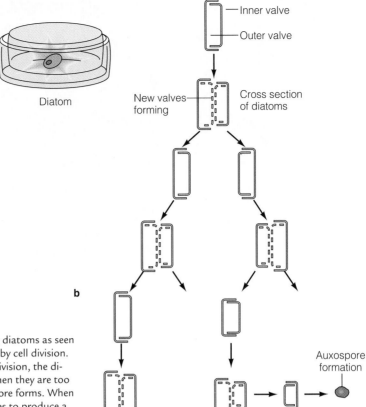

Diatom

Inner valve
Outer valve

New valves forming
Cross section of diatoms

Auxospore formation

b

FIGURE 14.4 How diatoms divide. (**a**) Dividing diatoms as seen with a light microscope. (**b**) Diatoms reproduce by cell division. Because a new inner valve is formed after each division, the diatoms become smaller with each generation. When they are too small to permit further reproduction, an auxospore forms. When conditions are suitable, the auxospore germinates to produce a full-sized diatom.

lightens the load. Not all diatoms need to float, however. Many nonplanktonic species lie on shallow bottoms, where light and nutrients are able to support photosynthesis. These benthic species are nearly always elongated (or *pennate*) in shape.

Like most single-celled organisms, diatoms reproduce by dividing in half and drifting apart (or, in the species of diatoms that form chains, remaining linked in long lines of cells). The cells may divide as rapidly as once each day. In most species the new valve is generated *within* the old one during division, so the average size of individuals in the population becomes smaller with time (**Figure 14.4**). Individuals reach a minimum of about one-fourth the size of the original cell. When the cells become too small—or, more accurately, when the cells have too high a ratio of glass to living tissue—they become too heavy to float in spite of their buoyant oils. The problem is solved by sexual reproduction, which generates an **auxospore,** a naked cell without valves. If conditions are still favorable for growth, the auxospore will expand to the diatom's original size, form two thin new valves, and begin the cycle anew. If growing conditions are unsuitable, the auxospore will become dormant and await an opportunity for growth—which may be weeks or months away. When diatoms die, their valves fall to the seafloor to accumulate as layers of siliceous ooze (see Chapter 5).

Dinoflagellates Most **dinoflagellates** are single-celled autotrophs (**Figure 14.5**). A few species live within the tissues of other organisms (the zooxanthellae of coral animals you will meet in Chapter 15, for example), but the great majority of di-

noflagellates live free in the water. Most have two whiplike projections, called **flagella,** in channels grooved in their protective outer cell wall of cellulose. One flagellum drives the organism forward, while the other causes it to rotate in the water (*dino* = whirling, *flagellum* = whip). Their flagella allow dinoflagellates to adjust their orientation and vertical position to make the best photosynthetic use of available light, or to move vertically in the water column to obtain nutrients.

Dinoflagellates are widely distributed, solitary organisms that reproduce by simple fission; they rarely form colonies. During reproduction, the cellulose covering that surrounds most species splits and the single cell divides in half. Each daughter subsequently replaces the missing portion of covering. Under favorable conditions the organisms can reproduce once a day, growing in number but not in size.

As you learned in this chapter's opener, a few common species of dinoflagellates are strongly bioluminescent. **Bioluminescence** is the process by which energy from a chemical reaction is transformed into light energy. The imaginatively named compound *luciferin* is oxidized by action of the enzyme *luciferase,* and in the process a blue-green light is emitted. The release of light is very efficient and thus is not accompanied by a release of heat, so the organism does not overheat in the oxidation process. Bioluminescence may have evolved as a sort of "intrusion alarm" (see Questions from Students #4).

Some species of dinoflagellates can become so numerous that the water turns a rusty red because light is reflecting from the accessory pigments within each cell. These species are

a — Tom Garrison

b — © Bill Curtsinger Photography

FIGURE 14.5 Dinoflagellates. (**a**) Two representatives of the genus *Ceratium,* a common dinoflagellate. Two flagella beat in opposing grooves in the armor-plated body: One groove is visible at the equator of the upper cell, between the two "horns." These specimens are about 0.5 millimeter (0.02 inch) across. (**b**) During a red tide, the presence of millions of dinoflagellates turns seawater brownish-red.

usually responsible for the phenomenon referred to as a "red tide" or (more generally) as a *harmful algal bloom* (HAB) (**Box 14.1**). During times of such rapid growth (usually in the spring), the concentration of microscopic planktonic organisms may briefly reach 6 to 8 million per liter (23 to 29 million per gallon).

Coccolithophores and Silicoflagellates Coccolithophores are small, single-celled autotrophs covered with disks of calcium carbonate (coccoliths) fixed to the outside of their cell walls (**Figure 14.6**). Coccolithophores live near the

FIGURE 14.6 Coccolithophores. (**a**) A rare light microscope photograph of *Emiliania huxleyi,* a coccolithophore. The tiny calcified plates (coccoliths) covering the cells are 6 micrometers (120 millionths of an inch) across. Photosynthetic pigments give the cells a golden or golden-brown color. The coccoliths (not the organic cells themselves) act like mirrors suspended in the water and can reflect a significant amount of the incoming sunlight. The reflectance from the blooms can be picked up by satellites in space, allowing the extent of the blooms of this species to be distinguished in fine detail. Figure 5.13c is a scanning electron microscope image of the same species that shows the individual coccoliths in more detail. (**b**) A coccolithophore bloom is clearly visible south of Ireland in this natural-color satellite image.

a — Ian Probert

b — SeaWIFS Project, NASA/Goddard Space Flight Center

BOX 14.1

HABs Can Be Trouble!

A harmful algal bloom (HAB) occurs when high concentrations of phytoplankton adversely affect the physiology of nearby organisms. The somewhat misleading name *red tide* is often used to describe patches of water turned a rusty red by the abundant growth of pigmented phytoplankton, usually certain dinoflagellate species (**Figure a**). But HABs don't always turn water red, nor are the organisms that cause them always visible. Factors thought to contribute to HABs include warm surface temperatures, reduced salinity, optimal nutrient and light conditions, and a mechanism that physically concentrates the dinoflagellates (such as gentle onshore winds).

Although the dinoflagellates responsible for most red tides are comparatively simple organisms, some have the ability to synthesize potent toxins as by-products of metabolism. Among the most effective poisons known, these toxins may affect nearby marine life if ingested (**Figure b**), or they may even indirectly poison humans through the food chain. Some of the toxins are similar in chemical structure to the muscle relaxant curare, but they are tens of times more powerful.

One of these compounds, a neurotoxin (nervous system poison), affects between 10,000 and 50,000 individuals annually—most of them on tropical and subtropical islands where herbivorous fishes make up much of the diet. The fishes graze on seaweeds, inadvertently ingesting dinoflagellates attached to the surface of the fronds. The toxin is stored in fat and so is most concentrated in the oldest and fattest (and thus most desirable) fish. Victims complain of abdominal pain, muscular aches, dizziness, anxiety, tingling in the hands and feet, and a curious inversion of the senses (hot seems cold, and vice versa); many die from the effects.

Clams, mussels, scallops, and oysters—invertebrates that filter seawater to obtain food—can also be dangerous to eat, and shellfish poisonings are increasingly common worldwide. Paralytic, diarrheic, neurotoxic, and amnesic shellfish poisonings have been reported by humans, but the filter-feeding invertebrate is insensitive or only marginally affected. A lethal dose of paralytic toxin kills by interfering with the normal function of heart muscle and affecting breathing. Neurotoxic compounds block communication between nerves and muscles; the first symptoms of intoxication are usually tingling lips and difficulty in swallowing. Diarrhetic shellfish poisoning causes diarrhea, nausea, and vomiting. Most mysterious is amnesic shellfish poisoning, a poorly understood disorder that results in the permanent loss of short-term memory. In 1987 three elderly people in eastern Canada died and 107 were affected by amnesic poisoning. They ingested blue mussels (*Mytilus edulis*) contaminated with domoic acid, a potent neurotoxin produced by the pennate diatom *Pseudonitzschia australis*. A 1998 *Pseudonitzschia* outbreak near California's Monterey Bay was responsible for the deaths of more than 400 sea lions and countless seabirds.

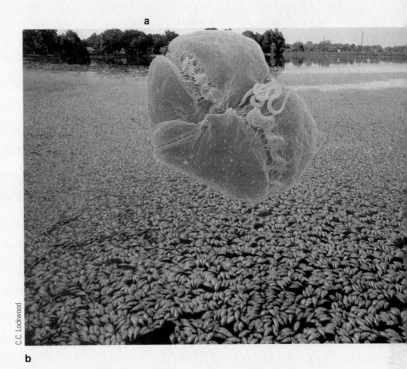

a

C.C. Lockwood

b

Red tides and toxic dinoflagellates. (**a**) A scanning electron micrograph of *Gymnodinium breve,* the dinoflagellate responsible for red tides along the Florida coast. Other dinoflagellate species produce red tides in other parts of the world. (**b**) A portion of a fish kill that resulted from a dinoflagellate bloom. (**c**) During a red tide, millions of dinoflagellates turn seawater brownish-red.

c

Suisan Aviation Ltd., Japan

In areas where fisheries are not regulated by governmental agencies, people should avoid eating filter feeders during summer months, when toxin-producing phytoplankton are most abundant. In the United States and many other countries, a governmental agency issues an advisory when shellfish from a particular area are unsafe. The advisory may remain in effect for six weeks or longer until the danger is past. Such warnings must be observed—a single clam can accumulate enough toxin to kill a human. There is no antidote; physicians can only treat the symptoms. Whales and seabirds are also at risk from dinoflagellate toxins.

People can be affected even if they don't eat seafood. Beachgoers in North Carolina complained of inflamed eyes and asthmalike symptoms when they inhaled dinoflagellates that had dried on the beaches and been blown inland by strong winds (**Figure c**).

The number and severity of HABs appear to be increasing. Perhaps this is not surprising: Coastal waters receive industrial, agricultural, and domestic wastes, rich in nitrogen and other plant nutrients that stimulate algal growth. Also, the long-distance transport of algal species in the ballast water of cargo vessels can introduce alien species into coastal waters, where they may thrive in the absence of the organisms that naturally consume them. Australia has recently issued strict guidelines for discharging ballast in the country's ports. In any case, seafood consumers should be aware of the source of their meal and the nature of poisoning symptoms.

ocean surface in brightly lit areas. Their translucent covering of coccoliths may act as a sunshade to prevent absorption of too much light. In areas of high coccolithophore productivity, most notably in the Mediterranean and Sargasso Seas, their numbers occasionally become so great that the water appears milky or chalky. Coccoliths can also build seabed deposits of ooze. The famous White Cliffs of Dover in England consist largely of fossil coccolith deposits uplifted by geological forces.

The **silicoflagellates** possess filamentous internal supporting structures of silica, the same material from which diatoms construct their valves. One or two flagella are always present. Compared to diatoms, however, the overall structure and biochemistry of silicoflagellates seem primitive. Little is known about their worldwide distribution or abundance.

Picoplankton Most other types of phytoplankton are very small and so are called **picoplankton** (*pico* = a trillionth). Recent evidence suggests that extraordinarily small picoplanktonic photosynthesizers may be much more numerous and contribute *much* more to overall plankton productivity than was previously thought, especially in open-ocean areas. **Box 14.2** explores this idea, which is of great interest to biological oceanographers.

Measuring Primary Productivity

We've noted that the phytoplankton are believed to contribute at least 40% of the food made by photosynthesis, but the actual numbers are not easy to measure. At first glance the problem might seem simple to solve: Collect all the organisms—large and small, autotrophs and heterotrophs—in an area and analyze their organic content and weight. Because all organisms are dependent on primary productivity, the mass of living tissue (biomass) in the area should be directly proportional to productivity.

This approach has drawbacks, though. A dense population of tiny, drifting autotrophs (a high biomass) would interfere with light penetration, so the autotrophs would manufacture carbohydrates slowly and productivity would be low. In contrast, a sparse population of drifting autotrophs (a low biomass) might encounter ideal conditions for photosynthesis and produce carbohydrates at a rapid rate. Small animals might immediately consume this production and keep the biomass at low levels, but productivity would be high.

What researchers need is a method of measuring the rate of productivity *directly*. Since we know the formula for photosynthesis (see again page 306), measuring any one component of photosynthesis will tell us about the others. For example, from a measurement of the carbon taken up by primary producers we can calculate the rate of carbohydrate production.

Researchers can do this calculation using atoms of carbon "tagged" by radioactivity. Though it is radioactive, carbon-14 (^{14}C) behaves chemically the same way as the much more common carbon-12 (^{12}C). And because it is radioactive, ^{14}C's progress through photosynthesis can be monitored. One of the ions formed when carbon dioxide dissolves in seawater is bicarbonate (HCO_3^-).[1] Scientists tag the carbon in biocarbonate and add known amounts of radioactive bicarbonate to two bottles of seawater. One bottle is exposed to light, and photosynthesis and respiration take place. The other bottle is shielded from light, so only respiration occurs. The amount of radioactive carbon incorporated into carbohydrates is measured when the organisms are filtered out of the samples. The radioactivity is measured, and productivity is calculated from the formula

$$\text{Rate of production} = \frac{(R_L - R_D) \times M}{R \times t}$$

where R is the total radioactivity added to the sample, t is the number of hours of incubation, R_L is the radioactive count in the "light" bottle sample, R_D is the count in the "dark" sample, and M is the total mass of all forms of carbon dioxide in the sample (in milligrams of carbon per cubic meter). Productivity

---

[1] This is shown in Figure 7.10.

BOX 14.2

Picoplankton!

In the mid-1970s biological oceanographers began to appreciate the contribution to oceanic productivity of extremely small phytoplankton called **picoplankton.** Picoplankton are so tiny that they were long mistaken for fine particles of suspended sediments. These organisms are often too small to be resolved by light microscopes, and they slip undetected through all but the finest filters. Their size, typically about 0.2 to 2 micrometers (4 to 40 millionths of an inch) across, is made up for by their abundance: an astonishing 100,000 cells per milliliter (3 million cells per ounce) in tropical and subtropical oceanic waters once thought to be all but devoid of phytoplankton.

The bacterium *Synechococcus* is typical of this newly recognized type of organism. It was discovered when individual cells— little more than naked photosynthetic machines—fluoresced bright orange when struck by ultraviolet light. Analysis of the fluorescence spectrum (living examples of *Synechococcus* are too small to study directly) revealed the presence of an odd chlorophyll variant previously known only in a rare mutant strain of corn. Structural details revealed by electron micrographs suggest that *Synechococcus* and its relatives are cyanobacteria.

Recent estimates suggest that picoplankton may account for up to 70% of all the photosynthetic activity in some parts of the open ocean! How could such a huge contribution to oceanic productivity have gone unnoticed for so long? In part because these autotrophs are so extremely small, and in part because they are efficiently grazed by microflagellates (protozoans). Additionally, the products of their photosynthetic activity are promptly used by even smaller heterotrophic bacteria in the immediate vicinity. Here is a complete microecosystem—a community operating on the smallest possible scale— that manufactures and consumes particulate and dissolved carbon in amounts almost beyond comprehension, a sort of ecological black market below the "official economy" of the relatively huge diatoms and dinoflagellates. As if they weren't busy enough, these heterotrophic bacteria also decompose organic material spilled into the water when phytoplankton are eaten by zooplankton, turn soluble organic materials released by zooplankton back into inorganic nutrients, and break down particulate organic matter into a dissolved form they can consume for their own growth. Biological oceanographers now believe that the greatest fraction of organic parti-

cles in the water column of the open ocean is composed of these metabolically active heterotrophic bacterial cells.

And what happens to the bacteria? Those that aren't consumed by microflagellates may become infected by viruses, which cause the cells to burst. Viruses that infect bacteria are referred to as *bacteriophages*; those that infect phytoplankton are called *phycoviruses*. Viruses are extremely small (usually 20 to 250 nanometers in diameter) and are fundamentally different from other forms of life. Viruses have no metabolism of their own and must rely on a host organism for energy-requiring processes, including reproduction. Although we have known about viruses for some time, only in the late 1980s were their abundances confirmed in a wide range of marine environments. How many are there? Between 10 and 100 million per milliliter of water—up to about 3 billion per ounce!

14.8

William P. Cochlan

Three bacteria—ellipsoidal, S-shaped, and C-shaped—from the Gulf of Bothnia (east of Sweden). The dark dots within the bacteria are viruses. The large hexagonal structure near the S-shaped bacterium is a relatively large virus with a diameter of 160 nanometers. The bacteria and viruses were photographed using a transmission electron microscope.

is expressed as the amount of carbon (in milligrams) bound— or *fixed*—in new carbohydrate per volume of water (in cubic meters) per unit time (per hour). The rate of production varies from zero to as much as about 80 milligrams of carbon fixed into carbohydrates per cubic meter per hour. These data can

be extrapolated to provide the amount of carbon fixed in the water column per square meter of surface per day ($gC/m^2/day$).

Another way to calculate the rate of carbohydrate production is the light-dark bottle technique. Researchers collect identical water samples from known depths, place the samples into pairs of identical transparent and opaque bottles, and then suspend an array of bottles from a buoy (**Figure 14.7**). The transparent bottles admit light; the opaque bottles block it. The difference in the uptake of carbon in each pair of bottles over time gives an indication of the gross rate of productivity at each depth. Light-dark bottle experiments with radioactive tracers may be conducted for various species and concentrations of producers at different light, nutrient, and temperature levels. Extrapolation from these experiments to other oceanic areas can then yield data on overall productivity.

Ocean conditions are complex and variable, however—rarely controlled. A new and promising method of gauging productivity may be the most effective and useful of all. Recent advances in remote sensing have made it possible to estimate the chlorophyll content of ocean water from orbiting satellites (as in Figures 13.4 and 14.9). Because the amount of chlorophyll present is directly related to the rate of photosynthesis, chlorophyll content can be a good indicator of productivity. As we have seen,

however, the biomass of drifting autotrophs—the amount of phytoplankton in an area—is not always a true indicator of how rapidly substances are cycling through that biomass.

Factors That Limit Productivity

As you may recall, a *limiting factor* is a physical or biological necessity whose presence in inappropriate amounts limits the normal actions of an organism—in this case, the production of carbohydrates. Photosynthetic autotrophs require four main ingredients to produce carbohydrates: water, carbon dioxide, inorganic nutrients, and sunlight. Obviously water is not a limiting factor in the ocean. Carbon dioxide is almost never a limiting factor either, because of its high solubility in water and because of the large quantity of carbon dioxide dissolved in the ocean. So, the two potential limiting factors in marine primary productivity are inorganic nutrients and light.

Availability of Nutrients Autotrophs require inorganic nutrients for two purposes: to construct the large organic molecules that make primary productivity possible and to construct their skeletons or protective shells. **Nonconservative nutrients** are nutrients that change in concentration with biological activity. After a period of rapid phytoplankton growth—a phenomenon known as a **plankton bloom**—ocean surface waters are often depleted of nonconservative nutrients such as nitrate, phosphate, iron, and silicate. These nutrients become parts of the producers or of the animals that have eaten the producers. Unfortunately for the producers, valuable nutrients incorporated into the bodies of dead organisms tend to sink below the sunlit zone. The resulting low nutrient concentrations are the most important factors limiting the growth of marine producers. Photosynthetic productivity cannot continue unless the upwelling of deep water returns these nutrients to the surface.

The opportunity for exchange of nutrient-depleted surface water and nutrient-rich deep water is relatively high where there is little or no thermocline. In the Antarctic the return of sunlight in the southern summer triggers tremendous phytoplankton blooms; Antarctic water sometimes becomes a rich planktonic soup as millions of diatoms compete for nutrients and sunlight. Indeed, because of the high nutrient levels from upwelling supplemented by iron suspended in glacial runoff from land, the summer waters between the Antarctic Convergence and the mainland of Antarctica are among the most productive on Earth. Phytoplankton blooms in the northern polar ocean are not usually as exuberant because the volume of upwelling water is much less there and glaciers don't contribute as much rock dust to the ocean.

Because of the stability of the horizontal layers, upwelling is uncommon in the tropical ocean, except in the equatorial Pacific or in areas where currents impinge on interrupting islands or continents. The clear blue of the tropical ocean is the sure signature of an oceanic desert in which deep upwelling rarely occurs. When nutrients are available or are tightly recycled, as in shallow coral reefs, the tropical ocean explodes into vigorous productivity.

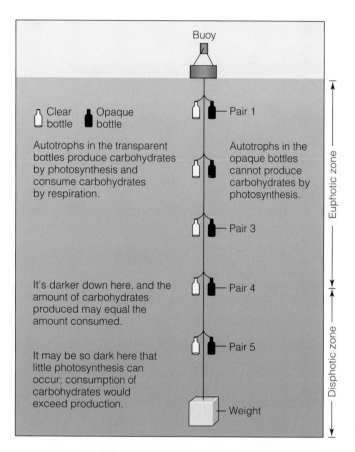

FIGURE 14.7 The light-dark bottle technique for estimating marine primary productivity. Clear and opaque bottles are filled with water from the area to be studied. Pairs of the bottles are lowered to depths at which 100%, 50%, 25%, 10%, 5%, and 1% of surface illumination is present, and photosynthesis and respiration are allowed to occur. Carbohydrate production by autotrophs in the bottles is measured by radioactively tagged bicarbonate ions added to the seawater in each bottle. The difference in the quantity of carbohydrates in each pair of bottles over time gives an indication of the productivity at each depth. Bottle pair 4 is near the compensation depth—the depth at which production of carbohydrates equals consumption (see Figure 14.8).

Availability of Light If adequate nutrients are present, primary productivity depends on illumination. Too little light is obviously limiting for photosynthesizers; very little photosynthesis proceeds below 100 meters (330 feet), and no photosynthesizers are known to function below 268 meters (879 feet). Too much light can also be inhibiting, however. You might think that productivity would be greatest right at the ocean surface, where light is brightest, but it isn't. Light there is often strong enough to overwhelm the photosynthetic chemistry of some photosynthesizers, especially phytoplankton.

Quantity is one important aspect of the light received by marine autotrophs. Quality, or color, is another. Chlorophyll is a green pigment and thus absorbs best in the red and violet wavelengths. Chlorophyll looks green because it reflects green light. But red and infrared light are effectively absorbed and converted to heat near the ocean surface; very little red light penetrates past 3 meters (10 feet). Phytoplankton stay near the surface to absorb red light, and primary productivity is thus highest in this top part of the euphotic zone.

Compensation Depth

So far, we've emphasized the importance of phytoplankton as producers, but remember that autotrophs respire as they photosynthesize; they use some of the carbohydrates and oxygen they produce. Carbohydrate production usually exceeds consumption, but not always. The deeper a phytoplankter's position, the less light it receives. At a certain depth, the production of carbohydrates and oxygen by photosynthesis through a day's time will exactly equal the consumption of carbohydrates and oxygen by respiration. This break-even depth is called the **compensation depth.** The compensation depth usually corresponds to the depth to which about 1% of surface light penetrates; it marks the bottom of the euphotic zone. If bottle pair 4 in Figure 14.9 were at compensation depth, the amount of carbohydrate produced in the transparent bottle would exactly equal the amount consumed, an indication of zero net productivity.

Figure 14.8 shows compensation depth graphically. Like the depth of greatest productivity, compensation depth changes with sun angle, turbidity, surface turbulence, and other factors. Remember that the compensation depth is always below the depth of greatest productivity and that producers can still "make a profit" between the depth of greatest productivity and the compensation depth. If a producer slips below its compensation depth for more than a few days, however, it will consume its carbohydrate reserves and die.

Because of their greater efficiency, diatoms have a deeper compensation depth than dinoflagellates. Open tropical seas have the deepest potential compensation depths, but these regions are not generally productive because of nutrient deficiencies.

Global Distribution of Plankton Productivity

Where is phytoplankton productivity the greatest? This question is among the most important in biological oceanography. Because phytoplankton form the base of nearly all oceanic food webs, the biological characteristics of any ocean area depend heavily on the presence and success of phytoplankton.

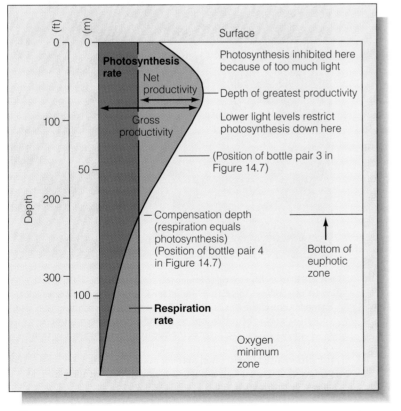

FIGURE 14.8 Compensation depth and its relationship to other aspects of productivity. Note the position of the bottom of the euphotic zone.

With some exceptions, the distribution of phytoplankton corresponds to the distribution of macronutrients. Because of coastal upwelling and land runoff, nutrient levels are highest near the continents. Plankton are most abundant there, and productivity is highest. The water above some continental shelves sustains productivity in excess of 1 gC/m²/*day*! But what of the open ocean? Where is productivity greatest away from land?

In the Tropics?

Water circulating in the great tropical gyres has abundant sunlight and CO_2 but is generally deficient in surface nutrients because the strong thermocline discourages the vertical mixing necessary to bring nutrients up from the depths. The tropical oceans away from land are therefore often considered oceanic deserts nearly devoid of visible plankton. The typical clarity of tropical water underscores this point. In most of the tropics, productivity rarely exceeds 30 gC/m²/yr, and seasonal fluctuation in productivity is low. But cyanobacteria and even tinier picoplanktonic producers may be binding carbon at rapid rates, and heterotrophic bacteria can be using it as fast as it is made. Even though the standing crop (biomass) of these producers is small, the exceedingly high rate of "throughput" of carbon dioxide and carbohydrate in these communities makes their contribution to general productivity very high. This "black market economy" (discussed in Box 14.2) is not available to fishes and other larger consumers because the larger zooplankton, which are their prey, are unable to separate these astonishingly small organisms from the surrounding water. Bacteria simply utilize the carbon and shuttle the metabolic products back to the tiny producers.

Tropical coral reefs are exceptions to this general rule. Reef areas, which account for less than 2% of the tropical ocean surface, are productive places because autotrophic dinoflagellates live *within* the tissues of the coral animals and don't drift as plankton. Nutrients are made available by coastal upwelling and by the coral's own metabolism. These nutrients are cycled tightly through the reef and are not lost to sinking.

In the Polar Regions?

At very high latitudes during the winter months, the low sun angle, reduced light penetration due to ice cover, and weeks or months of darkness severely limit productivity. At the height of summer, however, 24-hour daylight, a lack of surface ice, and the presence of upwelled nutrients can lead to spectacular plankton blooms. Such blooms cannot last, however, because the nutrients that support them are not quickly recycled and because the sun is above the critical angle for a few weeks at best, so the short-lived summer peak does not compensate for the long, unproductive winter months.

Average productivity at very high northern latitudes tends to be lower than at high southern latitudes, averaging less than 25 gC/m²/yr. The Arctic Ocean is almost surrounded by landmasses that limit water circulation and therefore nutrient upwelling, so nutrients are quickly depleted. The southern ocean, on the other hand, is enriched by water upwelling to replace sinking Antarctic Bottom Water. This rich mixture is stirred by the Antarctic Circumpolar Current. The Antarctic accounts for a much greater share of high-latitude production than the Arctic because its nutrients rarely become depleted.

In the Temperate and Subpolar Zones?

The tropics are generally out of the running because of nutrient deficiency, and the north polar ocean suffers from slow nutrient turnover and low illumination, so the overall productivity prize goes to the temperate and southern subpolar zones. Thanks to the dependable light and moderate nutrient supply, the annual production in the nearshore temperate and southern subpolar ocean areas is the greatest of any open-ocean area. Typical productivity in the temperate zone is about 120 gC/m²/yr. In ideal conditions southern subpolar productivity can approach 250 gC/m²/yr!

Figure 14.9 shows the levels of productivity in tropical, temperate, and northern polar ocean areas. Note that *nearshore* productivity is almost always higher than *open-ocean* productivity, even in the relatively productive temperate and south subpolar zones.

Curiously, the open-ocean area with the greatest annual productivity is an exception to the general picture developed in this section. Slender, cold fingers of high productivity pointing west from South America and Africa along the equator are a result of wind-propelled upwelling due to Ekman transport on either side of the geographical equator.

SeaWiFS Project, NASA/Goddard Space Flight Center

FIGURE 14.9 Measuring the concentration of chlorophyll in ocean water by satellite. This false-color image from the Coastal Zone Color Scanner aboard the *Nimbus 7* satellite shows average conditions in late spring in the Northern Hemisphere. It shows the concentration of chlorophyll in the upper layer of the ocean, with higher amounts indicated by green, orange, and red. Note the high phytoplankton concentrations induced by increased nutrient availability along the coasts. The centers of the oceanic gyres contain relatively few phytoplankton, as shown by their purple hue.

FIGURE 14.10 Variation in the biomass of phytoplankton by season and latitude. (The total area under each curve represents the productivity.)

Phytoplankton Productivity Through the Seasons

Figure 14.10 shows the relationship of phytoplankton biomass to season and latitude. The low, flat line representing annual tropical productivity contrasts with the high, thin peak representing the Arctic summer. The higher of the two peaks for the temperate zone indicates the plankton bloom of northern spring, caused by increasing illumination; the smaller temperate zone peak, representing the northern fall bloom, is caused by nutrients mixed toward the surface with increasing storm activity and less thermal stratification.

Zooplankton: The Heterotrophs

Heterotrophic plankton—the planktonic organisms that eat the primary producers—are collectively called **zooplankton** (*zoion* = animal). Zooplankters are the most numerous primary consumers of the ocean. They graze on the diatoms, dinoflagellates, and other phytoplankton at the bottom of the trophic pyramid much like cows grazing on grass. The mass of zooplankton is typically about 10% that of phytoplankton, which is reasonable because of the harvesting relationship that exists between them.

The variety of zooplankton is surprising; nearly every major animal group is represented. Each is expert at the painstaking concentration of food from water. The most abundant zooplankters, accounting for about 70% of individuals, are tiny shrimplike animals called **copepods** (**Figure 14.11**). Copepods are crustaceans, a group that also includes crabs, lobsters, and shrimp. Not all members of the zooplankton are small, however. Many are in the 1- to 2-centimeter ($\frac{1}{2}$- to 1-inch) size range. The largest drifters are giant jellyfish of the genus *Cyanea*; their bells may be larger than 3.5 meters (12 feet) in diameter! We have a special term for plankton larger than about 1 centimeter ($\frac{1}{2}$ inch) across: **macroplankton.**

FIGURE 14.11 A zooplanktonic copepod. Copepods probably are the most abundant and widely distributed animal in the world. The species shown here reaches a maximum size of about 0.5 millimeter (about 0.02 inch).

FIGURE 14.12 Krill (*Euphausia superba*). These shrimplike crustaceans, shown here about twice actual size, occur throughout the world ocean. They are particularly numerous in Antarctic seas.

Most zooplankton spend their whole lives in the plankton community, so we call them **holoplankton.** But some planktonic animals are the juvenile stages of crabs, barnacles, clams, sea stars, and other organisms that will later adopt a benthic or nektonic lifestyle. These temporary visitors are **meroplankton.** Most animal groups are represented in the meroplankton; even the powerful tuna serves a brief planktonic apprenticeship. These useful categories can be applied to phytoplankton as well as zooplankton. Holoplanktonic organisms are by far the more numerous forms of both phytoplankton and zooplankton.

One of the ocean's most important zooplankters is the pelagic arthropod known as **krill** (genus *Euphausia*, **Figure 14.12**), the keystone of the Antarctic ecosystem. This thumb-sized,

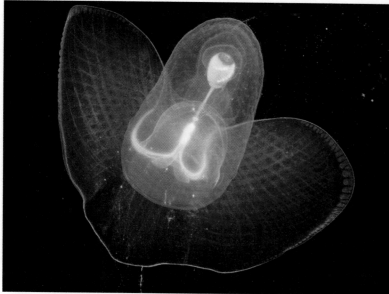

a

© Phillip Colla Photography

FIGURE 14.13 Jellies. (**a**) Unlikely as it may seem, the zooplankton includes gliding snails called pteropods. This example, genus *Corolla,* is about 8 centimeters (3 inches) across and feeds by capturing smaller drifting organisms in a mucus net. Pteropods contribute to calcareous oozes (see Chapter 5). (**b**) A crab of genus *Planes* hitches a ride on a string of primitive chordates (salps). Each salp (the yellow dots about the size of a pea) has constructed a jelly house for feeding and protection. (**c**) Light striking the feeding structures of this ctenophore (genus *Eurhamphaea*) diffracts into a rainbow of colors. The animal is about the size of a walnut.

b

Larry Madin/WHOI

Larry Madin/WHOI

c

shrimplike crustacean mostly grazes on the abundant diatoms of the southern polar ocean. In turn, krill are eaten in tremendous numbers by seabirds, squids, fishes, and whales. Some 500 to 750 million metric tons (550 to 825 million tons) of krill inhabit the Antarctic Ocean, with the greatest concentrations in the productive upwelling currents of the Weddell Sea. Krill travel in great schools that can extend over several square miles and collectively exceed the biomass of Earth's entire human population!

They behave more like schooling fish than planktonic crustaceans; their primary swimming mode is horizontal, not vertical.

The great diversity of members of the plankton community is perhaps best illustrated by the informally named "jellies." Common to all oceans, these diaphanous animals come from several taxonomic categories, including jellyfishes, pteropods, salps, and ctenophores (see Chapter 15). They range in size from microscopic to immense and use a wonderful range of

FIGURE 14.14 A foraminiferan. The word means "bearer of windows." Light streams through thin parts of the shell.

Wim van Egmond

adaptations to reduce weight, retard sinking, and snare prey. A few of these animals are shown in **Figure 14.13.**

Small but important, planktonic **foraminifera** are related to amoebas (**Figure 14.14**). Like amoebas, they extend long protoplasmic filaments to snare food. Most foraminifera have calcium carbonate shells. As we saw in Chapter 5, extensive white deposits of calcareous ooze have been built on the seabed from their skeletons. As was the case with some phytoplankton sediments, some of these layers have been uplifted and can be found on land.

The zooplankters depend on phytoplankton for both food and oxygen. The decomposition of falling biological debris and the activity of zooplankton often create an **oxygen minimum zone** below the well-lit surface zone: Oxygen is depleted by the animals there and not replaced by phytoplankton (see Figure 14.8). Some species of zooplankton and several kinds of small swimming animals make nightly migrations from the oxygen minimum zone toward the darkened surface layer to feed on the smaller organisms drifting there.

It is interesting to note that the largest marine animals, such as whale sharks (fish) and baleen whales (mammals), do not expend their energy tracking down and attacking big animals. Instead, these largest of all feeders concentrate zooplankton from the water and consume it in vast quantities. The zooplankton they eat are not usually the primary consumers but the somewhat larger secondary consumers, usually crustaceans such as krill that have themselves fed on the microscopic primary consumers. In this way whales and other large filter feeders can harvest energy closer to the source, gaining the advantage of efficiency and quantity.

Larger Marine Producers

We now shift our attention from drifters to attached autotrophs. Relatively large attached autotrophs (the protists we call seaweeds) account for between 2% and 5% of the ocean's total primary productivity.

Most marine autotrophs, large or small, are algae. **Algae** is a collective term for autotrophs possessing chlorophyll and capable of photosynthesis but lacking vessels to conduct sap. The single-celled diatoms and dinoflagellates discussed earlier are classified as **unicellular algae.** *Seaweed* is the informal name for large, marine **multicellular algae.** A few species of marine photosynthesizers are not algae but **angiosperms** (*angios* = covered, *sperma* = seed); these are flowering plants. The main groups of marine angiosperms are sea grasses and mangrove trees. Marine angiosperms are not considered seaweeds.

Algae are **nonvascular** organisms; that is, they do not have vessels. Algae require the same four ingredients for photosynthesis as vascular plants—carbon dioxide, sunlight, water, and inorganic nutrients—but in their case the ingredients are already present in one location, so vessels and flowing sap are not necessary. Either the alga is very small (perhaps a single cell) and lives in a moist spot on land where conditions are ideal, or it lives in water.

Multicellular marine algae occur in a great variety of sizes and shapes. Some types of algae form underwater forests; others grow in isolation. The largest can reach 62 meters (205 feet) in length. Their bodies are flexible, easily able to absorb shock, resistant to abrasion, streamlined to reduce water drag, and very strong. Their surfaces are often covered by a slick, mucilaginous material that lubricates them as they move, retards drying, and deters grazing animals. Algae do not grow below the euphotic zone because they all depend on photosynthesis to produce the energy-rich compounds necessary for life. Nearly 7,000 species of multicellular marine algae have been identified.

The marine lifestyle offers advantages. Large marine autotrophs suffer no droughts and nearly always have enough carbon dioxide for photosynthesis. Given suitable nutrient levels and a good foothold, only sunlight is required for productivity. Immersion in seawater brings the additional advantage of lightweight construction. A seaweed doesn't require strong support structures because it has nearly the same density as the surrounding seawater, so more of its bulk can be dedicated to photosynthesis. Indeed productivity in some wave-washed seaweed beds may be the highest per unit area of any autotrophic community on Earth.

The astonishingly high productivity of seaweeds is underscored by the fact that many species of large seaweeds are surprisingly leaky; carbohydrates and other products of photosynthesis diffuse from their blades and stipes like tea from a teabag. Up to half of all the organic matter they produce can be lost in this way! The foam visible in surf near kelp beds is produced in part by these substances. Sea urchins and other heterotrophs can absorb these molecules directly through their skin or outer membranes, "feeding" on kelp plants that may be tens or even thousands of meters away.

Accessory Pigments

How can a few species of tropical marine photosynthesizers live at depths greater than 250 meters (820 feet)? Because most marine autotrophs have evolved specialized accessory pigments. **Accessory pigments** (or *masking pigments*) are light-absorbing compounds closely associated with chlorophyll

b

FIGURE 14.15 (**a**) The thallus (body) of a typical multicellular alga. These organisms can grow at a rate of 50 centimeters (20 inches) per day and reach a length of 40 meters (132 feet). (**b**) *Macrocystis,* a brown alga (phaeophyte), showing a close-up view of the blades with gas bladders at the base of each. This fast-growing, productive alga is one of the dominant species found in the spectacular kelp forest of western North America.

molecules. Their presence in autotrophs greatly enhances photosynthesis because they absorb the dim blue light at depth and transfer its energy to the adjacent chlorophyll molecules. Accessory pigments may be brown, tan, olive green, or red; they give most marine autotrophs, especially seaweeds, their characteristic color. It is the combination of pigments in a photosynthesizer that determines its color and optimal depth distribution. The masking effect of accessory pigments is often so complete that an observer may be unaware that the organism contains any chlorophyll whatever, even when the seaweed is transparent. The absence of accessory pigments allows the bright green of chlorophyll to shine through.

Structure of Seaweeds

Common terms such as *leaf, stem,* and *root* are inappropriate for seaweeds because the definitions of those parts assume the presence of vessels. The structures in seaweeds that superficially resemble leaves are called **blades** (or fronds), the stemlike structures are termed **stipes,** and the root-shaped jumble at the base is appropriately named a **holdfast. Gas bladders** assist many species in reaching strongly illuminated surface water. Blades, stipes, and holdfast make up the body of the organism, the **thallus.** These parts are labeled in **Figure 14.15.**

A thallus may be large or small, branching or tufted, in sheet form or filamentous, encrusting or elongated, rounded or pointed—algal variety is tremendous. Algal blades are symmetrically equipped with photosynthesizing tissue. They absorb gases across their entire surface and even participate in reproduction. Stipes are strong, photosynthesizing, shock-absorbing links tying the blades (or sheets or filaments) into a unit. The holdfast does not take up water and nutrients from the substrate as does the vascular root it superficially resembles, but it does anchor the seaweed in place and can provide incidental shelter for a rich variety of animal life. The gas bladders range in size from tiny, grapelike bunches to single, volleyball-sized floats.

Classification of Seaweeds

Marine multicellular algae are classified in three divisions based on their pigments. The green algae, with their unmasked chlorophyll, are the **Chlorophyta** (*chloros* = green, *phyton* = plant); the brown algae, **Phaeophyta** (*phae* = tan, dusky); and the red algae, **Rhodophyta** (*rhodon* = rosy red).

The Chlorophytes The clear green color of chlorophytes is evidence of their lack of accessory pigments and suggests that they live at or near the surface, where red light is available. Land plants are thought to have evolved from chlorophyte ancestors. Only about 10% of the 7,000 species of Chlorophyta are marine, but some of those species are widely distributed and present in great numbers. Genus *Ulva* is a familiar, delicate, lettucelike edible seaweed able to tolerate the often impure waters of urban coasts. *Ulva* makes quick use of concentrated nutrients near sewage outfalls to grow and multiply.

Most other species of green algae prefer clean water and are branched or threadlike. Some encrust hard surfaces and some live on sand, but none reaches the size of the large brown seaweeds, the phaeophytes.

The Phaeophytes Nearly all of the 1,500 living species of phaeophytes are marine. Some species of these largest algae, which include the **kelps**, can reach lengths of 40 meters (132 feet)—the record length exceeds 60 meters (200 feet). To attain these dimensions the seaweed can grow at the spectacular rate of 50 centimeters (20 inches) per day! Part of the strategy of rapid growth is to reach bright surface water as soon as possible. Some brown algae are annuals; others live for up to seven years. The tan or brown color of phaeophytes comes from the accessory pigment *fucoxanthin*, which permits photosynthesis to proceed at greater depths than is possible for the unmasked chlorophytes. In ideal circumstances some larger brown algae can grow in water up to about 35 meters (115 feet) deep.

The Pacific Ocean's giant kelp forests, the world's largest, consist mostly of the magnificent genus *Macrocystis* (*macro* = large, *cyst* = bladder). Most brown algae live in temperate and polar habitats poleward of the 30° latitude lines, but a few live in the tropics. The worldwide distribution of kelp is shown in **Figure 14.16**.

The Rhodophytes Most of the world's seaweeds are red algae; there are more rhodophyte species—about 4,000—than all other major groups of algae combined. Rhodophytes tend to be smaller and more anatomically and biochemically complex than phaeophytes. Rhodophytes excel in dim light because of their sophisticated accessory pigments, reddish

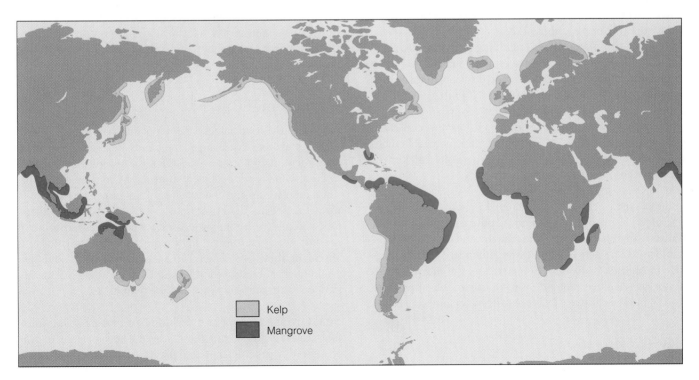

FIGURE 14.16 Worldwide distribution of kelp beds and mangrove communities.

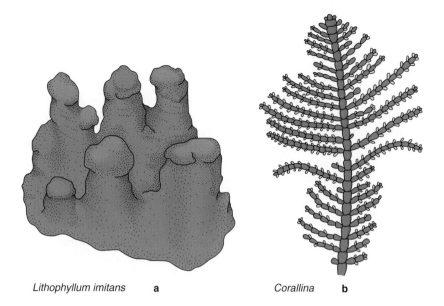

Lithophyllum imitans   **a**   Corallina   **b**

FIGURE 14.17 Two rhodophytes. (**a**) *Lithophyllum* is an encrusting species. (**b**) *Corallina* is an erect species.

proteins called *phycobilins*. These compounds absorb and transfer enough light energy to power photosynthetic activity at depths where human eyes cannot see light. The record depth for a photosynthesizer is held by a small rhodophyte discovered in 1984 at a depth of 268 meters (879 feet) on a previously undiscovered seamount in the clear, tropical Caribbean. The deepest rhodophytes grow very slowly and may be tens or even hundreds of years old.

Paradoxically, many red algae also live on rocks right at the water's surface. Surface light contains all colors of the visible spectrum, so the color-shifting capabilities of accessory pigments are not needed there. It is believed that the dark accessory pigments may act to shade the productive machinery from brilliant light. Most of the common surface-dwelling "reds" grow as a purple or pink film on rocks, shells, other seaweeds, and even glass or plastic. These coralline, or calcareous, algae look like a brilliant ceramic glaze, a bit of melting raspberry sherbet, or a knobby plastic membrane thrown carelessly over the surface. The name of one very common form, genus *Lithophyllum*, comes from the Latin terms for "stony" and "leaf," an apt description of an alga able to deposit calcium carbonate within its tissues (**Figure 14.17a**). Others grow in erect branching or bushy forms (**Figure 14.17b**). The gritty calcium in these seaweeds may deter grazing animals.

Another group of shallow rhodophytes is very important in the life of coral reefs. Like coral animals, encrusting coralline algae of the genus *Lithothamnium* remove large quantities of dissolved calcium carbonate from seawater and deposit it within their tissues. This activity helps cement the reef into a mass capable of resisting heavy surf. Some reefs in the Indian Ocean are not, as was once thought, of coral origin but rather were formed almost exclusively by the activity of coralline algae.

Seaweed Zonation

Zones or bands of dominant seaweeds are plainly visible on the sloping nearshore seafloor to the limit of light penetration. The vertical distribution of seaweeds depends in part on their pigments. Chlorophytes lack accessory pigments and are rarely found more than 10 meters (33 feet) below the surface. Phaeophytes can operate from the surface down to about 30 meters (100 feet). Rhodophytes, as we have seen, can function from the surface to the photosynthetic record depth of 268 meters (879 feet).

Factors other than the response of pigments to light are also important in determining where these algae will prosper, especially in the intertidal zone. Wave shock, varying salinity and pH, the presence of grazers from land and sea, the drying effects of sun and wind, abrasion by waterborne particles of sand—all these factors and many more affect the location of intertidal producers. The resulting effects of these factors on marine organisms can be seen at low tide by any observant beachcomber. Look for bands of different marine algae (and angiosperms) the next time you go to the shore (see Figure 16.7).

Marine Angiosperms

Angiosperms are advanced vascular plants that reproduce with flowers and seeds. Most large land plants are angiosperms. Relatively few angiosperms live in water; the advantages conferred by vessels and roots are largely unnecessary in an aquatic environment. However, a few species of angiosperms have colonized the ocean. All of these have descended from land ancestors, and all live in shallow coastal water. Angiosperms live at the surface, where the red light required for photosynthesis is abundant; they have no need for accessory pigments, and their chlorophyll is unmasked. The most conspicuous marine angiosperms are the sea grasses and the mangroves.

Sea Grasses Many people lump **sea grasses** in the informal seaweed group, but their resemblance to large marine algae is only superficial. These plants are not true grasses, but they do have leaves and stems as well as roots capable of extracting nutrients from the substrate. Extensive stands of sea

FIGURE 14.18 The bright green sea grass *Phyllospadix* in a tide pool. Seagrasses are vascular plants, not seaweeds.

grasses are found on the coasts of North America, on the Atlantic coast of Europe, in Eastern Asia, in temperate Australia, and in South Africa. They form broad gray or green submerged meadows, which support extraordinarily rich communities of heterotrophs. The life cycle of sea grasses is much like that of other angiosperms, but their stringy pollen is distributed by flowing water rather than by insects or wind. About 45 species of sea grasses are known.

The most common sea grass is eelgrass, genus *Zostera*, a common inhabitant of the muddy shallows of calm bays and estuar-

ies of the U.S. Atlantic and Pacific coasts. Similar habitats along the Gulf coast harbor stands of turtle grass (genus *Thalassia*) and manatee grass (genus *Syringodium*), angiosperms named after the animals that once shared their habitat and (in the case of the manatee) fed on them.

Perhaps the most beautiful sea grass is the vivid, emerald-green surf grass, genus *Phyllospadix*, with its seasonal flowers and fuzzy fruit (**Figure 14.18**). These hardy plants survive in the turbulent, wave-swept intertidal and subtidal zones of temperate East Asia and western North America.

Sea grasses can be remarkably productive. Some are capable of binding 1,000 gC/m²/yr, three to five times the nearshore average for phytoplankton. One explanation for this efficiency is the advantage conferred by roots. Anaerobic bacteria in subtidal mud have been shown to bind dissolved nitrogen into nitrates easily available to these plants.

Mangroves Low, muddy coasts in tropical and some subtropical areas are often home to tangled masses of trees known as **mangroves**. These large, flowering plants are never completely submerged, but because of their intimate association with the ocean we consider them to be marine plants (**Figure 14.19**). Mangroves thrive in the sediment-rich lagoons, bays, and estuaries of the Indo-Pacific, tropical Africa, and the tropical Americas. Their distribution depends on temperature, currents, and rainfall (see again Figure 14.16).

The sediment in which mangrove trees live must be covered with brackish or salt water for part or all of the day. Many mangroves avoid taking up salt ions from seawater, or they selectively remove salt from sap with salt-excreting cells. The fine coastal mud they colonize doesn't provide firm footing for these substantial plants, so an intricate network of arching prop roots is required for support. The strutlike prop roots are supplemented by many smaller roots equipped with breathing

FIGURE 14.19 A mangrove (*Rhizophora*) growing in salt water in Everglades National Park, Florida. The tangled roots descending from the main branch are called *prop roots* or *stilt roots*. They provide anchorage for the mangrove, trap sediment, and provide protection for small organisms.

pores and air passages. Atmospheric oxygen is conducted by these passages to the parts of the plant buried in oxygen-deficient mud. The root system also traps and holds sediments around the plant by interfering with the transport of suspended particles by currents. Mangrove forests thus assist in the stabilization and expansion of deltas and other coastal wetlands. The root complex also forms an impenetrable barrier and safe haven for organisms around the base of the trees.

The mangrove communities of south Florida, among the world's most widespread, consist primarily of the red mangrove (genus *Rhizophora*). The spreading leaves of the tree protect the residents from the tropical sun, and the tangled roots keep out large predators and also harbor huge numbers of fiddler crabs, worms, marine and terrestrial snails, fish, oysters, and red algae. Birds and insects inhabit the treetops, their droppings enriching the sediments below.

Mangrove seeds germinate on the trees. If the tide is out when the seed drops, the force of the fall plants the seed in the muck, where it will continue to grow. If the tide is in, the seed will drop into the water, suspend growth, float to a new location, and resume growth when it gets a foothold in the mud. The trees mature in 20 to 30 years and can attain a height of 8 to 10 meters (26 to 33 feet). New mangroves colonize mud and sandbars away from shore; they stabilize them and eventually create new land.

Commercial Importance

Chances are good that you have had some recent contact with marine autotrophs even if you haven't been in the ocean. The mucilaginous material that is so effective in making algal blades slick, in lowering friction, and in deterring grazers is also harvested and made into an important commercial product called *algin*. When separated and purified, its long, intertwining molecules are used to stiffen fabrics, make adhesives, suspend water and oil together in salad dressings, prevent the formation of gritty crystals in ice cream, clarify beer and wine, and manufacture shoe stains, soaps, and shaving cream. Fast-food restaurants now use carageenan, a similar seaweed extract, to replace some of the fat in newly popular healthier hamburgers. These substances also prevent fire-extinguishing foams from dispersing, and they permit chocolate milk to remain on the refrigerator shelf without separating and keep the abrasives in liquid car waxes from settling to the bottom of the bottle. In biological laboratories bacteria are cultured on agar made from seaweed extracts. Very likely even the ink that forms the letters you are now reading has an algin or carageenan component!

You can find out more about the uses of marine plants in the discussion of marine resources in Chapter 17.

Questions from Students

1. How does the productivity of marine plants compare with the productivity of land plants?

Look again at Figure 13.3c. Note that a rain forest, the most productive natural terrestrial community listed, is not as productive as coral reefs or kelp beds. Under ideal conditions, sea palms—a phaeophyte, genus *Postelsia*—can bind an astonishing 14,600 gC/m²/yr, the highest primary productivity recorded for any autotroph! Seaweeds are champion photosynthesizers. Unlike, say, an apple tree (which is composed mainly of nonproductive wood, bark, twigs, vessels, roots, and stems), nearly all of a seaweed's tissue is productive.

2. Do zooplankton have a compensation depth? If so, is it above, below, or at the same level as the compensation depth of most phytoplankters?

The concept of compensation depth is meaningless for zooplankton. Compensation depth applies to only *autotrophs*, organisms capable of both photosynthesis and respiration. Because animals don't make their own food—aren't autotrophic—productivity in them can *never* equal consumption.

3. Isn't it too cold for seaweeds to survive at high latitudes?

No. As with phytoplankton, seaweed productivity can be high regardless of water temperature. Indeed, kelp in some Canadian bays grows most rapidly when water is near 0°C (32°F). The important limiting factors are the few hours of daylight in winter months, the relative lack of nutrients in the north polar ocean, and the abrasive grinding of ice in nearshore areas.

4. What's this about bioluminescence in dinoflagellates having evolved, possibly, as an "intrusion alarm"?

Some dinoflagellates flash when they are attacked by zooplankton. The glow may startle the zooplankter and distract it from feeding. It may also alert larger organisms to the presence of that zooplankter and attract a predator toward the offending grazer. Behavioral studies suggest that dinoflagellate grazers react negatively to a dinoflagellate's blue flash, and perhaps this is why.

5. Phytoplanktonic organisms account for a significant percentage of world primary productivity. Why *are* phytoplankton so successful? Is there something about the planktonic lifestyle that lends itself to high efficiency?

Phytoplankton are successful because of their size. Because water supports them, small marine autotrophs don't require the elaborate support systems of large terrestrial plants. Their small size allows easy diffusion of required nutrients into their single cells and prompt transport of wastes out; vessels and sap are not needed. Nearly all of their tissue can be photosynthetic.

With transparent siliceous frustules to protect them, diatoms are especially successful. The relatively recent evolution of diatoms was a high point in the development of marine autotrophs, and the additional oxygen these organisms contributed to the atmosphere has greatly influenced the success of life on land. More plants have led to more animals. After a few hundred million years of evolution, the planktonic lifestyle is elegantly tuned and interlocked into the complex and productive system we now observe in the world as a whole.

6. You mentioned that krill can swim horizontally over considerable distances. Doesn't that disqualify them for inclusion in the plankton?

Japanese researchers, using two ships equipped with side-scan sonar, tracked a large school of swimming krill for 14 days across 278 kilometers (172 miles). This finding does indeed threaten krill's usual classification as zooplankton (animals unable to move consistently in one direction against waves or current flow). Traditions die hard, though, and I don't anticipate this keystone of the Antarctic ecosystem will lose its status as the premier zooplankter anytime soon.

Chapter in Perspective

In this chapter you learned that organisms that drift in the ocean are known collectively as plankton. The small plantlike organisms that make up phytoplankton are responsible for most of the ocean's primary productivity. (The larger marine producers we call seaweeds and relatively simple organisms that depend on chemosynthesis account for much of the rest.) Phytoplankton—and zooplankton, the small, drifting or weakly swimming animals that consume them—are usually the first links in oceanic food webs. Plankton are most common along the coasts, in the upper sunlit layers of the temperate zone, in areas of equatorial upwelling, and in the southern subpolar ocean.

Many physical and biological factors influence marine primary productivity, the most important being the availability of light and inorganic nutrients. Worldwide oceanic productivity is roughly equivalent to land productivity, but a *much* smaller mass of producers is responsible for productivity in the ocean than on land; marine producers are considerably more efficient in assembling glucose molecules.

The larger producers informally known as seaweeds are classified by color (that is, pigment composition) into three large groups: green, brown, and red algae. Some forms of brown algae, which we call kelp, grow in great underwater forests. Note that not all large marine autotrophs are algae; some are plants—sea grasses and mangroves.

In the next chapter you will learn more about the world of marine heterotrophs—animals. Freed from the need to make their own food, animals have evolved astonishing adaptations for grazing, predation, and parasitism.

Key Ideas

✓ The term *plankton* is a description of a lifestyle, not a collective natural category. Members of the plankton may be large or submicroscopic and include many plantlike species and nearly every major group of animals.
   Review, page 333; Figure 14.1, pages 334–335

✓ Phytoplankton—drifting photosynthetic autotrophs—are among the world's most important primary producers. Two groups, diatoms and dinoflagellates, account for most oceanic productivity.
   Review Figures 14.3–14.6, pages 336–338

✓ Primary productivity is measured in grams of carbon produced per square meter of ocean surface per year ($gC/m^2/yr$).
   Review Figures 13.2, 13.3, and 14.8, pages 307–308, 343

✓ Primary productivity is measured directly using radioactive tracers or light-dark bottle experiments.
   Review Figure 14.7, pages 340–342

✓ Many factors can limit primary productivity. Chief among these are lack of light and lack of nutrients.
   Review pages 342–343

✓ The compensation depth is the depth above which photosynthetic production of carbohydrate exceeds consumption of carbohydrate by the producer itself. The compensation depth defines the bottom of the euphotic zone.
   Review Figure 14.8, page 343

✓ On a total annual basis, primary productivity is greatest in the temperate zone near shore. The greatest productivity at any one time occurs around the Antarctic in local summer.
   Review Figures 14.9 and 14.10, pages 344–345

✓ Zooplankton—drifting heterotrophs—are the most abundant consumers in the ocean.
   Review Figures 14.11–14.14, pages 345–347

✓ Seaweeds, a form of marine algae, are efficient primary producers but collectively account for only a small percentage of total marine productivity.
   Review Figures 14.15–14.17, pages 347–350

✓ Seaweeds are classified by pigments into chlorophytes, phaeophytes, and rhodophytes. Algin, a substance derived from seaweeds, has considerable commercial importance.
   Review pages 349–350

✓ The most abundant marine angiosperms (true plants) are mangroves and sea grasses.
   Review Figures 14.18 and 14.19, pages 350–352

Terms and Concepts to Remember

accessory pigment, 347
algae, 347
angiosperm, 347
auxospore, 337
bioluminescence, 337
blade, 348
Chlorophyta, 349
coccolithophore, 338
compensation depth, 343
copepod, 345
diatom, 333
dinoflagellate, 337

flagellum, 337
foraminifera, 347
frustule, 333
gas bladder, 348
holdfast, 348
holoplankton, 345
kelp, 349
krill, 345
macroplankton, 345
mangrove, 351
meroplankton, 345
multicellular algae, 347

Study Questions

Reviewing What You've Learned

1. What do primary producers produce? How is productivity expressed?
2. What are plankton? How are plankton collected? How are members of the plankton community differentiated? How are zooplankton different from phytoplankton?
3. Describe the most abundant and important phytoplankters. Which are most efficient in converting solar energy to energy in chemical bonds?
4. How can primary productivity be measured? Which method is considered the most accurate?
5. What are algae? Are all algae seaweeds? How are seaweeds classified? Which seaweeds live at the greatest depths? Why?
6. Give examples of marine angiosperms. Are they vascular plants or nonvascular algae?
7. Of what commercial importance are marine plants?

Thinking Critically

1. What factors limit productivity? What methods have marine producers evolved to cope with the lack of red light needed by chlorophyll for photosynthesis?
2. What is compensation depth? What happens to phytoplankton below that depth? To zooplankton?
3. Where in the ocean is plankton productivity the greatest? Why?
4. How does a nonvascular alga differ from a vascular plant? Why are most marine autotrophs nonvascular?
5. **InfoTrac College Edition Project** Phytoplankton are the major primary producers of the ocean; all other life-forms depend on them, directly or indirectly. Accordingly, many marine scientists are concerned about changes that can alter phytoplankton populations. What would be the effects of a drastic drop in the ocean's phytoplankton? Research your answer to this question using InfoTrac College Edition.

Thinking Analytically

1. Is phytoplankton productivity highest at the ocean surface? What advantage would optimum productivity at a depth *below* the surface provide to phytoplankton?
2. Imagine a small tide-washed inlet on the coast of western Canada about the size of an Olympic swimming pool. Given optimal nutrients, a stable substrate, and the sunlight of summer, estimate *approximately* how many kilograms (wet weight) of a seaweed as productive as *Postelsia* could be produced in the inlet in one month. How does this compare with a field of alfalfa of the same surface area? Begin with a review of Figure 13.3 and the biology of seaweeds in this chapter.

Oceanography Now™

Assess your understanding of this chapter's topics with additional quizzing and conceptual-based problems at http://oceanography.brookscole.com/garrison5e

Study and Research on the Internet

Assess your understanding of this chapter's topics with additional quizzing and conceptual-based problems at:

 http://oceanography.brookscole.com/garrison5e

- **InfoTrac College Edition Research** leads you to critical thinking exercises that use InfoTrac College Edition® as a research tool.
- **Regional InfoTrac College Edition** articles are organized into East Coast, West Coast, and Gulf Coast regions, allowing you to study oceanography on a more local level.

    For more readings, go to InfoTrac College Edition, your online research library, at:

     http://infotrac.thomsonlearning.com

A wandering albatross (*Diomedea*) banks over the surface of the southern ocean in search of food. Named after Diomed, a warrior who found his way safely home after the Trojan Wars, this magnificent bird's wingspan is nearly 3.6 meters (12 feet).

© Paul Ward

MARINE ANIMALS

Masters of the Storm

The mild, supportive ocean brims with animal life. **Animals** are active multicellular organisms incapable of synthesizing their own food. Animals must obtain food from primary producers—or from other animals that have consumed primary producers—to ensure their own survival.

All animal species encounter the same basic problems—to find food, avoid predators, and reproduce—and all have ways to solve these problems. It is the *variety* of survival strategies—

adaptations to their environment—that makes marine animals so interesting. Each species, from sponge to whale, has a unique set of adaptations. It may sound like a circular argument, but by definition each living species is successful precisely because it is living: Its continuously evolving suite of adaptations has brought it through the rigors of food finding, predator avoidance, and reproduction time and time again.

The nature of these adaptations often amazes us. Consider the albatrosses. Foraging for months across the ocean in conditions of strong winds and high waves, seeking no shelter during storms, enduring tropical heat and polar sleet, soaring continuously through air with a gliding efficiency exceeding that of the most perfectly built human sailplane—the albatrosses are true masters of the sky.

The magnificent wandering albatrosses (genus *Diomedea*) are the largest of these birds. Their wingspan reaches 3.6 meters (12 feet), their weight 10 kilograms (22 pounds). The key to the albatross's success is its aerodynamically efficient wing—a very long, thin, narrow, cupped and pointed structure ideal for high-speed soaring and gliding. This specialized wing enables albatrosses to cover vast distances in search of food with very little expense of energy. Using the uplift from wind deflected by ocean waves to stay aloft, they soar in long looping arcs. Albatrosses rarely flap their wings, which lock at elbow and shoulder for nearly effortless transport. Flying is their natural state: an albatross's heart beats more slowly in flight than while it is sitting calmly on the ocean surface.

Satellite tracking data indicate that wandering albatrosses can cover 15,000 kilometers (9,300 miles) between visits to feed their chicks and reach speeds of 80 kilometers (50 miles)

per hour. High speeds over long distances are their specialty. One bird was observed to travel 808 kilometers (502 miles) at an *average* speed of 56 kilometers (35 miles) per hour.

Albatrosses can find schools of fish by the odor of fish oil wafted tens of kilometers downwind. They catch fish (and squid) by dipping their bill into the water during flight or while resting on the surface when the water is calm. Albatrosses take shore leave only during breeding season. They mate for life. Chicks are hatched and raised on remote islands to which albatrosses regularly migrate from nearly any point over the world ocean. They are astonishing animals, and no one who has seen one sweeping over the sea surface will ever forget the experience.

Animals like the albatross teach us an important lesson: *No matter how improbable a structure or behavior seems, it must benefit the species in some way—or it would not be present.* In the long run, every shape and movement helps an organism to succeed.

The history of animals begins far back in the history of the ocean. The path leading to today's formidable array of animals began with the availability of food.

The Origin of Animals

As you may recall from Chapter 2, the first organisms to evolve on Earth were probably tiny creatures adept at absorbing organic molecules that formed spontaneously in the ocean. Primitive life-forms could use the energy stored in these food molecules for growth and reproduction. Competition for food increased as these organisms grew more numerous. Had photosynthesis not evolved, life would probably have died out when the supply of usable energy-rich molecules in the environment was exhausted. Using sunlight, the first primary producers assembled their own food from inorganic molecules and then broke the food down to release energy. With the success and proliferation of these simple autotrophs, and with the abundance of oxygen they provided, the way was clear for the evolution of animals.

The first animal-like creatures were single-celled organisms. They began to prosper in the ocean during the **oxygen revolution**, a time of radical change in Earth's atmosphere. During the oxygen revolution, between about 2 billion and 400 million years ago, the activity of photosynthetic autotrophs changed the composition of the atmosphere from less than 1% free oxygen to its present oxygen-rich mixture of more than 20% (**Figure 15.1**). The growing abundance of free oxygen made aerobic respiration practical, speeding the disassembly of food molecules early animals obtained by eating the autotrophs. Ozone derived from this oxygen blocked most of the sun's dangerous ultraviolet radiation from reaching Earth's surface, permitting life to survive at the surface of the ocean and, later, on land.

Animals grew in complexity as they became more abundant. Instead of drifting apart after reproduction, some dividing cells stuck together and formed colonies. True animals evolved as these colonies distributed labor among specialized cells, eventu-

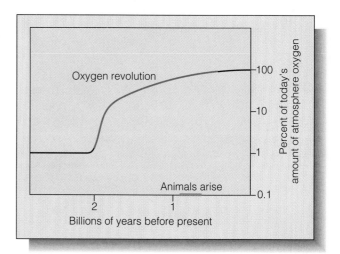

FIGURE 15.1 During the oxygen revolution, between about 2 billion and 400 million years ago, the activity of photosynthetic autotrophs—mostly bacteria—caused a rapid rise in the amount of oxygen in the air, making possible the evolution of animals. Animals are thought to have arisen between 900 and 700 million years ago.

ally increasing the interdependence among cells within the colony. The colonies ceased to be simple aggregations of individuals and began to take on specific architectures for specific tasks.

A group of animals that shares similar architecture, level of complexity, and evolutionary history is known as a **phylum** (plural, *phyla*; *phylon* = tribe). No one knows how many phyla of animals may have developed during the time of rapid animal proliferation that occurred near the end of the oxygen revolu-

a

b

c

FIGURE 15.2 Ancient marine animals. (**a**) From Australia's Ediacara Hills, the fossil of a segmented marine animal about 600 million years old. (**b**) From the Burgess Shale of British Columbia, a marine worm about 530 million years old. In basic form, it resembles species alive today. (**c**) A 500-million-year-old trilobite fossil. Trilobites were once abundant in the ocean, but all have died out.

tion. Small but fascinating "snapshots" of early marine life are preserved in fossils found in the Ediacara Hills of Australia, the Burgess Shale of British Columbia, and the Chengjiang beds of southwestern China. These sites were once parts of the warm, sediment-covered continental shelves of equatorial landmasses. When animals inhabiting these places were abruptly buried—perhaps by turbidity currents—their delicate features were preserved. The animal in **Figure 15.2a** is about 600 million years old and shows evidence of a segmented body plan. The 530-million-year-old worm in **Figure 15.2b** is similar in outline to species alive today.

Not all groups of early marine animals survived, however. Once abundant, trilobites like the one shown in **Figure 15.2c** have all perished. Indeed most of the animals in these ancient fossil beds are extinct and may represent failed branches in animal evolution, unique designs that were not suited to later environmental conditions.

Introducing the Invertebrates

More than 90% of all living and fossil animals, including all of the earliest multicellular animals, are categorized as invertebrates. **Invertebrates** are generally soft-bodied animals that lack a rigid internal skeleton for the attachment of muscles, but many invertebrates possess some sort of hard, protective outer covering, which can be continuous (like a snail shell) or segmented (like a lobster shell). The category is convenient though somewhat artificial, containing as it does a vast variety of different organisms in many diverse phyla that have little in common. Invertebrates range in size from microscopic worms to giant squids. Biologists currently recognize at least 33 invertebrate phyla, and almost every one of them has marine representatives. This chapter pre-

sents the nine most conspicuous invertebrate phyla in order of increasing complexity (for an overview, see **Table 15.1**).

Phylum Porifera

Sponges belong to the phylum **Porifera** (*porus* = holes, *ferre* = to bear), the most primitive true animals. Nearly all of the 10,000 species of these simple attached animals are marine. Sponges are widely distributed from intertidal zone to abyss and are found at all latitudes and in most benthic habitats. They range from the size of a bean to the size of a small automobile and come in a few basic shapes: branching, vaselike, and encrusting.

Table 15.1	Major Animal Phyla with Marine Examples
Phylum	**Marine Examples**
Invertebrates	
Porifera	Sponges
Cnidaria	Coral, jellyfish, sea anemones, siphonophores
Platyhelminthes	Flatworms, flukes, tapeworms
Nematoda	Roundworms
Annelida	Segmented worms
Mollusca	Chitons, snails, bivalves, squid, octopuses
Arthropoda	Crabs, shrimp, barnacles, copepods, krill
Echinodermata	Sea stars, sea urchins, sea cucumbers
Chordata[a]	Tunicates, salps, *Amphioxus*
Vertebrates	
Chordata	Fishes, reptiles, birds, mammals

[a] Phylum Chordata includes both vertebrate and invertebrate classes.

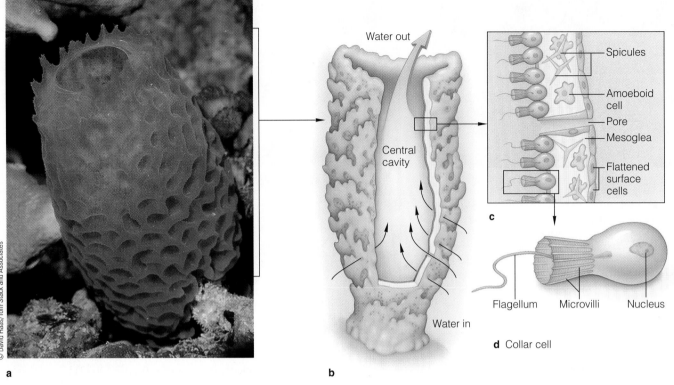

FIGURE 15.3 (a) An upright sponge. (b) The body plan of a simple sponge. (c) A section through a sponge body wall. (d) A type of collar cell.

All sponges are **suspension feeders:** They strain plankton and tiny organic food particles from the surrounding water. A large sponge may filter more than 1,500 liters (400 gallons) of water each day. **Figure 15.3** shows a cutaway diagram of a simple upright sponge. Water carrying food and oxygen enters the sponge through pores on its surface and is swept toward the exit opening by flagellated collar cells. The sticky collars snare food particles drifting past, and digestion begins. The captured nutrients are distributed to other cells of the organism by wandering amoeboid cells in the body of the sponge. Sponges have no digestive system, only individual digestive cells; they have no circulatory, respiratory, or nervous systems. Excretion and the movement of gases into and out of the animal occur by simple diffusion.

The flagellated cells of more elaborate sponges are concentrated in hundreds of tiny chambers. Their interiors resemble a Swiss cheese. A skeletal network of spicules (needles) of calcium carbonate or glassy silica prevents the internal chambers and canals from collapsing; a fibrous protein called *spongin* often serves the same purpose. Commercial natural bath sponges (not to be confused with the brightly colored synthetic sponges encrusting supermarket shelves) have been treated to retain only this spongin matrix; all the cells are gone.

Phylum Cnidaria

Jellyfish, sea anemones, and corals belong to the phylum **Cnidaria** (*knide* = nettle), which contains about 9,000 mostly marine species. (You may be more familiar with this phylum's old name, *Coelenterata*.) This group of carnivorous animals takes its name from the large, stinging cells called **cnidoblasts** (*knide* = nettle, *blastikos* = to shoot upward), deployed on tentacles that bend or retract toward the mouth. Each cnidoblast contains a capsule from which a coiled thread may be forcefully ejected (**Figure 15.4**). The thread can repel an aggressor or penetrate or entangle prey, often immobilizing the victim

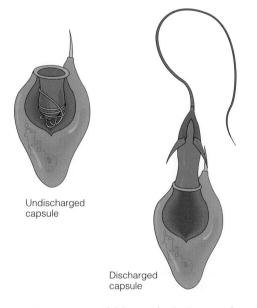

Undischarged capsule

Discharged capsule

FIGURE 15.4 A cnidoblast, with stinging capsule undischarged (left) and discharged (right).

with a toxin. Then the prey—which may include the larger zooplankters and small fish—is drawn into the mouth, leading to a saclike digestive cavity. Digested food is absorbed by cells of the inner layer and transported to other parts of the animal by migratory cells and by diffusion. Because the digestive cavity has only one opening, indigestible bones and other wastes are eliminated through the mouth.

Members of this group are built of two layers of cells. The inner layer (the gastrodermis) is responsible for digestion and reproduction, the outer layer (the epidermis) for capturing prey and protection from attack. The layers are connected by a jelly-like mesoglea (*mesos* = middle, *glia* = glue). Cnidarians exhibit **radial symmetry;** that is, their body parts radiate from a central axis like the spokes of a wheel. Some, such as sea anemones and corals, attach to rocks or other objects; others, such as jellyfish, swim freely in the water. No cnidarian possesses a definite head or concentration of sensory receptors, but a primitive network of nerves permits some species to respond to stimuli. These relatively simple animals depend on diffusion to move wastes and gases; they have no excretory or circulatory systems.

Cnidarians occur in two forms: medusae and polyps. Jellyfish are examples of the **medusa** body plan, named after a Greek mythological monster with a woman's face and hair that was a mass of writhing snakes. Medusae are predatory animals that swim by the rhythmic contraction of their bell-shaped bodies. They catch their prey with trailing tentacles armed with cnidoblasts. Some medusae are microscopically small, but the genus *Cyanea* can reach a bell diameter of 3.5 meters (12 feet), with tentacles 18 meters (60 feet) long! A smaller, more dangerous species is shown in **Figure 15.5.**

Sea anemones and corals are examples of the sedentary **polyp** (*polypous* = many-footed) body plan. Sea anemones have no skeleton and attach firmly to the substrate or burrow into it with a sticky basal disc on which they can slide slowly, like a snail. Corals contain a calcareous skeleton covered by living tissue, and they are permanently cemented in place. The anatomy of a coral polyp is shown in **Figure 15.6.**

Some corals are solitary animals with bodies up to 30 centimeters (12 inches) in diameter, but most of the more than 500 species are ant-sized organisms crowded into colonies called coral reefs (see Figures 12.26 and 12.27). The coral animals themselves construct the reefs by secreting hard skeletons of aragonite (a fibrous, crystalline form of calcium carbonate). The matrix of cup-shaped individual skeletons secreted by coral animals gives the colony its characteristic shape.

Tropical, reef-building corals are **hermatypic,** a term derived from the Greek word *hermatos,* meaning "mound-builder." Their bodies contain masses of single-celled, symbiotic dinoflagellates called **zooxanthellae** (*xanthos* = yellow), which carry on photosynthesis, absorb waste products, grow, and divide within their coral host. The coral animals provide a safe and stable environment and a source of carbon dioxide and nutrients; the zooxanthellae reciprocate by providing oxygen, carbohydrates, and the alkaline pH necessary to enhance the rate of calcium carbonate deposition. The coral occasionally absorbs one of the resident dinoflagellates, "harvesting" the organic compounds for its own use.

FIGURE 15.5 A sea wasp (*Chironix*), one of the most dangerous jellyfish. An inhabitant of tropical waters from Africa to northeastern Australia, it can kill a human within three minutes. The tentacles of a large specimen can be 15 meters (50 feet) long. *Chironix* has probably been responsible for more human deaths than sharks have.

Because of the needs of its zooxanthellae, hermatypic corals depend on light and warmth. Reef corals grow best in brightly lit water about 5 to 10 meters (16 to 33 feet) deep. Coral reefs can form to depths of 90 meters (300 feet), but growth rates decline rapidly past the optimum 5- to 10-meter depth. In ideal conditions coral animals grow at a rate of about 1 centimeter ($\frac{1}{2}$ inch) per year. They prefer clear water because turbidity prevents light penetration and because suspended inorganic particles interfere with feeding. The animals are protected from the harmful effects of bright sunlight by a mucous coating that contains an ultraviolet-blocking "suntan lotion" and by their habit of feeding at night; the polyps retract into their skeletal cups during the day to escape strong sunlight and predators.

Hermatypic corals also prefer water of normal or slightly elevated salinity. Coral animals are highly susceptible to osmotic shock, and exposure to fresh water is rapidly fatal; thus reefs growing in shallow water have a flat upper surface because rain is lethal. Fresh water and suspended sediments prevent reefs from forming near the mouths of rivers or in areas adjacent to islands or continents where rainfall is abundant.

The Cnidaria are a very successful group, but their simple architecture would not serve more advanced organisms. Their blind-sac digestive system allows only one batch of food to be processed at a time; feeding opportunities that arise during digestion cannot easily be accommodated. The lack of a distinct

a

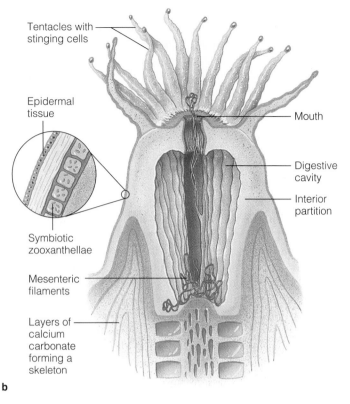

b

FIGURE 15.6 (**a**) A close-up of hermatypic coral, showing expanded polyps. (**b**) Anatomy of a reef coral polyp, with enlarged detail showing a cross section of the outer covering and tissue. The symbiotic photosynthetic zooxanthellae are crucial to the survival of this type of coral.

head with a concentration of sense organs is a drawback, as is the absence of circulatory, respiratory, and excretory systems. Having only two structural cell layers limits the complexity of systems and structures that can form within the organisms.

The Worm Phyla

15.6

A transition from relatively simple to more advanced organisms is made in the worm phyla, three of which are discussed here. The worm body plan exhibits **bilateral symmetry,** not radial symmetry; that is, the body has a left side and a right side that are mirror images of each another. Nearly all worms have some concentration of sensory tissue in what may be termed a head, and many have flow-through digestive systems and systems to circulate fluids and eliminate waste. Some are efficient parasites, but most are free-living. A few burrow in cavities; others roam the seabed or lurk under rocks.

The simplest worms are the well-named flatworms of phylum **Platyhelminthes** (*platys* = flat, *helmins* = worm). Some are parasitic of vertebrates, such as the tapeworms of fish and marine mammals. However, most marine flatworms are free-living predators and scavengers; they can be found on the shady underside of intertidal rocks or sharing colonies or burrows with other animals (**Figure 15.7**). Few examples exceed 3 centimeters ($1\frac{1}{4}$ inches) in length.

Flatworms are the most primitive organisms with a central nervous system. In some species a complex of nerve cells serving as a rudimentary brain connects the animal's simple nervous system to a pair of light-sensitive eyespots. The eyespots are small pigmented cups that can sense only the presence or absence of light—certainly a critical factor for an animal only a few cells thick that must avoid light to prevent overheating or detection.

FIGURE 15.7 An intertidal marine flatworm navigates the upper surface of an overturned rock. This primitive worm is very thin to allow quick diffusion of gases, nutrients, and waste.

FIGURE 15.8 Marine annelids of the class Polychaeta. (**a**) *Amphitrite,* a sedentary tube worm that extends its sticky tentacles across the seafloor to capture food. (**b**) *Eunice,* an active predator. (**c**) *Serpula,* which lives in stony tubes on the sides of rocks and captures prey with its grasping mouthparts.

FIGURE 15.9 A tube-dwelling polychaete with its "feather duster" appendages deployed. A mucous coating on the "duster" catches food, and the appendage itself works as a gill for gas exchange.

Larger flatworms are necessarily thin because they lack true respiratory and excretory systems. Gases must be exchanged and wastes eliminated by diffusion through the animal's surface, so no cell can be more than a few cell diameters from the outside.

The first animals in our phylum survey that possess a flow-through digestive system (a digestive tract with mouth and anus rather than a digestive cavity) are members of the plentiful phylum **Nematoda** (*nematos* = thread). Also called *round-worms,* these most successful of the worm phyla are present in nearly every imaginable terrestrial, aquatic, and marine habitat. Most of the 12,000 known species are free-living and microscopic, thriving in garden soil and marine sediments. Some make perfect parasites, however, and nearly all vertebrates and many invertebrates are parasitized by species of these long, thin worms. Readers who enjoy eating raw fish—sashimi—should read the information on fish parasites in **Box 15.1.**

The nematodes' true claim to marine fame is their astonishing numbers in some soft sediments. A Dutch researcher investigating shallow subtidal mud in the North Sea reported 4,420,000 tiny individuals in a sample 1 meter square (10.8 square feet) by 4 centimeters (1.57 inches) deep! Swimmers and gardeners have encountered these unobtrusive worms by the thousands without ill effects. Free-living worms cannot become parasitic.

Members of the phylum **Annelida** (*annelus* = ring) are the most evolutionarily advanced worms. Their bodies are divided into a number of similar rings or segments. **Metamerism,** as this segmentation is called, is a convenient strategy for increasing the size of an animal simply by adding nearly identical units. We see evidence of metamerism in most higher animals.

Each segment of an annelid can have its own circulatory, excretory, nervous, muscular, and reproductive systems, but some segments (such as those forming the head) are specialized for specific tasks. The familiar garden earthworm is an annelid.

The 5,400 species of class **Polychaeta** (*poly* = many, *chaetae* = bristles), the largest and most diverse class of annelids, are also the most important marine annelids (**Figure 15.8**). Polychaetes are often brightly colored or iridescent worms with pairs of bristly projections extending from each segment. They range in length from about 1 to 15 centimeters ($\frac{1}{2}$ to 6 inches). Some polychaetes burrow through and devour sediments or move freely over the bottom in search of food; others construct fixed parchment-like or calcareous tubes from which only parts of their heads emerge. Mobile polychaetes have well-developed heads with prominent sense organs, and they can function as efficient predators. Some skewer prey with their sharp mouthparts. Tube-dwelling polychaetes are a common sight to divers on most continental shelves. The "feather duster" tops of some of them exchange gases and remove food from the water (**Figure 15.9**). A few tropical forms have biting jaws at their tips, a surprise for the unwary student! All can instantly retract their feathery gills when disturbed.

The Advanced Invertebrates

Phylum Mollusca

The conspicuous phylum **Mollusca** (*molluscus* = soft bodied) contains 80,000 species, second in size only to the huge phylum

BOX 15.1   If It Moves, Don't Eat It!

People who enjoy sushi and sashimi (raw fish) run a slight risk of contracting herring worm disease or other maladies caused by marine parasites. More than 1,200 cases of humans infected by marine parasites have been reported, mostly from Japan but also from Europe and North America. About 50 cases of parasitic marine worm infections have been reported in the United States since 1980, probably because of an increase in the popularity of raw fish sushi, a Japanese delicacy.

The most dangerous parasites are nematodes of the genus *Anisakis*. These coiled worms—most often found in herring, red snapper, salmon, halibut, and some species of bass—are about 4 centimeters ($1\frac{1}{2}$ inches) long when unwound. They are usually coiled in balls the size of pinheads, either between groups of muscles or at the junction of muscles and organs. The worms look like tan or brown dots, often contrasting nicely with light-colored flesh.

Ingesting *Anisakis* can lead to much unpleasantness. Sometimes the parasite can be felt in the throat within 24 hours of the meal; coughing may dislodge it. Worms that have been swallowed may cause intense abdominal pain, nausea, vomiting, fever, and diarrhea. Symptoms of an infestation may resemble those of a peptic ulcer. A person will become dangerously ill if the worm penetrates the intestine. Japanese physicians see these infections rather frequently and recognize them, but American doctors often mistake the signs for an ulcer or appendicitis and treat them accordingly. In 1989 in the *New England Journal of Medicine*, a surprised surgeon reported removing a stricken patient's apparently normal appendix. Just as he was closing the incision he saw a 4.2-centimeter (1.65-inch) nematode crawl onto the surgical drapes.

Sushi lovers can greatly lessen the danger of *Anisakis* infection by patronizing specialty restaurants that employ experienced chefs. Sashimi from tuna or octopus is almost always free of the worms; salmon is potentially the most troublesome. Inspect each piece carefully, and *don't eat anything that moves!*

These parasites are killed by proper freezing or cooking. The fish must be frozen at $-20°C$ ($-4°F$) for more than 60 hours or cooked for at least 5 minutes at 60°C (140°F). (Microwave cooks please note: Quick microwaving of fish does not raise all parts of the portion to the same high temperature.) You can safely enjoy a hot fish dinner if the fish has been uniformly heated.

15.9

Arthropoda. The phylum Mollusca includes such diverse members as clams, snails, octopuses, and squid. Most mollusks are marine, and most have an external or internal shell. A few molluscan species possess acute sight and even considerable intelligence.

Mollusks and annelids probably shared a common origin—possibly a distantly ancestral segmented worm—and therefore share a few basic characteristics. Like annelids, mollusks are bilaterally symmetrical and generally have obvious heads, flow-through digestive tracts, and well-developed nervous systems. A few mollusks are segmented. Unlike annelids, however, some mollusks achieve great size, secrete beautifully fitted shells in which to take refuge, and exhibit great structural diversity.

We will briefly discuss three molluscan classes here: the class **Gastropoda**, the snails; the class **Bivalvia**, the clams, oysters, and mussels; and the class **Cephalopoda**, the nautiluses, octopuses, cuttlefish, and squid. Though very different in shape and habits, each class shows its link to a common ancient ancestor by sharing similar underlying parts. **Figure 15.10** shows some of the structural similarities in their body plans.

Gastropods (*gaster* = stomach, *pod* = foot) include the abalone, conch, limpet, and garden snail. Members of this largest class of mollusks usually inhabit relatively large shells, where they can seek refuge in case of danger. Some gastropods are grazers, some are suspension feeders, and some are preda-

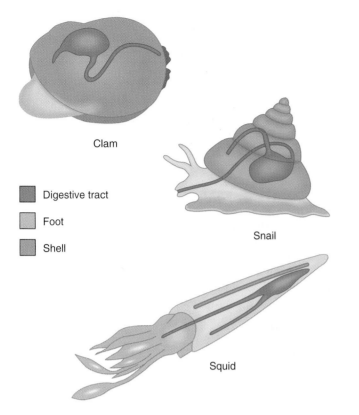

Clam

■ Digestive tract

□ Foot

■ Shell

Snail

Squid

FIGURE 15.10 The molluscan body plan as it has been modified in the major groups. (Source: From *Living Invertebrates* V. & J. Pearse/M. R. Buchsbaum, 1987, The Boxwood Press. Adapted with permission of the publisher.)

FIGURE 15.11 Gastropod shells are often aesthetically pleasing and have been prized by collectors for centuries.

tors. A few marine species—the pteropods and heteropods—are planktonic, but most marine gastropods are found wandering on rocky bottoms or other firm substrates.

Gastropod shells are often structures of great beauty (**Figure 15.11**). The animal adds to and enlarges the opening as it grows. The shell is frequently coiled to compress its mass and allow for easier maneuverability. A foot and head protrude from the shell while the snail moves about. The shell itself is secreted in three principal layers: a fibrous outer covering that may serve to distribute shock; a strong, crystalline layer of calcium carbonate ($CaCO_3$) to provide strength; and an inner layer of smooth $CaCO_3$ to provide nonabrasive surroundings for the resident. Not all gastropods have shells. Nudibranchs (*nudus* = naked, *branchia* = gills), also called *sea slugs*, are lovely (or even bizarre-looking) shell-less gastropods (**Figure 15.12**).

Although some snails use their foot to burrow, a gastropod foot cannot attach to sand or mud. The shell-and-foot configuration of gastropods is therefore not well suited to life on most of the ocean floor, which is covered by sediments. Evolution of the bivalves (*bi* = two, *valv* = door) admitted mollusks to this rich sedimentary habitat. Animals enclosed in twin shells (clams, oysters, mussels, and scallops), bivalves surrender mobility for protection, and they gather food by suspension feeding rather than pursuit (**Figure 15.13**). Burrowing species dig with a strong muscular foot and extend their siphons to the surface to obtain water and eject wastes. In other species the foot has different functions: In mussels it secretes the tough threads that attach the organisms to wave-swept rocks—a neat turnabout trick by which bivalves can invade a gastropod habitat.

The most highly evolved mollusks are the magnificent cephalopods (*cephalon* = head, *pod* = foot), a group of marine predators containing nautiluses, octopuses, and squid. These well-named animals have a head surrounded by a foot divided into tentacles. The nautiluses retain a large coiled external shell, but squid have only a thin vestige of the shell within their bodies, and octopuses have none at all. Cephalopods can move by creeping across the bottom, by swimming with special fins, or by squirting jets of water from an interior cavity.

Most cephalopods catch prey with stiff adhesive discs on their tentacles that function as suction cups, and they tear or bite the flesh with horny beaks. Nautiluses are an open-ocean group that hunts at considerable depths, their strong shells buoyed with gas-filled chambers (**Figure 15.14**). Little is known of their natural history. Squid and octopuses are more advanced cephalopods. They can confuse predators by secreting clouds of ink. Some kinds of squid eject a kind of "dummy" of coagulated ink that's a rough duplicate of their own size and shape—the squid is long gone by the time the attacker discovers the deception! At least one species of squid living below the euphotic zone produces sparkling luminous ink instead of black ink (which would be ineffective in the darkness).

FIGURE 15.12 A brightly colored reef nudibranch (*Ancula pacifica*) searches for food. The brilliant gill-like structures on its back assist in gas exchange. Although the nudibranch is usually in plain sight, its terrible taste seems to discourage animals from eating it.

Squid can grow to surprising sizes (**Figure 15.15**). The record length, including tentacles, is 18 meters (59 feet)! Most are much smaller.

Squid may be the largest invertebrates, but octopuses are the most intelligent. About as smart as puppies and with even better eyesight, some nearshore species of octopuses kept in captivity soon learn to recognize their keepers and forage at night through adjacent aquariums for tidbits. Octopuses use their visual acuity, ability to survive out of water for a short time, and intelligence to good advantage in their intertidal or subtidal homes, memorizing the positions of hiding places, escape holes, and good hunting locations.

Only about 450 species of cephalopods live today, a small percentage of the number of species known from fossils. The small number of species suggests that sophistication is no guarantee of biological success.

Phylum Arthropoda

The phylum **Arthropoda** (*arthron* = joint, *pod* = foot)—a group that includes the lobsters, shrimp, crabs, krill, and barnacles—is a phylum of superlatives. More than a million species of arthropods are now known! Arthropods are by far the most successful of Earth's animal phyla, occupying the greatest variety of habitats, consuming the greatest quantities of food, and existing in almost unimaginable numbers. As you read in the last chapter, one type of zooplanktonic arthropod—the krill—has the greatest biomass of any single species on Earth. The arthropod body plan is a variation on the basic annelid theme; bodies show clear segmentation with a pair (or pairs) of appendages per segment. All are bilaterally symmetrical.

Arthropods have not achieved the nervous system development of cephalopod mollusks, nor do they have the advantage of intelligence or extraordinary eyesight. They do, however, exhibit three remarkable evolutionary advances that have led to their great success:

■ An *exoskeleton*, a strong, lightweight, form-fitted external covering and support.

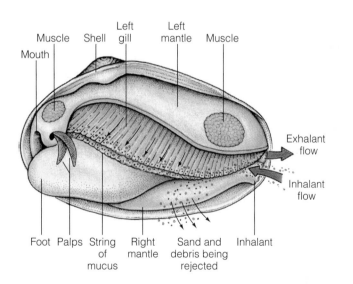

FIGURE 15.13 Bivalves are suspension feeders that make their living by filtering the water for edible particles. In this diagram (showing a bivalve with its left shell removed), water and tiny bits of food are swept into the animal by the movement of tracts of cilia on the gills. Food settles onto the gills and is then driven toward the mouth and swallowed.

FIGURE 15.14 A chambered nautilus (*Nautilus*) from the only living cephalopod group with an external shell. The animal occupies the outermost chamber; the other chambers are partially filled with gas to provide buoyancy. Chambers are added as the animal grows.

FIGURE 15.15 One of the largest specimens ever seen, this rare immature female giant squid measured 4.8 meters (16 feet) in length and weighed 150 kilograms (330 pounds) when recovered from a fisherman's net near New Zealand's remote Chatham Islands. Imagine the size of a full-grown adult!

- *Striated muscle*, a quick, strong, lightweight form of muscle that makes rapid movement and flight possible.
- *Articulation*, the ability to bend appendages at specific points. The appendages of more primitive phyla can usually bend anywhere (like wet spaghetti), but arthropod appendages bend at a joint. There are no ball-and-socket joints in arthropods; instead, each joint along an appendage moves through a different plane to ensure a full range of motion.

Most important of these advances is the **exoskeleton.** Unlike the often cumbersome shell of a gastropod, the exoskeleton of an arthropod fits and articulates like a finely tailored suit of armor. It is made in part of a tough, nitrogen-rich carbohydrate called **chitin,** which may be strengthened by calcium carbonate. Its three layers serve to waterproof the covering, tint it a protective color, and make it resilient and strong. Muscles within the animal are attached to the exoskeleton to move the appendages.

Such an arrangement sounds ideal, but the difficulties encountered by an organism with an exoskeleton are profound. How can muscle leverage be obtained? How can encrusting organisms be discouraged? How can ducts and feeding passages remain unblocked? And, perhaps most critically, how can the organism grow? That each of these problems has been solved is obvious by the group's overwhelming success, but the growth issue deserves a closer look.

We vertebrates grow by steadily adding length to the bones of our *internal* skeleton and bulk to our bodies. An *external* skeleton obviously limits growth and must be shed, or **molted,** at regular intervals. Arthropods do not have a steady growth pattern; instead, their external growth progresses in a series of

steplike jumps as the animal molts and replaces its exoskeleton (**Figure 15.16**). The arthropod grows without getting bigger between these jumps in size. An aquatic arthropod slowly substitutes body mass for water held in the tissues between molts. When molting, it suddenly takes on water from outside the body, expanding its tissues without growing in muscle mass. The shell splits and falls away, and, through a magnifi-

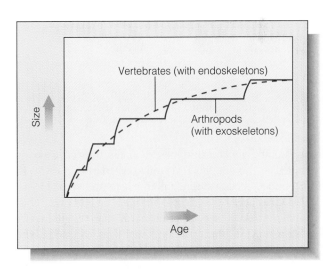

FIGURE 15.16 Generalized arthropod and vertebrate growth curves compared. The vertical segments of the arthropod curve represent molting periods. (Source: From James L. Sumich, *An Introduction to Biology of Marine Life,* 5/e. Copyright ©1992 Wm. C. Brown Communication, Inc. Dubuque, Iowa. Reprinted with permission of McGraw-Hill Companies.)

a

b

a

b

FIGURE 15.17 A molting arthropod. (**a**) A Dungeness crab (*Cancer magister*) backing out of the exoskeleton (right) that it is abandoning. (**b**) Clear of the old exoskeleton, the soft-bodied crab takes in water and expands. It immediately begins to secrete a new exoskeleton. Note the obvious increase in the animal's size.

FIGURE 15.18 Pentamerous (five-sided) symmetry in two classes of echinoderms. At some time in their lives, all members of this phylum possess a radially symmetrical body plan based on five sections or projections. (**a**) A sea star; (**b**) a close-up of the five-part jaws centered on the underside of a sea urchin.

cently orchestrated sequence of glandular secretions, the animal quickly regenerates a new exoskeleton one size larger (**Figure 15.17**).

The largest class of arthropods, the class Insecta, is poorly represented in the sea: There is only one marine genus with five known open-ocean species, all of which are water striders. The class **Crustacea** (*crustaceus* = having a shell or rind), however, includes 30,000 species of primarily marine, gill-breathing lobsters, crayfish, shrimp, crabs, water fleas, copepods, krill, amphipods, barnacles, and others. Their bodies usually have between 16 and 20 segments; the appendages may be specialized for sensing, food handling, walking, fighting, defense, and so forth. In numbers, about 70% of all zooplankton are copepods, minute crustaceans like those in Figure 14.11, which graze on diatoms and dinoflagellates. More familiar to most of us are the lobsters and crabs, including those we esteem for food. The largest crustacean is the king crab, which can reach a

legspan of 3.6 meters (12 feet). The heaviest individual, however, was a lobster caught off Chatham, Massachusetts, in 1949. The beast weighed 22 kilograms (48 pounds) and reportedly served ten people!

The activity and importance of billions of crustaceans as scavengers, predators, parasites, grazers, and general participants in oceanic biology are hard to overestimate. Large or small, obvious or retiring, crustaceans dominate the world of marine animals.

Phylum Echinodermata

The exclusively marine phylum **Echinodermata** (*echinos* = hedgehog, *derma* = skin) is an odd group sharply different from other members of the animal kingdom. The 6,000 species of echinoderms lack eyes or brains, have a radially symmetrical body plan based on five sections or projections (**Figure 15.18**), move slowly, and include only two known parasitic

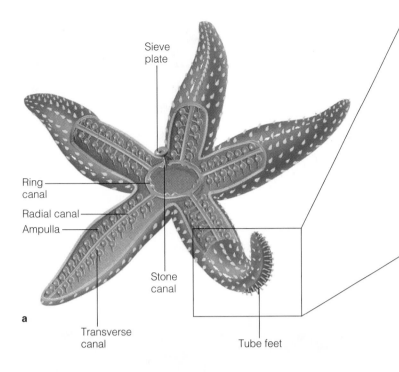

Sieve plate

Ring canal

Radial canal

Ampulla

Stone canal

Transverse canal

Tube feet

a

b

© Herve Chaumeton/Nature

FIGURE 15.19 (**a**) The water-vascular system in a sea star (shown in blue). Water enters the animal's body through a sieve plate, which excludes material that might clog the tubes and valves, and circulates through canals. (**b**) The tube feet lengthen, shorten, and bend because of changes in the distribution of water brought on by muscular contractions.

representatives. Living echinoderms are divided into five classes: The four most familiar are the class **Asteroidea,** the sea stars; the class **Ophiuroidea,** the brittle stars; the class **Echinoidea,** the sea urchins and sand dollars; and the class **Holothuroidea,** the sea cucumbers.

Nearly all asteroids (*aster* = star, *oidea* = resembling), or sea stars, are star-shaped echinoderms with five or more arms that are not completely delineated from the central disc (Figure 15.18a). The arms usually have spiny projections on top and delicate tube feet beneath. The tube feet work like suction cups and can grip objects; they also participate in gas exchange. Tube feet are part of a sea star's most striking feature: its unique **water-vascular system,** a complex of water-filled canals, valves, and projections used for locomotion and feeding (**Figure 15.19**). Operating like a hydraulic power system, the water-vascular system's plumbing can transmit forces generated by muscles at one side of the sea star to arms on the other side. Using this system, the tube feet of a sea star can grip a clam or mussel and exert a continuous pull (sometimes for hours) to force open its valves, even though one or more of the star's arms may tire. When the mussel finally opens, the sea star expels its stomach from its mouth, slips it between the mussel's shells, and digests the victim in place.

Delicate ophiuroids (*ophidion* = a snake) have long, slender arms. They are called *brittle stars* because of their unusual strategy for evading capture: If grasped by a predator, a brittle star will often detach its arm and escape. (The brittle star will later regenerate the arm.) Ophiuroids are perhaps the most widely distributed benthic marine animals; a few species are found in great numbers on the deep, sedimented seabeds of the world ocean (**Figure 15.20**). Many ophiuroid species also live beneath intertidal and subtidal rocks. Longitudinal grooves on the under-

side of each arm enable some species of brittle stars to locate food particles and transfer them by cilia to the mouth. Other species wave their arms through the water to capture plankton on sticky strands of mucus strung between adjacent arm spines.

Echinoids (*echinos* = hedgehog) are familiar to most coastal residents. The prickly appearance of a sea urchin or the smooth velvety surface of a sand dollar do not at first suggest the phylum's five-sided symmetry. Close observation, however, reveals

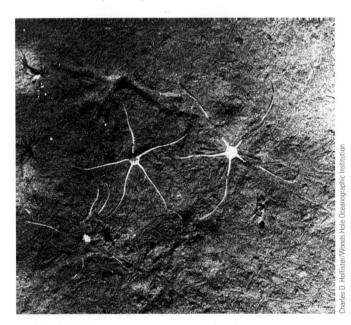

Charles D. Hollister/Woods Hole Oceanographic Institution

FIGURE 15.20 Ophiuroids (brittle stars) feed on edible particles in the surface layer of sediments on the continental slope off New England. Ophiuroids are among the most widely distributed of all benthic animals.

a steady internal temperature from metabolic heat; thus the internal body temperature of a fish is usually the same as that of the surrounding environment. About 40% of the 30,000-plus fish species live all or part of their lives in fresh water; 60% live exclusively in seawater. Fishes have evolved to fit almost every conceivable watery habitat, but they are most numerous on the bottom or in productive seawater over the continental shelves. Some species have a "sixth sense," an ability to detect small changes in the electrical field surrounding their bodies, that assists in the detection of prey or avoidance of predators. Some electric eels, catfish, and rays can use internally generated electricity for defense and offense.

The first fishes probably evolved in the ocean around 500 million years ago. These jawless animals were little more than motile sucking digestive tubes, but they did have the structural advantages of the chordate body plan. The earliest jawed fishes, with their grasping and crushing mouths, were far more successful at feeding on invertebrates with shells or exoskeletons than their jawless predecessors had been. Early jawed fishes were also equipped with paired fins, which stabilized their movements and minimized pitching and rolling when they attacked their prey.

The numbers and types of jawed fishes increased dramatically beginning about 410 million years ago. By the end of the Devonian period, the so-called Age of Fishes from 408 to 360 million years ago, jawed fishes had radiated into a vast number of aquatic and marine habitats in which their dominance has remained unchallenged. The ancestral jawed fishes gave rise to cartilaginous fishes (animals whose skeleton is made of stiff cartilage) and bony fishes. Only a few jawless forms, such as lampreys and hagfishes, have persisted to the present day.

The three living fish classes are as different from one another as they are different from amphibians, reptiles, birds, and mammals. We will look briefly at each in turn, beginning with the jawless fishes and ending with the advanced bony fishes.

Class Agnatha

15.17

Hagfishes and lampreys—members of the class **Agnatha** (*a* = lacking, *gnathos* = jaw)—lack jaws and have no paired appendages to aid in locomotion. Their thick, snakelike bodies are pierced by gill slits and (in some species) the openings of slime glands. Their round, sucking mouths are surrounded by organs sensitive to touch and smell. Agnathan eyes are degenerate and covered by thick skin. The body ends in a flattened tail that undulates to provide forward motion. The skin consists of resilient fibrous layers currently in high demand in South Korea for the manufacture of "eelskin" leather goods sold in upscale leather shops and department stores. Fewer than 50 species of agnathans are known.

The pinkish, soft hagfishes live in colonies on continental shelf sediments, where they burrow for polychaete worms or scavenge for weak or dead organisms. They prefer feeding on the soft inner flesh and internal organs of their prey, which they abrade with a rasping tongue. If a piece of food is too large, a hagfish will tie its flexible body into a loose knot, pass the knot toward the head, and brace against the prey for a better tearing

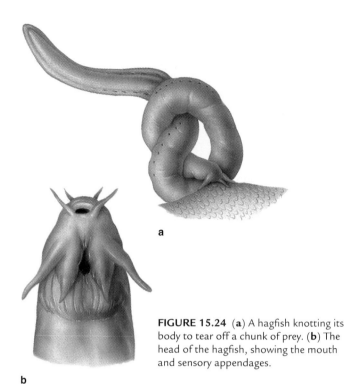

FIGURE 15.24 (**a**) A hagfish knotting its body to tear off a chunk of prey. (**b**) The head of the hagfish, showing the mouth and sensory appendages.

grip (**Figure 15.24**). A hagfish defends itself primarily by producing copious quantities of clinging slime from glands along the side of its body. This slippery, odorous mass deters all but the most determined would-be diner. The hagfish passes a knot down its body to shed the slime soon after danger has passed.

Lampreys, a group closely allied to hagfishes, possess a toothed, funnel-shaped mouth with which they can wear a passage into another vertebrate (**Figure 15.25**). Continuous sucking on this seeping wound provides nourishment for the lamprey. Prey are usually bony fish, but lampreys have been found feeding on several species of small whales. A lamprey will usually detach before it has killed the host. The largest hagfishes and lampreys can reach 1 meter (3.3 feet) in length, but most are smaller.

FIGURE 15.25 A lamprey pressing its toothed oral disc to the glass wall of an aquarium.

© Heather Angel

Class Chondrichthyes

Members of the group that includes sharks, skates, rays, and chimaeras have prowled the world ocean for at least 280 million years. Sharks have a history twice as long as that of dinosaurs. Fossilized teeth, fin spines, and shark's eggs found in marine sediments tell us that today's sharks and rays are not much different from their ancient ancestors.

All members of the class **Chondrichthyes** (*chondros* = cartilage, *ichthys* = fish) have a skeleton made of a tough, elastic tissue called **cartilage,** the material that gives your ears and nose their shape. Though there is some calcification in the cartilaginous skeleton, true bone is entirely absent from this group. Unlike the more primitive agnathans, these fish have jaws with teeth, paired fins, and often active lifestyles.

Only a small fraction of all fish species are members of class Chondrichthyes. About 350 species of sharks and 320 species of rays are known to exist. Sharks and rays tend to be larger than either agnathans or bony fishes, and except for some whales, sharks are the largest living vertebrates. Nearly all are marine, though a few species inhabit estuaries and a very few are permanent inhabitants of fresh water. Although there are many exceptions, sharks tend to favor swimming through open water, whereas rays tend to be found on or near the bottom. Chimaeras, such as ratfishes, are comparatively rare forms found mostly at and below middle depths.

Skates and rays have a flattened appearance, with spreading pectoral fins attached to their bodies to form a triangular or rounded winglike shape. They glide through the water with a slow, rhythmic flapping of their fins. Neither sharks nor rays have gas bladders, and both are slightly negatively buoyant: They will slowly sink if they stop swimming. The skin of rays is usually smooth and greatly variable in color. Sharks, in contrast, have a rough skin with thousands of tiny toothlike projections. One family of rays carries a defensive barb at the base of the tail capable of inflicting a serious wound. Another ray family contains members 7 meters (22 feet) across and weighing 1,700 kilograms (1.9 tons); an example is the giant manta (**Figure 15.26**). Still another produces a violent electric shock capable of stunning prey or disabling a human diver. The largest rays feed on plankton, but most smaller species crush mollusks and arthropods with smooth calcified plates in the mouth.

Sharks have an undeservedly bad reputation. Worldwide, sharks are responsible for about six known human fatalities each year. More people are killed in the United States each year by dogs than have been killed in the last century by sharks. For every human killed by a shark, humans kill more than 16 million sharks, mostly for food and medicines.

More than 80% of shark species are less than 2 meters (6.6 feet) long as adults, and only a few of the remaining 20% are aggressive toward humans. Sharks don't hold grudges or behave in the malignant ways so vividly portrayed in popular novels and movies. Still, some sharks are indeed dangerous to humans, and the great white sharks in the genus *Carcharodon* (*karcharos* = sharp, *odontos* = tooth) are perhaps the most dangerous of all (**Figure 15.27**). These swimmer's nightmares attain lengths of 7 meters (23 feet) and weigh up to 1,400 kilograms (3,000 pounds). (Reports of even larger great whites probably have resulted from misidentification of large, harmless species.) Great whites are not actually white but rather

FIGURE 15.26 A manta ray, sometimes called a devilfish because of the cartilaginous protuberances of the head, which are used to guide plankton into the mouth.

© Amos Nachoum/CORBIS

FIGURE 15.27 The great white shark (genus *Carcharodon*), a predator of seals, sea lions, and large fish. The great white shark is one of the most dangerous sharks encountered by human swimmers. In this extraordinary sequence taken off the coast of South Africa, a large great white shark rockets from the ocean with a freshly caught sea lion in its jaws.

grayish brown or blue above and creamy on the lower half. A dangerous relative, the mako shark, reaches lengths of 4 meters (13 feet) and is even known to attack small boats. These and other predatory species, such as tiger sharks and hammerhead sharks, are attracted to prey by vibrations in the water, which they detect with sensitive organs arrayed in lines beneath the surface of their skin. Smell also plays an important role in hunting their prey, usually fish and marine mammals.

Though most famous, the so-called man eaters are not the largest shark species. This honor goes to the immense, warm-water whale sharks in the genus *Rhineodon* (*rhine* = a rasp, a reference to the fish's rough skin), which reach sizes in excess of 18 meters (60 feet) and 41,000 kilograms (90,000 pounds) (**Figure 15.28**). Whale sharks and their somewhat smaller relatives, the basking sharks, are docile and present little threat to people. These greatest of fishes swim slowly near the surface with their huge mouths open, feeding on plankton. They may filter as much as 2,200 cubic meters (2,500 tons) of water per hour through a fine mesh of gill rakers. Accumulated plankton are periodically backflushed into the mouth, where they are concentrated for swallowing.

Class Osteichthyes

The 27,000-plus species of bony fishes, members of class **Osteichthyes** (*osteum* = bone, *ichthyes* = fish), owe much of their great success to the hard, strong, lightweight skeleton that supports them. These most numerous of fish—and most numerous and successful of all vertebrates—are found in almost every marine habitat, from tide pools to the abyssal depths. Their numbers include the air-breathing lungfishes and the lobe-finned coelacanths, whose ancient relatives broke from the path of fish evolution to establish the dynasties of land vertebrates.

About 90% of all living fishes are contained within the osteichthyan order **Teleostei** (*teleos* = perfect, *osteon* = bone), which contains the cod, tuna, halibut, perch, and other familiar species as well as less familiar ones (**Figure 15.29**). Within this large category are different fishes with gas-filled swim bladders to assist in maintaining neutral buoyancy, independently movable fins for well-controlled swimming and communication, great speed for pursuit or avoidance of predators, highly effective camouflage, social organization, the ability to cluster together in defensive schools, orderly patterns of migration, and other advanced features. Their economic importance is great: Some 70 million metric tons (77 million tons) of bony fishes are taken annually from the ocean to help satisfy the human demand for protein.

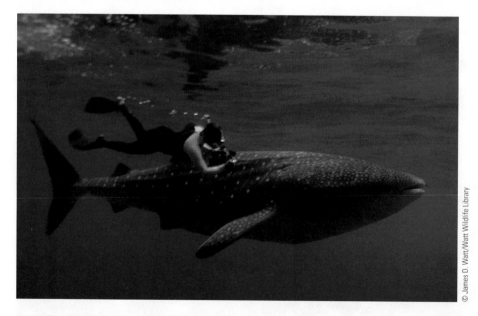

FIGURE 15.28 A diver hitches a ride on a whale shark. Unless he is struck by the fish's tail as he dismounts, the diver is in no danger. Divers are not part of the diet of this type of shark.

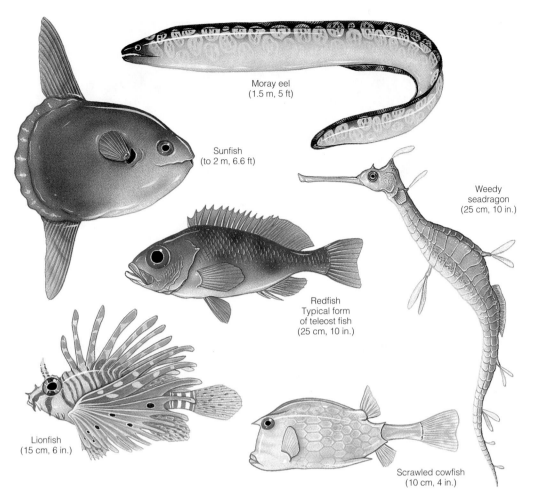

Moray eel
(1.5 m, 5 ft)

Sunfish
(to 2 m, 6.6 ft)

Weedy
seadragon
(25 cm, 10 in.)

Redfish
Typical form
of teleost fish
(25 cm, 10 in.)

Lionfish
(15 cm, 6 in.)

Scrawled cowfish
(10 cm, 4 in.)

FIGURE 15.29 Some of the diversity exhibited by teleost (bony) fishes. These fishes are not all drawn to the same scale.

The Problems of Fishes

What problems are unique to a fish? Seawater may seem to be an ideal habitat, but living in it does present difficulties. Water is about 1,000 times denser and 100 times more viscous than air, and it impedes motion more effectively at low speeds. How can a fish best move through it? How can a fish maintain its vertical position in the water column? What about breathing? Can oxygen and carbon dioxide be exchanged efficiently underwater? What about osmotic balance? How can some species of fish move from fresh water to salt water? How is salt balance maintained across the thin gill membranes? How can predators be thwarted? Can group behavior or camouflage or subterfuge be employed? There would seem to be many problems, but these most successful vertebrates have structures and behaviors to cope.

Movement, Shape, and Propulsion

Viscosity is a fluid's internal resistance to flow. Very small marine organisms are more directly affected by viscosity than larger ones are. Swimming is particularly difficult for tiny animals; the viscosity of water impedes their efforts at movement. For actively swimming marine organisms, the problem is **drag,** the resistance to movement of an organism caused by the fluid through which

it swims. The amount of drag depends on the viscosity of the water and the speed, shape, and size of the moving organism. Large swimming animals have a different problem with viscosity: the chaotic water movement known as **turbulence.** Water's resistance to flow results in turbulence around the swimmer and in the swimmer's trailing wake, both of which tend to slow the animal. An animal can minimize this frictional loss by streamlining. As an example, **Figure 15.30** shows three objects with identical frontal area moving through water. Streamlined shape **c** will have only about 6% of the drag of shape **a**'s disk. Note the teardrop shape of the fast-swimming tuna in Figure 15.30d.

A fish's forward thrust comes from the combined effort of body and fins. Muscles within slender, flexible fish (such as eels) cause the body to undulate in S-shaped waves that pass down the body from head to tail in a snakelike motion. The eel pushes forward against the water much as a snake pushes against the ground (**Figure 15.31a**). This type of movement is not very efficient, however. The body must wave back and forth across a considerable distance, exposing a large frontal area to the water, and the long body length (relative to width) needed to propagate the wave requires increased surface area, which increases drag. A more efficient swimming mechanism is shown in **Figure 15.31b.** More advanced fishes have a relatively inflexible body that undulates rapidly through a shorter distance; they also have

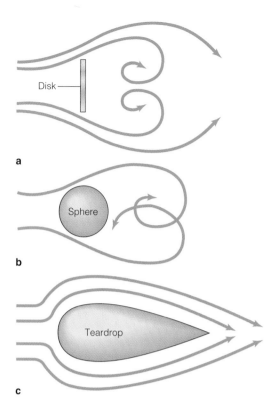

Disk

a

Sphere

b

Teardrop

c

d

© Timothy Reese

FIGURE 15.30 Turbulence and drag. At the same speed, with the same frontal area, shape **a** will have about 15 times more drag than shape **c**. Shape **b** shows only a small improvement in drag over the disk. (**d**) The fast-swimming yellowfin tuna is shaped like a teardrop to minimize turbulence and drag.

a hinged, scythelike tail to couple muscular energy to the water. The fish's body can be shorter and can face more squarely in the direction of travel, so the drag losses are lower.

Other unique adaptations have evolved in swimming species. Some fish secrete a small amount of friction-reducing mucus or oil onto their surface to minimize the formation of

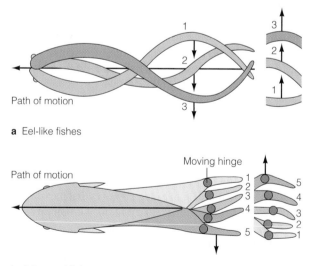

Path of motion

a Eel-like fishes

Moving hinge

Path of motion

b Advanced fishes

FIGURE 15.31 How a fish's body shape affects its swimming efficiency. The undulating movement of eel-like fishes (**a**) requires more energy to generate forward motion than the hinged-tail movement of more advanced fishes (**b**). (Source: From Pough et al., *Vertebrate Life*, 3/e © 1989. Reprinted by permission of Pearson Education, Inc. Upper Saddle River, NJ 07458.)

eddies, and some can tuck maneuvering fins into body recesses. In a sense, any reduction of drag is food saved, prey not needed, a greater ability to escape being eaten, and energy that can go to other concerns (such as reproduction).

How efficient are the best swimmers, and how fast can they swim? Estimates vary, but it is thought that in the fastest fish between 60% and 80% of muscle force delivered to the tail fin results in forward motion. Swordfish and marlin can reach 120 kilometers (75 miles) per hour in short bursts! Some of the fastest tuna (and a few swift sharks) sustain high speeds by maintaining their internal body temperature a few degrees above that of the surrounding water. This adaptation permits them to oxidize food more rapidly and generates greater muscle power per unit of weight.

Maintenance of Level

The density of seawater is nearly the same as the average density of living material, so the weight of a marine organism is usually counterbalanced by its buoyancy. Some parts of a fish are heavier than seawater and some parts lighter, but overall a marine fish has a density of around 1.07—about 5% denser than the same volume of seawater. To counteract the weight of heavy muscle and bone, many swimming fishes have gas-filled **swim bladders.** Gas is much lighter (less dense) than seawater, so only a small volume of gas is required.

Cartilaginous fishes have no swim bladders and must swim continuously to maintain their position in the water column. Sharks generate lift with an asymmetrical tail and fins that act like airplane wings. Bony fishes that appear to hover motion-

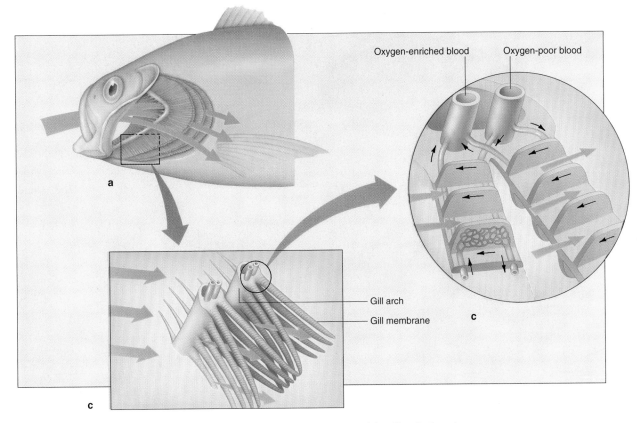

FIGURE 15.32 (**a**) A cutaway drawing of a mackerel, showing the position of the gills. The broad arrows in (**b**) and (**c**) indicate the flow of water over the gill membranes of a single gill arch. The small black arrows in (**c**) indicate the direction of blood flow through the capillaries of the gill filament in a direction opposite to that of the incoming water. This mechanism is called *countercurrent flow.*

less in the water usually have well-developed swim bladders just below their spinal columns. The quantity of gas is controlled both by secretion and absorption of gas from the blood and by muscular contraction of the swim bladder to compensate for temporary changes in depth.

A swim bladder may make life easier for a slow-moving teleost, but the fastest bony fishes lack swim bladders. Why? Fast, powerful predators such as tuna, mackerel, and swordfish must be able to chase prey between depths. The expansion and compression of gas in the bladder would vary rapidly with changing depth, and the risk of rupture would be too great. These speedy predators have power to spare and don't seem inconvenienced by their slight negative buoyancy.

Gas Exchange

How can fish breathe underwater? **Gas exchange,** the process of bringing oxygen into the body and eliminating carbon dioxide, is essential to all animals. At first glance the task may seem more difficult for water breathers than for air breathers, but air-breathing animals add an extra step. We air breathers must first dissolve gases in a thin film of water in our lungs before they can diffuse across a membrane.

Fish take in water containing dissolved oxygen at the mouth, pump it past fine **gill membranes,** and exhaust it through rear-facing gill slits. The higher concentration of free oxygen dissolved in the water causes oxygen to diffuse through the gill membranes into the animal; the higher concentration of CO_2 dissolved in the blood causes CO_2 to diffuse through the gill membranes to the outside. The gill membranes themselves are arranged in thin filaments and plates efficiently packaged into a very small space (**Figure 15.32**). Water and blood circulate in opposite directions—in a countercurrent flow—which increases transfer efficiency.

An active fish like a mackerel requires so much oxygen and generates so much waste CO_2 that its gill surface area must be ten times its body surface area. (Sedentary fishes have proportionally less gill area.) With their large gill area and countercurrent flow, active fish extract about 85% of the dissolved oxygen in water flowing past their gills. Air-breathing vertebrates, by contrast, extract only about 25% of the oxygen from air entering their lungs.

Osmotic Considerations

Marine invertebrates and primitive vertebrates such as agnathans have an internal salt concentration nearly identical to that of seawater. The body fluids of more advanced vertebrates generally are only about one-third as saline as seawater; they are hypotonic to their oceanic surroundings. The gill and intestinal membranes of fish are necessarily thin to allow the

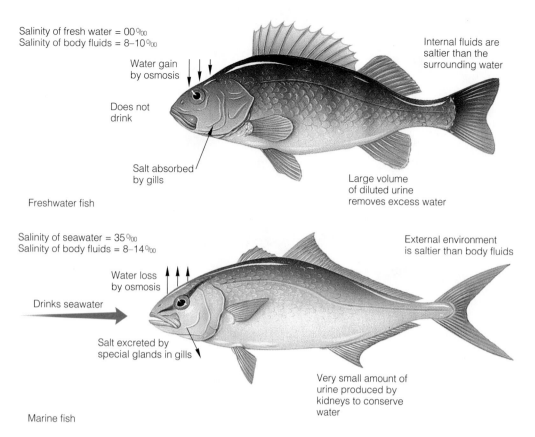

Salinity of fresh water = 00‰
Salinity of body fluids = 8–10‰

Water gain
by osmosis

Does not
drink

Salt absorbed
by gills

Freshwater fish

Internal fluids are
saltier than the
surrounding water

Large volume
of diluted urine
removes excess water

Salinity of seawater = 35‰
Salinity of body fluids = 8–14‰

Water loss
by osmosis

Drinks seawater

Salt excreted by
special glands in gills

Marine fish

External environment
is saltier than body fluids

Very small amount of
urine produced by
kidneys to conserve
water

FIGURE 15.33 Osmoregulation
in freshwater and marine fishes.

passage of gases and food molecules, but this thinness also makes the passage of water, or osmosis, inevitable.

As we saw in Chapter 13, osmosis moves water across membranes from regions of high water concentration to regions of low water concentration. Marine teleosts (bony fishes), with their higher relative internal concentration of water (and lower concentration of salts) per unit of fluid volume, continuously *lose* water to their environment; freshwater teleosts, with their lower relative internal concentration of water, constantly *absorb* water from their environment. If these fishes were incapable of **osmoregulation**—that is, if they had no active way of adjusting their internal salt concentration—they would quickly die of fluid imbalance.

Figure 15.33 summarizes the ways bony fishes cope with osmoregulatory difficulties. The skin of both freshwater and marine teleosts is nearly impermeable to water and salts. A freshwater fish (Figure 15.33a) does not drink water. It uses large kidneys to generate copious quantities of dilute urine to export the invading water, and it actively absorbs salts through its gills from both the surrounding water and its own urine. A marine fish (Figure 15.33b), on the other hand, makes only small quantities of urine, actively drinks seawater (some species drinking up to 25% of their body weight per day), and eliminates excess salts through special salt-secreting cells in the gills. Bony fish consume a substantial amount of energy in these essential osmoregulatory tasks.

Chondrichthyes (sharks, rays) employ a different strategy. Their internal fluids are supplemented with urea to produce a nearly neutral osmotic pressure across gill membranes. Although the proportion of dissolved solids in their body fluids is different from that in seawater, the quantity of dissolved solids is nearly identical, and the net flow of water into or out of the gills and intestinal membranes is minimal. Shark meat has an unpleasant, bitter taste (or tastes uncomfortably rich) unless it has been soaked in seawater to wash away the urea.

Feeding and Defense

Competitive pressure among the large number of fish species has caused a wonderful variety of feeding and defense tactics to evolve. Sight is very important to most fishes, enabling them to see their prey or avoid being eaten. Even some deep-water fishes that live below the photic zone have excellent eyesight for seeing luminous cues from potential mates or meals. Hearing is also well developed, as is the ability to detect low-frequency vibrations with the **lateral-line system.** This mechanism consists of a series of small canals in the skin and bones around the eyes, over the head, and down the sides of the body. The canals are richly supplied with nerves and connect to the surface through tiny pores in the skin. The nerves report changes in current direction, water pressure, or sonic environment to the brain. Predatory sharks use their lateral-line systems to detect prey.

Defense measures are well advanced in fishes. Some fishes such as sea horses (and their relatives the box fishes) depend on the simple expedient of armor plating for protection. Others, such as the puffer fish, inflate with water and erect bristly

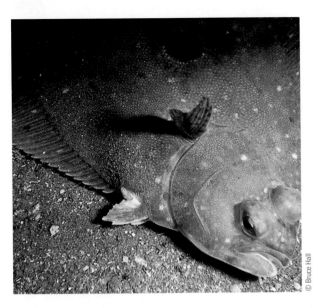

FIGURE 15.34 Active cryptic coloration in a turbot. The fish is able to blend into its surroundings because of chromatophores (skin cells that can change color). Though the fish is in plain view, the only obvious features are its round, black eyes.

FIGURE 15.35 Some fish avoid predators by disappearing completely—in one instant the predator sees a potential meal; in the next, it has left the ocean. This flying fish used specially modified pectoral fins to soar many meters across the surface. Some species lengthen the glide by extending a lower tail lobe into the water to re-accelerate.

spines to become less attractive as a snack. Actively swimming fish use a method of color blending called **countershading.** Their dark tops and silvery bottoms make the fish less obvious to predators above or below. More subtle means of offense or defense depend on trickery—looking like something you're not, or changing color to blend with the background. These kinds of **cryptic coloration** (*kryptos* = hidden) or camouflage may be active or passive. An example of active cryptic coloration is shown in **Figure 15.34.** The turbot—a bottom dweller—is capable of changing color to blend inconspicuously into its surroundings. This fish can reproduce the pattern of the background on its own surface. Passively cryptic animals look like nonthreatening objects (a seaweed-covered rock, for example) and can hide in plain sight. Unlike an actively cryptic animal, it cannot change its shape or color.

About a quarter of all bony fish species exhibit **schooling** behavior at some time during their life cycle. A fish school is a massed group of individuals of a single species and size class packed closely together and moving as a unit. There is no leadership in fish schools, and the movement of fish within them seems to be controlled automatically by direct interaction between lateral-line sensors and the locomotor muscles themselves. I can personally attest to the effectiveness of schooling as a means of defense. On a few diving trips I've noticed a large moving mass just beyond the limit of clear visibility. Is it a fish school, or is it a single large animal? Many predators might not stay around long enough to find out! Schools have the added benefits of reducing chance detection by a predator, providing ready mates at the appropriate time, and increasing feeding efficiency.

Perhaps the most effective way to avoid being eaten is simply to disappear. The surface-feeding flying fishes do this by ac-

celerating rapidly, leaping into the air, and spreading their fins to act as wings (**Figure 15.35**). Refractive differences between air and water prevent the surprised predator from seeing where its intended meal has gone.

Amphibians

Amphibians (frogs, salamanders, toads) are specialized vertebrates adapted to fresh water and to moist habitats on land. Only about 2,000 living species exist, and none is exclusively marine—although some large Southeast Asian frogs can tolerate water with a salinity of up to 28‰ for extended periods of time. Amphibians depend on the constant flow of water through their skins into their bodies to provide the fluid for the formation of urine to remove nitrogenous wastes. Placing an amphibian in seawater would cause water to flow through the skin in the opposite direction, dehydrating the animal. Their skin is too permeable to permit amphibians to colonize the marine environment.

Marine Reptiles

Each of the three main groups within the class **Reptilia** has marine representatives: turtles, sea snakes and marine lizards (iguanas), and marine crocodiles. Like all reptiles, marine reptiles are ectothermic, breathe air with lungs, are covered with scales and a relatively impermeable skin, and are equipped with special **salt glands** to concentrate and excrete excess salts from body fluids. Except for one widely ranging species of turtle, all marine reptiles require the warmth of tropical or subtropical waters.

Sea Turtles

15.28

The best-known and most successful living marine reptiles are the eight species of sea turtles. Unlike land turtles, from which they evolved, sea turtles have relatively small, streamlined shells without enough interior space to retract the head or limbs. The shell provides an effective passive defense, and adult sea turtles have no predators except humans. Their forelimbs are modified as flippers and provide propulsive power; their hind limbs act as rudders. The two species of green sea turtles (*Chelonia*) are the most abundant and widespread of living species (**Figure 15.36**). Green turtles range over great distances looking for the marine algae, turtle grass, and other plants on which they feed. The largest living turtle is the carnivorous Atlantic leatherback (*Dermochelys*), a streamlined animal with a soft, skin-covered "shell"; it reaches lengths in excess of 2 meters ($6\frac{1}{2}$ feet) and may weigh more than 600 kilograms (1,300 pounds). All sea turtle species are endangered.

Sea turtles are justly famous for their remarkable feats of navigation, which in some cases have been progressively increased by seafloor spreading. Sea turtles return at two-, three-, or four-year intervals to lay eggs on the beaches at which they were hatched. Homing behavior can be a great advantage to an animal; if the parent survived its earliest childhood at this location, it will probably be a suitable place for hatching the next generation. The navigation of green turtles to tiny Ascension Island, an emergent point of the Mid-Atlantic Ridge between Brazil and Africa, has been extensively studied. Researchers have found that the turtles use solar angle (to find latitude), wind wave direction, smell, and visual cues—first to find the island and then to discover the spot on the beach where they hatched perhaps 20 years before!

Marine Crocodiles

15.29

Fortunately, perhaps, there is only one living species of true marine crocodile (**Figure 15.37**). Marine crocodiles live in mangrove swamps and reef islands and on isolated mainland shores in the tropical western Pacific. They hunt in packs and sometimes move ashore to consume any large, warm-blooded animals they can find. They are extremely aggressive, extraordinarily fast in attack, and responsible for a few human deaths each year. Male marine crocodiles of northern Australia and Indonesia have attained lengths of 7 meters (23 feet) and may weigh over a ton! (The much smaller and less dangerous Florida and Louisiana crocodilians live in fresh, brackish, and salt waters.)

Marine Birds

15.30

Birds (class **Aves**) probably evolved from small, fast-running dinosaurs about 160 million years ago. Their reptilian heritage is clearly visible in their scaly legs and claws and in the configuration of their internal organs and skeletons. The success of the 8,600 living species of birds has resulted in large part from the evolution of feathers (derivatives of reptilian scales), which insulate the body and provide aerodynamic surfaces for flight. Birds (and mammals) are **endotherms**: They generate and regulate metabolic heat to maintain a constant internal temperature that is generally higher than that of their surroundings.

Flying birds have light, thin, hollow bones without fatty insulation; they have forsaken the heavy teeth and jaws of reptiles for a lightweight beak. Their highly efficient respiratory systems can accept great quantities of oxygen, and their large, four-chambered heart circulates blood under high pressure.

FIGURE 15.36 A green sea turtle covers her nest after laying eggs in the Heron Island preserve, Australia.

FIGURE 15.37 Compare the size of this marine crocodile to its trainer. Extremely aggressive, marine crocs are the world's largest, reaching a length of 7 meters (23 feet). They hunt in packs and can swim long distances between tropical islands.

All birds lay eggs on land, and most incubate them and provide care for the young. Some seabirds may stay at sea for years, but all must eventually return to land to breed.

Only about 270 kinds of birds, about 3% of known bird species, qualify as seabirds. Most seabirds live in the Southern Hemisphere. Like the marine reptiles, seabirds have special salt-excreting glands in their heads to eliminate the excess salt taken in with their food. Salty brine from these glands may sometimes be seen dripping from the tip of their beaks. Marine birds are voracious feeders and are often found wherever the ocean teems with life. True seabirds generally avoid land unless they are breeding; they obtain virtually all their food from the sea and seek isolated areas for reproduction.

Seabirds may be divided into four groups: tubenoses (albatrosses, petrels), pelicans and their relatives, the gull group, and penguins.

The Tubenoses

The 100 species of tubenoses of order Procellariiformes (*procella* = storm) are the world's most oceanic birds. Their prosaic common name does not convey any sense of their beauty and grace; it refers to the plumbing in their beak responsible for sensing air speed, detecting smells, and ducting saline water from the salt glands.

The largest of the tubenoses are the magnificent wandering albatrosses of genus *Diomedea*, which you met in this chapter's opener. These great birds can reach a wingspan of 3.6 meters (12 feet) and a weight of 10 kilograms (22 pounds). (Their ancestors were even larger; some had wingspans in excess of 5.5 meters, or 18 feet.) The key to these birds' success is their aerodynamically efficient wings (see the book's cover). Albatrosses can cover great distances in search of food with very little expense of energy, flying continuously for weeks or months at a time. Using the uplift from wind deflected by ocean waves to stay aloft, they soar in long looping arcs, but albatrosses rarely flap their wings.

The Pelicans and Their Relatives

Birds in the order Pelecaniformes (*pelekan* = pelican)—which includes cormorants, frigate birds, and boobies in addition to pelicans—all have throat pouches and webbed feet. These large birds, commonly seen off tropical and subtropical coasts, don't spend as much time over the open sea as their tubenose relatives do. Pelicans have broad, flat wings well adapted to slow flight. The wings are folded back when the pelican crashes into the water in pursuit of prey. In contrast, the aerobatic frigate birds can neither walk nor swim and so must feed on flying fish caught during flight, on small squid or fish obtained while hovering, or on regurgitated food stolen from other seabirds as they return to their rookeries from a day of hunting. These highly maneuverable birds are extraordinarily light and delicate; a frigate bird's skeleton weighs less than its feathers! The most oceanic of this group are the relatively small tropic birds, which are among the few seabirds

that regularly inhabit the relatively unproductive mid-oceanic gyres.

The Gulls

The order Charadriiformes (*charadra* = cleft dweller) is a shorebird group with both marine and freshwater representatives. A few of the 115-plus species of gulls and terns are familiar to anyone who has spent time near the shore. Some gulls are found far inland at lakes or at dumps (where they can indulge their skill at scavenging), but most are found along the coasts eating nearly anything available. They are extraordinarily maneuverable and efficient flyers, buoyant swimmers, and—thanks to long legs at the middle of the body—surprisingly good runners.

Terns, smaller and finer in shape than gulls, are more oceanic and may travel to sea for prolonged periods in search of food. Most terns are excellent divers and plunge for their prey. The arctic tern has the most extensive migratory route of any bird, completing a 24,000-kilometer (15,000-mile) round-trip each year. It avoids winter altogether by moving from high northern latitude to high southern latitude at just the right time.

The Penguins

Penguins have completely lost the ability to fly, but they use their reduced wings to swim for long distances and with great maneuverability. The name of their order, Sphenisciformes (*spheniskos* = little wedge), refers to the shortness and shape of their wings. Their flightlessness makes it practical to have fatty insulation, greasy peglike feathers, stubby appendages, and large size and weight; indeed such heat-conserving adaptations are critical to marine survival in very cold climates. Their neutral buoyancy is an advantage as they forage for food underwater. Emperor penguins, the largest of the living penguin species, may dive to depths of 265 meters (875 feet) and stay submerged for 10 minutes or longer. Penguins feed on fish, large zooplankters, bottom-dwelling mollusks or crustaceans, and squid. A few of the 18 species spend two uninterrupted years at sea between breedings.

Penguins are native only to the Southern Hemisphere and range from the size of a large duck to a height of more than 1 meter (3.3 feet) and a weight exceeding 36 kilograms (80 pounds). They are thought to consume about 86% of all food taken by birds in the southern ocean—about 34 million metric tons (37 million tons) per year, mostly larger zooplanktonic crustaceans. The small Galápagos penguin lives a comparatively easy life fishing the cold, nutrient-rich Humboldt Current at the equator, but its Antarctic relatives lead what must surely be the most rigorous existence of any seabird. For example, Emperor penguins breed and incubate during the bitterly cold Antarctic winter (**Figure 15.38**). Unlike most other birds, Emperors do not establish a territory, instead huddling together in tight crowds to conserve heat. The mass of penguins moves slowly around the breeding area as warm pen-

FIGURE 15.38 Emperor penguins and a maturing chick. One of the larger penguin species, Emperors subsist on a diet of mainly fish and squid.

guins from the center of the mob circulate to the outside to be replaced at the core by their chilled peripheral friends.

Marine Mammals

The class **Mammalia** (*mamma* = breast), to which humans belong, is the most advanced vertebrate group. About 4,300 species of mammals are known. The three living groups of marine mammals are the porpoises, dolphins, and whales of order **Cetacea;** the seals, sea lions, walruses, and sea otters of order **Carnivora;** and the manatees and dugongs of order **Sirenia.**

Each of these orders arose independently from land ancestors. They exhibit the mammalian traits of being homeothermic, breathing air, giving birth to living young that they suckle with milk from mammary glands, and having hair at some time in their lives. Unlike other mammals, however, these extraordinary creatures have become adapted to life in the ocean and are dramatically different in appearance and function both from other modern forms and from the terrestrial precursors from which they branched only about 50 million years ago.

All marine mammals share four common features:

■ Their *streamlined body shape* with limbs adapted for swimming makes an aquatic lifestyle possible. Efficient locomotion depends on minimum drag and maximum ability to transfer propulsive energy from the muscles to the water. Thin, stiff flippers and tail flukes situated at the rear of the animal drive it forward, and similarly shaped forelimbs act

as rudders for directional control. Drag is reduced by a slippery skin or hair covering.

■ They *generate internal body heat* from a high metabolic rate and conserve this heat with layers of insulating fat and, in some cases, fur. They also have relatively few surface capillaries, thus further reducing heat loss from the blood in the skin. Their large size gives them a favorable surface-to-volume ratio; with less surface area per unit of volume, they lose less heat through the skin. This is why no marine mammals are smaller than a sea otter; a small mammal would lose body heat too rapidly. These adaptations are critical to animals living in cold water where food is readily available.

■ The *respiratory system is modified* to collect and retain large quantities of oxygen. The air duct "plumbing" of marine mammals is typically much different from that of land mammals, and the lungs can be more thoroughly emptied before drawing a fresh breath. The biochemistry of blood and muscle is optimized for the retention of oxygen during deep, prolonged dives. Some whales can stay submerged for 90 minutes.

■ A number of *osmotic adaptations* free marine mammals from any requirement for fresh water. Unlike other marine vertebrates, the marine mammals do not have salt-excreting glands or tissues. They swallow little water during feeding (or at any other time), and their skin is impervious to water. This minimal seawater intake, coupled with their kidneys' ability to excrete a concentrated and highly saline urine, permit them to meet their water needs with the metabolic water derived from the oxidation of food.

Order Cetacea

The 79 living species of cetaceans (*ketos* = whale) are thought to have evolved from an early line of ungulates—hoofed land mammals related to today's horses and sheep—whose descendants spent more and more time in productive, shallow waters searching for food. Modern whales range in size from 1.8 meters (6 feet) to 33 meters (110 feet) in length and weigh up to 100,000 kilograms (110 tons). Their paddle-shaped forelimbs are used primarily for steering, and their hind limbs are reduced to vestigial bones that do not protrude from the body. They are propelled mainly by horizontal tail flukes that are moved up and down by powerful muscles at the animal's posterior end. A thick layer of oily blubber provides insulation, buoyancy, and energy storage. One or two nostrils at the top of the head have special valves to prevent the intake of water when submerged. Whales have large, deeply convoluted brains and are thought to form complex family and social groupings.

Modern cetaceans are further divided into two suborders. **Figure 15.39** shows representative whales in each division. Members of suborder **Odontoceti** (*odontos* = tooth), the toothed whales, are active predators and possess teeth to subdue their prey. Toothed whales have a high brain-weight-to-body-weight ratio, and though much of their additional brain tissue is involved in formulating and receiving the sounds on which they depend for feeding and socializing, many researchers believe them to be quite intelligent. Smaller whales in this group include the orca (killer whale) and the familiar dolphins and porpoises of oceanarium shows. The largest toothed whale is the 18-meter (60-foot) sperm whale, which can dive to at least 1,140 meters (3,740 feet) in search of the large squid that provide much of its diet.

Toothed whales search for prey using **echolocation,** the biological equivalent of sonar; they generate sharp clicks and other sounds that bounce off prey species and return to be recognized (**Figure 15.40**). Reflected sound is also used to build a "picture" of the animal's environment and to avoid hitting obstacles while swimming at high speed. Odontocete whales are now thought to use sound offensively as well. Recent research indicates that some odontocetes can generate sounds loud enough to stun, debilitate, or even kill their prey. In one experiment dolphins produced clicks as loud as 229 decibels, equivalent to a blasting cap exploding close to the target organism. Sperm whales, it has been calculated, may generate sounds exceeding 260 decibels! (The decibel scale is logarithmic; compare this figure to the 130-decibel noise of a military jet engine at full power 20 feet away!) How this prodigious noise is generated is not yet known, nor do we know how such energy is radiated from the whale without damaging the organs that produce and focus it.

Suborder **Mysticeti** (*mystidos* = unknowable), the whalebone or baleen whales, have no teeth and are thought to have branched from the line leading to toothed whales early in whale evolution. Filter feeders rather than active predators, these whales subsist primarily on shrimplike krill (see Figure 14.12). They do not dive deep but instead commonly feed a few meters below the surface. Their mouths contain interleaving triangular plates of bristly, hornlike **baleen** used to filter the zooplankton from great mouthfuls of water or from mud scooped from a shallow seabed (**Figure 15.41**). The plankton are concentrated as water is expelled, then swept from the baleen plates by the whale's tongue, compressed to wring out as much seawater as possible, and swallowed through a throat not much larger in diameter than a grapefruit. A great blue whale, largest of all animals, requires about 3 metric tons (6,600 pounds) of krill each day during the feeding season—about 1 million calories per day. When nursing, a baby blue whale gains weight at about 4 kilograms (9 pounds) an hour! The short, efficient food chain from phytoplankton to zooplankton to whale provides the vast quantity of food required for their survival. A feeding blue whale with its mouth hugely distended with seawater and krill is shown in **Figure 15.42.**

Mysticeti is an excellent name for these odd and wonderful animals. We know comparatively little of their social structure, intelligence, sound-producing abilities, navigational skills, or physiology. We do know that humpback whales use complex songs in group communication and that blue whales may use very low-frequency sound to communicate over tremendous distances. Recent studies of the lenses of the eyes of bowhead whales suggest they may more than 200 years old! Until recently our primary response to all whales has been to slaughter them in countless numbers for meat and oil, with little thought for their extraordinary abilities and assets. More of this depressing history may be found in the discussion of marine resources in Chapter 17.

Order Carnivora

The order Carnivora (*carnis* = flesh, *vorare* = to devour) includes land predators ranging from dogs and cats to bears and weasels, but the members of the carnivoran suborder **Pinnipedia**—the seals, sea lions, and walruses (**Figure 15.43**)—are almost exclusively marine. Unlike the cetaceans, the gregarious pinnipeds (*pinna* = wing, *pedalis* = foot) leave the ocean for varying periods of time to mate and raise their young. They appear to have evolved from the same stock as modern bears.

True seals have a smooth head with no external ear flaps, the external part of the ear having been sacrificed to further streamline the body. Seals are covered with a short, coarse hair without soft underfur. They are graceful swimmers that pursue small fish (their usual prey) with powerful side-to-side strokes of their hind limbs. These rear appendages are partially fused and always point back from the hind end of the body and are of very little use for locomotion on land. The elephant seal, named for its large size and long snout, holds the diving depth record for all air-breathing vertebrates: 1,560 meters (5,120 feet).

Sea lions, familiar to many as the performers in "seal" shows, have hind limbs with a greater range of motion and are thus more mobile on land. They have a streamlined head with small external ears and a pelt with soft underfur; unlike seals, they use their front flippers for propulsion.

Walruses are much larger than either seals or sea lions and may weigh as much as 1,800 kilograms (2 tons). Walruses use their tusks like sled runners to guide their sensitive "whiskers" just above the sediment surface, looking for clam siphons at depths up to 91 meters (300 feet). They dig up the clams with their mouths, crush the clam shell and remove the meat, then eject the inedible fragments. The large tusks are also useful for hauling their heavy bodies onto ice floes.

The suborder **Fissipedia** (*fissus* = split, *pedalis* = foot) has many members (including cats, dogs, raccoons, and bears) and two truly marine representatives: sea otters and polar bears, relative newcomers to the marine environment.

Sea otters, active and pleasing creatures, are the smallest marine mammals (**Figure 15.44**). Human demand for their fur, the densest and warmest of any animal, caused their near-extermination. The modern population of the Pacific sea otter descends from a very few individuals accidentally overlooked by fur hunters of the late nineteenth and early twentieth centuries. Playful and intelligent, otters rarely exceed 120 centimeters (4 feet) in length. They eat voraciously, consuming up to 20% of their body weight in mollusks, crustaceans, and echinoderms each day. Sometimes otters lie on their back in the water, balancing a rock on their chest. They hammer the shell of the prey against the rock until the shell cracks. Then they extract morsels of food with small nimble fingers and roll over in the water to clean away the debris.

Polar bears spend most of their lives stalking seals and stranded whales on the frozen northern polar ocean (**Figure**

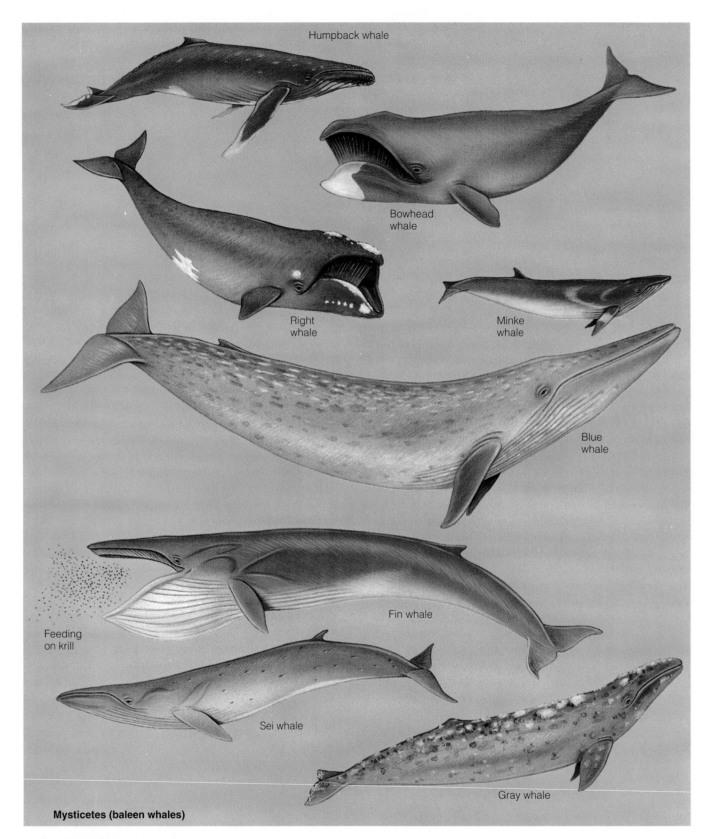

Humpback whale

Bowhead whale

Right whale

Minke whale

Blue whale

Feeding on krill

Fin whale

Sei whale

Gray whale

Mysticetes (baleen whales)

FIGURE 15.39 Some representatives of the order Cetacea. Note the differences between mysticetes and odontocetes.

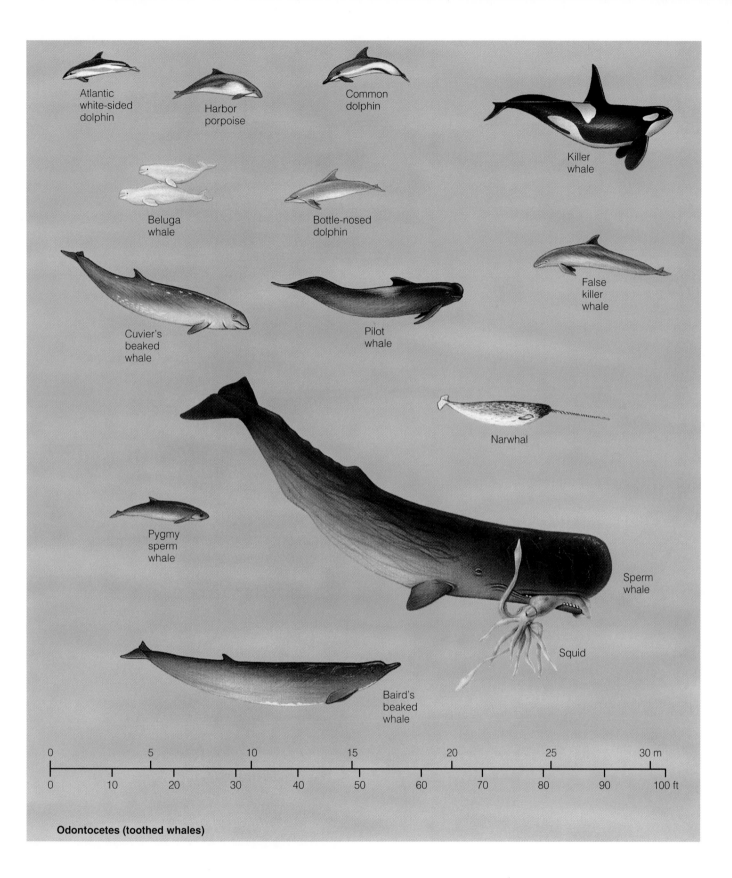

Atlantic
white-sided
dolphin

Harbor
porpoise

Common
dolphin

Killer
whale

Beluga
whale

Bottle-nosed
dolphin

False
killer
whale

Cuvier's
beaked
whale

Pilot
whale

Narwhal

Pygmy
sperm
whale

Sperm
whale

Squid

Baird's
beaked
whale

0	5	10	15	20	25	30 m

0	10	20	30	40	50	60	70	80	90	100 ft

Odontocetes (toothed whales)

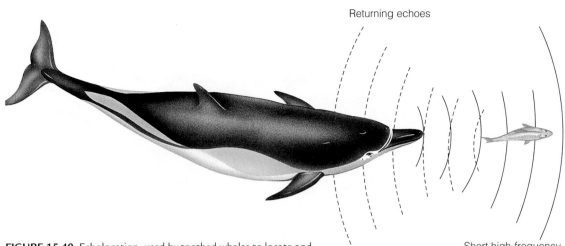

Returning echoes

Short high-frequency
sound pulses

FIGURE 15.40 Echolocation, used by toothed whales to locate and perhaps stun their prey.

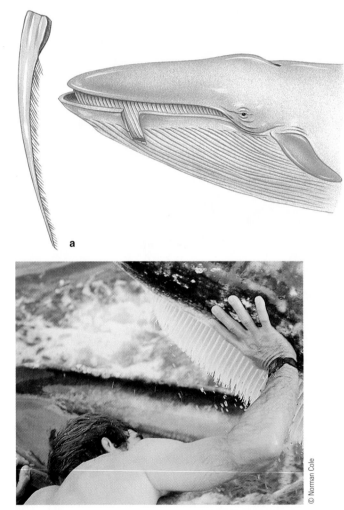

© Norman Cole

a

b

FIGURE 15.41 (**a**) A plate of baleen and its position in the jaw of a baleen whale. For clarity, the illustration shows an area of the mouth cut away. (**b**) A student and a live gray whale, up close and personal. The gray whale uses its stiff, coarse baleen plates to sieve crustaceans from shallow bottom mud.

© Richard Sears/MICS-Photo

FIGURE 15.42 A blue whale surfaces from a dive with throat pleats distended with water and food. The whale will filter the water in its mouth through plates of baleen, leaving behind krill, which it swallows. (For another view of a blue whale, see Figure 17.16.)

© Bruce Hall

a

Tom Garrison

b

FIGURE 15.43 Some representatives of the suborder Pinnipedia. (**a**) The California sea lion, the "seal" of seal shows. (**b**) Female elephant seals rest and molt on a central California beach.

15.45). They swim between ice floes with large, oarlike forepaws and can cross 100 kilometers (62 miles) of open water. The world's largest bears, male polar bears may grow to 2.5 meters (8.2 feet) and weigh 800 kilograms (1,800 pounds). They wander over enormous distances in search of food. One tagged bear moved 3,200 kilometers (2,000 miles) in a year, and bears have been seen in the vicinity of the pole. Apart from humans, polar bears have no enemies. Once an object of ex-

ploitation for their thick white fur, polar bears appear to be increasing in number under protection. As many as 40,000 are thought to exist today.

Order Sirenia

The bulky, lethargic, small-brained dugongs and manatees, collectively called sirenians (*siricis* = a mermaid), are the only

Courtesy of Friends of the Sea Otter/Richard Bucich

FIGURE 15.44 Two sea otters (*Enhydra lutris*) in a California kelp bed. Sea otters have the densest fur of any mammal.

FIGURE 15.45 A female polar bear with two cubs takes time out from seal hunting; the icy Arctic Ocean stretches beyond.

herbivorous marine mammals (**Figure 15.46**). Like the cetaceans, they appear to have evolved from the same ancestors as modern ungulates. They graze on sea grasses, marine algae, and estuarine plants in coastal temperate and tropical waters of North America, Asia, and Africa. Some species live in fresh water. The largest sirenians reach 4.5 meters (15 feet) in length and weigh 680 kilograms (1,500 pounds). They were first compared to mermaids by early Greeks, who noted the

FIGURE 15.46 A West Indian manatee (genus *Trichechus*), or sea cow.

manatee's habit of resting in an upright position in the water and holding a suckling calf to her breast. Sirenians have been hunted extensively, and only about 10,000 individuals are now thought to exist worldwide. Even though protected now in Florida, many are killed or wounded each year by the propellers of powerboats.

The animals we have discussed in this chapter do not exist in isolation but interact with each other and with plants and other autotrophs in complex marine communities. In the next chapter we will turn our attention to these groupings.

Questions from Students

1. Representatives of the phylum Mollusca are both the largest and the most intelligent invertebrates. Yet mollusks are *not* considered the most successful or the most highly evolved invertebrates. Why not?

The most often used measure of success in biology is the *number* of species and individuals within a group. The number of arthropod species and individuals greatly exceeds the number of molluskan species and individuals—arthropods are more successful.

The question of evolutionary position—that is, which group is most highly evolved—is a matter of some controversy, but consider this automotive analogy. Modern cars possess a number of sophisticated technological features, such as electronic fuel injection, turbochargers, independent suspension, high-speed radial tires, antilock brakes, aerodynamic enhancements, and nifty stereo systems. All of these innovations *could* be fitted to a 1959 Cadillac, but the older car's primitive chassis design would limit the performance of the total package. Cephalopod mollusks are an *old* chassis design. Excellent eyes and relatively high intelligence have been added to the ancient chassis, but the physical limitations of musculature and the lack of a jointed skeleton do not allow the class to exploit the full potential of these "options." Though well represented in the fossil record, fewer than 500 species of cephalopods exist today. They may represent an evolutionary dead end.

2. If krill are the most successful marine *invertebrate*, what is the most widely distributed and successful *vertebrate*?

Probably small pelagic bony fishes of genus *Cyclothone*. These *bristlemouths* are found just below the euphotic zone in virtually all the world ocean away from shore. Bristlemouths feed on plankton and on organisms that swim into their visual field at the limit of light penetration. Some ichthyologists believe there are more living individuals of *Cyclothone microdon* than of any other single vertebrate species on Earth.

3. I've heard that sharks don't get cancer. Could we find out what prevents cancer in sharks and synthesize it for human use?

In fact, sharks do develop cancer. The mistaken impression that shark cartilage may prevent or cure cancer in humans has contributed to sales of shark-derived supplements (derived mainly from hammerheads and spiny dogfish) worth more than $25 million in 1998. A large percentage of the 100 million sharks killed each year are taken for these ineffective medicines.

4. How do birds like the arctic tern and the wandering albatross navigate across the trackless ocean?

No one is certain. Experiments done at Cornell University with homing pigeons suggest that homing birds use a combination of magnetic and optical cues to return to their starting points. Even polarized light and the positions of certain stars might be involved. However it works, the behavior is not learned but is instinctive to the bird.

5. What's the difference between a dolphin fish and a dolphin mammal?

The name confusion began in Aristotle's time and has not abated. Dolphin mammals are small toothed whales; dolphin fish are teleosts (bony fish). The dolphin fish appears on restaurant menus and in seafood markets as mahi-mahi, dorado, and other names. Dolphin mammals are slaughtered in dismaying numbers in association with tuna fishing by some non-U.S. fishing fleets, but none of their meat appears on U.S. plates.

6. OK, how is a porpoise different from a dolphin?

Dolphin and *porpoise* are common names of two subtly different groups of odontocetes. Porpoises are the smaller members of the group; they have spade-shaped teeth, a triangular dorsal fin, and a smooth front end tapering to a point. Dolphins are usually larger and have an extended bottlelike jaw filled with sharp, round teeth. The small jumping whales in most oceanarium shows are dolphins. Indeed a killer whale is technically a dolphin and by far the largest member of that group. To make matters even more complicated, the common dolphin seen in ocean-themed amusement parks, *Tursiops truncatus*, is often referred to as a porpoise, even by show announcers. This confusion between common names points out how useful scientific names can be: The real name of the animal, *Tursiops*, is clear and unambiguous.

7. If whales are so intelligent, why haven't they learned to avoid the whalers' catcher boats?

There are many theories; some are presented in an interesting and curious book called *Mind in the Waters* by Joan McIntyre (see the bibliography). It may be that they are not as intelligent as was once thought. It may be that they have no concept of violence—a paradoxical thought considering the way odontocetes make their living. These animals perhaps should be considered intelligent for their native environment but should not be compared with terrestrial organisms, which have a radically different environment with radically different problems.

8. If some seals and whales dive to great depths, why don't they get the bends?

The bends is a painful and occasionally fatal condition brought on by nitrogen gas—the most plentiful component of air—leaving solution and forming bubbles in the blood. The condition is analogous to what happens in a newly opened soda bottle. Human divers take a source of air with them to depth, breathe the air under pressure, and dissolve excess nitrogen gas in their blood. If they have been down long enough, or deep enough, the release of pressure upon surfacing will be like taking the cap off the soda bottle. Whales don't use scuba tanks—they have no source of supplemental air at depth. They "tank up" at the surface by oxygenating their blood and tissues, but they have no excess gas to bubble from the blood at the end of their dives.

Chapter in Perspective

In this chapter you learned that animals must ultimately depend on primary producers (autotrophs) for nutrition. Animals could not exist on Earth until increasing levels of free oxygen in the atmosphere permitted them to metabolize food obtained from autotrophs. And remember, it was the photosynthetic autotrophs themselves that contributed huge quantities of oxygen to the environment. True multicellular animals arose between 900 and 700 million years ago, near the end of this oxygen revolution. Their variety is astonishing—a tribute to millions of years of complex interplay among environment, producer, and consumer.

Our survey of marine animals followed the course of their evolution. The complexity of animals increased as we moved from groups (phyla) whose basic structure seems to have solidified relatively early in the history of animals to groups that evolved more recently. Every marine animal has evolved effective adaptations for capturing prey, avoiding danger, maintaining thermal and fluid balance with their surroundings, and competing for space, and our survey of marine animals stressed these adaptations.

In the next chapter you will learn how these animals interact with one another and with their environment. The organisms you met in the last two chapters don't live alone. They are distributed throughout the marine environment in specific communities: groups of interacting producers, consumers, and decomposers that share a common living space. The types and variety of organisms found in a particular community depend on the physical and biological characteristics of that living space.

Key Ideas

✓ Animals are active multicellular organisms incapable of synthesizing their own food. Animals must obtain food from primary producers—or from other animals that have consumed primary producers—to ensure their own survival.
Review pages 355–357

✓ Animals arose near the end of the oxygen revolution— a time between about 2 billion and 400 million years

ago during which the activity of photosynthetic autotrophs changed the composition of the atmosphere to its present oxygen-rich mixture.
Review Figure 15.1, page 356

✓ A group of animals that shares similar architecture, level of complexity, and evolutionary history is known as a phylum.
Review page 356; Figure 13.23, page 328

✓ There were many more phyla early in the history of animals. The relatively few phyla present on Earth today represent the successful survivors.
Review Figure 15.2, page 357

✓ Each marine phylum has developed a set of adaptations—characteristics that have evolved to allow animals to capture prey, avoid danger, maintain thermal and fluid balance with their surroundings, and compete for space.
Review Figures 15.3, 15.6, 15.7, 15.10, 15.17, 15.18, 15.21, and 15.23, pages 358, 360, 363, 366, 368, 369

✓ Invertebrates are generally soft-bodied animals lacking a rigid internal skeleton, but many possess a hard outer protective covering. Almost every invertebrate phylum has marine representatives.
Review page 357

✓ All vertebrates are contained within the phylum Chordata. Vertebrates possess a bony spinal column as adults. The most abundant and diverse vertebrates are the fishes.
Review Figures 15.22 and 15.23, pages 368–369

✓ Marine mammals—specifically whales—are the largest animals ever to have lived on our planet.
Review Figures 15.39–15.46, pages 382–386

Terms and Concepts to Remember

Study Questions

Reviewing What You've Learned

1. What is an animal? How is an animal different from an autotroph?
2. Which animal phylum is most successful? How is success defined? What structural advances contribute most to that phylum's immense success?
3. Which phylum contains the largest representatives? The most intelligent?
4. Which phyla exhibit radial symmetry? Bilateral symmetry? No symmetry?
5. What are the classes of living fishes? Which is considered most primitive? Most advanced? Which class has the largest individuals? Which is the most economically important?
6. How do fishes swim? What problems are associated with swimming? How do fishes overcome these problems? How do fishes make defensive use of color, shape, and schooling behavior?
7. How do fishes "breathe" underwater? Do marine fishes experience any unusual problems by being immersed in seawater?
8. How are seabirds different from land birds? Are there more species of seabirds than land birds?
9. What characteristics are shared by all marine mammals?

Thinking Critically

1. When did the first true animals evolve? What atmospheric changes had to happen before animal life was possible? Are descendants of most of the early forms of animal life represented in the ocean today? Explain why.
2. How can an arthropod grow within a "tailored" shell? How can an animal grow without getting bigger, or get bigger without growing?

3. Are all chordates vertebrates? Are all vertebrates chordates? What distinguishes a vertebrate?

4. There are seven living classes of vertebrates, but only six are marine. List the seven classes. Which class has no permanent marine representatives? Why not?

5. How are odontocete (toothed) whales different from mysticete (baleen) whales? Which are the better known and studied?

6. **InfoTrac College Edition Project** The term *zooplankton* includes animals whose larval forms are small, such as crabs, squid, barnacles, and so on, as well as animals that remain small throughout their lives. Contrast the survival strategies of the two kinds of zooplankton. Would you expect the temporary larval members of the zooplankton community to have different ways of surviving compared to the lifelong residents? Use InfoTrac College Edition to research this question.

Thinking Analytically

1. Do you think there are more *species* of animals on land or in the ocean? What about *absolute numbers* of animals: Are there more animals on land or in the ocean?

2. There are more parasitic species of animals than all other sorts of animals combined. (If this sounds impossible, consider how many animals parasitize any animal you can think about, including ourselves.) Why do you think parasitism is such a runaway success? What are the drawbacks of being a parasite?

Study and Research on the Internet

Assess your understanding of this chapter's topics with additional quizzing and conceptual-based problems at:

http://oceanography.brookscole.com/garrison5e

- **InfoTrac College Edition Research** leads you to critical thinking exercises that use InfoTrac College Edition® as a research tool.
- **Regional InfoTrac College Edition** articles are organized into East Coast, West Coast, and Gulf Coast regions, allowing you to study oceanography on a more local level.

For more readings, go to InfoTrac College Edition, your online research library, at:

http://infotrac.thomsonlearning.com

CHAPTER 16

The hermit crab is eager to return to his tidal pool.

Tom Garrison

MARINE COMMUNITIES

The Resourceful Hermit

An enormous variety of organisms live in marine communities, but let's spend a moment with one of the more entertaining ones in an intertidal community, the hermit crab. These small, pleasant relatives of edible crabs and lobsters have engaged the attention of generations of seaside visitors. The hermits rush around sand-swept rocks or the floors of tidal pools, withdrawing into their borrowed shells at the slightest sign of danger. Their activity appears random, but their fighting, snooping, hiding, probing, and scuffling are purposeful. Like all animals, hermit crabs must struggle to eat, avoid predators, and mate. Hundreds of structural and behavioral adaptations contribute to their success. Sensors on the hermit's antennae and mouth parts alert him to the presence of food; good eyesight, muscular coordination, and a tough form-fitting covering usually foil fast-moving predators; and brilliant blue bands around the tips of his legs may signal his availability for mating.

One unique behavioral adaptation shared by all hermit crabs involves the selection of a temporary home. The front parts of a hermit crab—mostly pincers, antennae, and mouth parts—are formidable, but its hindquarters are delicate and subject to attack. To protect its flank, a hermit searches for any enclosed portable object to climb into, usually an unoccupied snail shell. A hermit crab's borrowed or stolen shell seems to be a source of both inordinate pride and perpetual difficulty. No shell is ever completely satisfactory. A hermit crab will carefully inspect any substitute dwelling, occupied or not, and consider whether to abandon its current digs in favor of the new candidate. House hopping proceeds fairly smoothly until the supply of suitably sized shells is exceeded by the number of potential occupants (as might happen when the crabs grow rapidly in times of abundant food). Then things get serious. Snails can be evicted from their self-made homes even before they're through with them—while the hermit gets a house *and* a meal in a single transaction. Two crabs may fight for hours or even days over one shell. Two others might simultaneously occupy the opposite ends of an abandoned worm tube, spending most of their time pulling each other in different directions.

Sometimes two crabs swap shells at a moment's notice. Renter's remorse sets in almost immediately, and they're off to see if they can find something even more suitable. At times human observers can't resist laughing at the all-too-human goings-on.

Marine Communities

Organisms are distributed throughout the marine environment in specific groups of interacting producers, consumers, and recyclers that share a common living space. These groups are called *communities*. A **community** is comprised of the many populations of organisms that interact at a particular location. A **population** is a group of organisms of the same species that occupy a specific area. The location of a community and the populations in it depend on the physical and biological characteristics of that living space (**Figure 16.1**).

The largest marine community—and also the most sparsely populated—lies within the uniform mass of permanently dark water between the sunlit surface and the deep bottom. Few animals live there because so little food is available, but those organisms that survive are among the strangest in the ocean. Opportunities for feeding in the deep open-ocean community are few and far between, and some animals are able to consume prey larger than themselves should the occasion arise. Because so few animals are present, mating is also a rare event; in a few species males and females become permanently bonded during their first encounter, with the male burrowing into the female's body for a lifelong free ride.

In contrast, the smallest obvious marine communities may be those established against solitary rocks on an otherwise flat, featureless seabed. Drifting larvae will colonize the place; the established community can seem an oasis of life and activity in an otherwise static sedimentary desert. Seaweeds will grow, worms will burrow, snails will scrape food from the hard surfaces, and small fishes will nestle among crevices. Hundreds of small plants and animals can live their lives within a meter of one another, interacting in a compact solitary community with no similar environment available for thousands of meters. The larvae of the next generation drift away with little chance of finding a suitable place to carry on their lives. Microscopic communities also exist; in fact, an interacting set of populations can exist on a single grain of sand or on one decomposing fish scale.

FIGURE 16.1 Mangrove roots support a diverse marine community that contains many populations of organisms.

NOAA

Organisms Within Communities

Communities are dependent on the availability of energy. As you learned in Chapter 13, living things cannot create new energy, but they can transform one kind of energy to a different kind. Using energy from the sun and the reactions of photosynthesis (or using energy from the chemical reduction of iron, sulfate, and manganese ions—chemosynthesis), primary producers assemble molecules of food (typically glucose) from atoms of carbon, hydrogen, and oxygen. The energy is passed from organism to organism in a food web (see Figure 13.6). These food webs and their related interactions usually define communities; the organisms in the web often share a common location and similar tolerances to the physical factors to which they are exposed.

There are many different places to live and many different "jobs" for organisms within even a simple community. A **habitat** is an organism's "address" within its community, its physical *location*. Each habitat has a degree of environmental uniformity. An organism's **niche** (*nidus* = nest) is its "occupation" within that habitat, its relationship to food and enemies, an expression of what the organism is *doing*. For example, the small fishes living among the coral heads in a coral reef community share the same habitat, but each species has a slightly different niche. Each population in the community has a different "job" for which its shape, size, color, behavior, feeding habits, and other characteristics particularly suit it.

Biologists also survey the *range* of species in a community. The variety of species in a given area is an expression of **biodiversity** (biological diversity). Communities with high biodiversity—a coral reef, for example—are characterized by complex interactions between and among residents.

The Influence of Physical and Biological Factors

The physical and biological factors in the environment determine the location and composition of a community. As we saw in Chapter 13, physical factors such as temperature, pressure, and salinity affect the success of an organism: Seaweeds adapted to cold water usually cannot survive prolonged exposure to abnormally warm water, and lower-than-normal salinity can dangerously disrupt the fluid balance of most marine invertebrates. Biological factors are influences on an organism by members of its own population or other populations, in its own or other communities. Biological factors include crowding, predation, grazing, parasitism, shading from light, generation of waste substances, and competition for limited oxygen.

A physical or biological factor that limits an organism's success in a community is a limiting factor for the organism, a factor that prevents the organism from feeding, growing, reproducing successfully, defending itself, sensing danger, or otherwise functioning successfully. As you may remember, some tropical fishes die in an unheated home aquarium because their physiology is optimized for a water temperature higher than the interior temperature of a typical house. If their

natural tropical environment temporarily dropped to the temperature of a cool room, some of the fishes would become too sluggish to catch fast-moving food or avoid larger predators. Temperature would *limit* their success. Other fishes can adapt relatively easily to changes in temperature, however. Temperature would be a limiting factor to a sensitive tropical fish species, but not to species with a wider temperature tolerance.

The prefixes *steno-* (meaning "narrow") and *eury-* (meaning "wide, broad") are sometimes used to describe those species, populations, or individuals that have narrow or wide tolerance to specific factors. **Stenothermal** species of tropical fish function better in a heated aquarium, but **eurythermal** fish do not need this extra attention. Likewise, **stenohaline** (*halos* = salt) marine organisms require a stable saline environment; they are unable to withstand relatively small fluctuations in salinity. **Euryhaline** species, however, can withstand a wide range of salinity and perhaps even tolerate some exposure to fresh water. Temperature and salinity are only two of the many physical factors that vary in a marine organism's native community. Light and pressure are two other examples.

Figure 16.2 provides an idealized look at the tolerance of organisms to a varying physical factor such as temperature. A eurythermal organism's optimal temperature range is wide (represented by a broad curve), but a stenothermal organism's optimal range is narrow (which would be represented by a steeper, tighter curve). The same reasoning applies to the salinity tolerance of stenohaline or euryhaline species, or to the pressure tolerance of stenobaric (*baros* = pressure) or eurybaric species.

An animal or plant is almost never exposed to fluctuations of only one physical or biological factor in its environment. An organism may die if subjected to changes in several environmental factors at once, even though each individual change is within its range of tolerance. Even small changes in temperature or salinity, survivable in themselves, might prove lethal if a particular species of tropical fish were exposed simultaneously to both.

A favorable balance of physical and biological factors is critical to each individual organism's success—and therefore to community success and longevity. The study of this balance, and of the relationships of organisms and interactions within communities, is called **ecology** (*oikos* = house, *logos* = study of). Marine ecologists are concerned with the types of organisms in marine communities as well as their habitats, niches, distribution, numbers, and reactions to variations in their environment.

Competition

The availability of resources such as food, light, and space in a community determines the number and composition of the populations of organisms within that community. Competition for the necessities of life may occur in the community between members of the *same* population or between members of *different* populations. Subtle swings in physical or biological factors may give one population the advantage for a time but then shift to favor another.

When members of the same population (all members of the same species) compete with each other, some individuals will

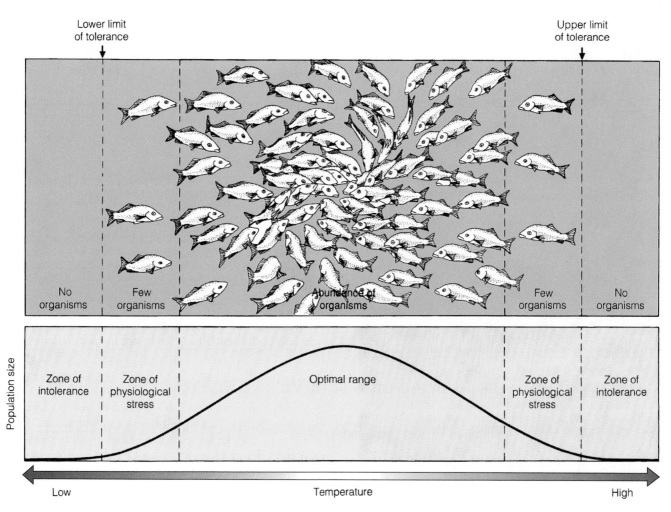

Lower limit of tolerance

Upper limit of tolerance

No organisms | Few organisms | Abundance of organisms | Few organisms | No organisms

Zone of intolerance | Zone of physiological stress | Optimal range | Zone of physiological stress | Zone of intolerance

Population size

Low — Temperature — High

FIGURE 16.2 Range of tolerance to a physical factor—in this case, temperature—for a population of organisms.

be larger, stronger, or more adept at gathering food, avoiding enemies, or mating. These animals tend to prosper, forcing their less successful relatives to emigrate, fight, or die in the course of competition. The most successful organisms have the most surviving offspring, so useful inheritable variations are passed along in greater quantity to the next generation. As we saw in our discussion of evolution, this kind of competition continually fine-tunes a population to its environment.

When members of different populations compete, one population may be so successful in its "job" that it eliminates competing populations. In a stable community, two populations cannot occupy the same niche for long. Eventually the more effective competitor overwhelms the less effective one. Extinction from this kind of head-to-head competition is probably uncommon, but restriction of a population because of competition between species is not. For example, the little barnacles of genus *Chthamalus* live on the uppermost rocks in many intertidal communities, while the larger limpets (*Collisella*) live lower on the rocks (**Figure 16.3**). Planktonic larvae of both species can attach themselves to rocks anywhere in the intertidal zone and begin to grow. In the lower zone the faster-growing limpets push the weaker barnacles off

the rocks, but at higher positions the limpets cannot survive because they are not as resistant to drying and exposure as the tough little barnacles are. At the top and bottom of their distribution, the two species do not compete for food or space. The competition at the intersection of their ranges prevents each species from occupying as much of the habitat as might otherwise be possible.

Growth Rate and Carrying Capacity

Organisms newly introduced into a favorable environment with no competitors for food or space will reproduce exponentially, tracing a J-shaped population growth curve. In nature very few populations reproduce at this maximal rate, however, because environmental conditions are rarely ideal and limiting factors in the environment quickly slow the rate of population growth. The sum of the effects of these limiting factors in the environment is called **environmental resistance**. Environmental resistance causes the actual population growth curve to be lower than the maximum potential growth curve. **Figure 16.4** shows the growth rate in number of indi-

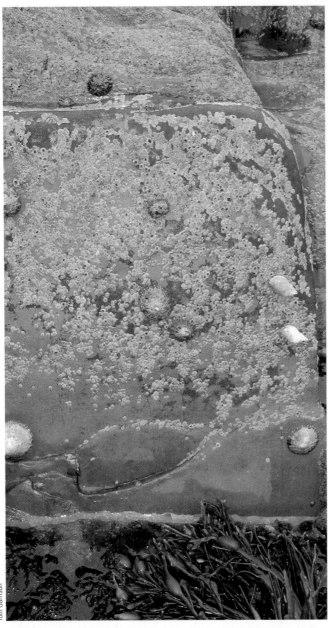

FIGURE 16.3 Competition between two species prevents either from occupying as much of the intertidal zone as might otherwise be possible. Small encrusting barnacles dominate most of this rock, but limpets at the bottom of the rock have probably prevented larval barnacles from gaining a foothold near the bottom.

viduals over time for both potential and actual situations. Note that the curve is S-shaped when limiting factors intrude; it gradually flattens toward an upper limit of the number of individuals in the population.

The final number of organisms oscillates around the **carrying capacity** of the environment for that species, the population size of each species that a community can support indefinitely under a stable set of environmental conditions. The carrying capacity changes if environmental conditions change. A marine population could *crash* if upwelling ceased, if new

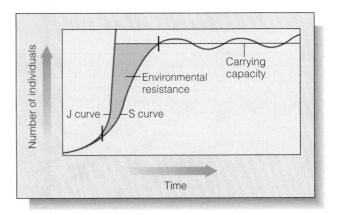

FIGURE 16.4 The J-shaped curve of population growth of a species is converted to an S-shaped curve when the population encounters environmental resistance. The physical or biological conditions responsible for the cessation of growth are called *limiting factors*.

predators were introduced, if climate varied, if food supplies dwindled, or if a new parasite infiltrated the population.

Distribution of Organisms

16.7

As we have seen, physical and biological factors affect the number and positions of organisms in a community. The number of individuals per unit area (or volume) is known as the **population density.** Rare individuals have a much lower population density than dominant ones. In general, more different species exist in benign habitats where physical factors stay near optimal values (like a coral reef or rain forest), and fewer species exist in rigorous habitats where physical factors range to extremes (like a beach or a desert). That is, "easy" habitats typically have high biodiversity—they contain more species in more niches within a given area—and harsh habitats usually have a lower species diversity. Relatively few species can cope with the stressful environment of the polar ocean surface, for example, but many species have adapted to the relatively benevolent environment of a tropical reef.

Individual organisms are almost never distributed randomly throughout their habitat. A **random distribution** implies that the position of one organism in a community *in no way influences* the positions of other organisms in the same community. Furthermore, a truly random distribution (such as that shown in **Figure 16.5a**) indicates that conditions are precisely the same throughout the habitat, an extremely unlikely situation except possibly in the unvarying benthic communities of abyssal plains.

The most common pattern for distribution of organisms is small patchy aggregations, or clumps. **Clumped distribution** (**Figure 16.5b**) occurs when conditions for growth are optimal in small areas because of physical protection (in cracks in an intertidal rock), nutrient concentration (near a dead body on the bottom), initial dispersal (near the position of a parent), or social interaction.

Uniform distribution (**Figure 16.5c**), with equal space between individuals (the arrangement of trees we see in or-

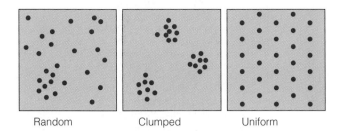

Random   Clumped   Uniform

FIGURE 16.5 Random, clumped, and uniform population distribution patterns. The clumped pattern is most common in nature; uniform is the rarest.

chards), is the rarest natural pattern of all. The distribution of some garden eels throughout their territories becomes almost uniform because each eel can extend from its burrow just far enough to hassle neighbor eels spaced at equal distance. But there is a break in the order of position. They don't line up row upon row like apple trees.

Change in Marine Communities

Like the organisms that constitute them, communities change through time, but marine communities do not evolve as rapidly as terrestrial communities. The slow changes associated with seafloor spreading, climate cycles, atmospheric composition, or newly evolved species have shaped this generally slow evolution. As on land, the species, community composition, and location of a marine community are changed by the environmental factors to which members of the community are exposed. Communities themselves can gradually modify the physical aspects of their environment. A coral reef is an extreme example: The massive accumulation of coral and sediments within the reef can alter current patterns, influence ocean temperature, and change the proportions of dissolved gases.

But rapid changes can occur in marine communities. A natural catastrophe—a volcanic eruption, a landslide that blocks a river, or the collision of an asteroid with Earth, for example—can disrupt a community. Similarly, human activities—such as altering an estuary by damming a river, dumping excess nutrients into a nearshore area, or stressing organisms with toxic wastes—can cause rapid, disruptive changes. Offshore communities change abruptly near new sewage outfalls, for example.

A stable, long-established community is known as a **climax community.** This self-perpetuating aggregation of species tends not to change unless disrupted by severe external forces, such as violent storms, significant changes in current patterns, epidemic diseases, or an influx of great amounts of fresh water or pollutants. A disrupted climax community can be reestablished through the process of **succession,** the orderly changes of a community's species composition from temporary inhabitants to long-term inhabitants. Disruption makes the environment more hostile to the original species, but destruction of species in the original community leaves open habitats and niches. A few highly tolerant species will move into the area, eventually

drawing in other species that depend on them. If the environment is permanently changed by the disruption, a climax community different from the previous one will be established.

Examples of Marine Communities

There are many distinct marine communities. We discussed the plankton community in Chapter 14; here we survey some other important examples.

Rocky Intertidal Communities

Anyone who spends time at the shore, especially a rocky shore, is soon struck by a curious contradiction. Although the rocky shore looks like a very difficult place for organisms to make a living, the **intertidal zone**—the band between the highest high-tide and lowest low-tide marks—is one of Earth's most densely populated areas. Hundreds of species and individuals crowd this junction of land and sea.

The problems of living in the intertidal zone are formidable. The tide rises and falls, alternately drenching and drying out the animals and plants. **Wave shock,** the powerful force of crashing waves, tears at the structures and underpinnings of the residents. Temperature can change rapidly as cold water hits warm shells, or as the sun shines directly on newly exposed organisms. In high latitudes ice grinds against the shoreline; in the tropics intense sunlight bakes the rocks. Predators and grazers from the ocean visit the area at high tide, and those from land have access at low tide. Too much fresh water can osmotically shock the occupants during storms. Annual movement of sediment onshore and offshore can cover and uncover habitats. Yet, astonishingly, the richness, productivity, and diversity of the intertidal rocky community—especially in the world's temperate zones—are matched by very few other places. There is intense competition for space. Life abounds.

One reason for the great diversity and success of organisms in the rocky intertidal zone is the large quantity of food available. The junction between land and ocean is a natural sink for living and once-living material. Minerals dissolved in water running off the land serve as nutrients for the inhabitants of the intertidal zone as well as for plankton in the area. The crashing surf and strong tidal currents keep nutrients stirred and ensure a high concentration of dissolved gases to support a rich population of autotrophs. Many of the larval forms and adult organisms of the intertidal community depend on plankton as their primary food source.

Another reason for the success of organisms here is the large number of habitats and niches to be occupied (**Figure 16.6**). The habitats of intertidal animals and plants vary from hot, high, salty splash pools to cool, dark crevices. These spaces provide hiding places, quiet places to rest, attachment sites, jumping-off spots, cracks from which to peer to obtain a surprise meal, footing from which to launch a sneak attack, secluded mating nooks, and darkness to shield a retreat. The niches of the creatures in this

a

FIGURE 16.6 A Pacific Coast tide pool and intertidal shore. (**a**) A diagrammatic view. (**b**) Key.

1 bushy red algae, *Endocladia*
2 sea lettuce, green algae, *Ulva*
3 rockweed, brown algae, *Fucus*
4 iridescent red algae, *Iridea*
5 encrusting green algae, *Codium*
6 bladderlike red algae, *Halosaccion*
7 kelp, brown algae, *Laminaria*
8 Western gull, *Larus*
9 intrepid marine biologist, *Homo*
10 California mussels, *Mytilus*
11 acorn barnacles, *Balanus*
12 red barnacles, *Tetraclita*
13 goose barnacles, *Pollicipes*
14 fixed snails, *Aletes*
15 periwinkles, *Littorina*
16 black turban snails, *Tegula*
17 lined chiton, *Tonicella*
18 shield limpets, *Collisella pelta*
19 ribbed limpet, *Collisella scabra*
20 volcano shell limpet, *Fissurella*
21 black abalone, *Haliotis*
22 nudibranch, *Diaulula*
23 solitary coral, *Balanophyllia*
24 giant green anemones, *Anthopleura*
25 coralline algae, *Corallina*
26 red encrusting sponges, *Plocamia*
27 brittle star, *Amphiodia*
28 common sea star, *Pisaster*
29 purple sea urchins, *Strongylocentrotus*
30 purple shore crab, *Hemigrapsus*
31 isopod or pil bug, *Ligia*
32 transparent shrimp, *Spirontocaris*
33 hermit crab, *Pagurus*, in turban snail shell
34 tide pool sculpin, *Clinocottus*

b

community are varied and numerous: Encrusting algae produce carbohydrates, snails scrape algae from rocks, hermit crabs scavenge for tidbits, and octopuses wait in ambush for likely meals.

The most obvious and important physical factor in intertidal communities is the rise and fall of the tides. Organisms living between the high-tide and low-tide marks experience very different conditions from those residing below the low-tide line. Within the intertidal zone itself, organisms are exposed to varying amounts of emergence and submergence. For example, **Figure 16.7a** plots the number of hours of exposure to air in a California intertidal zone through six months' time versus the tidal height. Because some organisms can tolerate many hours of exposure while others can tolerate only a very few hours per week or month, the animals and plants sort themselves into three or more horizontal bands, or subzones, within the intertidal zone. Each distinct zone is an aggregation of animals and plants best adapted to the conditions within that particular narrow habitat. The zones are often strikingly different in appearance, even to a person unfamiliar with shoreline characteristics. This zonation is clearly evident in the rocky shore of **Figure 16.7b.**

For intertidal areas exposed to the open sea, wave shock is a formidable physical factor. Fist-sized rocks have been thrown 100 meters (330 feet) into the air by the force of breaking waves. Large intertidal plants must be immensely strong, elastic, and slippery to avoid being shredded by wave energy. **Motile** animals move to protecting overhangs and crevices, where they cower during intense wave activity. Attached, or **sessile** (*sessilis* = sitting), animals hang on tightly, often gaining assistance from rounded or low-profile shells, which deflect the violent forces of rushing water around their bodies. Some sessile animals have a flexible foot that wedges into small cracks to provide a good hold; others (like mussels) form shock-absorbing cables that attach to something solid.

Desiccation (drying) by exposure to air and sunlight is another source of intertidal stress. Again, motile organisms have an advantage because they can move toward water left in tidal pools or muddy depressions by the retreating ocean. Attached animals and plants must await the water's return, huddled in low spots, moist pockets, or cracks in the rocks, or within tightly closed shells. Water trapped within a shell can keep gills moist for the needed exchange of gases. A protective mucous coating can retard evaporative water loss from exposed soft animal body parts or blades of seaweed. When the weather is too warm or when the tide is out for an unusually long time, a deceptive calm settles over the zone, to be relieved only by the returning ocean water.

Seaweed Communities 16.11

The shelter and high productivity of a kelp forest can help provide a near-ideal environment for animals. When light and nutrient conditions are optimal, large algae can make carbohydrate molecules so rapidly and in such quantity that sugars leak from their tissues like tea from a tea bag. Resident animals like sea urchins (**Figure 16.8**) are able to grow rapidly by collecting these molecules on their surfaces and transporting them directly into their bodies. As the algae weaken with age and

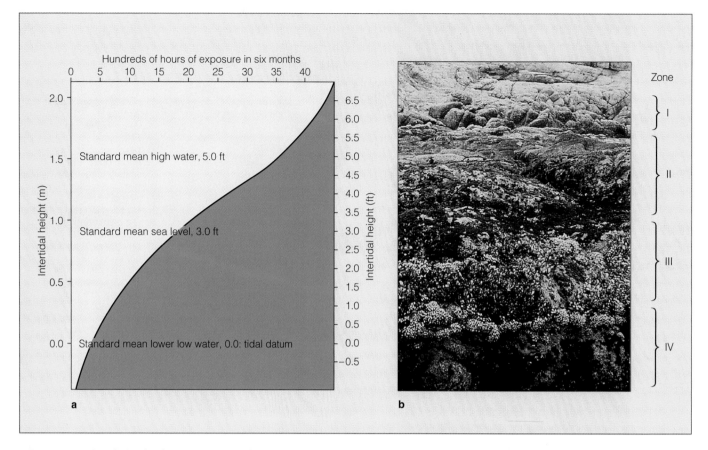

FIGURE 16.7 The relationship between amount of exposure and vertical zonation in a rocky intertidal community. (**a**) A graph of intertidal height versus hours of exposure. The 0.0 point on the graph, the tidal datum, is the height of mean lower low water. (**b**) Vertical zonation, showing four distinct zones. The uppermost zone (I) is darkened by lichens and cyanobacteria; the middle zone (II) is dominated by a dark band of the red alga *Endocladia*; the low zone (III) contains mussels and gooseneck barnacles; and the bottom zone (IV) is home to sea stars (*Pisaster*) and anemones (*Anthopleura*). The bands in the photograph correspond approximately to the heights shown in the graph. (From *Between Pacific Tides*, 5/e by Edward F. Ricketts et al. Revised by David W. Phillip. Stanford University Press, © 1985 by the Boards of Trustees of the Leland Stanford Junior University. Reprinted with permission.)

productivity declines, the urchins' sharp teeth can gnaw at the thalluses. Other animals graze on the blades, nestle within the holdfasts, and consume kelp flakes and debris. Sea otters may eventually move in to eat the urchins. As with all other communities, the kelp forest changes as its inhabitants come and go and as time passes.

Sand Beach and Cobble Beach Communities

Some intertidal areas are sandy, some are muddy, and others consist of gravel or cobbles. (A few shores combine all of these elements within a small area.) The usual rigors of the intertidal zone are intensified for organisms surviving on loose substrates. Indeed, it may surprise you to learn that in spite of its generally benign conditions, the ocean contains what may well be Earth's most hostile, rigorous, and dangerous environments for small living things: high-energy sand and cobble beaches.

As environments go, sand beaches don't seem particularly nasty places to us; many people consider the beach to be about the finest habitat around. Seals and sea lions spend a lot of time at the beach and seem to enjoy the experience as much as people do. In short, for organisms of about our size, the problems of living on a beach are manageable.

But for smaller organisms a beach is a forbidding place. Sand itself is the key problem. Many sand grains have sharp, pointed edges, so rushing water turns the beach surface into a blizzard of abrasive particles. Jagged grit works its way into soft tissues and wears away protective shells. A small organism's only real protection is to burrow below the surface, but burrowing is difficult without a firm footing. When the grain size of the beach is small, capillary forces can pin down small animals and prevent them from moving at all. If these organisms are trapped near the sand surface, they may be exposed to predation, to overheating or freezing, to osmotic shock from rain, or to crushing as heavy animals walk or slide on the beach.

As if this weren't enough, those organisms that survive must contend with the difficulty of separating food from swirling sand and the dangers of leaving telltale signs of their position for predators or being excavated by crashing waves. A few can run for their lives; some larger beach-dwelling crabs depend on their good eyesight and sprinting ability to outrace onrushing waves!

FIGURE 16.8 Sea urchins in a kelp bed. The urchins can absorb carbohydrates that leak from the algae, or gnaw the stipes and holdfasts with their teeth. Too many urchins can destroy a kelp forest by releasing the kelp from its holdfasts.

To these horrors must be added the usual problems of intertidal life discussed earlier. Not surprisingly, very few species have adapted to wave-swept sandy beaches! Some of the successful ones are shown in **Figure 16.9.** The few that have done so—mostly small, fast-burrowing clams, sand crabs, sturdy polychaetes, and other minute worms—consume a rich harvest of plankton and organic particles washed onto the beach and filtered from the water by the uppermost layer of sand.

Cobble beaches are even more uninviting. The rounded rocks clack and bump together as waves pound the shore; most small animals are crushed. Except for nimble, insectlike "beach hoppers" and a few species of scavenging terrestrial insects, most loose, rock-strewn shores are understandably void of anything much larger than microscopic organisms.

Surely the most difficult of all are the black sand beaches, derived from pulverized lava on tropical volcanic islands such as Hawaii. Besides all the other difficulties mentioned, we must add the lava's ability to store solar heat until temperatures approach 71°C (160°F) just below the surface of the sand. Almost nothing can tolerate these beaches for more than a few minutes—including human feet!

a

b

c

FIGURE 16.9 Sand beach organisms. (**a**) Dime-sized bean clams (*Donax*) lie at the surface awaiting a ride up the beach on an incoming wave. They will bury themselves in the loose sediment, push up their siphons, and filter the water for food. When the tide retreats, they will again pop to the surface and allow the waves to take them back down the beach. (**b**) A sand crab (*Emerita*), beloved of all beach-going children and beginning lab students, attempts to bury itself in anticipation of an onrushing wave. It gleans food from passing water with its feathery antennae. (**c**) A tardigrade, an example of an interstitial animal, an organism tiny enough to live in the spaces (or interstices) between sand grains and too small to be seen by the unaided eye.

a

FIGURE 16.12 The coral reef habitat. (**a**) A diagrammatic view. (**b**) Key. (**c**) A coral reef in the Red Sea.

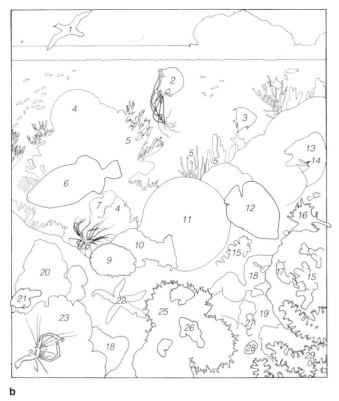

b

Key for coral reef habitat

1 black-capped petrel
2 sea nettle
3 angelfish
4 lobed corals
5 sea whips and soft corals
6 triggerfish
7 sea fans
8 tube anemone
9 orange stone coral
10 bryozoans
11 brain coral
12 butterfly fish
13 moray eel
14 cleaner fish
15 tube corals

16 muricid snail
17 nudibranch
18 sponges
19 colonial tunicate
20 giant clam
21 purple pseudochromid fish
22 cobalt sea star
23 soft corals
24 barber pole shrimp
25 sea anemones
26 clown fish
27 worm tubes
28 cowry
29 sea fan

c

© Peter Scoones

level of the last 3,000 to 10,000 years. Estuaries are probably more common today than they were at the height of the last glaciation, when sea level was lower.

Coral Reef Communities

Tropical coral reefs typically form in areas of high wave energy; indeed, reef organisms preferentially build into high-energy environments in an attempt to be the first to obtain dissolved and suspended material in the water. In most reefs there is an approximate balance between construction and destruction. The reef consists of actively growing coral colonies and fragments of material of different sizes from coral boulders worn down to fine sand.

Corals are by no means the only participants in reef life, however; they may account for only about half of the biomass in these areas. Other reef residents include calcareous algae whose secretions help "cement" the reef together as well as a bewildering array of encrusting, burrowing, producing, and consuming creatures ranging upward in size from the microscopic. Some tunnel into the coral or shatter it in search of food, contributing to the erosion of the reef. Fierce competition exists among reef organisms for food, living space, mates, and protection from predators. The bright colors, protective camouflage, spines, and various toxins and venoms common to tropical organisms are probably related to the intense struggle for existence that goes on in these beautiful but deceptively calm-looking places. A typical reef scene is depicted in **Figure 16.12.** More than a million species are thought to inhabit the ocean's coral reef ecosystems!

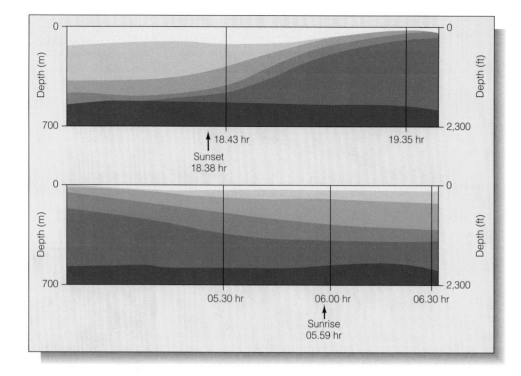

FIGURE 16.13 The movements of deep scattering layers, as recorded by an echo sounder. (**a**) Three distinct layers move toward the surface at sunset. (**b**) Before sunrise the layers move down again. This phenomenon is caused by organisms that migrate up and down with changing amounts of sunlight. (In both traces another scattering layer remains at a constant depth.)

The Open Ocean

About 83% of the ocean's total biomass is concentrated in its uppermost 200 meters (660 feet). Here we find most of the fishes and plankton. Of the ocean's biomass, only 0.8% is found below 3,000 meters (10,000 feet). Nearly all deep-ocean habitats are sparsely populated, but the few species of animals that have adapted to this impoverished place range through virtually all oceanic latitudes. There are no photosynthetic autotrophs in the deep ocean because there is no light. With the exception of rift communities (about which more in a moment), consumers at great depths must depend on the productivity of the water column above.

A peculiar pelagic community lives at the uppermost limits of the permanent darkness. Named after its ability to reflect sound pulses and appear to echo sounders as a false bottom, the **deep scattering layer** (**DSL**) is a relatively dense aggregate of fishes, squid, and other animals that usually migrate up and down in synchrony with daylight (**Figure 16.13**). Deep scattering layers—there are often more than one—are found in all ocean areas except the Arctic, and they are best developed in regions with high surface productivity. The DSL is most pronounced during daylight hours, when members of the community congregate at the lowest limit of light penetration. At nightfall many of the organisms migrate to the surface to feed on plankton. Most residents of the DSL have large, sensitive eyes, which permit them to feed by detecting the faint shadows of prey above. Some members of the community have built-in luminescent organs that cast dim blue light downward; this light masks their own shadows, so they have less chance of being detected and eaten (**Figure 16.14**).

Between the deep scattering layer and the bottom, in the bathypelagic zone, the ocean is nearly devoid of life. Rarely is

food available in this zone; tiny crumbs of organic material are usually broken down by microbes before reaching mid-depths, and the bodies of large organisms continue their fall to the seafloor. But recent research has shown that the deep-water environment is "patchy"; nutrient-rich zones caused by sinking remnants of phytoplankton blooms or the falling excretions of large fish populations can temporarily enrich a water parcel to the benefit of deep residents.

The few animals in this vast middle volume of ocean below the DSL are among Earth's most bizarre. Gulper eels have extendable jaws and a stomach capable of consuming prey larger than the eels themselves (**Figure 16.15**), an adaptation of great importance when one considers that a gulper eel may not encounter a feeding opportunity more often than once or twice a year! Bioluminescence, the biological production of light, is important here in both feeding and mate attraction. Some

FIGURE 16.14 *Diaphus elucens,* a common mesopelagic fish with large light organs on the body (the line of blue spots) and between the eyes. The light organs mask the fish's shadow and may identify it to potential mates. The fish is 5.5 centimeters (2 inches) long.

FIGURE 16.15 The deep-sea gulper eel (*Eurypharynx pelecanoides*), a bathypelagic species with a worldwide distribution beneath tropical waters. Its length is about 60 centimeters (24 inches). (Source: From J. C. Briggs, *Marine Zoogeography.* © 1974 by McGraw-Hill, Inc. Reprinted with permission.)

deep-swimming organisms attract their infrequent meals with a luminous lure (**Figure 16.16**). These animals also use patterns of glowing spots or lines to identify themselves to members of the same species, a necessary first step in mating. Some use flashes of light to dazzle or frighten potential predators.

The Deep-Sea Floor

Most of the deep-ocean floor is an area of endless sameness. It is eternally dark, almost always very cold, slightly hypersaline (to 36‰), and highly pressurized. Scientists once thought that

such rigors would limit the extent of communities there. Not so. In the 1980s researchers investigating bottoms at depths between 1,500 and 2,500 meters (5,000 to 8,000 feet) found an average of nearly 4,500 organisms per square meter. They took 21 samples (each 1 square meter) and recorded 798 species, 46 of which were new to science!

The feeding strategies of animals living on the deep-ocean floor are unique to this harsh environment. Tripod fish (**Figure 16.17**) use sensitive extensions of their fins and gill coverings to detect the movement of prey many meters away. Some organisms whose mouths blend with the natural contours of

FIGURE 16.16 Some species of deep-sea anglerfishes have bioluminescent lures. This fish is about 10 centimeters (4 inches) long.

FIGURE 16.17 A blind tripod fish, an abyssal benthic species. The long, curved projections on the fish's fins and gills are thought to aid in sensing the distant vibrations of prospective prey.

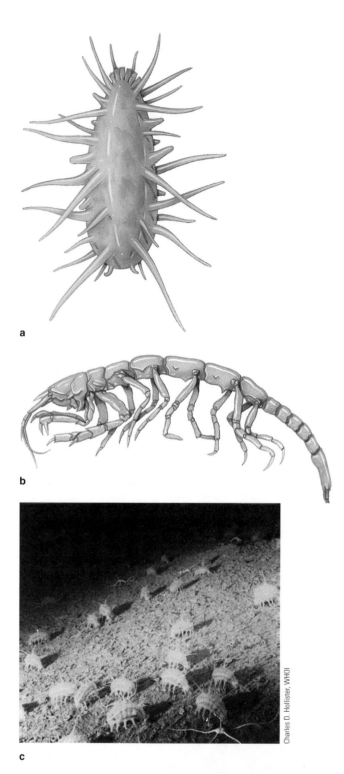

the ooze act as living caves into which small creatures crawl for protection. The predator need not even swallow to get the prey into its gut—back-pointing spines direct the victim along a one-way path to the stomach! Other species are capable of smelling large, sunken dead animals for many kilometers downcurrent and then spending weeks or months slowly following the scent to its source. The metabolic rate of organisms in cold water tends to be low, so most deep animals require relatively little food, move slowly, and live very long lives. Some may feed less than once a year and live to be hundreds of years old. Common deep benthic representatives are seen in **Figure 16.18.**

This deep, uniform environment, though rigorous, holds benefits for those few species adapted to it. Perhaps the best-adapted large organisms are the ubiquitous ophiuroids (brittle stars) that inhabit sedimentary bottoms, sometimes at arm-tip to arm-tip densities, at nearly all latitudes. They feed by collecting tiny nutritive particles—some of which will have fallen through the water for many months—along grooves beneath their arms and transporting them to their mouths. As you may recall from Chapter 15, brittle stars are among the most widely distributed of Earth's animals (see Figure 15.20).

The organisms in deep pelagic and benthic communities share some curious adaptations. Gigantism is a common characteristic: Individuals of representative families tend to be much larger in deep water than related individuals in the shallow ocean. Fragility is also common in the depths. Not only are heavy support structures unnecessary in the calm deep environment, but the relatively low water pH and high pressure discourage the deposition of calcium—and thus skeletal development. Some animals have slender legs or stalks to raise themselves above the sediment, and some come apart like warm gelatin at the slightest touch. Except for its influence on enzyme activity, hydrostatic pressure is not a problem for these animals. They live in balance with the great pressure; their internal pressure is precisely the same as the pressure outside their bodies.

Deep Rock Communities

Recent research has shown that seabed communities are not confined to the uppermost layer of sediments but are also found deep below the seafloor. What may prove to be the world's largest communities are only now being discovered. In the late 1970s researchers studying the quality of groundwater discovered that microorganisms could live in water-yielding rock formations. Because water coming from great depths can easily be contaminated by surface bacteria, a special drilling device was invented that could take uncontaminated samples of solid rock, even below the ocean bottom. Biologists were astonished to find microbial ecosystems in the pores between interlocking mineral grains of many rocks at drilling depths to 500 meters (1,650 feet) (**Figure 16.19**). Studies of cores from the Ocean Drilling Program show bacterial and archaean communities as deep as 800 meters (2,624 feet), and a Princeton geologist has recently extracted bacteria from water col-

Charles D. Hollister, WHOI

FIGURE 16.18 Abyssal benthic animals. (**a**) *Oneirophanta,* a 10-centimeter (4-inch) holothurian (sea cucumber) found on the abyssal plains of the North Atlantic. (**b**) *Apseudes galatheae,* a blind, thumb-sized crustacean found in the Kermadec Trench, north of New Zealand. (**c**) *Oneirophanta*—as in (**a**)—and brittle stars search for food on a continental slope at a depth of about 1,000 meters (3,300 feet). Brittle stars like these are among the world's most cosmopolitan organisms, found on nearly all deep sediments. (Source for Fig. 16.18a, b: From J. C. Briggs, *Marine Zoogeography.* © 1974 by McGraw-Hill, Inc. Reprinted with permission.)

FIGURE 16.19 Deep-living bacteria cultured from the minute spaces between mineral crystals in solid rock.

lected more than 3.2 kilometers (2 miles) beneath the South African coast!

The primitive organisms living at these depths can withstand very high temperatures and so are called **extremophiles** (*philos* = loving). In the oceanic crust the temperature rises roughly 15°C for every kilometer of depth. Some extremophiles are known to tolerate temperatures of 110°C; such organisms might be able to live as deep as 7 kilometers (4.4 miles) below the seafloor. What a vast habitat!

There may be as few as 100 or as many as 10 million bacteria in each gram of rock. What are they doing? What do they live on? Because their habitat receives no light, the autotrophs must be capable of chemosynthesis. The organisms chemically reduce iron, manganese, or sulfate ions, or generate methane from carbon dioxide, to supply the energy to build glucose molecules.[1] These primary producers are consumed by equally tiny primary consumers. Although a great many organisms are present, their metabolic rates appear to be very slow. As depth and pressure increase, the already microscopic pores in the rocks become smaller and the availability of chemosynthetic raw materials dwindles, but some researchers believe these well-hidden communities make up about one-third of Earth's total biomass!

---

[1] Reduction reactions involve the gain of electrons.

These aggregations of producers and consumers have been called SLIMES (for *subsurface lithoautotrophic microbial ecosystems*). New categories of bacteria have been found in their midst, including "ultramicrobacteria," dwarf bacteria apparently adapted to exceedingly small rock pores. It is tantalizing to think that these simple cells may be the remnants of Earth's first life-forms. Conditions on Earth at the time of the origin of life were hot and oxygen-free, and the genetic makeup of these bacteria and archaeans suggests they have evolved more slowly and in different directions than other forms of life here.

Hydrothermal Vent and Cold Seep Communities

Other organisms that can withstand extremes of temperature and pressure have also been found. The oceanographic world was excited in 1976 when scientists from the Scripps Institution of Oceanography discovered an entirely new type of marine community more than 3,000 meters (10,000 feet) below the surface. Using a towed camera platform, they were searching the seafloor along a spreading center 350 kilometers (220 miles) north and east of the Galápagos Islands. They found jets of water superheated to 350°C (660°F) blasting from rift vents in the young oceanic ridge. Seawater had percolated into fractures in the active crust and had been warmed by heat from nearby magma chambers at a depth of 1 to 3 kilometers (0.6 to 1.9 miles). Convection moved this water back to the seafloor, where it emerged as hot springs. The heated water dissolved minerals from the surrounding basaltic rock. As this water emerged and cooled, some inorganic sulfides precipitated, turning the water black. The term "black smoker" was quickly applied to these active vents (**Figure 16.20** and Figure 2.11).

Clustered around the vents were dense aggregations of large, previously unknown animals. Bottom water in the area was laden with hydrogen sulfide (H_2S), carbon dioxide, and oxygen, on which specialized archaea and bacteria were found to live. These chemosynthesizers form the base of a food chain that extends to the animals. Large crabs, clams, sea anemones, shrimp, and unusual worms in long parchment-like tubes were found in this warm oasis. These were termed hydrothermal vent communities.

Some of the so-called tube worms measured 3 to 4 meters (10 to 13 feet) in length and were the diameter of a human arm. They have been identified as pogonophorans, members of a small phylum of invertebrates also found in fairly shallow water; three species of the new genus *Riftia* have been found so far (**Figure 16.21a**). The tubes of these pogonophorans are flexible and capable of housing the length of the animal when it retracts. The animals extend tufts of tentacles from the openings of their tubes, but feeding was something of a puzzle because these animals have no mouth, digestive tract, or anus. The trunks of the worms were found to contain large feeding bodies tightly packed with bacteria similar to those seen in the water and on the bottom near the hydrothermal vents. The

a

FIGURE 16.20 The path of water associated with a hydrothermal vent. Seawater enters the fractured seabed near an active spreading center and percolates downward where it comes into contact with rocks heated by a nearby magma chamber. The warmed water expands and rises in a convection current. As it rises, the hot water dissolves minerals from the surrounding fresh basalt. When the water shoots from a weak spot in the seabed, some of these minerals condense to form a "chimney" up to 20 meters (66 feet) high and 1 meter (3.3 feet) in diameter. As the vented water cools, metal sulfides precipitate out and form a sedimentary layer downcurrent from the vent. Bacteria in the sediment, in the surrounding water, and within specialized organisms make use of the hydrogen sulfide (H_2S) in the water to bind carbon into glucose by chemosynthesis. This chemosynthesis forms the base of the food chains of vent organisms. (This illustration is not to scale.)

b

FIGURE 16.21 Some large organisms of hydrothermal vents. (**a**) *Riftia*, large tube worms (pogonophorans) that contain masses of chemosynthetic bacteria in special interior pouches. (**b**) A vent field dominated by the giant white clam *Calyptogena magnifica*. Each clam is about the size of a man's shoe and contains chemosynthetic bacteria within its gill filaments.

worms' tentacles absorb hydrogen sulfide from the water and transport it to the bacteria, which then use the hydrogen sulfide as an energy source to convert carbon dioxide to organic molecules. The ultimate source of the worms' energy (and the energy of most other residents in this community) is this energy-binding process—chemosynthesis—which replaces photosynthesis in the world of darkness.

The clams and shrimp of these vent communities are equally unusual. For example, the large, white clam *Calyptogena* grows among uneven basaltic mounds (**Figure 16.21b**). Each the size of a shoe, the clams shelter the same kinds of bacteria as *Riftia*. Though the clam retains its filter-feeding structures, it derives nutrition from the specialized bacteria embedded in the cells of its gill filaments. Small shrimp discovered at the vents in 1985 have been found to possess special organs that may allow them to sense heat from the vents. Such an adaptation would permit them to range away from the vents for food yet return to the warmth and richness of the home community. Hydrothermal vent communities have now been found off Florida and Louisiana, off California and Oregon, in the North Sea, east of Japan, and in at least 30 other locations.

Not all vent-type communities are on oceanic ridges, and not all are in areas spouting hot water, however. Cold seep communities are less dramatic but probably more widespread.

These areas are not always associated with the edges of tectonic plates, and about 25 large fields have been discovered so far. At cold seeps, hypersaline water rich in minerals, hydrogen sulfide, and sometimes methane percolates upward from beneath the seabed to emerge in broad fields at near-ambient temperatures. The slow upward seeping of cool, mineral-rich water encourages the growth of mats of chemosynthetic bacteria that are able to metabolize sulfur-containing compounds or methane. The methane, where present, appears to come from the decomposition of organic material in the sediment or underlying sedimentary rocks. The dominant large organisms at cold seeps are bivalve mollusks, pogonophoran worms, and a few species of sponges—but the heroes are the chemosynthetic bacteria and archaea that form the base of the food chains in these deep, mysterious places.

Studies of hydrothermal vent and cold seep communities suggest *many* questions. Do such communities occupy the active central rift valleys of a significant percentage of the 65,000 kilometers (41,000 miles) of Earth's oceanic ridges? Did the hot or cold vents—or even seemingly solid rock—serve as the birthplace of life on Earth? Perhaps the deep vent and seep communities will prove to be more important to overall marine productivity than has been previously supposed. Marine biologists are eager to continue their explorations.

Whale Fall Communities

You may wonder how the specialized organisms that inhabit vent and seep communities disperse over the great distances between vent systems. The problem becomes more acute with the recent knowledge that the lifetimes of the vents themselves may be measured in tens of years at most. How are such unique organisms recruited? How are they dispersed?

The answer may lie in the "stepping stones" provided by the fallen bodies of whales (**Figure 16.22**). Even though humans have greatly diminished the numbers of living whales, researchers estimate that whale carcasses may be spaced at roughly 25-kilometer (16-mile) intervals across areas like the North Pacific. Studies of fallen whale skeletons have shown the presence of sulfur-oxidizing chemosynthetic bacteria. As sulfide produced by these bacteria diffuses out of the bone, planktonic larvae of vent organisms might sense its presence, settle,

grow, and reproduce. With luck, their offspring might drift to another whale fall and repeat the process. After many steps the organisms would reach a new or newly active vent.

Symbiotic Interactions

Newcomers to biology are often surprised to learn that more than half of the animal species known to science are not free-living. Most are actively involved in close symbiotic relationships with at least one other life-form in their community. These relationships are often intricate and sometimes quite strange.

Symbiosis (*sym* = with, *bios* = life) is the biologist's term for the co-occurrence of two species in which the life of one is closely interwoven with the life of the other. The symbiotic bond is often so strong that one organism (the symbiont) is totally dependent on the other (the host).

There are three general types of symbioses: mutualism, commensalism, and parasitism. In **mutualism,** as the name implies, both the symbiont and the host—the larger organism with which the symbiont lives—benefit from the relationship. True mutualism is rare among marine organisms, but a few examples have been observed. One is the relationship between an anemone fish and its sea anemone. In this symbiosis a small, brightly colored anemone fish nestles within the stinging tentacles of a sea anemone (**Figure 16.23**). The mechanism that permits the little fish to do this without being stung is not well understood, but biologists believe that the fish gradually desensitizes the stinging cells of the anemone by using mucous secretions from its own skin. In return for the anemone's protection, the fish feeds the anemone scraps of food and may even lure prey within the anemone's reach.

Another example of mutualism is the relationship between certain cnidarians, such as coral animals, and the specialized dinoflagellates known collectively as *zooxanthellae* that live within their tissues. Both organisms benefit: The autotrophic dinoflagellates have a safe home and a ready source of carbon dioxide, and the animals have a handy, built-in source of

Craig Smith and Mike DeGruy

FIGURE 16.22 A whale fall community off the California coast. These communities may act as "stepping stones" for the specialized organisms that inhabit vent communities.

© Dave Breiter

FIGURE 16.23 Mutualism. Some species of sea anemones have a symbiotic relationship with anemone fish, in which the fish receives protection from predators and the anemone receives scraps of food from the fish.

carbohydrates. Without their resident dinoflagellates, the reef corals would be unable to deposit calcium, and the rich tropical reef communities we know today would not exist.

Perhaps the most striking examples of mutualism involve cleaning symbioses. In these relationships small organisms (usually fish or shrimp) establish a "cleaning station" at which they remove dead tissue and troublesome surface parasites from the skin, mouth, and gill coverings of larger animals (usually fish) that visit the site. The cleaner eats the dead tissue and parasites, so both animals benefit. The cleaned animals frequently defend the cleaning station and its cleaners from attack by would-be predators.

In **commensalism,** the symbiont benefits from the association, while its host neither benefits nor is harmed. For example, biologists once believed that the association between pilot fish and shark was mutualistic; the pilot fish was thought to guide the shark to a meal and in turn be permitted to dine on the scraps. We now know that the pilot fish is only an opportunistic commensal, taking what scraps it can from the shark's meal. (On the other hand, the mutualistic anemone fish/anemone partnership was thought to be commensal before the anemone fish was observed feeding the anemone.)

Some of the most curious commensals are those that enter their partner's habitat and, after a period of growth, cannot escape. Small pea crabs of genus *Fabia*, for example, live inside mussel shells, eating food particles that are brought inside by the normal feeding and respiratory actions of the mussel. After a time, the crab grows too large to fit through the gap between the mussel's shells.

Parasitism is the most highly evolved and by far the most common symbiotic relationship. The parasite lives in (or on) the host for at least part of its life cycle and obtains food at the host's expense. For obvious reasons, parasites do not usually kill their hosts, but they can seriously affect the host organism by reducing its feeding efficiency, depleting its food reserves, reducing its reproductive potential, lowering its resistance to disease, or otherwise sapping its energy. The host–parasite relationship is finely balanced and extraordinarily delicate. On one hand, the parasite must in some way be aware of the host's physical condition in order to avoid weakening the host so much that it dies. On the other hand, the parasite must take as much energy from the host as possible to ensure its own success.

All major phyla have parasitic marine representatives. However, the most widely distributed and successful parasitic marine animals are the roundworms of the phylum Nematoda. Like nearly all parasites, nematodes have a **species-specific relationship** with a host. A species-specific relationship is an exclusive relationship between two species; parasites can usually parasitize only one species of host. The reason for this interdependency is the delicacy of the biochemical feedback mechanisms informing the parasite that its activity may be overstressing the host. The feedback responses are, by necessity, tailored to specific host–parasite pairs. The parasite will not usually survive if it settles in or on a host for which it is not specifically "programmed."

More than one species of parasite *can* infect a single host, however. A significant percentage of the weight of many fishes may be in nematode worms and other parasites. The parasitic

FIGURE 16.24 Whale barnacles. These arthropods can be buried up to 3 centimeters ($1\frac{1}{4}$ inches) deep in the skin of certain whales. They derive part of their nutrition from the flesh and circulating fluids of the whale. Some individuals are up to 7.6 centimeters (3 inches) across.

burden of a normal, apparently healthy sea lion may exceed 2.3 kilograms (5 pounds) and 20 species! Parasites are usually small, but one species of nematode parasite of the fin whale reaches 7 meters (23 feet) in length and is the diameter of a pencil. And, by the way, parasites can have their own parasites!

These three types of symbioses (mutualism, commensalism, and parasitism) form a continuum in nature. There are few clearly mutualistic relationships that do not suggest at least a touch of commensalism, and few commensalistic relationships that do not hint of parasitism. The whale barnacles pictured in **Figure 16.24** glean most of their food directly from the ocean, but some of their nutrition is derived from the flesh and circulating fluids of their host. It's sometimes hard to tell where one kind of relationship stops and another begins.

Questions from Students

1. If the rocky, sandy, or muddy intertidal zones represent such a challenging mix of environmental factors, why do so many organisms live there?

Difficulty in biology is a relative term. It may seem a circular argument, but wherever organisms live, conditions for life at that place are biologically tolerable, food is available, and environmental conditions are not so extreme as to preclude success. Organisms live in abundance where energy is available. Where there is food, or sunlight, or biodegradable compounds, there is life. Natural selection has sorted out the ways that work in this zone from the ways that do not, and the adaptations that work give the organisms living in the intertidal zone's many niches access to a rich harvest of nutrients.

2. Is the species-specificity rule of parasitism ever broken?

Yes, sometimes with catastrophic results. As an example, the lung flukes that inhabit the respiratory tract of most sea lions will "abandon ship" if the sea lion is weakened or dying. A sea lion in this condition sometimes comes out of the water onto a beach to rest. If your pet dog should discover the animal there and sniff at its nose, some of the parasites could transfer from sea lion to dog and establish themselves in the dog's lungs. Because the dog is not the species-specific host for the lung parasites, the biochemical machinery that tells the parasite that it is weakening its host is missing. The dog may die a painful death in a few weeks from an overinfestation with lung flukes. The parasites will, of course, also die. The interaction will have been a failure in all respects.

3. Could any huge, undiscovered, Godzilla-type sea monsters live in the deep ocean?

Probably not, unless they can extract energy directly from water molecules! The deep pelagic feeding situation is simply not rich enough to support the energy needs of an active population of violent, aggressive, city-eating (*metrophagous?*) reptiles. Scientists never say never, but classic science fiction films aside, it doesn't look promising.

4. What are the historical foundations of ecology?

Many students confuse ecology with conservation efforts such as recycling or with environmental activism such as picketing large industrial polluters. But neither of these activities is ecology.

Ecology as a subdivision of natural science has a long history, which began with the writings of Pliny and other Romans who took an interest in a unified view of nature. Some European Renaissance scholars developed a similar outlook. In more recent times the French naturalists René-Antoine Réaumur (1663–1757) and Georges-Louis de Buffon (1707–88) and the great German zoologist and Darwinian Ernst Häckel, who coined the word *oekologie* in 1869, all helped to build the foundations of the modern science. Population studies were advanced by Charles Elton in his 1927 book *Animal Ecology*, and modern biometric analyses of populations and communities were pioneered by the Australian H. G. Andrewartha and his colleagues and contemporaries. The important concept of primary productivity was developed by two American freshwater biologists—E. Birge and C. Juday—in the 1930s.

Modern ecology came of age in 1942 when Robert L. Lindeman, an American, began the study of ecological energetics. His work detailing the flow of energy through ecosystems was among the first to consider an ecosystem as a unified whole. Most modern ecological studies can be traced to his insight.

Chapter in Perspective

In this chapter you learned that organisms are distributed throughout the marine environment in specific communities: groups of interacting producers, consumers, and decomposers that share a common living space. The types and variety of organisms found in a particular community depend on the physical and biological characteristics of that living space.

Any community is a dynamic place, growing and shrinking and changing its composition as residents respond to environmental fluctuations. The growth and distribution of organisms within communities depend on the often subtle interplay of physical and biological factors. The relative numbers of species and individuals in a community depend in part on whether its environment is relatively easy and free of stressors, or relatively hard and full of potential limiting factors. A few prominent marine communities were compared and contrasted in this chapter, and some representative adaptations of their residents were discussed. The main question: Why do organisms live where they do?

In the next chapter you will learn about the range of marine resources, from physical resources such as oil, natural gas, building materials, and chemicals; marine energy; and biological resources such as seafood, kelp, and pharmaceuticals to nonextractive resources like transportation and recreation. World economies are now dependent on oceanic resources, but we find that we cannot exploit those resources without damaging their source.

Key Ideas

✓ A community is all the living things in a defined area. The organisms of a community interact with and depend on one another, often in complex ways.
Review page 391

✓ An organism's habitat is its "address"; its niche is its "occupation."
Review page 392

✓ The distribution of organisms in communities is rarely random; clumped distribution is the most common.
Review Figure 16.5, page 395

✓ Though among Earth's most rigorous habitats, high-energy rocky and sandy shores are heavily populated by diverse organisms able to take advantage of the abundant nutrients found there.
Review Figures 16.6–16.8, pages 395–399

✓ Highly productive salt marshes and estuaries shelter a great variety of benthic life-forms and serve as nurseries for some pelagic organisms.
Review Figures 16.10 and 16.11, pages 400–401

✓ Coral reef communities are exceptions to the general rule that tropical oceans are unproductive. Closely cycled nutrients and zooxanthellae in coral organisms make high productivity and high biodiversity possible.
Review Figures 16.12, pages 402–403

✓ The open ocean below 3,000 meters (10,000 feet) is among the least densely populated habitats on Earth. Lack of food is the main limiting factor.
Review Figures 16.14–16.16, pages 404–405

✓ The deep-sea floor is the ocean's most uniform habitat. It is populated by large numbers of mostly small, highly specialized organisms.
Review Figures 16.17 and 16.18, pages 405–406

✓ Earth's largest community has only recently been discovered. It consists of bacteria and archaeans living in the minute spaces within solid rock and may extend to a depth of perhaps 7 kilometers (4.4 miles).
Review Figures 16.19 and 16.20, pages 407–408

✓ Deep vent communities near black smokers and atop cold seeps depend for energy on chemosynthesis by bacteria and archaeans.
Review Figures 16.20 and 16.21, pages 408–409

✓ Whale fall communities may act as "stepping stones" for organisms that colonize hydrothermal vents.
Review Figure 16.22, page 409

✓ Symbiosis is the co-occurrence of two species in which the life of one is closely interwoven with the life of the other. The types of symbiosis are mutualism, commensalism, and, the most common, parasitism.
Review Figures 16.23 and 16.24, pages 409–410

Terms and Concepts to Remember

biodiversity, 392	intertidal zone, 395
brackish, 400	motile, 397
carrying capacity, 394	mutualism, 409
climax community, 395	niche, 392
clumped distribution, 394	parasitism, 410
commensalism, 410	population, 391
community, 391	population density, 394
deep scattering layer (DSL), 404	random distribution, 394
	sessile, 397
desiccation, 397	species-specific relationship, 410
ecology, 392	
environmental resistance, 393	stenohaline, 392
estuary, 400	stenothermal, 392
euryhaline, 392	succession, 395
eurythermal, 392	symbiosis, 409
extremophile, 406	uniform distribution, 394
habitat, 392	wave shock, 395

Study Questions

Reviewing What You've Learned

1. How does a population differ from a community? A niche from a habitat?
2. Describe a typical growth curve for a population. What factors influence the shape and eventual height of the curve?
3. Which marine habitat is the most sparsely populated? Why?
4. Describe the residents of the deep ocean. How do inhabitants of the vent communities differ from other deep-floor organisms? What is the source of nutrition for vent communities?
5. What are the three types of symbioses? Give an example of each.

Thinking Critically

1. What is a limiting factor? Give a few examples.
2. In what ways can members of the same population compete with one another? How might members of different populations compete? Contrast the results of these kinds of competition.
3. What factors influence the distribution of organisms within a community? How are these distributions described? Why is random distribution so rare?
4. What problems confront the inhabitants of the intertidal zone? How do you explain the richness of the intertidal zone in spite of these rigors?
5. Why must the host–parasite relationship be so finely balanced? What would be the result of an imbalance?
6. **InfoTrac College Edition Project** *Pfiesteria piscicida* is a species of toxic algae that kills fish and makes people sick. In 1997 major outbreaks of this species occurred in Atlantic coastal estuaries and bays. Is the rapid growth of *Pfiesteria* caused by human activity? If so, what needs to be done? Research this question using InfoTrac College Edition.

Thinking Analytically

1. Why would larvae dispersing from hydrothermal vent communities need the assistance of whale fall "stepping stones" to colonize new vents? (*Hint:* Do you think hydrothermal vents are steady, relatively stable phenomena?)
2. What implications do you think the discovery of living things in deep rocks might have for discovering life on other planets?
3. What percentage of the total biosphere do you think deep extremophiles are? Which is the dominant mechanism of primary productivity on Earth: photosynthesis or chemosynthesis?

Oceanography ⬡ Now™
Assess your understanding of this chapter's topics with additional quizzing and conceptual-based problems at http://oceanography.brookscole.com/garrison5e

Study and Research on the Internet

Assess your understanding of this chapter's topics with additional quizzing and conceptual-based problems at:

● http://oceanography.brookscole.com/garrison5e

■ **InfoTrac College Edition Research** leads you to critical thinking exercises that use InfoTrac College Edition® as a research tool.
■ **Regional InfoTrac College Edition** articles are organized into East Coast, West Coast, and Gulf Coast regions, allowing you to study oceanography on a more local level.

For more readings, go to InfoTrac College Edition, your online research library, at:

● http://infotrac.thomsonlearning.com

CHAPTER 17

A dangerous way to make a living: Crab fishers work on deck in bad weather in the Bering Sea, Alaska. The 120-foot crab boats and crews are often buffeted by winter storms—winds over 100 miles per hour and seas to 50 feet are not unheard of. Large iron crab pots, some weighing more than 750 pounds, are moved by hand as rolling waves toss the boat like a cork in a bathtub. A crab fisher can make $20,000 per catch—for a successful trip lasting about six weeks. Commercial fishing is the most dangerous profession in the United States.

MARINE RESOURCES

An Endless Supply?

World economies began a period of unprecedented growth in the decade following World War II. Leaders looked to the sea to provide a greater proportion of the mineral and biological resources needed to sustain their economies and feed their citizens. The great ocean seemed an inexhaustible source of food, oil, and ore. It was comforting to think that as we learned more about the ocean, we would surely discover vast treasure troves available for the taking.

But resources that seemed limitless can be depleted in a few generations. Consider, for example, the history of the North Atlantic's most valuable resources, its fisheries. For hundreds of years, Georges Bank, one of the world's most productive fishing grounds, yielded seemingly endless supplies of cod,

413

haddock, flounder, and halibut. American colonists considered halibut a trash fish, unfit for human consumption. By the 1830s, however, halibut had become fashionable and demand skyrocketed. Individual halibut boats fishing Georges Bank brought in 10 tons of halibut per day! As the reproducing stock was removed, the fishery was rapidly and catastrophically depleted. Today catches of halibut are so rare in the northwestern Atlantic that regulators don't bother to keep statistics on them.

More recently, on the other side of the North Atlantic, fishers of at least seven European nations compete for cod and haddock whose historic abundance will soon be little more than a memory. Fishing pressure is now so great that less than 1% of one-year-old cod remain in the ocean long enough to spawn. Each year 60% of all the cod present are swept into nets. The hunt on the Dutch continental shelf is so intense that every square meter of seabed is trawled, on average, once or twice each year! The 1992 collapse of the northwestern cod fishery cost 35,000 jobs (18,000 of them in Newfoundland alone). The northeastern fishery is now teetering on the same brink.

Marine fish are the only wildlife still hunted on a large scale. Properly managed, fish and other biological resources can be a renewable resource, but minerals, oil, natural gas, and most other physical resources are nonrenewable. Because they are easier to exploit, physical resources on the land have many economic advantages over physical resources from the sea. Also, on land, valuable elements have been concentrated into ores by sedimentation, weathering, and other natural processes, and the variety of minerals that can be extracted far exceeds anything we have seen in the ocean. The most important physical resources from the sea are fluid resources such as petroleum (oil) and natural gas. These substances are probably as abundant in the continental shelves as on dry land, but obtaining them is very expensive and often dangerous.

With very few exceptions, the persistent prediction that the continued exploitation of the ocean will provide an increasing percentage of the material needs of a growing human population—of oceanic riches for everyone—cannot be realized. As we will see, marine resources grow more valuable as they become scarcer.

Marine Economics, Marine Resources

An economy is a system of production, distribution, and consumption of goods and services that satisfies people's wants or needs. In marine economics, individuals, businesses, and governments make economic decisions about what ocean-related goods and services to produce, how to produce them, how much to produce, and how to distribute and consume them.

In a free-market economic system (which, so far, exists only in theory), buying and selling are based on pure competition, and no seller or buyer can control or manipulate the market. Economic decisions are governed solely by supply, demand, and price; sellers and buyers have full access to information about the beneficial and harmful effects of goods and services to make informed decisions. Ideally, prices reflect all harmful costs of goods and services to the environment.

But ours is not a pure free-market economic system. Often prices do not reflect all harmful costs of goods and services to the environment. Consumers rarely have full access to information about the beneficial and harmful effects of goods and services to make informed decisions. With few exceptions, our present level of growth and exploitation of marine resources is unsustainable. As you read this chapter (and the next), remember the supply and demand nature of economic markets and think about the long-term implications of our growing dependence on oceanic resources.

We will discuss four groups of marine resources in this chapter:

- **Physical resources** result from the deposition, precipitation, or accumulation of useful substances in the ocean or seabed. Most physical resources are mineral deposits, but petroleum and natural gas, mostly remnants of once-living organisms, are included in this category. Fresh water obtained from the ocean is also a physical resource.

- **Marine energy resources** result from the extraction of energy directly from the heat or motion of ocean water.

- **Biological resources** are living animals and plants collected for human use.

- **Nonextractive resources** are uses of the ocean in place. Transportation of people and commodities by sea, recreation, and waste disposal are examples.

Marine resources can be classified as either renewable or nonrenewable.

- **Renewable resources** are naturally replaced by the growth of marine organisms or by natural physical processes.

- **Nonrenewable resources** such as oil, gas, and solid mineral deposits are present in the ocean in fixed amounts and cannot be replenished over time spans as short as human lifetimes.

Physical Resources

Physical resources from the ocean include hydrocarbon deposits (petroleum, natural gas, and methane hydrate), mineral deposits (sand and gravel, magnesium and its compounds, salts of various kinds, manganese nodules, phosphorites, and metallic sulfides), and fresh water.

Petroleum and Natural Gas

Offshore petroleum and natural gas generated nearly US$300 billion in worldwide revenues in 2001. Here the ocean makes a significant contribution to current world needs: About 35% of the crude oil and 26% of the natural gas produced in 2000 came from the seabed. About a third of known world reserves of oil and natural gas lie along the continental margins. Major U.S. marine reserves are located on the continental shelf of southern California, off the Texas and Louisiana Gulf coast, and along the North Slope of Alaska. The deep-sea floors probably contain little or no oil or natural gas.

Oil is a complex chemical soup containing perhaps a thousand compounds, mostly hydrocarbons. Petroleum is almost always associated with marine sediments, suggesting that the organic substances from which it was formed were once marine. Planktonic organisms or soft-bodied benthic marine animals are the most likely candidates. Their bodies apparently accumulated in quiet basins where the supply of oxygen was low and there were few bottom scavengers. The action of anaerobic bacteria converted the original tissues into simpler, relatively insoluble organic compounds that were probably buried—possibly first by turbidity currents, then later by the continuous fall of sediments from the ocean above. Further conversion of the hydrocarbons by high temperatures and pressures must have taken place at considerable depth, probably 2 kilometers (1.2 miles) or more beneath the surface of the ocean floor. Slow cooking under this thick sedimentary blanket for millions of years completed the chemical changes that produce oil.[1]

If the organic material cooked too long or at too high a temperature, the mixture turned to methane, the dominant component of natural gas. Deep sedimentary layers are older and hotter than shallow ones and have higher proportions of natural gas to oil. Very few oil deposits have been found below a depth of 3 kilometers (1.8 miles). Below about 7 kilometers (4.4 miles), only natural gas is found.

Oil is less dense than the surrounding sediments, so it can migrate from its source rock through porous overlying formations. It collects in the pore spaces of reservoir rocks when an impermeable overlying layer prevents further upward migration of the oil (**Figure 17.1**). When searching for oil, geologists use sound reflected off subsurface structures to look for the signature combination of layered sediments, depth, and reservoir structure before they drill.

Drilling for oil offshore is far more costly than drilling on land because special drilling equipment and transport systems are required. Most marine oil deposits are tapped from offshore platforms that rest in water less than 100 meters (330 feet) deep. As oil demand (and therefore price) continues to rise, however, deeper deposits farther offshore will be exploited from larger platforms (**Figure 17.2**). Currently the largest and heaviest platform is *Statfjord-B*, in position since 1981 northeast of the Shetland Islands in the North Sea. The tallest drilling platform is *Ursa*, a tension-leg structure deployed by the Shell Oil Company in 1998. *Ursa* floats in water more than 1,160 meters (3,800 feet) deep and is held in position off the Louisiana coast by 16 lengths of 81-centimeter

---

[1] How much plankton was needed to make a gallon of gasoline? Take a guess, and then read question 1 on page 434.

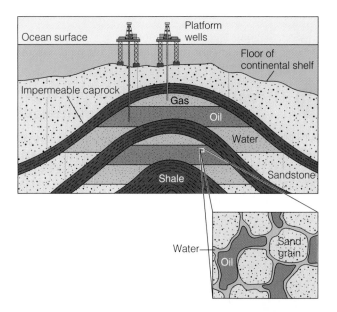

FIGURE 17.1 Oil and natural gas are often found together beneath a dome of impermeable caprock. Oil and gas are not found in great hollow reservoirs but within pore spaces in rock (inset).

(32-inch) steel cables anchored into the seabed and pulling against the semisubmerged platform's buoyancy. At full production the 14 wells on platform *Ursa* are projected to produce 30,000 barrels of oil and 80 million cubic feet of natural gas per day. The total cost of platform *Ursa* exceeded US$1.45 billion. Even larger tension-leg platforms are being planned.

Methane Hydrate

The largest known reservoir of hydrocarbons on Earth is not coal or oil but methane-laced ice crystals—methane hydrate—in the sediments of some continental slopes. Little is known about their formation, but methane hydrates exist in thin layers 200 to 500 meters (660 to 1,650 feet) below the seafloor, where they are stable and long-lived. Sediment rich in methane hydrate looks like green Play-Doh. When brought to the warm, low-pressure conditions at the ocean surface, the sediment fizzes vigorously as the methane escapes. It burns vigorously if ignited.

Though methane hydrate is abundant, exploitation of this resource would be very costly and dangerous. Even if engineers could bring the sediment to the surface before the methane disappeared, extracting the methane from the sediment and liquefying it for efficient use would be prohibitively expensive.

Could escape of methane from marine sediments have played a role in ancient climate change? Some researchers think so. Methane is a powerful greenhouse gas (see Chapter 18), and changes in ocean circulation that result in deep-ocean warming could release large quantities of methane. About 55 million years ago the deep ocean warmed by at least 4°C (7°F). The large-scale escape of methane from the seabed may have raised surface temperatures abruptly, melted surface ice, and lowered oxygen levels in the deep sea. If global warming continues today, will methane escaping from deep sediments exacerbate the problem?

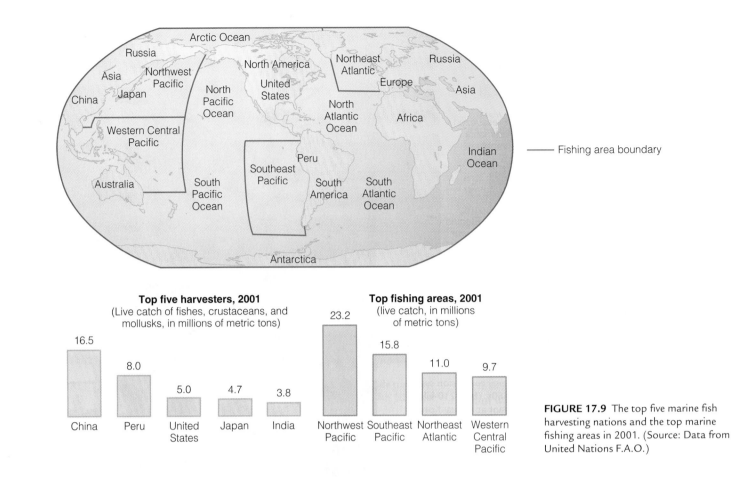

FIGURE 17.9 The top five marine fish harvesting nations and the top marine fishing areas in 2001. (Source: Data from United Nations F.A.O.)

FIGURE 17.10 A rich, diverse harvest of fresh marine life in a market in Hong Kong S.A.R., China.

Table 17.1	World Commercial Catch and Aquaculture Yield for 2001
Species Group	**Millions of Metric Tons, Live Weight**
Herring, sardines, anchovies	20.5
Carps, barbels, cyprinids	17.0
Cods, hakes, haddocks	9.2
Tunas, bonitos, billfishes	5.8
Salmons, trouts, smelts	2.6
Tilapias	2.1
Other fishes	44.9
Shrimps, crabs, lobsters, krill	5.7
Mollusks (oysters, scallps, squid, etc.)	18.6
Sea urchins, other echinoderms	0.1
Total for all marine sources, excluding marine mammals, marine algae, and aquatic plants	**130.2**

FIGURE 17.11 Orbimage's SeaStar Fisheries Information Service uses data from the Sea-Viewing Wide Field-of-View Sensor (SeaWiFS) on its OrbView-2 satellite to generate fish-finding charts like this one. This image, from information collected on 4 March 2003, shows high concentrations of plankton (green) off coastal Argentina, Uruguay, and Brazil. Computer models suggest that the red hatch-marked areas contain high concentrations of desirable fish.

a

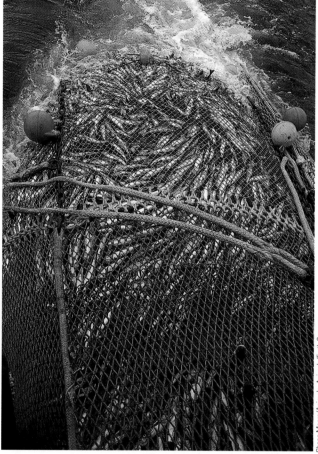

b

FIGURE 17.12 Stern trawler fishing. (**a**) After sonar on the trawler finds the fish, they're captured by a trawl net more than 122 meters (400 feet) wide. Boards angled to the water flow keep the net's mouth open. The largest nets extend about 0.8 kilometer ($\frac{1}{2}$ mile) behind the towing vessel and are large enough to hold a dozen 747 jetliners. (**b**) A haul of Alaskan pollock from the Bering Sea. (Pollock is a codlike fish popular in fish sticks and fast food.) About 60% of the fish landed in the United States is pulled from the Bering Sea, a resource worth US$1.1 billion before processing. Once thought inexhaustible, pollock stocks are plunging; about 25% of the pollock in the Bering Sea is caught each year.

100,000 workers each year.[2] Though estimates vary widely, the value in 2001 for the worldwide marine catch was thought to be around US$90 billion. Slightly more than half of the world's commercial marine catch is taken by the five countries shown in Figure 17.9. About 75% of the annual harvest is taken by commercial fishers who operate vast fleets working year-round, using satellite sensors, aerial photography, scouting vessels, and sonar to pinpoint the location of fish schools (**Figure 17.11**). Huge factory ships often follow the largest fleets to process, can, or freeze the animals on the run. Catching methods no longer depend on hooks and lines but rather on large trawl nets (**Figure 17.12**), purse seines, or gill nets. The living resources of the ocean are under furious assault: *Between 1950 and 1997 the commercial marine fish catch increased more than fivefold.*

The cost to obtain each unit of seafood has risen dramatically in spite of all this high-tech assistance. The increasing expense of fuel for the fishing fleets and processing plants, the cost of wages for the crews, and the greater distances that boats must cover to catch each ton of fish have all helped drive up the cost of seafood. In spite of greater efforts, the total marine catch leveled off in about 1970 and remained surprisingly stable until 1980, when greater demand and increasing prices began to drive the tonnage upward again. Harvests are now declining in spite of desperate attempts to increase yields. Since 1970 the world human population has grown, so the average *per capita* world fish catch has fallen alarmingly (**Figure 17.13**).

---

[2] Sebastian Junger's extraordinary 1997 book (and the 2000 movie) *The Perfect Storm* chillingly recalls these dangers.

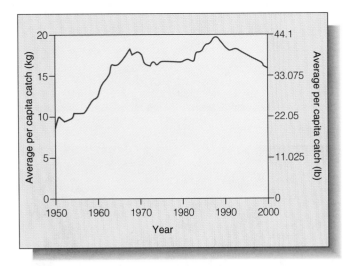

FIGURE 17.13 Average *per capita* world fish catch, 1950–2000. The per capita catch reached a peak in the late 1980s and has declined since. The decline is projected to continue in this century.

Fishery Mismanagement

Can we continue to take huge amounts food from the ocean year after year? The **maximum sustainable yield,** the maximum amount of each type of fish, crustaceans, and mollusks that can be caught without impairing future populations, probably lies between 100 and 135 million metric tons (110 and 150 million tons) annually. As can be seen in Table 17.1, the current yield now approaches the top figure. Fleets are obtaining fewer tons per unit of effort and are ranging farther afield in their urgent search for food. We may be perilously close to the catastrophic collapse of more fisheries, just as described at the beginning of the chapter for halibut and cod. As of 1997, 30% of recognized marine fisheries were overexploited or already depleted, and 50% more were at their presumed limit of exploitation. The U.S. National Marine Fisheries Service estimates that 45% of the fish stocks whose status is known are now **overfished**—so many fish have been harvested that not enough breeding stock is left to replenish the species. Recent trends may be inferred from **Figure 17.14.**

Even when faced with evidence of overfishing, the fishing industry seldom follows a rational course. The industry's dominant motivating force is quick financial return, even if it means depleting a stock and disrupting the equilibrium of a fragile ecosystem. Long-term stability is forsaken for short-term profit. When the catch begins to drop, the industry increases the number of boats and develops more efficient techniques for capturing the remaining animals. When the impending catastrophe is obvious, governments sometimes intervene to set limits or close a fishery altogether. In 1999 New England officials approved a plan to close a large section of the Gulf of Maine to fishing in an attempt to replenish once-abundant cod stocks. As many as 2,500 fishers and 700 boats were idled, and losses may exceed US$21 million a year. Given the choice between immediate profit and a long-term sustainable fishery, fishers made their priorities clear by

FIGURE 17.14 A century of dramatic decline in the fisheries of the North Atlantic. These data are for table fish (to be eaten directly by people), not for fish collected for oil or animal feed.

initiating a recall campaign against officials who had approved the ban!

Fishers sometimes abandon an exhausted fishery to search for new and profitable resources. New Zealand's orange roughy, a large, mild-tasting deep-water fish, caught on with chefs in the early 1980s. The orange roughy grows slowly, can live more than 100 years, and takes 25 to 30 years to become sexually mature. Using precision locating equipment and fishing techniques developed for deep water, fishers brought the orange roughy to the brink of **commercial extinction** in 13 years. (Commercial extinction is the depletion of a resource species to a point where it is no longer profitable to harvest.) With Pacific

FIGURE 17.15 Fish jam the rearmost part of a trawl net before it is hoisted from the water. Perhaps a third of these fish are unwanted species—bykill—that will be thrown over the side. (R. Watson, V. Christensen, and D. Pauly/Oceana. Reprinted with permission.)

orange roughy no longer in dependable supply, in 1996 interest turned to the Patagonian toothfish, another large, slow-growing deep-water fish. Prices have risen to US$7,000 per metric ton; boats from northern countries are in a race for the catch using 5,000 baited hooks each day on a 100-kilometer (62-mile) line. More than US$300 million worth of toothfish (called Chilean sea bass on restaurant menus) was taken in 1996, and twice that value in 1997. Few now remain; that fishery has also collapsed. The cycle repeats.

The intended organism is not the only victim. In some fisheries, **bykill**—animals unintentionally killed while collecting desirable organisms—sometimes greatly exceeds the target catch (**Figure 17.15**). Bottom trawling is especially devastating. The habitat itself is disturbed; slow-growing organisms and complex communities are ransacked. In 1995 the governor of Alaska said, "Last year's [Alaska bottom fishing] discards would have provided about 50 million meals." Worldwide bykill was around 27 million metric tons (30 million tons) in 1995, *a quantity nearly one-third of total landings!* In 2002 in the United States about 9.1 million metric tons (10 million tons) of nontarget fish is discarded each year—twice the catch of desired commercial and recreational fishing combined.

Some progress has been made to minimize bykill. Turtle exclusion devices (TEDs), chutes through which sea turtles are ejected from nets, have been mandated for shrimp fishing in U.S. territorial waters. As we shall see in our discussion of whaling, by forcing the redesign of tuna nets and the adoption of special ship maneuvers during netting, the "dolphin-safe" tuna campaign, an adjunct to the 1972 Marine Mammals Protection Act, has spared the lives of hundreds of thousands of dolphins.

There are other disruptive, mistargeted fishing techniques. One employs **drift nets**—fine, vertically suspended nylon nets as much as 7 meters (25 feet) high and 80 kilometers (50 miles) long. Drift net technology was pioneered by a United Nations agency to help impoverished Asian nations turn a profit from what had been subsistence fishing. Until 1993, Taiwanese, Korean, and Japanese vessels deployed some 48,000 kilometers (30,000 miles) of these "walls of death" each night—more than enough to encircle Earth! Drift nets caught the fish and squid for which they were designed, but they also entangled everything else that touched them, including turtles, birds, and marine mammals (**Figure 17.16**). Researchers at Duke University and St. Andrew's University (in the United Kingdom) estimate that about 800 small whales are killed each day in fishing nets.

Drift net fishing has been compared to strip-mining; the ocean is literally sieved of its contents. An estimated 30 kilometers (18 miles) of these fine, nearly invisible nets were lost each night—about 1,600 kilometers (1,000 miles) per season.

FIGURE 17.16 An unintentional but unavoidable consequence of drift net fishing.

FIGURE 17.17 An abandoned plastic fishing net drapes the intertidal zone at remote Clarion Island, Revillagigedo Islands, Mexico. Before being stranded here, this net may have entangled fish and marine mammals for years.

FIGURE 17.18 The ocean under furious assault. The Chinese and Korean fleets seen here in June 2001 are fishing for bait and squid across the South China Sea. From horizon to horizon, their brilliant lights pierce the dusk.

These remnants, made of nonbiodegradable plastic, become *ghost nets*, continuing to entangle fish and other animals perhaps for decades (**Figure 17.17**).

Although large-scale drift net fishing is now banned, pirate netting continues in the Pacific and has begun in the Atlantic. Completely unrestrained by regulation, these outlaw fishers operate where their activities cause maximum damage to valuable reproductive stock.

"Madhouse Economics"

The fishing industry spent US$124 billion to catch US$70 billion worth of fish and other marine life in 1995. Government subsidies made up most of the US$54-billion deficit. Artificial supports included fuel-tax exemptions, price controls, low-interest loans underwritten by government or state agencies, and grants of equipment. More than 1,600 large fishing vessels have been reflagged to maneuver around legal requirements in their original countries or to take advantage of subsidies. In an effort to preserve well-paying jobs, governments encourage construction of underutilized ships and research in new and often-destructive technology. The efficiency of the new ships reduces the number of fish available, thereby reducing the value of hulls and equipment on the market. Unable to sell their vessels, owners are forced to continue fishing to repay their loans. In an important 1995 paper in *Scientific American*, Carl Safina understandably described the situation as "madhouse economics."

Whaling

Since the 1880s, whales have been hunted to provide meat for human and animal consumption; oil for lubrication, illumination, industrial products, cosmetics, and margarine; bones for fertilizers and food supplements; and baleen for corset stays. **Figure 17.19** shows whales butchered for food and oil. An estimated 4.4 million large whales existed in 1900; today slightly more than 1 million remain. Eight of the 11 species of large whales once hunted by the whaling industry are commercially extinct. The industry pursued immediate profits despite obvious signs that most of the "fishery" was exhausted (**Figure 17.20**).

Substitutes exist for all whale products, but the harvest of most commercial species did not stop until whaling became uneconomical. In 1986 the International Whaling Commission, an organization of whaling countries established to manage whale stocks, placed a moratorium on the slaughter of large whales. Except for a suspiciously large harvest of whales taken by the Japanese for "scientific purposes," commercial whaling ceased in 1987. Fewer than 700 large whales were taken in 1988.

Now, however, the number being taken is rising. What protection there is may have come too late to save some species from extinction—and protection may be only temporary. Under intense pressure from its major fishing industry, Norway resumed whaling in 1993. Japan never stopped. Their target, the minke whale, is the smallest and most numerous of the great whale species (see Figure 15.39). The meat and blubber of this whale are prized in Japan as expensive delicacies. The minke whale population has been estimated at 1,200,000, a number that may withstand the present level of harvest. A decision to increase minke whaling, however, may doom the minke to the same fate as most other whale species. Between them, Japan and Norway caught 1,078 minkes in 1999. In 2000 Japan extended "scientific" whaling to sperm whales and Bryde's whales, and in 2003 to sei whales.

There is a glimmer of hope. In 1994 the International Whaling Commission voted overwhelmingly to ban whaling in about 21 million square kilometers (8 million square miles) around Antarctica, thus protecting most of the remaining large

FIGURE 17.19 Whales slaughtered for food and oil in the Faeroe Islands.

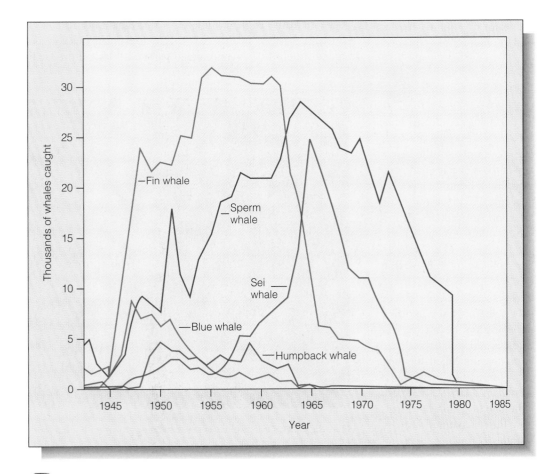

FIGURE 17.20 Commercial slaughter of the five largest whale species, 1940–1985. All five species included in this graph are commercially extinct, and the blue whale is still in danger of total extinction even though commercial blue whale hunting ended in 1964. At that time only about 1,000 blue whales were left, and experts fear that the population may have been too small to recover.

whales that feed in those waters. The sanctuary is often ignored, however. In their quest for minke (and other whales), Japanese and Norwegian whalers have entered the area and harpooned many animals. Chile, Peru, and North Korea are now considering joining the hunt. Pirate whalers based in other countries can also catch whales in the Antarctic and sell the flesh on the Japanese market. Conservation efforts can do some good. Although it was hunted nearly to extinction, the California gray whale has long been off limits to all but aboriginal hunters. Its numbers have grown, and it was removed from the endangered list in 1993.

Many more small whales have been killed than large ones, but not for food or raw materials. For reasons that are not well understood, dolphins (which are small whales) gather above schools of yellowfin tuna in the open sea. Fishers have learned to find the tuna by spotting the dolphins. Nets cast to catch the tuna also entangle the air-breathing dolphins, and the mammals drown. The dead dolphins are simply pitched over the side as waste. More than 6 million small whales have been killed in association with tuna fishing since 1971.

Passage of the U.S. Marine Mammals Protection Act in 1972 brought a drop in the number of dolphin deaths. From 1977 to 1987, however, the number of tuna boats in the U.S. fleet declined from more than 100 boats to 34 boats, while the foreign fleet grew from fewer than 10 to more than 70 boats. Foreign fishers do not always abide by the dolphin-saving provisions of the act, but the U.S. Congress voted in 1988 to ban importation of tuna not caught in accordance with new methods designed to reduce the kill. In 1990 American tuna-processing companies agreed to buy tuna only from fishers whose methods do not result in the deaths of dolphins. American commercial fishers have agreed to comply, and conservationists hope that the foreign fleets will follow their lead.

Fur-Bearing Mammals

Worldwide, between 300,000 and 450,000 seals and sea lions are taken annually for fur. Eight species of seals and one species of sea lion are of economic importance. At the beginning of the twentieth century a few of these species had been hunted nearly to extinction, but a system of hunting quotas allowed most of the populations to recover. A provision of the 1972 U.S. Marine Mammals Protection Act protects all fur-bearing marine mammals in U.S. territory except for the northern fur seal. Public pressure and an unfavorable economic climate contributed to the collapse of the Alaskan northern fur seal industry in 1986.

The harp seal, a species found in relatively great abundance on ice floes off the coasts of Canada in the Labrador and Barents Seas, has attracted much popular attention in the last decade. Newborn harp seal pups are covered with dense, pure white fur. Each spring, several thousand are killed for their pelts. Because these pups are so attractive and helpless, much effort has been made to stop the slaughter. Harp seals are not an endangered species, however; the rate at which harp seals are harvested is considerably below their maximum sustainable yield, and the population continues to grow.

Botanical Resources

Marine algae are also commercially exploited. The most important commercial product is **algin**, made from the mucus that slickens seaweeds. When separated and purified, algin's long, intertwining molecules are used to stiffen fabrics; to form emulsions such as salad dressings, paint, and printer's ink; to prevent the formation of large crystals in ice cream; to clarify beer and wine; and to suspend abrasives. The U.S. seaweed gel industry produces more than US$220 million worth of algin each year, and the annual worldwide value of products containing algin (and other seaweed substances) was estimated to be more than US$42 billion in 2000.

Seaweeds are also eaten directly, and some species are cultivated (**Figure 17.21**). The Japanese consume 150,000 metric tons (165,000 tons) of *nori* each year; seaweed and seaweed extracts are also eaten in the United States, Britain, Ireland, New Zealand, and Australia. Their mineral content and fiber are useful in human nutrition.

Aquaculture and Mariculture

Aquaculture is the growing or farming of plants and animals in any water environment under controlled conditions. Aquaculture production currently accounts for over one-quarter of all fish consumed by humans. Most aquaculture production occurs in China and the other countries of Asia. In 2001 more than 37 million metric tons (43 million tons) of fish—mostly freshwater fish—were produced worldwide. By 2010 fish aquaculture could overtake cattle ranching as a food source.

Mariculture is the farming of *marine* organisms, usually in estuaries, bays, or nearshore environments or in specially designed structures using circulated seawater. Mariculture facili-

FIGURE 17.21 Orderly plots of seaweed being grown for human food off an island east of Bali.

ties are sometimes placed near power plants to take advantage of the warm seawater flowing from their cooling condensers.

Worldwide mariculture production is thought to be about one-eighth that of freshwater aquaculture. Several species of fish, including plaice and salmon, have been grown commercially, and marine and brackish-water fish account for 67% of the total production. Shrimp mariculture is the fastest-growing and most profitable segment, with an annual global value exceeding US$6 billion in 1998.

In the United States annual revenue from mariculture exceeds US$150 million, with most of the revenue generated from salmon and oyster mariculture (**Figure 17.22**). About half of the oysters consumed in North America are cultured. Not all mariculture produces food; cultured pearls are an important industry in Japan, China, and northern Australia. Japan leads the world in mariculture.

Ranching, or open-ocean mariculture, is an interesting variation. Juveniles are grown in a certain area, released into the ocean, and expected to return when mature. The natural homing instinct of salmon makes salmon ranching quite successful—about 20% of the world supply of salmon is ranched. Yet only about 1 in 50 ranched juveniles returns to the area of release. In Japan recent attempts to extend ranching techniques to yellowtail and tuna have met with only limited success. Fish ranching is not without problems. Hatchery-raised fish can escape and breed with native populations, reducing the genetic diversity of the wild fish and their ability to survive. Fish in close quarters are more susceptible to disease, and their waste products (and feed) can contaminate the areas around their holding pens.

Mariculture is an expanding industry; it is growing at about 8% annually compared to a decline for the world marine fishery as a whole. Mariculture produces mostly luxury seafoods such as oysters and abalone, and it uses fish meal from so-called *trash fish* as feed. (Trash fish are edible but considered unappealing because they are bony or bad-tasting.) As world population increases and the demand for protein grows among the world's millions of undernourished people, however, today's trash fish may become more acceptable human food.

Drugs

The earliest recorded use of medicines derived from marine organisms appears in the *Materia Medica* of the emperor Shen Nung of China, 2700 B.C. Modern medical researchers estimate that perhaps 10% of all marine organisms are likely to yield clinically useful compounds. One medicine derived from a Caribbean sponge is already in use: Acyclovir, the first antiviral compound approved for humans, has been fighting herpes infections of the skin and nervous system since 1982. A class of anti-inflammatory drugs known as pseudopterosins, developed from the Caribbean sea whip by researchers at the University of California, has also been successful. Indeed pseudopterosins are among the university's top ten royalty-producing inventions.

Since 1986 more than 10,000 different species of marine invertebrates and marine algae have been collected by the Na-

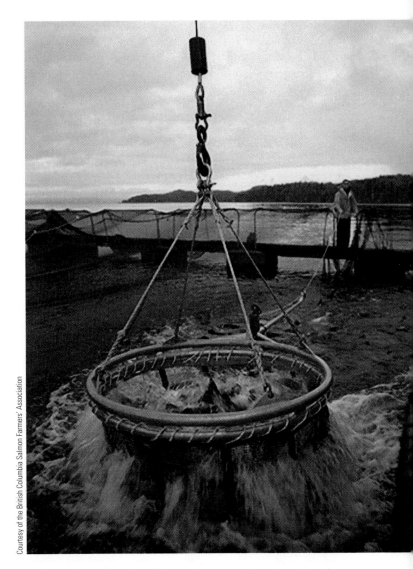

FIGURE 17.22 Harvesting fish at a salmon ranch in western Canada.

tional Institutes of Health for testing as drugs; they come mainly from the Indo-Pacific region, through contracts with the Harbor Branch Oceanographic Institute, the Australian Institute of Marine Science, the University of Canterbury (New Zealand), and the Coral Reef Research Foundation. A common bryozoan—a small encrusting invertebrate—has been found to produce a potent anticancer chemical that is now being tested in human volunteers. Extracts from 30% of all tunicate species investigated show antiviral and antitumor activity; one of these extracts, Didemnin-B, shows promise as a treatment for malignant melanoma, the deadliest form of skin cancer. Another tunicate derivative, Esteinascidin 743, has been found to be useful in the treatment of skin, breast, and lung cancers. A related compound found in the same organism, Aplidine, shows promise in shrinking tumors in pancreatic, stomach, bladder, and prostate cancer.

Cancer is not the only target. A new compound derived from cyanobacteria stimulated the immune system of test animals by 225% and cells in culture by 2,000%; the drug may be useful in treating AIDS. Vidabarin, another antiviral drug developed from sponges, may attack the AIDS virus directly.

Other useful chemicals derived from marine organisms include Padan, a powerful new insecticide derived from an annelid worm, and a new family of steroids that may prove to be useful in the regulation of human growth. In the not-too-distant future, drugs from marine organisms may be among our most valuable marine resources.

Nonextractive Resources

Transportation and recreation are the main nonextractive resources the oceans provide. People have been using the ocean for transportation for thousands of years. Through most of this time the transport of cargo has produced far more revenue than the movement of passengers. At present, oil tankers ship the greatest gross tonnage of any type of cargo (more than 285 million metric tons—314 million tons—annually). The capacity of the total oil tanker fleet is increasing at an annual rate of about 1.7%. Oil accounts for about 53% of the total value of world trade transported by sea; iron, coal, and grain make up 24% of the rest.

Nearly half of the world's crude oil production is transported to market by ships. Tankers are needed because very few of the major oil-drilling sites are close to areas where the demand for refined oil products is highest. The largest tankers are more than 430 meters (1,300 feet) long and 66 meters (206 feet) wide, and they carry more than 500,000 metric tons (3.5 million barrels) of oil (**Figure 17.23**).

Modern harbors are essential to transportation. Cargoes are no longer loaded and off-loaded piece by piece by teams of longshoremen. Today's harbors bristle with automated bulk terminals, high-volume tanker terminals (both offshore and dock-side), containership facilities (**Box 17.1**), roll-on–roll-off ports for automobiles and trucks, and passenger facilities required by the growing popularity of cruising. Most of this specialized construction has occurred since 1960. The Los Angeles–Long Beach port complex in southern California transfers more cargo than any other North American port. The port with the greatest transit volume is Shanghai, but the world's largest and busiest container port is in the Special Administrative Region of Hong Kong. In 1995 the Hong Kong facility was the first to move more than 1 million containers in a single month!

Transportation is sometimes combined with recreation. In the last decade the cruise industry has experienced spectacular growth. Passengers on luxurious liners can enjoy a few relaxing days on the ocean crossing the North Atlantic, visiting tropical islands, or touring places accessible to the public only by ship. (Indeed, tourism is now the world's largest industry.) Ocean-related leisure pursuits, including sport fishing, surfing, diving, day cruising, sunbathing, dining in seaside restaurants, and just plain relaxing, contribute to the economy. In addition to being important producers of revenue, public aquariums and marine parks (like Sea World) are centers of education, research, and captive breeding programs (**Figure 17.24**). Even public interest and curiosity about whales are a source of recreational revenue. In the United States in 2000, whale-watching trips generated an estimated annual direct revenue of US$210 million; indirect revenues including dolphin displays, whale artwork and books, and conservation group donations amounted to another US$400 million.

And don't forget about real estate. As coastal land values will attest, people enjoy living near the ocean. By 2015 about 165 million Americans will be living in coastal areas on property valued in excess of US$7 trillion.

The Law of the Sea

Prehistoric peoples living near the shore were the earliest users of marine resources. With the rise of nation-states and military establishments, the fight for control of marine resources began. Some nations assumed that the ocean belonged to all and endeavored to guarantee free right to passage and resources. Oth-

FIGURE 17.23 *Murex*, first of a series of five VLCCs (very large crude carriers) built for the Shell Oil Company by Daewoo Heavy Industries, South Korea. These ships, first of a new generation of tankers, are of double-hull design. Each can carry 2.15 million barrels of crude oil at a service speed of 28 kilometers (18 miles) per hour. Somewhat smaller than the largest tankers, *Murex* is 332 meters (1,089 feet) long and extends 22 meters (72 feet) below the surface when fully loaded. Built to high standards of reliability and safety, *Murex* is seen here high and unballasted on the day of her naming, 17 January 1995.

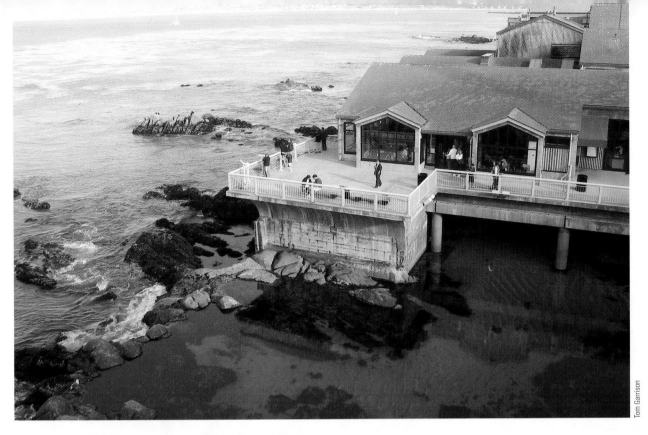

FIGURE 17.24 Public aquariums and marine parks are important resources. Monterey Bay Aquarium, seen here, and its associated Research Institute also conduct marine research.

ers decided that it belonged to none and tried to control access to ports and resources by force. In 1604 Hugo Grotius, a learned Dutch jurist, wrote *De Jure Praedae* (*On the Law of Prize and Booty*), a treatise justifying the action of a Dutch admiral who had successfully defended Dutch trading rights in a dispute with Portugal. One chapter of this work, in which Grotius defended free ocean access for all nations, was reprinted in 1609 under the title *Mare Liberum* (*A Free Ocean*). *Mare Liberum* formed the basis for all modern international **laws of the sea.**

About a century later, in 1703, the concept of territorial seas adjacent to land was recognized. A country's seaward boundary was set at about 5 kilometers (3 miles)—the distance a cannonball could be fired from shore. This 5-kilometer (3-mile) limit stood until 1945.

The United Nations and the International Law of the Sea

After World War II the technology became available to search for oil and natural gas on continental shelves. After U.S. oil companies found rich deposits beyond the 5-kilometer (3-mile) limit off Louisiana, President Harry Truman issued a proclamation annexing the physical and biological resources of the continental shelf contiguous to the United States. Other nations rushed to make similar claims.

The United Nations then became involved. A committee of the General Assembly began to formulate policy, which was later presented at the First United Nations Conference on Law of the Sea in 1958 in New York. Twenty-four years of effort by

delegates from many interested nations resulted in the 1982 Draft Convention on the Law of the Sea. In April 1982 the United Nations adopted the convention by a vote of 130 to 4, with 17 abstentions. (The United States, Turkey, Venezuela, and Israel voted against the convention.) By 1988 more than 140 countries had signed all or most parts of the treaty. It is now legally binding, but signatories have selectively chosen to respect or ignore its individual provisions.

Here are some important features of the 1982 Draft Convention:

- **Territorial waters** are defined as extending 12 miles (18.2 kilometers) from shore. A nation has the right to jurisdiction within its territorial waters. Straits used for international navigation are excluded from a nation's territorial waters in that any vessel has the right to innocent passage.

- The 200 nautical miles (370 kilometers) from a nation's shoreline constitute its **exclusive economic zone (EEZ)**. Nations hold sovereignty over resources, economic activity, and environmental protection within their EEZs.

- All ocean areas outside the EEZs are considered the **high seas.** In tradition, the high seas are common property to be shared by the citizens of the world. An International Seabed Authority was established to oversee the extraction of mineral resources from the deep sea.

- The values of protecting the ocean and preventing marine pollution were endorsed.

- Subject to some conditions, the freedom of scientific research in the ocean was encouraged.

Look at the labels on your clothing and shoes. Mine seem to come from everywhere—the Dominican Republic, Malaysia, Sri Lanka, Israel, Thailand, Bangladesh, and especially China. Our local supermarket stocks watermelon in the middle of winter, canned clams from Korea, fresh tuna from Thailand, and New Zealand apples. The citizens of Hong Kong consume more California oranges per capita than any people in the world. The electronics store at the local mall is brimming with DVD players and MP3 recorders, nearly all produced in the Far East.

This trade revolution is possible because of a revolution in shipping: the *container*. A container is a large, strong, internationally standardized packing box for cargo. Goods can be safely stowed and transported in a secure, locked space that is resistant to damage, theft, and pilferage. A container is designed for the most efficient use of space. It can easily be lifted and moved and may be carried by any type of transportation—road, rail, or sea.

Before the widespread use of containers (about 25 years ago), a large cargo ship could take two weeks to load, and the irregularly packaged cargo could easily be damaged or simply disappear. Now a modern containership can be loaded in two days and can cross the Pacific from China to Los Angeles in 16 days.

Standardization of container sizes makes loading the ships efficient. The basic unit, set by the International Organization of Standardization (ISO), is a 20-foot container. With dimensions of 20 feet in length, 8 feet in width, and 8 feet 6 inches in height, a container can be loaded with 15 to 20 metric tons of cargo. A larger container 40 feet in length has become more popular in the last decade; it can be loaded with up to 30 metric tons of cargo. A 45-foot container is used for special purposes. Additionally there are standardized designs for general-purpose containers, hardtop containers, flat-rack containers, bulk containers, and refrigerated containers. These efficiencies mean that imported goods are transported more quickly and at lower cost than ever before.

(**a**) A representative of a U.S. manufacturer of high-tech apparel inspects materials and negotiates the production of gloves to be made in a factory in Guangdong, China.

Specialized containerships, some nearly as large as the U.S. Navy's *Nimitz*-class aircraft carriers, are designed to transport up to 4,100 standard containers at a time. They operate safely with a small crew and are driven by efficient diesel or gas turbine engines.

The convention places about 40% of the world ocean under the control of the coastal countries, within the EEZs. The resources of the remaining 60%, the high seas, are to be shared by the citizens of the world.

The United States Exclusive Economic Zone

The United States did not sign the 1982 Draft Convention for a variety of reasons. Among these was concern that private

enterprise would be deprived of profits if it were made to share high seas resources with other countries. Instead, the United States unilaterally claimed sovereign rights and jurisdiction over all marine resources within its own 200-nautical-mile region, which it called the **United States Exclusive Economic Zone**. The proclamation—similar in most ways to the 1982 United Nations treaty but lacking the provision of shared high seas resources—was signed by President Ronald Reagan on 10 March 1983. The U.S. EEZ brings within national domain more than 10.3 million square kilometers (4 million square (**b**) Sealed safely in a container, the finished gloves are transported from factory to port through specialized economic zones in China. The gloves will depart from one of the busiest container ports in the world. In 1995, the state-owned facility on the Pearl River estuary in Hong Kong was the first to transfer one million containers in one month.

(**c**) After a 16-day journey across the Pacific, the apparel arrives at the world's largest container port near Los Angeles, California. *Ever Racer* is shown unloading.

A container cycle begins when a U.S. importer's representative approves the manufacture and transport of goods (**Figure a**). The containers are loaded and locked at a factory or production facility. They are moved by truck (**Figure b**) or rail to a container terminal, where cranes load them onto a containership for ocean transit (**Figure c**). Once unloaded, cleared through customs, and transported by rail and truck (**Figure d**) to the importer's central warehouse, the containers are unlocked and opened, and their contents are distributed. Container shipping is efficient, swift, and sure. Clothing made in Hong Kong can be worn in Kansas City less than a month after manufacture.

Containers that carried finished goods to the United States return to containerships and factories carrying raw materials (plastic pellets, fabric, computer chips, fruit, or even recycled cardboard). The cycle repeats.

International trade by sea is growing at an explosive rate. Look around you on the highways, and you'll see shipping containers on trucks and trains going everywhere. Almost all of them are part of the international container cycle.

17.30

(**d**) Containers are placed on trucks or trains for movement inland.

miles) of continental margins, an area 30% larger than the land area of the United States and a region of diverse geological and oceanographic settings (**Figure 17.25**).

The first step in exploring this new region has been to map the surface of the seafloor, a project that is still going on. The first bottom surveys were conducted off the Pacific coast of the United States because of the energy and mineral resources known to exist there. Massive deposits of various metallic sulfides are being explored at the hydrothermal vents on the Gorda and Juan de Fuca Ridges. More than 100 previously unknown volcanoes have been mapped within the U.S. EEZ off the Pacific coast. Huge faults, submarine landslides, seamounts, and details of the spreading crest of the oceanic ridges are a few of the features that have been discovered so far. Knowledge of the modern tectonic setting for ore formation has been valuable in locating new ore deposits on land.

Manganese nodules and crusts have been discovered within the U.S. EEZ off the Atlantic and Pacific coasts, Hawaii, and the Pacific Island territories. Cobalt, present in the manganese

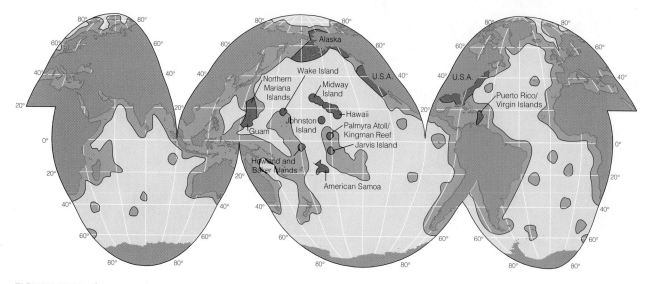

FIGURE 17.25 The U.S. Exclusive Economic Zone (EEZ), shown in red. Note how much of the EEZ is associated with distant possessions. Other Exclusive Economic Zones are shown in blue.

crusts, is being studied to determine how the deposit is formed and how it can be retrieved economically.

The EEZ project also includes research into meteorology, accurate weather forecasting, environmental studies, effects of plate movement on the ocean floor, effects of mining on seafloor organisms, and geohazards such as submarine landslides and earthquakes. It is hoped that this new era of marine exploration and research will lead to informed decisions about offshore activities that will benefit not only the United States but also other maritime nations. Maybe the treasure trove is really there, waiting for new developments in marine technology and international cooperation to tap the riches of the sea.

Questions from Students

1. How much plankton had to die to provide my car with a gallon of gasoline?

Jeffrey Dukes, an ecologist at the University of Massachusetts, recently wondered the same thing. He estimates that about 2% of phytoplankton fall to the seabed to be buried in sediments. Heat transforms about 75% of this mass into oil, but only a small fraction of the oil accumulates where humans can get at it. Depending on the quality of the oil pumped from the ground, about half can be refined into gasoline. Back-of-the-envelope calculations show that about 90 metric tons (99 tons) of dying phytoplankton went into each gallon of gasoline. Your tank can hold the remains of more than 1,000 metric tons of ancient diatoms, dinoflagellates, coccolithophores, and other planktonic producers. Suddenly the cost of a gallon of premium unleaded sounds like a bargain!

2. When will we run out of oil?

By the time the world's crude oil supplies have been substantially depleted, the total amount of oil extracted is expected to range from 1.6 to 2.4 trillion barrels. If this estimate is correct, consumption could continue at the present level until some time between 2025 and 2040, at which time it would drop quite rapidly. But, in fact, we will never run *completely* out of oil—there will always be some oil within Earth to reward great effort at extraction. The days of unlimited burning of so valuable a commodity are nearing an end, however. Future civilizations will surely look back in horror at the fact that their ancestors actually burned something as valuable as lubricating oil.

3. Where are petroleum geologists looking for new reserves of oil?

In 2000 a team of researchers at the U.S. Geological Survey raised its previous estimate of the world's remaining crude oil reserves by 20% to a total of 649 billion barrels. The largest reserves of undiscovered oil are believed to lie in existing fields in the Middle East; northeast of Greenland's continental shelf; in areas of Alaska, Siberia, coastal China, and the Caspian Sea; and in the deltas of the Niger and Congo Rivers in Africa. The shelf surrounding Antarctica may contain the greatest reserves of all. (The difficulties of Antarctic extraction would be extraordinary indeed!)

4. How does the sea salt I see in the health food stores differ from regular table salt?

Unlike the producers of regular table salt, the producers of commercial sea salt do not move the evaporating brine from pond to pond to isolate the different precipitates. Sea salt therefore contains the ocean's salts in their natural proportions, with NaCl making up 78% of the mix. Sea salt has a slightly bitter taste from the potassium and magnesium salts present. Some people believe that the variety of minerals in sea salt is beneficial, but most people obtain adequate amounts of these minerals from a normal diet.

5. What seafoods are the favorites in the United States?

In 2002 shrimp replaced canned tuna as the nation's favorite seafood. We ate a record 1.5 kilograms (3.4 pounds) of shrimp per person in that year. Canned tuna, long the reigning champ, plunged from 1.6 kilograms (3.5 pounds) per capita to 1.3 kilograms (2.9 pounds). Overall consumption of fish and shellfish dipped in 2002 by about 3% from the year before due to rising prices and decreased availability of favorite species.

6. I've heard that seafood contains mercury, a toxin. Can I still eat fish?

That depends. Mercury occurs naturally in the ocean, but most arrives as a by-product of coal burning and mining and other industries. Once in the water, mercury takes the form of methyl mercury, which is concentrated in the tissues of fishes high in the food chain—fishes like swordfish, large mackerel, and (in some cases) tuna. Methyl mercury is especially damaging to the developing brains of fetuses and children. The FDA advises pregnant women to avoid eating swordfish, shark, tilefish, and king mackerel, and to limit consumption of other fishes to 341 grams (12 ounces) a week. For information about an extreme case of mercury poisoning, see Box 18.1 in Chapter 18.

7. The problems of overfishing and bykill concern me. What seafood should I eat, and what should I avoid?

The Monterey Bay Aquarium in California maintains an up-to-date list of seafood obtained in deleterious ways or from endangered stocks. For the most recent information go to http://www.mbayaq.org/cr/cr_seafoodwatch/sfw_regional.asp.

Chapter in Perspective

In this chapter you saw two sides of humanity's use of the ocean. On the one hand, we find the ocean's resources useful, convenient, and essential. On the other, we cannot exploit those resources without damaging their source. World economies are now dependent on oceanic materials, and we are unwilling to abandon or diminish their use until we see clear signs of severe environmental damage.

Marine resources include physical resources such as oil, natural gas, building materials, and chemicals; marine energy; biological resources such as seafood, kelp, and pharmaceuticals; and nonextractive resources like transportation and recreation. The contribution of marine resources to the world economy has become so large that international laws now govern their allocation.

In spite of their abundance, marine resources provide only a fraction of the worldwide demand for raw materials, human food, and energy. Similar resources on land can usually be obtained more safely and at lower cost. The management of marine resources—especially biological resources—for long-term benefit has been largely unsuccessful.

In the next chapter you will learn that humanity has embarked on an unintentional global experiment in marine resource exploitation and waste management. We only hesitatingly adjust our course as we rush into the unknown.

Key Ideas

✓ Marine resources may be categorized as physical, energetic, biological, and nonextractive. They may be renewable or nonrenewable. The contribution of marine resources to the world economy has become so large that international laws now govern their allocation.
Review page 414

✓ Pollution can be a side-effect of resource extraction and use.
Review pages 414–434

✓ Though expensive to obtain, offshore oil and natural gas presently provide about a third and a quarter (respectively) of world needs. Sand and gravel are second in dollar value.
Review Figures 17.1 and 17.2, pages 415–416

✓ Except in a few locales, marine energy production is more experimental than practical.
Review Figures 17.6–17.8, pages 419–421

✓ Though fishery output continues to increase, *per capita* use of oceanic biological resources has fallen since about 1970. Biological resources are often taken without regard to replenishment time.
Review Figures 17.9–17.14, pages 422–424

✓ Worldwide bykill may amount to one-third of total biological resource utilization.
Review Figures 17.15 and 17.16, pages 425–426

✓ Whaling is legal and encouraged in some nations. Thousands of whales are still killed each year for food and oil or by accident in tuna nets.
Review Figure 17.19, pages 426–428

✓ Aquaculture—much of it freshwater—accounts for more than one-quarter of all fish consumed by humans. Shrimp mariculture is the most profitable and fastest-growing sector.
Review Figures 17.21 and 17.22, pages 428–429

✓ Promising drugs are being extracted from marine resources; some are approved for human use.
Review pages 429–430

✓ Nonextractive resources, which include transportation and recreation, are among the fastest growing and potentially among the most valuable ocean resources.
Review pages 430–431

✓ World economics strongly depends on marine transport. Most international cargo is transported in standard containers aboard containerships.
Review Figures 17.23 and 17.24 and Box 17.1, pages 430–433

✓ International agreements, collectively termed laws of the sea, govern resource allocation and trade.
Review Figure 17.25, pages 430–434

Terms and Concepts to Remember

Study Questions

Reviewing What You've Learned

1. Distinguish between physical and biological resources, and between renewable and nonrenewable resources.
2. What are the three most valuable physical resources? How does the contribution of each to the world economy compare to the contribution of that resource derived from land?
3. What are the sources of metals mined or extracted (or potentially mined or extracted) from the sea?
4. Distinguish between mariculture and aquaculture. Does freshwater or seawater aquaculture provide more human food, worldwide? What are the prospects for mariculture?
5. What is a nonextractive resource? Give some examples.

Thinking Critically

1. How are oil and natural gas thought to be formed? How can these substances be extracted from the seabed? Why are the physical characteristics of the surrounding rock important?
2. What method of ocean energy extraction has been most practical? Which has the greatest potential? Why has the latter not been exploited on a large scale?
3. Does the ocean provide a substantial percentage of all protein needed in human nutrition? Of all animal protein? What is the most valuable biological resource? The fastest-growing fishery?
4. What are the signs of overfishing? How does the fishing industry often respond to these signs? What is the usual result? What is bykill?
5. What are the advantages and disadvantages of a proclaimed EEZ? Do you think the United States was justified in proclaiming its own EEZ separate from the provisions of the 1982 United Nations Convention?
6. **InfoTrac College Edition Project** Offshore oil drilling requires large drilling and storage structures. These structures can become obsolete, however, and their disposal can become problematic. Oil companies' proposed solution has been to sink the structures; one example is Shell Oil's Brent spar oil storage buoy off the coast of Scotland. This solution has been strongly opposed by environmental groups. Do you think sinking is an acceptable method for disposing of obsolete oil rigs? Why or why not? Do some research using InfoTrac College Edition to back up your position.

Thinking Analytically

1. Imagine a conversation between the owner of a fishing fleet and a government official responsible for managing the fishery. List five talking points that each person would bring to a conference table. What would be the likely outcome of the resulting discussion?
2. Review the information about primary productivity in Chapter 14. Estimate the surface area of a wheat or alfalfa field that would produce energy equivalent to the 90 metric tons of phytoplankton required to produce a gallon of gasoline.

 Oceanography Now™

Assess your understanding of this chapter's topics with additional quizzing and conceptual-based problems at http://oceanography.brookscole.com/garrison5e

Study and Research on the Internet

Assess your understanding of this chapter's topics with additional quizzing and conceptual-based problems at:

http://oceanography.brookscole.com/garrison5e

- **InfoTrac College Edition Research** leads you to critical thinking exercises that use InfoTrac College Edition® as a research tool.
- **Regional InfoTrac College Edition** articles are organized into East Coast, West Coast, and Gulf Coast regions, allowing you to study oceanography on a more local level.

 For more readings, go to InfoTrac College Edition, your online research library, at:

http://infotrac.thomsonlearning.com

Giant statues look to the horizon on Easter Island (Rapa Nui). The civilization that built the statues had all but destroyed its world when the first Europeans arrived in 1722.

ENVIRONMENTAL CONCERNS

A Cautionary Tale

Let me tell you a story.

Easter Island was home to a culture that rose to greatness amid abundant resources, attained extraordinary levels of achievement, and then died suddenly and alone, terrified, in the empty vastness of the Pacific. The inhabitants had destroyed their own world.

Europeans first saw Easter Island in 1722. The Dutch explorer Jacob Roggeveen encountered the small volcanic speck on Easter morning while on a scouting voyage. The island, dotted with withered grasses and scorched vegetation, was populated by a few hun-dred skittish, hungry, ill-clad Polynesians who lived in caves. During his one-day visit, Roggeveen was amazed to see more than 200 massive stone statues standing on platforms along the coast. At least 700 more statues were later found partially completed, lying in the quarries as if they had just been abandoned by workers. Roggeveen immediately recognized a problem: "We could not comprehend how it was possible that these people, who are devoid of thick heavy timber for making machines, as well as strong ropes, nevertheless had been able to erect such images." The islanders had no wheels, no powerful animals, and no resources to accomplish this artistic, technical, and organizational feat. And there were too few of them—600 to 700 men and about 30 women.

When Captain James Cook visited in 1774, the islanders paddled to his ships in canoes "put together with manifold small planks . . . cleverly stitched together with very fine twisted threads." Cook noted the natives "lacked the knowledge and materials for making tight the seams of the canoes, they are accordingly very leaky, for which reason they are compelled to spend half the time in bailing." The canoes held only one or two people each, and they were less than 3 meters (10 feet) long. Only three or four canoes were seen on the entire island, and Cook estimated the human population at less than 200. By the time of Cook's visit, nearly all the statues had been overthrown, tipped into pits dug for them, often onto a spike placed to shatter their faces upon impact.

Archaeological research on Easter Island has revealed a chilling story. Only 165 square kilometers (64 square miles) in extent and the easternmost outpost of Polynesia, Easter Island is one of the most isolated places on Earth. Voyagers from the Marquesas first reached Easter Island about A.D. 350, possibly after being blown off course by storms. The place was a miniature paradise. In its fertile volcanic soils grew dense forests of palms, daisies, grasses, hauhau trees, and toromino shrubs. Large oceangoing canoes could be built from the long, straight, buoyant Easter Island palms and strengthened with rope made from the hauhau trees. Toromino firewood cooked the fish and dolphins caught by the newly arrived fishermen, and forest tracts were cleared to plant crops of taro, bananas, sugarcane, and sweet potatoes. The human population thrived.

By the year 1400 the population had blossomed to between 10,000 and 15,000 people. Gathering, cultivating, and distributing the rich bounty for so many inhabitants required complex political control. As numbers grew, however, stresses began to be felt: The overuse and erosion of agricultural land caused crop yields to decline, and the nearby ocean was stripped of benthic organisms so the fishermen had to sail greater distances in larger canoes. As resources became inadequate to support the growing population, those in power appealed to the gods. They redirected community resources to carve worshipful images of unprecedented size and power. The people cut down more large trees for ropes and rolling logs to place the heads on huge carved platforms.

By this time the island's seabirds had been consumed and no new birds came to nest. The rats that had hitchhiked to the island on the first canoes were raised for food. Palm seeds were prized as a delicacy. All available land was under cultivation. As resources continued to shrink, wars broke out over dwindling food and space. By about 1550 no one could venture offshore to harpoon dolphins or fishes because the palms needed to construct seagoing canoes had all been cut down. The trees used to provide rope lashings were extinct, their wood burned to cook what food remained. The once-lush forest was gone. Soon the only remaining ready source of animal protein was being utilized: The people began to hunt and eat each other.

Central authority was lost and gangs arose. As tribal wars raged, the remaining grasses were burned to destroy hiding places. The rapidly shrinking population retreated into caves from which raids were launched against enemy forces. The vanquished were consumed, their statues tipped into pits and destroyed. Around 1700 the human population crashed to less than a tenth of its peak numbers. No statues stood upright when Cook arrived.

Even if the survivors had wanted to leave the island, they could not have done so. No suitable canoes existed; none could be made. What might the people have been thinking as they chopped down the last palm? Generations later, their successors died, wondering what the huge stone statues had been looking for.

As you read this chapter, you may be struck by similarities between Easter Island and Earth. In this last chapter we will survey the pollutants, the habitat destruction, the mismanagement of living resources, and the global changes that are currently stressing the planet's environment. It's a depressing list, but at the end you'll find a glimmer of hope. After all, the Easter Islanders had no books and no histories of other doomed societies. We might be able to learn from their mistakes.

Marine Pollution

The ocean's great volume and relentless motion dissipate and distribute natural and synthetic substances. For this reason, we humans have long used the sea as a dump for our wastes. But the ocean's ability to absorb is not inexhaustible. **Marine pollution** is the introduction into the ocean by humans of substances or energy that changes the quality of the water or affects the physical, chemical, or biological environment. It is not always easy to identify a pollutant; some materials labeled as pollutants are produced in large quantities by natural processes. For example, a volcanic eruption can produce immense quantities of carbon dioxide, methane, sulfur compounds, and oxides of nitrogen. Excess amounts of these substances produced by human activity may cause global warming and acid rain. For this reason we need to distinguish between *natural pollutants* and *human-generated* ones.

No one knows to what extent we have contaminated the ocean. By the time the first oceanographers began widespread testing, the Industrial Revolution was well under way and changes had already occurred. Traces of synthetic compounds have now found their way into every oceanic corner. It is sad to consider that we will never know what the natural ocean was like or what remarkable plants and animals may have vanished as a result of human activity. Our limited knowledge of pristine conditions is gleaned from small seawater samples recovered from deep within the polar ice pack and from tiny air bubbles trapped in glaciers. Few undisturbed habitats are left to study, and few marine organisms are completely free of the effects of ocean pollutants.

Characteristics of a Pollutant

About three-quarters of the pollution entering the ocean comes from human activities on land (**Figure 18.1**). A **pollutant** causes damage by interfering directly or indirectly with the biochemical processes of an organism. Some pollution-induced changes may be instantly lethal; other changes may weaken an organism over weeks or months, alter the dynamics of the population of which it is a part, or gradually unbalance the entire community.

In most cases an organism's response to a particular pollutant depends on its sensitivity to the combination of *quantity* and *toxicity* of that pollutant. Some pollutants are toxic to organisms in tiny concentrations. For example, the photosynthetic ability of some species of diatoms is diminished when chlorinated hydrocarbon compounds are present in parts-per-trillion quantities. Other pollutants may seem harmless, as when fertilizers flowing from agricultural land stimulate plant growth in estuaries. Still other pollutants may be hazardous to some organisms but not to others. For example, crude oil interferes with the delicate feeding structures of zooplankton and coats the feathers of birds, but it simultaneously serves as a feast for certain bacteria.

Pollutants also vary in their *persistence*; some reside in the environment for thousands of years, while others last for only a few minutes. Some pollutants break down into harmless substances spontaneously or through physical processes (like the shattering of large molecules by sunlight). Sometimes pollutants are removed from the environment through biological activity. For example, some marine organisms escape permanent damage by metabolizing hazardous substances to harmless ones. Indeed many pollutants are ultimately **biodegradable**—that is, able to be broken down by natural processes into simpler compounds. Many pollutants resist attack by water, air, sunlight, or living organisms, however, because the synthetic compounds of which they are composed resemble nothing in nature.

The ways in which pollutants are changing the ocean and the atmosphere are often difficult for researchers to determine. Environmental impact cannot always be predicted or explained. As a result, marine scientists vary widely in their opinions about what pollutants are doing to the ocean and atmosphere and what to do about it. Environmental issues are frequently emotional, and media reports tend to sensationalize short-term incidents (like oil spills) rather than more serious, long-term problems (like atmospheric changes or the effects of long-lived chlorinated hydrocarbon compounds).

Oil

Oil is a natural part of the marine environment. Oil seeps have been leaking large quantities of oil into the sea for millions of years. The amount of

FIGURE 18.1 Sources of marine pollution. *(Source: Joint Group of Experts on the Scientific Aspects of the Marine Environment, The State of the Marine Environment, UNEP Regional Seas Reports and Studies No. 115, Nairobi: U.N. Environment Programme, 1990.)*

oil entering the ocean has increased greatly in recent years, however, because of our growing dependence on marine transportation for petroleum products, offshore drilling, nearshore refining, and street runoff carrying waste oil from automobiles (**Figure 18.2**).

In the late 1990s people were using nearly 6.8 billion metric tons (7.5 billion tons) of crude oil each year, slightly more than half of which was transported to market in large tankers. In 1998 about 6 million metric tons (6.6 million tons) of oil entered the world ocean. Natural seeps accounted for only about 10% of this annual input, about 600,000 metric tons. About a third of the total was associated with marine transportation. Some of this

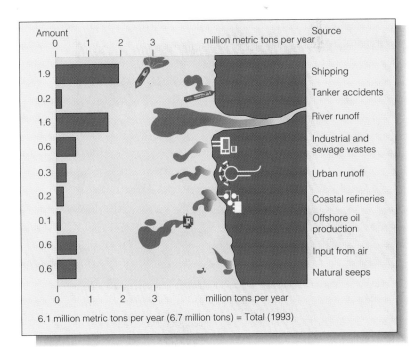

FIGURE 18.2 Sources and volume of ocean oil pollution.

FIGURE 18.3 A bird coated with crude oil. Such birds die unless the oil is removed with a detergent solution. Many will die of cold or toxic effects even when the oil is removed; more will die because of suppressed immune systems. About half a million seabirds die each year from the effects of oil pollution.

oil was not spilled in well-publicized tanker accidents but was released during the loading, discharging, and flushing of tanker ships. Between 150,000 and 450,000 marine birds are killed each year by oil released from tankers (**Figure 18.3**).

Much more oil reaches the ocean in runoff from city streets or as waste oil dumped down drains, poured into dirt, or hidden in trash destined for a landfill. Every year more than 900 million liters (about 240 million gallons) of used motor oil—about 22 times the volume of the *Exxon Valdez* spill—finds its way to

the sea (**Figure 18.4**). This oil is much more toxic than crude or newly refined oil because it has developed carcinogenic and metallic components from the heat and pressure within internal combustion engines. It's no wonder that a sea surface completely free of an oil film is quite rare (**Figure 18.5**).

It is difficult to generalize about the effects that a concentrated release of oil—an oil spill from a tanker, coastal storage facility, or drilling platform—will have on the marine environment. The consequences of a spill vary according to several factors: its location and proximity to shore; the quantity and composition of the oil; the season of the year, currents, and weather conditions at the time of release; and the composition and diversity of the affected communities. Intertidal and shallow-water subtidal communities are most sensitive to the effects of an oil spill.

Spills of *crude* oil are generally larger in volume and more frequent than spills of refined oil. Most components of crude oil do not dissolve easily in water, but those that do can harm the delicate juvenile forms of marine organisms even in minute concentrations. The remaining insoluble components form sticky layers on the surface that prevent the free diffusion of gases, clog adult organisms' feeding structures, kill larvae, and decrease the sunlight available for photosynthesis. Even so, crude oil is not highly toxic, and it is biodegradable. Though crude oil spills look terrible and generate great media attention, most forms of marine life in an area recover from the effects of a moderate spill within about five years. For example, the 907 million liters (240 million gallons) of light crude oil released into the Persian Gulf during the 1991 Gulf War (**Table 18.1**) dissipated relatively quickly and will probably cause little long-term biological damage.

Spills of *refined* oil, especially near shore where marine life is abundant, can be more disruptive for longer periods of time. The refining process removes and breaks up the heavier com-

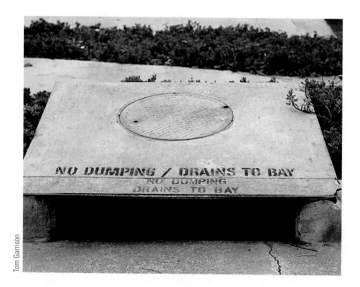

FIGURE 18.4 Storm drains empty directly into rivers, bays, and ultimately the ocean. Each year about 240 million gallons of used motor oil—about 22 times the volume of the 1989 *Exxon Valdez* tanker spill—is dumped down drains, poured into dirt, or concealed in trash headed for landfills. Much of this finds its way to the ocean.

FIGURE 18.5 On 19 November 2002 a tanker carrying 76,000 cubic meters (20 million gallons) of oil split open off the northwestern coast of Spain. Three days later the Canadian *RADARSAT* satellite imaged the resulting oil slick. The coastal city of Lar Coruña is located just above the 3-mile mark overlay.

Table 18.1 Largest Marine Oil Spills Since 1960

Rank	Date	Incident and Location	Size of Spill (millions of gallons)
1	26 Jan 1991	Discharged into Persian Gulf during Gulf War	240.0
2	3 June 1979	Well spill into the Bay of Campeche, Mexico	140.0
3	19 July 1979	Tanker *Atlantic Empress* wreck and subsequent tow, Caribbean NE of Trinidad and Tobago	83.8
4	4 Feb 1983	Well spill at Nowruz, Iran, Persian Gulf	80.0
5	6 Aug 1983	Fire aboard tanker *Castillo de Bellver*, South Africa	78.5
6	16 March 1978	Tanker *Amoco Cadiz* aground off Portsall, Brittany, France	68.7
7	10 Nov 1988	Tanker *Odyssey*, breakup NE of St. Johns, Newfoundland, Canada	43.1
8	11 April 1991	Tanker *Haven*, Mediterranean Sea off Genoa, Italy	42.3
9	18 Mar 1967	Tanker *Torrey Canyon* aground at Land's End, southwestern England	38.2
10	19 Dec 1972	Tanker *Sea Star*, Gulf of Oman, Oman	37.9
.			
.			
.			
46	24 March 1989[a]	Tanker *Exxon Valdez*, Prince William Sound, Alaska	11.0

[a] The *Exxon Valdez* spill of March 1989 was the third worst in U.S. history. But when ranked with other massive marine oil spills involving both shipping and well accidents worldwide, it falls to number 46.
Source: International Oil Spill Database, Cutter Information Corporation.

ponents of crude oil and concentrates the remaining lighter, more biologically active ones. Components added to oil during the refining process also make it more deadly. Spills of refined oil are of growing concern because the amount of refined oil transported to the United States rose dramatically through the 1980s and 1990s. A 1969 spill of refined oil at West Falmouth, Massachusetts, killed countless marine organisms and caused grave environmental damage that is still apparent in the reduced species diversity and fertility. In a similar accident in Mexico, the tanker *Tampico* went aground and dumped her load of refined gasoline into an unspoiled Baja California bay. The area still shows biological scars after nearly three decades.

The volatile components of any oil spill eventually evaporate into the air, leaving the heavier tars behind. Wave action causes the tar to form into balls of varying sizes. Some of the tar balls fall to the bottom, where they may be assimilated by bottom organisms or incorporated into sediments. Bacteria will eventually decompose these spheres, but the process may take years to complete, especially in cold polar waters. This oil residue—especially if derived from refined oil—can have long-lasting effects on seafloor communities. The fate of spilled oil is summarized in **Figure 18.6.**

The methods used to contain and clean up an oil spill sometimes cause more damage than the oil itself. Detergents used to disperse oil are especially harmful to living things. Cleanup of the 1967 *Torrey Canyon* accident off the southern coast of England, one of the first large tanker accidents, did much more environmental damage than the 100,000 metric tons (110,000 tons) of crude oil released. Some resort beaches in the south of England were closed for two seasons, not because of oil residue but because of the stench of decaying marine life killed by the chemicals used to make the shore look clean.

Even the more sophisticated methods that were used to deal with the *Exxon Valdez* disaster, the third worst oil spill in U.S. history (and the 46th worst spill ever; see Table 18.1), seem to have done more harm. When the supertanker *Exxon Valdez* ran aground in Alaska's Prince William Sound on 24 March 1989, more than 40 million liters (almost 11 million gallons; 29,000 metric tons) of Alaskan crude oil—about 22% of her cargo—escaped from the crippled hull. Only about 17% of this oil was recovered by a work crew of more than 10,000 people using containment booms, skimmer ships, bottom scrapers, and absorbent sheets. About 35% of the oil evaporated, 8% was burned, 5% was dispersed by strong detergents, and 5% biodegraded in the first five months. The rest of the oil, some 30% of the spill, formed oil slicks on Prince William Sound and fouled more than 450 kilometers (300 miles) of coastline.

Recent analysis of the affected parts of Prince William Sound shows the cleaned areas to be in generally worse shape than the areas left alone. Most of the small animals that make up the base of the food chain in these areas were cooked by the 65°C

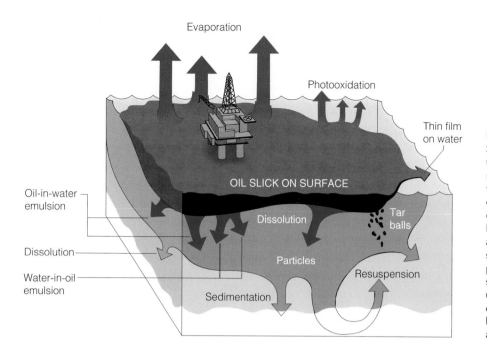

Evaporation

Photooxidation

Oil-in-water emulsion

Dissolution

Water-in-oil emulsion

Thin film on water

OIL SLICK ON SURFACE

Dissolution

Tar balls

Particles

Resuspension

Sedimentation

FIGURE 18.6 The fate of oil spilled at sea. Smaller molecules evaporate or dissolve into the water beneath the slick. Sunlight and atmospheric oxygen cause oxidation of the surface layer. Within a few days, water motion coalesces the oil into tar balls and semisolid emulsions of water-in-oil and oil-in-water. Tar balls and emulsions may persist for months after their formation. Though bonding with sediment particles may sink oil droplets, suspended oil does not enter benthic subtidal sediments in large quantities. If crude oil is left undisturbed, bacterial activity will eventually consume it. Refined oil can be more toxic, however, and natural cleaning processes take a proportionally longer time to complete.

(150°F) water used to blast oil from between the rocks. Others were smothered when the high-pressure jets rearranged sand and mud (**Figure 18.7**). It appears that an overambitious cleanup program can be counterproductive. Sylvia Earle, chief on-site scientist of the National Oceanic and Atmospheric Administration (NOAA), has said, "Sometimes the best, and ironically the most difficult, thing to do in the face of an ecological disaster is to do nothing." The biological cost of the spill will not be known for many years. The cost of the cleanup has exceeded US$3.5 billion. The legal costs could be even higher: In 1994 a federal jury awarded US$5 billion in punitive damages to 14,000 plaintiffs including fishers, native villagers, and fish processors. Exxon appealed the verdict; litigation continues.

Of course the best way to deal with oil pollution is to prevent it from happening in the first place. Tanker design is being modified to limit the amount of oil intentionally released in transport. Legislation is being considered that would limit new tanker construction to stronger, double-hull designs (although shipping experts are uncertain whether even a double-hull design could have prevented the *Exxon Valdez* spill). Perhaps most important, crew testing and training are being upgraded.

Heavy Metals

Small quantities of heavy metals are capable of causing damage to organisms by interfering with normal cell metabolism.

FIGURE 18.7 Cleaning up the March 1989 *Exxon Valdez* oil spill in Prince William Sound, Alaska. Ironically, nearly as much environmental damage was done in the cleanup process as in the original spill.

© Tony Dawson Photography

For the people of Minamata, Japan, who were poisoned by mercury released into the ocean from a nearby factory, heavy metal pollution is a continuing horror story. Between 1953 and 1960 more than 100 people who ate shellfish taken from Minamata Bay were afflicted by a form of mercury poisoning now called Minamata disease (**figure**). The symptoms included kidney damage, neuromuscular deterioration, birth defects, insanity, and eventually death.

The source of the mercury was a plastics plant that was dumping waste mercuric chloride into the bay. Bacterial action changed this chemical into a form that could be taken up and concentrated by benthic organisms. Villagers who ate clams and oysters, especially pregnant women and young children, were seriously affected.

The discharge of mercury has been stopped, but the bay (which had been a source of food for the poor) is still completely unusable for fishing and clamming, even after more than 40 years. This situation will not change for generations; mercury is not easily removed from sediments and has a long residence time.

A mother bathes her daughter, a victim of Minamata disease.

In 1997 the last surviving victims of the poisoning ended their 40-year battle for compensation. Each accepted a cash settlement of $24,200 from the Chisso Corporation.

18.6

Among the dangerous heavy metals being introduced into the ocean are mercury, lead, copper, and tin. Human activity releases about 5 times as much mercury and 17 times as much lead into the ocean as is derived from natural sources, and the incidences of mercury and lead poisoning—major causes of brain damage and behavioral disturbances in children—have increased dramatically over the last two decades. Lead particles from industrial wastes, landfills, and gasoline residue reach the ocean through runoff from land during rains, and the lead concentration in some shallow-water, bottom-feeding species is increasing at an alarming rate. Consumers should be wary of seafoods taken near shore in industrialized regions.

Copper, another heavy metal, is so effective in killing marine organisms that it has long been used in marine antifouling paints. The ship *Pac Baroness*, a freighter carrying 21,000 metric tons (23,000 tons) of finely powdered copper, sank in 448 meters (1,480 feet) of water off the coast of central California after a collision in 1987. Her toxic cargo was scattered over the seabed, and a plume of copper-tainted water was detected 40 kilometers (24 miles) downcurrent from the wreck. Marine life was significantly disrupted in the area, which is a major fishing zone for Dover sole and rock cod. A similar incident off Holland in 1965 killed more than 100,000 fish and destroyed commercial mussel beds.

Tributyl tin, an antifouling paint additive banned in the United States in the late 1980s, is still accumulating in dol- phins. Though the compound has been removed from store shelves, it persists in the food chain. Tributyl tin is a potent suppressor of the immune system and may damage the animals' ability to fight off bacterial and viral diseases. Tin accumulations are unusually high in small whales found in strandings, and biologists suspect a relationship may exist.

Canadian and Norwegian researchers have found that coal combustion, electric utilities, steel and iron manufacturing, fuel oils, fuel additives, and incineration of urban refuse are the major sources of oceanic and atmospheric contamination by heavy metals. They now believe that the toxicity of heavy metals is a greater health problem than the toxicity of all organic and radioactive wastes. Wellness-conscious consumers see fish as a safe and healthful food. But with the ocean still receiving heavy metal–contaminated runoff from the land, a rain of pollutants from the air, and the fallout from shipwrecks, we can only wonder how much longer most seafood will be safe to eat.

18.7

Synthetic Organic Chemicals

Heavy metals are not the only poisons that may contaminate seafood. Many different synthetic organic chemicals also enter the ocean and become incorporated into its organisms. Some of these synthetic organic substances are listed in **Table 18.2**. Ingestion of even small amounts of these compounds can cause illness or even death. Rachel Carson's alarming discussion of the effects

Table 18.2 Synthetic Organic Chemicals That Have Been Detected in the Ocean

Name	Major Health Effects
Aldicarb (Temik)	High toxicity to the nervous system
Benzene	Chromosomal damage, anemia, blood disorders, and leukemia
Carbon tetrachloride	Cancer; liver, kidney, lung, and central nervous system damage
Chloroform	Liver and kidney damage; suspected cancer
Dioxin	Skin disorders, cancer, genetic mutations
Ethylene dibromide (EDB)	Cancer and male sterility
Polychlorinated biphenyls (PCBs)	Liver, kidney, and lung damage
Trichloroethylene (TCE)	In high concentrations, liver and kidney damage, central nervous system depression, skin problems, and suspected cancer and mutations
Vinyl chloride	Liver, kidney, and lung damage; lung, cardiovascular, and gastrointestinal problems; cancer and suspected mutations

Source: Miller, 1997.

FIGURE 18.8 Rachel Carson, author of the influential 1962 book *Silent Spring* on the misuse of synthetic pesticides. This work is generally credited with beginning the environmental movement in the United States.

of these compounds in her 1962 book *Silent Spring* began the environmental movement (**Figure 18.8**).

Halogenated hydrocarbons—a class of synthetic hydrocarbon compounds that contain chlorine, bromine, or iodine—are used in pesticides, flame retardants, industrial solvents, and cleaning fluids. The concentration of **chlorinated hydrocarbons**—the most abundant and dangerous halogenated hydrocarbons—is so high in the water off New York State that officials have warned women of childbearing age and children under 15 not to consume more than half a pound of local bluefish a week. (They are told *never* to eat striped bass caught in the area.) One administrator of the U.S. Environmental Protection Agency has written, "Anyone who eats the liver from a lobster taken from an urban area is living dangerously."

The level of synthetic organic chemicals in seawater is usually very low, but some organisms at higher levels in the food chain can concentrate these toxic substances in their flesh. This **biological amplification** is especially hazardous to top carnivores in a food web.

The damage caused by biological amplification of DDT, a chlorinated hydrocarbon pesticide, is particularly instructive. In the early 1960s brown pelicans began to produce eggs with thin shells containing less than normal amounts of calcium carbonate. The eggs broke easily, no chicks were hatched, and the nests were eventually abandoned. The pelicans were disappearing. The trail led investigators to DDT. Plankton absorbed DDT from the water; fishes that fed on these microscopic organisms accumulated DDT in their tissues; and the birds that fed on the fishes ingested it, too (**Figure 18.9**). The whole food chain was contaminated, but because of biological amplification the top carnivores were most strongly affected. A chemical interaction between DDT and the birds' calcium-depositing tissues prevented the formation of proper eggshells. DDT was eventually banned for use in the United States, and the pelican and osprey populations are recovering.

Biological amplification of other chlorinated hydrocarbons has also affected other species. **Polychlorinated biphenyls (PCBs)**, fluids once widely used to cool and insulate electrical devices and to strengthen wood or concrete, may be responsible for the behavior changes and declining fertility of some populations of seals and sea lions on islands off the California coast. PCBs have also been implicated in a deadly viral epidemic among dolphins in the western Mediterranean. Ingestion of the chemical may have severely weakened the dolphins' immune system and made it impossible for them to fight the infection—as many as 10,000 dolphins died during the summer of 1990. Even more alarming to biologists is the recent discovery that the nearshore dolphins off U.S. coasts are intensely contaminated. The concentration of chlorinated hydrocarbons in these animals exceeds 6,900 parts per million

FIGURE 18.9 The concentration of the pesticide DDT in the fatty tissues of organisms was biologically amplified approximately 10 million times in this food chain of an estuary adjacent to Long Island Sound, near New York City. Dots represent DDT, and arrows show small losses of DDT through respiration and excretion.

In the figure the following labels appear:

DDT in fish-eating birds (pelicans) 25 ppm

DDT in large fish (needlefish) 2 ppm

Concentration has increased 10 million times

DDT in small fish (minnows) 0.5 ppm

DDT in small zooplankton 0.04 ppm

DDT in water 0.000003 ppm

(ppm), concentrations high enough to disrupt the dolphins' immune systems, hormone production, reproductive success, neural function, and ability to stave off cancers.[1] These levels vastly exceed the 50-ppm limit the U.S. government considers hazardous for animals, and the 5 ppm considered the maximum acceptable level for humans! Investigations are continuing, as are probes into the effects of dioxin and other synthetic organic poisons accumulating in the oceanic sink.

Marine animals are even being killed by illicit drug trafficking. In 1997, 42 dolphins and at least three large whales were found dead off Mexico's west coast, victims of the cyanide-based chemicals used by drug merchants to mark ocean drop-off sites. Ships approach the shore at night, drop bales of illegal drugs overboard, mark them with toxic luminescent compounds, and retreat into international waters. Pick-up crews in small boats retrieve the drugs. Squid and fishes are also drawn to the light, encounter the poisonous chemicals, and die. They are in turn eaten by larger animals, with disastrous results.

Figure 18.10 indicates the locations of U.S. coastal areas degraded by synthetic and processed organic chemicals. The production of such chemicals subject to biological amplification in food chains currently exceeds 91 million metric tons (100 million tons) each year. Even greater expanses of ocean will be affected if the predicted 1% of that production reaches the sea.

Eutrophication

Not all pollutants kill organisms. Some dissolved organic substances act as nutrients or fertilizers that speed the growth of marine autotrophs, causing eutrophication. **Eutrophication** (*eu* = good, well; *trophos* = feeding) is a set of physical, chemical, and biological changes that take place when excessive nutrients are released into the water. Too much fertility can be as destructive as too little. Eutrophication stimulates the growth of some species to the detriment of others, destroying the natural biological balance of an ocean area. The extra nutrients

---

[1] If you drag a beached porpoise into the ocean, you could theoretically receive a $10,000 fine for improper disposal of polluted materials!

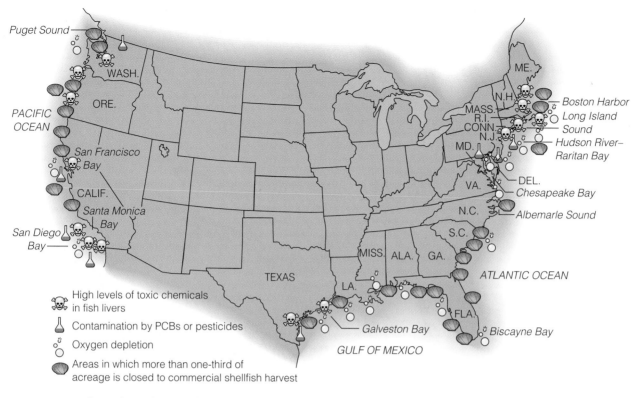

FIGURE 18.10 Chemical assault on major U.S. coastal areas.

come from wastewater treatment plants, factory effluent, accelerated soil erosion, or fertilizers spread on land. They usually enter the ocean from river runoff and are particularly prevalent in estuaries. Eutrophication is occurring at the mouths of almost all the world's rivers.

The Mississippi provides an example. In 1999 a zone of low or no oxygen covered 20,000 square kilometers (7,700 square miles) of the continental shelf south and west of the Mississippi delta (**Figure 18.11**). Fishes, shrimp, and other animals suffocated if they could not flee. The primary cause of this "dead zone" is the nitrogen fertilizer washed into the gulf from millions of farm fields across the Mississippi Basin. A spring drought in 2000 reduced the nutrient flow fueling the algal blooms that deplete oxygen; the low-oxygen zone shrunk to just 4,400 square kilometers (1,700 square miles) in July 2000.

As river flow increases, however, so will the size of the "dead zone." In the future government officials hope to reduce fertilizer runoff to minimize the problem.

The most visible manifestations of eutrophication are the red tides, yellow foams, and thick green slimes of vigorous plankton blooms (see Box 14.1). These blooms typically consist of one dominant phytoplankter that grows explosively and overwhelms other organisms. Huge numbers of algal cells can choke the gills of some animals and (at night, when sunlight is unavailable for photosynthetic oxygen production) deplete the free oxygen content of surface water. (Note the symbols for oxygen depletion in Figure 18.10.)

In nearshore waters, low oxygen content is now thought to cause more mass fish deaths than any other single agent, including oil spills. It is a leading threat to commercial shellfisheries.

FIGURE 18.11 The "dead zone," an area of anoxic conditions on the continental shelf off southern Louisiana. Algal blooms fueled by nutrients running off farmlands and into the Mississippi River deplete the oxygen in this area. The "dead zone" shrank in 2000 because of decreased river flow during a spring drought. Unless nutrient runoff is limited, the zone will grow. (Source: *Science*, 289 (August 4, 2000), p. 723. Reprinted by permission.)

FIGURE 18.12 "Black water" in a Florida bay, February 2002. Visibility in the usually clear water dropped to less than 3 meters (10 feet). A plankton bloom caused by an abundance of sewage-related nutrients in the water is the likely cause. Such blooms have economic consequences; normally abundant fish are almost completely absent in this area.

FIGURE 18.13 Plastic mounds on an isolated shore of Niihau, one of the Hawaiian islands. This material floated from river mouths and was jettisoned by fishers and cruise ship crews.

Toxic substances released from the abundant algae can sicken or kill other species. In 1978 a phytoplankton bloom was linked to the catastrophic die-off of fish near the German coast of the North Sea. In 1988 a similar fate befell dolphins off the U.S. Atlantic coast. As many as a million fluke and flounder were killed in the summer of 1989 when an algal bloom depleted the oxygen in New Jersey's Raritan Bay. During some of these blooms, beaches were covered with foul-smelling foam more than 1 meter (3.3 feet) thick.

These exceptional algae blooms appear to be increasing in number and intensity (**Figure 18.12**). There is little mention of foam events before about 1930, but since 1978 at least one has occurred every year. A similar pattern has been reported for red tides. The threat is especially great in countries that are experiencing rapid economic growth. After two decades of breathless development, severe harmful algal blooms (HABs) have begun to plague China's southeastern coastline. In 1998 a massive bloom killed up to 75% of Hong Kong's fish farms, wiping out many local fishers. The Chinese government estimates US$240 million in direct economic losses from 45 HAB incidents from 1997 to 1999.

Solid Waste

Not all pollutants enter the ocean in a dissolved state; much of the burden arrives in solid form. Some solid waste is ultimately biodegradable, but plastic, which now makes up almost 7% of the solid waste stream, is not. Scientists estimate that some kinds of synthetic materials—plastic six-pack holders, for example—will not decompose for about 400 years!

Americans generate 120 million metric tons of plastic waste, about 500 kilograms (1,100 pounds) per person each year. Plas-

tics now account for more than 10% of all solid waste. Because the ocean is treated as the ultimate dump, some of this waste plastic finds its way to the sea. In 1997 more than 4,500 volunteers scoured the New Jersey–New York coast to collect debris. More than 75% of the 209 tons recovered was plastic (paper and glass accounted for another 15%). The dunes of plastic debris dumped overboard from cruise ships or swept to sea in flooding rivers can build to surreal proportions (**Figure 18.13**).

A 1987 survey by the Woods Hole Oceanographic Institution found that each square mile of ocean surface off the northeastern coast of the United States had more than 46,000 pieces of plastic floating on the surface. This material included ropes, fishing line and nets, plastic sheeting and bags, and granules of broken plastic cups. The problem has become worse in the intervening years—a staggering 100,000 marine mammals and 2 million seabirds die *each year* after ingesting or being caught in plastic debris! Sea turtles mistake plastic bags for their jellyfish prey and die from intestinal blockages. Seals and sea lions starve after becoming entangled in nets (**Figure 18.14**) or muzzled by six-pack rings. The same kinds of rings strangle fish and seabirds. Adding ingredients to plastics that would hasten their decomposition would add only 5% to 7% to their cost, but this price increase is currently unacceptable to industry.

What should we do with plastic and other solid wastes such as glass and paper, disposable diapers, scrap metal, building debris, and all the rest? Dumping it into the ocean is clearly unacceptable, yet places to deposit this material are becoming scarce. In 1999 the average New Yorker threw away more than a ton of waste annually. California's Los Angeles and Orange Counties generate enough solid waste to fill Dodger Stadium every eight days. Transportation of waste to sanitary landfills becomes more expensive as nearby landfills reach capacity.

FIGURE 18.14 Sea lions (seen here) and seals die by the hundreds each year after becoming entangled in plastic debris, especially discarded and broken fishing nets.

Is incineration the answer? We currently burn about 10% of our trash. By contrast, Sweden burns half its solid wastes. Unfortunately, even with air pollution control devices, incinerators still emit great quantities of tiny particles, heavy metal residue, and immense amounts of carbon dioxide. Dumping the residual ash is also a problem; ash from Philadelphia's garbage incinerators has been turned away by states as far away as South Carolina, and one shipload of the stuff was refused entry at an African port.

Is recycling the answer? The Japanese currently recycle about 50% of their solid waste and are importing even more; scrap metal and waste paper headed for Japan are the two biggest exports from the Port of New York. Americans are buying back their own refuse in the form of appliances, automobiles, and the cardboard boxes that hold their MP3 players and compact disc recorders. Massachusetts and California have set a goal of recycling 25% of their waste; the city of Seattle is now approaching 30%. The direct savings to consumers as well as the environmental rewards to ocean and air will be significant.

The *best* solution is a combination of recycling and reducing the amount of debris we generate by our daily activities. We will soon have no other choice.

Sediment

Runoff from mining, farming, forestry, and other land uses often contains large amounts of sediment. This material can cloud the water, impede photosynthesis, and clog the gills of marine organisms. Coastal ecosystems have been smothered and buried by soils and sands washed into the ocean after strong rains. The problem is particularly apparent in areas of rapidly growing human populations and in the vicinity of coastal mining and dredging operations. In Louisiana, coastal marshes are vanishing as sediments from upstream bury them at the rate of 130 to 155 square kilometers (50 to 60 square miles) a year. The sediments come from poor agricultural practices and uncontrolled commercial development. Sediments washing from local rivers have killed 75% of Costa Rica's Caribbean reefs. Philippine reefs are similarly endangered. By the early 1980s about 75% of the reefs surrounding the Philippines had been damaged beyond recovery by the effects of sedimentation and erosion. The economic loss of Philippine reefs is tremendous; more than 100,000 jobs and US$80 million in fish catches are lost each year. Because of lowered reef productivity, 5 million Filipinos are unable to catch the fish and crustaceans they require for proper nutrition. Half of all children living in coastal regions of the Philippines are malnourished.

Sewage

Until 1992 almost 2 billion liters (half a billion gallons) of partially treated **sewage** poured into Boston Harbor every day. A new sewage treatment plant opened in that year, 22 years after the deadline mandated by the U.S. Clean Water Act. About the same amount of treated sewage pours from outfall pipes off Los Angeles. Treatment plants in southern California are sometimes overwhelmed after heavy rainstorms, and raw sewage enters the ocean in large quantity. Rain or shine, areas around the entrance to San Diego Harbor are often so contaminated with sewage from the city of Tijuana, Mexico, that anyone who enters the water runs the risk of bacterial or viral infection.

Not only human sewage is involved. In June 1995, 26 million gallons of hog feces and urine spilled into a narrow stream feeding North Carolina's New River estuary. Thousands of fish died from direct effects, and hundreds of thousands more perished as algae grew and consumed oxygen.

About 98% of sewage is water; the rest is a mixture of organic matter containing bacteria and viruses, toxic metal compounds, synthetic organic chemicals, and other debris. Sewage treatment separates the fluid component from the solids, treats the water to kill disease organisms and reduce the levels of nutrients, and releases it into a river or the ocean. The remaining semisolid sewage sludge is digested, thickened, dried, and shipped to landfills, burned to generate electricity, or dumped into the ocean. The liquid effluent wanders from its outlet and circulates with currents, but sludge and other insoluble residues may stay near the outfall or dump site for years. The amount of wastewater and sewage sludge increased by 60% in the 1990s.

Sludge may be an even greater problem than raw sewage. The water just south of Long Island, New York, has been one of the most intensely used ocean dumping sites in the world. The area is covered with sludge, which creates an oxygen-poor environment in which few animals can survive. During storms some of this material washes up on local beaches, contaminates shellfish beds, and routinely causes disease outbreaks among people consuming raw oysters and clams from the area. After a public outcry, a new dumping site was selected beyond the edge of the continental shelf. Ten million tons of wet sludge produced by treatment plants in New York and New Jersey since March 1986 have been dropped there by huge barges. The plume from this sludge now contaminates the Gulf Stream.

Both liquid effluent and sludge contribute to eutrophication. As we've seen, some forms of algae thrive in nutrient-rich wastewater, where they multiply to prodigious numbers, secrete toxins, or deplete oxygen in the surrounding water and cause the death of fish and crustaceans.

Waste Heat

Shoreside electrical generating plants use seawater to cool and condense steam. The seawater is returned to the ocean about 6°C (10°F) warmer, a difference that may overstress marine organisms in the effluent area. **Figure 18.15** shows how widely the heated water spreads out around a coastal power installation. Some recent power plant designs minimize the environmental impact by pumping colder water from farther offshore, warming it to the temperature of the seawater surrounding the plant site, and then releasing it. This method minimizes the impact on the surrounding communities, but it still shocks those eggs, larvae, plankton, and other organisms that are sucked through the power plant with the cooling water.

Introduced Species

Several thousand species are in transit every day in the ballast water of tankers and other ships. Juvenile forms of marine organisms can easily hitch rides across otherwise insurmountable oceanic barriers and set up housekeeping at distant shores. These foreign organisms—called **introduced species** or exotic species—sometimes outcompete native species and reduce biological diversity in their new habitats. New marine diseases can also be introduced in this way. Even canals and fishery enhancement projects can introduce potentially destabilizing new species.

The Chinese mitten crab (genus *Eriocheir*, **Figure 18.16**) and a common decorative aquarium seaweed (genus *Caulerpa*) are good examples of exotic marine species that can wreak destruction. Probably brought to the western United States and the Atlantic coast of Europe as larval forms in ships' ballast water, the mitten crab is an energetic crustacean that burrows into riverbanks and levees, causing them to collapse. Sometimes reaching densities of 30 animals per square meter (2.7 per square foot), the crabs clog water supply pipes and, if frustrated in migration, wander onto streets and into people's backyards and homes. The aquarium alga may be even more of a menace. It spreads through fragmentation. After escaping into the Mediterranean in 1984, this Caribbean native now chokes more than 4,050 hectares (10,000 acres) of ocean floor off Spain, France, Italy, and Croatia. It has recently turned up in southern California bays, where it is outcompeting local organisms and greatly reducing biodiversity.

The problems caused by introduced species will grow. There are about 250 known exotic species in California's San

FIGURE 18.15 Thermal effluent from power plants and sewage outfalls, shown in a thermograph of the Quinault River as it enters the Pacific at Taholah, Washington. In this infrared photograph, white areas indicate warmer water.

FIGURE 18.16 Chinese mitten crabs, exotic (introduced) species that burrow into riverbanks and levees, causing them to collapse.

Francisco Bay, the most invaded estuary in the world—more inadvertent global experimentation.

The Costs of Pollution

In 1998 government and industry in the United States spent about US$220 billion—an average of US$800 for each American—on the control of atmospheric, terrestrial, and marine pollution. This figure was equivalent to about 1.6% of the gross national product, or 2.7% of capital expenditures by U.S. business. That same year the United States lost 4% of its gross national product through environmental damage. Clearly the financial costs of pollution will continue to increase.

But there are other costs. Failure to control pollution will eventually threaten our food supply (marine and terrestrial), destroy whole industries, produce a greater disparity between have and have-not nations, and cause a decline in the health of all the planet's inhabitants. To these costs must be added the aesthetic costs of an ocean despoiled by pollution; few of us look forward to sharing the beach with oiled birds, jettisoned diapers, or clumps of medical waste.

Habitat Destruction

The pollution processes we have discussed don't affect individual organisms alone. They influence whole habitats, especially the most complex and biologically sensitive shallow-water habitats.

Bays and Estuaries

The hardest-hit habitats are estuaries, the hugely productive coastal areas at the mouths of rivers where fresh water and seawater meet. Pollutants washing down rivers enter the ocean at estuaries, and estuaries often contain harbors, with their potentials for oil spills. As little as 1 part of oil for every 10 million parts of water is enough to seriously affect the reproduction and growth of the most sensitive bay and estuarine species. Some of the estuaries along Alaska's Prince William Sound, site of the 1989 *Exxon Valdez* accident, were covered with oil to a depth of 1 meter (3.3 feet) in places. The spill's effects on the US$150-million-a-year salmon, herring, and shrimp fishery will be felt for years to come.

Other Pacific coast habitats have been polluted in different ways. Bottom sediments in Seattle's Elliott Bay are contaminated with a poisonous mix of lead, arsenic, zinc, cadmium, copper, and PCBs. Tumors on the livers of English sole, which dwell on the sediments, have been linked to these compounds. Pollutants are so abundant in southern San Francisco Bay that clams and mussels contain near-lethal concentrations of heavy metals. Birds migrating from Central America to the Arctic Circle run a risk of being poisoned by stopping there to feed. There is also a risk to humans who eat ducks shot in this area because of the high concentrations of pollutants they contain. Similar warnings apply to all fish caught in and near southern California estuaries between Santa Barbara and Ensenada, Mexico.

Estuaries and bays along the U.S. Gulf coast, one of the most polluted bodies of water on Earth, are also severely stressed. About 40% of the nation's most productive fishing grounds, including its most valuable shrimp beds, are found in the Gulf. Nearly 60% of the Gulf's oyster and shrimp harvesting areas—about 13,800 square kilometers (5,300 square miles)—either are permanently closed or have restrictions placed on them because of rising concentrations of toxic chemicals and sewage. Half of Galveston Bay, once classed as the second most productive estuary in the United States, is off limits to oyster fishers because of sewage discharges.

Estuaries along the Atlantic coast are also threatened. From southern Florida to central Georgia, more than 325 square kilometers (125 square miles) of sea grasses (which act as nurseries for a great variety of marine life) have been killed by a virus. Scientists speculate that pollutants from urban and agricultural runoff have weakened the plants' resistance. Fishers to the north in Chesapeake Bay have been mystified by a sudden decline in the abundance of fish and crustaceans, a change marine scientists attribute to increasing pollution in the area. Lobstermen in New England have noticed an alarming increase in the incidence of tumors in lobsters' tail and leg joints. Could these changes also be caused by toxic wastes? The beluga whale population of Canada's St. Lawrence estuary collapsed in the early 1980s. High levels of PCBs, DDT, and heavy metals—substances biologically amplified in the whales' food—were blamed for the tumors, ulcers, respiratory ailments, and failed immune systems discovered during the autopsies of 72 dead whales.

People also "develop" estuaries into harbors and marinas. In the 1960s and 1970s California led the world in the acreage of bays and estuaries filled for recreational marinas. Harbors grew smaller as more of their area was filled for docks and storage facilities. One hundred and fifty years ago, San Francisco Bay covered 1,131 square kilometers (437 square miles). Today only 463 square kilometers (179 square miles) remains—the rest has been filled in. Filling of estuaries is just as threatening to the natural reproductive cycles of shrimp and fish as poisoning by toxic wastes.

Some states control the development of coastal regions. A citizens' initiative passed by Californians in 1972 limited development of that state's coastal zone. Massachusetts laws make it illegal to fill any marsh or estuarine region, even areas that are privately owned. Similar legislation is pending in a few other coastal states.

Coral Reefs

Pollution may also be damaging coral reefs. Marine biologists have been baffled by recent incidents of coral bleaching—corals expelling their symbiotic zooxanthellae—in the Caribbean and tropical Pacific. A 1989 bleaching event in the Caribbean was just the most recent ecological disruption of the area in a series that included a massive fish kill in 1980, a die-off of about 95% of the individuals of one sea urchin species in 1983–84, and a slowly spreading coral malady known as white-band disease. Biologists do not know for certain what caused these disturbances, but increasing pollution is thought to be partially responsible.

Some chemical pollution is intentional. Especially damaging to tropical reefs has been the practice of using cyanide to collect tropical fish. Fishers squirt a solution of sodium cyanide over the reef to stun valuable species. Many fish die; those that survive are sent to collectors all over the world. At the same time the invertebrate populations of the sensitive coral reef communities are decimated.[2]

Not all coral reef pollutants are chemicals. Fishers in Indonesia and Kenya dynamite reefs to kill fish that hide among the coral branches. Reefs throughout the world are mined for construction material, for ornamental pieces, or for their calcium carbonate (to make plaster and concrete). About 27% of the world's coral reefs have already disappeared. Researchers fear that 70% of the reefs close to large population centers will disappear in the next 50 years.

Other Habitats

Between 1963 and 1977, about half of India's extensive mangrove forests were cut down. About a third of Ecuador's mangroves have been converted to ponds used in shrimp mariculture. Agricultural expansion is expected to wipe out all Philippine mangroves in ten years. Not even the calm, cold communities of the abyssal plains are safe from disruption: Imagine the effect manganese nodule mining will have on the delicate organisms of the deep bottom.

Marine Sanctuaries

Beginning in 1972, the federal government has established a dozen national marine sanctuaries (**Figure 18.17**). These areas are intended as safe havens for marine life. They vary in size, but collectively cover about 46,000 square kilometers (18,000 square miles) of coral reefs, whale migration corridors, under-

---

[2] For more on this troubling practice, see Questions from Students on page 460.

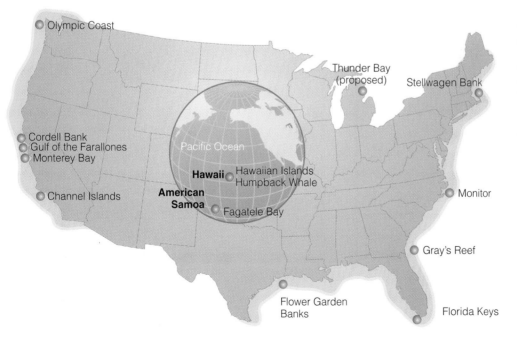

FIGURE 18.17 The 12 U.S. marine sanctuaries. Since 1972, a dozen national marine sanctuaries have been designated in the coastal waters of the United States and American Samoa. Despite their name, most sanctuaries are not off limits to commercial exploitation. (Source: *Popular Science* (September 2000). Reprinted with permission from *Popular Science* magazine, copyright 2000, Times Mirror Magazines, Inc.)

sea archaeological sites, deep canyons, and zones of extraordinary beauty and biodiversity.

Despite their name, these sanctuaries are not off limits to commercial fishing, trawling, or dredging. In May 2000 President Bill Clinton issued an executive order directing federal agencies to establish a national framework for managing sanctuaries, wildlife refuges, and other protected areas that together cover about 1% of U.S. territorial waters. The new framework was to include ecological reserves where "consumptive uses" of marine resources would be prohibited. As of January 2004, the framework had not been fully implemented.

Global Changes

The ocean and the atmosphere are extensions of each other, and human activity has changed the atmosphere as it has changed the ocean. Pollutants injected into the air can have global consequences for the ocean and for all of Earth's inhabitants. Potentially the most destructive atmospheric problems are depletion of the ozone layer, global warming, and acid rain.

Ozone Layer Depletion

Ozone is a molecule formed of three atoms of oxygen. Ozone occurs naturally in the atmosphere. A diffuse layer of ozone mixed with other gases—the **ozone layer**—surrounds the world at a height of about 20 to 40 kilometers (12 to 25 miles).

Seemingly harmless synthetic chemicals released into the atmosphere—primarily **chlorofluorocarbons** (**CFCs**) used as cleaning agents, refrigerants, fire-extinguishing fluids, spray-can propellants, and insulating foams—are converted by the energy of sunlight into compounds that attack and partially deplete Earth's atmospheric ozone. Ozone levels in the stratosphere have decreased by about 3% over some of the United States since 1969. A 4% drop has been noted over Australia and New Zealand, and a 50% decrease has been observed near the North and South Poles (**Figure 18.18**). The amount of depletion varies with latitude (and with the seasons) because of variations in the intensity of sunlight. In September 2000 the largest ozone depletion ever recorded was detected over Antarctica; about 11 million square kilometers (4 million square miles) was affected.

This decline in ozone alarms scientists because stratospheric ozone intercepts some of the high-energy ultraviolet radiation coming from the sun. Ultraviolet radiation injures living things by breaking strands of DNA and unfolding protein molecules. Species normally exposed to sunlight have evolved defenses against average amounts of ultraviolet radiation, but increased amounts could overwhelm those defenses. Land plants such as soybeans and rice would be subjected to sunburn that decreases their yields. Even plankton in the uppermost 2 meters (6.5 feet) of ocean would be affected. Recent research in fact indicates a decrease in phytoplankton primary productivity in the coastal waters around Antarctica (**Figure 18.19**). In 1997 biologists at the U.S. Palmer Research Station in Antarctica confirmed lethal ultraviolet damage to seastar embryos drifting near the sea surface.

Our own species would not escape: A 1% decrease in atmospheric ozone would probably be accompanied by a 5% to 7% increase in the incidence of human skin cancer (**Figures 18.20** and **18.21**). According to medical researchers' estimates, the 4% ozone depletion over Australia and New Zealand will cause at least a 20% increase in cases of human skin cancers over the next two decades. Strong ultraviolet light can also sup-

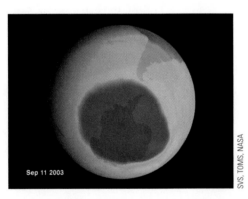

ACTIVE FIGURE 18.18 The seasonal ozone "hole" above Antarctica, as recorded by the *Earth Probe* satellite in September 2003. The lowest ozone values (the "hole") are indicated by magenta and purple. This hole is smaller than in recent years, in part because of international efforts to reduce the use of ozone-damaging chemicals and in part because of warmer than normal air in the surrounding stratosphere.

Oceanography ⌇ Now™ | At the Active Figures link at http://oceanography.brookscole.com/garrison5e you can view the formation of the ozone hole.

−3.0   −2.5   −2.0   −1.5   −1.0   −0.5   0.0   0.5   1.0

Change in primary productivity August–December, 1992

FIGURE 18.19 The depletion of stratospheric ozone (the ozone hole) over the Antarctic allows more ultraviolet radiation to reach the ocean surface, which lowers phytoplankton productivity.

FIGURE 18.20 A malignant melanoma in an early stage. The outer discolored area is about the size of a dime. Left untreated, this most dangerous form of skin cancer could spread and cause the death of the patient. Skin cancer appears to be associated with exposure to the sun. If you spend much time in the sun, you increase your risk of skin cancer. Be alert for changes in moles or for any pigmented lesion that is asymmetrical, has an irregular border, is an unusual color, grows, or begins to itch. If you notice any of these characteristics, see a physician.

FIGURE 18.21 If ozone depletion continues, the "perfect tan" may have to be none at all. (Source: © The New Yorker Collection 1989. Stuart Leeds from cartoonbank.com. All rights reserved.)

press the immune system and cause eye cataracts. The quantity of photochemical smog shrouding our cities will also increase as ozone levels fall.

In June 1990 representatives of 53 nations agreed to ban major production and use of ozone-destroying chemicals by the year 2000. Legislation has been proposed to phase out production of all forms of CFC in the United States, and a sense of great urgency surrounds ongoing research to find safe substitutes for these substances. Recent data indicate that these measures are having an effect. CFC concentrations peaked near the beginning of 1994 (**Figure 18.22**). Researchers hope to see a decrease in the sizes of polar ozone holes by 2010.

Oceanography⊜Now™ | Click Oceanography Interactive to work through an activity on ozone hole through weather and climate.

Global Warming

The surface temperature of Earth fluctuates slowly over time. The global temperature trend has been generally upward in the 18,000 years since the last ice age, but the *rate* of increase has recently accelerated. This rapid warming is probably the result of an enhanced **greenhouse effect,** the trapping of heat by the atmosphere. Glass in a greenhouse is transparent to light but not to heat. The light is absorbed by objects inside the greenhouse, and its energy is converted into heat. The temperature inside a greenhouse rises because the heat is unable to escape. On Earth **greenhouse gases**—carbon dioxide, water vapor, methane, CFCs, and others—take the place of glass.

Heat that would otherwise radiate away from the planet is absorbed and trapped by these gases, causing a rise in surface temperature. **Figure 18.23** shows this mechanism.

The greenhouse effect is necessary for life; without it, Earth's average atmospheric temperature would be about −18°C (0°F). Earth has been kept warm by natural greenhouse gases. The sources of these gases are volcanic and geothermal processes, the decay and burning of organic matter, and respiration and other biological sources. The removal of these gases by photosynthesis and absorption by seawater appears to prevent the planet from overheating. But the human demand for quick energy to fuel industrial growth, especially since the beginning of the Industrial Revolution, has injected unnatural amounts of new carbon dioxide into the atmosphere from the combustion of fossil fuels. Carbon dioxide is now being produced at a greater rate than it can be absorbed by the ocean. **Figure 18.24** shows how much carbon dioxide concentrations have increased in the atmosphere since about 1750. The atmosphere's carbon dioxide content now rises at the rate of 0.4% each year. Though CFC levels are now declining, methane levels continue to increase. (The primary sources of methane are intestinal gases from cows and termites and the decay of refuse and vegetation cleared from land.)

There has been a 5°C (9°F) rise in global temperature from the end of the last ice age until today. Carbon dioxide and other human-generated greenhouse gases produced since 1880 are thought to be responsible for about half of that increase. (The other half is thought to be caused by an increase in solar output over the past century.) A new survey of 300 bore-hole temperatures on four continents has confirmed that Earth is getting

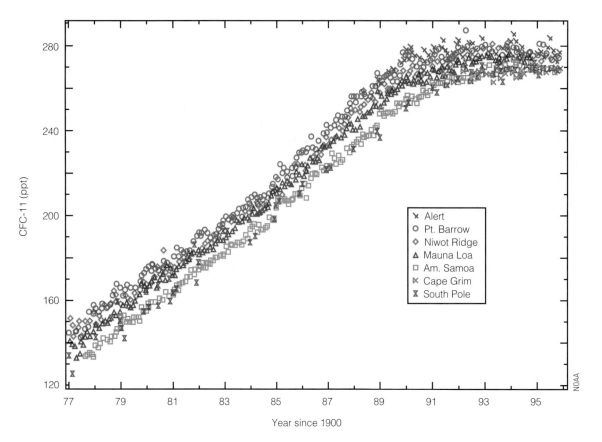

FIGURE 18.22 The accumulation of chlorofluorocarbon-11 (CFC-11) in the atmosphere has leveled off as a result of voluntary and mandated emission reductions. Monthly averages are reported as dry mixing ratios in parts per trillion (ppt) for CFC-11 at ground level for four NOAA/CMDL stations (Pt. Barrow, Alaska; Mauna Loa, Hawaii; Cape Matatula, American Samoa; and South Pole) and three cooperative stations [Alert, Northwest Territories, Canada (Atmospheric Environment Service); Niwot Ridge, Colorado (University of Colorado); and Cape Grim Baseline Air Pollution Station, Tasmania, Australia (Commonwealth Scientific and Industrial Research Organization)]. (Courtesy of NOAA/CMDL.)

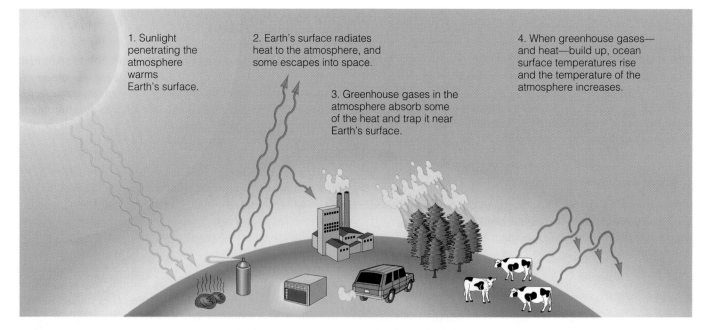

FIGURE 18.23 How the greenhouse effect works. Since the beginning of the Industrial Revolution, emissions of carbon dioxide, methane, and other greenhouse gases have increased. Most researchers believe that these gases have contributed to a general warming of Earth's atmosphere and ocean.

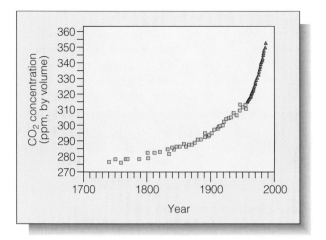

FIGURE 18.24 The rise in atmospheric carbon dioxide concentration since 1750. Measurements up to about 1960 (■) came from samples of air trapped in the Antarctic ice; later measurements (▲) were made directly at Mauna Kea, Hawaii. (After Watson et al., 1990.)

warmer and that the rate of warming has been increasing since 1900. Subsurface temperatures confirm that the average global temperature has increased 0.4° to 0.8°C (0.7° to 1.4° F) over the last century (**Figure 18.25**). The El Niño year 1998 was the warmest on record; 2001 was the second warmest. Experts who met in Shanghai in January 2001 estimated that if greenhouse emissions continue to rise, the average temperature will increase another 5.8°C (10.5°F) over the next century.[3]

The south polar ice sheets are already shrinking in the warmth. Imagine the effect of a significant rise in sea level on the harbors, coastal cities, river deltas, and wetlands where one-third of the world's people now live. As **Figures 18.26** and 12.1 suggest, the costs to society would be enormous.

Other problems are associated with global warming. Thought to be due in part to warmer ocean water and diminished winds to provide the light dusting of terrestrial iron

---

[3] This estimate is based on projections of data, not on models.

FIGURE 18.25 Global annual surface temperature relative to the 1951–80 mean based on surface air measurements at meteorological stations and satellite measurements of sea surface temperature. (NASA–Goddard, after Hansen et al., 2002.)

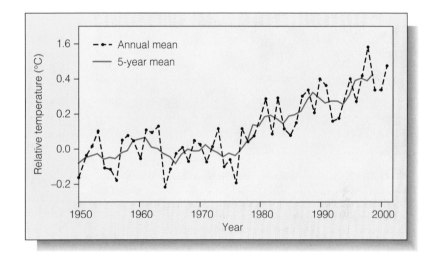

FIGURE 18.26 For island nations such as the Maldives, even a small rise in sea level could spell disaster. Strung out across 880 kilometers (550 miles) of the Indian Ocean, 80% of this island nation of 263,000 people lies less than 1 meter (3.3 feet) above sea level. Of the country's 1,180 islands, only a handful would survive the median estimate of sea level rise by 2100. Most of the population lives in fishing villages on low islands like the one shown here, where the effects of this century's sea-level rise of 10 to 25 centimeters (4 to 10 inches) have already been felt.

FIGURE 18.27 Vapor trails left by commercial airliners and smoke trails from ships trigger clouds in the atmosphere over the northern Pacific on 29 April 2002. If conditions are right, these "triggered" clouds can block incoming sunlight and outgoing heat, causing less of a difference between day and night temperatures. How will these increasingly common clouds affect weather and climate? No one knows.

NASA/Tom Garrison

needed for their metabolism, phytoplankton productivity in the last 20 years has dropped by about 9% in the North Pacific and nearly 7% in the North Atlantic. Worldwide, researchers at the Goddard Space Flight Center found a decline in worldwide phytoplankton productivity of more than 6% in that length of time. Less phytoplankton means less carbon dioxide uptake and significant changes in oceanic ecosystems. Warming may also shift the strength and position of ocean surface currents. What would happen to the economy of Europe if the Gulf Stream were to alter course? All too often, as **Figure 18.27** shows, we have no clear idea of the effects of our activities.

Unfortunately, it will be exceedingly difficult to curtail global warming by decreasing the production of carbon dioxide and methane. In the last hundred years, industrial production has increased fiftyfold; we have burned roughly 1 billion barrels of oil, 1 billion metric tons of coal, and 10 billion cubic meters of natural gas. Carbon dioxide is a major product of combustion for these hydrocarbon compounds. The world's energy demand is projected to increase 3.5 times between now and 2025, with carbon dioxide emissions 65% higher than today. China alone plans to add 18,000 megawatts of coal-fired electric generating capacity each year, equal to Louisiana's entire power grid.

At a meeting in Kyoto, Japan, in 1997, leaders and representatives of 160 countries established carbon dioxide emission targets for each developed country. The United States, for example, would reduce its carbon dioxide emissions to 7% less than 1990 levels by 2012. This goal is thought to be economically untenable, and the Kyoto Protocol has not been ratified by the U.S. Senate. In any case, those levels would only slow the eventual effects on Earth's climate.

Alternatives to fossil fuel must be found if we are to maintain world economies and prevent an increase in global temperature with all its uncertainties. The only alternative source of energy that currently produces significant amounts of power is **nuclear energy,** which now generates about 17% of the electricity produced in the United States. Despite much publicity to the contrary, these pressurized water reactors have good records of dependable power production and safety.[4] The problem with nuclear power lies not so much in the everyday operation of the reactors but in disposing of the nuclear wastes they produce. By 1992 about 55,000 highly radioactive spent-fuel assemblies were in temporary storage in deep pools of cooled water; they must be stored for 10,000 years before their levels of radioactivity will be low enough to pose no environmental hazard. Radioactive substances emit **ionizing radiation,** a form of energy that is able to penetrate and permanently damage cells. It is mandatory that artificial sources of ionizing radiation be isolated from the environment.

Will citizens of industrial countries (and countries that wish to become industrialized) agree to lessen the danger of increased global warming by slowing their economic growth, decreasing their dependence on fossil fuel combustion, and developing safe alternative sources of energy? Some insight may be gained from the behavior of ranchers and industrialists in the rain forests of New Guinea, the Philippines, and Brazil. The Amazon rain forest of Brazil is being burned at a rate of about 12 square kilometers (almost 5 square miles) *per hour,* equivalent to the area of West Virginia every year. Huge stands of trees that should be nurtured to absorb excess carbon dioxide are being destroyed. The cleared land is used for farms, cattle ranches, roads, and cities. The priorities are clear.

What Can Be Done?

In a pivotal paper published in 1968, biologist Garrett Hardin examined what he termed "The Tragedy of the Commons." Hardin's title was suggested by his study of societies in which some agricultural areas were held *in common*—that is, jointly owned by all residents. Citizens of these societies owned small homes, plots of land, and perhaps a cow that was put to pasture

---

[4] The Soviet reactor at Chernobyl that exploded in 1986 had a much different design.

C. Mayhew & R. Simmon (NASA/GSFC), NOAA/NGDC, DMSP Digital Archive

FIGURE 18.28 Earth at night. Human-made light shows developed or populated areas on the surface. This composite image emphasizes the present extent of industrialization and resource use.

on the commons. Each farmer *kept* the milk and cheese given by his cow but *distributed* the costs of cow ownership—overgrazing of the commons, cow excrement, fouled drinking water, and so on—among all the citizens. This arrangement worked well for centuries because wars, diseases, and poaching kept the numbers of people and cows well below the carrying capacity of the land. But eventually political stability and relative freedom from disease allowed the human (and cow) population to increase. Farmers pastured more cows on the commons and gained more benefits. Soon the overstressed commons could no longer sustain the growing numbers of cows, and the area held in common was ruined. Eventually no cows could survive there.

The lesson applies to our present situation. Hardin noted that in our social system each individual tends to act in ways that maximize his or her material gain. Each of us gladly keeps the *positive* benefit of work but willingly distributes the *costs* among all. For example, this morning I drove to my college office; the benefit to me was one trip to my office. A cost of this short drive was the air pollution generated by the fuel combustion in my car's engine. Did I route the exhaust fumes through a hose to a mask held tightly over my nose and mouth? That is, did I reserve the environmental costs of my actions for my own use, just as I had reserved for myself the benefit of my ride to work? No. I shared those fumes with my fellow Californians, just as you shared your morning's sewage with your fellow citizens, or just as the factory down the road shared its carbon dioxide with all the world. Indeed, *the world itself* is our commons. The modern tragedy of the commons rests on these kinds of actions.

The carrying capacity of the whole Earth-commons may already have been exceeded. Births now exceed deaths by about 3 people per second, 10,400 per hour. *Each year* there are 95 million more of us, nearly one-third the population of the United States. The number of people tripled in the twentieth century and is expected to double again before reaching a plateau sometime in this century. Another billion humans will join the world population in the next 10 years, 92% of them in underdeveloped countries (**Figure 18.29**).[5] One-fifth of the world's people already suffer from abject poverty and hunger.

This exploding population is not content with using the same proportion of resources used today. Citizens of the world's least developed countries are influenced by education and advertising to demand a developed-world standard of living. They look with misguided envy at the United States, a country with 5% of the planet's population that consumes 35% of its raw-material resources and 25% of its energy, while generating 30% of industry-related carbon dioxide. Human demand has exceeded Earth's ability to regenerate resources since at least the early 1980s. Since 1961, human demand on Earth's organisms and raw materials has doubled and now exceeds Earth's replacement capacity by at least 20% (**Figure 18.30**).

Can the world support a population whose expectations are rising as rapidly as their numbers? In Garrett Hardin's words, "We can maximize the number of humans living at the lowest possible level of comfort, or we can try to optimize the quality of life for a smaller population." The burgeoning human

---

[5] In the United States alone, the population is growing by the equivalent of four Washington D.C.s every year, another New Jersey every 3 years, another California every 12.

Time	Estimated World Population	Doubling Time (Years)
8000 B.C.	5,000,000	1,500
A.D. 1650	500,000,000	200
1850	1,000,000,000	80
1930	2,000,000,000	45
1975	4,000,000,000	35
2000	6,000,000,000	25
2025	8,000,000,000 ?	?

Domestication of plants, animals 11,000 years ago

Agriculturally based urban societies

Beginning of industrial, scientific revolutions

Bubonic plague

14,000 13,000 12,000 11,000 10,000 9000 8000 7000 6000 5000 4000 3000 2000 1000 | 1000 2000

B.C. | A.D.

FIGURE 18.29 Growth curve for the human population. The diagram's vertical axis represents the world population, in billions. (The slight dip between the years 1347 and 1351 represents the 25 million deaths in Europe from the bubonic plague.) The growth pattern over the past two centuries has been exponential, sustained by agricultural revolutions, industrialization, and improvements in health care. The table in the box tells how long it took for the human population to double in size at different times in history. The number of people on Earth now exceeds 6 billion. The last billion was added in only 12 years.

population is the greatest environmental problem of all. The last Easter Islanders would have understood completely.

We cannot expect science to solve the problem for us. Most of the decisions and necessary actions fall outside pure science in the areas of values, ethics, morality, and philosophy. *The solution to environmental problems, if one exists, lies in education and action.* Each of us is obliged to become informed on issues that affect Earth, its ocean, and its air—to learn the arguments and weigh the evidence. Once informed, we must act in rational ways. Chaining yourself to an oil tanker is not rational, but selecting well-designed, long-lasting, recyclable products made by responsible companies with minimal environmental impact (and encouraging others to do so) certainly is.

FIGURE 18.30 A thick haze lingers over the eastern reaches of the world's most populous country. Factory floor to the world, China is home to nearly 1.3 billion people. China's gross national product has grown by an average of 9.8% per year in real terms since 1980, and China will exceed the United States as the world's largest economy in about 2020. If present trends continue, China will surpass the United States as the number one emitter of greenhouse gases about five years later. The new entrepreneurial spirit of the Chinese people and the successes of their enterprises are clearly evident on the labels of your clothing, shoes, appliances and housewares, and high-technology hardware. This true-color image was taken by the *OrbView* satellite in January 2002. India will surpass China in population in about 2030.

Sea WiFS/NASA/ORBIMAGE

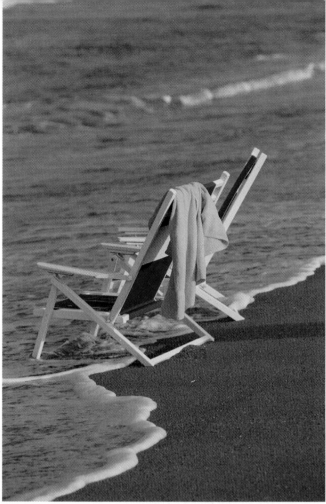

FIGURE 18.31 A clean, calm corner of the world ocean.

Obvious answers and quick solutions are often misleading; a great deal of research and work are needed to give reliable insight into the many difficult questions that confront us. The present trade-off between financial and ecological considerations is often strongly tilted in favor of immediate gain, short-term profit, and immediate convenience. Education may be the only way to modify these destructive behaviors. Garrett Hardin suggests that absolute freedom in a commons brings ruin to all.

Humanity is part of the natural world, not its master. We may be able to learn to live in harmony with this small, beautiful, blue world. We need not relive the Tragedy of the Commons on a planetary scale. True convenience and true progress depend on several things: the preservation of open space, serious and sustained attention to population control, conversion to a steady-state economy (instead of one that must grow to stay alive), business incentives for preservation, the use of renewable resources, and, above all, *public education in environmental issues.* We must ask ourselves difficult questions: What is the optimal quality of life? How can I achieve balance between my material needs and the needs of the planet? What do I want to leave for my children? How can I reserve the quiet, renewing ocean for myself, for all species, and for the future? Our cities are crowded and our tempers are short. Times of turbulent change lie before us. The trials ahead will be severe.

Each of us, individually, needs to take a stand. We must preserve the sunsets and fog, the waves to ride, the cold, clean windblown spray on our faces—our one world ocean (**Figure 18.31**). Margaret Mead summarized our potential for making a difference: "Never doubt that a small group of thoughtful, committed citizens can change the world. Indeed it is the only thing that ever has." We need to start now.

Questions from Students

1. What can an individual do to minimize his or her impact on the ocean and atmosphere?

Remember that Earth and all its millions of life-forms are interconnected. There are no true consumers, only users: Nothing can truly be thrown away (there is no "away"). We must abandon the pollute-and-move-on ethic that has guided the actions of most humans for thousands of years. *Our task is not to multiply and subdue Earth;* we must work toward a society more in harmony with the fundamental rhythms of life that sustain us. It may not be too late to change our ways. We need to act individually to effect change. We should think globally and act locally.

2. Is pollution always a bad thing?

Some forms of pollution bring temporary benefits. For example, some of Florida's once-endangered manatees are thriving at the warm outfalls of coastal power stations. On a larger scale, if increased greenhouse warming does develop, some computer models indicate increased rainfall, longer growing seasons, and increased crop yields over broad latitude bands in the temperate zones of both hemispheres. On the whole, though, the less human intervention in complex natural systems, the better.

3. What are the most dangerous threats to the environment, overall?

The underlying causes of the problems discussed in this chapter are human population growth and a growth-dependent economy. Stanford professor Paul Ehrlich said, "Arresting global population growth should be second in importance on humanity's agenda only to avoiding nuclear war." The present world population, now more than 6.1 billion, seems doomed to reach at least 10 billion before leveling off. And what if everybody wants to live in first-world comfort?

4. What role does public perception play in pollution issues?

A large role, indeed! Until recently, relatively small-scale but highly visible insults have claimed most of the public's attention and have driven us to action. The messy breakup of the oil tanker *Torrey Canyon* off the southern coast of England in 1967 galvanized world opinion and set the stage for the present environmental movement. More recently, in the summer of 1988 beachgoers were horrified to discover that more than 80 kilometers (50 miles) of northern New Jersey and Long Island beaches had been temporarily closed because of medical debris littering the shore. Some of the dozens of vials of blood, syringes, stained bandages, and surgical sutures tested positive for the viruses that cause AIDS and hepatitis B. Similar incidents occurred in Rhode Island and Massachusetts.

Appalling and visible though such incidents are, their long-term effects on the ocean as a whole are negligible. Public attention has recently turned to issues of larger consequence. The drought and heat of the past few summers brought terms like *greenhouse effect* and *ozone layer* to local newspapers and dinner table conversation. Weekend fishers worry about eating their catch. They wonder about the unseen threats as much as the obvious ones. Perceptions are changing.

5. How about aquarium fishes? Are we wrecking reef ecosystems by buying them?

In the recent past, most reef fishes were collected indiscriminately by people with no interest in preserving the species or environment for the future. People stunned fishes by squirting a solution of cyanide into reef crevices. The reef would be poisoned, and the fishes would often die of liver failure shortly after arriving in their new homes. That's changing in many places, in part because aquarium fishes are now big business and consumers are becoming aware of the methods used in their collection.

More than 20 million fishes from 1,471 species, and another 10 million invertebrates, are caught every year for the trade, now worth about US$330 million annually. For countries like Sri Lanka, protecting and preserving the "fishery" provide steady employment for 50,000 low-income people. The Marine Aquarium Council, a trade association, is beginning a certification process to provide information to consumers about the methods used to collect any fishes or other reef animals they may want to purchase. This could force countries like the Philippines and Indonesia, whose collectors rely heavily on cyanide, to abide by sustainable collection methods or convert to mariculture.

Chapter in Perspective

In this chapter you learned that our species has always exercised its capacity to consume resources and pollute its surroundings, but only in the last few generations have our efforts affected the ocean and atmosphere on a planetary scale. The introduction into the biosphere of unnatural compounds (or natural compounds in unnatural quantities) has had—and will continue to have—unexpected detrimental effects. The destruction of marine habitats and the uncontrolled harvesting of the ocean's living resources have also disturbed delicate ecological balances. We have embarked on a time of inadvertent global experimentation, and we find ourselves in difficult situations for which solutions do not come easily.

This is an unpleasant chapter—the only one I don't enjoy working on. Nonetheless, I urge you to think carefully about what you've seen, then read the Afterword that follows. The problems are yours to solve. You'll find some hope there.

Key Ideas

✓ A pollutant causes damage by interfering directly or indirectly with the biochemical processes of an organism. **Review pages 438–439**

✓ Spills of crude oil are generally larger in volume and more frequent than spills of refined oil. Given equal volumes, refined or used oil is more damaging than crude oil to the marine environment. **Review Figures 18.1–18.6, pages 439–442**

✓ Though it generated much media attention, the *Exxon Valdez* oil spill was only the 46th worst spill since 1960. Americans flush more oil down drains every year than was spilled by *Exxon Valdez*. **Review Figure 18.7 and Table 18.1, pages 442, 444**

✓ Synthetic organic chemicals and heavy metals are persistent environmental poisons and can sometimes be concentrated by biological amplification through food chains. **Review Figures 18.9 and 18.10, Table 18.2, and Box 18.1, pages 443–446**

✓ Uncontrolled algal growth through artificially high nutrient levels—eutrophication—can be dangerous to organisms and community diversity. **Review Figures 18.11 and 18.12, pages 446–447**

✓ Introduced (exotic) species are becoming more of a problem as marine transportation grows. **Review Figure 18.16, pages 449–450**

✓ Human activity is destroying habitats. Coral reefs are especially hard hit, with an estimated 70% of the world's reefs at risk. **Review pages 450–451**

✓ Depletion of the atmosphere's ozone layer exposes surface organisms to higher levels of ultraviolet radiation, and consequently to a greater risk of cancer. **Review Figures 18.18–18.22, pages 452–454**

✓ Changes in the composition of the atmosphere can influence global climate, ocean temperatures, and sea level. Experts estimate that at least half of recent global warming is human-caused. **Review Figures 18.23–18.27, pages 453–456**

✓ The solution to environmental problems, if one exists, lies in education and rational action. **Review pages 456–459**

Terms and Concepts to Remember

biodegradable, 439
biological amplification, 444
chlorinated
  hydrocarbons, 444

chlorofluorocarbons
  (CFCs), 452
eutrophication, 445
greenhouse effect, 453

Study Questions

Reviewing What You've Learned

1. What is pollution? What factors determine how dangerous a pollutant is?
2. What is eutrophication? How can "good eating" be hazardous to marine life?
3. What parts of the marine environment are hardest hit by human activities?
4. What synthetic chemicals appear to be causing the depletion of Earth's protective ozone layer? What is the likely result?
5. How is waste heat dangerous to marine life?
6. What introduced species affect your area? Are they dangerous? Costly?
7. What was the largest oil spill? How does this compare with the routine introduction of oil into the marine environment by intentional release and leakage from landfills?

Thinking Critically

1. Why is refined oil more hazardous to the marine environment than crude oil? Which is spilled more often? What happens to oil after it enters the marine environment?
2. What heavy metals are most toxic? How do these substances enter the ocean? How do they move from the ocean to marine organisms and people?
3. Few synthetic organic chemicals are dangerous in the very low concentrations in which they enter the ocean. How are these concentrations increased? What can be the outcome when these substances are ingested by organisms in a marine food chain?
4. What is the greenhouse effect? Is it always detrimental? What gases contribute to the greenhouse effect? Why do most scientists believe that Earth's average surface temperature will increase over the next few decades? What may result?
5. What is the tragedy of the commons? Do you think Garrett Hardin was right in applying the old idea to modern times? What will you do to minimize your negative impact on the ocean and atmosphere?
6. How might global warming or a decrease in stratospheric ozone *directly* affect the ocean?
7. **InfoTrac College Edition Project** Debris and pollution in the ocean can be a deadly problem for marine life as well as an unsightly and dangerous addition to beaches.

In some cases, however, cleanup may do more harm than good. Explore the question of marine pollution and its cleanup. In what cases should we emphasize prevention rather than remediation? Can there be a positive side to disposing of manufactured materials in the sea? Find information for your answer using InfoTrac College Edition.

Thinking Analytically

1. The cost of pollution and habitat mismanagement, over time, will be higher than the cost of doing nothing. But the cost *now* is cheaper. Arguing only from practical standpoints (that is, avoiding an appeal to emotion), how could you convince the executive board of a first-world industrial corporation dependent on an ocean resource to reduce or eliminate the negative effects of its activities?
2. How would you convince the board of a corporation in a developing country (say, China)?
3. If a pollutant has effects at a dilution of one part-per-billion by weight, how much seawater is contaminated by the release of 1 metric ton of the material into a bay? Consider either Chesapeake Bay or San Francisco Bay in your calculations.
4. You and your wife spend a two-week vacation in Hawaii during which you eat top marine carnivores (bluefin and albacore tuna, swordfish, marlin) for most lunches and every dinner. A month after your return, you and your wife are pleased to discover you're pregnant. How concerned should you be about the health of your child?

Oceanography Now™
Assess your understanding of this chapter's topics with additional quizzing and conceptual-based problems at http://oceanography.brookscole.com/garrison5e

Study and Research on the Internet

Assess your understanding of this chapter's topics with additional quizzing and conceptual-based problems at:

 http://oceanography.brookscole.com/garrison5e

- **InfoTrac College Edition Research** leads you to critical thinking exercises that use InfoTrac College Edition® as a research tool.
- **Regional InfoTrac College Edition** articles are organized into East Coast, West Coast, and Gulf Coast regions, allowing you to study oceanography on a more local level.

For more readings, go to InfoTrac College Edition, your online research library, at:

http://infotrac.thomsonlearning.com

Afterword

The marine sciences are at the threshold of a new age. The recent revolutions in biology and geology are being assimilated, and the road ahead seems clearer. A revolution in the design of sampling devices, robot submersible vehicles, and data processing has brought new vigor to oceanography. Satellite-borne sensors can provide in an instant data that would have taken years to collect using surface ships. Shipboard technology has become so sophisticated that Wyville Thomson or Fridtjof Nansen would hardly recognize our sensors and sampling devices.

The tools may be different, but the spirits of those who use them remain the same. Today's marine scientists are like all the men and women who have gone before: *We want to know about the ocean.* We eagerly search our mailboxes for journals bearing the latest research news, scan the Internet daily for discoveries, inspect new samples with the enthusiasm of little kids, and share our insights with anyone at the drop of a hat. I am personally delighted that you have traveled with me this far. Those of us who enjoy an oceanographic background (and this now includes you) look at Earth with greater understanding than we did before we began. The whole concept of an ocean world appeals to us, gives us profound pleasure, and sobers us with a deep sense of responsibility. In no other field of science do so many ideas interweave to form so rich a tapestry.

Our journey together is over, but before we go our separate ways, I have four last ideas to share:

- *Change* has been a recurrent theme of this book. Earth's climate has changed with time, as has its atmospheric composition, its ocean chemistry, the size and positions of its continents, and its life-forms. Our Earth may seem a calm and stable home, but it is really a violent place for inhabitation by such seemingly delicate objects as living things. Even so, life and the ocean have grown old together. The story of Earth is the story of change and chance; its history is written in the rocks, the water, and the genes of the millions of organisms that have evolved here. We are survivors.

  That survival may now be in question. Change is now progressing at an unnatural rate, and these human-induced changes are imposing stress on natural systems. What we do *with* and *to* the ocean is literally of planetary consequence. In the last century we have developed the physical, chemical, and biological machinery to destroy or rejuvenate the world ocean and all of its life. A painful time of inadvertent global experimentation lies just ahead.

  All of us who love the colors and textures of this small wet world need to act to moderate the negative effects of the looming environmental crisis. In Chinese the written character for the word *crisis* has two components: danger and opportunity. Informed citizens will express their concern, discuss this concern with others, and act whenever possible to minimize the threats and take advantage of new opportunities. Intelligence and beauty must triumph; we have no other rational alternative.

- Appreciation of the ocean doesn't come exclusively from the realm of science. Philosophers, artists, composers, and poets have had much to say about the sea. Read Homer's description of the ocean in *The Odyssey* (try Books IV, X, and XI). See how Lord Byron's feeling for the ocean colors his poetry (see, for instance, *Childe Harold's Pilgrimage*, stanzas 183 and 184). Read modern poet Robinson Jeffers's powerful *Continent's End*. Share Prospero's marine magic in Shakespeare's *The Tempest*. See what Chinese philosopher Lao Tsu has to say about our need to find solace in the natural world. Enjoy some of the evocative woodcuts of Rockwell Kent and the impressionistic ocean paintings of English artist J. M. W. Turner. Listen to Benjamin Britten's *Four Sea Interludes* from *Peter Grimes*, and Ralph Vaughan-Williams's *Sea Symphony* and *Sinfonia Antarctica* (but take care not to blow out your sound system). Read the ocean novels of Herman Melville and Jack London, and try reading the journals and accounts of the famous explorers and scientists you have met in this book. Sit on a quiet beach at night with the stars of the Milky Way shining softly overhead. The pervasive inspiration of the wave-breathing ocean is never far away.

- Don't let your involvement stop here. Lifelong learning is the truest joy, a pleasure that does not diminish with age, a source of wisdom and calm. We can learn much about patience, hope, and optimism from the ocean. We can learn much about the world—and about ourselves—by looking for the oceanic connections among things. I hope your interest in learning about the ocean has just been kindled. There is much good in the world. Go and add to it.

- Share your insight with family and friends. *Use* your new knowledge—make it your own. You don't need to be a college professor to talk to people about the beauty, history, and future of the ocean. My wife has always been patient with and receptive to my oceanic tilt. Our children and grandchildren are my tolerant built-in students. Our daughter, son, son-in-law, and granddaughter are participants in Boxes 15.1 and 17.1. Along with your children, they will inherit the world.

Measurements and Conversions

Other than the United States, only two countries in the world—Liberia and Myanmar—do not use metric measurements. The metric system, a contribution of the French Revolution, conquered Europe along with Napoleon. It is based on a decimal system, a system familiar to Americans because of our decimal money system: 10 cents to a dime, 10 dimes to a dollar.

The first move toward a rational system of measurement was made in 1670 by Gabriel Mouton, the vicar of St. Paul's Church in Lyon, France. Instead of the then-prevalent measurement system based on the width of the king's hand, or the length of his outstretched right arm, or the weight of a particular basket of stones kept in the palace, Mouton suggested a length measure based on the arc of 1 minute of longitude, to be subdivided decimally. Other measurements would follow from this unit of length. His proposal contained the three major characteristics of the metric system: using Earth itself as a basis for measurement, subdividing decimally (by 10s), and using standard prefixes (*kilo, centi, milli,* and so on). These ideas were debated for 125 years before being implemented by a commission appointed by Louis XVI in one of his last official acts before being imprisoned during the French Revolution. One ten-millionth of the distance from the North Pole to the equator (on the line of longitude passing through Paris) was selected as the standard unit of length, the meter. A new unit of weight was derived from the weight of 1 cubic meter of pure water. Temperature was to be based on pure water's boiling and freezing points. A list of prefixes for decimal multiples and submultiples was proposed. In 1795 a firm decision was made to establish the system throughout France, and in 1799 the metric system was implemented "for all people, for all time."

At first, people objected to the changes, but the government insisted that old measurements be included side by side with the equivalent new (metric) ones. In everyday competition, the advantages of the metric system proved decisive; in 1840 it was declared a legal monopoly in France. The French public had been won over to the new, simple, rational system of measurement. All of Europe—and, eventually, virtually all other countries—followed.

Not the United States, however. Though Ben Franklin proposed that the country convert in the eighteenth century, the people of the United States have continued to insist that the metric system—now known as the Système International (SI)—is too difficult to learn and work with. The federal government has urged conversion to metric units to increase opportunities for international trade. In August 1988, President Ronald Reagan signed the Omnibus Trade and Competitiveness Act. This act amended the 1975 Metric Conversion Act, stating that by 1992 all federal agencies must, wherever feasible, use the metric (SI) system in their purchases, grants, and other business. (It should be noted that Canada began to convert in 1970 and has been metric since 1980.)

The government may be making the change, but the public clings tenaciously to inches, pints, and pounds. Why? Is it really simpler to add $\frac{1}{16}$ of an inch, $\frac{1}{32}$ of an inch, and $\frac{3}{8}$ of an inch to cut a bookshelf to length? Can you remember how many cups to a quart? How many pints to a gallon? How many miles to a league? The reason we continue to use the old English Imperial system (which, of course, the English have long since abandoned) is that it is familiar to us. We know how long 5 inches is, and how much a quart is, and what 72° Fahrenheit represents. Perhaps by following the French example—by having measurements expressed everywhere in English *and* metric measurements—we may be able to make a complete conversion within a generation or two. That's why American and metric measurements are used together throughout this book. The process has already begun, of course: You use 35mm film, 2-liter soft drink containers, 750-milliliter wine bottles, 100-watt light bulbs—and you might run a 10-K (10-kilometer) race on Saturday.

The conversion factors listed here will give you an idea of how American and metric (SI) units are equivalent. Don't panic—the system is as rational and logical as it has always been. Note that 1 meter equals 100 centimeters and that 1 centimeter equals 10 millimeters. Note that 2.54 centimeters equals 1 inch. (See the table of conversion factors if you wish to convert from one system to the other.) Some numerical oceanographic data are included in supplemental tables.

Geological Time

As we saw in Chapter 1, astronomers and geologists have determined that Earth originated about 4.6 billion years ago. They have divided Earth's age into eras, roughly corresponding to major geological and evolutionary changes that have taken place, as shown in the figure below. Note that the time spans of the different eras are not shown to scale; if they were, the chart would run off the page.

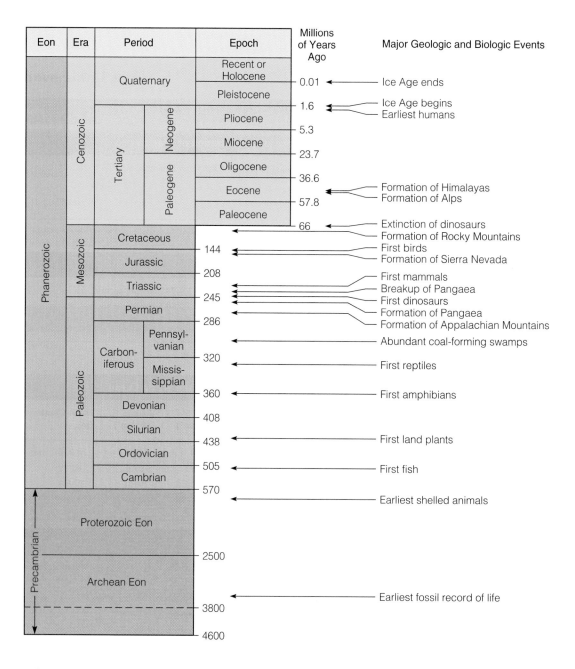

Eon	Era	Period		Epoch	Millions of Years Ago	Major Geologic and Biologic Events
Phanerozoic	Cenozoic	Quaternary		Recent or Holocene	0.01	Ice Age ends
				Pleistocene	1.6	Ice Age begins / Earliest humans
		Tertiary	Neogene	Pliocene	5.3	
				Miocene	23.7	
			Paleogene	Oligocene	36.6	
				Eocene	57.8	Formation of Himalayas / Formation of Alps
				Paleocene	66	Extinction of dinosaurs / Formation of Rocky Mountains
	Mesozoic	Cretaceous			144	First birds / Formation of Sierra Nevada
		Jurassic			208	First mammals / Breakup of Pangaea
		Triassic			245	First dinosaurs / Formation of Pangaea
	Paleozoic	Permian			286	Formation of Appalachian Mountains
		Carboniferous	Pennsylvanian		320	Abundant coal-forming swamps / First reptiles
			Mississippian		360	First amphibians
		Devonian			408	
		Silurian			438	First land plants
		Ordovician			505	First fish
		Cambrian			570	Earliest shelled animals
Precambrian		Proterozoic Eon			2500	
		Archean Eon			3800	Earliest fossil record of life
					4600	

APPENDIX III
Latitude and Longitude, Time, and Navigation

The ocean is large and easy to get lost in. A backyard, like that shown in **Figure 1,** is smaller, but we can still be lost in it if we don't have a frame of reference. Note that the yard is framed by a fence. We can refer to this frame to establish our position—in this case, at the intersection of perpendicular lines drawn from fence posts 2 and C. Many towns are arranged in this way: Fourth and D Streets intersect at a precise spot based on the municipal frame of reference.

But the World Is Round: Spherical Coordinates

If the world were flat, a simple scheme of rectangular coordinates would serve all mapping purposes—a rectangle, like the yard in Figure 1, has four sides from which to measure. A sphere has no edges, no beginnings or ends, so what shall we use as a frame of reference for Earth? Because Earth turns, the poles—the axis of rotation—are the only absolute points of reference. We can draw an imaginary line equidistant from the North and South Poles, a line that *equates* the globe into northern and southern halves: the equator. Other lines, drawn parallel to the equator, further divide the sphere north and south of the equator. These lines, or parallels, are lines of **latitude** (**Figure 2**).

We can further subdivide Earth by drawing lines at regular intervals through both poles. Note that unlike the parallels, these lines, called meridians, are all equally long. Meridians are lines of **longitude** (**Figure 3**).

If you travel north from the equator, you can count the parallels (lines of latitude) that cross your path to find out how far you have gone. Likewise, if you travel east from a reference meridian, you can count the meridians (lines of longitude) that cross your path to find out how far you have gone. Just as a football player on the field knows his distance from the goal line by the yard lines that cross his run, so you know how far north or east you have gone by the lines that have crossed your path.

Because there are no continuous lines of fence posts on the spherical Earth, our reference frame for latitude and longitude must be marked from the equator and poles by some other means. This is done by degrees.

Why Degrees?

Degrees measure fractions of a circle. We need to know what fraction of Earth's circumference separates us from the equator and from the reference meridian to have a definite idea of our location.

FIGURE 1

FIGURE 2

FIGURE 3

FIGURE 4

Babylonian astronomers first divided the circle into 360 degrees (°). Why 360? The moon cycles around Earth every 30 days. It takes about 12 months ("moonths") to make a year. Thus, 30 × 12 = 360, the number of days they supposed was in a year. Circles were divided the same way. As we saw in Chapter 2, the Greek librarian Hipparchus applied this division to the surface of Earth.

In **Figure 4** we have marked the position of Sydney, Canada. A line drawn from Sydney to the center of Earth intersects the plane of the equator at an angle of 46° to the north. That is its latitude.

The reference meridian, the meridian from which all others are marked, is known as the prime meridian. Unlike the equator, there is no earthly reason the prime meridian should pass through any particular place. It passes through Greenwich, England, because an international agreement signed in 1884 decreed it so. The meridian on which Sydney, Canada, lies intersects the plane of the prime meridian at an angle of 60°. The angular distance of Sydney from the prime meridian is 60° to the west. That is its longitude. So its position is 46°N 60°W.

We can do this for each hemisphere. A line drawn to the center of Earth from Sydney, Australia, intersects the plane of the equator at an angle of 34° south latitude. Sydney, Australia, lies 151° east of the prime meridian. Thus, its position is 34°S 151°E. (Note that the greatest possible longitude is 180°; once you pass 180°, the line opposite the prime meridian, you begin to come around the other side of Earth and the angle to Greenwich decreases.)

What Does Time Have to Do with This?

Meridians are often numbered from the prime meridian in 15° increments. Earth takes 24 hours to complete a 360° rotation.

Divide 360° by 24 hours and you get 15, the number of degrees the sun moves across the sky in 1 hour. Meridians on a globe are often spaced to represent 1 hour's turning of Earth toward or away from the sun, toward or away from the moon.

You can use this fact to find your east-west position, your longitude. Imagine that you have a radio that can tell you the precise time of noon at Greenwich.[1] If your local noon comes *before* Greenwich noon, you are east of Greenwich. For instance, if the sun is highest in your sky at 10 A.M. Greenwich time, you are 2 hours before—30° east—of Greenwich. Earth must turn 2 more hours before the sun will shine directly above the Greenwich meridian. If your local noon is *after* Greenwich noon, you are west of Greenwich. Suppose that the sun is at high noon and your chronometer, set at Greenwich time, says 6 P.M. That means that Earth has been turning 6 hours since noon at Greenwich, and 6 hours times 15° per hour is 90°. That's your longitude relative to Greenwich: 90°W.

Navigation

Longitude is half the problem. To find latitude and obtain a position, we need to measure the angle north or south of the equator. But we can't use the time difference between local noon and Greenwich noon to determine latitude because the sun moves from east to west and we want to measure north-south position. Instead, we use the angle of the North Star

[1] Any shortwave radio will do. Tune it to 2.5, 5, 10, 15, or 20 mHz for radio stations WWV (Colorado) or WWVH (Hawaii). These stations broadcast time signals giving a measure of coordinated universal time, an international time standard based on the time at Greenwich. For a telephone report of coordinated universal time, call WWV at (303) 499-7111 or go to www.time.gov on the Internet.

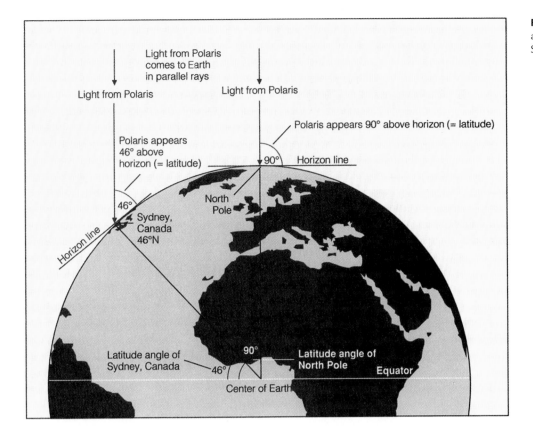

above the horizon. Polaris, the current North Star, lies almost exactly above the North Pole. If we were standing at the North Pole, the North Star would appear almost directly overhead; ideally, the angle from the horizon to the star would be 90°, the same as the latitude of the North Pole (**Figure 5**). At Sydney, Canada, the angle from the horizon to the star would be about 46°—again, the same as the latitude. What would the angle of Polaris be at the equator, 0° latitude? (If you enjoyed your high school geometry course, you might try to prove that the angle from the horizon to Polaris is equal to the latitude at any position in the Northern Hemisphere.)

Polaris is not visible in the Southern Hemisphere, so how can we find south latitude? By finding the angle above the horizon of other stars. In practice, navigators in both hemispheres use a sextant to measure angles from the horizon to selected stars, planets, the moon, and the sun. The time of the observation is carefully noted. The navigator takes these readings to his or her stateroom, consults a series of mathematical tables, does some relatively simple calculations to compensate for observational errors, and comes up to the pilothouse with the vessel's latitude and longitude, accurate (in the best of circumstances) to within $\frac{1}{2}$ mile, marked on a small slip of paper. The daily results are always entered into the ship's log.

New Tricks

Discovering position by measuring the angular positions of heavenly bodies—celestial navigation—is a dying art. Global positioning satellites, loran-C, inertial platforms, radar, and other electronic wonders have largely replaced the romance of a navigator standing on the bridge squinting through a sextant. The slip of paper has been supplanted by the glow of backlit liquid-crystal readouts or a chart with an X marking the ship's position, accurate to within about 2 meters (6 feet), feeding out of a slot. Still, when the power fails, the human navigator becomes the most popular person on board.

Maps and Charts

It is easier to draw a diagram to show someone how to get to a place than to describe the process in words. For centuries, travelers have made special diagrams—maps and charts—to jog their own memories and to show others how to reach distant destinations. A **map** is a representation of some part of Earth's surface, showing political boundaries, physical features, cities and towns, and other geographical information. A **chart** is also a representation of Earth's surface, but it has been specially designed for convenient use in navigation. It is intended to be worked on, not merely looked at. A **nautical chart** is primarily concerned with navigable water areas. It includes information such as coastlines and harbors, channels, obstructions, currents, depths of water, and the positions of aids to navigation.

Any flat map or chart is necessarily a distortion of the spherical Earth. If we roll a flat sheet of paper around a globe to form a cylinder, the paper will contact the globe only along one curve. Let's assume that it's the equator. If the lines of latitude and longitude on the globe are covered with ink, only the equator will contact the paper and print an exact replica of itself. Unroll the cylinder, and that part of the new map will be a perfect representation of Earth. To include areas north and south of the equator, we will have to "throw them forward" onto the paper; we need to *project* them in some way.

Now imagine our globe to be a translucent sphere. If we place a bright light at its center, we can project the lines of latitude and longitude onto the rolled paper cylinder (**Figure 1**). Careful tracing of these lines will result in a map, but the areas away from the equator will be distorted: The farther from the equator, the greater the distortion. A useful modification of this projection—one that does not distort high latitudes as dramatically—was devised by Gerhardus Mercator, a Flemish cartographer who published a map of the world in 1569. Though landmasses and ocean areas are not depicted as accurately in a Mercator projection as they would be on a globe, such a map is still useful because it enables mariners to steer a course over long distances by plotting straight lines.

The distortion in Mercator projections has led generations of schoolchildren to believe that Greenland is the same size as South America (**Figure 2**). Mercator charts can distort our perceptions of the ocean as well: The area of the continental

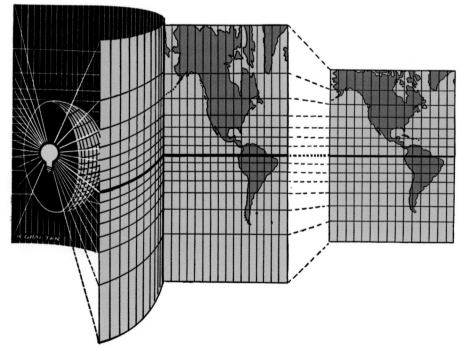

FIGURE 1 Central projection of a globe upon a cylinder, and a modified map structure, the Mercator, made to the same scale along the equator.

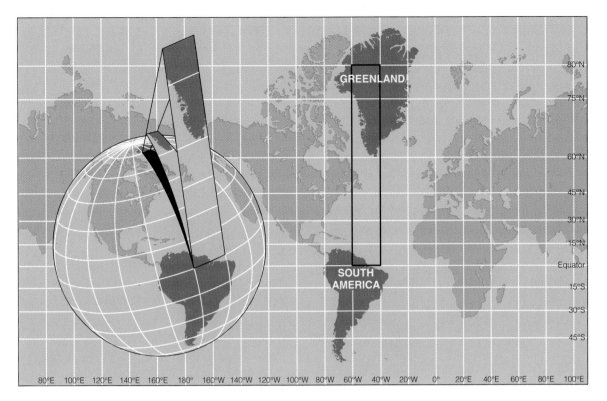

FIGURE 2 A gore of the globe peeled and projected according to the scheme devised by Gerhardus Mercator. This is the projection used on modern sailing charts. Note that this projection's distortion at high latitudes makes Greenland and South America appear about the same size. Next time you're near a globe, check their real sizes.

shelves at high latitudes, the amount of primary productivity in the polar regions, and the importance of ocean currents at the northerly and southerly extremes of an ocean basin may be exaggerated if presented in Mercator projection. The projection used in this book—a further modification of the Mercator projection known as the Miller projection—was chosen for its more accurate representation of surface area at high latitudes.

Mapmakers have invented other projections, each with advantages and disadvantages for particular uses. Some are conical projections: a flat sheet of paper wrapped into a cone with its edge touching the globe at a line of latitude north (or south) of the equator and the point of the cone above the North (or South) Pole. Conical projections do not distort high-latitude areas in the same way a Mercator projection does and, if drawn for the ocean area in which a mariner is sailing, can be used to draw great circle routes as straight lines. However, the distortions inherent in a conical projection prevent it from being used to represent more than about one-third of the globe on a single sheet of paper. Other projections, like the point-contact projection shown in **Figure 3,** try to minimize distortion around a specific location. All map and chart projections are distorted in some way; a sphere cannot be flattened onto a plane without deformation. Marine scientists necessarily become familiar with various chart projections and are careful to use the proper chart for the intended purpose.

Figure 4 is a Mercator projection of the world. On it are indicated areas of interest discussed in this book.

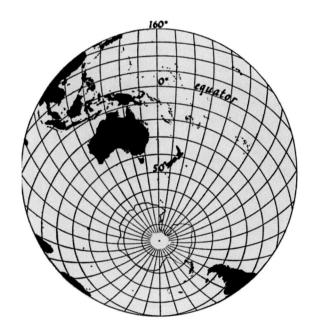

FIGURE 3 A Lambert equal-area projection (a type of point-contact projection), centered at 50°S 160°E.

FIGURE 4 A map of the world, with many oceanic features labeled.

For Further Study

Herring, T. 1996. "The Global Positioning System." *Scientific American*, 274 (no. 2): 44–50.

Krause, G., and M. Tomczak. 1995. "Do Marine Scientists Have a Scientific View of the Earth?" *Oceanography* 8 (no. 1): 11–16. Chart distortions often distort our interpretation of data, as this well-illustrated paper demonstrates.

Wilford, J. N. 1998. "Revolution in Mapping." *National Geographic* 193 (no. 2): 6–39. The usual thorough treatment of a rapidly changing topic.

The Coriolis Effect

Newton's first law states that an object in motion will naturally move in a straight line, but an object at rest on Earth's surface is moving in a curved path. Because the motion is not straight, it must be experiencing some acceleration, which may be written as v^2/R (R equals the distance between Earth's center and that of the object; it is, for all practical purposes, Earth's radius). The force is written $F = v^2/R$. This centrifugal inertia varies with latitude and is written, per unit mass, as

Force = mass × (velocity)² × radius at that latitude
× cosine of latitude
$$F = Mv^2R\phi(\cos \phi)$$

Thus, the centrifugal acceleration is

$$F/M = v^2R\phi(\cos \phi)$$

The Coriolis effect (F_c) (per unit mass) is a function of an object's velocity across Earth's surface (horizontal velocity v) and the vertical component of Earth's rotational velocity. It is mathematically stated as $F_{CF} = 2wv \sin \phi$, where w is Earth's angular velocity and ϕ is the latitude. This deflecting force is always at right angles to the velocity and, of course, performs no work. The Coriolis effect is greatest at the poles (90° latitude) and zero at the equator (0° latitude). Mathematically, this is due to the fact that the sine is a maximum (1.00) at 90° and zero at 0°. In physical terms, however, the Coriolis effect vanishes at the equator because there the vertical component of Earth's rotation vanishes; the vertical component is greatest at the poles. When an object is at rest, its centrifugal force and deflecting force are equal, but when the object has a horizontal motion across Earth's surface, it has, depending on the direction, an excess or deficit of centrifugal inertia, which is termed the *Coriolis effect*.

It is by no means a foregone conclusion that the deflection due to the Coriolis effect on a moving object will be greater at higher latitudes, because the velocity of the object can vary. One would expect a large deflection if such an object moves very rapidly. The Coriolis effect would indeed be large, and the distance traversed by the object would also be large per unit time, but in that time period Earth would rotate only a small angle distance, and the object's actual deflection would be small. In the same period of time, a slow-moving object would cover a short distance along its path of travel (velocity = distance/time, so distance = velocity × time), producing a large deflection but a small Coriolis acceleration.

While the deflection force is not a physical force, it is necessarily real to an observer within a rotating system and must be taken into consideration when describing motion, especially for movements of fluids such as water or of air, which move at relatively low velocities and experience appreciable deflection over long distances.

In addition, the equation $C_F = 2wv \sin \phi$ actually expresses the magnitude of the Coriolis acceleration and would, for a unit mass, numerically equal the Coriolis effect. The product of the object's mass and the Coriolis acceleration is the Coriolis effect, $C_F = ma$ (Newton's second law). For objects in *vertical* motion, the Coriolis acceleration is greatest at the equator and least at the poles.

Source: Ingmanson and Wallace, 1995, courtesy of Wadsworth Publishing Co.

APPENDIX VI
Taxonomic Classification of Marine Organisms

Exclusively nonmarine phyla generally have been omitted, along with most extinct phyla and classes.

KINGDOM BACTERIA: Single-celled prokaryotes with a single chromosome that reproduce asexually and exhibit high metabolic diversity.

KINGDOM ARCHAEA: Superficially similar to bacteria, but with genes capable of producing different kinds of enzymes. Often live in extreme environments.

KINGDOM PROTISTA: Eukaryotic single-celled, colonial, and multicellular autotrophs and heterotrophs.

PHYLUM CHRYSOPHYTA. Diatoms, coccolithophores, silicoflagellates.

PHYLUM PYRROPHYTA. Dinoflagellates, zooxanthellae.

PHYLUM CRYPTOPHYTA. Some "microflagellates"; cryptomonads.

PHYLUM EUGLENOPHYTA. A few "microflagellates"; mostly freshwater.

PHYLUM ZOOMASTIGINA. Nonphotosynthesizing flagellated protozoa.

PHYLUM SARCODINA. Amoebas and their relatives.

Class Rhizopodea. Foraminiferans.
Class Actinopodea. Radiolarians.

PHYLUM CILIOPHORA. Ciliated protozoa.

PHYLUM CHLOROPHYTA. Multicellular green algae.

PHYLUM PHAEOPHYTA. Brown algae, kelps.

PHYLUM RHODOPHYTA. Red algae, encrusting and coralline forms.

KINGDOM FUNGI: Fungi, mushrooms, molds, lichens; mostly land, freshwater, or highest supratidal organisms; heterotrophic.

KINGDOM PLANTAE: Photosynthetic autotrophs.

DIVISION ANTHOPHYTA. Flowering plants (angiosperms). Most species are freshwater or terrestrial. Marine eelgrass, manatee grass, surfgrass, turtle grass, salt marsh grasses, mangroves.

KINGDOM ANIMALIA: Multicellular heterotrophs.

PHYLUM PLACOZOA. Amoeba-like multicellular animals.

PHYLUM MESOZOA. Wormlike parasites of cephalopods.

PHYLUM PORIFERA. Sponges.

PHYLUM CNIDARIA. Jellyfish and their kin; all are equipped with stinging cells.

Class Hydrozoa. Polyplike animals that often have a medusa-like stage in their life cycle, such as Portuguese man-of-war.
Class Scyphozoa. Jellyfish with no (or reduced) polyp stage in life cycle.
Class Cubozoa. Sea wasps.
Class Anthozoa. Sea anemones, coral.

PHYLUM CTENOPHORA. "Sea gooseberries," comb jellies; round, gelatinous, predatory, common.

PHYLUM PLATYHELMINTHES. Flatworms, tapeworms, flukes; many free-living predatory forms, many parasites.

PHYLUM NEMERTEA. Ribbon worms.

PHYLUM GNATHOSTOMULIDA. Microscopic, wormlike; live between grains in marine sediments.

PHYLUM GASTROTICHA. Microscopic, ciliated; live between grains in marine sediments.

PHYLUM ROTIFERA. Ciliated; common in fresh water, in plankton, and attached to benthic objects.

PHYLUM KINORYNCHA. Small, spiny, segmented, wormlike; live between grains in marine sediments; all marine.

PHYLUM ACANTHOCEPHALA. Spiny-headed worms; all parasitic in vertebrate intestines.

PHYLUM ENTOPROCTA. Polyplike, small, benthic suspension feeders.

PHYLUM NEMATODA. Roundworms. Common, free-living, parasitic.

PHYLUM BRYOZOA. Common, small, encrusting colonial marine forms.

PHYLUM PHORONIDA. Shallow-water tube worms; suspension feeders; a few centimeters long; all marine.

PHYLUM BRACHIOPODA. Lampshells; bivalve animals, superficially like clams; scarce, mainly in deep water.

PHYLUM MOLLUSCA. Mollusks.
    Class Monoplacophora. Rare deep-water forms with limpetlike shells.
    Class Polyplacophora. Chitons.
    Class Aplacophora. Shell-less, sand burrowing.
    Class Gastropoda. Snails, limpets, abalones, sea slugs, pteropods.
    Class Bivalvia. Clams, oysters, scallops, mussels, shipworms.
    Class Cephalopoda. Squid, octopuses, nautiluses.
    Class Scaphopoda. Tooth shells.

PHYLUM ARTHROPODA.
    Subphylum Crustacea. Copepods, barnacles, krill, isopods, amphipods, shrimp, lobsters, crabs.
    Subphylum Chelicerata. Horseshoe crabs, sea spiders.
    Subphylum Uniramia. Insects, centipedes, millipedes; one genus and five species in the ocean.

PHYLUM PRIAPULIDA. Small, rare, wormlike, subtidal.

PHYLUM SIPUNCULA. Peanut worms; all marine.

PHYLUM ECHIURA. Spoon worms.

PHYLUM ANNELIDA. Segmented worms; includes polychaetes such as feather duster worms and some oligochaete deep-sea bristle worms.

PHYLUM TARDIGRADA. "Water bears"; tiny, eight-legged animals with the ability to survive long periods of hibernation.

PHYLUM PENTASTOMA. Tongue worms; parasites of vertebrates.

PHYLUM POGONOPHORA. Beard worms; no digestive system; deep-water tube worms; all marine.

PHYLUM ECHINODERMATA. Spiny-skinned, benthic, radially symmetrical, most with a water-vascular system.

    Class Asteroidea. Sea stars.
    Class Ophiuroidea. Brittle stars, basket stars.
    Class Echinoidea. Sea urchins, sand dollars, sea biscuits.
    Class Holothuroidea. Sea cucumbers.
    Class Crinoidea. Sea lilies, feather stars.
    Class Concentricycloidea. Sea daisies.

PHYLUM CHAETOGNATHA. Arrowworms; stiff-bodied, planktonic, predaceous, common.

PHYLUM HEMICHORDATA. Acorn worms; unsegmented burrowers.

PHYLUM CHORDATA.
    Subphylum Urochordata. Sea squirts, tunicates, salps.
    Subphylum Cephalochordata. Lancelets, *Amphioxus*.
    Subphylum Vertebrata.
        Class Agnatha: Jawless fishes: lampreys, hagfishes; cartilaginous skeleton.
        Class Chondrichthyes. Sharks, skates, rays, sawfish, chimaeras; cartilaginous skeleton.
        Class Osteichthyes. Bony fishes.
        Class Amphibia. Frogs, toads, salamanders; no marine species.
        Class Reptilia. Sea snakes, turtles, one species of crocodile.
        Class Aves. The birds.
            Order Sphenisciformes. Penguins.
            Order Procellariformes. Albatrosses, petrels.
            Order Charadriiformes. The gulls.
            Order Pelecaniformes. The pelicans.
    Class Mammalia. Warm-blooded, with hair and mammary glands.
        Order Cetacea. Whales, porpoises, dolphins.
        Order Sirenia. Manatees.
        Order Carnivora. Two marine families.
            Suborder Pinnipedia. Seals, sea lions, walruses.
            Suborder Fissipedia. Sea otters.
        Order Primates. One family that regularly enters the ocean.
            Family Hominidae. Humans.

APPENDIX VII

Calculating the Tide-Generating Force

We can calculate the strength of the tide-generating force as follows: Let the distance between the center of Earth and that of the moon be R, let the radius of Earth be r, and let m be the mass of the moon divided by the mass of Earth. The average gravitational attraction, which is equal to the centrifugal acceleration, is then proportional to m/R^2 because the gravitational attraction is proportional to the mass and inversely proportional to the square of the distance.

Consider a point on Earth closest to the moon. The distance of this point from the center of the moon is $R - r$, and so the gravitational attraction is proportional to

$$\frac{m}{(R - r)^2}$$

The tide-generating force, which is equal to the difference between the gravitational attraction and the centrifugal force, is proportional to

$$\frac{m}{(R - r)^2} - \frac{m}{R^2} = \frac{mR^2 - m(R - r)^2}{(R - r)^2 R^2}$$

Because r is small compared to R, the equation is approximately equal to $2mr/R^3$. In contrast, in the same units, the gravitational attraction of Earth is $1/r^2$. Thus, the ratio of the tide-generating force to the acceleration of gravity is

$$\frac{Tidal\ force}{a_g} = \frac{2mr^3}{R^3} = 1.176 \times 10^{-7}$$

Because $m = 1/81.45$, $R = 60/34r$.

APPENDIX VIII
Periodic Table of the Elements

Group

Noble Gases

IA(1)   (18)

Atomic number → 11
Symbol → Na
Atomic mass → 22.99

Atomic masses are based on carbon-12. Numbers in parentheses are mass numbers of most stable or best known isotopes of radioactive elements.

Period	IA(1)	IIA(2)	IIIB(3)	IVB(4)	VB(5)	VIB(6)	VIIB(7)	(8)	(9) VIII	(10)	IB(11)	IIB(12)	IIIA(13)	IVA(14)	VA(15)	VIA(16)	VIIA(17)	(18)
1	1 H 1.008																	2 He 4.003
2	3 Li 6.941	4 Be 9.012											5 B 10.81	6 C 12.01	7 N 14.01	8 O 16.00	9 F 19.00	10 Ne 20.18
3	11 Na 22.99	12 Mg 24.31											13 Al 26.98	14 Si 28.09	15 P 30.97	16 S 32.06	17 Cl 35.45	18 Ar 39.95
4	19 K 39.10	20 Ca 40.08	21 Sc 44.96	22 Ti 47.90	23 V 50.94	24 Cr 52.00	25 Mn 54.94	26 Fe 55.85	27 Co 58.93	28 Ni 58.7	29 Cu 63.55	30 Zn 65.38	31 Ga 69.72	32 Ge 72.59	33 As 74.92	34 Se 78.96	35 Br 79.90	36 Kr 83.80
5	37 Rb 85.47	38 Sr 87.62	39 Y 88.91	40 Zr 91.22	41 Nb 92.91	42 Mo 95.94	43 Tc 98.91	44 Ru 101.1	45 Rh 102.9	46 Pd 106.4	47 Ag 107.9	48 Cd 112.4	49 In 114.8	50 Sn 118.7	51 Sb 121.8	52 Te 127.6	53 I 126.9	54 Xe 131.3
6	55 Cs 132.9	56 Ba 137.3	57* La 138.9	72 Hf 178.5	73 Ta 180.9	74 W 183.9	75 Re 186.2	76 Os 190.2	77 Ir 192.2	78 Pt 195.1	79 Au 197.0	80 Hg 200.6	81 Tl 204.4	82 Pb 207.2	83 Bi 209.0	84 Po (210)	85 At (210)	86 Rn (222)
7	87 Fr (223)	88 Ra 226.0	89** Ac (227)	104 Unq (261)	105 Unp (262)	106 Unh (263)	107 Uns (262)	108 Uno (265)	109 Une (266)									

← Transition Elements →

Inner Transition Elements

Lanthanide Series 6*	58 Ce 140.1	59 Pr 140.9	60 Nd 144.2	61 Pm (145)	62 Sm 150.4	63 Eu 152.0	64 Gd 157.3	65 Tb 158.9	66 Dy 162.5	67 Ho 164.9	68 Er 167.3	69 Tm 168.9	70 Yb 173.0	71 Lu 175.0
Actinide Series 7**	90 Th 232.0	91 Pa 231.0	92 U 238.0	93 Np 237.0	94 Pu (244)	95 Am (243)	96 Cm (247)	97 Bk (247)	98 Cf (251)	99 Es (252)	100 Fm (257)	101 Md (258)	102 No (259)	103 Lr (260)

APPENDIX IX
Working in Marine Science

Working in the marine sciences is wonderfully appealing to many people. They sometimes envision a life of diving in warm, clear water surrounded by tropical fish, or descending to the seabed in an exotic submersible outfitted like Captain Nemo's fictional submarine in *20,000 Leagues Under the Sea*, or living with intelligent dolphins in a marine life park. Then reality sets in. There are rewards from working in the marine sciences, but they tend to be less spectacular than the first dreams of students looking to the ocean for a life's work.

A marine science worker is paid to bring a specific skill to a problem. If that problem lies in warm, tropical water or in a marine park, fine. But more likely, the problem will yield only to prolonged study in an uncomfortable, cold, or dangerous environment. The intangible rewards can be great; the physical rewards are often slim. Having said that, let me add that no endeavor is more interesting or exciting, and few are more intellectually stimulating. Doing marine science is its own reward.

Training for a Job in Marine Science

Marine science is, of course, science. And science requires mathematics—you need math to do the chemistry, physics, measurements, and statistics that lie at the heart of science. Your first step in college should be to take a math placement test, enroll in an appropriate math class, and spend time doing math. *Math is the key to further progress in any area of marine science.*

With your math skills polished, start classes in chemistry, physics, and basic biology. Surprisingly, except for one or two introductory marine science classes, you probably won't take many marine science courses until your junior year. These introductory classes will be especially valuable because a balanced survey of the marine sciences can aid you in selecting an appealing specialty. Then, with a good foundation in basic science, you can begin to concentrate in that specialty.

Other skills are important, too. The ability to write and speak well is crucial in any science job. Also critical is computer literacy. Expertise in photography or foreign languages or the ability to field-strip and rebuild a diesel engine or hydraulic winch will put you a step above the competition at hiring time. Certification as a scuba diver is almost mandatory; you can never have too much diving experience. (Remember, though, diving is only a tool, a way to deliver an informed set of eyes and an educated brain to a work site.) You should be in good health. Indeed, good aerobic fitness is essential in most marine science jobs; stamina is often a crucial factor in long experiments under difficult conditions at sea. It is also desirable to be physically strong—marine equipment is heavy and often bunglesome. And it helps greatly if you are not prone to seasickness.

Deciding what school to attend will depend on your skills. Readers of this book will probably be enrolled in a general oceanography course in a college or university. The first step would be to discuss your interests with your professor (or his or her teaching assistants). You'll need to attend a four-year college or university to complete the first phase of your training. If you're attending a two-year institution, picking a specific transfer institution can come later, but keep a few things in mind: No matter where you take your first two years of training, you need thorough preparation in basic science. You should attend an institution with strengths in the area of your specialty (such as geology, biology, and marine chemistry). And you should be reasonable in your expectations of acceptance if you're a transfer student (that is, don't try for Stanford or Yale with a B average).

Another thing: Most marine scientists have completed a graduate degree (a master's degree or doctorate). Most graduate students hold teaching or research assistantships (that is, they get paid for being grad students). In all, progress to a final degree is a long road, but the journey is itself a pleasure.

If the thought of four or more years of higher education does not appeal to you, does that mean there's no hope? Not at all. Many students begin a program with the goal of becoming a marine technician, animal trainer at a marine life park, marina or boatyard employee or manager, or crew member on a private yacht. Those jobs don't always require a bachelor's degree. Jobs at Sea World and other marine theme parks do require athletic ability, extreme patience, public speaking skills, a love of animals, and usually diving experience. Few positions are available, but there is some turnover in the ranks of junior trainers, and being hired is certainly possible.

Becoming a marine technician is an especially attractive alternative to the all-out chemistry-physics-math academic route. For every highly trained marine scientist, there are perhaps five technical assistants who actually do the experiments, maintain the equipment, work daily with organisms, and build special apparatus. Marine technicians tend to spend more time at hands-on tasks than marine scientists. Most of these folks (including the author of the letter that ends this appendix) have the equivalent of a two-year technical degree, usually from a community college.

Don't quit your job, burn your bridges, leave your family, sell your possessions, and dedicate yourself monklike to marine science. Do some investigation. Nothing is as valuable as *actually going out and talking to people who do things that you'd like to do.*

Ask them if they enjoy their work. Is the pay OK? Would they start down the same road if they had it to do all over again? You may decide to expand your involvement in marine science in a more informal way by becoming a volunteer; joining the Sierra Club, Audubon Society, Greenpeace, or other environmental group; working for your state's fish and game office as a seasonal aide; or attending lectures at local colleges and universities.

If you decide to continue your education, don't be discouraged by the time it will take. Have a general view of the big picture, but proceed one semester at a time. Again, remember that the educational journey is itself a great pleasure. Don Quixote reminds us of the joys of the road, not the inn.

The Job Market

Marine science is very attractive to the general public. People are naturally drawn to thoughts of working in the field. Unfortunately, there aren't a great many jobs in the marine sciences. But there will always be some jobs, and people will fill them. Those people will be the best prepared, most versatile, and most highly motivated of those who apply. Perhaps not surprisingly, marine biology is the most popular marine science specialty. Unfortunately, it is also the area with the smallest number of nonacademic jobs. Museums, aquariums, and marine theme parks employ biologists to care for animals and oversee interpretive programs for the public. A few marine biologists are employed as monitoring specialists by water management agencies like sanitation districts, which discharge waste into the ocean. Electrical utilities that use seawater to cool the condensers in power-generating plants almost always have a handful of marine biologists on staff to watch the effects of discharged heat on local marine life and to write the reports required by watchdog agencies. State and federal agencies employ marine biologists to read and interpret those documents and to set standards. Relatively small businesses, like private shipyards, agricultural concerns, and chemical plants, can't afford their own staff biologists, so private consulting firms staffed by marine biologists and other specialists have arisen to assist in the preparation of the environmental impact reports required of businesses under various legislation.

There are more jobs in physical oceanography: marine geology, ocean engineering, and marine chemistry and physics. Thousands of marine geologists work for oil and mineral companies; indeed, with the increasing emphasis on offshore resources, the market for these people may be increasing. Marine engineers are needed to design, construct, and maintain offshore oil rigs, ships, and harbor structures. Marine chemists are hard at work figuring ways to stop corrosion and to extract chemicals from seawater. Physicists are vitally interested in the transmission of underwater sound and light, in the movement of the ocean, and in the role the ocean plays in global weather and climate. Economists, lawyers, writers, and mathematicians also work in the marine science field.

Many biological and physical oceanographers are teachers and professors. Indeed, there are nearly as many marine scientists employed in the academic world as there are in private industry and government. If you like the idea of teaching, you might consider this avenue. The demand for science teachers at all educational levels is already great and is expected to increase.

Four factors will be significant in influencing your employability:

1. *Experience.* Employers are favorably impressed by experience, especially work experience related to the duties of the position for which you are applying. Volunteer work counts.

2. *Grades.* Good grades are important, especially for positions in government agencies. A grade point average of 3.0 or higher in all college work increases your chances of employment and should give you a higher starting salary.

3. *Geographical availability.* Don't restrict yourself geographically. Not everyone can work in Hawaii or California, but four out of ten marine scientists work in just three states: California, Maryland, and Virginia.

4. *Diversification.* Again, mastery of more than one specialty gives you an employment edge. Being a plankton connoisseur and also able to repair a balky computer while ordering in-port supplies over a radiotelephone in Spanish makes a lasting impression.

Report from a Student

Students in marine science programs graduate, get jobs, and move on. One of the pleasures of being a professor is hearing from them. One of our former students, an employee of the Marine Science Institute at the University of California, Santa Barbara, reported his activities as part of a team using the submersible *Alvin* to investigate plumes of warm water issuing from hydrothermal vents along the southern Juan de Fuca Ridge. The nature of his work—and his enthusiasm for it—is clearly evident in this excerpt. Dan Dion writes:

> The buoyant plume experiment wasn't going very well. The chemistry dives were pushed back because of technical difficulties and poor weather (rough seas cut two dives). The first two buoyant plume dives ended in failure. The first one because of mechanical/electrical problems, the second because of a computer crash. Everyone worked around the clock to get things in order for dive 2440. I was scheduled to go down with John Trefrey, from the Florida Institute of Technology. Cindy Van Dover was our pilot. We launched *Alvin* right on schedule at 0800, and descended from the glacier blue water into the bioluminescent snowstorm of the euphotic zone. During the hour and a half descent we listened to the music of Enya in the soft light of the sub as we busily prepared ourselves for the experiment: booting up the computers, loading film in the cameras, tapes in the recorders, etc. We had three laptop computers to deal with in the cramped spaces of the sub. I was in charge of two of them, one that plotted our in-sub navigation (from transponders), and the other that controlled and recorded data from the [continuous temperature-depth-conductivity probe]. The third laptop was connected to the chemical analyzer and John was in control of that. All of

FIGURE 1 Dan Dion attaching hydraulic actuators to water-sampling bottles, in preparation for a dive by *Alvin*. The bottles were part of a sampling program that included measurements of water conductivity, transmissivity, temperature, and iron and manganese ion content near a hydrothermal vent. This information was later merged with data from transponder navigation to obtain a three-dimensional map of plume structure and chemistry.

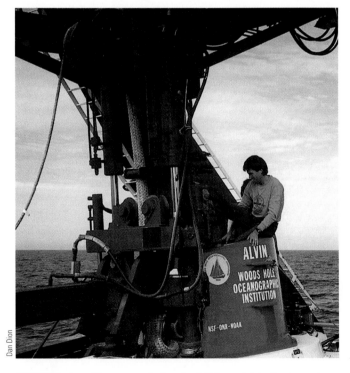

FIGURE 2 Dan Dion enters the hatch of *Alvin* to visit hydrothermal vents 2,261 meters (7,416 feet) below the surface off the coast of Oregon.

the instruments were operating perfectly. I periodically saved the computer file in the event of another crash. We reached the bottom right on target; Monolith Vent was in sight, 2261 meters below the surface. We did a video survey of the vent, especially a chimney that was rapidly growing back after the geologists had decapitated it just a few days earlier. We ascended to 55 meters-off-bottom and began our drive-throughs. To me, the navigator, it was the ultimate video game. From the computer screen I would guide *Alvin* through a dark abyss, calling out headings that would maneuver us into a "lawn mower" pattern crisscrossing the plume. It was quite visible, and even beautiful; wispy, intricate patterns of "smoke" which seemed to dance like graceful ghosts. We completed passes at 35, 20, 10, and 5 meters above the bottom, then one last one at 45 meters. Eight hours of sub time went by so quickly! Our dive was a huge success; in addition to all the samples we obtained, we generated over 25 megabytes of data. I used everything I learned . . . from computer skills to navigation and marlinspike seamanship (and, of course, chemistry!).

Marine science is equipment training sessions, long cruises, seminars and lectures, visiting experts, hot sand volleyball games, and chilly labs with classical music. Marine science is a long and demanding road, but it is, quite honestly, great fun. Captain Nemo and his sub never had it this good!

For More Information

Two useful Internet sites

This Web site, hosted by Joe Wible, Hopkins Marine Station of Stanford University, contains links to many career-related sites:

www.marine.stanford.edu/HMSweb/careers.html

This Web site lists the member institutions of the Joint Oceanographic Institutions, universities at the forefront of oceanographic research:

www.joi-odp.org/JOI/Members.html

Organizations

American Society of Limnology and Oceanography A non-profit, professional scientific society that seeks to promote the interests of limnology and oceanography and related sciences and to further the exchange of information across the range of aquatic science disciplines.

www.aslo.org

Association for Women in Science A nonprofit association dedicated to achieving equity and full participation for women in science, mathematics, engineering, and technology.

www.awis.org

Center for Marine Conservation A nonprofit membership organization dedicated to protecting marine wildlife and its habitats and to conserving coastal and ocean resources.

www.cmc-ocean.org

Marine Advanced Technology Training Center For training of marine technicians—specialists in the deployment and maintenance of tools used in marine science.

www.marinetech.org

Marine Technology Society An international, interdisciplinary society devoted to ocean and marine engineering science and policy.

www.mtsociety.org

National Marine Educators Association An organization that brings together those interested in the study and enjoyment of the world of water—both fresh and salt.

www.marine-ed.org

National Oceanic and Atmospheric Administration A government agency that guides the use and protection of our oceans and coastal resources, warns of dangerous weather, charts the seas and skies, and conducts research to improve our understanding and stewardship of the environment.

www.noaa.gov

Oceanic Engineering Society An organization that promotes the use of electronic and electrical engineers for instrumentation and measurement work in the ocean environment and the ocean/atmosphere interface.

www.oceanicengineering.org

The Oceanography Society A professional society for scientists in the field of oceanography.

www.tos.org

The Society for Marine Mammalogy A professional organization that supports the conservation of marine mammals and the educational, scientific, and managerial advancement of marine mammal science.

www.marinemammalogy.org

Joint Oceanographic Institutions The Joint Oceanographic Institutions (JOI) is a consortium of U.S. academic institutions that brings to bear the collective capabilities of the individual oceanographic institutions on research planning and management of the ocean sciences.

www.joi-odp.org

Note: The organizations listed here are but a sampling of the resources that are available to help you learn more about careers in the marine sciences. Each contact you make in your search for information will lead you to more contacts and more information.

Glossary

absolute dating Determining the age of a geological sample by calculating radioactive decay and/or its position in relation to other samples.

absorption Conversion of sound or light energy into heat.

abyssal hill Small sediment-covered inactive volcano or intrusion of molten rock less than 200 meters (650 feet) high, thought to be associated with seafloor spreading. Abyssal hills punctuate the otherwise flat abyssal plain.

abyssal plain Flat, cold, sediment-covered ocean floor between the continental rise and the oceanic ridge at a depth of 3,700 to 5,500 meters (12,000 to 18,000 feet). Abyssal plains are more extensive in the Atlantic and Indian Oceans than in the Pacific.

abyssal zone The ocean between about 4,000 and 5,000 meters (13,000 and 16,500 feet) deep.

accessory pigment One of a class of pigments (such as fucoxanthin, phycobilin, and xanthophyll) that are present in various photosynthetic plants and that assist in the absorption of light and the transfer of its energy to chlorophyll; also called *masking pigment*.

accretion An increase in the mass of a body by accumulation or clumping of smaller particles.

acid A substance that releases a hydrogen ion (H^+) in solution.

acid rain Rain containing acids and acid-forming compounds such as sulfur dioxide and oxides of nitrogen.

acoustical tomography A technique for studying ocean structure that depends on pulses of low-frequency sound to sense differences in water temperature, salinity, and movement beneath the surface.

active margin The continental margin near an area of lithospheric plate convergence; also called *Pacific-type margin*.

active sonar A device that generates underwater sound from special transducers and analyzes the returning echoes to gain information of geological, biological, or military importance.

active transport The movement of molecules from a region of low concentration to a region of high concentration through a semipermeable membrane at the expense of energy.

adaptation An inheritable structural or behavioral modification. A favorable adaptation gives a species an advantage in survival and reproduction. An unfavorable adaptation lessens a species' ability to survive and reproduce.

adhesion Attachment of water molecules to other substances by hydrogen bonds; wetting.

Agnatha The class of jawless fishes: hagfishes and lampreys.

ahermatypic Describing coral species lacking symbiotic zooxanthellae and incapable of secreting calcium carbonate at a rate suitable for reef production.

air mass A large mass of air with nearly uniform temperature, humidity, and density throughout.

algae Collective term for nonvascular plants possessing chlorophyll and capable of photosynthesis. (Singular, alga.)

algin A mucilaginous commercial product of multicellular marine algae; widely used as a thickening and emulsifying agent.

alkaline Basic. See also *base*.

amphidromic point A "no-tide" point in an ocean caused by basin resonances, friction, and other factors around which tide crests rotate. About a dozen amphidromic points exist in the world ocean. Sometimes called a *node*.

angiosperm A flowering vascular plant that reproduces by means of a seed-bearing fruit. Examples are sea grasses and mangroves.

angle of incidence In meteorology, the angle of the sun above the horizon.

animal A multicellular organism unable to synthesize its own food and often capable of movement.

Animalia The kingdom to which multicellular heterotrophs belong.

Annelida The phylum of animals to which segmented worms belong.

Antarctic Bottom Water The densest ocean water (1.0279 g/cm^3), formed primarily in Antarctica's Weddell Sea during Southern Hemisphere winters.

Antarctic Circle The imaginary line around Earth parallel to the equator at 66°33'S, marking the southernmost limit of sunlight at the June solstice. The Antarctic Circle marks the northern limit of the area within which, for one day or more each year, the sun does not set (around 21 December) or rise (around 21 June).

Antarctic Circumpolar Current The current driven by powerful westerly winds north of Antarctica. The largest of all ocean currents, it continues permanently eastward without changing direction.

Antarctic Convergence Convergence zone encircling Antarctica between about 50° and 60°S, marking the boundary between Antarctic Circumpolar Water and Subantarctic Surface Water.

Antarctic Ocean An ocean in the Southern Hemisphere bounded to the north by the Antarctic Convergence and to the south by Antarctica.

aphelion The point in the orbit of a satellite where it is farthest from the sun; opposite of *perihelion*.

aphotic zone The dark ocean below the depth to which light can penetrate.

apogee The point in the orbit of a satellite where it is farthest from the main body; opposite of *perigee*.

Aqua A NASA satellite designed to obtain data on Earth's water cycle.

aquaculture The growing or farming of plants and animals in a water environment under controlled conditions. Compare *mariculture*.

Arctic Circle The imaginary line around Earth parallel to the equator at 66°33'N, marking the northernmost limit of sunlight at the December solstice. The Arctic Circle marks the southern limit of the area within which, for one day or more each year, the sun does not set (around 21 June) or rise (around 21 December).

Arctic Convergence Convergence zone between Arctic Water and Subarctic Surface Water.

Arctic Ocean An ice-covered ocean north of the continents of North America and Eurasia.

Arthropoda The phylum of animals that includes shrimp, lobsters, krill, barnacles, and insects. The phylum Arthropoda is the world's most successful.

artificial system of classification A method of classifying an object based on attributes other than its reason for existence, its ancestry, or its origin. Compare *natural system of classification*.

Asteroidea The class of the phylum Echinodermata to which sea stars belong.

asthenosphere The hot, plastic layer of the upper mantle below the lithosphere, extending some 350 to 650 kilometers (220 to 400 miles) below the surface. Convection currents within the asthenosphere power plate tectonics.

atmosphere The envelope of gases that surround a planet and are held to it by the planet's gravitational attraction.

atmospheric circulation cell Large circuit of air driven by uneven solar heating and the Coriolis effect. Three circulation cells form in each hemisphere. See also *Ferrel cell*; *Hadley cell*; *polar cell*.

atoll A ring-shaped island of coral reefs and coral debris enclosing, or almost enclosing, a shallow lagoon from which no land protrudes. Atolls often form over sinking, inactive volcanoes.

atom The smallest particle of an element that exhibits the characteristics of that element.

authigenic sediment Sediment formed directly by precipitation from seawater; also called *hydrogenous sediment*.

autotroph An organism that makes its own food by photosynthesis or chemosynthesis.

auxospore A naked diatom cell without valves; often a dormant stage in the life cycle following sexual reproduction.

Aves The class of birds.

backshore Sand on the shoreward side of the berm crest, sloping away from the ocean.

backwash Water returning to the ocean from waves washing onto a beach.

bacteria Single-celled prokaryotes, organisms lacking membrane-bound organelles.

baleen The interleaved, hard, fibrous, hornlike filters within the mouth of baleen whales.

barrier island A long, narrow, wave-built island lying parallel to the mainland and separated from it by a lagoon or bay. Compare *sea island*.

barrier reef A coral reef surrounding an island or lying parallel to the shore of a continent, separated from land by a deep lagoon. Coral debris islands may form along the reef.

basalt The relatively heavy crustal rock that forms the seabeds, composed mostly of oxygen, silicon, magnesium, and iron. Its density is about 2.9 g/cm³.

base A substance that combines with a hydrogen ion (H⁺) in solution.

bathyal zone The ocean between about 200 and 4,000 meters (700 and 13,000 feet) deep.

bathybius Thomas Henry Huxley's name for an artifact of marine specimen preservation he thought was a remnant of the "primeval living slime."

bathymetry The discovery and study of submerged contours.

bathyscaphe Deep-diving submersible designed like a blimp, which uses gasoline for buoyancy and can reach the bottom of the deepest ocean trenches. From the Greek *batheos* ("depth") and *skaphidion* ("a small ship").

bay mouth bar An exposed sandbar attached to a headland adjacent to a bay and extending across the mouth of the bay.

beach A zone of unconsolidated (loose) particles extending from below the water level to the edge of the coastal zone.

beach scarp A vertical wall of variable height marking the landward limit of the most recent high tides; corresponds with the berm at extreme high tides.

benthic zone The zone of the ocean bottom. See also *pelagic zone*.

berm A nearly horizontal accumulation of sediment parallel to shore; marks the normal limit of sand deposition by wave action.

berm crest The top of the berm; the highest point on most beaches; corresponds to the shoreward limit of wave action during most high tides.

big bang The hypothetical event that started the expansion of the universe from a geometric point; the beginning of time.

bilateral symmetry Body structure having left and right sides that are approximate mirror images of each other. Examples are crabs and humans. Compare *radial symmetry*.

biodegradable Able to be broken down by natural processes into simpler compounds.

biodiversity The variety of different species within a habitat.

biogenous sediment Sediment of biological origin. Organisms can deposit calcareous (calcium-containing) or siliceous (silicon-containing) residue.

biogeochemical cycle Natural processes that recycle nutrients in various chemical forms from the nonliving environment to living organisms and then back to the nonliving environment.

biological amplification Increase in the concentration of certain fat-soluble chemicals such as DDT or heavy-metal compounds in successively higher trophic levels within a food web.

biological factor A biologically generated aspect of the environment, such as predation or metabolic waste products, that affects living organisms. Biological factors usually operate in association with purely physical factors such as light and temperature.

biological resource A living animal or plant collected for human use; also called a *living resource*.

bioluminescence Biologically produced light.

biomass The mass of living material in a given area or volume of habitat.

biosynthesis The initial formation of life on Earth.

Bivalvia The class of the phylum Mollusca that includes clams, oysters, and mussels.

Bjerknes, Vilhelm (1862–1951) Pioneering Norwegian physicist and discoverer of the nature and formation of extra-tropical cyclones, which cause most mid-latitude weather.

blade Algal equivalent of a vascular plant's leaf; also called a *frond*.

bond See *chemical bond*.

brackish Describing water intermediate in salinity between seawater and fresh water.

breakwater An artificial structure of durable material that interrupts the progress of waves to shore. Harbors are often shielded by a breakwater.

buffer A group of substances that tends to resist change in the pH of a solution by combining with free ions.

buoyancy The ability of an object to float in a fluid by displacement of a volume of fluid equal to it in mass.

bykill Animals unintentionally killed when desirable organisms are collected.

$C = \sqrt{gd}$ Relationship of velocity (C), the acceleration due to gravity (g), and water depth (d) for shallow-water waves.

$C = L/T$ Relationship of velocity (C), wavelength (L), and period (T) for deep-water waves; velocity increases as wavelength increases. Typically measured in meters per second.

caballing Mixing of two water masses of identical densities but different temperatures and salinities, such that the resulting mixture is denser than its components.

calcareous ooze Ooze composed mostly of the hard remains of organisms containing calcium carbonate.

calcium carbonate compensation depth The depth at which the rate of accumulation of calcareous sediments

equals the rate of dissolution of those sediments. Below this depth, sediment contains little or no calcium carbonate.

calorie The amount of heat needed to raise the temperature of 1 gram (0.035 ounce) of pure water by 1°C (1.8°F).

capillary wave A tiny wave with a wavelength of less than 1.73 centimeters (0.68 inch), whose restoring force is surface tension; the first type of wave to form when the wind blows.

carbon cycle The movement of carbon from reservoirs (sediment, rock, ocean) through the atmosphere (as carbon dioxide), through food webs, and back to the reservoirs.

Carnivora The order of mammals that includes seals, sea lions, walruses, and sea otters.

carrying capacity The size at which a particular population in a particular environment will stabilize when its supply of resources—including nutrients, energy, and living space—remains constant.

cartilage A tough, elastic tissue that stiffens or supports.

cartographer A person who makes maps and charts.

catastrophism The theory that Earth's surface features are formed by catastrophic forces such as the biblical flood. Catastrophists believe in a young Earth and a literal interpretation of the biblical account of Creation.

celestial navigation The technique of finding one's position on Earth by reference to the apparent positions of stars, planets, the moon, and the sun.

cell The basic organizational unit of life on this planet.

Cephalopoda The class of the phylum Mollusca that includes squid, octopuses, and nautiluses.

Cetacea The order of mammals that includes porpoises, dolphins, and whales.

CFCs See *chlorofluorocarbons*.

***Challenger* expedition** The first wholly scientific oceanographic expedition, 1872–76; named for the steam corvette used in the voyage.

chart A map that depicts mostly water and the adjoining land areas.

chemical bond An energy relationship that holds two atoms together as a result of changes in their electron distribution.

chemical equilibrium In seawater, the condition in which the proportion and amounts of dissolved salts per unit volume of ocean are nearly constant.

chemosynthesis The synthesis of organic compounds from inorganic compounds using energy stored in inorganic substances such as sulfur, ammonia, and hydrogen. Energy is released when these substances are oxidized by certain organisms.

chitin A complex nitrogen-rich carbohydrate from which parts of arthropod exoskeletons are constructed.

chiton A polyplacophoran mollusk.

chlorinated hydrocarbons The most abundant and dangerous class of halogenated hydrocarbons, synthetic organic chemicals hazardous to the marine environment.

chlorinity A measure of the content of chloride, bromine, and iodide ions in seawater. We derive salinity from chlorinity by multiplying by 1.80655.

chlorofluorocarbons (CFCs) A class of halogenated hydrocarbons thought to be depleting Earth's atmospheric ozone. CFCs are used as cleaning agents, refrigerants, fire-extinguishing fluids, spray-can propellants, and insulating foams.

chlorophyll A pigment responsible for trapping sunlight and transferring its energy to electrons, thus initiating photosynthesis.

Chlorophyta Green algae.

Chondrichthyes The class of fishes with cartilaginous skeletons: the sharks, skates, rays, and chimaeras.

Chordata The phylum of animals to which tunicates, Amphioxus, fishes, amphibians, reptiles, birds, and mammals belong.

chromatophore A pigmented skin cell that expands or contracts to affect color change.

chronometer A very consistent clock. It doesn't need to tell accurate time, but its rate of gain or loss must be constant and known exactly so that accurate time may be calculated.

clamshell sampler A sampling device used to take shallow samples of the ocean bottom.

classification A way of grouping objects according to some stated criteria.

clay Sediment particle smaller than 0.004 millimeter in diameter; the smallest sediment size category.

climate The long-term average of weather in an area.

climax community A stable, long-established community of self-perpetuating organisms that tends not to change with time.

clockwise Rotation around a point in the direction that clock hands move.

clumped distribution Distribution of organisms within a community in small, patchy aggregations, or clumps; the most common distribution pattern.

Cnidaria The phylum of animals to which corals, jellyfish, and sea anemones belong.

cnidoblast Type of cell found in members of the phylum Cnidaria that contains a stinging capsule. The threads that evert from the capsules assist in capturing prey and repelling aggressors.

coast The zone extending from the ocean inland as far as the environment is immediately affected by marine processes.

coastal cell The natural sector of a coastline in which sand input and sand outflow are balanced.

coastal upwelling Upwelling adjacent to a coast, usually induced by wind.

coccolithophore A very small planktonic alga carrying discs of calcium carbonate, which contributes to biogenous sediments.

cohesion Attachment of water molecules to each other by hydrogen bonds.

colligative properties Those characteristics of a solution that differ from those of pure water because of material held in solution.

Columbus, Christopher (1451–1506) Italian explorer in the service of Spain who discovered islands in the Caribbean in 1492. Although traditionally credited as the discoverer of America, he never actually sighted the North American continent.

commensalism A symbiotic interaction between two species in which only one species benefits and neither is harmed.

commercial extinction Depletion of a resource species to a point where it is no longer profitable to harvest the species.

community The populations of all species that occupy a particular habitat and interact within that habitat.

compass An instrument for showing direction by means of a magnetic needle swinging freely on a pivot and pointing to magnetic north.

compensation depth The depth in the water column at which the production of carbohydrates and oxygen by photosynthesis exactly equals the consumption of carbohydrates and oxygen by respiration. The break-even point for autotrophs. Generally a function of light level.

compound A substance composed of two or more elements in a fixed proportion.

condensation theory Premise that stars and planets accumulate from contracting, accreting clouds of galactic gas, dust, and debris.

conduction The transfer of heat through matter by the collision of one atom with another.

conservative constituent An element that occurs in constant proportion in seawater; for example, chlorine, sodium, and magnesium.

constructive interference The addition of wave energy as waves interact, producing larger waves.

consumer A heterotrophic organism.

continental crust The solid masses of the continents, composed primarily of granite.

continental drift The theory that the continents move slowly across the surface of Earth.

continental margin The submerged outer edge of a continent, made of granitic crust; includes the continental shelf and continental slope. Compare *ocean basin*.

continental rise The wedge of sediment forming the gentle transition from the outer (lower) edge of the continental slope to the abyssal plain; usually associated with passive margins.

continental shelf The gradually sloping submerged extension of a continent, composed of granitic rock overlain by sediments; has features similar to the edge of the nearby continent.

continental slope The sloping transition between the granite of the continent and the basalt of the seabed; the true edge of a continent.

contour current A bottom current made up of dense water that flows around (rather than over) seabed projections.

convection Movement within a fluid resulting from differential heating and cooling of the fluid. Convection produces mass transport or mixing of the fluid.

convection current A single closed-flow circuit of rising warm material and falling cool material.

convergence zone The line along which waters of different density converge. Convergence zones form the boundaries of tropical, subtropical, temperate, and polar areas.

convergent evolution The evolution of similar characteristics in organisms of different ancestry; the body shape of a porpoise and a shark, for instance.

convergent plate boundary A region where plates are pushing together and where a mountain range, island arc, and/or trench will eventually form; often a site of much seismic and volcanic activity.

Cook, James (1728–1779) Officer in the British Royal Navy who led the first European voyages of scientific discovery.

copepod A small planktonic arthropod, a major marine primary consumer.

coral Any of more than 6,000 species of small cnidarians, many of which are capable of generating hard calcareous (aragonite, $CaCO_3$) skeletons.

coral reef A linear mass of calcium carbonate (aragonite and calcite) assembled from coral organisms, algae, mollusks, worms, and so on. Coral may contribute less than half of the reef material.

core The innermost layer of Earth, composed primarily of iron, with nickel and heavy elements. The inner core is thought to be a solid 6,000°C (11,000°F) sphere, the outer core a 5,000°C (9,000°F) liquid mass. The average density of the outer core is about 11.8 g/cm^3, and that of the inner core is about 16 g/cm^3.

Coriolis, Gaspard Gustave de (1792–1843) The French scientist who in 1835 worked out the mathematics of the motion of bodies on a rotating surface. See *Coriolis effect*.

Coriolis effect The apparent deflection of a moving object from its initial course when its speed and direction are measured in reference to the surface of the rotating Earth. The object is deflected to the right of its anticipated course in the Northern Hemisphere and to the left in the Southern Hemisphere. The deflection occurs for any horizontal movement of objects with mass and has no effect at the equator.

cosmogenous sediment Sediment of extraterrestrial origin.

counterclockwise Rotation around a point in the direction opposite to that in which clock hands move; also called *anticlockwise*.

countercurrent A surface current flowing in the opposite direction from an adjacent surface current.

countershading A camouflage pattern featuring a dark upper surface and a lighter bottom surface.

covalent bond A chemical bond formed between two atoms by electron sharing.

crest See *wave crest*.

crust The outermost solid layer of Earth, composed mostly of granite and basalt; the top of the lithosphere. The crust has a density of 2.7–2.9 g/cm^3 and accounts for 0.4% of Earth's mass.

Crustacea The class of phylum Arthropoda to which lobsters, shrimp, crabs, barnacles, and copepods belong.

cryptic coloration Camouflage; may be active (under control of the animal) or passive (an unalterable color or shape).

Curie point The temperature above which a material loses its magnetism.

current Mass flow of water. (The term is usually reserved for horizontal movement.)

cyclone A weather system with a low-pressure area in the center around which winds blow counterclockwise in the Northern Hemisphere and clockwise in the Southern Hemisphere. Not to be confused with a tornado, a much smaller weather phenomenon associated with severe thunderstorms. See also *extratropical cyclone; tropical cyclone*.

Darwin, Charles (1809–1882) An English biologist and the co-discoverer (with Alfred Russell Wallace) of evolution by natural selection.

deep scattering layer (DSL) A relatively dense aggregation of fishes, squid, and other mesopelagic organisms capable of reflecting a sonar pulse that resembles a false bottom in the ocean. Its position varies with the time of day.

deep-water wave A wave in water deeper than one-half its wavelength.

deep zone The zone of the ocean below the pycnocline, in which there is little additional change of density with increasing depth; contains about 80% of the world's water.

degree An arbitrary measure of temperature. One degree Celsius (°C) = 1.8 degrees Fahrenheit (°F).

delta The deposit of sediments found at a river mouth, sometimes triangular in shape (named after the Greek letter Δ).

denitrifying bacteria Bacteria capable of converting nitrite or nitrate to gaseous nitrogen.

density The mass per unit volume of a substance, usually expressed in grams per cubic centimeter (g/cm^3).

density curve A graph showing the relationship between a fluid's temperature or salinity and its density.

density stratification The formation of layers in a material, with each deeper layer being denser (weighing more per unit of volume) than the layer above.

dependency A feeding relationship in which an organism is limited to feeding on one species or, in extreme cases, on one size phase of one species.

deposition Accumulation, usually of sediments.

depositional coast A coast in which processes that deposit sediment exceed erosive processes.

desalination The process of removing salt from seawater or brackish water.

desiccation Drying.

destructive interference The subtraction of wave energy as waves interact, producing smaller waves.

diatom Earth's most abundant, successful, and efficient single-celled phytoplankton. Diatoms possess two interlocking valves made primarily of silica. The valves contribute to biogenous sediments.

diffusion The movement—driven by heat—of molecules from a region of high concentration to a region of low concentration.

dinoflagellate One of a class of microscopic single-celled flagellates, not all of which are autotrophic. The outer covering is often of stiff cellulose. Planktonic dinoflagellates are responsible for "red tides."

dispersion Separation of wind waves by wavelength (and therefore wave speed) as they move away from the fetch (the place of their formation). Dispersion occurs because waves with long wavelengths move more rapidly than waves with short wavelengths.

disphotic zone The lower part of the photic zone, where there is insufficient light for photosynthesis.

dissolution The dissolving by water of minerals in rocks.

dissolved organic carbon (DOC) Organic (carbon-containing) molecules dissolved in water.

dissolved organic nitrogen (DON) Nitrogen-containing organic molecules dissolved in seawater. Most DON is in the form of protein.

disturbing force The energy that causes a wave to form.

diurnal tide A tidal cycle of one high tide and one low tide per day.

divergent evolution Evolutionary radiation of different species from a common ancestor.

divergent plate boundary A region where plates are moving apart and where new ocean or rift valley will eventually form. A spreading center forms the junction.

doldrums The zone of rising air near the equator known for sultry air and variable breezes. See also *intertropical convergence zone (ITCZ)*.

downwelling Circulation pattern in which surface water moves vertically downward.

drag The resistance to movement of an organism induced by the fluid through which it swims.

drift net Fine, vertically suspended net that may be 7 meters (25 feet) high and 80 kilometers (50 miles) long.

drumlin A streamlined hill formed by a glacier.

DSL See *deep scattering layer*.

dynamic theory of tides Model of tides that takes into account the effects of finite ocean depth, basin resonances, and the interference of continents on tide waves.

earthquake A sudden motion of Earth's crust resulting from waves in Earth caused by faulting of the rocks or by volcanic activity.

eastern boundary current Weak, cold, diffuse, slow-moving current at the eastern boundary of an ocean (off the west coast of a continent). Examples include the Canary Current and the Humboldt Current.

ebb current Water rushing out of an enclosed harbor or bay because of the fall in sea level as a tide trough approaches.

Echinodermata The phylum of exclusively marine animals to which sea stars, brittle stars, sea urchins, and sea cucumbers belong.

Echinoidea The class of the phylum Echinodermata to which sea urchins and sand dollars belong.

echo sounder A device that reflects sound off the ocean bottom to sense water depth. Its accuracy is affected by the variability of the speed of sound through water.

echolocation The use of reflected sound to detect environmental objects. Cetaceans use echolocation to detect prey and avoid obstacles.

ecology Study of the interactions of organisms with one another and with their environment.

ectotherm An organism incapable of generating and maintaining steady internal temperature from metabolic heat and therefore whose internal body temperature is approximately the same as that of the surrounding environment; a cold-blooded organism.

eddy A circular movement of water usually formed where currents pass obstructions, or between two adjacent currents flowing in opposite directions, or along the edge of a permanent current.

EEZ See *exclusive economic zone*.

Ekman spiral A theoretical model of the effect on water of wind blowing over the ocean. Because of the Coriolis effect, the surface layer is expected to drift at an angle of 45° to the right of the wind in the Northern Hemisphere and 45° to the left in the Southern Hemisphere. Water at successively lower layers drifts progressively to the right (N) or left (S), though not as swiftly as the surface flow.

Ekman transport Net water transport, the sum of layer movement due to the Ekman spiral. Theoretical Ekman transport in the Northern Hemisphere is 90° to the right of the wind direction.

El Niño A southward-flowing nutrient-poor current of warm water off the coast of western South America, caused by a breakdown of trade-wind circulation.

electron A tiny negatively charged particle in an atom responsible for chemical bonding.

element A substance composed of identical atoms that cannot be broken down into simpler substances by chemical means.

endotherm An organism capable of generating and regulating metabolic heat to maintain a steady internal temperature. Birds and mammals are the only animals capable of true endothermy. A warm-blooded organism.

energy The capacity to do work.

ENSO Acronym for the coupled phenomena of El Niño and the Southern Oscillation. See also *El Niño; Southern Oscillation*.

entropy A measure of the disorder in a system.

environmental resistance All the limiting factors that act together to regulate the maximum allowable size, or carrying capacity, of a population.

epicenter The point on Earth's surface directly above the focus of an earthquake.

epipelagic zone The lighted, or photic, zone in the ocean.

equator See *geographical equator; meteorological equator*.

equatorial upwelling Upwelling in which water moving westward on either side of the geographical equator tends to be deflected slightly poleward and replaced by deep water often rich in nutrients. See also *upwelling*.

equilibrium theory of tides Idealized model of tides that considers Earth to be covered by an ocean of great and uniform depth capable of instantaneous response to the gravitational and inertial forces of the sun and the moon.

Eratosthenes of Cyrene (276–192 B.C.) Greek scholar and librarian at Alexandria who first calculated the circumference of Earth about 230 B.C.

erosion A process of being gradually worn away.

erosional coast A coast in which erosive processes exceed depositional ones.

estuary A body of water partially surrounded by land where fresh water from a river mixes with ocean water, creating an area of remarkable biological productivity.

euphotic zone The upper layer of the photic zone in which net photosynthetic gain occurs. Compare *photic zone*.

euryhaline Describing an organism able to tolerate a wide range in salinity.

eurythermal Describing an organism able to tolerate a wide variance in temperature.

eurythermal zone The upper layer of water, where temperature changes with the seasons.

eustatic change A worldwide change in sea level, as distinct from local changes.

eutrophication A set of physical, chemical, and biological changes brought about when excessive nutrients are released into water.

evaporite Deposit formed by the evaporation of ocean water.

evolution Change; the maintenance of life under constantly changing conditions by continuous adaptation of successive generations of a species to its environment.

excess volatiles A compound found in the ocean and atmosphere in quantities greater than can be accounted for by the weathering of surface rock. Such compounds probably entered the atmosphere and ocean from deep crustal and upper mantle sources through volcanism.

exclusive economic zone (EEZ) The offshore zone claimed by signatories to the 1982 United Nations Draft Convention on the Law of the Sea. The EEZ extends 200 nautical miles (370 kilometers) from a contiguous shoreline. See also *United States Exclusive Economic Zone*.

exoskeleton A strong, lightweight, form-fitted external covering and support common to animals of the phylum Arthropoda. The exoskeleton is made partly of chitin and may be strengthened by calcium carbonate.

experiments Tests that simplify observation in nature or in the laboratory by manipulating or controlling the conditions under which observations are made.

extratropical cyclone A low-pressure mid-latitude weather system characterized by converging winds and ascending air rotating counterclockwise in the Northern Hemisphere and clockwise in the Southern Hemisphere. An extratropical cyclone forms at the front between the polar and Ferrel cells.

extremophile An organism capable of tolerating extreme environmental conditions, especially temperature or pH level.

fault A fracture in a rock mass along which movement has occurred.

Ferrel, William (1817–1891) The American scientist who discovered the mid-latitude circulation cells of each hemisphere.

Ferrel cell The middle atmospheric circulation cell in each hemisphere. Air in these cells rises at 60° latitude and falls at 30° latitude. See also *westerlies*.

fetch The uninterrupted distance over which the wind blows without a significant change in direction, a factor in wind-wave development.

Fissipedia The carnivoran suborder that includes sea otters.

fjord A deep, narrow estuary in a valley originally cut by a glacier.

fjord estuary An estuary in a fjord, a steep, submerged, U-shaped valley.

flagellum A whiplike structure used by some small organisms and gametes to move through the environment. (Plural, flagella.)

float method A method of current study that depends on the movement of a drift bottle or other free-floating object.

flood current Water rushing into an enclosed harbor or bay because of the rise in sea level as a tide crest approaches.

flow method A method of current study that measures the current as it flows past a fixed object.

food General term for organic molecules capable of providing energy to heterotrophs when combined with oxygen during biochemical respiration.

food web A group of organisms associated by a complex set of feeding relationships in which the flow of food energy can be followed from primary producers through consumers.

foraminiferan One of a group of planktonic amoeba-like animals with a calcareous shell, which contributes to biogenous sediments.

forced wave A progressive wave under the continuing influence of the forces that formed it.

Forchhammer's principle See *principle of constant proportions*.

foreshore Sand on the seaward side of the berm, sloping toward the ocean, to the low-tide mark.

fracture zone Area of irregular, seismically inactive topography marking the position of a once-active transform fault.

Franklin, Benjamin (1706–1790) Published the first chart of an ocean current in 1769.

free wave A progressive wave free of the forces that formed it.

freezing point The temperature at which a solid can begin to form as a liquid is cooled.

fringing reef A reef attached to the shore of a continent or island.

front The boundary between two air masses of different density. The density difference can be caused by differences in temperature and/or humidity.

frontal storm Precipitation and wind caused by the meeting of two air masses, associated with an extratropical cyclone. Generally, one air mass will slide over or under the other, and the resulting expansion of air will cause cooling and consequently rain or snow.

frustule The siliceous external cell wall of a diatom consisting of two interlocking valves fitted together like the halves of a box.

fucoxanthin A brown or tan accessory pigment found in many species of brown algae and some species of diatoms.

fully developed sea The theoretical maximum height attainable by ocean waves given wind of a specific strength, duration, and fetch. Longer exposure to wind will not increase the size of the waves.

galaxy A large rotating aggregation of stars, dust, gas, and other debris held together by gravity. There are perhaps 50 billion galaxies in the universe and 50 billion stars in each galaxy.

gas bladder In multicellular algae, an air-filled structure that assists in flotation.

gas exchange Simultaneous passage, through a semipermeable membrane, of oxygen into an animal and carbon dioxide out of it.

Gastropoda The class of the phylum Mollusca that includes snails and sea slugs.

geographical equator 0° latitude, an imaginary line equidistant from the geographical poles.

geostrophic Describing a gyre or current in balance between the Coriolis effect and gravity; literally, "turned by Earth."

gill membrane The thin boundary of living cells separating blood from water in a fish's (or other aquatic animal's) gills.

Global Positioning System (GPS) Satellite-based navigation system that provides a geographical position—longitude and latitude—accurate to less than 1 meter.

GPS See *Global Positioning System*.

granite The relatively light crustal rock—composed mainly of oxygen, silicon, and aluminum—that forms the continents. Its density is about 2.7 g/cm^3.

gravimeter A sensitive device that measures variations in the pull of gravity at different places on Earth's surface.

gravity wave A wave with wavelength greater than 1.73 centimeters (0.68 inch), whose restoring forces are gravity and momentum.

greenhouse effect Trapping of heat in the atmosphere. Incoming short-wavelength solar radiation penetrates the atmosphere, but the outgoing longer-wavelength radiation is absorbed by greenhouse gases and reradiated to Earth, causing a rise in surface temperature.

greenhouse gases Gases in Earth's atmosphere that cause the greenhouse effect; include carbon dioxide, methane, and CFCs.

groin A short, artificial projection of durable material placed at a right angle to shore in an attempt to slow longshore transport of sand from a beach; usually deployed in repeating units.

group velocity Speed of advance of a wave train; for deepwater waves, half the speed of individual waves within the group.

Gulf Stream The strong western boundary current of the North Atlantic, off the Atlantic coast of the United States.

guyot A flat-topped, submerged inactive volcano.

gyre Circuit of mid-latitude currents around the periphery of an ocean basin. Most oceanographers recognize five gyres plus the Antarctic Circumpolar Current.

habitat The place where an individual or population of a given species lives; its "mailing address."

hadal zone The deepest zone of the ocean, below a depth of 5,000 meters (16,500 feet).

Hadley, George (1685–1768) A London lawyer and philosopher who worked out the overall scheme of wind circulation in an effort to explain the trade winds.

Hadley cell The atmospheric circulation cell nearest the equator in each hemisphere. Air in these cells rises near the equator because of strong solar heating there and falls because of cooling at about 30° latitude. See also *trade winds*.

half-life Time required for one-half of all the unstable radioactive nuclei in a sample to decay.

halocline The zone of the ocean in which salinity increases rapidly with depth. See also *pycnocline*.

Harrison, John (1693–1776) British clockmaker who invented the modern chronometer in 1760.

heat A form of energy produced by the random vibration of atoms or molecules.

heat budget An expression of the total solar energy received on Earth during some period of time and the total heat lost from Earth by reflection and radiation into space through the same period.

heat capacity The heat, measured in calories, required to raise 1 gram of a substance 1° Celsius. The input of 1 calorie of heat energy raises the temperature of 1 gram of pure water by 1° C.

Henry the Navigator (1394–1460) Prince of Portugal who established a school for the study of geography, seamanship, shipbuilding, and navigation.

hermatypic Describing coral species possessing symbiotic zooxanthellae within their tissues and capable of secreting calcium carbonate at a rate suitable for reef production.

heterotroph An organism that derives nourishment from other organisms because it is unable to synthesize its own food molecules.

hierarchy Grouping of objects by degrees of complexity, grade, or class. A hierarchical system of nomenclature is based on distinctions within groups and between groups.

high-energy coast A coast exposed to large waves.

high seas That part of the ocean past the exclusive economic zone that is considered common property to be shared by the citizens of the world; about 60% of the ocean area.

high tide The high-water position corresponding to a tidal crest.

holdfast A complex branching structure that anchors many kinds of multicellular algae to the substrate.

holoplankton Permanent members of the plankton community. Examples are diatoms and copepods. Compare *meroplankton*.

Holothuroidea The class of the phylum Echinodermata to which sea cucumbers belong.

horse latitudes Zones of erratic horizontal surface air circulation near 30°N and 30°S latitudes. Over land, dry air falling from high altitudes produces deserts at these latitudes (for example, the Sahara).

hot spot A surface expression of a plume of magma rising from a stationary source of heat in the mantle.

hurricane A large tropical cyclone in the North Atlantic or eastern Pacific, whose winds exceed 118 kilometers (74 miles) per hour.

hydrogen bond A relatively weak bond formed between a partially positive hydrogen atom and a partially negative oxygen, fluorine, or nitrogen atom of an adjacent molecule.

hydrogenous sediment A sediment formed directly by precipitation from seawater; also called *authigenic sediment*.

hydrostatic pressure The constant pressure of water around a submerged organism.

hydrothermal vent A spring of hot, mineral- and gas-rich seawater found on some oceanic ridges in zones of active seafloor spreading.

hypertonic Referring to a solution having a higher concentration of dissolved substances than the solution that surrounds it.

hypothesis A speculation about the natural world that may be verified or disproved by observation and experiment.

hypotonic Referring to a solution having a lower concentration of dissolved substances than the solution that surrounds it.

ice age One of several periods (lasting several thousand years each) of low temperature during the last million years. Glaciers and polar ice were derived from ocean water, lowering sea level at least 100 meters (328 feet). (See Appendix II, "Geological Time.")

ice cap Permanent cover of ice; formally limited to ice atop land, but informally applied also to floating ice in the Arctic Ocean.

iceberg A large mass of ice floating in the ocean that was formed on or adjacent to land. Tabular icebergs are tablelike or flat; pinnacled icebergs are castellated, or jagged. Southern icebergs are often tabular; northern icebergs are often pinnacled.

inlet A passage giving the ocean access to an enclosed lagoon, harbor, or bay.

insolation rate The amount of solar energy reaching Earth's surface per unit time.

interference Addition or subtraction of wave energy as waves interact; also called *resonance*. See also *constructive interference; destructive interference*.

internal wave A progressive wave occurring at the boundary between liquids of different densities.

intertidal zone The marine zone between the highest high-tide point on a shoreline and the lowest low-tide point. The intertidal zone is sometimes subdivided into four separate habitats by height above tidal datum, typically numbered 1 to 4, land to sea.

intertropical convergence zone (ITCZ) The equatorial area at which the trade winds converge. The ITCZ usually lies at or near the meteorological equator; also called the *doldrums*.

introduced species A species removed from its home range and established in a new and foreign location; also called *exotic species*.

invertebrate Animal lacking a backbone.

ion An atom (or small group of atoms) that becomes electrically charged by gaining or losing one or more electrons.

ionic bond A chemical bond resulting from attraction between oppositely charged ions. These forces are said to be "electrostatic" in nature.

ionizing radiation Fast-moving particles or high-energy electromagnetic radiation emitted as unstable atomic nuclei disintegrate. The radiation has enough energy to dislodge one or more electrons from atoms it hits to form charged ions, which can react with and damage living tissue.

island arc Curving chain of volcanic islands and seamounts almost always found paralleling the concave edge of a trench.

isostatic equilibrium Balanced support of lighter material in a heavier, displaced supporting matrix; analogous to buoyancy in a liquid.

isotonic Referring to a solution having the same concentration of dissolved substances as the solution that surrounds it.

ITCZ See *intertropical convergence zone*.

Jason-1 A follow-on satellite mission to *TOPEX/Poseidon*.

kelp Informal name for any species of large phaeophyte.

kingdom The largest category of biological classification. Five kingdoms are presently recognized.

knot A speed of 1 nautical mile per hour. See also *nautical mile*.

krill *Euphausia superba*, a thumb-size crustacean common in Antarctic waters.

La Niña An event during which normal tropical Pacific atmospheric and oceanic circulation strengthens and the surface temperature of the eastern South Pacific drops below average values; usually occurs at the end of an ENSO event. See also *ENSO*.

lagoon A shallow body of seawater generally isolated from the ocean by a barrier island. Also, the body of water enclosed within an atoll, or the water within a reverse estuary.

land breeze Movement of air offshore as marine air heats and rises.

Langmuir circulation Shallow, wind-driven circulation of water in horizontal, spiral bands.

latent heat of fusion Heat removed from a liquid during freezing (or added to a solid during thawing) that produces a change in state but not a change in temperature. For pure water, 80 calories per gram at 0°C (32°F).

latent heat of vaporization Heat added to a liquid during evaporation (or released from a gas during condensation) that produces a change in state but not a change in temperature. For pure water, 585 calories per gram at 20°C (68°F).

lateral-line system A system of sensors and nerves in the head and midbody of fishes and some amphibians that functions to detect low-frequency vibrations in water.

latitude Regularly spaced imaginary lines on Earth's surface running parallel to the equator.

law A large construct explaining events in nature that have been observed to occur with unvarying uniformity under the same conditions.

law of the sea Collective term for laws and treaties governing the commercial and practical use of the ocean.

Library of Alexandria The greatest collection of writings in the ancient world, founded in the third century B.C. by Alexander the Great; could be considered the first university.

light Electromagnetic radiation propagated as small, nearly massless particles that behave like both a wave and a stream of particles.

limiting factor A physical or biological environmental factor whose absence or presence in an inappropriate amount limits the normal actions of an organism.

Linnaeus, Carolus Carl von Linné (1707–1778). Swedish "father" of modern taxonomy.

lithification Conversion of sediment into sedimentary rock by pressure or by the introduction of a mineral cement.

lithosphere The brittle, relatively cool outer layer of Earth, consisting of the oceanic and continental crust and the outermost, rigid layer of mantle.

littoral zone The band of coast alternately covered and uncovered by tidal action; the intertidal zone.

longitude Regularly spaced imaginary lines on Earth's surface running north and south and converging at the poles.

longshore bar A submerged or exposed line of sand lying parallel to shore and accumulated by wave action.

longshore current A current running parallel to shore in the surf zone, caused by the incomplete refraction of waves approaching the beach at an angle.

longshore drift Movement of sediments parallel to shore, driven by wave energy.

longshore trough Submerged excavation parallel to shore adjacent to an exposed sandy beach; caused by the turbulence of water returning to the ocean after each wave.

low-energy coast A coast only rarely exposed to large waves.

low tide The low-water position corresponding to a tidal trough.

low-tide terrace The smooth, hard-packed beach seaward of the beach scarp on which waves expend most of their energy. Site of the most vigorous onshore and offshore movement of sand.

lower mantle The rigid portion of Earth's mantle below the asthenosphere.

lunar tide Tide caused by gravitational and inertial interaction of the moon and Earth.

macroplankton Animal plankters larger than 1 to 2 centimeters ($\frac{1}{2}$ to 1 inch). An example is the jellyfish.

Magellan, Ferdinand (c. 1480–1521) Portuguese navigator in the service of Spain who led the first expedition to circumnavigate Earth, 1519–22. He was killed in the Philippines.

magma Molten rock capable of fluid flow; called *lava* above ground.

magnetometer A device that measures the amount and direction of residual magnetism in a rock sample.

Mahan, Alfred Thayer An American naval officer and strategist; the influential author of *The Influence of Sea Power upon History, 1660–1783*.

Mammalia The class of mammals.

mangrove A large flowering shrub or tree that grows in dense thickets or forests along muddy or silty tropical coasts.

mantle The layer of Earth between the crust and the core, composed of silicates of iron and magnesium. The mantle has an average density of about 4.5 g/cm^3 and accounts for about 68% of Earth's mass.

mantle plume Ascending columns of superheated mantle originating at the core–mantle boundary.

map A representation of Earth's surface, usually depicting mostly land areas. See also *chart*.

mariculture The farming of marine organisms, usually in estuaries, bays, or nearshore environments or in specially designed structures using circulating seawater. Compare *aquaculture*.

marine energy resource Any resource resulting from the direct extraction of energy from the heat or movement of ocean water.

marine pollution The introduction by humans of substances or energy into the ocean that changes the quality of the water or affects the physical and biological environment.

marine science The process (or result) of applying the scientific method to the ocean, its surroundings, and the life-forms within it; also called *oceanography* or *oceanology*.

masking pigment See *accessory pigment*.

mass A measure of the quantity of matter.

mass extinction A catastrophic, global event in which major groups of species perish abruptly.

Maury, Matthew (1806–1873) "Father" of physical oceanography. Probably the first person to undertake the systematic study of the ocean as a full-time occupation, and probably the first to understand the global interlocking of currents, wind flow, and weather.

maximum sustainable yield The maximum amount of fish, crustaceans, and mollusks that can be caught without impairing future populations.

mean sea level The height of the ocean surface averaged over a few years' time.

medusa Free-swimming body form of many members of the phylum Cnidaria.

membrane A complex structure of proteins and lipids that forms boundaries around and within the cell. It is usually semipermeable, allowing some kinds of molecules to pass through but not others.

meroplankton The planktonic phase of the life cycle of organisms that spend only part of their life drifting in the plankton.

mesosphere The rigid inner mantle, similar in chemical composition to the asthenosphere.

metabolic rate The rate at which energy-releasing reactions proceed within an organism.

metamerism Segmentation; repeating body parts.

***Meteor* expedition** German Atlantic expedition begun in 1925; the first to use an echo sounder and other modern optical and electronic instrumentation.

meteorological equator The irregular imaginary line of thermal equilibrium between hemispheres. It is situated about 5° north of the geographical equator, and its position changes with the seasons, moving slightly north in northern summer. Also called the *thermal equator*.

meteorological tide A tide influenced by the weather. Arrival of a storm surge will alter the estimate of a tide's height or arrival time, as will a strong, steady onshore or offshore wind.

metrophagy Tendency for large reptiles to eat entire cities.

microtektite Translucent oblong particles of glass, a component of cosmogenous sediment.

Milky Way galaxy The name of our galaxy; sometimes applied to the field of stars in our home spiral arm, which is correctly called the Orion arm.

mineral A naturally occurring inorganic crystalline material with a specific chemical composition and structure.

mixed layer See *surface zone*.

mixed tide A complex tidal cycle, usually with two high tides and two low tides of unequal height per day.

mixing time The time necessary to mix a substance through the ocean, about 1,600 years.

mixture A close intermingling of different substances that still retain separate identities. The properties of a mixture are heterogeneous; they may vary within the mixture.

molecule A group of atoms held together by chemical bonds. The smallest unit of a compound that retains the characteristics of the compound.

Mollusca The phylum of animals that includes chitons, snails, clams, and octopuses.

molt To shed an external covering.

monsoon A pattern of wind circulation that changes with the season. Also, the rainy season in areas with monsoon wind patterns.

moon tide See *lunar tide*.

moraine Hills or ridges of sediment deposited by glaciers.

motile Able to move about.

multicellular Consisting of more than one cell.

multicellular algae Algae with bodies consisting of more than one cell. Examples are kelp and *Ulva*.

mutation A heritable change in an organism's genes.

mutualism A symbiotic interaction between two species that is beneficial to both.

Mysticeti The suborder of baleen whales.

Nansen bottle A water-sampling instrument perfected early in this century by the Norwegian scientist and explorer Fridtjof Nansen.

natural selection A mechanism of evolution that results in the continuation of only those forms of life best adapted to survive and reproduce in their environment.

natural system of classification A method of classifying an organism based on its ancestry or origin.

nautical chart A chart used for marine navigation.

nautical mile The length of 1 minute of latitude, 6,076 feet, 1.15 statute miles, or 1.85 kilometers. (See Appendix I.)

neap tide The time of smallest variation between high and low tides occurring when Earth, moon, and sun align at right angles. Neap tides alternate with spring tides, occurring at two-week intervals.

nebula Diffuse cloud of dust and gas.

nekton Drifting organisms.

Nematoda The phylum of animals to which roundworms belong.

neritic Of the shore or coast; refers to continental margins and the water covering them, or to nearshore organisms.

neritic sediment Continental shelf sediment consisting primarily of terrigenous material.

neritic zone The zone of open water near shore, over the continental shelf.

niche Description of an organism's functional role in a habitat; its "job."

nitrifying bacteria Bacteria capable of fixing gaseous nitrogen into nitrite, nitrate, or ammonium ions.

nitrogen cycle The cycle in which nitrogen moves from its largest reservoir (the atmosphere) through the ocean, ocean sediments, and food webs, and then back to the atmosphere.

node The line or point of no wave action in a standing pattern. See also *amphidromic point*.

nodule Solid mass of hydrogenous sediment, most commonly manganese or ferromanganese nodules and phosphorite nodules.

nonconservative constituent An element whose proportion in seawater varies with time and place, depending on biological demand or chemical reactivity. An element with a short residence time; for example, iron, aluminum, silicon, trace nutrients, dissolved oxygen, and carbon dioxide.

nonconservative nutrient A compound or ion that is needed by autotrophs for primary productivity and that changes in concentration with biological activity.

nonextractive resource Any use of the ocean in place, such as transportation of people and commodities by sea, recreation, or waste disposal.

nonrenewable resource Any resource that is present on Earth in fixed amounts and cannot be replenished.

nonvascular Describing photosynthetic autotrophs without vessels for the transport of fluid. Examples are algae.

nor'easter (northeaster) Any energetic extratropical cyclone that sweeps the eastern seaboard of North America in winter.

North Atlantic Deep Water Cold, dense water formed in the Arctic that flows onto the floor of the North Atlantic ocean.

notochord Stiffening structure found at some time in the life cycle of all members of the phylum Chordata.

nuclear energy Energy released when atomic nuclei undergo a nuclear reaction such as the spontaneous emission of radioactivity, nuclear fission, or nuclear fusion. About 17% of the electrical power generated in the United States is provided by the nuclear fission of uranium in civilian power reactors.

nucleus (physics) The small, dense, positively charged center of an atom that contains the protons and neutrons.

nutrient Any needed substance that an organism obtains from its environment except oxygen, carbon dioxide, and water.

ocean (1) The great body of saline water that covers 70.78% of the surface of Earth. (2) One of its primary subdivisions, bounded by continents, the equator, and other imaginary lines.

ocean basin Deep-ocean floor made of basaltic crust. Compare *continental margin*.

oceanic crust The outermost solid surface of Earth beneath ocean floor sediments, composed primarily of basalt.

oceanic ridge Young seabed at the active spreading center of an ocean, often unmasked by sediment, bulging above the abyssal plain. The boundary between diverging plates. Often called a mid-ocean ridge, though less than 60% of the length exists at mid-ocean.

oceanic zone The zone of open water away from shore, past the continental shelf.

oceanography The science of the ocean. See also *marine science*.

oceanus Latin form of *okeanos*, the Greek name for the "ocean river" past Gibraltar.

Odontoceti The suborder of toothed whales.

oolite sand Hydrogenous sediment formed when calcium carbonate precipitates from warmed seawater as pH rises, forming rounded grains around a shell fragment or other particle.

ooze Sediment of at least 30% biological origin.

ophiolite An assemblage of subducting oceanic lithosphere scraped off (obducted) onto the edge of a continent.

Ophiuroidea The class of the phylum Echinodermata to which brittle stars belong.

orbit In ocean waves, the circular pattern of water particle movement at the air–sea interface. Orbital motion contrasts with the side-to-side or back-and-forth motion of pure transverse or longitudinal waves.

orbital inclination The 23°27' "tilt" of Earth's rotational axis relative to the plane of its orbit around the sun.

orbital wave A progressive wave in which particles of the medium move in closed circles.

osmoregulation The ability to adjust internal salt concentration.

osmosis The diffusion of water from a region of high water concentration to a region of lower water concentration through a semipermeable membrane.

Osteichthyes The class of fishes with bony skeletons.

outgassing The volcanic venting of volatile substances.

overfishing Harvesting so many fish that there is not enough breeding stock left to replenish the species.

oxygen minimum zone A zone in which oxygen is depleted by animals and not replaced by phytoplankton.

oxygen revolution The time span, from about 2 billion to 400 million years ago, during which photosynthetic autotrophs changed the composition of Earth's atmosphere to its current oxygen-rich mixture.

ozone O_3, the triatomic form of oxygen. Ozone in the upper atmosphere protects living things from some of the harmful effects of the sun's ultraviolet radiation.

ozone layer A diffuse layer of ozone mixed with other gases surrounding the world at a height of about 20 to 40 kilometers (12 to 25 miles).

P wave Primary wave; a compressional wave that is associated with an earthquake and that can move through both liquid and rock.

Pacific Ring of Fire The zone of seismic and volcanic activity that encircles the Pacific Ocean.

paleoceanography The study of the ocean's past.

paleomagnetism The "fossil," or remanent, magnetic field of a rock.

Pangaea Name given by Alfred Wegener to the original "protocontinent." The breakup of Pangaea gave rise to the Atlantic Ocean and to the continents we see today.

Panthalassa Name given by Alfred Wegener to the ocean surrounding Pangaea.

parasitism A symbiotic relationship in which one species spends part or all of its life cycle on or within another, using the host species (or food within the host) as a source of nutrients; the most common form of symbiosis.

partially mixed estuary An estuary in which an influx of seawater occurs beneath a surface layer of fresh water flowing seaward. Mixing occurs along the junction.

passive margin The continental margin near an area of lithospheric plate divergence; also called *Atlantic-type margin*.

passive sonar A device that detects the intensity and direction of underwater sounds.

PCBs See *polychlorinated biphenyls*.

pelagic Of the open ocean; refers to the water above the deep-ocean basins, sediments of oceanic origin, or organisms of the open ocean.

pelagic sediment Sediments of the slope, rise, and deep-ocean floor that originate in the ocean.

pelagic zone The realm of open water. See also *benthic zone*.

perigee The point in the orbit of a satellite where it is closest to the main body; opposite of *apogee*.

perihelion The point in the orbit of a satellite where it is closest to the sun; opposite of *aphelion*.

period See *wave period*.

pH scale A measure of the acidity or alkalinity of a solution; numerically, the negative logarithm of the concentration of hydrogen ions in an aqueous solution. A pH of 7 is neutral; lower numbers indicate acidity, and higher numbers indicate alkalinity.

Phaeophyta Brown multicellular algae, including kelps.

photic zone The thin film of lighted water at the top of the world ocean. The photic zone rarely extends deeper than 200 meters (660 feet). Compare *euphotic zone*.

photon The smallest unit of light energy.

photosynthesis The process by which autotrophs bind light energy into the chemical bonds of food with the aid of chlorophyll and other substances. The process uses carbon dioxide and water as raw materials and yields glucose and oxygen.

phycobilin A reddish accessory pigment found in red algae.

phylum One of the major groups of the animal kingdom whose members share a similar body plan, level of complexity, and evolutionary history (see Appendix VI). (Plural, phyla.) (The major groups of the plant kingdom are called divisions.)

physical factor An aspect of the physical environment that affects living organisms, such as light, salinity, or temperature.

physical resource Any resource that has resulted from the deposition, precipitation, or accumulation of a useful nonliving substance in the ocean or seabed; also called a *nonliving resource*.

phytoplankton Plantlike, usually single-celled members of the plankton community.

picoplankton Extremely small members of the plankton community, typically 0.2 to 2 micrometers (4 to 40 millionths of an inch) across.

Pinnipedia The carnivoran suborder that contains the seals, sea lions, and walruses.

piston corer A seabed-sampling device capable of punching through up to 25 meters (80 feet) of sediment and returning an intact plug of material.

planet A smaller, usually nonluminous body orbiting a star.

plankter Informal name for a member of the plankton community.

plankton Drifting or weakly swimming organisms suspended in water. Their horizontal position is to a large extent dependent on the mass flow of water rather than on their own swimming efforts.

plankton bloom A sudden increase in the number of phytoplankton cells in a volume of water.

plankton net Conical net of fine nylon or Dacron fabric used to collect plankton.

Plantae The kingdom to which multicellular vascular autotrophs belong.

plate One of about a dozen rigid segments of Earth's lithosphere that move independently. The plate consists of continental or oceanic crust and the cool, rigid upper mantle directly below the crust.

plate tectonics The theory that Earth's lithosphere is fractured into plates that move relative to each other and are driven by convection currents in the mantle. Most volcanic and seismic activity occurs at plate margins.

Platyhelminthes The phylum of animals to which flatworms belong.

plunging wave A breaking wave in which the upper section topples forward and away from the bottom, forming an air-filled tube.

polar cell The atmospheric circulation cell centered over each pole.

polar front Boundary between the polar cell and the Ferrel cell in each hemisphere.

polar molecule A molecule with unbalanced charge. One end of the molecule has a slight negative charge, and the other end has a slight positive charge.

pollutant A substance that causes damage by interfering directly or indirectly with an organism's biochemical processes.

Polychaeta The largest and most diverse class of phylum Annelida. Nearly all polychaetes are marine.

polychlorinated biphenyls (PCBs) Chlorinated hydrocarbons once widely used to cool and insulate electrical devices and to strengthen wood or concrete. PCBs may be responsible for the changes in and declining fertility of some marine mammals.

Polynesia A large group of Pacific islands lying east of Melanesia and Micronesia and extending from the Hawaiian Islands south to New Zealand and east to Easter Island.

Polynesians Inhabitants of the Pacific islands that lie within a triangle fromed by Hawaii, New Zealand, and Easter Island.

polynya A gap in polar pack ice at which liquid water contacts the atmosphere.

polyp One of two body forms of Cnidaria. Polyps are cup-shaped and possess rings of tentacles. Coral animals are polyps.

poorly sorted sediment A sediment in which particles of many sizes are found.

population A group of individuals of the same species occupying the same area.

population density The number of individuals per unit area.

Porifera The phylum of animals to which sponges belong.

potable water Water suitable for drinking.

precipitate (1) A solid substance formed in an aqueous reaction. (2) The process by which a solute forms in and falls from a solution. The falling of water or ice from the atmosphere.

precipitation Liquid or solid water that falls from the air and reaches the surface as rain, hail, or snowfall.

pressure Force per unit area.

prey An organism consumed by a predator.

primary consumer Initial consumer of primary producers. The consumers of autotrophs; the second level in food webs.

primary forces The forces that induce and maintain water flow in ocean current systems: thermal expansion, wind friction, and density differences.

primary producer An organism capable of using energy from light or energy-rich chemicals in the environment to produce energy-rich organic compounds; an autotroph.

primary productivity The synthesis of organic materials from inorganic substances by photosynthesis or chemosynthesis; expressed in grams of carbon bound into carbohydrate per unit area per unit time (g C/m²/yr).

Prince Henry the Navigator Established a center at Sagres, Portugal, for the study of marine science and navigation in the mid-1450s.

principle of constant proportions The proportions of major conservative elements in seawater remain nearly constant, though total salinity may change with location; also called *Forchhammer's principle*.

progressive wave A wave of moving energy in which the wave form moves in one direction along the surface (or junction) of the transmission medium (or media).

Protista The kingdom of single-celled nucleated organisms to which protozoa, diatoms, and dinoflagellates belong; also called *Protoctista*.

proton A positively charged particle at the center of an atom.

protostar A tightly condensed knot of material that has not yet attained fusion temperature.

pteropod A small planktonic mollusk with a calcareous shell, which contributes to biogenous sediments.

pycnocline The middle zone of the ocean in which density increases rapidly with depth. Temperature falls and salinity rises in this zone.

radial symmetry Body structure in which the body parts radiate from a central axis like spokes from a wheel. An example is a sea star. Compare *bilateral symmetry*.

radioactive decay The disintegration of unstable forms of elements, which releases subatomic particles and heat.

radiolarian One of a group of usually planktonic amoeba-like animals with a siliceous shell, which contributes to biogenous sediments.

radiometric dating The process of determining the age of rocks by observing the ratio of unstable radioactive elements to stable decay products.

random distribution Distribution of organisms within a community whereby the position of one organism is in no way influenced by the positions of other organisms or by physical variations within that community; a very rare distribution pattern.

reef A hazard to navigation; a shoal, a shallow area, or a mass of fish or other marine life.

refraction Bending of light or sound waves as they move at an angle other than 90° between media of different optical or acoustical densities. See also *wave refraction*.

refractive index The degree of refraction from one medium to another expressed as a ratio. The higher the ratio (refractive index), the greater the bending of waves between media.

refractometer A compact optical device that determines the salinity of a water sample by comparing the refractive index of the sample to the refractive index of water of known salinity.

relative dating Determining the age of a geological sample by comparing its position to the positions of other samples.

renewable resource Any resource that is naturally replaced on a seasonal basis by the growth of living organisms or by other natural processes.

Reptilia The class of reptiles, including turtles, crocodiles, iguanas, and snakes.

residence time The average length of time a dissolved substance spends in the ocean.

respiration Release of stored energy from chemical bonds in food; carbon dioxide and water are formed as by-products. (Respiration is a biochemical process and is not the same as the mechanical process of breathing.)

restoring force The dominant force trying to return water to flatness after formation of a wave.

reverse estuary An estuary along a coast in which salinity increases from the ocean to the estuary's upper reaches because of evaporation of seawater and a lack of freshwater input.

Rhodophyta Red, multicellular algae.

Richter scale A logarithmic measure of earthquake magnitude. A great earthquake measures above 8 on the Richter scale.

rip current A strong, narrow surface current that flows seaward through the surf zone and is caused by the escape of excess water that has piled up in a longshore trough.

rogue wave A single wave crest much higher than usual, caused by constructive interference.

S wave Secondary wave; a transverse wave that is associated with an earthquake and that cannot move through liquid.

salinity A measure of the dissolved solids in seawater, usually expressed in grams per kilogram or parts per thousand by weight. Standard seawater has a salinity of 35‰ at 0°C (32°F).

salinometer An electronic device that determines salinity by measuring the electrical conductivity of a seawater sample.

salt gland Specialized tissue responsible for concentration and excretion of excess salt from blood and other body fluids.

salt wedge estuary An estuary in which rapid river flow and small tidal range cause an inclined wedge of seawater to form at the mouth.

sand Sediment particle between 0.062 and 2 millimeters in diameter.

sand spit An accumulation of sand and gravel deposited downcurrent from a headland. Sand spits often curl at their tips.

sandbar A submerged or exposed line of sand accumulated by wave action.

saturation State of a solution in which no more of the solute will dissolve in the solvent. The rate at which molecules of the solute are being dissolved equals the rate at which they are being precipitated from the solution.

scattering The dispersion (or "bounce") of sound or light waves when they strike particles suspended in water or air. The amount of scatter depends on the number, size, and composition of the particles.

schooling Tendency of small fish of a single species, size, and age to mass in groups. The school moves as a unit, which confuses predators and reduces the effort spent searching for mates.

science A systematic way of asking questions about the natural world and testing the answers to those questions.

scientific method The orderly process by which theories explaining the operation of the natural world are verified or rejected.

scientific name The genus and species name of an organism.

sea Simultaneous wind waves of many wavelengths forming a chaotic ocean surface. Sea is common in an area of wind wave origin.

sea breeze Onshore movement of air as inland air heats and rises.

sea cave A cave near sea level in a sea cliff cut by processes of marine erosion.

sea cliff A cliff marking the landward limit of marine erosion on an erosional coast.

sea grass Any of several marine angiosperms. Examples are *Zostera* (eelgrass) and *Phyllospadix* (surfgrass). Sea grasses are not seaweeds.

sea ice Ice formed by the freezing of seawater.

sea island An island whose central core was connected to the mainland when sea level was lower. Rising ocean separates these high points from land, and sedimentary processes surround them with beaches. Compare *barrier island*.

sea level The height of the ocean surface. See also *mean sea level*.

sea power The means by which a nation extends its military capacity onto the ocean.

sea state Ocean wave conditions at a specific place and time, usually stated in the Beaufort scale.

seafloor spreading The theory that new ocean crust forms at spreading centers, most of which are on the ocean floor, and pushes the continents aside. Power is thought to be provided by convection currents in Earth's upper mantle.

seamount A circular or elliptical projection from the seafloor, more than 1 kilometer (0.6 mile) in height, with a relatively steep slope of 20° to 25°.

SEASTAR Satellite capable of measuring the distribution of chlorophyll at the ocean surface, a measure of marine productivity.

seaweed Informal term for large marine multicellular algae.

second law of thermodynamics Disorder (entropy) in a closed system must increase over time. If disorder decreases, it does so at the expense of energy. Because the universe as a whole may be considered a closed system, it follows that an increase in order in one part must result in a decrease in order in another.

secondary consumer Consumer of primary consumers.

sediment Particles of organic or inorganic matter that accumulate in a loose, unconsolidated form.

seiche Pendulum-like rocking of water in an enclosed area; a form of standing wave that can be caused by meteorological or seismic forces, or that may result from normal resonances excited by tides.

seismic Referring to earthquakes and the shock of earthquakes.

seismic sea wave Tsunami caused by displacement of earth along a fault. (Earthquakes and seismic sea waves are caused by the same phenomenon.)

seismic wave A low-frequency wave generated by the forces that cause earthquakes. Some kinds of seismic waves can pass through Earth. See also *P wave*; *S wave*.

seismograph An instrument that detects and records earth movement associated with earthquakes and other disturbances.

semidiurnal tide A tidal cycle of two high tides and two low tides each lunar day, with the high tides of nearly equal height.

sensible heat Heat whose gain or loss is detectable by a thermometer or other sensor.

sessile Attached; nonmotile; unable to move about.

sewage Waste water with a significant organic content, usually frm domestic or industrial sources.

sewage sludge Semisolid mixture of organic matter, microorganisms, toxic metals, and synthetic organic chemicals removed from wastewater at a sewage treatment plant.

shadow zone (1) The wide band at Earth's surface 105° to 143° away from an earthquake in which seismic waves are nearly absent. P waves are absent because they are refracted by Earth's liquid outer core; S waves are absent from this band and the zone immediately opposite the earthquake site because they are absorbed by the outer core. (2) In sonar, the volume of ocean from which sound waves diverge and in which a submarine may hide.

shallow-water wave A wave in water shallower than $\frac{1}{20}$ its wavelength.

shelf break The abrupt increase in slope at the junction between continental shelf and continental slope.

shore The place where ocean meets land. On nautical charts, the limit of high tides.

side-scan sonar A high-resolution sound-imaging system used for geological investigations, archaeological studies, and the location of sunken ships and airplanes.

siliceous ooze Ooze composed mostly of the hard remains of silica-containing organisms.

silicoflagellate A tiny, single-celled phytoplankter with a siliceous skeleton.

silt Sediment particle between 0.004 and 0.062 millimeter in diameter.

Sirenia The order of mammals that includes manatees, dugongs, and the extinct sea cows.

slack water A time of no tide-induced currents that occurs when the current changes direction.

sofar *Sound fixing and ranging.* An experimental U.S. Navy technique for locating survivors on life rafts, based on the fact that sound from explosive charges dropped into the layer of minimum sound velocity can be heard for great distances. See also *sofar layer.*

sofar layer Layer of minimum sound velocity in which sound transmission is unusually efficient for long distances. Sounds leaving this depth tend to be refracted back into it. The sofar layer usually occurs at mid-latitude depths around 1,200 meters (4,000 feet).

solar nebula The diffuse cloud of dust and gas from which the solar system originated.

solar system The sun together with the planets and other bodies that revolve around it.

solar tide Tide caused by the gravitational and inertial interaction of the sun and Earth.

solstice One of two times of the year when the overhead position of the sun is farthest from the equator. The time of the solstice is midway between equinoxes.

solute A substance dissolved in a solvent. See also *solution.*

solution A homogeneous substance made of two components, the solvent and the solute.

solvent A substance able to dissolve other substances. See also *solution.*

sonar *Sound navigation and ranging.*

sound A form of energy transmitted by rapid pressure changes in an elastic medium.

sounding Measurement of the depth of a body of water.

Southern Oscillation A reversal of airflow between normally low atmospheric pressure over the western Pacific and normally high pressure over the eastern Pacific; the cause of El Niño. See also *El Niño.*

speciation The formation of new species. Charles Darwin suggested that this is accomplished through isolation and natural selection.

species Any group of actually or potentially interbreeding organisms reproductively isolated from all other groups and capable of producing fertile offspring. (Note: The word *species* is both singular and plural.)

species diversity Number of different species in a given area.

species-specific relationship An exclusive relationship between two species. Parasites are usually species-specific; that is, they can usually parasitize only one species of host.

spilling wave A breaking wave whose crest slides down the face of the wave.

spreading center The junction between diverging plates at which new ocean floor is being made; also called *spreading zone.*

spring tide The time of greatest variation between high and low tides occurring when Earth, moon, and sun form a straight line. Spring tides alternate with neap tides throughout the year, occurring at two-week intervals.

standing wave A wave in which water oscillates without causing progressive wave forward movement. There is no net transmission of energy in a standing wave.

star A massive sphere of incandescent gases powered by the conversion of hydrogen to helium and other heavier elements.

state An expression of the internal form of matter. Water exists in three states: solid, liquid, and gas. A solid has a fixed volume and fixed shape, a liquid has a fixed volume but no fixed shape, and a gas has neither fixed volume nor fixed shape.

stenohaline Describing an organism unable to tolerate a wide range in salinity.

stenothermal Describing an organism unable to tolerate wide variance in temperature.

stipe Multicellular algal equivalent of a vascular plant's stem.

Stokes drift A small net transport of water in the direction a wind wave is moving.

storm Local or regional atmospheric disturbance characterized by strong winds often accompanied by precipitation.

storm surge An unusual rise in sea level as a result of the low atmospheric pressure and strong winds associated with a tropical cyclone. Onrushing seawater precedes landfall of the tropical cyclone and causes most of the damage to life and property.

stratigraphy The branch of geology that deals with the definition and description of natural divisions of rocks; specifically, the analysis of relationships of rock strata.

subduction The downward movement into the asthenosphere of a lithospheric plate.

subduction zone An area at which a lithospheric plate is descending into the asthenosphere. The zone is characterized by linear folds (trenches) in the ocean floor and strong deep-focus earthquakes; also called a *Wadati–Benioff zone.*

sublittoral zone The ocean floor near shore. The inner sublittoral extends from the littoral (intertidal) zone to the depth at which wind waves have no influence; the outer sublittoral extends to the edge of the continental shelf.

submarine canyon A deep, V-shaped valley running roughly perpendicular to the shoreline and cutting across the edge of the continental shelf and slope.

subsidence Sinking, often of tectonic origin.

Subtropical Convergence Convergence zone marking the boundary between Central Water and either Subarctic or Subantarctic Surface Water. The northern Subtropical Convergence lies at about 45°N in the Pacific and 60°N in the Atlantic; the southern Subtropical Convergence lies at 40° to 50°S.

succession The changes in species composition that lead to a climax community.

sun tide See *solar tide.*

supernova The explosive collapse of a massive star.

superplume A very large mantle plume.

supralittoral zone The splash zone above the highest high tide; not technically part of the ocean bottom.

surf The confused mass of agitated water rushing shoreward during and after a wind wave breaks.

surf beat The pattern of constructive and destructive interference that causes successive breaking waves to grow, shrink, and grow again over a few minutes' time.

surf zone The region between the breaking waves and the shore.

surface current The horizontal flow of water at the ocean's surface.

surface-to-volume ratio A physical constraint on the size of cells. As a cell's linear dimensions grow, its surface area does not increase at the same rate as its volume. As the surface-to-volume ratio decreases, each square unit of outer membrane must serve an increasing interior volume.

surface zone The upper layer of ocean in which temperature and salinity are relatively constant with depth. Depending on local conditions, the surface zone may reach to 1,000 meters (3,300 feet) or be absent entirely. Also called the *mixed layer.*

surging wave A wave that surges ashore without breaking.

suspension feeder An animal that feeds by straining or otherwise collecting plankton and tiny food particles from the surrounding water.

sverdrup (sv) A unit of volume transport named in honor of oceanographer Harald U. Sverdrup: 1 million cubic meters of water flowing past a fixed point each second.

swash Water from waves washing onto a beach.

swell Mature wind waves of one wavelength that form orderly undulations of the ocean surface.

swim bladder A gas-filled organ that assists in maintaining neutral buoyancy in some bony fishes.

symbiosis The co-occurrence of two species in which the life of one is closely interwoven with the life of the other; mutualism, commensalism, or parasitism.

synoptic sampling Simultaneous sampling at many locations.

taxonomy In biology, the laws and principles covering the classification of organisms.

tektite A small, rounded, glassy component of cosmogenous sediments, usually less than 1.5 millimeters ($\frac{1}{20}$ inch) in length; thought to have formed from the impact of an asteroid or meteor on the crust of Earth or the moon.

Teleostei The osteichthyan order that contains the cod, tuna, halibut, perch, and other species of bony fishes.

telepresence The extension of a person's senses by remote sensors and manipulators.

temperate zone The mid-latitude area between the Tropic of Cancer and the Arctic Circle and between the Tropic of Capricorn and the Antarctic Circle.

temperature The response of a solid, liquid, or gas to the input or removal of heat energy. A measure of the atomic and molecular vibration in a substance, indicated in degrees.

temperature–salinity (T-S) diagram A graph showing the relationship of temperature and salinity with depth.

terrane An isolated segment of seafloor, island arc, plateau, continental crust, or sediment transported by seafloor spreading to a position adjacent to a larger continental mass; usually different in composition from the larger mass.

terrigenous sediment Sediment derived from the land and transported to the ocean by wind and flowing water.

territorial waters Waters extending 12 miles from shore and in which a nation has the right to jurisdiction.

thallus The body of an alga or other simple plant.

theory A general explanation of a characteristic of nature consistently supported by observation or experiment.

thermal equator See *meteorological equator.*

thermal equilibrium The condition in which the total heat coming into a system (such as a planet) is balanced by the total heat leaving the system.

thermal inertia Tendency of a substance to resist change in temperature with the gain or loss of heat energy.

thermocline The zone of the ocean in which temperature decreases rapidly with depth. See also *pycnocline.*

thermohaline circulation Water circulation produced by differences in temperature and/or salinity (and therefore density).

thermostatic property A property of water that acts to moderate changes in temperature.

tidal bore A high, often breaking wave generated by a tide crest that advances rapidly up an estuary or river.

tidal current Mass flow of water induced by the raising or lowering of sea level owing to passage of tidal crests or troughs. See also *ebb current; flood current.*

tidal datum The reference level (0.0) from which tidal height is measured.

tidal range The difference in height between consecutive high and low tides.

tidal wave The crest of the wave causing tides; another name for a tidal bore; not a tsunami or seismic sea wave.

tide Periodic short-term change in the height of the ocean surface at a particular place, generated by long-wavelength progressive waves that are caused by the interaction of gravitational force and inertia. Movement of Earth beneath tide crests results in the rhythmic rising and falling of sea level.

tombolo Above-water bridge of sand connecting an offshore feature to the mainland.

top consumer An organism at the apex of a trophic pyramid, usually a carnivore.

TOPEX/Poseidon Joint French–U.S. satellite carrying radars that can determine the height of the sea surface with unprecedented accuracy. Other experiments in this five-year program included sensing water vapor over the ocean, determining the precise location of ocean currents, and determining wind speed and direction.

tornado Localized, narrow, violent funnel of fast-spinning wind, usually generated when two air masses collide; not to be confused with a cyclone. (The tornado's oceanic equivalent is a waterspout.)

trace element A minor constituent of seawater present in amounts of less than 1 part per million.

trade winds Surface winds within the Hadley cells, centered at about 15° latitude, that approach from the northeast in the Northern Hemisphere and from the southeast in the Southern Hemisphere.

transform fault A plane along which rock masses slide horizontally past one another.

transform plate boundary Places where crustal plates shear laterally past one another. Crust is neither produced nor destroyed at this type of junction.

transverse current East-to-west or west-to-east current linking the eastern and western boundary currents. An example is the North Equatorial Current.

trench An arc-shaped depression in the deep-ocean floor with very steep sides and a flat sediment-filled bottom coinciding with a subduction zone. Most trenches occur in the Pacific.

trophic level A feeding step within a trophic pyramid.

trophic pyramid A model of feeding relationships among organisms. Primary producers form the base of the pyramid; consumers eating one another form the higher levels, with the top consumer at the apex.

Tropic of Cancer The imaginary line around Earth parallel to the equator at 23°27′N, marking the point where the sun shines directly overhead at the June solstice.

Tropic of Capricorn The imaginary line around Earth parallel to the equator at 23°27′S, marking the point where the sun shines directly overhead at the December solstice.

tropical cyclone A weather system of low atmospheric pressure around which winds blow counterclockwise in the Northern Hemisphere and clockwise in the Southern Hemisphere. It originates in the tropics within a single air mass, but may move into temperate waters if the water temperature is high enough to sustain it. Small tropical cyclones are called *tropical depressions*, larger ones *tropical storms*, and great ones *hurricanes*, *typhoons*, or *willi-willis*, depending on location.

tropics The area between the Tropic of Cancer and the Tropic of Capricorn.

trough See *wave trough*.

tsunami Long-wavelength, shallow-water wave caused by rapid displacement of water. See also *seismic sea wave*.

tunicate A type of suspension-feeding invertebrate chordate.

turbidite A terrigenous sediment deposited by a turbidity current; typically, coarse-grained layers of nearshore origin interleaved with finer sediments.

turbidity current An underwater "avalanche" of abrasive sediments thought responsible for the deep sculpturing of submarine canyons and a means of transport for sediments accumulating on abyssal plains.

turbulence Chaotic fluid flow.

ultraplankton Extremely small plankton, smaller than nanoplankton.

undercurrent A current flowing beneath a surface current, usually in the opposite direction.

unicellular Consisting of a single cell.

unicellular algae Algae with bodies consisting of a single cell. Examples are diatoms and dinoflagellates.

uniform distribution Distribution of organisms within a community characterized by equal space between individuals (the arrangement of trees in an orchard); the rarest natural distribution pattern.

uniformitarianism The theory that all of Earth's geological features and history can be explained by processes occurring today and that these processes must have been at work for a very long time.

United States Exclusive Economic Zone The region extending seaward from the coast of the United States for 200 nautical miles, within which the United States claims sovereign rights and jurisdiction over all marine resources.

United States Exploring Expedition The first U.S. oceanographic research voyage, launched in 1838.

upwelling Circulation pattern in which deep, cold, usually nutrient-laden water moves toward the surface. Upwelling can be caused by winds blowing parallel to shore or offshore.

valve In diatoms, each half of the protective silica-rich outer portion of the cell. The complete outer covering is called the *frustule*.

vascular plant Plant having vessels for transport of fluid through leaves, stems, and roots. Examples are sea grasses, mangroves, and maple trees.

velocity Speed in a specified direction.

vertebrate A chordate with a segmented backbone.

Vikings Seafaring Scandinavian raiders who ravaged the coasts of Europe around A.D. 780–1070.

viscosity Resistance to fluid flow. A measure of the internal friction in fluids.

voyaging Traveling (usually by sea) with a specific purpose.

Wadati–Benioff zone See *subduction zone*.

water mass A body of water identifiable by its salinity and temperature (and therefore its density) or by its gas content or another indicator.

water vapor The gaseous, invisible form of water.

water-vascular system System of water-filled tubes and canals found in some representatives of the phylum Echinodermata and used for movement, defense, and feeding.

wave Disturbance caused by the movement of energy through a medium.

wave crest Highest part of a progressive wave above average water level.

wave-cut platform The smooth, level terrace sometimes found on erosional coasts that marks the submerged limit of rapid marine erosion.

wave diffraction Bending of waves around obstacles.

wave frequency The number of waves passing a fixed point per second.

wave height Vertical distance between a wave crest and the adjacent wave troughs.

wave period The time it takes for successive wave crests to pass a fixed point.

wave reflection The reflection of progressive waves by a vertical barrier. Reflection occurs with little loss of energy.

wave refraction Slowing and bending of progressive waves in shallow water.

wave shock Physical movement, often sudden, violent, and of great force, caused by the crash of a wave against an organism.

wave steepness Height-to-wavelength ratio of a wave. The theoretical maximum steepness of deep-water waves is 1:7.

wave train A group of waves of similar wavelength and period moving in the same direction across the ocean surface. The group velocity of a wave train is half the velocity of the individual waves.

wave trough The valley between wave crests below the average water level in a progressive wave.

wavelength The horizontal distance between two successive wave crests (or troughs) in a progressive wave.

weather The state of the atmosphere at a specific place and time.

Wegener, Alfred (1880–1930) German scientist who proposed the theory of continental drift in 1912.

well-mixed estuary An estuary in which slow river flow and tidal turbulence mix fresh and salt water in a regular pattern through most of its length.

well-sorted sediment A sediment in which particles are of uniform size.

West Wind Drift Current driven by powerful westerly winds north of Antarctica. The largest of all ocean currents, it continues permanently eastward without changing direction. See *Antarctic Circumpolar Current.*

westerlies Surface winds within the Ferrel cells, centered around 45° latitude, that approach from the southwest in the Northern Hemisphere and from the northwest in the Southern Hemisphere.

western boundary current Strong, warm, concentrated, fast-moving current at the western boundary of an ocean (off the east coast of a continent). Examples include the Gulf Stream and the Japan (Kuroshio) Current.

westward intensification The increase in speed of geostrophic currents as they pass along the western boundary of an ocean basin.

Wilson, John Tuzo (1908–1993) Canadian geophysicist who proposed the theory of plate tectonics in 1965.

wind The mass movement of air.

wind duration The length of time the wind blows over the ocean surface, a factor in wind wave development.

wind-induced vertical circulation Vertical movement in surface water (upwelling or downwelling) caused by wind.

wind strength Average speed of the wind, a factor in wind wave development.

wind wave Gravity wave formed by transfer of wind energy into water. Wavelengths from 60 to 150 meters (200 to 500 feet) are most common in the open ocean.

world ocean The great body of saline water that covers 70.78% of Earth's surface.

xanthophyll A yellow or brown accessory pigment that gives some marine autotrophs a yellow or tan appearance.

zone Division or province of the ocean with homogeneous characteristics.

zooplankton Animal members of the plankton community.

zooxanthellae Unicellular dinoflagellates that are symbiotic with coral and that produce the relatively high pH and some of the enzymes essential for rapid calcium-carbonate deposition in coral reefs.

Acknowledgments

This page constitutes an extension of the copyright page. We have made every effort to trace the ownership of all copyrighted material and to secure permission from copyright holders. In the event of any question arising as to the use of any material, we will be pleased to make the necessary corrections in future printings. Thanks are due to the following authors, publishers, and agents for permission to use the material indicated.

Text and Illustration Credits

Chapter 3
Figure 3.2 From *The Earth: An Introduction to Physical Geology*, 4/e by Edward J. Tarbuck and Frederick K. Lutgens. ©1993 Macmillan College Publishing. Reprinted by permission of Pearson Education, Inc. Upper Saddle River, NJ 07458. **Figure 3.3a, b** From *The Earth's Dynamic Systems*, 9/e by Hamblin and Christiansen. Reprinted by permission of Pearson Education, Inc., Upper Saddle River, NJ 07458. **Figure 3.3c, d** From *Understanding Earth*, 3/e by Frank Press and Raymond Siever. © 1994, 1998, and 2001 by W. H. Freeman and Company. Used with permission. **Box 3.1c** B. Bolt, "Earthquakes and Geological Discover," Scientific American Books, © 1993. Reprinted with permission of W. H. Freeman and Company. **Figures 3.13 and 3.27** From *Understanding Earth*, 3/e by Frank Press and Raymond Siever. © 1994, 1998, and 2001 by W. H. Freeman and Company. Used with permission. **Figure 3.30** From *Earth's Dynamic Systems*, 8/e by Hamblin & Christiansen, ©1998. Reprinted by permission of Pearson Education, Inc., Upper Saddle River, NJ 07458. **Figure 3.31** After R. Montelli, "Seismic tomography beyond ray theory," Ph.D. Thesis, Princeton University, 2003. Used with permission. **Figure 3.37** Reprinted by permission of Dr. Roger Billham, University of Colorado.

Chapter 6
Figure 6.4 Fig 2.5, p. 29 from *Physical Geology* by Arthur N. Stahler. Copyright © 1981 by Arthur N. Stahler. Reprinted by permission of Pearson Education, Inc. **Figure 6.13** From G. P. Kuiper, ed., *The Earth as a Planet*, © 1954 The University of Chicago Press. Reprinted by permission.

Chapter 8
Figure 8.4 From Levin Danielson, *Meteorology*, Fig 3.4, p. 73. © 1998 McGraw-Hill Companies. Reprinted with permission. **Figure 8.10** *Calvin and Hobbes.* ©1990 Universal Press Syndicate. Reprinted with permission of Universal Press Syndicate. **Figure 8.14** *Meteorology*, 5/e by Moran and Morgan. ©1997 Prentice-Hall, Inc. Reprinted with permission. **Figure 8.16** From *Earth in Crisis: An Introduction to the Earth Sciences*, 2/e, Thomas L. Burrus, Herbert J. Spiegel, 1980. C. V. Mosby Co. Reprinted by permission of Thomas L. Burrus. **189:** Reprinted with permission of Alan D. Iselin.

Chapter 9
Figure 9.5 From *Laboratory Exercises in Oceanography* by Pipkin, Gorslin, Casey and Hammond. ©1987 by W. H. Freeman and Company. Reprinted with permission of W. H. Freeman and Company. **Table 9.1** From M. Grant Gross, *Oceanography: A View of the Earth*, 5/e, p. 173, ©1990. Prentice-Hall. Reprinted with permission.

Chapter 10
Figure 10.31 Research by Philip L. F. Liv, Seung Nam Seo, and Sung Bum Yoon, and Civil and Environmental Engineering, Cornell University. Visualization by Catherine Devine, Cornell Theory Center. Used by permission.

Chapter 12
Figure 12.31 Used by permission of Dr. William Galloway, University of Texas, Austin and the Houston Geological Survey. **Figure 12.31** Used by permission of Glen Wheless.

Chapter 15
Figure 15.10 From *Living Invertebrates*, V. & J. Pearse/M. & R. Buchsbaum, 1987, The **Box**wood Press. Adapted with permission of the publisher. **Figure 15.16** From James L. Sumich, *An Introduction to Biology of Marine Life*, 5/e. Copyright ©1992 Wm. C. Brown Communication, Inc., Dubuque, Iowa. Reprinted with permission of McGraw-Hill Companies. **Figure 15.21** From *Living Invertebrates*, V. & J. Pearse/M. & R. Buchsbaum, 1987, The **Box**wood Press. Adapted with permission of the publisher. **Figure 15.21** From Pough et al., *Vertebrate Life*, 3/e ©1989. Reprinted by permission of Pearson Education, Inc., Upper Saddle River, NJ 07458.

Chapter 16
Figure 16.7 From *Between Pacific Tides*, 5/e by Edward F. Ricketts, et al. Revised by David W. Phillip. Stanford University Press, ©1985 by the Board of Trustees of the Leland Stanford Junior University. Reprinted with permission. **Figures 16.15 and 16.18** From J. C. Briggs, *Marine Zoogeography*. ©1974 by McGraw-Hill, Inc. Reprinted with permission.

Chapter 17
Figure 17.9 Data from United Nations F.A.O. **Figure 17.14** R. Watson, V. Christensen, and D. Pauly/Oceana. Reprinted with permission.

Chapter 18
Figure 18.11 *Science*, 289 (August 4, 2000), p. 723. Reprinted by permission. **Figure 18.17** *Popular Science* (September 2000). Reprinted with permission from *Popular Science* magazine, copyright 2000, Times Mirror Magazines, Inc. **Figure 18.21** © The New Yorker Collection 1989 Stuart Leeds from cartoonbank.com. All rights reserved.

Appendixes
Appendix III, Figures 1–4 and **Appendix IV, Figures 1 and 3** from David Greenhood, *Down to Earth: Mapping for Everybody.* **Appendix IX, Figures 1–2 and excerpt pages 480–481** reprinted by permission of Dan Dion.

Photo Credits

Preface
Page xviii Tom Garrison. **Page xix** NASA. **Page xx** NASA/ORBIMAGE.

Chapter 1
Chapter opening photo NASA/GSFC. **Figure 1.3** © Pal Hermansen/The Image Bank/Getty Images. **Figure 1.4a** Scripps Institution of Oceanography. **Figure 1.4b** Tom Garrison. **Figure 1.6** © Taschen Verlag GMBH/de Espona Inforgraficia SL. **Figure 1.8** © Thomas Hartwell/Corbis Saba. **Figure 1.10** Tom Garrison. **Figure 1.11** © 1991 Herbert Kawainui Kane. **Figure 1.12** Reprinted with permission from Doubleday, a division of Bantam, Doubleday, Dell Publishing Group, Inc., from *Sailing Ships* by Bjørn Landstrøm, 1978. **Figure 1.13** © 1991 Herbert Kawainui Kane. **Figure 1.15** James Ford Bell Library, University of Minnesota. **Figure 1.16a** The Granger Collection, New York. **Figure 1.17** © National Maritime Museum, Greenwich. **Figure 1.18** © 1991 Herbert Kawainui Kane. **Figure 1.19a** Tom Garrison. **Figure 1.19b** National Maritime Museum, London. **Figure 1.20** Tom Garrison. **Figure 1.21** United States Naval Academy Museum. **Figure 1.22** U.S. Naval Observatory Library. Artist, Beverly Stautz. **Figure 1.23** Library of Congress. **Figure 1.25** Reprinted by permission of the Royal Geographical Society, London, archives from the journal of Lt. Pelham Aldrich kept aboard *H.M.S. Challenger*, 1872–1875. **Figure 1.27** Archive, Scripps Institution of Oceanography. **Figure 1.28a** Hulton Deutsch Collection Limited/Getty Images. **Figure 1.28b** Alda Amundsen, Scott Polar Research Institute. **Figure 1.28c** Courtesy of Dale Chayes, Lamont-Doherty Earth Observatory of Columbia University. **Box 1.2a** AP/Wide World Photos. **Box 1.2b** © 2001 Friends of the Hunley. All Rights Reserved. **Figure 1.29** JAMSTEC. **Figure 1.30** Musée océanophiuque de Monaco, Photo F.P. **Figure 1.31a, b** Tom Garrison. **Figure 1.32** NASA.

Chapter 2

Chapter opening photo GSFC/NASA. **Figure 2.1** European Southern Observatory. **Figure 2.2** Image from TRACE (Transition Region and Coronal Explorer), a mission of the Stanford-Lockheed Institute for Space Research (a joint program of Lockheed-Martin Advanced Technology Center's Solar and Astrophysics Laboratory and Stanford's Solar Observatories Group), and part of the NASA Small Explorer program. **Figure 2.3** Rerendered from *Living in the Universe* by Bennett, Shostak, Jakosky. Copyright © 2003 Pearson Education, Inc., publishing as Addison Wesley, San Francisco, California. **Figure 2.4** European Southern Observatory. **Figure 2.5** Hubble Space Telescope/STSci. **Figure 2.7** © William K Hartmann. **Figure 2.8** Excerpted from *The History of Earth.* Copyright © 1991 by William Hartmann and Ron Miller. Used by permission of Workman Publishing Co., Inc., New York. All Rights Reserved. **Figure 2.9a** © William K Hartmann. **Figure 2.9b** © Don Dixon. **Figure 2.11** © Ralph White/CORBIS. **Figure 2.12** Department of Earth and Space Sciences/UCLA. **Figure 2.13** Hubble Space Telescope/STSci. **Box 2.1a, b** NASA/Robert Sullivan. **Box 2.1c** Maria Zuber/NASA/GSFC, M.I.T. **Box 2.1d, e** Malin Space Science Systems/MGS/JPL/NASA. **Box 2.1f** Bruce MacIntosh, Seran Gibbard, Don Gavel, and Claire Max/Lawrence Livermore National Laboratory. **Box 2.1g** © Don Davis.

Chapter 3

Chapter opening photo Magnús Tumi Guomundsson, Science Institute, University of Iceland. **Box 3.1 left** National Portrait Gallery, London. **Box 3.1 right** © National Survey & Cadastre, Denmark. **Figure 3.4** Wilbur Garrett/NGS Image Collection. **Figure 3.7** Dorothy L. Stout. **Figure 3.10** © Alfred Wegener Institute for Polar and Marine Research. **Figure 3.18d** V. Courtillot. **Figure 3.26a** Fundamental Photographs. **Figure 3.33** Dr. Peter Sloss, NOAA/National Geophysical Data Center. **Figure 3.35** Hans-Ulrich Schmincke.

Chapter 4

Chapter opening photo (a) © Perry Thorsvik/National Geographic Society; **(b)** JAMSTEC. **Figure 4.1** Archive, Scripps Institution of Oceanography. **Figure 4.2b** Robert Profeta. **Figure 4.3a** Marie Tharp and Charles D. Hollister, Woods Hole Oceanographic Institution. **Figure 4.3b** Lamont-Doherty Earth Observatory of Columbia University. **Figure 4.4b** Andrew Goodwillie, Scripps Institution of Oceanography. **Figure 4.5a** U.S. Department of the Navy. **Figure 4.5c** Karen Marks, NOAA. **Figure 4.12, 4.13, 4.15** William Haxby, Lamont-Doherty Earth Observatory of Columbia University. **Figure 4.17** Ray Sterner/The Johns Hopkins University, Applied Physics Laboratory. **Figure 4.18** Society for Sedimentary Geology. **Figure 4.19** U.S. Department of the Navy. **Figure 4.20** Charles D. Hollister, Woods Hole Oceanographic Institution. **Figure 4.22a** ALCOA, Inc. **Figure 4.22b** Lamont–Doherty Earth Observatory of Columbia University. **Figure 4.22c** Charles D. Hollister, Woods Hole Oceanographic Institution. **Figure 4.23** Dr. Jian Lin, Woods Hole Oceanographic Institution. **Figure 4.25** Woods Hole Oceanographic Institution. **Figure 4.27** Charles D. Hollister. **Figure 4.28b** Stanley Hart, Woods Hole Oceanographic Institution. **Figure 4.31** Dr. Peter Sloss, NOAA/National Geophysical Data Center. **Box 4.1a** Dr. Don Walsh, International Maritime, Inc. **Box 4.1b** Naval Historical Center. **Box 4.1c** Dr. Andreas Rechnitzer. **Figure 4.32** OAR/NURP/JAMSTEC.

Chapter 5

Chapter opening photo (a) Don Davis/NASA; **(b)** © Richard Norris/Scripps Institution of Oceanography. **Figure 5.1–5.4** Charles D. Hollister. **Figure 5.7a** Liam Gumley, University of Wisconsin-Madison. **Figure 5.7b** Department of Geology, University of Delaware. **Figure 5.7c** SeaWiFS Project, NASA/GSFC and ORBIMAGE. **Figure 5.8** Alan Hildebrand, University of Calgary. **Figure 5.9** NGDC/NOAA. **Figure 5.11** Mark Drinkwater. **Figure 5.13a** Howard Spero. **Figure 5.13b** Wim van Egmond. **Figure 5.13c** Jeremy Young, University of California. **Figure 5.15a** Wim van Egmond. **Figure 5.15b** Greta Fryxell. **Figure 5.16a, b** Susumu Honjo/Woods Hole Oceanographic Institution. **Figure 5.17a, b** Unocal Research. **Figure 5.18a** Scripps Institution of Oceanography/UCSD. **Figure 5.19a** Tom Garrison. **Figure 5.20** Joint Oceanographic Institutes. **Figure 5.21** Deep Sea Drilling Project, Texas A&M University. **Box 5.1a, b** Dr. James Broda, Woods Hole Oceanographic Institution. **Figure 5.24** NASA.

Chapter 6

Chapter opening photo © Kaz Moari/The Image Bank/Getty Images. **Figure 6.7** © Donovan Reese/Stone/Getty Images. **Box 6.1** Tom Garrison. **Figure 6.11** Josefino Corniso/NASA. **Figure 6.12** NASA/MODIS. **Figure 6.19** Gunther Konnen, by permission of Bob Greenler. **Figure 6.20c, d**

Bruce Hall. **Box 6.2a** Brian Dushaw/University of Washington. **Figure 6.25b** EG&G Ocean Products. **Figure 6.25c** WESMAR.

Chapter 7

Chapter opening photo Jacques Louis David, 1788. The Metropolitan Museum of Art. Purchase, Mr. and Mrs. Charles Wrightsman Gift, in honor of Everett Fahy, 1977 (1977.10). Copyright © 1989 by the Metropolitan Museum of Art. **Figure 7.5b** Woods Hole Oceanographic Institution. **Figure 7.6** Photo courtesy of Dr. William Cochlan, Hancock Institute for Marine Science, University of Southern California. **Figure 7.7** David Breiter. **Box 7.1** © Rommel/Masterfile. **Figure 7.9** Art, Lisa Starr.

Chapter 8

Chapter opening photo (a) © PA News/CORBIS KIPA; **(b)** AP/Wide World Photos. **Figure 8.2b** Stephen Gassman/Index Stock Imagery. **Figure 8.6** NOAA/Tom Garrison. **Figure 8.15** NASA. **Figure 8.20** NASA and Tom Garrison. **Figure 8.21** © Bettmann Archive/CORBIS. **Figure 8.22** SeaWiFS Project, NASA/Goddard Space Flight Center, and ORBIMAGE. **Figure 8.23b** © M. Laca/Weatherstock. **Figure 8.24** JPL/NASA. **Figure 8.25b** NOAA/National Weather Service. **Box 8.1** NOAA. **Figure 8.26** Jim Bell, Cornell University, Astronomy Department.

Chapter 9

Chapter opening photo © Collections/Mike Pooler. **Figure 9.7c, 9.8a** NASA/JPL. **Figure 9.10** NASA/ORBIMAGE. **Box 9.1** © Bettmann/CORBIS. **Figure 9.12a** O. Brown, R. Evans, M. Carle/University of Miami Rosenstiel School of Marine and Atmospheric Science. **Figure 9.12b** NASA/ORBIMAGE. **Figure 9.15b** William Hsieh. **Figure 9.16b** SeaWiFS Project, NASA/Goddard Space Flight Center and ORBIMAGE. **Figure 9.18** Mobil Oil Corporation. **Figure 9.19b, d** NASA. **Figure 9.20a–e** TOPEX/Poseidon Team, CNES, NASA. **Figure 9.28b** Tom Garrison. **Figure 9.28c** Philip Richardson/Woods Hole Oceanographic Institution. **Figure 9.28e** RD Instruments. **Figure 9.28f** Webb Research.

Chapter 10

Chapter opening photo © Ron Dahlquist. **Figure 10.9** Tom Garrison. **Box 10.1** Raychel Ciemma. **Figure 10.12** NASA. **Figure 10.15** U.S. Navy. **Figure 10.16** Dr. W. C. Stillwell. Reprinted by permission of Suzanne M. Stillwell. **Figure 10.18a, b** © Dennis Junor/Creation Captured. **Figure 10.19** From Erich Lessing, Art Resource, NY. Crane, Walter (1845–1915). *Neptune's Horses.* Oil on Canvas. Neue Pinakothek, Munich, Germany. Reprinted by permission of Art Resource. **Figure 10.20b** NASA. **Figure 10.21** Tom Garrison. **Figure 10.22** University of Hawaii Press. **Figure 10.25b, c** NASA. **Figure 10.26b** AP/Wide World Photos. **Figure 10.26a** Tom Garrison. **Figure 10.26b** Thames Barrier Visitors Centre. **Figure 10.29** William F. Haxby & Lincoln F. Pratson. **Figure 10.32a–c** USGS. **Figure 10.33a** D. Millsap. **Figure 10.33b** USGS. **Figure 10.34** Bishop Museum, Hawaii. **Figure 10.35** Kyoto Photo Service. **Figure 10.37a** Tom Garrison. **Figure 10.37b** AP/Wide World Photos.

Chapter 11

Chapter opening photo Photography by J. Eric Jones. Used with permission. **Figure 11.1 (both)** Scott Walking Adventures, Halifax, Nova Scotia. **Figure 11.16** Richard Ray/Space Geodesy branch, NASA/GSFC. **Figure 11.19 (both)** © Jeff Greenberg/Photo Researchers, Inc. **Box 11.1a** Arthur Rackham. **Box 11.1b** BØDØ Arrangement, Norway. **Figure 11.20** Gary Florin and Cabrillo Marine Aquarium. **Figure 11.21** Photothèque EDF.

Chapter 12

Chapter opening photo Tom Garrison. **Figure 12.1a, b** Courtesy of William Haxby, Lamont/Columbia. **Figure 12.2** © Ken W. Davis/Tom Stack & Associates. **Figure 12.3b** © Steve Vidler/SuperStock, Inc. **Figure 12.3d** © John S. Shelton. **Figure 12.5** NASA. **Figure 12.6** Digital Vision/Getty Images. **Figure 12.7** USGS/Hawaii Volcanoes National Park. **Figure 12.8** © George P. Lafaker, USGS. **Figure 12.9** NASA. **Figure 12.10** © Alan Hoelzle. **Figure 12.13a, b** © John S. Shelton. **Figure 12.14a, c** © Dr. William Wallace. **Figure 12.14b** Tom Garrison. **Figure 12.14d** © Dante de la Rosa. **Figure 12.15b** © John S. Shelton. **Figure 12.16** Steve Elgar, Woods Hole Oceanographic Institution. **Figure 12.17b** NASA. **Figure 12.19** © Jack Mertz. **Figure 12.20** NASA. **Box 12.1a** © Billy E. Barnes/Stock, Boston/PictureQuest. **Box 12.1b** AP/Wide World Photos. **Figure 12.22a, b** © Gerald G. Kuhn. **Figure 12.22c** © W. Demetrakas, O. C. Camera. **Figure 12.23a** USGS. **Figure 12.23b** USGS/EROS Data Center. **Figure 12.26** Great Barrier Reef Marine Park Authority. Queensland, Australia. **Figure 12.27d** © Douglas Faulkner, Science Source/Photo Researchers, Inc.

Figure 12.28 © Brian Parker/Tom Stack and Associates. **Figure 12.32** © John S. Shelton. **Figure 12.33a, b** The Fairchild Aerial Photography Collection at Whittier College. **Figure 12.35** AP/Wide World Photos.

Chapter 13
Chapter opening photo © Cathy Church/Picturesque/PictureQuest. **Figure 13.1** © Linda E. Tway. **Figure 13.2** Wim van Egmond. **Figure 13.4** SeaWiFS Project, NASA/Goddard Space Flight Center and ORBIMAGE. **Figure 13.7a** © Micrographica/micrographia.com. **Figure 13.7b** © Bruce Hall. **Figure 13.7c** © Graham Munro/photolibrary/PictureQuest. **Figure 13.9** © James Watt/Visuals Unlimited. **Figure 13.15** M. Sheetz, R. Painter, and S. Singer, *Journal of Cell Biology*, 70:193 (1976) by permission of The Rockefeller University Press. **Figure 13.18** L. K. Townsend. **Box 13.1a, b** NASA. **Box 13.1c** © Don Davis. **Box 13.1d** Dr. Virgil Sharpton/University of Alaska–Fairbanks. **Figure 13.20 (top to bottom)** Frank Lane Picture Agency Limited; Field Museum of Chicago; © SuperStock; © Ed Degginger.

Chapter 14
Chapter opening photo © Frank Borges Llosa/frankly.com. **Figure 14.2b** Dennis Kelly, Orange Coast College. **Figure 14.3a, c** Wim van Egmond. **Figure 14.3b, 14.4a** From C. Shih and R. G. Kessel, *Living Images* © 1982 Sudbury, MA: Jones & Bartlett Publishers. www.jbpub.com. Reprinted with permission. **Figure 14.5a** Tom Garrison. **Figure 14.5b** © Bill Curtsinger Photography. **Figure 14.6a** Ian Probert. **Figure 14.6b** SeaWiFS Project, NASA/Goddard Space Flight Center and ORBIMAGE. **Box 14.1a** Florida Fish & Wildlife Conservation Commission, Florida Marine Research Institute, St. Petersburg. **Box 14.1b** C. C. Lockwood/Cactus Clyde Productions. **Box 14.1c** Suisan Aviation Ltd., Japan. **Box 14.2** Dr. William P. Cochlan, Hancock Institute for Marine Sciences/University of Southern California. **Figure 14.9** SeaWiFS Project, NASA/Goddard Space Flight Center. **Figure 14.11** Wim van Egmond. **Figure 14.12** © Uwe Kils. **Figure 14.13a** © Phillip Colla Photography. **Figure 14.13b, c** Larry Madin/Woods Hole Oceanographic Institution. **Figure 14.14** Wim van Egmond. **Figure 14.15b** Bruce & Valerie Hall. **Figure 14.18** Tom Garrison. **Figure 14.19** National Park Service.

Chapter 15
Chapter opening photo © Paul Ward. **Figure 15.2a** The South Australian Museum. **Figure 15.2b, c** Dr. Chip Clark. **Figure 15.3a** © David Haas/Tom Stack and Associates. **Figure 15.5** © Ben Cropp/Norbert Wu Productions, www.norbertwu.com. **Figure 15.6a** Pat Mason. **Figure 15.7** Tom Garrison. **Figure 15.9** © Bruce Hall. **Box 15.1** Tom Garrison. **Figure 15.11** Tom Garrison. **Figure 15.12** © Mark McMahon. **Figure 15.14** © 1992 Sea World of California, Inc. **Figure 15.15** Kim Griggs/GAMMA. **Figure 15.17a, b** Washington State Department of Fisheries. **Figure 15.18a, b** © Bruce Hall. **Figure 15.19b** © Herve Chaumeton/Nature. **Figure 15.20** Charles D. Hollister/Woods Hole Oceanographic Institution. **Figure 15.22** © Herve Chaumeton/Nature. **Figure 15.25** © Heather Angel. **Figure 15.26** © Amos Nachoum/CORBIS. **Figure 15.27** Chris Fallows/www. apexpredators.com. **Figure 15.28** © James D. Watt/Watt Wildlife Library. **Figure 15.30** © Timothy Reese. **Figure 15.34** © Bruce Hall. **Figure 15.35** © OSF, Debbie Stone/Animals Animals. **Figure 15.36** © Kevin

Schafer/CORBIS. **Figure 15.37** Tom Garrison. **Figure 15.38** © SuperStock, Inc. **Figure 15.41b** © Norman Cole. **Figure 15.42** © Richard Sears/MICS-Photo. **Figure 15.43a** © Bruce Hall. **Figure 15.43a** Tom Garrison. **Figure 15.44** Courtesy of Friends of the Sea Otter/Richard Bucich. **Figure 15.45** © Leonard Rue/Visuals Unlimited. **Figure 15.46** © Tom & Therisa Stack/Tom Stack & Associates.

Chapter 16
Chapter opening photo Tom Garrison. **Figure 16.1** NOAA. **Figure 16.3** Tom Garrison. **Figure 16.8** © Brad Sheard/SuperStock. **Figure 16.9a** Tom Garrison. **Figure 16.9b** © Stan Elms/Visuals Unlimited. **Figure 16.9c** © Andrew Syred/Photo Researchers, Inc. **Box 16.1** © Ed Ricketts, Jr. **Figure 16.10, 16.11** Tom Garrison. **Figure 16.12c** © Peter Scoones. **Figure 16.14** © E. Widder/HBOI, 1999. **Figure 16.16** © 2003 Norbert Wu. **Figure 16.17, 16.18** Charles D. Hollister/Woods Hole Oceanographic Institution. **Figure 16.19** © Henry Aldrich, University of Florida. **Figure 16.21a** © Science VU/Visuals Unlimited. **Figure 16.21b** Woods Hole Oceanographic Institution. **Figure 16.22** Craig Smith and Mike DeGruy. **Figure 16.23** © Dave Breiter. **Figure 16.23** © Norman Cole.

Chapter 17
Chapter opening photo © Dan Parrett/Alaska Stock. **Figure 17.2a, b** Courtesy of Shell Oil Company. **Figure 17.3** Stan Celestian, Glendale Community College. **Figure 17.4** Courtesy of Cargill Salt, Newark, California. **Figure 17.5** George Skene/Orlando Sentinel. **Figure 17.6** © Adam Schmedes/Loke Film. **Figure 17.7** Illustration by Diavision. Courtesy of Hammerfest Strøm. **Figure 17.8b** National Energy Laboratory of Hawaii. **Figure 17.10** Tom Garrison. **Figure 17.11** Orbimage © Oribital Imaging Corportion. **Figure 17.12b** © Shane Moore/Animals Animals, Earth Scenes. **Figure 17.15** © Bill Curtsinger Photography. **Figure 17.16** © Tom Campbell's Photographic. **Figure 17.17** Anthony R. Ramirez. **Figure 17.18** Tom Garrison. **Figure 17.20** © Wedigo Ferchi/Bruce Coleman, Inc./PictureQuest. **Figure 17.21** © Yann Arthus Bertrand/CORBIS. **Figure 17.22** Courtesy of the British Columbia Salmon Farmers Association. **Figure 17.23** Shell Oil Company. **Figure 17.24** Tom Garrison. **Box 17.1a, b** Tom Garrison. **Box 17.1c** John Wilson Cramer IV. **Box 17.1d** Port of Long Beach.

Chapter 18
Chapter opening photo Hiroyuki Matsumoto/Black Star Publishing/PictureQuest. **Figure 18.3** Paul Hanna/Reuters/Landov. **Figure 18.4** Tom Garrison. **Figure 18.5** NASA/JPL. **Figure 18.7** © 1989 Tony Dawson Photography. **Box 18.1** © W. Eugene Smith/Black Star Publishing. **Figure 18.8** © Erich Hartmann/Magnum Photos. **Figure 18.12** SeaWiFS/NASA/ORBIMAGE. **Figure 18.13** © Ken Sakamoto/Black Star Publishing. **Figure 18.14** © Tom Campbell's Photographic. **Figure 18.15** NOAA. **Figure 18.16** State of California. **Figure 18.18** SVS, TOMS, NASA. **Figure 18.19** Kevin Arrigo/Standard University. **Figure 18.20** Courtesy of Dr. Ronald D. Barr, UC-Irvine Medical Center. **Figure 18.22** NOAA. **Figure 18.26** Reprinted by permission of the Los Angeles Times Syndicate. **Figure 18.27** NASA/Tom Garrison. **Figure 18.28** C. Mayhew & R. Simmon (NASA/GFSC), NOAA, DMSP Digital Archive. **Figure 18.30** SeaWiFS/NASA/ORBIMAGE. **Figure 18.31** © ThinkStock/SuperStock.

Index

Note: Page references in italics refer to figures.